Modern Reduction Methods

Edited by
Pher G. Andersson and
Ian J. Munslow

Related Titles

Yamamoto, H., Ishihara, K. (eds.)

Acid Catalysis in Modern Organic Synthesis

2008
ISBN: 978-3-527-31724-0

Roberts, S. M.

Catalysts for Fine Chemical Synthesis V 5 – Regio and Stereo-Controlled Oxidations and Reductions

2007
Online Book Wiley Interscience
ISBN: 978-0-470-09024-4

de Vries, J. G., Elsevier, C. J. (eds.)

The Handbook of Homogeneous Hydrogenation

2007
ISBN: 978-3-527-31161-3

Torii, S.

Electroorganic Reduction Synthesis

2006
ISBN: 978-3-527-31539-0

de Meijere, A., Diederich, F. (eds.)

Metal-Catalyzed Cross-Coupling Reactions

2004
ISBN: 978-3-527-30518-6

Bäckvall, J.-E. (ed.)

Modern Oxidation Methods

2004
ISBN: 978-3-527-30642-8

Modern Reduction Methods

Edited by
Pher G. Andersson and Ian J. Munslow

WILEY-VCH Verlag GmbH & Co. KGaA

The Editors

Prof. Dr. Pher G. Andersson
Uppsala University
Department of Organic Chemistry
Husargatan 3
751 23 Uppsala
Sweden

Dr. Ian J. Munslow
Uppsala University
Department of Biochemistry and
Organic Chemistry
Husargatan 3
751 23 Uppsala
Sweden

■ All books published by Wiley-VCH are carefully produced. Nevertheless, authors, editors, and publisher do not warrant the information contained in these books, including this book, to be free of errors. Readers are advised to keep in mind that statements, data, illustrations, procedural details or other items may inadvertently be inaccurate.

Library of Congress Card No.:
applied for

British Library Cataloguing-in-Publication Data
A catalogue record for this book is available from the British Library.

Bibliographic information published by the Deutsche Nationalbibliothek
Die Deutsche Nationalbibliothek lists this publication in the Deutsche Nationalbibliografie; detailed bibliographic data are available on the Internet at <http://dnb.d-nb.de>.

© 2008 WILEY-VCH Verlag GmbH & Co. KGaA, Weinheim

All rights reserved (including those of translation into other languages). No part of this book may be reproduced in any form – by photoprinting, microfilm, or any other means – nor transmitted or translated into a machine language without written permission from the publishers. Registered names, trademarks, etc. used in this book, even when not specifically marked as such, are not to be considered unprotected by law.

Composition   SNP Best-set Typesetter Ltd., Hong Kong
Printing   betz-druck GmbH, Darmstadt
Binding   Litges & Dopf GmbH, Heppenheim
Cover Design   Schulz Grafik-Design, Fußgönheim

Printed in the Federal Republic of Germany
Printed on acid-free paper

ISBN: 978-3-527-31862-9

Contents

Preface *XV*
List of Contributors *XVII*

Part One Alkene Reductions

1	**Reduction of Functionalized Alkenes** *3*	
	Jean-Pierre Genet	
1.1	Introduction *3*	
1.2	Asymmetric Hydrogenation of Dehydroamino Acids *4*	
1.2.1	Rh-Catalyzed Reactions *4*	
1.2.1.1	Hydrogenation with Chiral Bisphosphine Ligands *4*	
1.2.1.2	Mechanism of the Asymmetric Hydrogenation with Rhodium Catalysts *8*	
1.2.1.3	Rh-Catalyzed Hydrogenation with Monophosphorus Ligands *9*	
1.2.2	Ruthenium- and Iridium-Catalyzed Reactions *10*	
1.2.2.1	Ruthenium *10*	
1.2.2.2	Mechanism of the Ruthenium-Catalyzed Asymmetric Hydrogenation *12*	
1.2.2.3	Iridium *13*	
1.3	Simple Enamides *13*	
1.4	Hydrogenation of β-(Acylamino) Acrylates *16*	
1.5	Hydrogenation of Unsaturated Carboxylic Acids and Esters *18*	
1.5.1	Mechanistic Aspects of the Ru-(BINAP)-Catalyzed Hydrogenation of Carboxylic Acids *21*	
1.6	Hydrogenation of Unsaturated Esters, Lactones, Amides and Ketones *22*	
1.7	Hydrogenation of Unsaturated Alcohols *25*	
1.7.1	Diastereoselective Hydrogenation with Rh and Ir Catalysts *25*	
1.7.2	Enantioselective Hydrogenation with Chiral Ru and Ir Catalysts *26*	
1.8	Synthesis of Pharmaceutical Intermediates *28*	
1.9	Conclusion *32*

Modern Reduction Methods. Edited by Pher G. Andersson and Ian J. Munslow.
Copyright © 2008 WILEY-VCH Verlag GmbH & Co. KGaA, Weinheim
ISBN: 978-3-527-31862-9

2 Hydrogenation of Unfunctionalized Alkenes 39
Jarle S. Diesen and Pher G. Andersson

2.1 Introduction 39
2.2 Iridium Catalysis 39
2.2.1 Catalysts 41
2.2.1.1 Ligands 41
2.2.1.2 Anion 41
2.2.2 Substrates 44
2.2.3 Mechanism 52
2.3 Ruthenium and Rhodium Catalysis 53
2.4 Chiral Metallocene Catalysts 57
2.4.1 Titanium and Zirconium Systems 57
2.4.2 Lanthanide Systems 60
2.5 Conclusion 62

3 The Development and Application of Rhodium-Catalyzed Hydroboration of Alkenes 65
Anthony G. Coyne and Patrick J. Guiry

3.1 Introduction 65
3.2 Mechanism 66
3.3 Selectivity of Metal-Catalyzed Hydroboration 67
3.3.1 Regioselectivity 67
3.3.2 Stereoselectivity 69
3.3.2.1 Chiral P,P Ligands 69
3.3.2.2 Chiral P,N Ligands 74
3.4 Recent Applications in Synthesis 82
3.5 Conclusion 84

4 Alkene Reduction: Hydrosilylation 87
Penelope A. Mayes and Patrick Perlmutter

4.1 Introduction 87
4.2 Isolated Alkenes 87
4.2.1 Palladium 87
4.2.1.1 Aromatic Alkenes 87
4.2.1.2 Nonaromatic Alkenes 93
4.2.2 Metals Other Than Palladium 94
4.3 Conjugated Alkenes 95
4.3.1 Acyclic 1,3-Dienes 95
4.3.2 Cyclic 1,3-Dienes 98
4.3.3 Enynes 98
4.4 α,β-Unsaturated Systems 99
4.4.1 Copper 99
4.4.2 Tandem Processes 102
4.5 Conclusions 103

Part Two Carbonyl Reactions

5 Carbonyl Hydrogenation *109*
Christian Hedberg
5.1 Introduction *109*
5.2 Asymmetric Hydrogenation of Activated Ketones and β-Keto Esters *109*
5.2.1 α-Keto Ester Hydrogenation *112*
5.2.2 1,3-Diketones *112*
5.2.3 Hydrogenation of β-Keto Ester Analogues *114*
5.2.4 Mechanism *115*
5.2.5 Catalyst Preparation *117*
5.2.6 Dynamic Kinetic Resolution (DKR) of β-Keto Esters *118*
5.3 Ketone Hydrogenation *120*
5.3.1 Mechanism *121*
5.3.2 Aryl Alkyl Ketones *124*
5.3.3 Dialkyl Ketones *127*
5.3.4 Diaryl and Aryl Heteroaryl Ketones *127*
5.3.5 Phosphine-free Hydrogenation of Alkyl Aryl Ketones *128*
5.3.6 α,β-Unsaturated Ketones *132*

6 Reduction of Carbonyl Compounds by Hydrogen Transfer *135*
Serafino Gladiali and Rossana Taras
6.1 Introduction *135*
6.2 Historical Overview *135*
6.3 General Background *136*
6.4 Hydrogen Donors *137*
6.5 Catalysts *138*
6.6 Mechanisms *139*
6.7 Ligands *143*
6.8 Substrates *148*
6.8.1 Ketones *148*
6.8.2 Functionalized Carbonyl Compounds *150*
6.8.3 Aldehydes *151*
6.8.4 Conjugated Carbonyl Compounds *152*

7 Carbonyl Hydroboration *159*
Noriyoshi Arai and Takeshi Ohkuma
7.1 Introduction *159*
7.2 Recent Topics in Diastereoselective Reduction *159*
7.3 Enantioselective Reduction *163*
7.3.1 Reagents (Introduction) *163*
7.3.2 Simple Ketones *167*
7.3.2.1 Aromatic Ketones *167*

7.3.2.2	Aliphatic Ketones	168
7.3.3	α,β-Unsaturated Ketones	169
7.3.4	α-Hetero Substituted Ketones	171
7.3.5	Keto Esters	171
7.3.6	Diketones	173
7.4	Synthetic Applications	173
7.4.1	Reduction of Chiral Ketones with Chiral Reducing Agents	173
7.4.2	Application to Natural Product Synthesis	175

8 Diverse Modes of Silane Activation for the Hydrosilylation of Carbonyl Compounds *183*
Sebastian Rendler and Martin Oestreich

8.1	Introduction	183
8.2	Metal-Catalyzed Hydrosilylations	185
8.2.1	Silane Activation by Oxidative Addition	185
8.2.2	Silane Activation by σ-Bond Metathesis	188
8.2.3	Silane Activation by High-valent Oxo Complexes	193
8.3	Transition-metal-free Hydrosilylations	197
8.3.1	Brønsted Acid-promoted Hydrosilylations	197
8.3.2	Lewis Acid-catalyzed Hydrosilylations	198
8.3.3	Lewis Base-catalyzed Hydrosilylations	200
8.4	Closing Remarks	202

9 Enzyme-catalyzed Reduction of Carbonyl Compounds *209*
Kaoru Nakamura and Tomoko Matsuda

9.1	Introduction	209
9.1.1	Differences between Chemical and Biological Reductions	209
9.1.1.1	Selectivity	209
9.1.1.2	Safety of the Reaction	210
9.1.1.3	Natural Catalysts	210
9.1.1.4	Catalyst Preparation	210
9.1.1.5	Large-scale Synthesis and Space–Time Yield	210
9.1.2	Reaction Mechanism	211
9.2	Hydrogen Sources	211
9.2.1	Alcohol as a Hydrogen Source for Reduction	212
9.2.2	Sugars as Hydrogen Sources for Reduction	212
9.2.3	Formate as a Hydrogen Source for Reduction	213
9.2.4	Molecular Hydrogen as a Hydrogen Source for Reduction	213
9.2.5	Light Energy as a Hydrogen Source for Reduction	214
9.2.6	Electric Power as a Hydrogen Source for Reduction	214
9.3	Methodology for Stereochemical Control	215
9.3.1	Screening of Biocatalysts	215
9.3.2	Modification of Biocatalysts by Genetic Methods	216
9.3.2.1	Engineered Yeast	216
9.3.2.2	Overexpression	217

9.3.2.3   Modification of Biocatalysts: Directed Evolution *218*
9.3.3   Modification of Substrates *218*
9.3.4   Modification of Reaction Conditions *219*
9.3.4.1   Acetone Treatment of the Cell *219*
9.3.4.2   Selective Inhibitors *220*
9.4   Medium Engineering *221*
9.4.1   Organic Solvent *221*
9.4.1.1   Water-soluble Organic Solvent *221*
9.4.1.2   Aqueous–Organic Two-Phase Reaction *221*
9.4.2   Use of Hydrophobic Resin *222*
9.4.3   Supercritical Carbon Dioxide *223*
9.4.4   Ionic Liquid *224*
9.5   Synthetic Applications *225*
9.5.1   Reduction of Aldehydes *225*
9.5.2   Reduction of Ketones *225*
9.5.3   Dynamic Kinetic Resolution and Deracemization *227*
9.5.3.1   Dynamic Kinetic Resolution *227*
9.5.3.2   Deracemization through Oxidation and Reduction *230*
9.6   Conclusions *231*

Part Three Imino Reductions

10   Imine Hydrogenation *237*
           Carmen Claver and Elena Fernández
10.1   Recent Advances in the Asymmetric Hydrogenation of Imines *237*
10.1.1   Iridium Catalysts *238*
10.1.1.1   Iridium / P-P Ligands *238*
10.1.1.2   Iridium / Phosphine–Phosphite Ligands *241*
10.1.1.3   Iridium / Diphoshite, Diphosphinite and Phosphinite–Phosphite Ligands *241*
10.1.1.4   Iridium / P,N-Ligands *243*
10.1.1.5   Iridium / N-Ligands *245*
10.1.1.6   Other Iridium / Phosphorous Systems *246*
10.1.2   Rhodium and Palladium Catalysts *247*
10.1.3   Ruthenium Catalysts *248*
10.1.4   Titanium and Zirconium Catalysts *248*
10.1.5   Gold Catalysts *249*
10.2   Green Approaches *249*
10.2.1   Aqueous–Organic Two-Phase Solvent Systems *249*
10.2.2   Catalyst Immobilization on Insoluble Materials *252*
10.2.3   Carbon Dioxide / Ionic Liquid Media *255*
10.3   Mechanistic Insights *257*
10.3.1   Homolytic and Heterolytic H_2-Metal Activation *258*
10.3.2 H^-/H^+ Transfer to the Imine in the Inner or Outer Coordination Sphere *259*

10.3.3	Neutral or Ionic Mechanisms	*261*
10.3.4	Ligand-assisted Mechanisms	*264*
10.3.5	Enantiodifferentiation Steps	*264*

11 Imino Reductions by Transfer Hydrogenation *271*
Martin Wills

11.1	History and Background	*271*
11.2	Mechanisms of C=N Bond Reduction by Transfer Hydrogenation	*271*
11.3	Asymmetric Reduction of C—N Bonds: Catalysts, Mechanisms and Results	*273*
11.3.1	Organometallic Catalysts Based on Ru, Rh, and Ir	*273*
11.3.1.1	Mechanistic Discussion	*278*
11.3.2	Asymmetric Reductive Aminations Using Ammonium Formate to Give the Primary Amine Directly (the Leuchart–Wallach Reaction)	*279*
11.3.3	MPV Type Reductions	*280*
11.3.4	Carbene-based Catalysts	*281*
11.3.5	Organocatalytic Methods	*281*
11.4	Specific Synthetic Applications	*283*
11.4.1	Sultams	*283*
11.4.2	Tetrahydroisoquinolines and Tetrahydro-β-carbolines	*285*
11.5	Conclusion	*291*

12 Hydroboration and Diboration of Imines *297*
Stephen A. Westcott and R. Thomas Baker

12.1	Introduction	*297*
12.2	Uncatalyzed Reactions	*298*
12.2.1	Imines	*298*
12.2.2	Diimines	*303*
12.2.3	Tosylhydrazones	*304*
12.2.4	Nitriles	*305*
12.3	Catalyzed Reactions	*307*
12.3.1	Transition Metals	*310*
12.4	Conclusions	*312*

13 Hydrosilylation of imines *321*
Olivier Riant

13.1	Introduction	*321*
13.2	Rh, Ir, Ru Based Catalysts	*322*
13.3	Titanium-based Catalysts	*324*
13.4	Zinc-, Copper-, and Rhenium-based Catalysts	*328*
13.5	Lanthanide-based Catalysts	*330*
13.6	Tin-based Catalysts	*331*

13.7	Chiral Lewis Bases as Catalysts	*333*
13.8	Miscellaneous Methods	*334*
13.9	Conclusion	*335*

Part Four Miscellaneous Reductions

14   Alkene and Imino Reductions by Organocatalysis *341*
       Hans Adolfsson
14.1   Introduction *341*
14.2   Reducing Agents *342*
14.2.1   N-Heterocyclic Hydrogen Donors *342*
14.2.2   Silanes *343*
14.3   Alkene Reduction *343*
14.3.1   Alkene Reduction by Transfer Hydrogenation of α,β-Unsaturated Aldehydes and Ketones *344*
14.3.2   Alkene Reduction in Organocatalytic Tandem Processes *348*
14.4   Imine Reductions *351*
14.4.1   Enantioselective Reductions of Ketimines Using Trichlorosilane as Reducing Agent *351*
14.4.2   Enantioselective Reductions of Ketimines Using Hantzsch Esters as Reducing Agents *354*
14.4.3   Organocatalytic Reductive Amination of Aldehydes and Ketones *357*
14.5   Concluding Remarks *358*

15   Alkyne Reductions *363*
       Ian J. Munslow
15.1   Introduction *363*
15.2   Hydrogenation *363*
15.2.1   Semi-hydrogenation *364*
15.3   Hydroboration *366*
15.3.1   Catalysis *368*
15.3.1.1 Palladium *368*
15.3.1.2 Rhodium *368*
15.3.1.3 Zirconium *370*
15.3.1.4 Titanium *371*
15.3.2   Mechanism *371*
15.4   Hydrosilylation *373*
15.4.1   Terminal Alkynes *374*
15.4.2   Internal Alkynes *378*
15.4.3   Mechanism *381*
15.5   Conclusions *382*

16   **Metal-Catalyzed Reductive Aldol Coupling**   *387*
  Susan A. Garner and Michael J. Krische
16.1   Introduction – Reductive Generation of Enolates from Enones   *387*
16.2   The Reductive Aldol Reaction   *389*
16.2.1   Rhodium   *390*
16.2.2   Cobalt   *398*
16.2.3   Iridium   *401*
16.2.4   Ruthenium   *401*
16.2.5   Palladium   *403*
16.2.6   Copper   *403*
16.2.7   Nickel   *406*
16.2.8   Indium   *407*
16.3   Conclusion   *408*

17   **Dissolving Metals**   *419*
  Miguel Yus and Francisco Foubelo
17.1   Introduction   *419*
17.2   Reduction of Compounds with C=X Bonds   *420*
17.2.1   Reduction of Carbonyl Compounds   *420*
17.2.2   Reduction of Imines   *422*
17.3   Reduction of Carboxylic Acids and Their Derivatives   *423*
17.4   Reduction of functional groups bearing N, O and S   *424*
17.4.1   Reduction of Sulfoxides   *424*
17.4.2   Reduction of Nitro Compounds   *425*
17.4.3   Reduction of Compounds with N—X Bonds (X = N, O, S)   *425*
17.5   Reduction of C=C and C≡C Bonds   *426*
17.5.1   Reduction of C=C Bonds   *426*
17.5.2   Reduction of C≡C Bonds   *430*
17.6   Partial Reduction of Aromatic and Heteroaromatic Rings   *431*
17.6.1   The Birch Reduction of Aromatic Compounds   *431*
17.6.2   Partial Reduction of Heteroaromatic Rings   *433*
17.7   Reduction of Compounds with C—X Bonds   *434*
17.7.1   Reduction of α-Functionalized Carbonyl Compounds   *434*
17.7.2   Reduction of C—Hal to C—H Bonds   *435*
17.7.3   Reduction of C—O to C—H Bonds   *436*
17.7.4   Reduction of C—N to C—H Bonds   *438*
17.7.5   Reduction of C—S to C—H Bonds   *439*
17.7.6   Reduction of C—C to C—H Bonds   *440*

18   **Hydrometallation of Unsaturated Compounds**   *447*
  Usein M. Dzhemilev and Askhat G. Ibragimov
18.1   Introduction   *447*
18.2   Thermal Hydroalumination   *448*
18.2.1   Alkenes   *448*
18.2.2   Dienes (Unconjugated)   *452*

18.2.3	Dienes (Conjugated)	*454*
18.2.4	Alkynes	*456*
18.3	Catalytic Hydroalumination	*456*
18.3.1	Alkenes	*456*
18.3.2	Dienes	*467*
18.3.3	Alkynes	*467*
18.4	Catalytic Hydromagnesiation	*472*
18.4.1	Alkenes	*472*
18.4.2	Dienes	*475*
18.4.3	Alkynes	*479*
18.5	Summary	*482*

Index *491*

Preface

The reduction of organic compounds is perhaps the most common reaction type in organic synthesis today. The development in this field has been tremendous, progressing from the use of stoichiometric reagents to the modern reactions that use cheaper, "greener" catalytic methods. The importance of the field has also been highlighted by the awarding of the 2001 Nobel Prize in Chemistry to Knowles and Noyori for their work on asymmetric hydrogenation, a key reaction among reductions.

The rapid development of this large field renders the task of editing a book on the topic difficult – how does one condense into a single book a subject that could easily fill multiple volumes? Although there are many ways to organize the material, we thought it would be helpful to do so according to the functional group being reduced. Needless to say, we also had to place severe constraints on the number of chapters in the book, as well as the number of methods for reducing certain functionalities. The aim of the book has been to present the reader with comprehensive material that covers the field and focuses on recent advances in the area.

We would like to thank the authors of each chapter for their willingness to contribute their expertise. We are also grateful for all the help we have received from VCH, and especially from Manfred Köhl and Stefanie Volk.

In Part 1, three methods for the reduction of alkenes are discussed: the asymmetric hydrogenations of both functionalized and unfunctionalized olefins are reviewed in Chapters 1 and 2, the hydroboration of alkenes is discussed in Chapter 3, and Chapter 4 covers advances in the hydrosilylation of alkenes that were published after 1998.

Chapter 5 reviews the hydrogenation of carbonyl compounds and Chapter 6 discusses their reduction via transfer hydrogenation. Recent advances in the diastereo- and enantioselective hydroborations of carbonyl compounds are reviewed in Chapter 7. Current state-of-the-art hydrosilylations of carbonyls are summarized in Chapter 8, which covers transition-metal-catalyzed and transition metal-free methods, and even metal-free procedures. The fundamentals of and new methodology for improving the reactivity and selectivity of enzymatic carbonyl reductions are explained in Chapter 9.

Modern Reduction Methods. Edited by Pher G. Andersson and Ian J. Munslow.
Copyright © 2008 WILEY-VCH Verlag GmbH & Co. KGaA, Weinheim
ISBN: 978-3-527-31862-9

In Chapter 10 several new strategies for the hydrogenation of imines are described. This chapter deals first with iridium-based catalytic systems, then considers other metals (Rh, Ru, Ti, Zr and Au). The transfer hydrogenation of imines, focusing on more recent synthetic applications, is reviewed in Chapter 11. Chapter 12 highlights some of the latest developments in the use of boron hydrides and related diboron compounds for the reduction of imines. The hydrosilylations of imines that have been discovered and developed during the past 10–15 years are the focus of Chapter 13.

Chapter 14 discusses a field that has gained an enormous attention recently: the use of small organic compounds as catalysts for the reduction of alkenes and imine functionalities. Reviewed in Chapter 15 are three methods for the reduction of alkynes: hydrogenation, hydroboration, and hydrosilylation. Recent advances in metal-catalyzed reductive aldol couplings are reviewed in Chapter 16. Dissolving metals have been extensively used as reducing agents for more than a century, and Chapter 17 discusses recent interest in their use for the selective reduction of specific polar functional groups (such as hindered cyclic ketones) and the reductive cleavage of some activated bonds. Chapter 18 focuses on the transition-metal-catalyzed hydroalumination and hydromagnesiation of alkenes, dienes and alkynes.

We hope that this book will be of value to chemists involved in organic synthesis in both academic and industrial research, and that it will stimulate further development in this important field.

Uppsala, January 2008

Pher G. Andersson
Ian J. Munslow

List of Contributors

Hans Adolfsson
Stockholm University
Department of Organic Chemistry
The Arrhenius Laboratory
6th floor
106 91 Stockholm
Sweden

Pher G. Andersson
Uppsala University
Department of Biochemistry and
Organic Chemistry
Husargatan 3
751 23 Uppsala
Sweden

Noriyoshi Arai
Faculty and Graduate School of
Engineering
Hokkaido University
N13, W8, Kita-ku
Sapporo 060-8628
Japan

R. Thomas Baker
Los Alamos National Laboratory
Chemistry Division
MS J582
Los Alamos, NM 87545
USA

Carmen Claver
Universitat Rovira i Virgili
Facultat de Química
c/Marcel li Domingo s/n
43007 Tarragona
Spain

Anthony G. Coyne
University College Dublin
School of Chemistry and
Chemical Biology
Centre for Synthesis and
Chemical Biology
Belfield
Dublin
Ireland

Jarle S. Diesen
Uppsala University
Department of Biochemistry and
Organic Chemistry
Husargatan 3
751 23 Uppsala
Sweden

Usein M. Dzhemilev
Russian Academy of Sciences
Institute of Petrochemistry and
Catalysis
141 Prospekt Oktyabrya, Ufa
450075, Bashkortostan
Russia

Modern Reduction Methods. Edited by Pher G. Andersson and Ian J. Munslow.
Copyright © 2008 WILEY-VCH Verlag GmbH & Co. KGaA, Weinheim
ISBN: 978-3-527-31862-9

Elena Fernández
Universitat Rovira i Virgili
Facultat de Química
c/Marcel li Domingo s/n
43007 Tarragona
Spain

Francisco Foubelo
Universidad de Alicante
Instituto de Síntesis Orgánica
Apolo. 99
03080 Alicante
Spain

Susan A. Garner
University of Texas at Austin
Department of Chemistry and
Biochemistry
1 University Station, A5300
Austin, TX 78712-1167
USA

Jean-Pierre Genet
Ecole Nationale Supérieure
de Chimie de Paris
Laboratoire de Synthèse Sélective
Organique et Produits Naturels
UMR CNRS 7573
11 rue P.& M. Curie
75231 Paris Cedex 05
France

Serafino Gladiali
Università di Sassari
Dipartimento di Chimica
Via Vienna 2
07100 Sassari
Italy

Patrick J. Guiry
University College Dublin
School of Chemistry and
Chemical Biology
Centre for Synthesis and
Chemical Biology
Belfield
Dublin
Ireland

Christian Hedberg
Max Planck Institute of Molecular
Physiology
Department of Chemical Biology
Otto-Hahn-Strasse 11
44227 Dortmund
Germany

Askhat G. Ibragimov
Russian Academy of Sciences
Institute of Petrochemistry and
Catalysis
141 Prospekt Oktyabrya, Ufa
450075, Bashkortostan
Russia

Michael J. Krische
University of Texas at Austin
Department of Chemistry and
Biochemistry
1 University Station, A5300
Austin, TX 78712-1167
USA

Tomoko Matsuda
Tokyo Institute of Technology
Department of Bioengineering
4259 Nagatsuta
Midori-ku
Yokohama 226-8501
Japan

Penelope A. Mayes
Monash University
School of Chemistry
Box 23, Victoria 3800
Australia

Ian J. Munslow
Uppsala University
Department of Biochemistry and
Organic Chemistry
Husargatan 3
751 23 Uppsala
Sweden

Kaoru Nakamura
Kyoto University
Institute for Chemical Research
Uji
Kyoto 611-0011
Japan

Martin Oestreich
Westfälische Wilhelms-Universität
Münster
Organisch-Chemisches Institut
Corrensstrasse 40
48149 Münster
Germany

Takeshi Ohkuma
Faculty and Graduate School of
Engineering
Hokkaido University
N13, W8, Kita-ku
Sapporo 060-8628
Japan

Patrick Perlmutter
Monash University
School of Chemistry
Box 23
Victoria 3800
Australia

Sebastian Rendler
Westfälische Wilhelms-Universität
Münster
Organisch-Chemisches Institut
Corrensstrasse 40
48149 Münster
Germany

Olivier Riant
Unité de chimie organique et
médicinale
Département de chimie, Bt Lavoisier
Place Louis Pasteur 1
1348 Louvain la Neuve
Belgium

Rossana Taras
Università di Sassari
Dipartimento di Chimica
Via Vienna 2
07100 Sassari
Italy

Stephen A. Westcott
Mount Allison University
Department of Chemistry
Sackville, NB E4L 1G8
Canada

Martin Wills
The University of Warwick
Department of Chemisty
Coventry, CV4 7AL
United Kingdom

Miguel Yus
Universidad de Alicante
Instituto de Síntesis Orgánica
Departamento de Química Orgánica
Apolo. 99
03080 Alicante
Spain

Part One
Alkene Reductions

1
Reduction of Functionalized Alkenes
Jean-Pierre Genet

1.1
Introduction

While a variety of methods is now available for the stereoselective reduction of olefins, catalytic hydrogenation continues to be the most useful technique for addition of hydrogen to various functional groups. Catalytic hydrogenations can be carried out under homogeneous or heterogeneous conditions, both employing a similar range of metals. Heterogeneous catalysts have had a strong impact on the concept of catalysis. They have provided powerful tools to the chemical industry and organic chemistry, allowing the chemo-, regio- and stereo-selective reduction of a wide range of functional groups and generally easy catalyst separation [1]. Homogeneous catalysts have found applications in a number of special selectivity problems or where enantioselectivity is the most important. Today, highly selective catalysts have revolutionized asymmetric synthesis. For two decades, homogeneous asymmetric hydrogenation has been dominated by rhodium(I)-based catalysts of prochiral enamides. Knowles and Horner initiated the development of homogeneous asymmetric hydrogenation in the late 1960s using modified Wilkinson's catalysts [2]. An important improvement was introduced when Kagan and Dang demonstrated that the biphosphine DIOP, having the chirality located within the carbon skeleton, was superior to a monophosphine in Rh-catalyzed asymmetric hydrogenation of dehydroamino acids [3]. Knowles made a significant discovery of a C_2-symmetric chelating P*-stereogenic biphosphine DIPAMP that was employed with rhodium(I) for the industrial production of L-DOPA [4].

In the 1990s, the next breakthrough was Noyori's demonstration that the well-designed chiral complex containing Ru(II)-BINAP catalyzes asymmetric hydrogenation of prochiral olefins and keto groups to produce enantiomerically enriched compounds with excellent enantioselectivity [5]. Not surprisingly, such versatile Rh- and Ru-based systems have had significant industrial impact and have been widely investigated in modern organometallic laboratories. For these beautiful achievements W. S. Knowles [6] and R. Noyori [7] were awarded the 2001 Nobel Prize in chemistry. Today, asymmetric hydrogenation is a core technology [8];

Modern Reduction Methods. Edited by Pher G. Andersson and Ian J. Munslow.
Copyright © 2008 WILEY-VCH Verlag GmbH & Co. KGaA, Weinheim
ISBN: 978-3-527-31862-9

thousands of very efficient chiral ligands with diverse structures have been developed [9]. Combinatorial approaches combined with high-throughput screening techniques have facilitated the discovery of new catalysts and increased the cost-effectiveness of a given process [10]. The catalysts used in asymmetric hydrogenation are not limited to those with Rh or Ru metals. Catalysts derived from other transition metals such as Ir, Pd, Ti, Pt are also effective. Asymmetric hydrogenation of functionalized olefins with Rh, Ru and Ir will be the predominant topics of this chapter.

1.2
Asymmetric Hydrogenation of Dehydroamino Acids

1.2.1
Rh-Catalyzed Reactions

Since the invention of the well-designed Rh-complexes containing chiral biphosphines for the asymmetric hydrogenation of dehydroamino acids and the synthesis of L-DOPA [4], this reaction has become the model reaction to evaluate the efficiency of new chiral ligands. Indeed, a wide range of chiral phosphorus ligands with great structural diversity have been found to be effective for the synthesis of (R)- or (S)-enantiomers (Scheme 1.1).

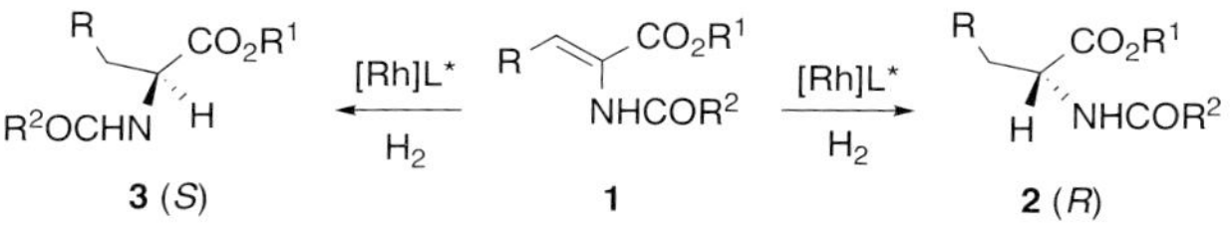

a: R = Ph, R^1 = H, R^2 = Ph
b: R = Ph, R^1 = H, R^2 = Me
c: R = p-FPh, R^1 = H, R^2 = Me
d: R = Ph, R^1 = Me, R^2 = Me
e: R = H, R^1 = Me, R^2 = Me
f: R = C_3H_7, R^1 = Me, R^2 = H
g: R = R^1 = H, R^2 = Me

Scheme 1.1 Rh-catalyzed reduction of dehydroamino acids.

1.2.1.1 Hydrogenation with Chiral Bisphosphine Ligands
For the Rh-catalyzed hydrogenation of 2-(acetamido)acrylic acid derivatives and (Z)-2-(acetamido) cinnamic acids and esters, cationic Rh catalysts and low hydrogen pressure are generally used. Examples are shown in Scheme 1.2. The reaction of (Z)-α-(acetamido)cinnamic acid in the presence of preformed [Rh-(S)-BINAP(MeOH)$_2$]ClO$_4$ produces (S)-N-acetylphenylalanine with nearly 100% ee [11]. The Rh-tetra-Me-BITIOP and Rh-CyP-PHOS catalysts developed by Sannicolo and Chan respectively were also found efficient in this reaction [12]. Pye and Rossen have developed a ligand based on a paracyclophane backbone, [2,2]-

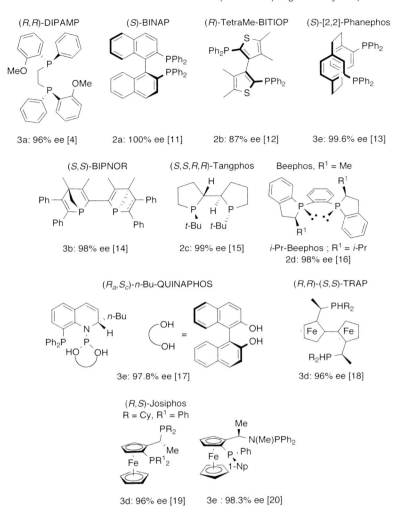

Scheme 1.2 Selected examples of Rh-asymmetric hydrogenation of dehydroamino acids (**1**).

Phanephos, which has shown high enantioselectivity, up to 99% ee, in rhodium-catalyzed hydrogenation [13].

Twenty years after the discovery of DIPAMP by Knowles [4], several new generations of P*-chiral bisphosphines have been developed. Mathey and coworkers have designed BIPNOR, a bisphosphane with two chiral non-racemizable bridgehead phosphorus centers. BIPNOR has shown good enantioselectivities up to 98% ee in the hydrogenation of α-(acetamido) cinnamic acids [14]. A rigid P*-chiral bispholane ligand Tangphos has been reported by Zhang. This readily accessible ligand is very efficient for asymmetric hydrogenation of (acylamino)acrylic acid [15]. A new class of bisphosphine bearing one or two benzophospholanes has been designed by Saito (1,2-bis-(2-isopropyl-2,3-dihydro-1H-phosphindo-1-1-yl) benzene (i-Pr-BeePHOS)). i-Pr-Beephos has been found to provide high enantioselectivity in the Rh-catalyzed asymmetric hydrogenation of α-dehydroamino acids [16]. Leitner has developed a series of mixed phosphoramidite-phosphine (Quinaphos) [17a], phosphinite-phosphine [17b, c] and Ito the TRAP ligands [18] that have obtained excellent efficiency in Rh-asymmetric hydrogenation of dehydroalanine derivatives. Since the discovery in 1994 of Josiphos, a ferrocene-based ligand devised by Togni and Spindler [19a], this class of bisphosphines including Taniaphos, Mandyphos families [19b], phosphinoferrocenyl phosphines [20] and (iminophosphoranyl) ferrocenes [21] have also shown excellent enantioselectivities in Rh-catalyzed reactions. As shown in Scheme 1.2, the biphosphines are highly efficient in the asymmetric hydrogenation of (Z)-dehydroamino acid derivatives with very high enantioselectivity. However, there are many reactions of interest where catalysts bearing these phosphines perform poorly in terms of enantioselectivity and efficiency. In particular, the hydrogenation of the (E)-isomeric substrates gives poor enantioselectivities and proceeds at a much lower rate. Interestingly, a new class of C_2-symmetric bisphospholane ligands has been prepared by Burk et al. and used in rhodium-catalyzed asymmetric hydrogenation. The Rh-Duphos catalyst provides high enantioselectivities for both (E)- and (Z)-dehydroamino acid derivatives [22] as shown in Scheme 1.3. In these hydrogenations, no separation of (E)- and (Z)-isomeric substrates is necessary. The hydrogenation of α,β-dienamides with Rh-Duphos proceeds chemoselectively, only one alkene function being reduced to give chiral γ,γ-unsaturated amino acids with both high regioselectivity (>98%) and ee (99%).

A short and efficient synthesis of optically pure (R)- and (S)-3-(hetero) alanines has been developed from isomerically pure (Z)-α-amino-α,β-dehydro-t-butyl esters using Rh-MeDuphos (Equation 1.1) [23b].

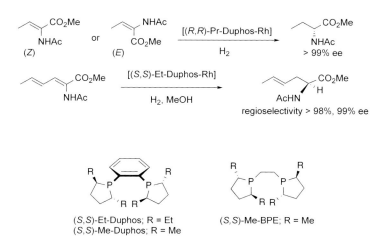

Scheme 1.3 Asymmetric hydrogenation of (1) with Rh-Duphos catalysts.

1.2 Asymmetric Hydrogenation of Dehydroamino Acids

$$\begin{array}{c} \text{Ar} \diagdown \text{C} = \text{C} \diagup \text{CO}_2\text{-}t\text{-Bu} \\ \text{NHCO}_2\text{CH}_2\text{Ph} \end{array} \xrightarrow[\text{(ii) CF}_3\text{CO}_2\text{H}]{\text{(i) [Rh] (}S,S\text{)-Me-Duphos}\atop 60\text{ bar H}_2,\ 40\ ^\circ\text{C, MeOH}} \begin{array}{c} \text{Ar} \diagdown \diagup \text{CO}_2\text{H} \\ \text{H} \\ \text{NHCO}_2\text{CH}_2\text{Ph} \end{array}$$

ee >99%   (1.1)

Ar = 3-furyl, 3-thienyl, N-methyl-2-pyrrolyl

Hydrogenation of β,β-disubstituted-α-dehydroamino acids remains a relatively difficult problem. Duphos ligands and analogues provide excellent enantioselectivity up to 99% ee for a wide range of substrates [23a]. The modular nature of these ligands, which allows simple adjustment of their steric and electronic properties, can be achieved through the ability to modify both the phospholane core and the R-substituents. Since their useful application in Rh-asymmetric hydrogenation of olefins [22c], many structural modifications of the phospholane core have been reported. The conformationally rigid and bulky bisphosphane Penphos has been designed by Zang [24]. A series of modified Duphos and DPE ligands containing hydroxyl ether and ketal groups at C3 and C4 of the phospholane have been reported [25]. The ligands (**4**) with four hydroxyl groups enabled hydrogenation to be carried out in aqueous solution while maintaining the high efficacy of Duphos and DPE ligands [25a]. The fully functionalized enantiopure bisphospholane Rophos is highly effective for the hydrogenation of unsaturated phosphonate (Equation 1.2).

$$\begin{array}{c} \text{Ph} \diagdown \text{C}=\text{C} \diagup \text{P(O)(OMe)}_2 \\ \text{NHCOMe} \end{array} \xrightarrow[1\text{ bar H}_2]{\text{[Rh]Rophos}} \begin{array}{c} \text{R}\diagdown \diagup \text{P(O)(OMe)}_2 \\ \text{H} \quad \text{NHCOMe} \end{array} \qquad \text{Rophos}$$

99.5% ee   (1.2)

The synthesis of a new class of chiral bisphosphetane ligands related structurally to the Duphos and DPE ligands (Scheme 1.4) such as 1,2-bis(phosphetano)

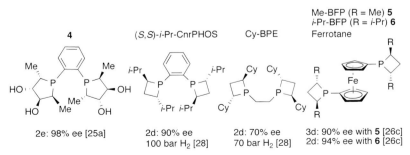

Scheme 1.4 Rh-asymmetric hydrogenation of (**1**) with (**4**) and bisphosphetane ligands.

benzenes CnrPHOS, 1,2-bis(phosphetano) ethanes BPE and 1,1-bis(phosphetano) ferrocene has been reported by Marinetti and Genet [26]. Later in 2000, Burk *et al.* reported a similar synthesis of these interesting 1,1-bis(phosphetano)ferrocene ligands named FerroTANE [27]. Interestingly, (S,S)-*i*-Pr-CnrPHOS provides moderate ee of (R)-methyl-N-acetylphenylalanine at 5 bar (500 kPa) of hydrogen (74%). However, at 100 bar of H_2 higher optical yields are observed (up to 90% ee) [28]. This nonconventional stereochemical issue can be related to the electron-rich nature of the phosphetane ligands.

Imamoto and coworkers have developed a series of electron-rich P*-chiral bisphosphanes [29] such as BisP*, Miniphos and 1,1'-di-*t*-butyl-2,2'-dibenzophosphenetyl. The Miniphos ligand leads to highly strained C_2-symmetric chelates when bound to a metal center. These ligands having both a conformational rigidity and an ideal chiral environment have shown significant enantioselectivities up to 99% ee in the hydrogenation of α-dehydroamino acids as shown in Scheme 1.5.

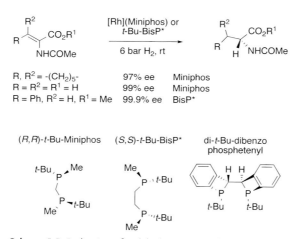

Scheme 1.5 Reduction of α-dehydroamino acids with electron-rich P*-chiral bisphosphanes.

1.2.1.2 Mechanism of the Asymmetric Hydrogenation with Rhodium Catalysts

The practical importance of asymmetric hydrogenation has stimulated a great interest in the mechanistic aspects of this reaction. Over the last 30 years, the mechanism of the rhodium-catalyzed asymmetric hydrogenation has been actively investigated. Success in this field is evident from theoretical and experimental studies [8d, 30]. The acylamino substituent plays a crucial role in the enantioselection. The amide carbonyl group provides an additional binding site for the catalyst, placing the substrate precisely within the coordination sphere of rhodium (Scheme 1.6) giving rise to two diastereomeric catalyst–substrate complexes competing for H_2 addition. Previously, it has been well accepted that the "unsaturated-alkene" mechanism (Halpern-Brown), pathway (a), Scheme 1.6, via **(7)** was

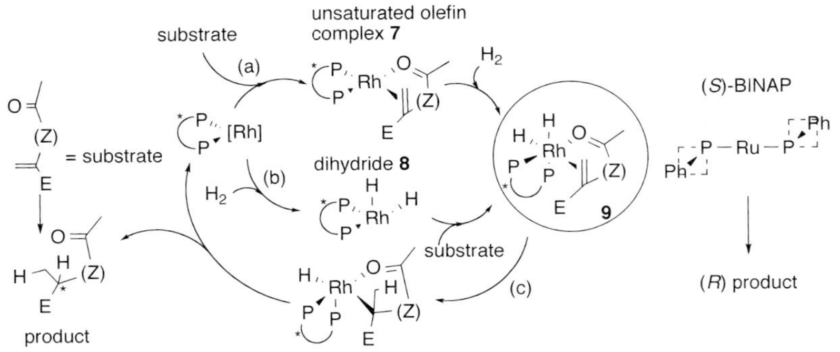

Scheme 1.6 General of the aspects mechanism of the Rh-catalyzed hydrogenation of (1).

operating with a wide range of phosphines, the Rh-(S)-BINAP giving (R) configured α-amino acids.

Gridnev and Imamoto, through experimental and computational studies, have established that the Rh-catalyzed asymmetric hydrogenation with electron-rich P*-stereogenic ligands such as Miniphos proceeds with a different mechanism. A "dihydride" mechanism (pathway (b), Scheme 1.6) via (8) is proposed [31]. However, it is suggested that the differences in these mechanisms are not significant in the stereoselection since they join at a single pathway (c), forming the common intermediate (9) before stereoselection occurs. They also suggested a new approach for the prediction of the sense of enantioselectivity [31].

1.2.1.3 Rh-Catalyzed Hydrogenation with Monophosphorus Ligands

The high degree of enantioselectivity resulting from chiral biphosphanes in the Rh-catalyzed asymmetric hydrogenation of α-dehydroamino acids can be explained as a result of decreased rotational freedom in the postulated metallacycle of the catalytic pathway [3]. However, isolated cases have been reported in which this long-standing theory is not as general as usually assumed [32]. More recently, it has been recognized that chiral phosphite (10) [33], phosphoramidites such as Monophos [34], monodentate phospholane (11) [35] and phosphinane (12) [36], are excellent ligands in this hydrogenation reaction as shown in Scheme 1.7. Zhou introduced a novel monodentate phosphorus ligand containing 1,1'-spirobiindane backbone which was found to be particularly effective in the hydrogenation of methyl-2-acetamidocinnamate at 1 bar (10^5 Pa) pressure of H_2, with up to 97.8% ee at 0 °C [37]. Interestingly, enantioselectivities up to 99% were achieved in the rhodium-catalyzed asymmetric hydrogenation of the N-formyl dehydroamino ester using Monophos ligand where a mixture of E and Z isomers was reduced with excellent ee values using 2 mol% of catalyst [34c]. Ding has recently reported an efficient system of heterogenization of Ferringa's catalyst by a self-supported strategy that can be recycled several times for the enantioselective hydrogenation of dehydroalanine derivatives [38].

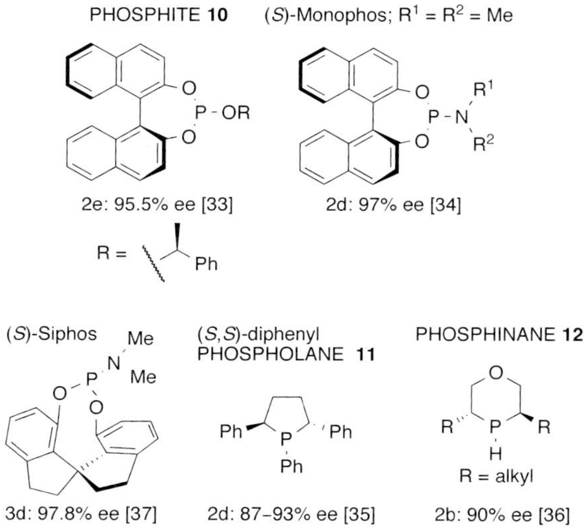

PHOSPHITE 10   (S)-Monophos; R¹ = R² = Me

2e: 95.5% ee [33]   2d: 97% ee [34]

R = CH(Me)Ph

(S)-Siphos   (S,S)-diphenyl PHOSPHOLANE 11   PHOSPHINANE 12

3d: 97.8% ee [37]   2d: 87–93% ee [35]   2b: 90% ee [36]
R = alkyl

Scheme 1.7 Hydrogenation of dehydroamino acids with chiral Rh-monophosphorus ligand catalysts.

1.2.2
Ruthenium- and Iridium-Catalyzed Reactions

1.2.2.1 Ruthenium

The development of chiral ruthenium-BINAP complexes has considerably enhanced the scope of enantioselective hydrogenation of olefins and keto groups [5, 7]. The axially 1,1′-binaphthyl moiety generally displays very high chiral recognition properties. The optically active bis-(triarylphosphines), BIPHEMP and MeO-BIPHEMP, containing the axially chiral biphenyl core have been developed by Roche [39]. In this context, several structural variations of the BINAP and MeO-BIPHEMP have been designed by many research groups and applied extensively to the reduction of olefins and keto groups.

Unlike the Rh-based hydrogenation of α-(arylamino) acrylates, the corresponding Ru-chemistry has not been studied extensively. The first two examples were described 20 years ago by Ikariya-Saburi using Ru-(S)-BINAP and by James and coworkers with Ru-(S,S)-Chiraphos. These complexes catalyze the hydrogenation of (Z)-α-(acylamino) cinnamates giving (S)-phenylalanine derivatives with 92% [40] and 97% ee respectively [41]. More recently several chiral Ru-(bisphosphine) catalysts have been used in this reaction as shown in Table 1.1. The [Ru-(R,R)-DIPAMP(Br₂)] and [Ru-(R,R)-DIPAMP(2-methylallyl)₂] [42] complexes also catalyze the asymmetric hydrogenation of N-acetyldehydroalanine giving (R)-N-acetylalanine with 35–38% ee The non-C_2 symmetric biaryl (bisphosphino)-MeO-NAPhe-PHOS and TriMe-NAPhePHOS (R = R¹ = H) ligands designed by Genet and Marinetti have been used in Ru-catalyzed asymmetric hydrogenation of dehydroamino acids giving 70% ee [43]. These results are comparable to those obtained with MeO-BIPHEP (68%).

1.2 Asymmetric Hydrogenation of Dehydroamino Acids | 11

Table 1.1 Ruthenium-catalyzed hydrogenation of (Z)-α-(acylamino) cinnamates.

Substrate	Catalyst	H_2 (bar)	ee%	Product	Reference
1b	Ru(S)-BINAP	2	86	2b(S)	[40]
1e	Ru(S)-BINAP	2	87	2e(S)	[40]
1b	Ru(S,S)-Chiraphos	1	97	2b(S)	[41]
1g	Ru(R,R)-DIPAMP	12	35	2g(R)	[42b]
1e	Ru(S)-BIPHEMP	2	80	2e(S)	[36]
1d	Ru(R)-MeO-NaPhePHOS	10	70	3d(R)	[43b]
1d	Ru(R)-MeO-BIPHEP	10	69	2d(R)	[46b]
1b	$RuCl_2(R)$-13-$(DMF)_2$	4	87	3b(R)	[44]
1a	Ru(S)-Synphos	5	86	2a(S)	[45c]
1d	Ru(R)-P-Phos	1	88	3(R)	[46]

The bis-steroidal atropisomeric phosphine (13) has been developed from equilenine and gives up to 87% ee in the hydrogenation of (Z)-α-acetamidocinnamic acid [44]. A new atropisomeric ligand containing a bis-benzodioxane core Synphos was reported independently by Genet and Vidal [45] and by Chan [46]. A Ru-(S)-Synphos complex catalyzes asymmetric hydrogenation of (Z)-(acylamino)cinnamate giving (S)-protected phenylalanine in 86% ee [45c]. Chan and coworkers have developed a P-PHOS ligand containing a dipyridyl unit that is highly efficient for the asymmetric hydrogenation of (Z)-2-acetamidocinnamate obtaining 88% ee at 1 bar (10^5 Pa) of H_2 [47]. Lemaire and coworkers have reported a Ru-(5,5′)-

perfluoroalkylated BINAP which is a useful catalyst in the asymmetric hydrogenation of methyl-2-acetamidoacrylate in supercritical carbon dioxide with full conversion and high enantioselectivity (74% ee) [48].

The Ru-complexes of the atropisomeric family exhibit slight differences in enantioselectivity toward the same substrate. This may be attributed to the electronic and stereoelectronic properties of these ligands. The optical purities are significantly reduced in comparison with those of the Rh-catalyzed reaction. More interestingly, the Rh and Ru hydrogenation catalysts consisting of the same chiral biphosphines exhibit an opposite sense of asymmetric hydrogenation of dehydroamino acids.

1.2.2.2 Mechanism of the Ruthenium-Catalyzed Asymmetric Hydrogenation

In contrast to the Rh-catalyzed reaction involving a Rh-dihydride intermediate, the Ru-hydrogenation proceeds via a Ru monohydride species. Bergens and coworkers using (Z)-methyl-α-acetamidocinnamate (MAC) and [Ru-(R)-(BINAP)(H)(MeCN)(solv)$_2$]BF$_4$ (14) (solv = MeOH or THF) [49] have found that the stoichiometric reaction between MAC and the Ru-catalyst resulted in rapid formation of a predominant species in solution. As shown in Scheme 1.8, the reaction of MAC with H$_2$ gas resulted in formation of MACH$_2$ and [Ru(R)-(BINAP)(H)(η^6-MACH$_2$)]$^+$ (17). MACH$_2$ was then liberated in refluxing acetonitrile with an ee of up to 94%. They also have established that at −40 °C the Si-olefin catalyst adduct (15) is formed which is the immediate precursor of (16) [49c].

Bergen's investigations paved the way for a deeper understanding of the mechanism of Ru-catalyzed asymmetric hydrogenation of enamides. Noyori and cowork-

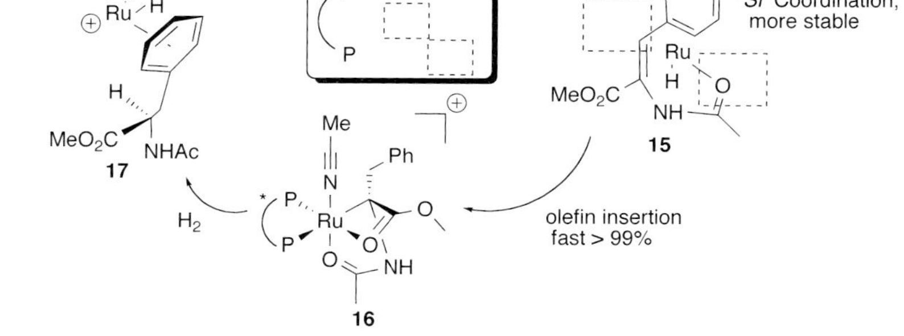

Scheme 1.8 Mechanism of [Ru-(R)-(BINAP)(H)(MeCN)]BF$_4$-catalyzed hydrogenation of MAC. Anion is $^-$BF$_4$.

ers have investigated the hydrogenation of α-(acylamino) acrylic esters using [Ru-(S)-BINAP(CH$_3$CO$_2$)$_2$] by means of kinetic studies, deuterium labeling experiments, isotope effect NMR measurements and X-ray analysis of Ru complexes. They have established that the Ru-BINAP diacetate-catalyzed reaction occurs via a monohydride-unsaturated pathway for both the major and minor enantiomers, and they form a short-lived enamide complex that delivers the hydride at C3 giving a five-membered metallacycle. Actually, the enantioselection is controlled by the stability of diastereoisomeric RuH/olefin complexes [50] as shown in Scheme 1.9 and is opposite to that of the hydrogenation using Rh(I) complexes with the same BINAP ligand.

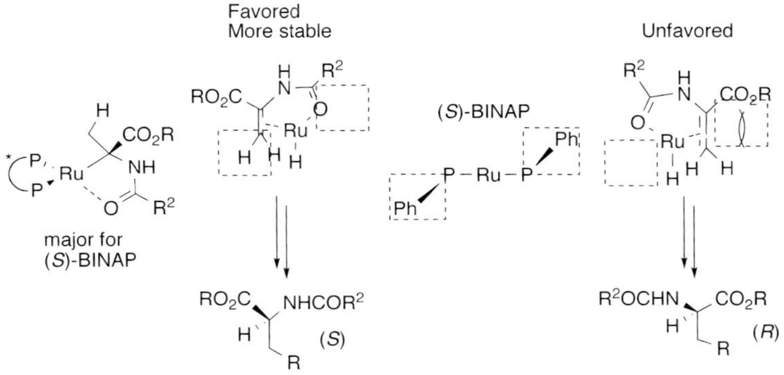

Scheme 1.9 Enantioselection control via diastereoisomeric RuH/olefin complexes.

1.2.2.3 Iridium

The Ir-catalyzed hydrogenation of unsaturated enamides to amino acid derivatives has been much less studied. One successful example has been reported by Knochel [51] using a new type of P,N ligand derived from camphor. The hydrogenation of methyl (Z)-α-(acetamido)cinnamate in the presence of [Ir(L*)(COD)Cl] produces (S)-N-acetyl phenylalanine with 96.5% ee (Equation 1.3).

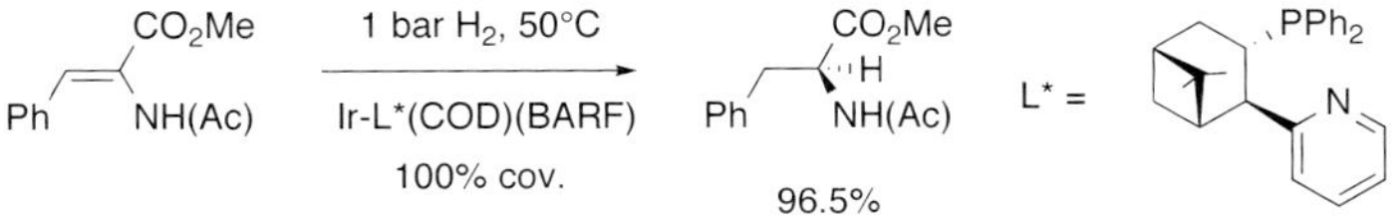

(1.3)

1.3
Simple Enamides

Rh-catalyzed hydrogenation of enamides has attracted much attention recently opening up an enantioselective route to chiral protected amines. Noyori has reported a general and straightforward method for synthesizing enantiomerically pure tetrahydroisoquinoline alkaloids [5a]. Ru-(S)-BINAP and Ru-(S)-BIPHEMP complexes give almost perfect enantioselectivities in enamide hydrogenation for a wide array of tetrahydroquinolines (Equation 1.4) [52]. The present reaction provides access to morphinic and synthetic morphinans and benzomorphans analogues.

$$\text{MeO-ArCH=CR-NCHO} \xrightarrow[\text{MeOH}]{\text{1-4 bar H}_2, \text{ Ru-}(S)\text{-BINAP}} \text{MeO-Ar-CH}_2\text{R-NCHO} \quad ee > 99\% \tag{1.4}$$

R = H, 4-MeO-C$_6$H$_4$, 3,4-(MeO)$_2$-C$_6$H$_3$

An efficient asymmetric synthesis of *N*-Boc-(*R*)-3-amino-2,3,4,5-tetrahydro-1*H*-[1]benzazepin-2-one, an important intermediate for the preparation of an angiotensin-converting enzyme inhibitor, based on asymmetric hydrogenation of an acyclic enamide, has been reported by Merck (Equation 1.5) [53].

$$\text{benzazepinone-NHBoc (enamide)} \xrightarrow[\text{(S)-BINAP RuCl}_2 \text{ 80\% yield}]{\text{3.4 bar H}_2} \text{benzazepinone-NHBoc} \quad 82\% \text{ ee} \tag{1.5}$$

The most general method for the hydrogenation of simple enamides has been reported by Burk and coworkers using Rh-Duphos or DPE ligand. *E/Z* mixtures of β-substituted enamides can be hydrogenated without purification since both are reduced to acyl amines with high enantioselectivities ranging from 74.8% to 98.5% ee (Equation 1.6) [54].

$$\text{Ar-CR=CH-NHAc} + \text{Ar-CH=CR-NHAc} \xrightarrow[\text{4 bar H}_2]{\text{Rh-}(S,S)\text{-Me-Duphos}} \text{Ar-CHR-CH}_2\text{-NHAc} \quad 74.8\text{-}98\% \text{ ee} \tag{1.6}$$

R = H, Me, Et, CH$_2$Ph, Cy

A series of 2-substituted *N*-acetylindoles can be reduced to optically active indolines by using the Rh-Ph-TRAP catalyst with enantioselectivities up to 94% (Equation 1.7) [55].

$$\text{2-R-N-Ac-indole} \xrightarrow[\text{i-PrOH, 60°C}]{\substack{\text{5 bar H}_2, \\ \text{Rh(acac)(col)} \\ (S,S)\text{-}(R,R)\text{-PhTRAP}}} \text{2-R-N-Ac-indoline} \quad 94\% \text{ ee} \tag{1.7}$$

R = Bu, *i*-Bu, CO$_2$Me

Table 1.2 Rhodium-catalyzed reduction of enamides.

Substrate	Ligand	H_2 (bar)	ee	Configuration	Reference
R = H, Ar = Ph	Manniphos R = Me	10	99.5	(R)	[56]
R = H, Ar = p-CF_3Ph	Tangphos	1.4	99	(R)	[15]
R = H, Ar = p-NO2Ph	DiSquare P*	2	99	(R)	[57]
R = H, Ar = p-ClPh	Morphos	55	99	(R)	[58]
R = H, Ar = Ph	Aaphos	10	87	(R)	[59]
R = H, Ar = Ph	(17)	10	93	(R)	[60]
R = H, Ar = m-CO_2MePh	t-Bu-BisP*	3	97	(S)	[29d]
R = H, Ar = Ph	(18)	20.6	96.5	(S)	[61]

Monophosphites derived from D-mannitol (Manniphos) have been reported very recently and are highly efficient for Rh-catalyzed asymmetric hydrogenation of simple acyclic and endocyclic enamides [56]. Very high ee values have been obtained with Rh-catalysts using the P*-chiral ligand Tangphos [15], t-Bu-BisP* [29d], DiSquare P* [57], phosphoramidite (Morphos) mixture of phosphates, phosphonites and phosphines [58], chiral planar cyrhetrenes [59] and ferrocenylethylamine-monophosphoramidites (17) [60]. The air stable fluorinated ferrocenylphosphine aminophosphine ligand (18) has also been applied and allows efficient hydrogenation of enamides. These reactions afford N-acylated aryl amines with high enantioselectivities up to 96.5% ee (Table 1.2) [61].

The o-Ph-hexane-BIPHEP and o-phenyl-MeO-BIPHEP, Rh-catalysts have been applied successfully to hydrogenation of cyclic enamides (Equation 1.8) [62].

(1.8)

The hydrogenation of enamides bearing an endocyclic tetra substituted carbon–carbon double has also been reported at 100 bar (10 MPa) of H_2 using a Rh-Duphos catalyst generated in situ from [Ru(COD)(methylallyl)$_2$] and HBF_4 with high diastereoselectivity and acceptable ee up to 72% [63].

1.4
Hydrogenation of β-(Acylamino) Acrylates

The synthesis of optically active β-amino acids and their derivatives has gained much attention in view of their pharmacological activities and usefulness as chemical building blocks. In recent years, good to excellent enantioselectivities have been obtained by employing chiral monodentate and bidentate phosphorus-containing ligands for Rh- or Ru-catalyzed asymmetric hydrogenation of β-(acylamino) acrylates. Some selected examples are shown in Table 1.3.

For example, BINAP [64], t-BuBisP* [65], phenyl P-Phos [66] and mixtures of two different chiral monodentate ligands, phosphites (19) and phosphonites (20) [67], were all found to be efficient in the hydrogenation of (E)-β-alkyl-β-(acylamino) acrylates. However, the hydrogenation of (Z)-isomers, especially (Z)-β-aryl-β-(acylamino) acrylates, were less selective, though significant improvements have been made using Rh-catalyzed hydrogenation with Duphos [68], BICP [69] or Ferrotane [70]. Recently, a Rh(I) catalyst containing the hybrid ferrocenyl phosphine phosphoramidite ligand (17) has been used in the hydrogenation of both (E)- and (Z)-isomers under identical conditions (10 bar (1 MPa) of H_2) affording β-amino acid derivatives with excellent enantioselectivities of 92–99% ee [71]. Hoge and coworkers have reported efficient Rh-trichickenPHOS and Rh-P* chirogenic bisphosphine catalysts for reductions both (E)- and (Z)-β-acetamido dehydroamino acids [72].

A significant drawback to this reaction is the requirement of an acyl group on nitrogen. This group is considered essential to satisfy the chelation requirements between the substrate and the metal (21). The first general and highly enantioselective method for hydrogenation of unprotected β-enamine esters

1.4 Hydrogenation of β-(Acylamino) Acrylates

Table 1.3 Ruthenium- and rhodium-catalyzed reduction of β-(acylamino) acrylates.

	Substrate	Catalyst	H_2 (bar)	ee %	Configuration	Reference
(E)	R = R^1 = Me	Ru (R)-BINAP	1	96	(S)	[64]
(Z)	R = R^1 = Me	Ru (R)-BINAP	4	5		[64]
(E)	R = R^1 = Me	Rh (S,S)-t-BuBISP*	3	98.7	(R)	[65]
(E)	R = Me, R^1 = i-Pr	Rh (R)-P-PHOS	4	97.4	(R)	[66]
(Z)	R = Et, R^1 = i-Pr	Rh (R,R)-BICP	10	90.7	(R)	[69]
(Z)	R = R^1 = Me	Rh (S,S)-Et-Duphos	1	86.7	(S)	[68]
(E)	R = Me, R^1 = Ph	Rh (R,R)-Et-Ferrotane	1	>99	(S)	[70]
(Z)	R = Ph, R^1 = Et	Rh-(17)		98	(R)	[71]
(E) or (Z)	R = Et, R^1 = i-Pr	Rh (R)-Trichickenphos	1.5	99	(R)	[72]

has been reported using a Rh-ferrocenophosphine complex under relatively mild conditions, proceeding with high enantioselectivity up to 96.1% ee (Equation 1.9) [73].

$$(1.9)$$

Enantioselective hydrogenation of tetrasubstituted olefins of cyclic or acyclic β-(acylamino) acrylates is much more difficult to accomplish. The asymmetric hydrogenation of (E)-α,β-bis(N-acylamino) acrylates using Rh-(R,R)-(S,S)-PrTRAP provides optically active (2S,3R)-2,3-bis(N-acylamino)carboxylates with 79–82% ee (Equation 1.10) [74].

$$\underset{\text{NHAc}}{\overset{\text{CbzHN}}{\underset{R}{\bigvee}}}\text{CO}_2\text{Me} \xrightarrow[\text{(R,R),(S,S)-TRAP}]{\text{H}_2, \text{[Rh(COD)}_2\text{]BF}_4 \text{ (1\%)}} \underset{R \quad \text{NHAc}}{\overset{\text{CbzHN} \quad \text{CO}_2\text{Me}}{\bigvee}} \quad (1.10)$$

R = Me, Et, t-Bu

Recently Zhang and coworkers have been successful in the hydrogenation of tetrasubstituted olefins [75] using monomeric [Ru(COD)(2-methylallyl)$_2$] and HBF$_4$ in the presence of a chiral phosphorus ligand such as (S)-C$_3$-Tunaphos or biaryl ligands providing high cis selectivity and enantioselectivities up to 99% ee with cyclopentenyl and cyclohexenyl derivatives (Equation 1.11). However, lower ee values were obtained in the hydrogenation of cycloheptenyl and cyclooctenyl substrates (44–80% ee). This procedure for the generation of Ru-catalysts was previously reported for the industrial production of dehydrojasmonate [76]; see Section 1.6.

$$\quad (1.11)$$

n = 1 R = Et 99% ee
n = 2 R = Me 92% ee

(S)-C$_3$-Tunaphos

1.5
Hydrogenation of Unsaturated Carboxylic Acids and Esters

Noyori provided a breakthrough in this area with the discovery of Ru-BINAP dicarboxylate complexes [5a]. The ruthenium-BINAP dicarboxylate complexes afford high enantioselectivities in the hydrogenation of α,β-unsaturated acids. However, the catalytic efficiencies are highly sensitive to the substitution pattern and reaction conditions, particularly the hydrogen pressure [8a]. Interestingly, in the asymmetric hydrogenation of geranic acid, only the double bond closest to the carboxyl group is reduced (Equation 1.12) [5b].

$$\quad (1.12)$$

81% ee

Other Ru-atropisomeric complexes such as [Ru-BIPHEMP-(2-methylallyl)$_2$] or [Ru-(R)-MeO-BIPHEP(Br$_2$)] [42a, 77] [Ru-BINAP-(allyl)(acac-F$_6$)] [78] can also be used in this transformation. Asymmetric hydrogenation of α-aryl-substituted acrylic acids has been extensively studied due to the pharmaceutical importance of the resulting products. The anti-inflammatory drugs (S)-naproxen and (S)-ibuprofen can be efficiently obtained by high-pressure hydrogenation of 2-(6′-methoxynapthyl-2′-yl) acid using [Ru-(S)-BINAP(OCOCH$_3$)$_2$] [79a, b]. [Ru-(R)-(P-Phos)(acac)$_2$] was also found to be an economically attractive system for the synthesis of (R)-

naproxen with up to 96% ee obtained (Equation 1.13) [79c]. A Ru-sulfonated BINAP catalyst absorbed in a polar solvent phase onto a porous glass bead has been designed by Davis [79d, e], this system can be efficiency recycled and the efficiency rivals that of solution chemistry (95% ee). Roche has reported the efficient synthesis of a pharmaceutical intermediate toward Mibefradil, (S)-2-(4-fluorophenyl)-3-methylbutanoic acid (Equation 1.14) [80].

$$\text{naproxen precursor} \xrightarrow[\text{or }(R)\text{-P-Phos-Ru}]{\substack{H_2 \\ (R)\text{-BINAP-Ru} \\ \text{or }(R)\text{-sulfonated BINAP-Ru}}} \text{naproxen, 95% ee} \quad (1.13)$$

$$\text{mibefradil precursor} \xrightarrow[60 \text{ bar } H_2]{(R)\text{-MeO-BIPHEP-Ru}} \text{mibefradil precursor, 92% ee} \quad (1.4)$$

The hydrogenation of tiglic acid in supercritical CO_2 and ionic liquids catalyzed by the H_8-BINAP-Ru(II) complex proceeds cleanly with cis stereochemistry affording 2-methylbutanoic acid with 89–93% ee and over 99% yield [81]. 2-2-Dimethylidenesuccinic acid [82] is hydrogenated by Ru-(R)-BINAP complex at 100 bar (10 MPa) giving (2S,3S)-dimethyl succinic acid as the major product (96%) and a minor amount of the meso isomer (1.2%) (Equation 1.15).

$$\xrightarrow[(R)\text{-BINAP-Ru}]{H_2} \quad (1.15)$$

A trifluoromethyl-substituted unsaturated acid was hydrogenated by Chemi on a large scale (340 kg) with {Ru(p-cymene)[(−)tetraMe-Bitiop]I$_2$} to the corresponding fluorinated saturated acid with 92% ee (Equation 1.16) [12, 83a]. Recently Zhou designed a new spirofluorene-biphosphane ligand SFDPs with a high rigidity and a large dihedral angle, with a 3,4,5-trimethylphenyl group. This ruthenium complex is highly efficient for asymmetric hydrogenation of a wide range of methylhydrocinnamic acids (Equation 1.16) [83b].

$$R\text{-CH=C(CH}_3)\text{COOH} \xrightarrow[\text{Ru-TetraMe-BITIAMP}]{50 \text{ bar } H_2} F_3C\text{-CH(CH}_3)\text{COOH}$$

(R)-TetraMe-BITIOP   SFDPs   R = CF$_3$ 92% ee   [83a]
                                       R = Ph, p-OMe, 90–94% ee [83b]
                                       2-napht etc ...   (1.16)

Ar = 3,4,5,-Me$_3$-C$_6$H$_2$

Very recently Gladiali has developed an interesting system for the reduction of α,β-unsaturated acids and esters using the Ph-BINEPINE ligand [84] by hydrogen transfer using formic acid as the reducing agent in the presence of a Rh(I) complex. A facile preparation of a series of chiral α-aryloxy carboxylic acids via asymmetric hydrogenation of the corresponding unsaturated acids using [Ru-(S)-BINAP(Cl)] and NEt$_3$ in methanol has been reported (Equation 1.17) [85].

$$\text{Ar-O-C(=CHR}^1\text{)-CO}_2\text{H} \xrightarrow[\text{6.2 bar H}_2,\ 25°C]{\text{Ru (S)-BINAP}} \text{Ar-O-CH(CH}_2\text{R}^1\text{)-CO}_2\text{H} \quad R^1 = \text{Me, }n\text{-C}_5\text{H}_{11} \quad R = \text{H, 2-Me, 2-F, 4-NO}_2 \quad (1.17)$$

ee 32–96%

Very recently [4-(4-chlorobenzyl)-7-fluoro-5-(methylsulfonyl)-1,4-dihydrocyclopenta[b]indol-3-(2H)ylidene] acetic acid was reduced enantioselectively (up to 92% ee) with the [Ru-(S)-BINAP-(p-cymene)Cl$_2$] catalyst giving the prostaglandin D$_2$ (PGD2) receptor antagonist (Equation 1.18) [86].

$$\text{(substrate)} \xrightarrow[\text{50°C, MeOH, 15 h}]{\substack{(S)\text{-BINAP [Ru(}p\text{-cymene)Cl}_2]_2 \\ 0.3–1.2\ \text{mol\%, 2.4 bar H}_2}} \text{(product)} \quad (1.18)$$

Hydrogenations of itaconic acids and derivatives have been studied extensively (Table 1.4) [8a]. Reactions using a cationic-Rh-BICHEP complex proceed at 5 bar (500 kPa) of H$_2$ giving 2-methylsuccinic acid in 93% ee [87]. High reactivity is generally observed with bisphosphinamidite electron-rich phosphane ligands such as Et-Duphos [88] and Tangphos [89]. A biphosphite [90a], derived from dianhydro-D-mannite [90b] and biphosphonite, spirophosphoramidite [91] and t-BuBisP* [92] ligands are also suitable for this reaction. The Ru-catalyzed hydrogenation of itaconate derivatives has recently been reported using the two atropisomeric ligands (S)-Difluorophos [93] and (S)-Synphos having a narrow dihedral angle [45]. Under 4 bar (400 kPa) at 50 °C the hydrogenation of dimethyl itaconate afforded the (S) product in 92% ee (Synphos) and 85% ee (Difluorophos). These differences in selectivity are probably due to the steric and electronic properties of the biaryl backbone of these ligands [93b].

The hydrogenation of β-substituted itaconic acids and derivatives is more difficult. However, Burk and coworkers using a chiral 1,1′-biphosphetanylferrocene ligand (Et-Ferrotane) [27] have developed a very efficient Rh-catalyst for the hydrogenation of β-substituted monoamido itaconates derived from morpholine (Equation 1.19).

1.5 Hydrogenation of Unsaturated Carboxylic Acids and Esters

Table 1.4 Reduction of itaconic acids.

Substrate	Catalyst	H$_2$ (bar)	ee %	Configuration	Ref.
R^1 = R = H	Rh(R)-(bichep)(nbd)X	1	96	(R)	[87]
R^1 = i-Pr	[Rh(S,S)-Et-Duphos(cod)]BF$_4$	5.5	99	(R)	[88]
R^1 = R = H	Rh(Tangphos)	1.5	99	(S)	[89]
R^1 = H, R = Me	Rh(cod)$_2$BF$_4$-(22)		96	(R)	[90]
R^1 = H, R = Me	Rh t-BuBisP*	2	98.6		[92]
R^1 = H, R = Me	Ru(S)-Difluorphos/(S)-Synphos	4	85–92	(S)	[93]

R = Ph, p-F-C$_6$H$_4$, thienyl, n-Bu, t-Bu

$$\text{(1.19)}$$

99% ee

An interesting example is shown in Equation 1.20 with Rh-Me-Duphos in which a trisubstituted olefin is reduced to afford a chiral succinate [94]. The Ru-MeO-BIPHEP catalyst is also very efficient in this process, providing access to the large-scale production of this intermediate for the synthesis of candoxatril (see Section 1.8.)

$$\text{(1.20)}$$

99% ee

1.5.1
Mechanistic Aspects of the Ru-(BINAP)-Catalyzed Hydrogenation of Carboxylic Acids

Several groups, most notably those of Noyori [95], Halpern et al. [96], Brown et al. [97] and Chan [98], have reported on the mechanism of enantioselective

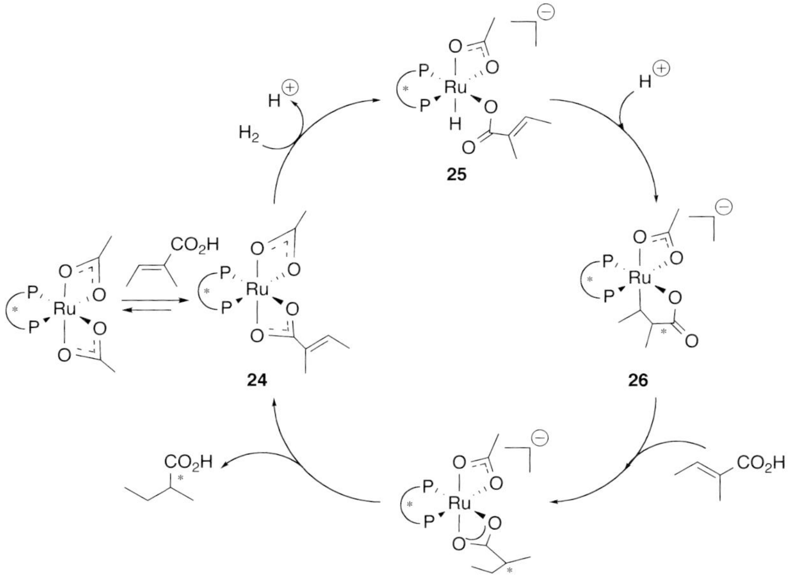

Scheme 1.10 Halpern's mechanism.

hydrogenation of α,β-unsaturated acids/esters in alcohols using ruthenium-bis(phosphine)-dicarboxylate species as catalyst. These studies have revealed (Scheme 1.10) that the mechanism follows four main steps: first, a rapid equilibration between the catalyst and tiglic acid by carboxylate ligand exchange **(24)**; second, heterolytic cleavage of dihydrogen generating a monohydride ruthenium intermediate **(25)**; third, olefin-hydride insertion giving the 5-membered heterometallacycle **(26)**; and fourth, solvolysis of the Ru–C bond to complete the cycle.

Very recently Bergens and coworkers [99] using [Ru-(R)-BINAP(H)(solv)$_{3-n}$(MeCN)$_n$]BF$_4$ (n = 0–3) have examined the hydrogenation of tiglic and angelic acids. The catalyst enters the catalytic cycle by reaction with the substrate to generate the carboxylate compound **(27)** and dihydrogen. Isotope labeling studies support Halpern's mechanism, heterolytic cleavage of H$_2$ resulting in formation of monohydride **(28)** and rapid H–D exchange between **(28)** and methanol-d_4 gives **(29)**. Olefin hydride or deuteride insertion leads to the five-membered metallacycle intermediate **(30)**; solvolysis of the ruthenium–carbon bond completes the catalytic cycle (Scheme 1.11).

1.6
Hydrogenation of Unsaturated Esters, Lactones, Amides and Ketones

There are several highly selective catalysts for the hydrogenation of α-(acylamino) acrylic acids, enamides, α,β-unsaturated carboxylic acids, and allylic and homoal-

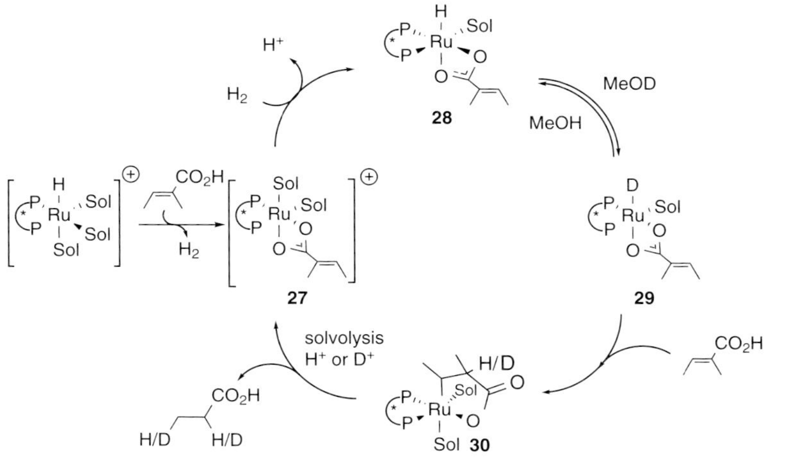

1.6 Hydrogenation of Unsaturated Esters, Lactones, Amides and Ketones | 23

Scheme 1.11 Proposed catalytic cycle for ruthenium bis(phosphine) catalysts.

lylic alcohols. In contrast, reports of efficient enantioselective hydrogenation of olefins with aprotic oxygen or nitrogen functionalities such as esters, ketones and lactones are limited.

However, Takaya and coworkers have established that several Ru-BINAP systems are efficient for the hydrogenation of alkylidene ketones or lactones [100]. Using [RuCl-(S)-BINAP-benzene]Cl, [Ru-(S)-BINAP(OAc)$_2$] or [Ru$_2$-(S)-BINAP(Cl)$_4$]$_2$(NEt$_3$), alkylidene ketones or lactones were reduced to their corresponding chiral products with high selectivity (Equations 1.21 and 1.22). Interestingly, (R)-4-methyl-2-oxetanon was prepared in 85% yield and with good enantioselectivity (92%) (Equation 1.23). In this reaction, chelation of the substrate to a Ru(II) species also seems to be important for high selectivity. Exocyclic C=C olefins are particularly good substrates; in contrast, endo compounds gave poor enantioselectivity.

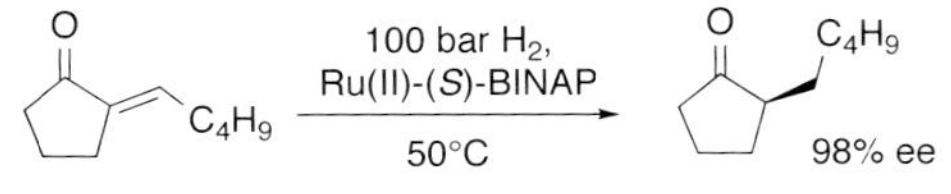

(1.21)

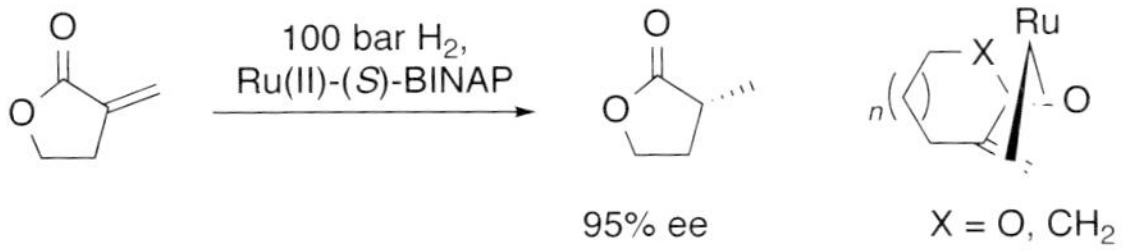

(1.22)

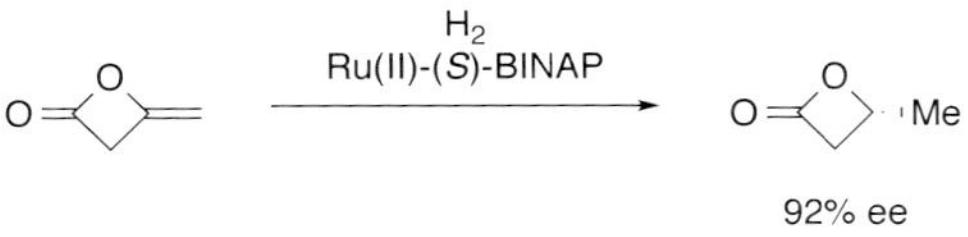

(1.23)

1 Reduction of Functionalized Alkenes

Bruneau and Dixneuf have described the enantioselective hydrogenation of α-methylene-1,3-dioxolane-2-ones catalyzed by chiral biphosphine ruthenium [101]. The use of [Ru-(S)-BINAP(O$_2$CF$_3$)$_2$] leads to optically active cyclic carbonates with enantioselectivies up to 95% ee (Equation 1.24). The excellent enantioselectivities observed may be due to the formation of the chelate intermediate **(31)**.

$$(1.24)$$

R = Me, R-R spirocylopentyl, spirohexenyl

Consiglio *et al.* at Roche used a cationic ruthenium containing the bulky electron-rich ligand (6,6′-dimethoxybiphenyl-2,2′diyl)bis[3,5-di(*t*-butyl-phenylphosphine)], have described the chemoselective and enantioselective hydrogenation of various substituted 2-pyrones to the corresponding 5,6-dihydropyrones (ee values up to 97%) (Equation 1.25) [102].

R = C$_{11}$H$_{23}$, R^1 = *n*-C$_6$H$_{13}$ 94% (99% selectivity) (*R*)

$$(1.25)$$

During the course of a collaboration with Firmenich, a considerable breakthrough in the hydrogenation of tetrasubstituted carbon–carbon double bonds was realized by Genet and Rautenstrauch, using [Ru(P*P)(H)(η^6-COT)]PF$_6$ (COT = cyclocta-1,3,5,7-triene) prepared *in situ* from [Ru(COD)η^3-(methylallyl)$_2$] with HBF$_4$ and various P*P = (BINAP, DIPAMP, Duphos) ligands in a weakly coordinating solvent [76]. This new type of complex is highly active and has demonstrated its utility in the enantioselective hydrogenation of difficult to reduce β-oxoesters. Thus, the enantioselective hydrogenation of the tetrasubstituted "dehydrodione" with [Ru(P*P)(H)(η^6-COT)]PF$_6$, (P*P) = (*R*,*R*)Me-Duphos or Josiphos, leads directly to the (+)-*cis*-methyl dehydrojasmonate with up to 90% ee (Equation 1.26), a commercially important perfume component. Since 1998, Firmenich has developed a process on multi tonnes/year scale for this compound named paradisone.

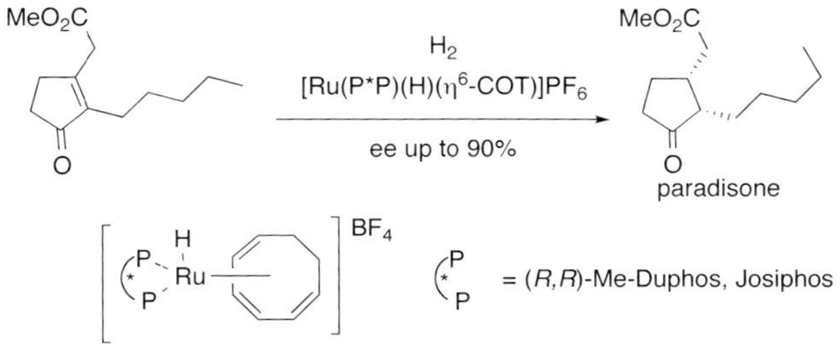

$$\text{(1.26)}$$

Scheme shows hydrogenation with $[\text{Ru}(P^*P)(H)(\eta^6\text{-COT})]PF_6$, ee up to 90%, producing paradisone. $P^*P = (R,R)$-Me-Duphos, Josiphos.

The success of this system may be attributed (Scheme 1.12), firstly, to the formation under hydrogen pressure of the highly unsaturated ruthenium intermediate **(32)**, which acts as a pre-catalyst. Secondly, the chelation of the oxoester entails metal binding to the double bond to give **(33)**, followed by olefin insertion leading to **(34)**, which by hydrogenolysis delivers paradisone with regeneration of the catalyst [9b].

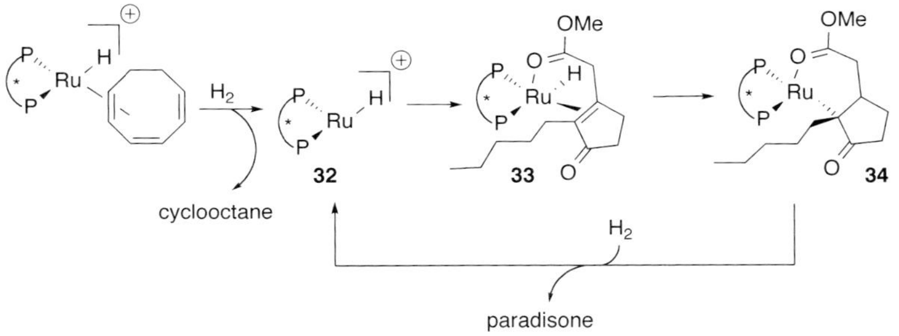

Scheme 1.12 Ru-catalyzed synthesis of paradisone.

1.7
Hydrogenation of Unsaturated Alcohols

1.7.1
Diastereoselective Hydrogenation with Rh and Ir Catalysts

Substrate-directing chemical reactions are a major challenge to organic chemists. Several functional groups and reagents have proved to be highly efficient, giving exceptional levels of stereoselection [103]. The directed hydrogenation of functional olefins employing cationic rhodium or iridium catalysts has considerable potential [104]. The most common directing group, the hydroxyl, acts as a donor to the metal center in the catalytic cycle. Wilkinson's catalyst requires formation of the alkoxide, whereas cationic Rh(I) and Crabtree's catalyst [Ir(COD)PCy$_3$(Pyr)]PF$_6$ have emerged as the most commonly used catalysts in the hydrogenation of allylic and homoallylic alcohols [105].

For example the cationic rhodium complex is effective for the stereoselective of allylic and homoallylic alcohols (35) and (36) [106, 107], while the cationic iridium catalyst is less effective in the hydrogenation of 36. In contrast Crabtree's catalyst is highly efficient in the directed reduction of a number of diverse cyclic olefinic alcohols 37, 38 [108]. The hydrogenation with iridium gives predominately the *trans*-indanone, Table 1.5 shows some selected examples.

Table 1.5 Diasteroselection of unsaturated alcochols.

Substrate	Major product	Catalyst	Selectivity	Reference
PhO$_2$S—C(=CH$_2$)—CH(Me)—OH **35**	PhO$_2$S—CH(Me)—CH(Me)—OH	[Rh]$^+$	99.5 : 0.5	[106]
HO—CH$_2$—CH=C(Me)—CH(Me)—CH$_2$—OTBS **36**	HO—CH$_2$—CH(Me)—CH(Me)—CH$_2$—OTBS	[Rh]$^+$ [Ir]$^+$	95 : 5 73 : 27	[107]
iPr,,,—OH (methylcyclohexene) **37**	iPr,,,—OH (cyclohexane)	[Ir]$^+$	>99.9	[105]
OH (bicyclic enone) **38**	OH (bicyclic ketone, H)	[Ir]$^+$	27 : 1	[108]

[Rh]$^+$ = Ph$_2$P—(CH$_2$)$_n$—PPh$_2$ Rh(nbd)$^+$ BF$_4^-$ or CF$_3$SO$_3^-$

[Ir]$^+$ = (Cy)$_3$P—Ir(py)(cod)$^+$ PF$_6^-$

1.7.2
Enantioselective Hydrogenation with Chiral Ru and Ir Catalysts

Allylic and homoallylic alcohols can be reduced with high selectively using Ru(BINAP) dicarboxylate, reported by Noyori [109]. For example, geraniol and nerol are quantitatively and chemoselectively converted to citronellol with up to 99.9% ee (Equations 1.27 and 1.28). Only the allylic alcohol is hydrogenated, leaving the isolated C6—C7 double bond. The complex tetra(−)-Me-BITIAMP-Ru(OAc)$_2$ is also effective in reducing geraniol to (R)-citronellol (94% ee) [12a].

1.7 Hydrogenation of Unsaturated Alcohols

$$\text{geraniol} \xrightarrow[\text{or } (-)\text{TetraMe-BITIAMP-Ru}]{(S)\text{-BINAP-Ru(II)}, 30\text{--}100 \text{ bar } H_2} (R)\text{-citronellol } 100\% \text{ (94--99\% ee)} \quad (1.27)$$

TetraMe-BITIANP

$$\text{nerol} \xrightarrow[\text{or } (-)\text{-}(S)\text{-MeO-BIPHEP-Ru}]{(S)\text{-BINAP-Ru(II)}, 30\text{--}100 \text{ bar } H_2} (S)\text{-citronellol } 100\% \quad (1.28)$$

Takasago carries this process out on a 300 t/scale, while Roche has developed a similar process using Ru-(S)-MeO-BIPHEMP or Ru-(S)-BIPHEMP for (S)-citronellol and for the elaboration of a vitamin E intermediate (Equation 1.29) [110].

$$\xrightarrow[60 \text{ bar } H_2, 20°C]{(S)\text{-MeO-BIPHEP-Rh(II)}} \quad \text{de} > 96\% \quad (1.29)$$

Deslongchamps and coworkers have described a synthetic approach toward the macrocyclic dienone **(41)**, a key intermediate in the synthesis of (+)-chatancin. Major synthetic steps include two sequential enantioselective hydrogenations of the two allylic alcohols **(39)** and **(40)** using a Ru(BINAP) catalyst (Scheme 1.13) [111].

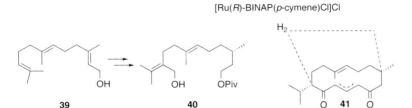

Scheme 1.13 Synthetic approach toward (+)-chatancin.

A chiral moiety close to the olefinic bond in an allylic alcohol profoundly affects the stereochemistry of the Ru(II)-catalyzed hydrogenation; Scheme 1.14 shows an interesting application to the synthesis of carbapenem. The extremely high diastereoselectivity in the hydrogenation of (42) to the β-product (43) with (R)-RuTolBINAP can be explained by the cooperation of the efficient Ru-catalyst, olefin and center of chirality.

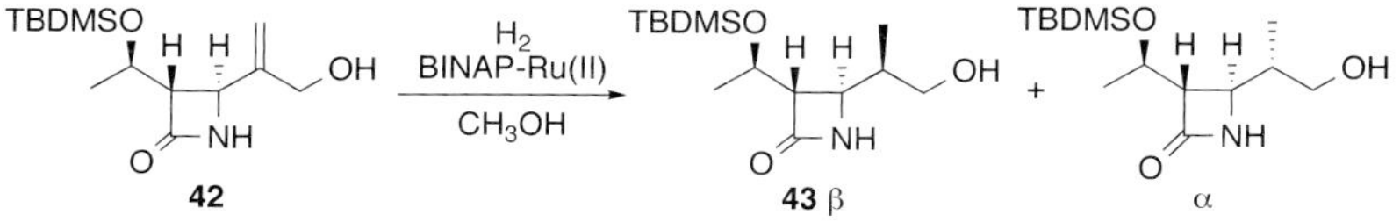

	43 β	α
Ru(OAc)$_2$(R)-TolylBINAP	99.9	0.1
Ru(OAc)$_2$(S)-TolylBINAP	22	78

Scheme 1.14 Synthesis of carbapenem.

The major stereoisomer has been used the synthesis of the antibiotic precursor 1-β-methylcarbapenem [112]. Homogeneous hydrogenation with (R)- or (S)-BINAP diacetate complex allows kinetic resolution of allylic secondary alcohols [113].

Recently iridium catalysts containing chiral phosphine-oxazoline [114] and phosphine thiazole [115] ligands have been reported to catalyze the asymmetric hydrogenation of trisubstituted acrylic allylic alcohols (44) with high enantioselectivity up to 99% ee (Scheme 1.15).

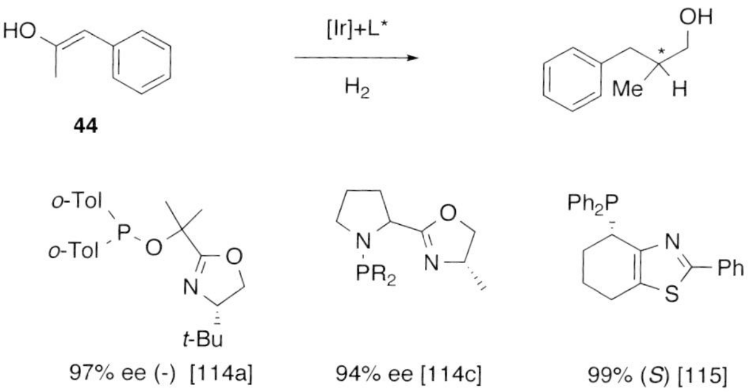

Scheme 1.15 Ir-catalyzed reduction of allylic alcohols.

1.8
Synthesis of Pharmaceutical Intermediates

The production of enantiomerically pure drugs has brought asymmetric synthesis to the forefront of drug discovery [116]. Asymmetric hydrogenation of functionalized alkenes has been very successful, given the vast array of efficient catalysts now available, and this technique has been employed in the synthesis of many pharmaceutical intermediates. Some selected examples using asymmetric hydrogenation are shown in Schemes 1.16–1.19.

The synthesis of pregabalin, (S)-3-aminomethyl-5-methylhexanoic acid, a marketed anticonvulsant, has been achieved by the Rh-catalyzed asymmetric hydrogenation of 3-cyano-5-methylhex-3-enoic salt **(45)**. Burk and coworkers (Dowpharma, Chirotech) used [(R,R)-Et-Duphos-Rh(COD)]BF$_4$ to provide the desired (S)-3-cyano-5-methylhexanoate in very high ee (97.4%); the reaction takes place at 6.2 bar (620 kPa) of H$_2$ in 15 min in methanol [117]. Hoge designed two new enantioselective Rh-catalysts using a P-chirogenic, 1,2-bisphospholanoethane **(46)** [118] and (S)-trichikenPHOS [119].

In connection with the synthesis of a thrombin inhibitor (CR220), Jendralla and coworkers (HMR/Aventis) described the enantioselective hydrogenation of dehydroamino acid **(47)** using Rh-(S,S)-BPPM with 96% ee. Precipitation from the reaction mixture increased the selectivity to >99.8% ee; this process has been scaled up to 10 kg (Scheme 1.16) [120].

Faucher and coworkers at Boehringer-Ingelheim have prepared ethyl-(S)-acetamido-8-nonenoate, one of the building blocks required for the synthesis of BILN 2061, an HCV protease inhibitor. This compound was obtained by enantioselective (99% ee) and chemoselective hydrogenation of the corresponding dehydroamino ester **(48)** using Rh-(S,S)-Et-Duphos (Scheme 1.17) [121].

To synthesize the endopeptidase C inhibitor Candoxatril (Pfizer), Challenger devised an enantioselective route [122] using Noyori's catalyst. However, the double bond isomerization limited the chemical yield. As shown previously, Burk *et al.* at Chirotech improved the process using the Rh(I)-Duphos catalyst [94]. Asymmetric hydrogenation of the trisubstituted olefin **(49)** using the procedure of Genet was successfully optimized and scaled up by Bulliard and coworkers at PPG-SIPSY, with the (COD)(methylallyl)$_2$Ru + HBr, MeO-BIPHEP catalyst system (231 kg/batch size). The corresponding chiral succinate was prepared with 96.5% ee as a crude product (>99% ee after recrystallization). This procedure using

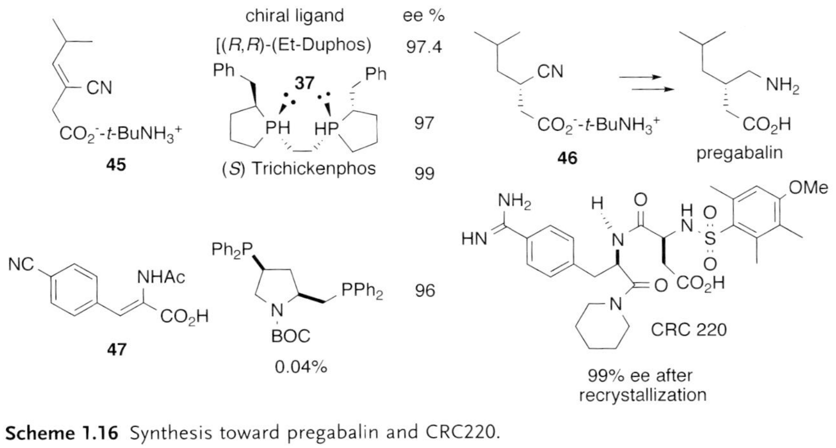

Scheme 1.16 Synthesis toward pregabalin and CRC220.

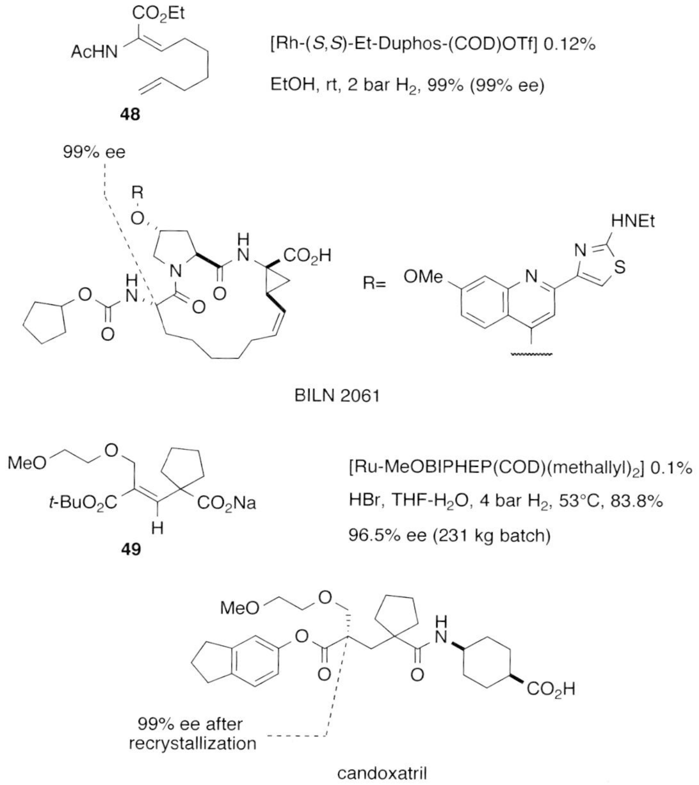

Scheme 1.17 Synthesis toward BILN2061 and candoxatril.

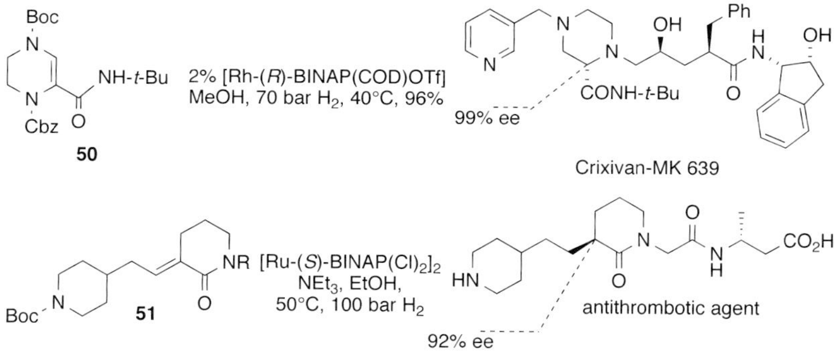

Scheme 1.18 Synthesis toward Crixivan-MK 639 and an anti thrombotic agent.

ruthenium metal is a cost-efficient process and has been used to manufacture more than two tonnes of candoxatril intermediate for phase III clinical trials [123].

Rossen and coworkers (Merck) have developed the preparation of the piperazine subunit of the anti-HIV drug Crixiran MK639. The hydrogenation of **(50)** using a

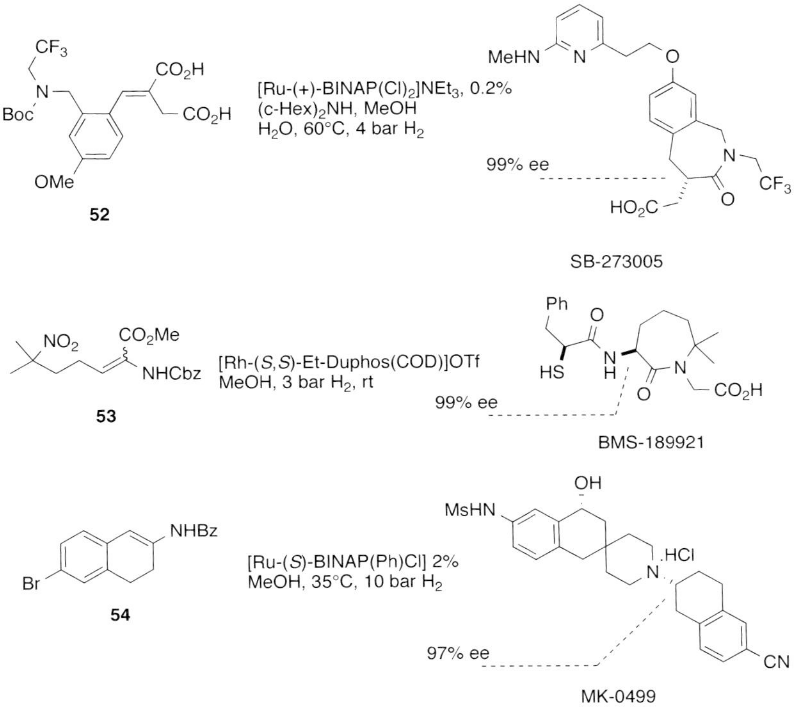

Scheme 1.19 Synthesis toward SB-273005, BMS-189921 and MK-0499.

Rh-(R)-BINAP catalyst at high pressure (70 bar) delivered the target compound in 99% ee (Scheme 1.18) [124]. Another application developed by Merck and reported by Chung and coworkers describes the enantioselective preparation of an antithrombotic agent. The asymmetric hydrogenation of a 3-alkylidene-2-piperidone **(51)** catalyzed by [Ru-(S)-BINAP-Cl$_2$]$_2$NEt$_3$ at high pressure (100 bar (10 MPa)) leads to the chiral 3-alkyl piperidinone in 92% ee [125].

The large-scale synthesis of SB-273005, a vitronectin receptor antagonist, was developed by Wallace and coworkers (GSK). The enantioselective hydrogenation of an itaconic derivative **(52)** was screened with several rhodium and ruthenium catalysts. The optimization of the catalyst system using [Ru$_2$Cl$_4$(R)-BINAP)]NEt$_3$ in the presence of dicyclohexylamine proved to be the most consistent method to obtain high selectivities (94% ee; 99% after recrystallization) [126]. To prepare the potent vasopeptidase inhibitor (BMS-189921), Singh and coworkers devised a highly enantioselective catalytic hydrogenation of an E/Z mixture of the dehydroamino acid **(53)**. Based on the literature the reaction was achieved using [Rh-(S,S)-EtDuphos(COD)] and as expected excellent ee values (99%) were obtained (Scheme 1.19) [127].

Tschaen and coworkers (Merck) have described the enantioselective hydrogenation of a cyclic enamide **(54)** with a Ru-(S)-BINAP catalyst that provided the desired chiral building block in 97% ee for the synthesis of the potent anti-arrhythmic agent (MK-0499) [128].

1.9
Conclusion

Asymmetric hydrogenation of olefins has been the subject of interest of many research groups in academia and industry. The successful hydrogenation reactions described herein proceed generally with both high enantioselectivities and conversions. During the past two decades the field of asymmetric hydrogenation has been growing very rapidly. A wide range of efficient enantioselective methods is now available. As a result, this technology is being more widely practiced and applied to large-scale asymmetric synthesis of fragrances, pharmaceutical and agrochemicals intermediates. It is obvious that many industrial process chemists have integrated these catalytic methods for the development of cost-effective processes. Issues such as development and catalyst activity are now at least as important as selectivity. In this regard, high-throughput and parallel-screening methods offer great assistance in catalyst optimization. Thus, there remain a variety of problems to be solved, as the development of more reactive catalysts is highly desirable. From a synthetic viewpoint, asymmetric hydrogenation has a bright future, particularly in the field of the synthesis of bioactive compounds.

References

1 (a) Siegel, S. (1991) *Comprehensive Organic Synthesis*, Vol. **8** (eds B.M. Trost and I. Flemming), Pergamon, Oxford, pp. 417–41. (b) Harada, K. and Munegumi, T. (1991) *Comprehensive Organic Synthesis*, Vol. **8** (eds B.M. Trost and I. Flemming), Pergamon, Oxford, pp. 139–58.

2 (a) Horner, L., Siegel, H. and Büthe, H. (1968) *Angewandte Chemie (International Ed. in English)*, **7**, 942. (b) Knowles, W.S. and Sabacky, M.J. (1968) *Chemical Communications*, **22**, 1445.

3 (a) Dang, T.P. and Kagan, H.B. (1971) *Chemical Communications*, **10**, 481. (b) Kagan, H.B. and Dang, T.P. (1972) *Journal of the American Chemical Society*, **94**, 6429.

4 (a) Vineyard, B.D., Knowles, W.S., Sabacky, M.J., Bachman, G.L. and Weinkauff, D.J. (1977) *Journal of the American Chemical Society*, **99**, 5946. (b) Knowles, W.S. (1986) *Journal of Chemical Education*, **63**, 222.

5 (a) Noyori, R., Ohta, M., Hsiao, Y., Kitamura, M., Ohta, T. and Takaya, H. (1986) *Journal of the American Chemical Society*, **108**, 7117. (b) Ohta, T., Takaya, H., Kitamura, M., Nagai, K. and Noyori, R. (1987) *Journal of Organic Chemistry*, **52**, 3176. (c) Kitamura, M., Ohkuma, T., Inoue, S., Sayo, N., Kumobayashi, H., Akutagawa, S., Ohta, R., Takaya, H. and Noyori, R. (1988) *Journal of the American Chemical Society*, **110**, 629. (d) Noyori, R., Ohkuma, T., Kitamura, M., Takaya, H., Sayo, N., Kumobayashi, H. and Akutagawa, S. (1987) *Journal of the American Chemical Society*, **109**, 5856.

6 Knowles, W.S. (2002) (Nobel Lecture), *Angewandte Chemie (International Ed. in English)*, **41**, 1998.

7 Noyori, R. (2002) (Nobel Lecture), *Angewandte Chemie (International Ed. in English)*, **41**, 2008.

8 For some recent reviews: (a) Noyori, R. (1994) Homogeneous Asymmetric Hydrogenation. *Asymmetric Catalysis in Organic Synthesis*, John Wiley & Sons, Inc., New York. (b) Genet, J.P. (1996) New developments in chiral ruthenium(II) catalysts for asymmetric hydrogenations and synthetic applications. *Reduction in Organic Synthesis*, ACS Symposium Series 641, Chapter 2 (ed. A.F. Abdel Magid), American Chemical Society, Washington DC. (c) Noyori, R. and Ohkuma, T. (2001) Asymmetric catalysis by architectural and functional molecular engineering: practical chemostereoselective hydrogenation of ketones. *Angewandte Chemie (International Ed. in English)*, **40**, 40. (d) Brown, J.M. (1999) Hydrogenation of functionalized carbon-carbon double bond. *Comprehensive Asymmetric Catalysis*, Vol. **1** (eds E.N. Jacobsen, A. Pfaltz and H. Yamamoto), Springer, Berlin, p. 122. (e) Ohkuma, T. and Noyori, R. (1999) Hydrogenation of carbonyl groups. *Comprehensive Asymmetric Catalysis*, Springer, Berlin, p. 199. (f) Ohkuma, T., Kitamura, M. and Noyori, R. (2000) Asymmetric Hydrogenation. *Catalytic Asymmetric Synthesis*, 2nd edn (ed. I. Ojima), Springer, Berlin pp. 1–110.

9 (a) McCarthy, M. and Guiry, P. (2001) *Tetrahedron*, **57**, 3809. (b) Genet, J.P. (2003) *Accounts of Chemical Research*, **36**, 908. (c) Tang, W. and Zhang, X. (2003) *Chemical Reviews*, **103**, 3029.

10 (a) Kagan, H.B. (1998) *Journal of Organometallic Chemistry*, **567**, 3. (b) Shimizu, K.D., Snapper, M.L. and Hoveyda, A.H. (1998) *Chemistry – A European Journal*, **4**, 1885. (c) Jakel, C. and Paciello, R. (2006) *Chemical Reviews*, **106**, 2912.

11 (a) Miyashita, A., Yasuda, A., Takaya, H., Toriumi, K., Ito, T., Souchi, T. and Noyori, R. (1980) *Journal of the American Chemical Society*, **102**, 7932. (b) Noyori, R. and Takaya, H. (1990) *Accounts of Chemical Research*, **23**, 345.

12 (a) Benincori, T., Brenna, E., Sannicolo, F., Trimarco, L., Antognazza, P., Cesarotti, E., Demartin, F. and Pilati, T. (1996) *Journal of Organic Chemistry*, **61**, 6244. (b) Benincori, T., Cesarotti, E., Piccolo, O. and Sannicolo, F. (2000) *Journal of Organic Chemistry*, **65**, 2043. (c) Wu, J., An Yeung, T.L., Kwok, W.H., Ji, J.X., Zhou, Z., Yeung, C.H. and Chan, A.S.C. (2005) *Advanced Synthesis and Catalysis*, **347**, 507.

13 Pye, P.J., Rossen, K., Reamer, R.A., Tsou, N.T., Volante, R.P. and Reider, P.J. (1997) *Journal of the American Chemical Society*, **119**, 6207.

14 (a) Robin, F., Mercier, F., Ricard, L., Mathey, F. and Spagnol, M. (1997) *Chemistry – A European Journal*, **3**, 1365. (b) Mathey, F., Mercier, F., Robin, F. and Ricard, L. (1998) *Journal of Organometallic Chemistry*, **577**, 117.

15 Tang, W. and Zhang, X. (2002) *Angewandte Chemie (International Ed. in English)*, **41**, 1612.

16 Shimizu, H., Saito, T. and Kumobayashi, H. (2003) *Advanced Synthesis and Catalysis*, **345**, 185.

17 (a) Francio, G., Farone, F. and Leitner, W. (2000) *Angewandte Chemie (International Ed. in English)*, **39**, 1428. (b) Yan, Y., Chi, Y. and Zhang, X. (2004) *Tetrahedron: Asymmetry*, **16**, 2173.

18 (a) Sawamura, M., Kuwano, R. and Ito, Y. (1995) *Journal of the American Chemical Society*, **117**, 9602. (b) Kuwano, R., Vemura, T., Saitoh, M. and Ito, Y. (2004) *Tetrahedron: Asymmetry*, **15**, 2263.

19 (a) Togni, A., Breutel, C., Schnyder, A., Spindler, F., Landert, H. and Tijani, A. (1994) *Journal of the American Chemical Society*, **116**, 4062. (b) Spindler, F., Malan, C., Lotz, M., Kesselgruber, M., Pittelkow, U., Rivas-Nass, A., Briel, O. and Blaser, H.U. (2004) *Tetrahedron: Asymmetry*, **15**, 2299.

20 (a) Boaz, N.W., Mackenzie, E.B., Debenham, S.D., Large, S.E. and Ponasik, J.A. Jr (2005) *Journal of Organic Chemistry*, **70**, 1872. (b) Chen, W., Mbafor, W., Roberts, S. and Whittall, J.

(2006) *Journal of the American Chemical Society*, **128**, 3922.
21 Co, T.T., Shim, S.C., Cho, C.S. and Kim, T.J. (2005) *Organometallics*, **24**, 4824.
22 (a) Burk, M.J., Feaster, J.E., Nugent, W.A. and Harlow, R.L. (1993) *Journal of the American Chemical Society*, **115**, 10125. (b) Burk, M.J., Allen, J.G. and Kiesman, W.G. (1998) *Journal of the American Chemical Society*, **120**, 657. (c) Review: Burk, M.J. (2000) *Accounts of Chemical Research*, **33**, 363.
23 (a) Burk, M.J., Gross, M.F. and Martinez, J.P. (1995) *Journal of the American Chemical Society*, **117**, 9375. (b) Masquelin, T., Broger, E., Müller, K., Schmid, R. and Obrech, D. (1994) *Helvetica Chimica Acta*, **77**, 1395.
24 Jiang, Q., Xiao, D., Zhang, Z., Cao, P. and Zhang, X. (1999) *Angewandte Chemie (International Ed. in English)*, **38**, 516.
25 (a) Rajanbabu, T.V., Yan, Y.Y. and Shin, S. (2001) *Journal of the American Chemical Society*, **123**, 10207. (b) Li, W., Zhang, Z., Xiao, D. and Zhang, X. (2000) *Journal of Organic Chemistry*, **65**, 3489. (c) Holz, J., Quirmach, M., Schmidt, U., Heller, D., Stürmer, R. and Borner, A. (1998) *Journal of Organic Chemistry*, **63**, 8031.
26 (a) Marinetti, A., Kruger, V. and Buzin, F. (1997) *Tetrahedron Letters*, **38**, 2947. (b) Marinetti, A., Genet, J.P., Jus, S., Blanc, D. and Ratovelomanana-Vidal, V. (1999) *Chemistry – A European Journal*, **5**, 1160. (c) Marinetti, A., Labrue, F. and Genet, J.P. (1999) *Synlett: Accounts and Rapid Communications in Synthetic Organic Chemistry*, **12**, 1975. (d) Marinetti, A., Jus, S., Genet, J.P. and Ricard, L. (2001) *Journal of Organometallic Chemistry*, **624**, 162.
27 Berens, U., Burk, M.J., Gerlach, A. and Hems, W. (2000) *Angewandte Chemie (International Ed. in English)*, **39**, 1981.
28 Marinetti, A., Jus, S. and Genet, J.P. (1999) *Tetrahedron Letters*, **40**, 8365.
29 (a) Imamoto, T., Watanabe, J., Wada, J., Masuda, H., Yamada, H., Tsuruta, H., Matsukawa, S. and Yamaguchi, K.

(1998) *Journal of the American Chemical Society*, **120**, 1635. (b) Imamoto, T., Oohara, N. and Takahashi, H. (2004) *Synthesis*, **9**, 1353. (c) Yamamoi, Y. and Imamoto, T. (1999) *Journal of Organic Chemistry*, **64**, 2988. (d) Gridnev, I.D., Yasutake, M., Higashi, N. and Imamoto, I. (2001) *Journal of the American Chemical Society*, **123**, 5268. (e) Crépy, K.V. and Imamoto, T. (2003) *Advanced Synthesis and Catalysis*, **345**, 79. (f) Imamoto, T., Crepy, K.V. and Katagire, K. (2004) *Tetrahedron: Asymmetry*, **15**, 2213.
30 (a) Chan, A.S.C. and Halpern, J. (1980) *Journal of the American Chemical Society*, **102**, 838. (b) Brown, J.M. and Chaloner, P.A. (1980) *Chemical Communications*, **8**, 344. (c) Landis, C.R. and Feldgus, S. (2000) *Angewandte Chemie (International Ed. in English)*, **39**, 2863.
31 (a) Gridnev, I.D., Higashi, H., Asakura, K. and Imamoto, T. (2000) *Journal of the American Chemical Society*, **122**, 7183. (b) Gridnev, I.D. and Imamoto, T. (2004) *Accounts of Chemical Research*, **37**, 633.
32 (a) Lacasse, F. and Kagan, H.B. (2000) *Chemical and Pharmaceutical Bulletin*, **48**, 315. (b) Jerphagnon, T., Renaud, J.L. and Bruneau, C. (2004) *Tetrahedron: Asymmetry*, **15**, 2101.
33 Reetz, M.T. and Mehler, G. (2000) *Angewandte Chemie (International Ed. in English)*, **39**, 3889.
34 (a) Minnaard, M., van den Berg, A.J., Schudde, E.P., van Esch, J., de Vries, A.H.M., de Vries, J.G. and Feringa, B.L. (2000) *Journal of the American Chemical Society*, **112**, 11539. (b) Minnaard, M., van den Berg, A.J., Haak, R.M., Leeman, M., Schudde, E.P., Meetsma, A., Feringa, B.L., de Vries, A.H.M., Maljaars, C.E.P., Willians, C.E., Hyett, D., Boogers, J.A.F., Henderickx, H.J.W. and de Vries, J.G. (2003) *Advanced Synthesis and Catalysis*, **345**, 308. (c) Panella, L., Aleixandre, A.M., Kruidhof, G.J., Robertus, J., Feringa, B.L., de Vries, J.G. and Minnaard, A.J. (2006) *Journal of Organic Chemistry*, **71**, 2026.
35 (a) Galland, A., Dobrota, C., Toffano, M. and Fiaud, J.C. (2006) *Tetrahedron: Asymmetry*, **17**, 2354. (b) Dobrato, C., Toffano, M. and Fiaud, J.C. (2004) *Tetrahedron Letters*, **45**, 8153. (c) Guillen,

F., Rivard, M., Toffano, M., Legros, J.Y., Daran, J.C. and Fiaud, J.C. (2002) *Tetrahedron*, **58**, 5895.
36 Ostermeier, M., Prieß, J. and Helmchen, G. (2002) *Angewandte Chemie (International Ed. in English)*, **41**, 612.
37 Fu, Y., Xie, J.H., Hu, A.G., Zhou, H., Wang, L.X. and Zhou, Q.L. (2002) *Chemical Communications*, **5**, 480.
38 Wang, X. and Ding, K. (2004) *Journal of the American Chemical Society*, **126**, 10524.
39 (a) Schmid, R., Cereghetti, M., Heiser, B., Schönholzer, P. and Hansen, H.J. (1988) *Helvetica Chimica Acta*, **71**, 897. (b) Schmid, R., Foricher, J., Cereghetti, M. and Schönholzer, P. (1991) *Helvetica Chimica Acta*, **74**, 370.
40 (a) Ikariya, T., Ishii, Y., Kawano, H., Arai, T., Saburi, M., Yoshikawa, S. and Akutagawa, S. (1985) *Chemical Communications*, **13**, 922. (b) Kawano, H., Ikariya, T., Ishii, Y., Saburi, M., Yoshikawa, S., Uchida, Y. and Kumobayashi, H. (1989) *Journal of the Chemical Society, Perkin Transactions I*, **9**, 1571.
41 James, B.R., Pacheco, A., Rettig, S.J., Thorburn, I.S., Ball, R.G. and Ibers, J.A. (1987) *Journal of Molecular Catalysis*, **41**, 147.
42 (a) Genet, J.P., Pinel, C., Ratovelomanana-Vidal, V., Mallart, S., Pfister, X., Bischoff, L., Caño de Andrade, M.C., Darses, S., Galopin, C. and Laffitte, J.A. (1994) *Tetrahedron: Asymmetry*, **5**, 675. (b) Genet, J.P., Pinel, C., Mallart, S., Juge, S., Cailhol, N. and Laffitte, J.A. (1992) *Tetrahedron Letters*, **33**, 5343.
43 (a) Michaud, G., Bulliard, M., Ricard, L., Genet, J.P. and Marinetti, A. (2002) *Chemistry – A European Journal*, **8**, 3327. (b) Madec, J., Michaud, G., Genet, J.P. and Marinetti, A. (2004) *Tetrahedron: Asymmetry*, **15**, 2253.
44 Enev, V., Ewers, Ch.L.J., Harre, M., Nickisch, K. and Mohr, J.T. (1997) *Journal of Organic Chemistry*, **62**, 7092.
45 (a) Duprat de Paule, S., Champion, N., Ratovelomanana-Vidal, V., Dellis, P. and Genet, J.P. French patent 0,112,499, N° PCT/FR02/03146. World patent WO 03,029,259 (b) Duprat de Paule, S., Jeulin, S., Ratovelomanana-Vidal, V., Genet, J.P. Champion, N. and Dellis, P. (2003) *Tetrahedron Letters*, **44**, 823. (c) Duprat de Paule, S., Jeulin, S., Ratovelomanana-Vidal, V., Genet, J.P. Champion, N. and Dellis, P. (2003) *European Journal of Organic Chemistry*, **10**, 1931.
46 Pai, C.C., Lin, Y.M., Zhou, Z.Y. and Chan, A.S.C. (2002) *Tetrahedron Letters*, **43**, 2789.
47 Wu, J., Pai, C.C., Kwok, W.H., Guo, R.V., Au-Yeung, T.T.L., Yeung, G.H. and Chan, A.S.C. (2003) *Tetrahedron: Asymmetry*, **14**, 987.
48 Berthod, M., Mignani, G. and Lemaire, M. (2004) *Tetrahedron: Asymmetry*, **15**, 1121.
49 (a) Wiles, J.A., Bergens, S.H. and Young, U.G. (1997) *Journal of the American Chemical Society*, **119**, 2940. (b) Wiles, J.A. and Bergens, S.H. (1998) *Organometallics*, **17**, 2228. (c) Wiles, J.A. and Bergens, S.H. (1999) *Organometallics*, **18**, 3709.
50 Kitamura, M., Tsukamoto, M., Bessho, Y., Yoshimura, M., Kobs, U., Widhalm, M. and Noyori, R. (2002) *Journal of the American Chemical Society*, **124**, 6649.
51 Bunlaksananusorn, T., Polborn, K. and Knochel, P. (2003) *Angewandte Chemie (International Ed. in English)*, **42**, 3941.
52 (a) Kitamura, M., Hsiao, Y., Ohta, M., Tsukamoto, M., Ohta, T., Takaya, H. and Noyori, R. (1994) *Journal of Organic Chemistry*, **59**, 297. (b) Heiser, B., Broger, E.A. and Cramiri, Y. (1991) *Tetrahedron: Asymmetry*, **2**, 51.
53 Armstrong, J.D., III, Eng, K.K., Keller, J.L., Purick, R.M., Hartner, F.W., Jr, Choi, W.B., Askin, D. and Volante, R.P. (1994) *Tetrahedron Letters*, **35**, 3239.
54 Burk, M.J., Wang, Y.M. and Lee, J.R. (1996) *Journal of the American Chemical Society*, **118**, 5142.
55 Kuwano, R., Sato, K., Kurokawa, T., Karube, D. and Ito, Y. (2000) *Journal of the American Chemical Society*, **122**, 7614.
56 (a) Huang, H., Zheng, Z., Luo, H., Bai, C., Hu, X. and Chen, H. (2004) *Journal of Organic Chemistry*, **69**, 2355. (b) Huang,

H., Liu, X., Chen, S., Chen, H. and Zheng, Z. (2004) *Tetrahedron: Asymmetry*, **15**, 2011.

57 Imamoto, T., Oohara, N. and Takahashi, H. (2004) *Synthesis*, **9**, 1353.

58 (a) Hoen, R., Bernsmann, M., van den Berg, H., Minnaard, A.J., de Vires, J.G. and Feringa, B.L. (2004) *Organic Letters*, **6**, 1433. (b) Bernsmann, H., Hoen, M., van den Berg, R., Minnaard, A.J., Mehler, G., Reetz, M.T., Vires, J.G. and Feringa, B.L. (2005) *Journal of Organic Chemistry*, **70**, 943. (c) Reetz, M.T., Mehler, G. and Meiswinkel, A. (2004) *Tetrahedron: Asymmetry*, **15**, 2165.

59 Stemmler, R.T. and Bolm, C. (2005) *Journal of Organic Chemistry*, **70**, 9925.

60 Zeng, Q.H., Hu, X.P., Duan, Z.C., Liang, X.M. and Zheng, Z. (2006) *Journal of Organic Chemistry*, **71**, 393.

61 Li, X., Jia, X., Xu, L., Kok, S.H.L., Yip, C.W. and Chan, A.S.C. (2005) *Advanced Synthesis and Catalysis*, **347**, 1904.

62 (a) Tang, W., Chi, Y. and Zhang, X. (2002) *Organic Letters*, **4**, 1695. (b) Wu, S., He, M. and Zhang, X. (2004) *Tetrahedron: Asymmetry*, **15**, 2177.

63 Dupau, P., Bruneau, C. and Dixneuf, P.H. (2001) *Advanced Synthesis and Catalysis*, **343**, 331.

64 Lubell, W.D., Kitamura, M. and Noyori, R. (1991) *Tetrahedron: Asymmetry*, **2**, 543.

65 Yasutake, M., Gridnev, I.D., Higashi, N. and Imamoto, T. (2001) *Organic Letters*, **3**, 1701.

66 (a) Wu, J., Chen, X., Guo, R., Yeung, C.H. and Chan, A.S.C. (2003) *Journal of Organic Chemistry*, **68**, 2490. (b) Qiu, L., Wu, J., Chan, S., Au-Yeung, T.T.L., Ji, J.X., Guo, R., Pai, C.C., Zhou, Z., Li, X., Fan, Q.H. and Chan, A.S.C. (2004) *Proceeding of the National Academy of Sciences of the U. S. A*, **101**, 5815. (c) Qiu, L., Kwong, F.Y., Wu, J., Lam, W.H., Chan, S., Yu, W.Y., Lin, Y.M., Guo, R., Zhou, Z. and Chan, A.S.C. (2006) *Journal of the American Chemical Society*, **128**, 5955. (d) Wu, J. and Chan, A.S.C. (2006) *Accounts of Chemical Research*, **39**, 711.

67 Reetz, M.T. and Li, X. (2004) *Tetrahedron*, **60**, 9709.

68 Heller, D., Holz, J., Drexler, H.J., Lang, J., Drauz, K., Krimmer, H.P. and Börner, A. (2001) *Journal of Organic Chemistry*, **66**, 6816.

69 Zhu, G., Chen, Z. and Zhang, X. (1999) *Journal of Organic Chemistry*, **64**, 6907.

70 You, J., Drexler, H.J., Zhang, S., Fischer, C. and Heller, D. (2003) *Angewandte Chemie (International Ed. in English)*, **42**, 913.

71 Hu, X.P. and Zheng, Z. (2005) *Organic Letters*, **7**, 419.

72 (a) Wu, H.P. and Hoge, G. (2004) *Organic Letters*, **6**, 3645. (b) Hoge, G. and Samas, B. (2004) *Tetrahedron: Asymmetry*, **15**, 2155. (c) Jia, X., Li, X., Lam, W.S., Kok, S.H.L., Xu, L., Lu, G., Yeung, C.H. and Chan, A.S.C. (2004) *Tetrahedron: Asymmetry*, **15**, 2273.

73 Hsiao, Y., Rivera, N.R., Rosner, T., Krska, S.W., Njolito, E., Wang, F., Sun, Y., Armstrong, J.D., III, Grabowski, E.J.J., Tillyer, R.D., Spindler, F. and Malan, C. (2004) *Journal of the American Chemical Society*, **126**, 9918.

74 Kuwano, R., Okuda, S. and Ito, Y. (1998) *Tetrahedron: Asymmetry*, **9**, 2773.

75 Tang, W., Wu, S. and Zhang, X. (2003) *Journal of the American Chemical Society*, **125**, 9570.

76 Dobbs, D.A., Vanhessche, K.P.M., Brazi, E., Rautenstrauch, V., Lenoir, J.Y., Genet, J.P., Wiles, J. and Bergens, S.H. (2000) *Angewandte Chemie (International Ed. in English)*, **39**, 1992.

77 Genet, J.P., Mallart, S., Pinel, C., Juge, S. and Laffitte, J.A. (1991) *Tetrahedron: Asymmetry*, **2**, 43.

78 Alcock, N.W., Brown, J.M., Rose, M. and Wienand, A. (1991) *Tetrahedron: Asymmetry*, **2**, 47.

79 (a) Mashima, K., Kusano, K.H., Ohta, T., Noyori, R. and Takaya, H. (1989) *Chemical Communications*, **17**, 1208. (b) Uemura, T., Zhang, X., Matsumura, K., Sayo, N., Kumobayashi, H., Ohta, T., Nozaki, K. and Takaya, H. (1996) *Journal of Organic Chemistry*, **61**, 5110. (c) Pai, C.C., Lin, C.W., Lin, C.C., Chen, C.C. and Chan, A.S.C. (2000) *Journal of the American Chemical Society*, **122**, 11513. (d) Review: Wu, J. and Chan, A.S.C. (2006) *Accounts of Chemical Research*, **39**, 711.

(e) Wan, K.T. and Davis, M.E. (1994) *Nature*, **370**, 449. (f) Wan, K.T. and Davis, M.E. (1993) *Tetrahedron: Asymmetry*, **4**, 2461.

80 Crameri, Y., Foricher, J., Scalone, M. and Schmid, R. (1997) *Tetrahedron: Asymmetry*, **8**, 3617.

81 (a) Xiao, J., Nefkens, S.C.A., Jessop, P.G., Ikariya, T. and Noyori, R. (1996) *Tetrahedron Letters*, **37**, 2813. (b) Jessop, P.G., Stanley, R.R., Brown, R.A., Eckert, C.A., Liotta, C.L., Ngo, T.T. and Pollet, P. (2003) *Green Chemistry*, **5**, 123.

82 Saburi, M., Takeuchi, H., Ogasawara, M., Tsukahara, T., Ishii, Y., Ikariya, T., Takahashi, T. and Uchida, Y. (1992) *Journal of Organometallic Chemistry*, **428**, 155.

83 (a) Benincori, T., Rizzo, S., Sannicolo, F. and Piccolo, O. (2000) Proceedings of Chiral Source Symposium The Catalyst Group, Spring House, USA. (b) Cheng, X., Zhang, Q., Xie, J.H., Wang, L.X. and Zhou, Q.L. (2005) *Angewandte Chemie (International Ed. in English)*, **44**, 1118. (c) For a Rh(I) catalyst system using (RN2) ligand see Yamada, I., Yamaguchi, M. and Yamagishi, T. (1996) *Tetrahedron: Asymmetry*, **7**, 3339.

84 Alberico, E., Nieddu, I., Taras, R. and Gladiali, S. (2006) *Helvetica Chimica Acta*, **89**, 1716.

85 Maligres, P.E., Krska, S.W. and Humphrey, G.R. (2004) *Organic Letters*, **6**, 3147.

86 Tellers, D.M., McWilliams, J.C., Humphrey, G., Journet, M., DiMichele, L., Hinksmon, J., McKeown, A.E., Rosner, T., Sun, Y. and Tillyer, R.D. (2006) *Journal of the American Chemical Society*, **128**, 17063.

87 Chiba, T., Miyashita, A., Nohira, H. and Takaya, H. (1991) *Tetrahedron Letters*, **32**, 4745.

88 Burk, M.J., Bienewald, F., Harris, M. and Zanotti-Gerosa, A. (1998) *Angewandte Chemie (International Ed. in English)*, **37**, 1931.

89 Tang, W., Liu, D. and Zhang, X. (2003) *Organic Letters*, **5**, 205.

90 (a) Reetz, M.T. and Neugebauer, T. (1999) *Angewandte Chemie (International Ed. in English)*, **38**, 179. (b) Reetz, M.T., Gosberg, A., Goddard, R. and Kyung, S.H. (1998) *Chemical Communications*, **19**, 2077.

91 (a) Wu, S., Zhang, W., Zhang, Z. and Zhang, X. (2004) *Organic Letters*, **6**, 3565. (b) Lin, C.W., Lin, C.C., Lam, L.F.L., Au-Yeung, T.T.L. and Chan, A.S.C. (2004) *Tetrahedron Letters*, **45**, 7379.

92 (a) Gridnev, I.D., Yamanoi, Y., Higashi, N., Tsuruta, H., Yasutake, M. and Imamoto, T. (2001) *Advanced Synthesis and Catalysis*, **343**, 118. (b) Liu, D., Li, W. and Zhang, X. (2004) *Tetrahedron: Asymmetry*, **15**, 2181.

93 (a) Jeulin, S., Duprat de Paule, S., Ratovelomanana-Vidal, V., Genet, J.P., Champion, N. and Dellis, P. (2004) *Angewandte Chemie (International Ed. in English)*, **43**, 320. (b) Jeulin, S., Duprat de Paule, S., Ratovelomanana-Vidal, V., Genet, J.P., Champion, N. and Dellis, P. (2004) *Proceeding of National Academy of Sciences of the U. S. A.*, Special Feature issue on Asymmetric Catalysis, **101** (*16*), 5799.

94 Burk, M.J., Bienewald, F., Challenger, S., Derrick, A. and Ramsden, J.A. (1999) *Journal of Organic Chemistry*, **64**, 3290.

95 Ohta, T., Takaya, H. and Noyori, R. (1990) *Tetrahedron Letters*, **31**, 7189.

96 Ashby, M.T. and Halpern, J. (1991) *Journal of the American Chemical Society*, **113**, 589.

97 (a) Brown, J.M., Rose, M., Knight, F.I. and Wienand, A. (1995) *Recueil des Travaux Chimiques des Pays-Bas*, **114**, 242. (b) Brown, J.M. (1993) *Chemical Society Reviews*, **22**, 25.

98 Chan, A.S.C., Chen, C.C., Yang, T.K., Huang, J.H. and Lin, Y.C. (1995) *Inorganica Chimica Acta*, **234**, 95.

99 Daley, C.J.A., Wiles, J.A. and Bergens, S.H. (2006) *Inorganica Chimica Acta*, **359**, 2760.

100 Ohta, T., Miyake, T., Seido, N., Kumobayashi, H. and Takaya, H. (1995) *Journal of Organic Chemistry*, **60**, 357.

101 Le Gendre, P., Braun, T., Bruneau, C. and Dixneuf, P.H. (1996) *Journal of Organic Chemistry*, **61**, 8453.

102 Fehr, M.J., Consiglio, G., Scalone, M. and Schmid, R. (1999) *Journal of Organic Chemistry*, **64**, 5768.

103 Hoveyda, A.H., Evans, D.A. and Fu, G.C. (1993) *Chemical Reviews*, **93**, 1307.
104 Brown, J.M. (1987) *Angewandte Chemie (International Ed. in English)*, **26**, 190.
105 Crabtree, R.H. and Davis, M.W. (1983) *Organometallics*, **2**, 681.
106 Ando, D., Bevan, C., Brown, J.M. and Price, D.W. (1992) *Chemical Communications*, **8**, 592.
107 Evans, D.A., Morrissey, M.M. and Dow, R.L. (1985) *Tetrahedron Letters*, **26**, 6005.
108 Stork, G. and Kahne, D.E. (1983) *Journal of the American Chemical Society*, **105**, 1072.
109 Takaya, H., Ohta, T., Sayo, N., Kumobayashi, H., Kutagawa, S.A., Inoue, S.I., Kasahara, I. and Noyori, R. (1987) *Journal of the American Chemical Society*, **109**, 1596.
110 Schmid, R. and Scalone, M. (1999) *Comprehensive Asymmetric Catalysis*, Vol. III (eds E.N. Jacobsen, A. Pfaltz and H. Yamamoto), Springer, Berlin, pp. 1439–49.
111 Toro, A., l'Heureux, A. and Deslongchamps, P. (2000) *Organic Letters*, **2**, 2737.
112 Galland, J.C., Roland, S., Malpart, J., Savignac, M. and Genet, J.P. (1999) *European Journal of Organic Chemistry*, **3**, 621.
113 Kitamura, M., Kasahara, I., Manabe, K., Noyori, R. and Takaya, H. (1988) *Journal of Organic Chemistry*, **53**, 710.
114 (a) Smidt, S.P., Menges, F. and Pfaltz, A. (2004) *Organic Letters*, **6**, 2023. (b) Drury, W.J., III, Zimmermann, N., Keenan, M., Hayashi, M., Kaiser, S., Goddard, R. and Pfaltz, A. (2004) *Angewandte Chemie (International Ed. in English)*, **43**, 70. (c) Xu, G. and Gilbertson, S.R. (2003) *Tetrahedron Letters*, **44**, 953.
115 Hedberg, C., Källström, K., Brandt, D., Hansen, L.K. and Andersson, P.G. (2006) *Journal of the American Chemical Society*, **128**, 2995.
116 Reviews: (a) Farina, V., Reeves, J.T., Senanayake, C.H. and Song, J.J. (2006) *Chemical Reviews*, **106**, 2734. (b) Blaser, H.U., Spindler, F. and Studern, M. (2001) *Applied Catalysis A: General*, **221**, 119.
117 Burk, M.J., de Koning, P.D., Grote, T.M., Hoekstra, M.S., Hoge, G., Jennings, R.A., Kissel, W.S., Le, T.V., Lennon, I.C., Mulhern, T.A., Rasmden, J.A. and Wade, R.A. (2003) *Journal of Organic Chemistry*, **68**, 5731.
118 Hoge, G., Wu, H.P., Kissel, W.S., Pflum, D.A., Greene, D.J. and Bao, J. (2004) *Journal of the American Chemical Society*, **126**, 5966.
119 Hoge, G. (2003) *Journal of the American Chemical Society*, **125**, 10219.
120 Jendralla, H., Seuring, B., Herchen, J., Kulitzscher, B., Wunner, J., Stüber, W. and Koschinsky, R. (1995) *Tetrahedron*, **51**, 12047.
121 Faucher, A.M., Bailey, M.D., Beaulieu, P.L., Brochu, C., Duceppe, J.S., Ferland, J.M., Ghiro, E., Gorys, V., Halmos, T., Kawai, S.H., Poirier, M., Simoneau, B., Tsantrizos, Y.S. and Llinàs-Brunet, M. (2004) *Organic Letters*, **6**, 2901.
122 Challenger, S., Derrick, A., Masson, C.P. and Silk, T.V. (1999) *Tetrahedron Letters*, **40**, 2187.
123 Bulliard, M., Laboue, B., Lastennet, J. and Roussiasse, S. (2001) *Organic Process Research and Development*, **5**, 438.
124 Rossen, K., Pye, P.J., DiMichele, L.M., Volante, R.P. and Reider, P.J. (1998) *Tetrahedron Letters*, **39**, 6823.
125 Chung, J.Y.L., Zhao, D., Hugues, D.L., McNamara, J.M., Grabowski, E.J.J. and Reider, P.J. (1995) *Tetrahedron Letters*, **36**, 7379.
126 Wallace, M.D., McGuire, M.A., Yu, M.S., Goldfinger, L., Liu, L., Dai, W. and Shilcrat, S. (2004) *Organic Process Research and Development*, **8**, 738.
127 Singh, J., Kronenthal, D.R., Schwinden, M., Godfrey, J.D., Fox, R., Vawter, E.J., Zhang, B., Kissick, T.P., Patel, B., Mneimne, O., Humora, M., Papaioannou, C.G., Szymanski, W., Wong, M.K.Y., Chen, C.K., Heikes, J.E., DiMarco, J.D., Qiu, J., Deshpande, R.P., Gougoutas, J.Z. and Mueller, R.H. (2003) *Organic Letters*, **5**, 3155.
128 Tschaen, D.M., Abramson, L., Cai, D., Desmond, R., Dolling, U.H., Frey, L., Karady, S., Shi, Y.J. and Verhoeven, T.R. (1995) *Journal of Organic Chemistry*, **60**, 4324.

2
Hydrogenation of Unfunctionalized Alkenes
Jarle S. Diesen and Pher G. Andersson

2.1
Introduction

Although the enantioselective hydrogenation of alkenes is one of the most widely investigated methods in asymmetric catalysis [1], the majority of the reported substrates contain a functionality that can coordinate to a metal. These include enamides, allylic or homoallylic alcohols and α,β- or β,γ-unsaturated carboxylic acids. These chelating olefins have a long and successful history in ruthenium- and rhodium-catalyzed asymmetric hydrogenations [1]. However, there are many classes of substrates toward which these catalysts perform without satisfactory control. These unfunctionalized, or largely unfunctionalized [2], alkenes still represent a challenging class of substrates.

The most common catalysts for hydrogenating unfunctionalized alkenes are perhaps the heterogeneous catalysts Pd/C and PtO_2 (Adam's catalyst), which have found widespread use in various applications [3]. However, relatively few homogeneous catalysts are capable of effecting the same transformations, especially when either tri- or tetrasubstituted alkenes need to be hydrogenated asymmetrically.

2.2
Iridium Catalysis

Crabtree reported the first homogeneous iridium catalyst for olefin hydrogenation in 1977 [4]. This achiral complex, often called "Crabtree's catalyst," **(1)** (Figure 2.1) was able to reduce a range of unfunctionalized olefins, including tri- and tetrasubstituted ones.

Crabtree's catalyst showed impressive turnover frequencies (TOFs) for a number of substituted olefins (Table 2.1); however, only low conversions for tri- or tetrasubstituted olefins were obtained (entries 3 and 4).

It was also noted that a competitive deactivation reaction took place during the catalysis, especially in the presence of hindered alkenes [5]. Crabtree reported the

Modern Reduction Methods. Edited by Pher G. Andersson and Ian J. Munslow.
Copyright © 2008 WILEY-VCH Verlag GmbH & Co. KGaA, Weinheim
ISBN: 978-3-527-31862-9

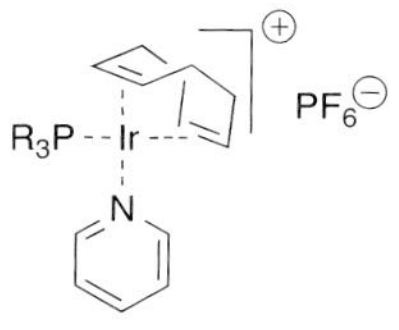

R = Cy, **1a**; *i*-Pr, **1b**

Figure 2.1 Crabtree's catalyst (**1**).

Table 2.1 Alkene hydrogenation with (**1b**).[a]

Entry	Alkene	Conversion (%)	TOF (h^{-1})
1	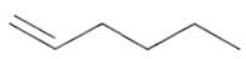	100	6400
2		100	4500
3	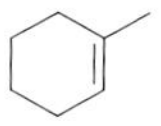	35	3800
4	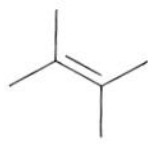	<40	4000

a   Conditions: (**1b**) 0.1 mol%, 0.83 bar H_2, CH_2Cl_2, 0 °C.

formation of, and subsequently fully characterized, a catalytically inactive iridium trimer (Equation 2.1).

$$3\,[Ir(COD)(PCy_3)Py][PF_6] + 10\,H_2 \longrightarrow [\{IrH_2Py_3(PCy_3)\}_3(\mu_3\text{-}H)][PF_6] + HPF_6 + 3\,COA \qquad (2.1)$$

Twenty-one years after Crabtree's initial report, Pfaltz and coworkers reported the first successful Ir catalyst for asymmetric hydrogenation of unfunctionalized alkenes ((**2**), Figure 2.2) [6], employing a previously reported phosphinooxazoline ligand [7].

In this report they showed that (**2b**), which had a similar coordination environment to (**1**), gave encouraging results but was limited in terms of substrate scope and catalyst deactivation. Importantly, most trisubstituted substrates were reduced in over 90% ee and even the tetrasubstituted alkene 2-methyl-3-phenylbut-2-ene was quantitatively reduced in high ee (81%). Under strictly anhydrous and anaerobic conditions quantitative reductions could still be obtained with catalyst loadings of only 0.5 mol%, though catalyst deactivation remained an issue. However, upon changing the PF_6^- counterion to the very weakly coordinating tetrakis[3,5-bis(trifluoromethyl)phenyl]borate ion (BAr_F^-, **3**, Figure 2.2) the Ir complexes became more

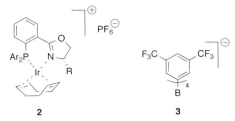

Figure 2.2 Pfaltz's first-generation Ir complex (**2**). The counterion BAr$_F^-$ (**3**).

robust; effective reductions could be achieved using catalyst loadings as low as 0.02 mol%.

2.2.1
Catalysts

Most of the successful Ir-based catalysts that have been reported to date adopt a similar coordination environment to Crabtree's catalyst, possessing a cationic Ir center, a coordinating, bidentate *N,P*-ligand, a 1,5-cyclooctadiene (COD) ligand and the counterion BAr$_F^-$.

2.2.1.1 Ligands

The phosphinooxazoline (**4**), used in Pfaltz original catalyst, has evolved into many different classes of successful ligands (Figure 2.3).

Extensive variations of the ligand backbone, oxazoline and phosphine groups have proved successful, although common to all these ligands is an sp^2-hybridized coordinating nitrogen atom. Phosphorus coordination occurs most commonly via a phosphine group ((**4**) [6, 8], (**5**) [9], (**6**) [10], (**7**) [11], (**11**) [12], (**12**) [13], (**15**) [14], (**17**) [15], (**20**) [16], (**21**) [17] and (**22**) [18]), though ligands with phosphinites ((**8**) [19], (**13**) [20], (**14**) [20, 21], (**16**) [22], (**18**) [23] and (**19**) [24]) and phosphites ((**9**) and (**10**) [25]) have also successfully been developed. More recently, phosphorus coordination has been replaced by the use of electron-rich *N*-heterocyclic carbenes [26] ((**23**) [27], (**24**) and (**25**) [28] and (**26**) [29]) (Figure 2.4).

Nitrogen coordination has typically been through five-membered rings, such as the oxazoline, imidazoline or thiazole moieties ((**4**)–(**17**)) and ((**22**)–(**26**)), though more recent reports have taken advantage of pyridine ((**18**)–(**20**)) and imine (**21**) functionalities. Schenkel and Ellman reported the chiral nonracemic ligand (**21**) in 2004. Although it is not the most efficient ligand for asymmetric Ir-catalyzed hydrogenations – 94% ee for the reduction of α-methyl-*trans*-stilbene (cf. entry 1, Table 2.2) – it is interesting because it incorporates heteroatom chirality [17].

2.2.1.2 Anion

The role of the anion has been studied, and experiments show that coordinating anions such as halides result in Ir catalysts of lower activity than those with weakly

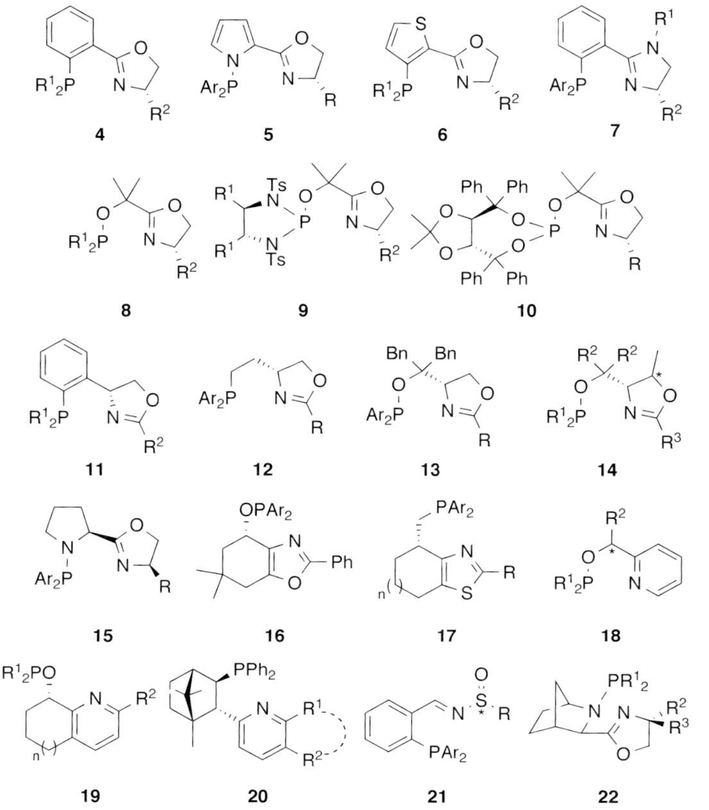

Figure 2.3 An overview of ligands utilized in Ir-catalyzed asymmetric hydrogenations.

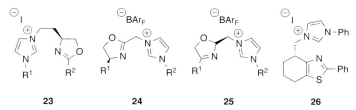

Figure 2.4 Imidazolium precursors to electron-rich N-heterocyclic carbenes.

coordinating anions [30]. A recent report by Pfaltz and coworkers showed a large decrease in the reaction rate of catalysts with the following series of weakly coordinating anions: $[Al\{OC(CF_3)_3\}_4]^- > BAr_F^- > [B(C_6F_5)_4]^- > PF_6^- \gg BF_4^- > CF_3SO_3^-$ [31]. Similarly to Crabtree's catalyst, complexes with the PF_6^- anion suffer from extreme water sensitivity and deactivation via the formation of inactive trimeric Ir-hydride clusters, which have been isolated and structurally characterized [32]

Table 2.2 A summary of some of the best results obtained in the iridium-catalyzed asymmetric hydrogenation of standard substrates.[a]

Entry	Alkene	Ligands	Conversion (%)	ee (%)
1	Ph-C(Ph)=CHMe	(5): (Ar = o-Tol, R = t-Bu) (6): (R^1 = Cy, R^2 = t-Bu) (11): (R^1 = Cy, R^2 = t-Bu) (16): (Ar = o-Tol) (23): (R^1 = 2,6-(i-Pr)$_2$C$_6$H$_3$, R^2 = Ad)	>99	99
2	(E)-4-MeO-C$_6$H$_4$-C(Me)=CHMe	(14): (R^1 = R^3 = Ph, R^2 = Bn)	>99	99
3	(Z)-4-MeO-C$_6$H$_4$-C(Me)=CHMe	(14): (R^1 = R^3 = Ph, R^2 = Bn)	>99	92
4	4-MeO-C$_6$H$_4$-C(Et)=CH$_2$	(16): (Ar = o-Tol)	99	97
5	4-MeO-C$_6$H$_4$-C(Me)=CMe$_2$	(4): (R^1 = Ph, R^2 = i-Pr) (18): (R^1 = t-Bu, R^2 = t-Bu)	>99	81
6	6-MeO-1-methyl-3,4-dihydronaphthalene	(8): (R^1 = o-Tol, R^2 = t-Bu)	>99	99
7	Ph-C(Me)=CH-OH	(17): (Ar = R = Ph)	99	99
8	Ph-C(Me)=CH-OAc	(16): (Ar = o-Tol)	99	99
9	Ph-C(Me)=CH-CO$_2$Et	(17): (Ar = R = Ph)	99	95

a Conditions: cat. 0.01–2 mol%, 50 bar H$_2$, 2 h, CH$_2$Cl$_2$.

from the reaction of (2) with H$_2$ [6]. The formation of these inactive trimers is much slower for catalysts containing the BAr$_F^-$ counterion than for those using PF$_6^-$. However, unlike complexes having the PF$_6^-$ anion, complexes with weakly coordinating anions such as [Al{OC(CF$_3$)$_3$}$_4$]$^-$ and BAr$_F^-$ are relatively moisture- and air-stable, frequently give TOFs of over 5000 h^{-1} and turnover numbers (TONs) in the range of 2000 to 5000, and also remain active after all substrate has been consumed [31].

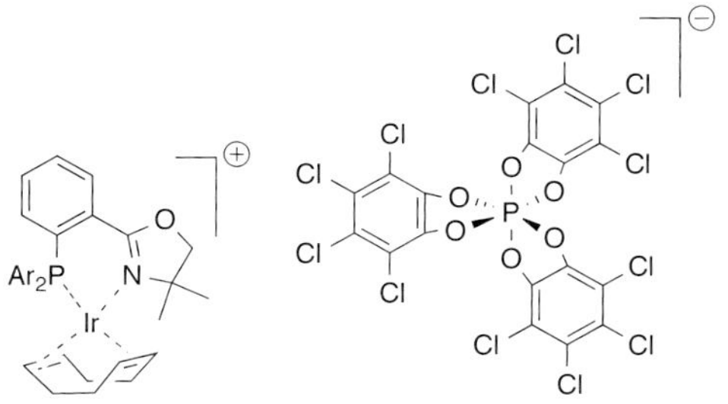

Figure 2.5 Complex **(27)**, an analogue of Pfaltz's catalyst that is only chiral in the anion, Δ-TRISPHAT.

Pfaltz has also examined the use of the chiral anion Δ-TRISPHAT (Figure 2.5) [33]. A racemic product was obtained when the Ir complex **(27)** (Figure 2.5) was used in the hydrogenation of *trans*-α-methylstilbene. However, the complex [Ir**(4)**(COD)][Δ-TRISPHAT] (R^1 = Ph, R^2 = *i*-Pr) hydrogenated *trans*-α-methyl stilbene with essentially the same optical purity as did the corresponding $BArF^-$ complex [6, 31], implying that the anion plays no part in the enantioselectivity of the reaction.

2.2.2
Substrates

Most of the compounds shown in Figure 2.2, when utilized as ligands in the Ir-catalyzed hydrogenation of unfunctionalized (*E*)-trisubstituted alkenes under similar conditions (cat. 0.5 mol%, rt, 50 bar H_2, 2 hours, CH_2Cl_2), perform excellently, typically giving full conversion and high (≥90%) ee (Equation 2.2). Table 2.2 displays the best results obtained for a range of *standard* substrates. Not all ligands have been reported with all substrates, and therefore the reader is encouraged to consult each individual publication for a complete description of the results.

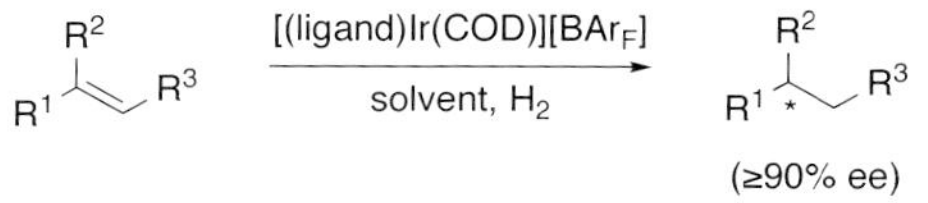

$$\text{R}^1\underset{}{\overset{\text{R}^2}{\diagup\!\!\!\diagdown}}\text{R}^3 \xrightarrow[\text{solvent, H}_2]{[(\text{ligand})\text{Ir}(\text{COD})][\text{BArF}]} \text{R}^1\underset{*}{\overset{\text{R}^2}{\diagup\!\!\!\diagdown}}\text{R}^3 \quad (\geq 90\% \text{ ee}) \tag{2.2}$$

Terminal alkenes (entry 4) have proved difficult to reduce with high selectivities. Reported selectivities for the reduction of 2-(4-methoxyphenyl)-1-butene range from 60% to 97% ee, ligands **(4)** (R^1 = *o*-Tol, R^2 = *t*-Bu) and **(16)** (Ar = *o*-Tol) respectively [6, 22]. Interestingly, temperature and pressure play crucial roles in the selectivity of asymmetric hydrogenation of this substrate. Burgess reported that, at 25 °C, decreasing the pressure from 85 to 2 bar had the dramatic effect of increasing selectivity from 25% to 90% ee, using **(23)** (R^1 = 2,6-(i-Pr)$_2$-Ph, R^2 = 1-Ad). However, in sharp contrast, at 85 bar and −15 °C the product having the opposite absolute configuration was obtained in 60% ee [27a].

Although Crabtree's achiral catalyst **(1)** [5] is highly efficient in reducing tetrasubstituted alkenes, only limited success has been reported with asymmetric versions (entry 5) [6]. The only other catalysts that have reacted with both high conversion and selectivity are the cationic zirconocene complexes (see below) reported by Buchwald [34].

More interesting substrates are the *weakly* coordinating functionalized alkenes such as the allylic alcohols, α,β-unsaturated acetates and *trans*-β-cinnamic esters (entries 7–9), because they allow the possibility for further functionalization. Allylic alcohols have been only briefly investigated and the substrate range is highly limited; only Ar = Ph and p-(t-Bu)-C$_6$H$_4$ have been reported [6, 35]. The highest reported ee value for *trans*-2-methyl-3-phenyl-prop-2-ene-1-ol (entry 7) is 99%, by Andersson and coworkers, applying the thiazole-based ligand **(17)** (Ar = Ph, o-Tol, R = Ph). However, other ligands perform with similar levels of selectivity and conversion [9, 19, 21, 23].

These *standard* substrates are what many researchers commonly use in their investigations; however, many less common compounds have also been reported as substrates (Table 2.3).

Pfaltz has reported the asymmetric hydrogenation of a range of substrates with heteroaromatic substituents. These were reduced with excellent enantioselectivities (Figure 2.6) [35]. These are interesting substrates due to the application of heterocycles in industry. Although the catalytic activity of Ir complexes with N,P-ligands can be substantially reduced in the presence of coordinating groups, thiophene, furan and pyrrole substrates were all successfully reduced, in excellent ee and yield, although pyridyl substrates inhibited the reaction.

More recently, Pfaltz has expanded the scope to include the asymmetric hydrogenation of furans (Figure 2.7) [24b]. These reductions permit the enantioselective synthesis of tetrahydrofuran and benzodihydrofuran systems, which are found in many natural products and biologically active compounds.

Dienes and polyenes are an interesting substrate class as their asymmetric reduction has the potential to generate multiple enantiomeric centers with control of enantio- and diastereoselectivities (Equation 2.3).

$$\underset{}{\text{R}}\underset{\text{R}}{\overset{\text{R}}{\diagup}}\xrightarrow[\text{H}_2]{[\text{Ir}]} \underset{\text{ent}}{\text{R}\diagup\text{R}} + \underset{\text{meso}}{\text{R}\diagup\text{R}} \tag{2.3}$$

Prior to the work of Burgess and coworkers, few research groups were engaged in this field [36–38]. Burgess investigated the asymmetric hydrogenation of a

Table 2.3 The best results of Ir-catalyzed asymmetric hydrogenation of various substrates.[a]

Entry	Alkene	Ligands	Conversion (%)	ee (%)
1	2-naphthyl C(=CH₂)Ph	(23): (R^1 = 2,6-(i-Pr)$_2$-C$_6$H$_3$, R^2 = 1-Ad)	90	93
2	p-MeO-C$_6$H$_4$-C(Et)=CHPh	(4): (R^1 = o-Tol, R^2 = t-Bu)	97	95
3	p-MeO-C$_6$H$_4$-C(Me)=CHEt	(23): (R^1 = 2,6-(i-Pr)$_2$-C$_6$H$_3$, R^2 = 1-Ad)	95	84
4	p-MeO-C$_6$H$_4$-C(Et)=C(Me)Ph	(4): (R^1 = o-Tol, R^2 = t-Bu)	97	95
5	p-MeO-C$_6$H$_4$-CH=C(Me)Et	(23): (R^1 = 2,6-(i-Pr)$_2$-C$_6$H$_3$, R^2 = 1-Ad)	58	49
6	p-MeO-C$_6$H$_4$-C(Me)=CH(i-Pr)	(23): (R^1 = 2,6-(i-Pr)$_2$-C$_6$H$_3$, R^2 = 1-Ad)	>99	97
7	p-MeO-C$_6$H$_4$-C(Me)=CH(i-Pr)	(23): (R^1 = 2,6-(i-Pr)$_2$-C$_6$H$_3$, R^2 = 1-Ad)	92	84
8	Ph-CH$_2$-C(Me)=CH-CO$_2$Et	(19): (n = 1, R^1 = o-Tol, R^2 = Ph)	99	88
9	Ph-CH$_2$-C(Me)=CH-CO$_2$Et	(19): (n = 1, R^1 = t-Bu, R^2 = Ph)	99	90
10	Ph-CH$_2$-furanone	(4): (R^1 = Ph, R^2 = i-Pr)	58	12
11	Cl-C$_6$H$_4$-C(=CH$_2$)-CH$_2$OH	(4): (R^1 = Ph, R^2 = i-Pr)	99	88
12	Ph-CH$_2$-C(Me)=CH-CH$_2$OH	(14): (R^1 = Cy, R^2 = Ph, R^3 = Bn)	100	82

a Conditions: cat. 0.01–2 mol%, 50 bar H$_2$, 2 h, CH$_2$Cl$_2$.

number of dienes using the [Ir]-**(23)** (R^1 = 2,6-(i-Pr)$_2$-C$_6$H$_3$, R^2 = Ad) complex (Table 2.4).

Enantioselectivity of the major product, in most cases, was very good at 70–99% (except entry 3), though diastereoselectivity in favor of the *ent* form was generally poor except for entries 7 and 8.

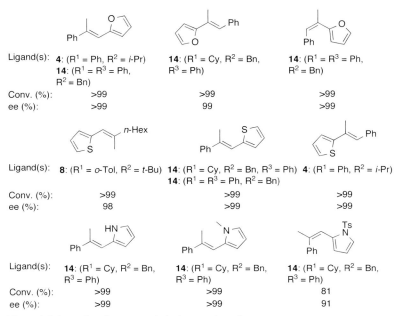

Figure 2.6 Ir-catalyzed asymmetric hydrogenation of substrates with heteroaromatic substituents.

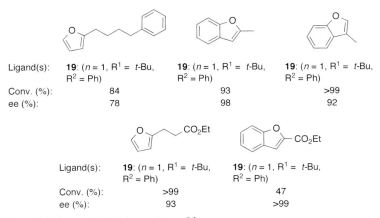

Figure 2.7 Asymmetric hydrogenation of furans.

All the substrates discussed so far have contained at least one aryl substituent close to the C—C double bond. This narrow substrate scope limits the potential usefulness of the catalysts developed; there is a need for catalysts that are able to reduce purely alkyl-substituted alkenes. This reaction has not been investigated

Table 2.4 Asymmetric hydrogenation of dienes.[a]

Entry	Alkene	Conversion (%)	ent:meso	ee[b] (%)
1	Ph, Ph (1,3-diene)	>99	1.0:2.9	87
2	Ph, Ph	>99	2.1:1.0	86
3	Ph, Ph	88	1.0:2.9	24
4	Ph, Ph	>99	1.3:1.0	98
5	(m-tolyl)-diene-(m-tolyl)	>99	1.3:1.0	97
6	(5-Me-furyl)-diene-(5-Me-furyl)	>99	1.8:1.0	99
7	Ph, Ph	93	14:1.0	98
8	(p-tolyl)-diene-(p-tolyl)	>99	20:1.0	99
9	(5-Me-furyl)-diene-(5-Me-furyl)	>99	1.2:1.9	70
10	Ph, Ph	>99	5.8:1.0	99

a Conditions: cat. 1 mol%, 50 atm H$_2$, 24 h, CH$_2$Cl$_2$.
b Major product.

extensively, except for the use of Ir-**(19)** complex ($n = 1$, R^1 = t-Bu, R^2 = Ph). Using Ir-**(19)**, Bell *et al.* were able to reduce 2-cyclohexylbutene in 92% ee [24a]. This rather simple substrate served as a proof of concept, and the authors continued their investigations. They hydrogenated useful natural product precursors such as γ-tocotrienyl acetate with great success (Equation 2.4).

>98% R,R,R: <0.5% R,S,R: <0.5% R,S,S

>99% Conv.

(2.4)

Andersson and coworkers have recently reported the hydrogenation of a range of vinyl silanes using [Ir(**17**)(COD)][BAr$_F$] (Table 2.5) [39]. Vinyl silanes are useful building blocks that can be used in a number of synthetic processes, such as the Tamao–Fleming oxidation [40]. Conversions were excellent in all cases, though apart from one substrate (entry 2, 98% ee) only moderate enantioselectivities (28–58% ee) were obtained.

Table 2.5 Ir-catalyzed hydrogenation of vinyl silanes.[a]

Entry	Alkene	Conversion (%)	ee (%)
1	Ph, TMS alkene	>99	28
2	Ph, TMS alkene	>99	98
3	Ph, TMS alkene	>99	58
4	tetralin-TMS	>99	48
5	Cl-alkyl, TMS alkene	>99	55
6	Ph, SiMe$_2$Ph alkene	>99	55

a Conditions: cat. 0.5 mol%, 30 bar H$_2$, 12 h, CH$_2$Cl$_2$.

Table 2.6 Ir-catalyzed hydrogenation of enol phosphinate esters.[a]

Entry	Alkene	Conversion (%)	ee (%)
1	Ph, O-P(=O)(Ph)Ph	>99	95
2	(4-MeC₆H₄), O-P(=O)(Ph)Ph	97	96
3[b]	(4-MeOC₆H₄), O-P(=O)(Ph)Ph	48	98
4	(4-BrC₆H₄), O-P(=O)(Ph)Ph	>99	>99
5	(4-F₃CC₆H₄), O-P(=O)(Ph)Ph	>99	99
6	(4-O₂NC₆H₄), O-P(=O)(Ph)Ph	>99	92
7	Cy, O-P(=O)(Ph)Ph	>99	92
8	t-Bu, O-P(=O)(Ph)Ph	>99	>99

a Conditions: cat. 0.5 mol%, 30 bar H$_2$, CH$_2$Cl$_2$.
b Reaction performed in the presence of poly(4-vinylpyridine) resin (10 mg), cat. 2 mol% and 50 bar H$_2$.

More recently, Andersson and coworkers reported the asymmetric hydrogenation of a range of enol phosphinate esters and tri- and tetrasubstituted fluorinated olefins using [Ir(**22**)(COD)][BAr$_F$] (R^1 = o-Tol, R^2 = H, R^3 = i-Pr) [41, 42]. Aryl enol phosphinate esters bearing electron-withdrawing groups at the *para* position (entries 4–6, Table 2.6) were more reactive (full conversion in 0.5–1 hours) than those bearing electron-donating groups (entries 2 and 3) (full conversion in 3–4 hours) [41]. Of particular interest was the reduction of alkyl enol phosphinates (entries 7 and 8), as purely aliphatic substrates are typically difficult to reduce selectively. Andersson and coworkers obtained excellent selectivities and conversions for these substrates. The resulting alkyl phosphinate esters were easily converted to the corresponding alcohols, with no loss of ee, via treatment with n-BuLi.

Andersson and coworkers used [Ir(**22**)(COD)][BAr$_F$] (R^1 = o-Tol, R^2 = H, R^3 = i-Pr) to reduce a number of tri- and tetrasubstituted fluorinated olefins with ee values up to 90% (Table 2.7) [42]. They increased enantioselectivities up to >99% (entries 2 and 3) using [Ir(**28**)(COD)][BAr$_F$] (Figure 2.8). Importantly, they were able to

Table 2.7 Hydrogenation of tri- and tetrasubstituted fluorinated olefins.^{a}

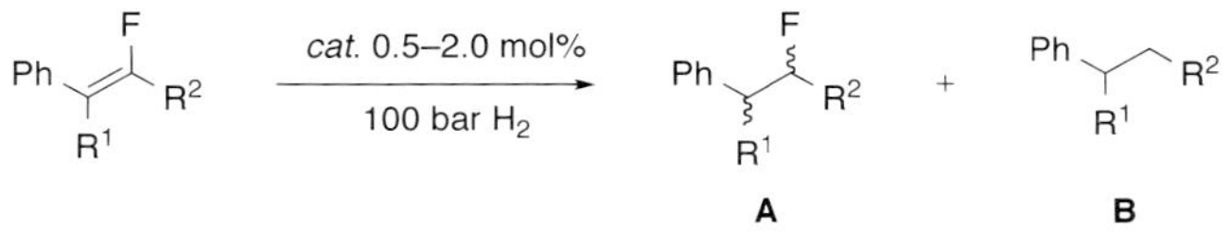

Entry	Alkene	cat.	Conversion (%)	ee (%)	Ratio A:B
1	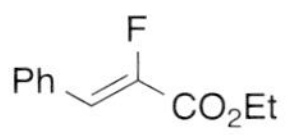 Ph–C(F)=C(H)–CO$_2$Et	[Ir(**22**)(COD)][BAr$_F$]	99	29	98:2
2	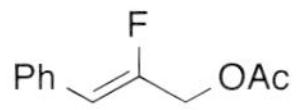 Ph–C(F)=C(H)–OAc	[Ir(**28**)(COD)][BAr$_F$]	82	>99	95:5
3	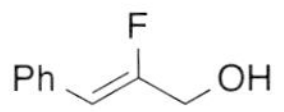 Ph–C(F)=C(H)–OH	[Ir(**28**)(COD)][BAr$_F$]	97	99	100:0
4	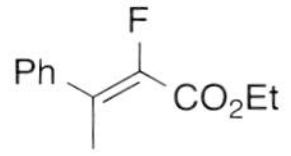 Ph–C(Me)=C(F)–CO$_2$Et	[Ir(**22**)(COD)][BAr$_F$]	21	57	100:0
5	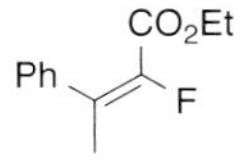 Ph–C(Me)=C(CO$_2$Et)–F	[Ir(**22**)(COD)][BAr$_F$]	25	74	100:0
6	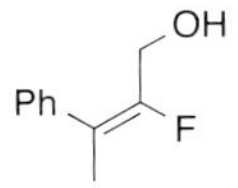 Ph–C(Me)=C(OH)–F	[Ir(**22**)(COD)][BAr$_F$]	24	90	71:29

a Conditions: cat. 0.5–2.0 mol%, RT–40 °C, 20–100 bar H$_2$, CH$_2$Cl$_2$.

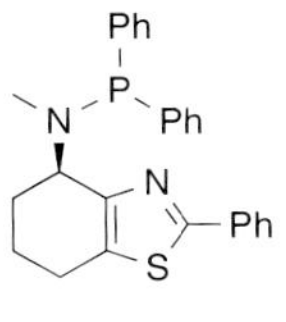

Figure 2.8 Ligand (**28**) used in the Ir-catalyzed hydrogenation of fluorinated olefins.

hydrogenate olefins having two prochiral carbons (entries 4–6) with iridium for the first time.

However, as shown in Table 2.7, defluorination was found to be an issue. Subsequent experiments showed that defluorination occurred prior to hydrogenation of the carbon–carbon double bond.

2.2.3
Mechanism

Recent studies have suggested a different mechanism from the original proposal for Ir-catalyzed hydrogenation of olefins [43]. However, these experimental and computational studies each arrived at different conclusions about the catalytic cycle.

Studies have shown that different pathways are likely to operate depending on hydrogen pressure, temperature, substrate and catalyst [27a, 44]. In addition, for a number of substrates, deuterium studies have shown significant levels of deuterium incorporation at the allylic positions of the carbon–carbon double bond, indicating the formation of Ir-allyl intermediates and/or double bond migration [45].

Brandt *et al.* reported an experimental and theoretical study from which they proposed an Ir(III)–Ir(V) catalytic cycle (Scheme 2.1, pathway **A**) [46]. Following COD hydrogenation and dissociation, the available coordination sites are occupied with two solvent molecules and two hydride ligands giving a solvated Ir-dihydride complex. Molecular hydrogen and an olefin replace the two solvent molecules, followed by the rate-determining migratory insertion of the olefin into an Ir-hydride bond. Subsequent reductive elimination of the alkane completes the catalytic cycle.

Burgess and Hall have also proposed a similar Ir(III)–Ir(V) catalytic cycle (Scheme 2.1, pathway **B**) [47]. However, their studies suggested that the olefin reacts metathetically with the coordinated molecular hydrogen, in preference to a hydride, giving a σ-alkyl-Ir(V) complex. This is followed by fast reductive elimination of the alkane. Burgess has further postulated the existence of an Ir(V) intermediate from a kinetic study of the hydrogenation of 2,3-diphenylbutadiene [36].

However, Chen *et al.* studied the Ir-catalyzed hydrogenation of styrene, using electrospray ionization tandem mass spectrometry, and interpreted the results as discounting an Ir(III)–Ir(V) mechanism [48]. Analyzing catalytically active solutions under H_2 pressure, they observed three new peaks whose masses corresponded to the composition [Ir(**4**)(styrene)]$^+$, [Ir(**4**)(styrene)(H_2)]$^+$ and [Ir(**4**)(styrene)(H_2)$_2$]$^+$ ((**4**); Ar = Ph, R = *i*-Pr). Although the latter species could arise from either of the mechanisms discussed, gas-phase experiments suggested that it played no active role in the catalytic cycle. When they bombarded [Ir(**4**)(styrene)]$^+$ with D_2, products corresponding to incorporation of one or of two deuteriums were observed [Ir(**4**)(styrene)]$^+$-D_1 and [Ir(**4**)(styrene)]$^+$-D_2 but not of three. Under their reaction conditions they argued that if an Ir(III)–Ir(V) cycle was active, then a species of the corresponding to [Ir(**4**)(styrene)(D_2)$_2$]$^+$ would have the opportunity

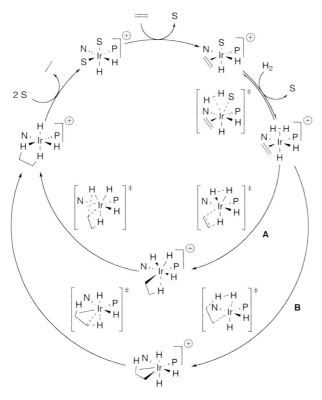

Scheme 2.1 Catalytic cycles for Ir-catalyzed hydrogenation proposed by Brandt (pathway **A**) and Burgess and Hall (pathway **B**).

to form, and so "given the reversibility of elementary steps in the gas-phase reaction, even a transient Ir(V) intermediate with three chemically equivalent deuterides (and an alkyl group with one deuterium atom) would produce at least partial incorporation of more than two deuterium atoms into the styrene substrate." Based on the data obtained, they instead proposed that the mechanism proceeds via an Ir(I)–Ir(III) dihydride cycle (Scheme 2.2).

2.3
Ruthenium and Rhodium Catalysis

Although Wilkinson's catalyst [RhCl(PPh$_3$)$_3$] [49–51] is an ineffective catalyst for the hydrogenation of tri- or tetrasubstituted alkenes, the cationic complex **(29)** (Figure 2.9) has been shown to be as effective as Crabtree's catalyst [52]. Recently

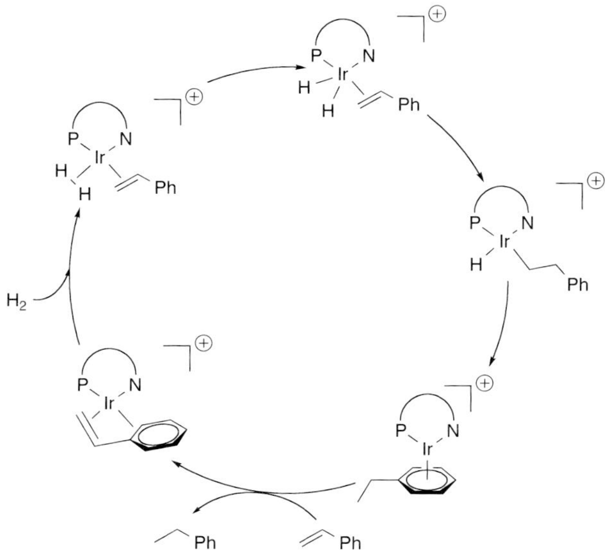

Scheme 2.2 Chen's proposed mechanism for iridium-catalyzed hydrogenation.

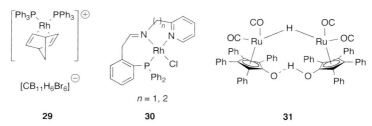

Figure 2.9 Achiral rhodium and ruthenium catalysts for olefin hydrogenation.

Osborn and coworkers have reported that the complex (**30**) is active in the hydrogenation of 1-methylcyclohexene [53]. Shvo's catalyst (**31**), typically used in the hydrogenation of aldehydes and ketones, has also been shown to catalyze the hydrogenation of 1-methylcyclohexene quantitatively at high temperature and pressure [54].

Whereas Ru and Rh complexes are the catalysts of choice for functionalized alkenes, there are no real practical reports of their use in asymmetric hydrogenations of unfunctionalized tri- or tetrasubstituted alkenes. Their chemistry with 1,1-disubstituted alkenes is more substantial, though the substrate scope is small and only moderate enantioselectivities have been reported. A number of Rh catalysts with chiral phosphine ligands have been reported (Figure 2.10), and although they have been investigated in the hydrogenation of styrene derivatives, few studies concern substrates other than 2-phenyl-1-butene [55].

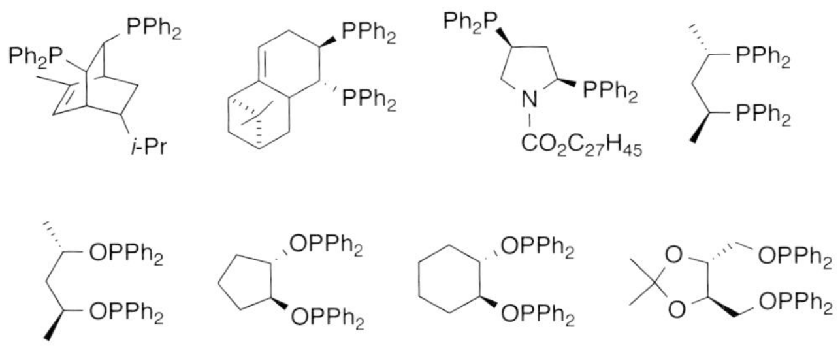

PPh2
Ph2P
i-Pr
PPh2
PPh2
Ph2P
N
PPh2
CO2C27H45
PPh2
PPh2
OPPh2
OPPh2
OPPh2
OPPh2
OPPh2
OPPh2
O
O
OPPh2
OPPh2

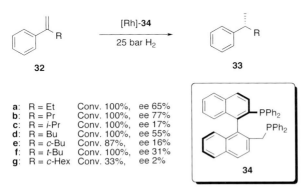

Figure 2.10 Chiral phosphine ligands tested in Ru- and Rh-catalyzed hydrogenations.

R
[Rh]-34
25 bar H2
R
32
33
a: R = Et Conv. 100%, ee 65%
b: R = Pr Conv. 100%, ee 77%
c: R = i-Pr Conv. 100%, ee 17%
d: R = Bu Conv. 100%, ee 55%
e: R = c-Bu Conv. 87%, ee 16%
f: R = t-Bu Conv. 100%, ee 31%
g: R = c-Hex Conv. 33%, ee 2%
PPh2
PPh2
34

Scheme 2.3 Rhodium catalyzed hydrogenation of some of disubstituted alkenes.

Two reports from Takaya and coworkers stand out as exceptional among rhodium-catalyzed asymmetric hydrogenations. They evaluated the hydrogenation of 2-phenyl-1-butene derivatives catalyzed by $[RhCl(COD)]_2$ and **(34)** (Scheme 2.3) [56]. Substrates bearing linear alkyl chains were smoothly hydrogenated with moderate enantioselection; the best ee value was 77% for substrate (**32b**). However, substrates with either secondary or tertiary alkyl groups were only hydrogenated with only poor selectivities (**32c**, **32e**, **32f** and **32g**).

They also reported the use of several Rh and Ru complexes in the asymmetric hydrogenation of various terminal alkenes [57]. The best results (80% ee and 81% conversion; entry 3, Table 2.8) were obtained with $[RhI(COD)]_2$/(*R*)-BINAP. Using the catalyst system $[RhX(COD)]_2$/(*R*)-BINAP (X = I, Br, Cl, OAc and OPh), they investigated the effects of H_2 pressure and the Rh source on the reaction, and found that enantioselectivities depended heavily on the anionic ligand used, with selectivity decreasing in the order I > Br > Cl, OAc, OPh. Although the H_2 pressure had little effect on enantioselectivity, enhanced reaction rates were observed at higher pressures.

Takaya also investigated a number of Ru-BINAP complexes with these substrates, and with $[Ru(R)\text{-BINAP}(OAc)_2]$ obtained results similar to those for $[RhI(COD)]_2$/(*R*)-BINAP (Table 2.8). The effects of changing the Ru source and H_2

Table 2.8 Asymmetric hydrogenation of various terminal alkenes by Takaya and coworkers.[a]

Entry	Alkene		[Rh]		[Ru]	
			Conversion (%)	ee (%)	Conversion (%)	ee (%)
1	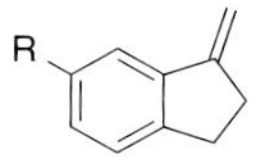	R = H	100	66	100	78
2		R = OMe	100	43	100	45
3	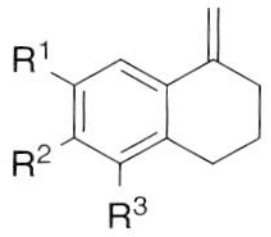	R^1 = H, R^2 = H, R^3 = H	81	80	90	69
4		R^1 = OMe, R^2 = H, R^3 = H	85	71	100	65
5		R^1 = H, R^2 = OMe, R^3 = H	53	77	100	75
6		R^1 = H, R^2 = H, R^3 = OMe	100	47	100	61
7	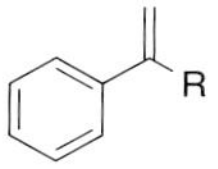	R = Et	100	29	100	9
8		R = i-Pr	70[b]	35[b]	70	16
9		R = t-Bu	22	7	11	30

a Conditions: [Rh] = [RhI(COD)]$_2$/(R)-BINAP, CH$_2$Cl$_2$, 25 bar H$_2$, 30 °C; [Ru] = [Ru(R)-BINAP(OAc)$_2$], 100 bar H$_2$, 30 °C.
b [RhCl(COD)]$_2$ was used as the rhodium source.

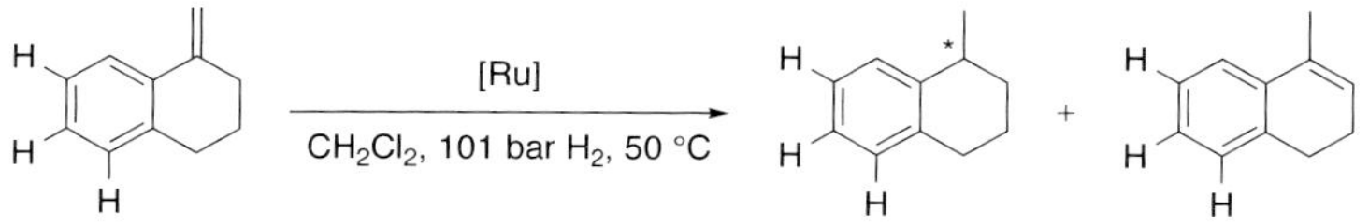

[Ru] source	35(ee%):36
Ru(OAC)$_2$ / (R)-BINAP	89(66):11
Ru$_2$Cl$_4$ / (R)-BINAP / NEt$_3$	49(65):51
[RuI(R)-BINAP(p-cymene)]I	0(-):100

Scheme 2.4 Chemoselectivity of terminal olefin hydrogenation by asymmetric ruthenium catalysts.

pressure were also investigated. The enantioselectivity of hydrogenation was mostly independent of H$_2$ pressure and the Ru source, but chemoselectivity depended strongly on the Ru source (Scheme 2.4). No isomerization was observed in the absence of H$_2$, leading the authors to propose that it was caused by a metal-hydride species.

Another ruthenium catalyst that catalyses the hydrogenation of unfunctionalized alkenes is [RuCl$_2$(R,R)-Me-Duphos](dmf)$_n$ [58]. Ruthenium complexes of the

2.4 Chiral Metallocene Catalysts

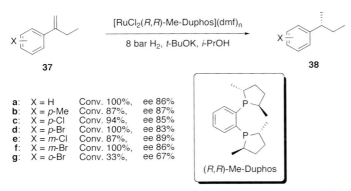

a:	X = H	Conv. 100%,	ee 86%
b:	X = p-Me	Conv. 87%,	ee 87%
c:	X = p-Cl	Conv. 94%,	ee 85%
d:	X = p-Br	Conv. 100%,	ee 83%
e:	X = m-Cl	Conv. 87%,	ee 89%
f:	X = m-Br	Conv. 100%,	ee 86%
g:	X = o-Br	Conv. 33%,	ee 67%

Scheme 2.5 [RuCl$_2$(R,R)-Me-Duphos](dmf)$_n$-catalyzed hydrogenation of disubstituted olefins.

type [RuCl$_2$(diphosphine)](dmf)$_n$ generally have a low reactivity toward the hydrogenation of styrenes [58]. However, Noyori and coworkers reported that an alkali base, such as t-BuOK, in 2-propanol remarkably enhanced reaction rates (Scheme 2.5). The combination of catalyst with t-BuOK and 2-propanol proved crucial for the success of the reaction. Other phosphines and alcohols tested resulted in low yields and enantioselectivities, and in the absence of a strong base the initial reaction rate was around 10 times slower and the opposite sense of enantioselection was observed [58].

The term "unfunctionalized alkene" is a misnomer [2]; many reports include weakly coordinating substrates such as α,β-unsaturated esters, diketenes, enol ethers and anhydrides in this category (Figure 2.11). These substrates have been reduced with either Rh or Ru catalysts, and this section briefly illustrates some of the substrates used and the results obtained. However, for a more thorough discussion, one should turn to a text dealing with reduction of functionalized alkenes.

Ruthenium catalysts have also been evaluated for other substrates, such as cyclic lactones and ketones, giving ee values ranging from moderate to excellent (Scheme 2.6) [59, 60].

A number of 2-pyrones and enones have also successfully been hydrogenated by Rh-(R,R)-Me-Duphos [61–63] or Ru-(**34**) [64] catalysts.

2.4 Chiral Metallocene Catalysts

2.4.1 Titanium and Zirconium Systems

Early results by Kagan and coworkers showed the possibility of using chiral non-racemic titanocene complexes (Figure 2.12) in the asymmetric reduction of simple

2 Hydrogenation of Unfunctionalized Alkenes

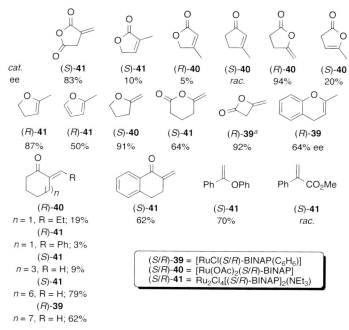

Figure 2.11 Some substrates evaluated with various Ru-BINAP complexes [59]. [a] 0.9 NEt$_3$ added.

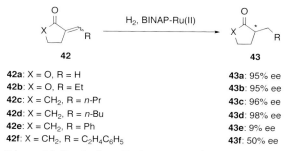

42a: X = O, R = H
42b: X = O, R = Et
42c: X = CH$_2$, R = n-Pr
42d: X = CH$_2$, R = n-Bu
42e: X = CH$_2$, R = Ph
42f: X = CH$_2$, R = C$_2$H$_4$C$_6$H$_5$

43a: 95% ee
43b: 95% ee
43c: 96% ee
43d: 98% ee
43e: 9% ee
43f: 50% ee

Scheme 2.6 Results from BINAP-Ru(II)-catalyzed hydrogenation of five membered cyclic ketones and lactones [59].

olefins. For example, the prochiral olefin 2-phenyl-1-butene was reduced in 15% ee [65]. Vollhardt showed that when the titanocene precatalyst **(44)** was activated using n-BuLi optical yields of up to 33% could be achieved in the reduction of 2-phenyl-1-butene [66]. A number of other research groups have prepared various chiral nonracemic group (IV) metallocene precatalysts, which generally use an alkyl lithium, such as n-BuLi, as a cocatalyst for activation [67]. Although the best

R* = (−)-menthyl, **44a**; (+)-neomenthyl, **44b**   (−)-menthyl   (+)-neomenthyl

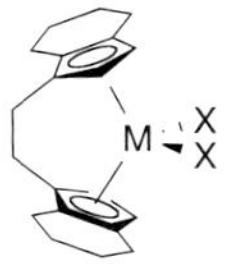

M = Ti, **45**; Zr, **46**

Figure 2.12 Chiral nonracemic group (IV) metallocenes.

enantioselectivities were high, 96% ee [67e], only a small number of prochiral 1,1-disubstituted alkenes have been reported.

It was not until Buchwald and coworkers reported the use of reduced forms of Brintzinger's complexes **(45)** (X_2 = binaphtholate) [68] and **(46)** (X = Me) [34] that group (IV) metallocenes (Figure 2.12) became interesting synthetically. The titanocene precatalyst **(45)** (X_2 = binaphtholate), activated with n-BuLi and stabilized with phenylsilane, was used in the asymmetric hydrogenation of trisubstituted alkenes with great success; many substrates were reduced in >90% ee (entries 1–8, Table 2.9) [68]. Of potentially greater interest was the reduction of a number of tetrasubstituted alkenes (entries 9–16) with the precatalyst **(46)** (X = Me) [69]. This time Buchwald activated the zirconocene precatalyst **(46)** with an ammonium tetrakis(pentafluorophenyl)borate [$PhMe_2HN$][$B(C_6F_5)_4$] cocatalyst. In general the substrates were limited to cyclic alkenes (entries 10–16); however, the alkanes were obtained with high ee, 52–99%. Interestingly, in the case of 2-methyl-3-phenylindene (entry 15), the sense of enantioselection was found to be pressure-dependent: at 6 bar of H_2, the major enantiomer was the opposite of that obtained at 138 bar.

It was proposed that the mechanism proceeded via formation of a metallocene-hydride and subsequent alkene insertion to give a metal alkyl. Hydrogenolysis would then regenerate the metal hydride and form the product. This model correctly predicted the known absolute configurations of the products, though it was not supported by any other evidence [68].

Drawbacks such as high catalyst loadings (generally 5 mol%), high pressures of H_2 (up to 138 bar) and the highly air-sensitive nature of the metal complexes mean that although these group (IV) metallocenes perform well across a wide range of substrates and give the best results to date for tetrasubstituted alkenes, they are unlikely to find widespread application.

Table 2.9 Asymmetric hydrogenation using (**45**) and (**46**).

Entry	Alkene	Yield (%)	ee (%)	Entry	Alkene	Yield (%)	ee (%)
1	Ph-CH=CH-Ph	94	>99	9	F-C₆H₄-C(Me)=CHMe (isopropylidene)	77	96
2	MeO-C₆H₄-C(=CHPh)-	79	95	10	1-methylindene	87	93
3	1-methyl-dihydronaphthalene	70	83	11	1-methyl-3-n-Bu-indene	96	92
4	6-MeO-1-methyl-dihydronaphthalene	70	93	12	1-methyl-3-Ph-indene	89	99
5	6-MeO-1-Ph-dihydronaphthalene	87	87	13	1-n-Bu-3-Ph-indene	94	98
6	6-MeO-3-methyl-dihydronaphthalene	77	92	14	1-methyl-3-Et-indene	95	52
7	Ph-C(Me)=CH-CH₂-NBn₂	74	95	15	1-Ph-indene	94	78
8	Ph-C(Me)=CH-CH₂-OMe	86	94	16	1-methyl-dihydronaphthalene	91	92

Entries 1–8: (**45**), 1 bar H$_2$. Entries 9, 10 and 12–16: (**46**), 69–138 bar H$_2$. Entry 11: (**46**), 6 bar H$_2$.

2.4.2
Lanthanide Systems

Marks and coworkers published the synthesis and use of a chiral nonracemic lanthanide catalysts, (**47**) and (**48**), in the asymmetric hydrogenation of unfunctionalized alkenes (Figure 2.13) [70–73].

High TOFs were observed (up to 26 000 h^{-1} at 25 °C in the reduction of 2-phenylbut-1-ene with (**48**) (Ln = Sm, R* = (+)-neomenthyl)). The best ee value reported for this substrate was 98%, using the precatalyst 70/30 (S/R)-(**48**) (M = Sm, R* = (−)-menthyl), at −80 °C. It was observed that enantioselection was sensitive to temperature and to the size of the Ln metal, with ee value decreasing markedly with decreasing ionic radius of Ln.

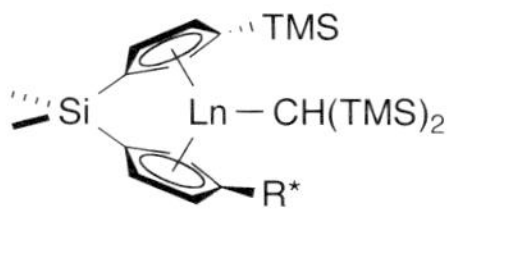

47
Ln = Y or Lu
R* = (+)-neomenthyl of (–)-menthyl

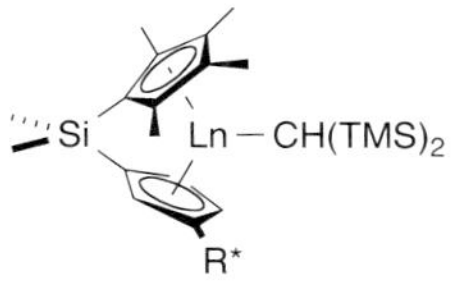

48
Ln = La, Nd, Sm, Y or Lu
R* = (+)-neomenthyl, (–)-menthyl or (–)-phenylmenthyl

Figure 2.13 Lanthanide-based catalysts for the asymmetric hydrogenation of unfunctionalized olefins.

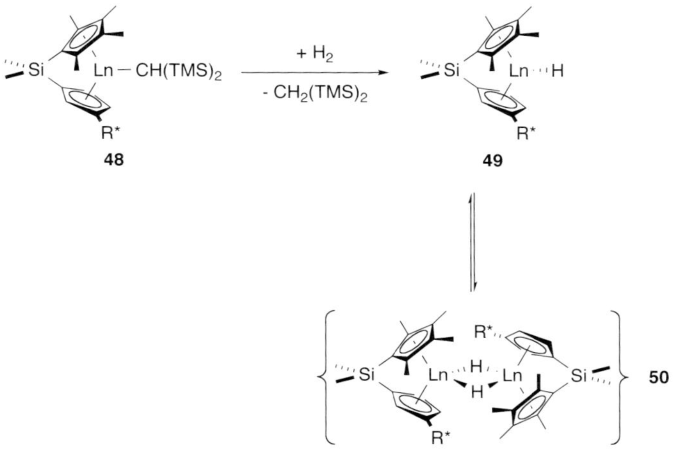

Scheme 2.7 Formation of Ln-H **(49)** and Ln-H dimer **(50)**.

Although few substrates were reported, Marks and coworkers performed a highly detailed study of the reaction mechanism [72], and found it to be both ligand- and substrate-dependent. They proposed that in the case of **(48)** (M = Sm) initial hydrogenolysis of the $LnCH(TMS)_2$ bond formed the transitory Ln-H **(49)**, which subsequently underwent rapid olefin insertion followed by another hydrogenolysis giving the product and reforming **(49)**. The Ln-H dimer **(50)**, in equilibrium with **(49)**, was considered too unreactive to play any major role in the catalysis (Scheme 2.7).

Although asymmetric hydrogenation by these lanthanide catalysts is mechanistically interesting, proceeding via a pathway similar to the one outlined by Buchwald for group (IV) metallocene catalysts (see above), there are issues regarding their synthetic applicability. The Ln complexes are extremely air-sensitive and difficult to prepare, and were demonstrated only for the asymmetric hydrogenations of two 1,1-disubstituted alkenes and the asymmetric deuterations of styrene and 1-pentene.

2.5 Conclusion

The hydrogenation of unfunctionalized alkenes is still a challenging reaction. Limited substrate scope and difficult handling and preparation have limited the practical applications of the chiral metallocene catalysts reported to date. Ru- and Rh-based hydrogenations of unfunctionalized alkenes have achieved only moderate success with a range of 1,1-disubstituted alkenes. However, for substrates with weakly coordinating functionalities, good enantioselectivities are possible.

The current focus in the hydrogenation of unfunctionalized alkenes is on iridium catalysis, which has been based on the pioneering work of Crabtree and, more recently, Pfaltz. Since the work of Pfaltz, ligand design and evolution have been the driving forces in chiral iridium catalysis for asymmetric hydrogenations. Only recently has the development of substrate scope been reported. However, there still remain many substrate classes, such as 1,1-disubstituted and tetrasubstituted alkenes, for which only limited examples have been reported with satisfactory results. There also exists no generally accepted mechanism for the iridium-catalyzed hydrogenation of unfunctionalized alkenes.

It may be hoped that further catalyst development will allow chemists to move beyond the standard substrates, which have dominated the field to date, to more practically interesting substrates.

References

1 Brown, J.M. (1999) *Comprehensive Asymmetric Catalysis*, Vol. 1 (eds E.N. Jacobsen, A. Pfaltz and H. Yamamoto), Springer-Verlag, Berlin, pp. 121–82.
2 Cui, X. and Burgess, K. (2005) *Chemical Reviews*, **105**, 3272.
3 (a) Hershberg, E.B., Oliveto, E.P., Gerold, C. and Johnson, L. (1951) *Journal of the American Chemical Society*, **73**, 5073. (b) Rylander, P.N. (1979) *Catlytic Hydrogenation in Organic Synthesis*, Academic Press, New York.
4 Crabtree, R.H., Felkin, H. and Morris, G.E. (1977) *Journal of Organometallic Chemistry*, **141**, 205.
5 Crabtree, R.H. (1979) *Accounts of Chemical Research*, **12**, 331.
6 Lightfoot, A., Schnider, P. and Pfaltz, A. (1998) *Angewandte Chemie (International Ed. in English)*, **37**, 2897.
7 Sprinz, J. and Helmchen, G. (1993) *Tetrahedron Letters*, **34**, 1769.
8 Blackmond, D.G., Lightfoot, A., Pfaltz, A., Rosner, T., Schnider, P. and Zimmermann, N. (2000) *Chirality*, **12**, 442.
9 Cozzi, P.G., Zimmermann, N., Hilgraf, R., Schaffner, S. and Pfaltz, A. (2001) *Advanced Synthesis and Catalysis*, **343**, 450.
10 Cozzi, P.G., Menges, F. and Kaiser, S. (2003) *Synlett*, **XX** 833.
11 Menges, F., Neuburger, M. and Pfaltz, A. (2002) *Organic Letters*, **4**, 4713.
12 Liu, D., Tang, W. and Zhang, X. (2004) *Organic Letters*, **6**, 513.
13 Hou, D.-R., Reibenspies, J.H., Colacot, T.J. and Burgess, K. (2000) *Chemistry – A European Journal*, **7**, 5391.
14 Xu, G. and Gilbertson, S. (2003) *Tetrahedron Letters*, **44**, 953.
15 Hedberg, C., Källström, K., Brandt, P., Hansen, L.K. and Andersson, P.G. (2006) *Journal of the American Chemical Society*, **128**, 2995.
16 Bunlaksananusorn, T., Polborn, K. and Knochel, P. (2003) *Angewandte Chemie (International Ed. in English)*, **42**, 3941.

17 Schenkel, L.B. and Ellman, J.A. (2004) *Journal of Organic Chemistry*, **69**, 1800.
18 Trifonova, A., Diesen, J.S., Chapman, C.J. and Andersson, P.G. (2004) *Organic Letters*, **6**, 3825.
19 (a) Smidt, S.P., Menges, F. and Pfaltz, A. (2004) *Organic Letters*, **6**, 2023. (b) Smidt, S.P., Menges, F. and Pfaltz, A. (2004) *Organic Letters*, **6**, 3653.
20 Blankenstein, J. and Pfaltz, A. (2001) *Angewandte Chemie (International Ed. in English)*, **40**, 4445.
21 Menges, F. and Pfaltz, A. (2002) *Advanced Synthesis and Catalysis*, **344**, 40.
22 Källström, K., Hedberg, C., Brandt, P., Bayer, A. and Andersson, P.G. (2004) *Journal of the American Chemical Society*, **126**, 14308.
23 Drury, W.J., III, Zimmermann, N., Keenan, M., Hayashi, M., Kaiser, S., Goddard, R. and Pfaltz, A. (2004) *Angewandte Chemie (International Ed. in English)*, **43**, 70.
24 (a) Bell, S., Wüstenberg, B., Kaiser, S., Menges, F., Netscher, T. and Pfaltz, A. (2006) *Science*, **311**, 642. (b) Kaiser, S., Smidt, S.P. and Pfaltz, A. (2006) *Angewandte Chemie (International Ed. in English)*, **45**, 5194.
25 (a) Hilgraf, R. and Pfaltz, A. (1999) *Synlett: Accounts and Rapid Communications in Synthetic Organic Chemistry*, **XX**, 1814. (b) Hilgraf, R. and Pfaltz, A. (2005) *Advanced Synthesis and Catalysis*, **347**, 61.
26 For recent reviews see reference 2 and: (a) Perry, M.C. and Burgess, K. (2003) *Tetrahedron: Asymmetry*, **14**, 951. (b) Hermann, W.A. (2002) *Angewandte Chemie (International Ed. in English)*, **41**, 1290. (c) Hermann, W.A. and Köcher, C. (1997) *Angewandte Chemie (International Ed. in English)*, **36**, 2162.
27 (a) Perry, M.C., Cui, X., Powell, M.T., Hou, D.-R., Reibenspies, J.H. and Burgess, K. (2003) *Journal of the American Chemical Society*, **125**, 113. (b) Powell, M.T., Hou, D.-R., Perry, M.C., Cui, X. and Burgess, K. (2001) *Journal of the American Chemical Society*, **123**, 8878.
28 Nanchen, S. and Pfaltz, A. (2006) *Chemistry – A European Journal*, **12**, 4550.
29 Källström, K. and Andersson, P.G. (2006) *Tetrahedron Letters*, **47**, 7477.
30 Krossing, I. and Raabe, I. (2004) *Angewandte Chemie (International Ed. in English)*, **43**, 2066.
31 Smidt, S.P., Zimmermann, N., Studer, M. and Pfaltz, A. (2004) *Chemistry – A European Journal*, **10**, 4685.
32 Smidt, S.P., Pfaltz, A., Martínez-Viviente, E., Pregosin, P.S. and Albinati, A. (2003) *Organometallics*, **22**, 1000.
33 Lacour, J., Ginglinger, C., Grivet, C. and Bernardinelli, G. (1997) *Angewandte Chemie (International Ed. in English)*, **36**, 608.
34 Grossman, R.B., Doyle, R.A. and Buchwald, S.L. (1991) *Organometallics*, **10**, 1501.
35 Pfaltz, A., Blankenstein, J., Hilgraf, R., Hörmann, E., McIntyre, S., Menges, F., Schönleber, M., Smidt, S.P., Wüstenberg, B. and Zimmermann, N. (2003) *Advanced Synthesis and Catalysis*, **345**, 33.
36 Cui, X. and Burgess, K., (2003) *Journal of the American Chemical Society*, **125**, 14212.
37 Cui, X., Fan, Y., Hall, M.B. and Burgess, K. (2005) *Chemistry – A European Journal*, **11**, 6859.
38 Cui, X., Ogle, J. and Burgess, K. (2005) *Chemical Communications*, **XX**, 672.
39 Källström, K., Munslow, I.J., Hedberg, C. and Andersson, P.G. (2006) *Advanced Synthesis and Catalysis*, **348**, 2575.
40 (a) Fleming, I., Henning, R. and Plaut, H. (1984) *Journal of the Chemical Society, Chemical Communications*, **XX**, 29. (b) Tamao, K., Ishida, N., Tanaka, T. and Kumada, M. (1983) *Organometallics*, **2**, 1694.
41 Cheruku, P., Gohil, S. and Andersson, P.G. (2007) *Organic Letters*, **9**, 1659.
42 Engman, M., Diesen, J.S., Paptchikhine, A. and Andersson, P.G. (2007) *Journal of the American Chemical Society*, **129**, 4536.
43 Taqui Khan, M.M., Taqui Khan, B. and Begum, S. (1986) *Journal of Molecular Catalysis*, **34**, 9.
44 McIntyre, S., Hörmann, E., Menges, F., Smidt, S.P. and Pfaltz, A. (2005) *Advanced Synthesis and Catalysis*, **347**, 282.
45 Hou, D.-R., Reibenspies, J.H. and Burgess, K. (2001) *Journal of Organic Chemistry*, **66**, 206.
46 Brandt, P., Hedberg, C. and Andersson, P.G. (2003) *Chemistry – A European Journal*, **9**, 339.

47 Fan, Y., Cui, X., Burgess, K. and Hall, M.B. (2004) *Journal of the American Chemical Society*, **126**, 16688.
48 Dietiker, R. and Chen, P. (2004) *Angewandte Chemie (International Ed. in English)*, **43**, 5513.
49 Hussey, A.S. and Takeuchi, Y. (1969) *Journal of the American Chemical Society*, **91**, 672.
50 Osborn, J.A., Jardine, F.H., Young, J.F. and Wilkinson, G. (1966) *Journal of the Chemical Society*, **XX**, 1711.
51 Burgess, K. and Donkin, W.A. (1995) *Encyclopedia of Reagents for Organic Synthesis*, Vol. 2 (ed. L.A. Paquette), John Wiley & Sons, Inc., New York.
52 Rifat, A., Patmore, N.J., Mahon, M.F. and Weller, A.S. (2002) *Organometallics*, **21**, 2856.
53 Masson, J.-P., Bahsoun, A.A., Youinou, M.-T. and Osborn, J.A. (2002) *Comptes Rendus des Seances de l'Academie de Sciences (Sciences Chimiques)*, **5**, 303.
54 Blum, Y., Czarkie, D., Rahamin, Y. and Shvo, Y. (1985) *Organometallics*, **4**, 1459.
55 (a) Horner, L., Siegel, H. and Büthe, H. (1968) *Angewandte Chemie*, **80**, 1034. (b) Hayashi, T., Tanaka, M. and Ogata, I. (1977) *Tetrahedron Letters*, **18**, 295. (c) Tanaka, M. and Ogata, I. (1975) *Journal of the Chemical Society. Chemical Communications*, **XX**, 735. (d) Bakos, J., Toth, I., Heil, B. and Marko, L. (1985) *Journal of Organometallic Chemistry*, **279**, 23. (e) Achiwa, K. (1977) *Tetrahedron Letters*, **18**, 3735.
56 Inagaki, K., Ohta, T., Nozaki, K. and Takaya, H. (1997) *Journal of Organometallic Chemistry*, **531**, 159.
57 Ohta, T., Ikegami, H., Miyake, T. and Takaya, H. (1995) *Journal of Organometallic Chemistry*, **502**, 169.
58 Forman, G.S., Ohukuma, T., Hemms, W.P. and Noyori, R. (2000) *Tetrahedron Letters*, **41**, 9471.
59 Ohta, T., Miyake, T., Seido, N., Kumobayashi, H. and Takaya, H. (1995) *Journal of Organic Chemistry*, **60**, 357.
60 Le Gendre, P., Braun, T., Bruneau, C. and Dixneuf, P.H. (1996) *Journal of Organic Chemistry*, **61**, 8453.
61 Robinson, A., Li, H.-Y. and Feaster, J. (1996) *Tetrahedron Letters*, **37**, 8321.
62 Wiles, J.A., Bergens, S.H., Vanhessche, K.P.M., Dobbs, D.A. and Rautenstrauch, V. (2001) *Angewandte Chemie (International Ed. in English)*, **40**, 914.
63 Dobbs, D.A., Vanhessche, K.P.M., Brazi, E., Rautenstauch, V., Lenoir, J.-Y., Genet, J.-P., Wiels, J. and Bergens, S.H. (2000) *Angewandte Chemie (International Ed. in English)*, **39**, 1992.
64 Fehr, M.J., Consiglio, G., Scalone, M. and Schmid, R. (1999) *Journal of Organic Chemistry*, **64**, 5768.
65 Cesarotti, E., Ugo, R. and Kagan, H.B. (1979) *Angewandte Chemie (International Ed. in English)*, **18**, 779.
66 Halterman, R.L. and Vollhardt, K.P.C. (1988) *Organometallics*, **7**, 883.
67 (a) Cesarotti, E., Ugo, R. and Kagan, H.B. (1979) *Angewandte Chemie (International Ed. in English)*, **18**, 779. (b) Carlini, C., Chiellini, E. and Solaro, R. (1980) *Journal of Polymer Science*, **27**, 5599. (c) Cesarotti, E., Ugo, R. and Vitello, R. (1981) *Journal of Molecular Catalysis*, **12**, 63. (d) Paquette, L.A., McKinney, J.A., McLaughlin, M.L. and Rheingold, A.L. (1986) *Tetrahedron Letters*, **27**, 5599. (e) Halterman, R.L., Vollhardt, K.P.C., Welker, M.E., Blaser, D. and Boese, R. (1987) *Journal of the American Chemical Society*, **109**, 8105. (f) Waymouth, R. and Pino, P. (1990) *Journal of the American Chemical Society*, **112**, 4911. (g) Paquette, L.A., Sivik, M.R., Bzowej, E.I. and Stanton, K.J. (1995) *Organometallics*, **14**, 4865.
68 Broene, R.D. and Buchwald, S.L. (1993) *Journal of the American Chemical Society*, **115**, 12569.
69 Troutman, M.V., Appella, D.H. and Buchwald, S.L. (1999) *Journal of the American Chemical Society*, **121**, 4916.
70 Conticello, V.P., Brard, L., Giradello, M.A., Tsuji, Y., Sabat, M., Stern, S.L. and Marks, T.J. (1992) *Journal of the American Chemical Society*, **114**, 2761.
71 Girardello, M.A., Conticello, V.P., Brard, L., Gagné, M.R. and Marks, T.J. (1994) *Journal of the American Chemical Society*, **116**, 10241.
72 Haar, C.M., Stern, C.L. and Marks, T.J. (1996) *Organometallics*, **15**, 1765.
73 Roesky, P.W., Denninger, U., Stern, C.L. and Marks, T.J. (1997) *Organometallics*, **16**, 4486.

3
The Development and Application of Rhodium-Catalyzed Hydroboration of Alkenes

Anthony G. Coyne and Patrick J. Guiry

3.1
Introduction

Hydroboration, the addition of a boron–hydrogen bond across an unsaturated moiety, was first discovered by H. C. Brown in 1956 [1]. Usually the reaction does not require a catalyst and the borane reagent, most commonly diborane (B_2H_6) or a borane adduct ($BH_3 \cdot THF$), reacts rapidly at room temperature to afford, after oxidation, the *anti*-Markovnikov alkene hydration product. However, when the boron of the hydroborating agent is bonded to heteroatoms, as is the case in catecholborane (1,3,2-benzodioxaborole) the electron density at boron is increased and elevated temperatures are needed for hydroboration to occur [2, 3].

The development of a catalytic hydroboration process was aided by the observation of Kono and Ito in 1975 that Wilkinson's catalyst [Rh(PPh$_3$)$_3$Cl] undergoes oxidative addition when treated with catecholborane (**2**), Scheme 3.1 [4]. Subsequently, Westcott *et al.* reported the isolation of the oxidative addition adduct with the tri-isopropylphosphine derivative and its characterization by X-ray crystallography [5]. However, it was another decade before the idea of developing a rhodium-catalyzed olefin hydroboration process came to fruition with the first examples reported by Männig and Nöth [6].

The conversion of an alkene (**1**) into an organoborane intermediate (**3**) has made this a valuable synthetic technique, particularly since the development of enantioselective variants [7, 8]. They serve as synthons for numerous functional groups [9] and are often subject to a consecutive carbon–oxygen [10], carbon–carbon [11–13] boron–carbon [14], boron–chlorine [5], carbon–nitrogen [15] and other bond-forming reactions, Scheme 3.2.

These and other approaches toward extending the synthetic scope of catalytic asymmetric hydroboration have been the subject of recent excellent reviews [16–18]. Many metals including nickel [19], ruthenium [20], iridium [21], lanthanum [22], titanium [23] and zirconium [24] have been employed in this transformation with varying degrees of success, although rhodium has remained the metal of choice for transition metal hydroboration.

Modern Reduction Methods. Edited by Pher G. Andersson and Ian J. Munslow
Copyright © 2008 WILEY-VCH Verlag GmbH & Co. KGaA, Weinheim
ISBN: 978-3-527-31862-9

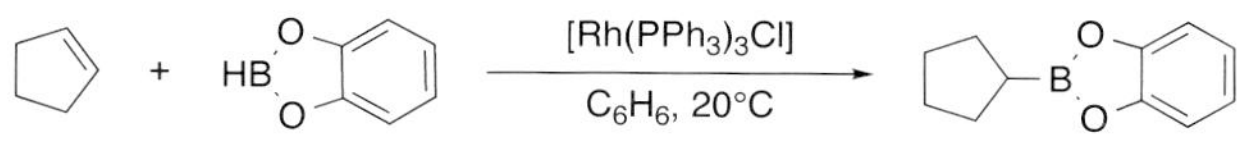

Scheme 3.1 Rhodium-catalyzed hydroboration of an alkene using catecholborane.

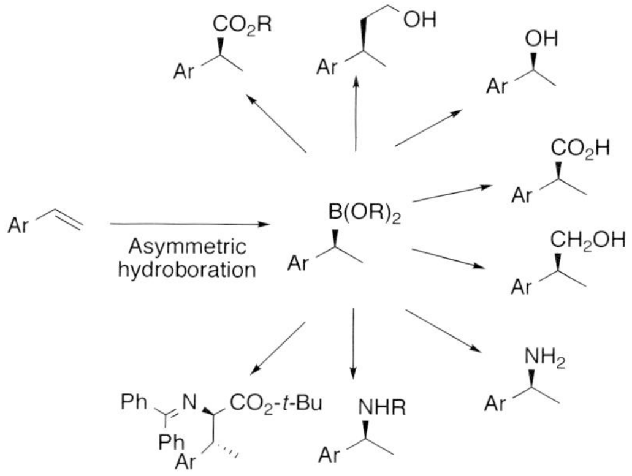

Scheme 3.2 Transformations of the carbon–boron bond.

3.2
Mechanism

The mechanism of rhodium-catalyzed hydroboration is thought to depend on the nature of the substrate, the catalyst, the ligand used and the reaction conditions employed, Scheme 3.3 [25].

The dissociation of a triphenylphosphine ligand from Wilkinson's catalyst (**4**) generates the catalytically active species (**5**) to which the B–H bond of the borane reagent oxidatively adds to give (**6**). The analogous complex with $P(i\text{-}Pr)_3$ rather than PPh_3 has been isolated and structurally characterized by Westcott and coworkers [5]. Coordination of the alkene *trans* to chlorine generates (**7**). The hydride and boryl ligands are *trans* in the reactive form of this complex [26]. Subsequent migratory insertion of the alkene into the rhodium-hydride bond produces the regioisomeric alkyl boronate esters (**8**) and (**9**). Upon reductive elimination these give the *anti*-Markovnikov product (**10**) or the Markovnikov product (**11**), respectively, and the catalytic species (**5**) is regenerated. Supporting evidence for this last step comes from stoichiometric studies of osmium boryl complexes by Roper and Wright [27]. Theoretical studies have suggested that reductive elimination is the slowest step in the original mechanism proposed by Männig and Nöth [6], later supported by Evans and Fu [28], and is a dissociative mechanism analogous to that established for the corresponding hydrogenation process. After oxidative addition of the borane, coordination of the alkene takes place with simultaneous dissociation of

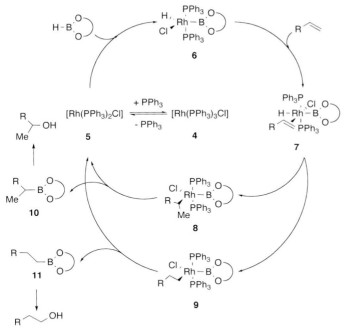

Scheme 3.3 Mechanism for hydroboration.

one of the PPh_3 ligands. Burgess and coworkers favored an alternative associative mechanism [29] in which the olefin and both PPh_3 ligands are bound to a six-coordinate Rh species. Hydroboration has also been studied by theoretical methods [26]. In addition, the nature of the catalytic cycle has been addressed experimentally and by means of quantum chemistry methods, and this has been reviewed recently [30]. Dorigo and Schleyer [31] conducted an ab initio study of the dissociative mechanism and categorically excluded the possibility of an associative mechanism.

3.3 Selectivity of Metal-Catalyzed Hydroboration

3.3.1 Regioselectivity

Männig and Nöth first demonstrated that the use of a catalyst could direct the course of the hydroboration reaction toward a different chemoselectivity than the uncatalyzed variant [6]. There have been considerable efforts to find alternative hydroborating agents other than catecholborane. These include 4,4,6-trimethyl-1,3,2-dioxaborinane [6] **(12)** and pinacolborane [32] **(13)**, with emphasis on boron hydrides bearing boron–oxygen bonds, as well as borazine [33] **(14)** (Figure 3.1). Despite these endeavors, catecholborane **(2)** is still the most useful borane and rhodium complexes are the most useful catalysts for catalytic hydroboration.

The uncatalyzed hydroboration-oxidation of an alkene usually affords the *anti*-Markovnikov product while the catalyzed version can be induced to produce either Markovnikov **(15)** or *anti*-Markovnikov **(16)** products (Scheme 3.4).

The regioselectivity obtained with a catalyst has been shown to depend on the ligands attached to the metal and also on the steric and electronic properties of the reacting alkene [34]. In the case of monosubstituted alkenes, except for vinyl arenes, the *anti*-Markovnikov alcohol **(16)** is obtained as the major product in either the presence or absence of a metal catalyst. However, the difference is that the metal-catalyzed reaction with catecholborane proceeds to completion within minutes at room temperature, while extended heating at 90 °C is required for the uncatalyzed transformation [21]. It should be noted that there is a reversal of regioselectivity from Markovnikov B—H addition in unfunctionalized terminal olefins to the *anti*-Markovnikov manner in monosubstituted perfluroalkenes, both in the achiral and chiral versions [35, 36].

Brown examined the asymmetric hydroboration of unsymmetrical *trans*-stilbenes **(17)** and the reaction was found to proceed with both high regioselectivity and enantioselectivity (in terms of enantiomeric excess (ee)) [37]. Rhodium and iridium QUINAP complexes give the same regioisomer **(18)** but opposite enantiomers (Scheme 3.5).

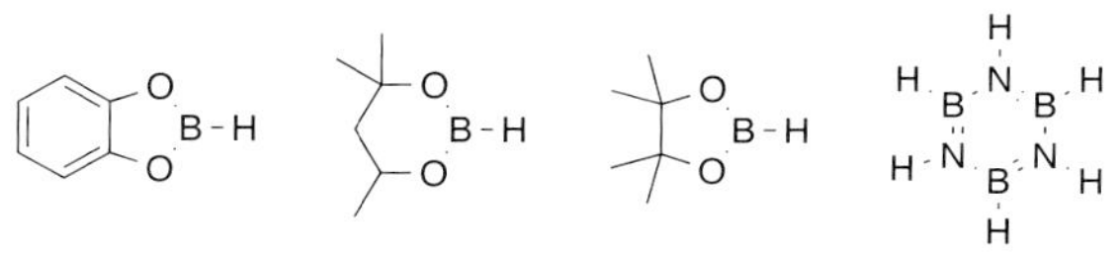

Figure 3.1 Hydroborating reagents.

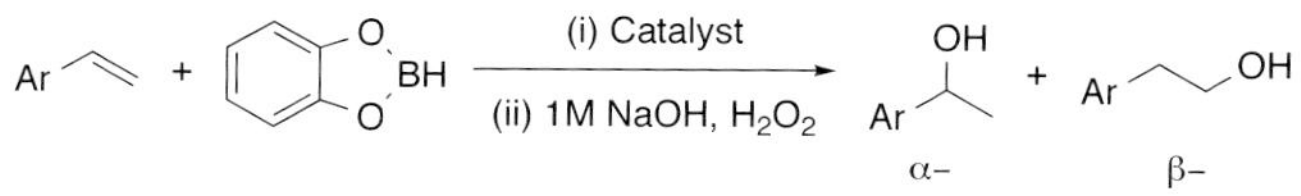

Scheme 3.4 Hydroboration of styrene showing differing additions.

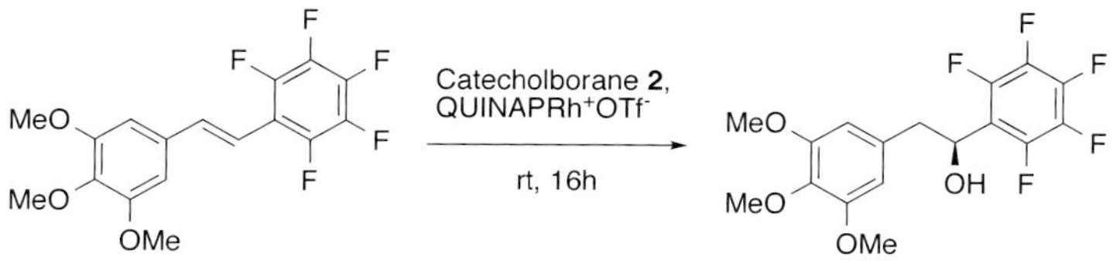

Scheme 3.5 Hydroboration of unsymmetrical stilbenes using Rh-QUINAP catalyst.

The hydroboration of unsymmetrical stilbene **(17)** gave the single regioisomer **(18)** in a 77% yield and with an ee of 88%.

3.3.2
Stereoselectivity

The development of an asymmetric variant of rhodium-catalyzed hydroboration was another significant milestone in the evolution of this transformation. The hydroboration-oxidation of substituted alkenes to regioselectively give the Markovnikov product can be used to introduce a chiral center. In fact, the vast majority of catalytic hydroborations have been applied to the formal synthesis of enantiomerically enriched alcohols by oxidation of the initial catecholborane adduct with basic hydrogen peroxide [15], although this synthetic scope has been extended to other functionalities [17, 18]. The attractive feature of the catecholborane hydroboration-oxidation sequence is the facile removal of the catechol by-product by simple extraction with aqueous base.

Two methods have been used to incorporate enantiodiscrimination into rhodium-catalyzed hydroboration. One such method, reported by Brown in 1990, involved the use of chiral hydroborating agents derived from ephedrine and pseudoephedrine in conjunction with an achiral catalyst [38]. The hydroboration of *para*-methoxystyrene **(19)** with a pseudoephedrine-derived borane **(20)** produced the most promising result: an enantiomeric excess of 76% after oxidation, with the α:β ratio 82% in favor of the secondary alcohol **(21)**. In other cases, enantioselectivities close to 90% were observed using chiral boron sources, but the general utility of this procedure was limited by poor regiochemical control. Neither did the use of chiral ligands in conjunction with the chiral borane reagents lead to significantly enhanced ee values (Scheme 3.6).

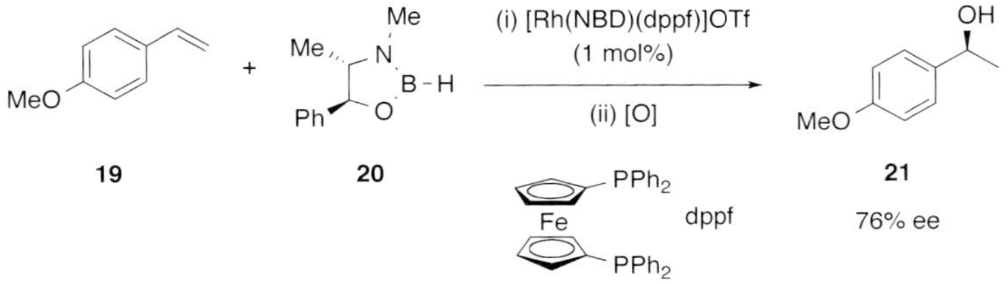

Scheme 3.6 Enantioselective hydroboration of *p*-methoxystyrene with pseudoephedrine-derived borane.

Instead, by far the most common approach toward catalytic asymmetric hydroboration is to use a chiral catalyst and an achiral borane source. Both homobidentate *P,P* and heterobidentate *P,N* ligand classes have been employed in this transformation with varying degrees of success.

3.3.2.1 Chiral P,P Ligands
The development of chiral catalysts for use in enantioselective rhodium-catalyzed hydroborations was pioneered by Burgess [39], Suzuki [40] and Hayashi [41]. Initial

probings into enantioselective hydroboration by Burgess [39] centered on norbornene **(22)** as substrate where regioselectivity was not an issue and the results obtained were promising (Scheme 3.7).

Exo-norborneol **(23)** was consistently afforded in excellent yields (>90%) upon treatment of norbornene **(22)** with catecholborane **(2)** in the presence of a catalytic amount (1 mol%) of cationic Rh(I)-complexes of centrally chiral (*R,R*)-DIOP **(24)** or axially chiral (*R*)-BINAP **(25)** [42, 43].

Suzuki found that the rate of hydroboration depended both on the substitution pattern of the alkene and on the chiral ligands employed. In general, 1,1-disubstituted alkenes gave a lower yield of *anti*-Markovnikov alcohol after oxidation (<30%) than did internal alkenes: 58–93% of Markovnikov product, compared with (*S,S*)-DIOP **(24)**. All chiral diphosphine ligands employed were effective, irrespective of the type of chirality present. However, (*S,S*)-DIOP **(24)** was found to induce the highest asymmetry: 58% ee for the hydroboration-oxidation of the cyclic alkene indene. This was later optimized to an enantioselectivity of 74% when the reaction was carried out at −30 °C for 120 hours.

Nonetheless, with these encouraging results, the real breakthrough occurred when Hayashi carried out enantioselective hydroboration of vinyl arenes with cationic diphosphine–rhodium complexes [40]. The most remarkable feature about Hayashi's work was the complete reversal in regioselectivity observed in the catalyzed hydroboration relative to the uncatalyzed variant. Irrespective of the substitution pattern of the styrene, the regioselectivity of the alcohol after oxidation overwhelmingly (>99/1) favored the Markovnikov alkene hydration product in a complementary regioselectivity to the uncatalyzed process. Using the same chiral diphosphine ligands as Suzuki, Hayashi first investigated the enantioselective hydroboration of styrene **(28)** with catecholborane **(2)** (Scheme 3.8).

(*R*)-BINAP **(25)** was found to be the most effective chiral ligand, giving a high yield of 1-phenylethanol **(29)** in 57% ee after 30 minutes at room temperature. Other diphosphine ligands, (*S,S*)-Chiraphos **(26)** and (*R,S*)-BPPFA **(27)** were all

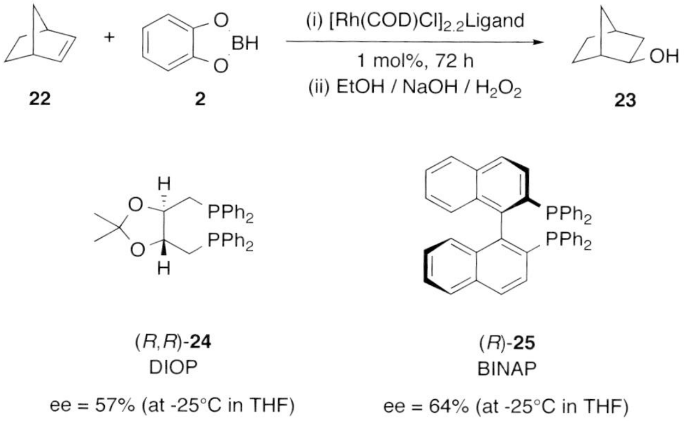

Scheme 3.7 Hydroboration of norbornene using DIOP and BINAP ligands.

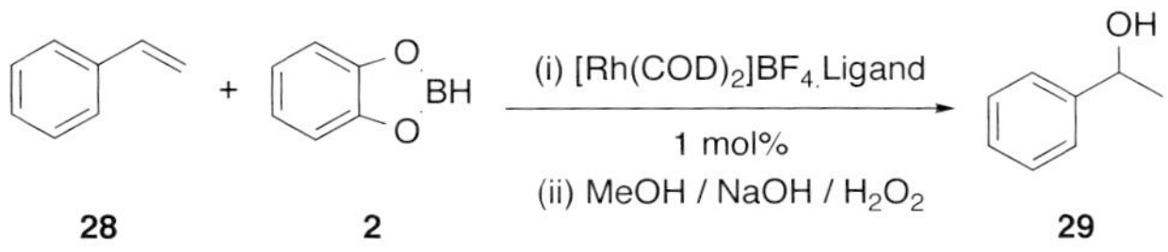

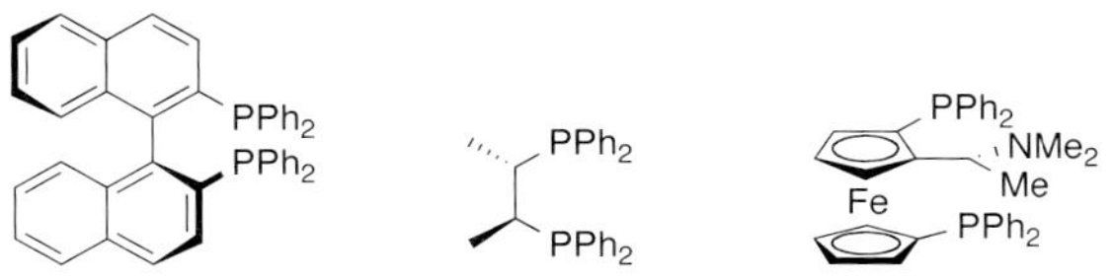

3.3 Selectivity of Metal-Catalyzed Hydroboration | 71

Scheme 3.8 Hayashi exploration of enantioselective hydroboration of styrene using various ligands.

less enantioselective for this reaction (<23% ee), despite the comparable catalytic activity of their cationic rhodium/ligand complexes to that of (*R*)-BINAP **(25)**. The cationic rhodium/(*R*)-BINAP **(25)** complex was highly active as a catalyst, the hydroboration of styrene **(28)** being completed in 30 minutes even at −30 °C with 1 mol% of the catalyst. Like Burgess [39], Hayashi found that there was an inverse relationship between the asymmetry induced and the reaction temperature. Reactions at −78 °C were carried out in Dimethoxyethane (DME), since catecholborane **(2)** was frozen and hence insoluble in tetrahydrofuran (THF) at this temperature. (*R*)-BINAP **(25)** was found to be the most effective chiral ligand at −78 °C, giving a high yield of 1-phenylethanol **(29)** in 94% ee.

With the proven catalytic activity of the cationic rhodium/(*R*)-BINAP **(25)** complex, Hayashi extended his study of asymmetric hydroboration-oxidation to substituted styrene derivatives **(19, 30–38b)** (Table 3.1).

Irrespective of their electronic properties, functional groups in the *para*-position (substrates **19, 30** and **31**) did not appear to influence the enantioselectivity of the product alcohols obtained after oxidation (89–94% ee). Similarly, the reaction of *meta*-substituted styrene **(32)** also proceeded in high enantiomeric excess, although more sterically hindered *ortho*-substituted derivatives **(33)** and **(34)** showed somewhat lower enantioselectivity. In all cases, the near-perfect regioselectivity (>99/1) for the Markovnikov alkene hydration product was maintained. β-Substituted styrenes **(35)** and **(36)** were unreactive at low temperatures and required room-temperature conditions for a reasonable rate of reaction to occur, but this compromise may in part have been responsible for the poor to modest asymmetry (16–42% ee) induced in their product alcohols. α-Substituted styrene **(37)** also required reaction at room temperature, but the *anti*-Markovnikov alkene hydration product was formed preferentially in both lower yield and enantiomeric purity than Suzuki had achieved with a [RhCl(ethene)$_2$]$_2$·2(*S*)-BINAP **(25)** catalyst system (38% ee).

Table 3.1 Hydroboration of various substituted alkenes using the BINAP ligand system.

	30	31	19	32	33	34	35	36	37	38
R	4-Cl	4-Me	4-MeO	3-Cl	2-Cl	2-MeO	H	H	Ph	Indene
R'	H	H	H	H	H	H	H	H	Me	

Entry[a]	Substrate	Temperature (°C)	Solvent	Time (h)	Yield (%)	% ee (configuration)
1[b]	(30)	−78	DME	6	98	91 (R)
2[b]	(31)	−78	DME	2	77	94 (R)
3[b]	(19)	−78	THF/DME[c]	6	54	89 (R)
4[b]	(32)	−78	DME	2	99	85 (R)
5	(33)	−50	THF	1	30	72 (R)
6	(34)	−30	THF	0.5	84	82 (R)
7	(E)-(35)	25	THF	34	65	42 (S)
8	(Z)-(35)	25	THF	48	58	18 (S)
9	(E)-(36)	25	THF	50	48	16 (S)
10	(37)	25	THF	3.5	27[d]	19 (S)
11	(38)	25	THF	3	65[e]	13 (S)

a Regioselectivity (α/β) > 99/1 in all cases with the exception of entries 10 and 11.
b Reaction with 2 mol% of the catalyst.
c THF/DME = 1/3.
d Regioselectivity (α/β) 39/61.
e Regioselectivity (α/β) 93/7.

Hydroboration-oxidation of indene (**38**), to give the Markovnikov product (**36**) also occurred in significantly lower selectivity (65% yield, 13% ee) to Suzuki's result with DIOP (**24**) (91% yield, 74% ee), a result which merely serves to highlight the many control factors at play in catalytic asymmetric hydroboration.

Burgess studied the possibility of a correlation between ligand chelate-ring size and hydroboration product distribution [44]. The hydroboration of 2-phenylpropene (**37**) to give, after oxidation, the Markovnikov product (**39**) or the anti-Markovnikov product (**40**), was investigated with a broader range of diphosphine ligands. As well as (R)-BINAP (**25**), an expanded class of C_2-symmetric centrally chiral diphosphine ligands was employed (Table 3.2) Among these were (S,S)-BDPP (**41**), DIOP analogues (R,R)-2-MeO-DIOP (**42**) and (R,R)-3-MeO-DIOP (**43**), all of which have an asymmetric carbon backbone, and (R,R)-DIPAMP (**44**), which is chiral at phosphorus [45].

(S,S)-Chiraphos (**26**) and (R,R)-DIPAMP (**44b**) (entries 1 and 2, Table 3.2), form five-membered chelate-ring structures. The ligand (S,S)-BDPP (**41**) (entry 3) is

3.3 Selectivity of Metal-Catalyzed Hydroboration | 73

Table 3.2 Hydroboration of 2-phenylpropene using different P,P ligands.

Entry	Ligand	Rh : Ligand	(39) : (40)	% ee of (40)
1	(S,S)-(26)	1:1	1:19	25[a]
2	(R,R)-(44)	1:1	1:6	0
3	(S,S)-(41)	1:1	1:1	27
4	(R)-(25)	1:1	1:6	25
5	(R,R)-(24)	1:1	1:9	27
6	(R,R)-(24)	1:2	8:1	[b]
7	(R,R)-(42)	1:1	[b]	15
8	(R,R)-(43)	1:1	4:1	15

a Absolute configuration (R) in all cases.
b Not determined.

structurally analogous to (S,S)-Chiraphos (26) except that it chelates the metal in a six-membered ring. Both (R)-BINAP (25) and (R,R)-DIOP (24) (entries 4 and 5) give seven-membered chelate rings, as do (R,R)-2-MeO-DIOP (42) and (R,R)-3-MeO-DIOP (43) (entries 7 and 8). These results indicate that there is little correlation between chelate-ring size and product distribution. However, combined with data from the hydroboration-oxidation of norbornene (22) and indene (38) with the same seven diphosphine ligands as in Table 3.2, some general pointers for the rational design of more effective chiral ligands were garnered. In general, ligands that gave five-membered chelate rings were less enantioselective than those that gave six- or seven-membered chelates; for example, (S,S)-Chiraphos (26) and (R,R)-DIPAMP (44) gave essentially racemic products for the hydroboration of norbornene (22). However, this reaction proceeded in 82% and 69% ee, respectively, for (R,R)-2-MeO-DIOP (42) and (R,R)-3-MeO-DIOP (43), highlighting that subtle stereoelectronic modifications of existing chiral ligands could have a pronounced effect on the degree of asymmetry induced.

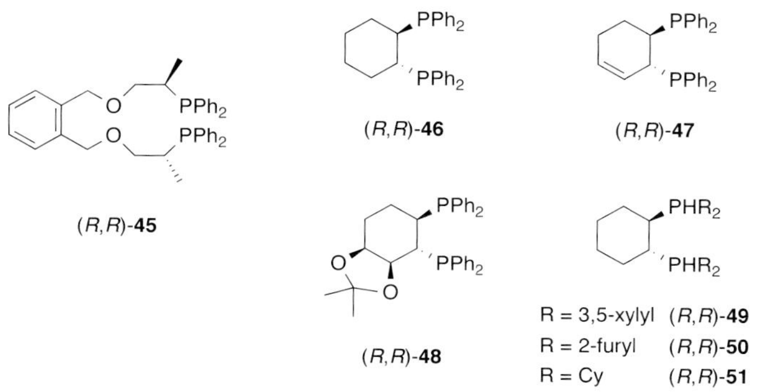

Figure 3.2 Diphosphine ligands for the hydroboration of alkenes.

As a greater understanding has emerged of the control factors at play in enantioselective rhodium-catalyzed hydroborations, an increased number of novel chiral diphosphine ligands have been reported in the chemical literature, the majority of these within the last seven years. Given the proven success of BINAP (25) in catalytic asymmetric hydroborations, it is perhaps surprising that so few BINAP analogues have been reported.

Whereas the majority of phosphine ligands that had been tested in catalytic asymmetric synthesis were electron-rich, (more so than triphenylphosphine), among them DIOP (24), DIPAMP (44) and BINAP (25b), there was nonetheless literature precedent for enhancing catalyst performance by reducing the basicity of the phosphine [18].

However, the vast majority of novel chiral diphosphines reported have possessed elements of either atom-centered or planar chirality. Included in the former class is the C_2-symmetric diphosphine (45) from the group of Flor (Figure 3.2) [46].

This ligand may be regarded as a diphosphine containing two intramolecular solvent molecules, due to the two hemilabile ether donor groups. The cationic Rh(I)-complex of (R,R)-(45) was tested in the asymmetric hydroboration of styrene (28) with catecholborane (2), at both 20 °C and −25 °C The catalyst system showed high activity and chemoselectivity (>96%) at both temperatures. The regioselectivity for formation of the branched product, 1-phenylethanol (29), was 82% and 95% at 20 °C and −25 °C respectively. Despite the high chemoselectivity and regioselectivity, enantioselectivity was very poor: <1% ee at 20 °C and an optimum 10% ee at the lower temperature.

Of greater success were Knochel's C_2-symmetric diphosphine ligands (46–51) (Figure 3.2), which were employed in the hydroboration of styrene and substituted styrenes [47]. The corresponding 1-arylethanols were furnished with excellent regioselectivity (>99:1) and variable enantioselectivity after oxidation.

3.3.2.2 Chiral P,N Ligands

The first successful axially chiral phosphinamine ligand in asymmetric catalysis was QUINAP (52) reported by Brown in 1993 and the original synthesis has since been modified (Figure 3.3) [48–50]. The donor nitrogen atom is incorporated in an isoquinoline unit to form a six-membered chelate ring.

Brown and coworkers tested cationic Rh(I)-complexes of QUINAP **(52)** in the enantioselective hydroboration-oxidation of vinyl arenes [51] which proceeded with excellent regioselectivities, in most cases >95%. The product configuration induced was the same as for BINAP **(25)**; rhodium complexes of the (*R*)-ligand gave rise to the (*R*)-secondary alcohol.

In contrast to BINAP **(25)**, the Rh/QUINAP **(52)** catalyst system was effective at ambient temperature (Scheme 3.9). In fact, while the enantioselectivities

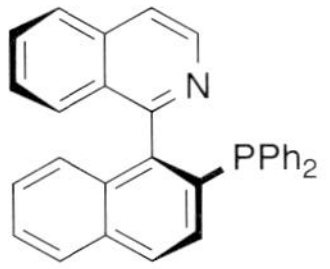
N
PPh2

(*S*)-**52**

QUINAP

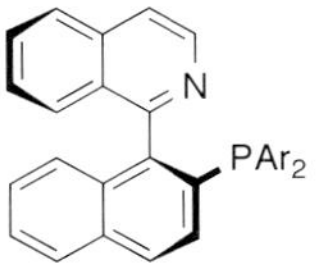
N
PAr2

53 Ar = 3-MeC_6H_4
54 Ar = 3,5-$(Me)_2C_6H_3$
55 Ar = 2-furyl
56 Ar = biphenyl

Figure 3.3 Axially chiral *P,N* ligands.

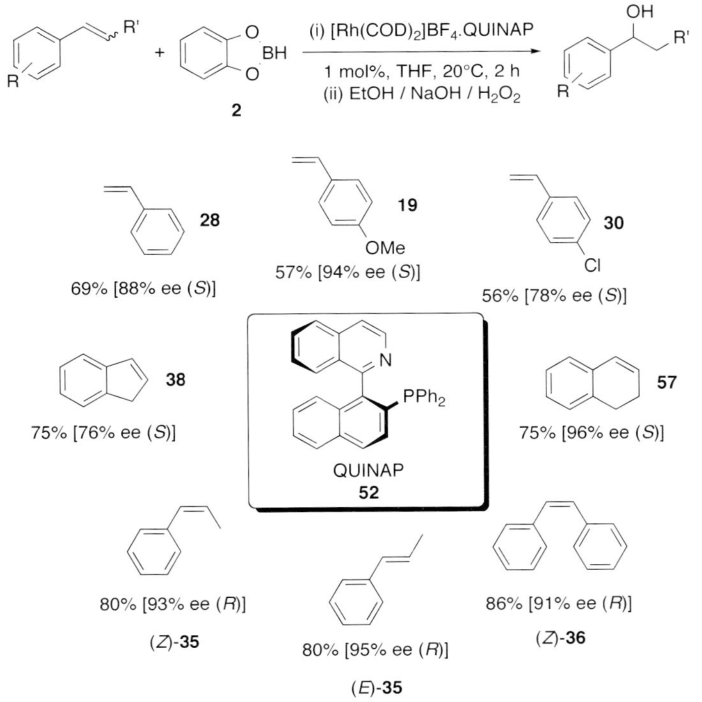
R'
R
O
BH
O
2
(i) [Rh(COD)2]BF4.QUINAP
1 mol%, THF, 20°C, 2 h
(ii) EtOH / NaOH / H2O2
OH
R'
R
28
69% [88% ee (S)]
19
OMe
57% [94% ee (S)]
30
Cl
56% [78% ee (S)]
38
75% [76% ee (S)]
N
PPh2
QUINAP
52
57
75% [96% ee (S)]
80% [93% ee (R)]
(Z)-35
80% [95% ee (R)]
(E)-35
86% [91% ee (R)]
(Z)-36

Scheme 3.9 Hydroboration of various substituted vinyl arenes using QUINAP.

unusually showed little change between 0 °C and 40 °C, the effect of further cooling was found to be deleterious. For best results, the rhodium complex was reprecipitated before use (THF/light petroleum ether) and the vinyl arene substrates were purified by distillation prior to reaction. Both the regioselectivity and enantioselectivity were sensitive to the ligand-to-rhodium ratio, a finding also observed with diphosphine ligands [52].

High asymmetries were induced with styrene (28) (88% ee) and styrene derivative (19) with electron-donating substituents in the *para*-position (94% ee), but electron-withdrawing substituents as in 4-chlorostyrene (30) gave lower ee values (78%). Of importance, however, was the finding that cyclic vinyl arenes, such as indene (38) and dihydronaphthalene (57), also gave high enantioselectivities (76 and 96%, ee respectively). The hydroboration-oxidation of such sterically demanding substrates was far less successful when complexes of BINAP (25) were employed (19% ee for indene (38); dihydronaphthalene (57) has not been reported). In general, hydroborations catalyzed by rhodium complexes of QUINAP (52) were tolerant of vinyl arenes that had (*E*)- or (*Z*)-β-substituents but not α-substituents. This represented a considerable expansion of substrate scope compared with previous *P*,*P* (or *P*,*N*) complexes employed, which were effective only for monosubstituted vinyl arenes. Specifically, the hydroboration of (*Z*)- and (*E*)-1-phenyl-1-propene (37) proceeded with 93% and 95% ee, respectively, with QUINAP (52). The corresponding results with BINAP (25) were 18% and 42% ee in reactions run at ambient temperature. The high yields unexpectedly afforded by both isomers of (35) suggested a high possibility of a rhodium-hydride-driven *cis–trans* isomerization, and certainly the reaction with *trans*-stilbene (36) (45 turnovers in 20 hours (85% ee)) was much slower than that of *cis*-stilbene (36).

QUINAP (52) is amenable to structural variation at several points. Among these are the aryl groups on phosphorus, which were systematically varied to examine the effect on the efficiency and enantiodifferentiating ability of the ligand. Thus the analogues (53–56) (Figure 3.3) were prepared and resolved in a similar manner to QUINAP (52).

Their rhodium complexes were subsequently applied in hydroboration-oxidation of vinyl arenes with comparable regioselectivities to QUINAP (52). Enantioselectivities of up to 93% were observed, compared with 97% for the parent catalyst system. The key finding from this study was that the parent diphenylphosphino ligand QUINAP (52b) gave superior results for electron-rich vinyl arenes, whereas by making the donor phosphorus atom less electron-rich, as in the difurylphosphino ligand (55), superior results for electron-poor vinyl arenes were obtained.

It became apparent during a mechanistic investigation of the allylic alkylation process with QUINAP (52), involving both solution ^{1}H NMR and solid-state studies, that the 3-H of the isoquinoline unit takes up a position in space near the metal that leads to critical ligand–reactant steric interactions thought to be significant for asymmetric induction [53]. This finding led to the design by Brown of the vaulted analogue PHENAP (58) (Figure 3.4), where the donor nitrogen atom is part of a phenanthridine unit.

phenanthridine N donor quinazoline N donor pyridine N donor

(*R*)-**58** (*R*)-**65**

PHENAP (*S*)-2-R-Quinazolinap Pyphos

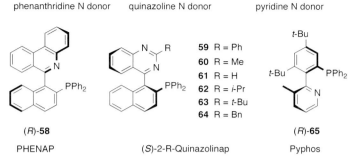

Figure 3.4 *P,N* ligands used in hydroboration.

(*aS,R*)-**66** (*aS,S*)-**66**

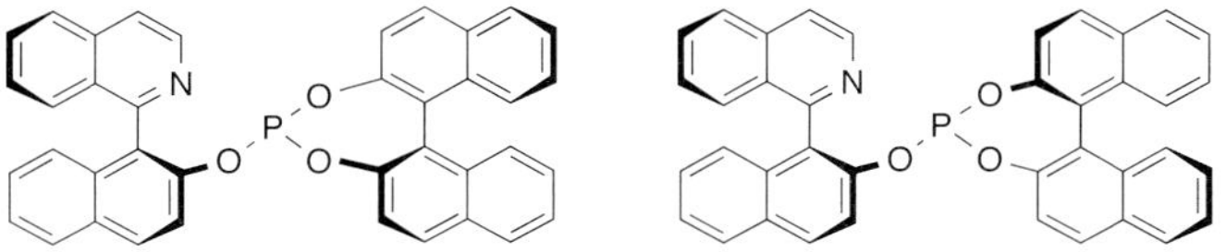

Figure 3.5 QUINAP-derived phosphite ligands used in hydroboration of alkenes.

PHENAP **(58)** was prepared and resolved in a similar manner to QUINAP **(52)** and tested in asymmetric rhodium-catalyzed hydroboration-oxidations [54, 55]. Regioselectivities were on a par with QUINAP **(52)**. Impressive enantioselectivities were obtained and the sterically demanding cyclic substrates were hydroborated in 64–84% ee, as well as β-substituted vinyl arene (*E*)-**(35)** in a similar enantioselectivity as with QUINAP **(52)**.

For QUINAP complexes the *P,N* chelate is a pronounced boat and there is reduced steric demand in the region of space around the isoquinoline nitrogen, certainly when compared with the aryl residues of BINAP **(25)** [51]. This structural difference, coupled with the intrinsic electronic features of the *P,N* ligand, explains why the asymmetric induction is higher with chiral QUINAP **(52)** than with chiral BINAP **(25)** [51].

In a departure in the quest for QUINAP-type ligands with an electron-deficient phosphorus atom, a novel QUINAP-derived triarylphosphite ligand **(66)** (Figure 3.5) was developed by Brown ten years after the parent ligand **(52)** was first reported [56].

Ligand **(66)** was prepared directly from a single enantiomer of the corresponding naphthol of QUINAP **(52)**, an early intermediate in the original synthesis, and both enantiomers of BINOL. Application in hydroboration found that, in practice, only one of the cationic rhodium complexes of the diastereomeric pair proved effective, (*aS,S*)-**(66)**. While (*aS,S*)-**(66)** gave 68% ee for the hydroboration of styrene **(28)** (70% yield), the diastereomer (*aS,R*)-**(66)** afforded the product alcohol after oxidation in an attenuated 2% ee (55% yield) and the same trend was apparent in the hydroboration of electron-poor vinyl arenes.

A series of axially chiral 2-substituted quinazoline-containing phosphinamine ligands, the "Quinazolinaps," **(59–64)** (Figure 3.4), has been prepared and resolved by our research group [57–60]. The naphthalene-quinazoline pivot was chosen since it would be inert to racemization [58]. In light of the mechanistic observations on related ligand systems, the 2-position of the quinazoline (equivalent to the 3-position of QUINAP **(52)**) is believed to be important for asymmetric induction. Therefore, it was of interest to vary the substituent at the 2-position in an effort to investigate the effect of steric demand on the degree of enantioselection observed. Of interest also was the reduced basicity of the Quinazolinap donor nitrogen relative to QUINAP **(52)**. Cationic rhodium complexes of these ligands were prepared and applied in the enantioselective hydroboration-oxidation of a range of vinyl arenes [60, 61], carefully chosen to highlight the effect on reactivity and enantioselectivity of different aryl substituents and β-substitution. A selection of these results, with the emphasis on the ligand that gave the highest enantioselectivity for each substrate, is shown in (Scheme 3.10). Like QUINAP **(52)** and PHENAP **(58)**, the (S)-ligand gave rise to the (S)-secondary alcohol.

The 2-substituted-Quinazolinap-derived rhodium complexes proved extremely efficient catalysts for the hydroboration-oxidation of vinyl arenes. For styrene

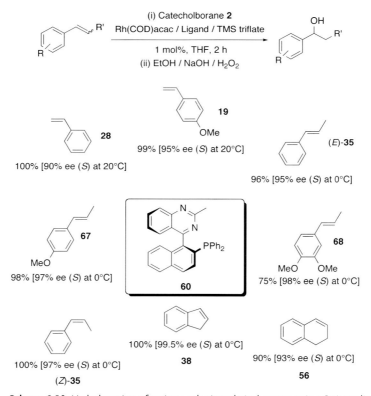

Scheme 3.10 Hydroboration of various substituted vinyl arenes using Quinazolinenap.

derivatives, in most cases quantitative conversions were obtained after just 2 hours at the relevant temperature. Higher enantioselectivities were afforded with a 4-methoxy substituent **(19)** (up to 95% ee), compared with the 4-chloro **(30)** or unsubstituted **(28)** analogues, a trend also observed in hydroboration with rhodium complexes of QUINAP **(52)**. This highlights that both the electronic nature of the substrate and the inherent steric properties of the catalyst are important for high asymmetric induction. It is noteworthy that in most cases optimum enantioselectivities were afforded by the less sterically demanding 2-methyl **(60)** and 2-unsubstituted analogues **(61)**, although the steric bulk of the 2-benzyl ligand **(64)** proved most effective for the hydroboration of styrene **(28)** at 0 °C In the Quinazolinap series as a whole, it was found that an electron-releasing substituent, as in **(19)**, had an adverse effect on regioselectivity compared with **(28)** and **(30)** [60].

Gratifyingly, a marked increase in regioselectivity and enantioselectivity was observed when β-substituted substrates were employed. On the whole, the hydroboration of (Z)-isomers proceeded with better conversions, regioselectivities and enantioselectivities than the hydroboration of the corresponding (E)-isomers. This is epitomized by the hydroborations of (E)- and (Z)-1-phenyl-1-propene **(37)**, and *cis*- and *trans*-stilbene **(36)**. For the hydroboration of (Z)-1-phenyl-1-propene **(37)**, optimum enantiomeric excesses of 96% and 97% at room temperature and 0 °C, respectively, were obtained with (S)-2-methyl-Quinazolinap **(60)** and (S)-2-isopropyl-Quinazolinap **(62)**, the best results reported to date for this substrate. These two ligands were again to the fore for the hydroboration of *trans*-anethole **(67)**, although, in comparison with (E)-1-phenyl-1-propene **(37)**, regioselectivities were lower due to the additional 4-methoxy aromatic substituent. Surprisingly, a more favorable regiochemistry was obtained with *trans*-3,4-dimethoxy-β-methylstyrene **(68)** relative to *trans*-anethole **(67)**, enantioselectivity was also excellent. Not for the first time, the steric influence of the 2-methyl analogue **(60)** proved optimal. The highest enantioselectivity to date for the hydroboration of *cis*-stilbene (Z)-**(36)**, 99% ee, was achieved with the rhodium catalyst derived from (S)-2-isopropyl-Quinazolinap **(62)**. The 2-methyl and 2-*tert*-butyl ligands **(60)** and **(63)** were also highly effective in this transformation. Furthermore, the cationic Quinazolinap/rhodium complexes proved excellent catalysts for the hydroboration of indene **(38)** and dihydronaphthalene **(32)**. While (S)-Quinazolinap **(61)** afforded 98% ee in the hydroboration of indene **(38)**, the effect of introduction of a 2-methyl group into the ligand framework was striking. As well as quantitative conversion and complete regioselectivity, an exceptional enantiomeric excess of 99.5% was observed; this remains the best result reported yet for the hydroboration of indene **(38)**.

The application of the diphosphine Josiphos **(69)** to the asymmetric hydroboration-oxidation of styrene **28** afforded 92% ee and excellent regioselectivity (>99:1) but this ligand was only effective at −78 °C and only for this substrate, Figure 3.6 [62]. By way of nucleophilic substitution with a pyrazole functionality at the *pseudo*-benzylic phosphine, Togni prepared a series of pyrazole-containing ferrocenyl ligands (Figure 3.6). These planar chiral ligands also possess an element of atom-centered chirality in the *pseudo*-benzylic position of the side-chain. The simple

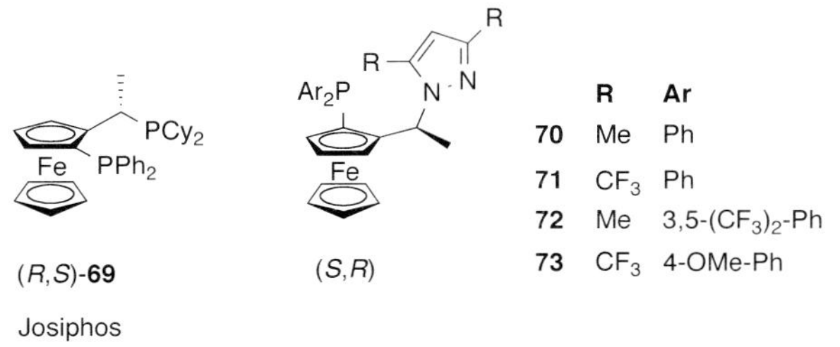

Figure 3.6 Ferrocene ligands used in the asymmetric hydroboration of alkenes.

Table 3.3 Ferrocene ligands in the asymmetric hydroboration of styrene.

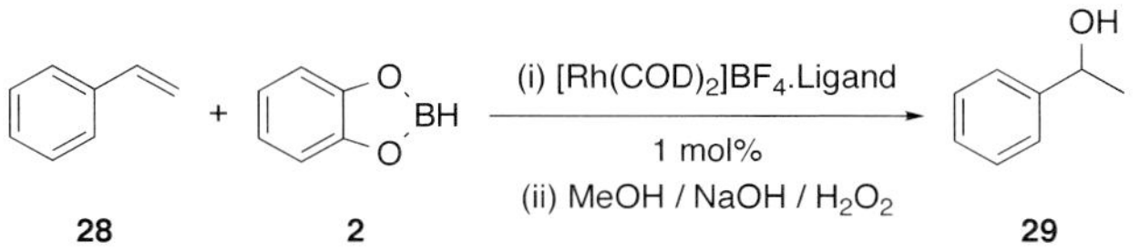

Entry[a]	Ligand	Yield (%)	α:β	% ee (configuration)
1	(S,R)-(70)	91	66:34	95 (R)
2	(S,R)-(71)	78	61:39	33 (R)
3	(R,S)-(72)	63	[b]	98.5 (S)
4	(R,S)-(73)	28	[b]	5 (S)

a Typical conditions: catecholborane (2), 1.0 mol% [Rh(COD)$_2$]BF$_4$·Ligand, THF, 20 °C 3–5 hours.
b Not quoted.

synthetic approach allows for the preparation of a wide variety of analogues with easy modification of their stereoelectronic properties.

Rhodium complexes (1 mol%) prepared in situ from [Rh(COD)$_2$]BF$_4$ and the respective ligands (70–73) were employed in the enantioselective hydroboration-oxidation of styrene (28) (Table 3.3)[63, 64].

In contrast to the application of Josiphos (69) to the same transformation, these phosphinamine ligands were highly effective at ambient temperature. Ligand (70b), containing the 3,5-dimethylpyrazolyl fragment, afforded an excellent 95% ee for the hydroboration of styrene (28) (entry 1). However, the ligands are limited in terms of regiochemical control (entries 1 and 2). Increasing the size of the pyrazole substituent from methyl to isopropyl resulted in a slightly lower 92% ee [63], showing that there was some dependence of the enantioselectivity on the steric properties of the ligand. However, the single most dominant influence was the electronic nature of the ligand. The different electronic properties of the pyrazole and phosphine moieties exert opposite influences and high asymmetries are induced when the combination exists such that nitrogen is a good σ-donor (electron-rich pyrazole) and phosphorus a good π-acceptor (electron-poor phosphine) [63].

To this end, while essentially conserving the steric bulk, replacement of the pyrazole methyl groups in **(70)** with the electron-withdrawing trifluoromethyl groups in **(72)** gave rise to a dramatic decrease in enantioselectivity, although the regioselectivity was largely unaffected. However, when the CF_3 groups were placed on the phosphine instead in ligand **(72)**, thereby rendering the phosphine electron deficient, an exceptional 98.5% ee was obtained. This remains the best result reported yet for the hydroboration of styrene **(28)**. In a striking illustration of the interplay of electronic effects and enantioselectivity, the combination of an electron-poor pyrazole and an electron-rich phosphine as in ligand **(73)** afforded only 5% ee. The possibility that the low enantioselectivities in the case of electron-poor pyrazole ligands **(71)** and **(73)** was due to partial dissociation of the ligand from rhodium during catalysis was found to be unlikely and the pronounced effects on enantioselectivity were therefore deemed largely electronic in nature [63].

In terms of chiral ligands that incorporate both planar and atom-centered chirality, the ferrocenyl framework is unrivalled in its versatility. Moreover, the synthetic approach toward this ligand class allows for tailoring of the ligand through steric and electronic effects. The latter was in evidence in the electronic asymmetry of Togni's *P*,*N* ligands **(70–73)**. Recently, the group of Knochel has reported a set of nine new chiral ferrocenyl phosphinamine ligands **(74–82)** (Figure 3.7). These ligands were tested in the rhodium-catalyzed asymmetric hydroboration-oxidation of styrene **(28)**, and representative results are shown in Table 3.4 [65].

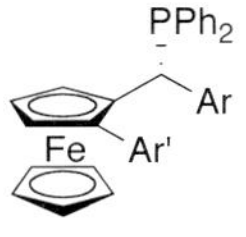

Ar'	Ar		
	Phenyl	o-Tolyl	3,5-Xylyl
2-Pyrimidyl	74	75	76
2-Pyridyl	77	78	79
2-Quinolyl	80	81	82

Figure 3.7 Ferrocene *P*,*P* ligands.

Table 3.4 Ferrocene *P*,*P* ligands in the asymmetric hydroboration of styrene.

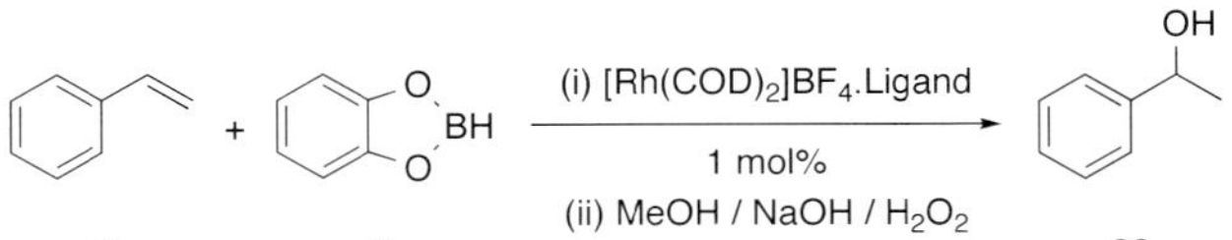

Entry[a]	Ligand	Time (h)	Conversion (%)	α:β	% ee
1	(74)	19	74	97:3	57[b]
2	(78)	14	54	84:16	80
3	(80)	16	>99	64:36	92

a Typical conditions: catecholborane **(2)**, THF, −45 °C 1.0 mol% $[Rh(COD)_2]BF_4 \cdot 2Ligand$.
b Absolute configuration (*S*) in all cases.

Ligands **(74–76)** with a pyrimidyl group showed excellent regioselectivity but only moderate conversion and enantioselectivity (entry 1). In contrast, rhodium complexes of ligands **(80–82)** bearing a quinolyl group were highly efficient catalysts, inducing excellent enantioselectivity but at the expense of the regioselectivity (entry 3). The performance of the 2-pyridyl ligands **(77–79)** lay between these two extremes (entry 2). Optimal results were obtained with ligand **(80)** which gave 92% ee and almost quantitative conversion for the hydroboration of styrene **(28)**, although with modest regioselectivity (entry 3).

3.4
Recent Applications in Synthesis

Among recent examples that highlight the synthetic utility of transition metal-catalyzed hydroborations are its direction toward a formal synthesis of the nonsteroidal anti-inflammatory agents ibuprofen **(83)** and naproxen **(84)** [11, 32, 66] as well as the antidepressant sertraline **(85)** (Figure 3.8) [67]. In the majority of cases rhodium-catalyzed hydroboration is utilized and the rhodium(I) source is generally Wilkinson's catalyst $RhCl(PPh_3)_3$ **(4)**.

Grieco in the total synthesis of (−)-epothilone B **(86)** used a rhodium-catalyzed hydroboration as a key step in the synthesis of the macrocyclic ring (Figure 3.9) [68]. Completion of the synthesis of the C(3)–C(12) fragment was carried out using a rhodium-catalyzed hydroboration as the key step.

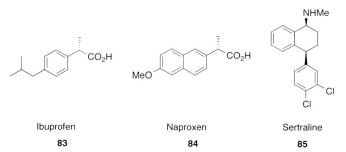

Figure 3.8 Common drugs potentially prepared by asymmetric hydroboration.

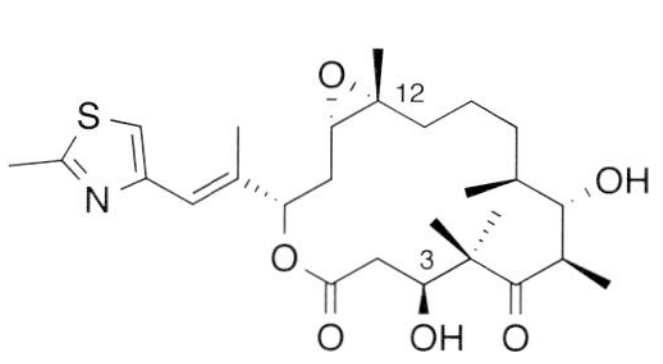

Figure 3.9 (−)-Epothilone.

Roush used a rhodium-catalyzed hydroboration to construct a component **(88)** of the natural product (−)-bafilomycin A$_1$ **(89)** (Scheme 3.11) [69]. The hydroboration was carried out in 87% yield.

Trost used a rhodium(I)-catalyzed hydroboration to obtain a key intermediate **(91)** in 90% yield in the synthesis of the pyrone ring of the natural product (−)-malyngolide **(92)** (Scheme 3.12) [70].

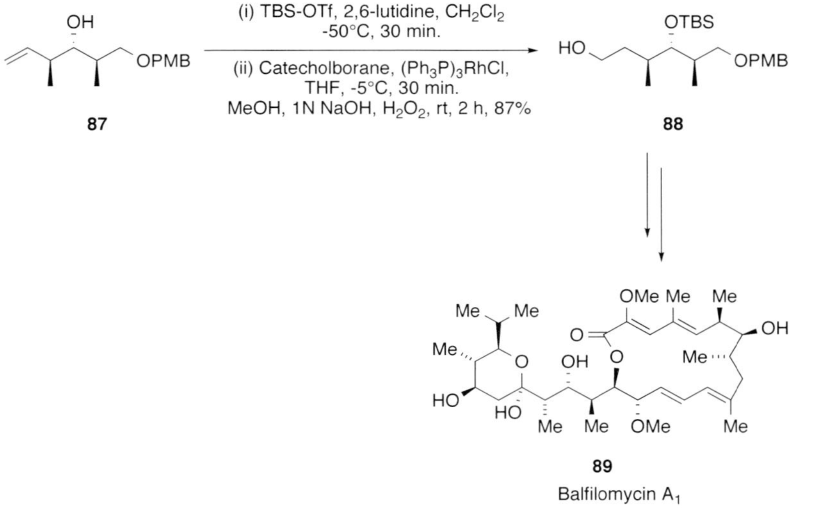

Scheme 3.11 Synthesis of balfilomycin A$_1$ with hydroboration as a key step.

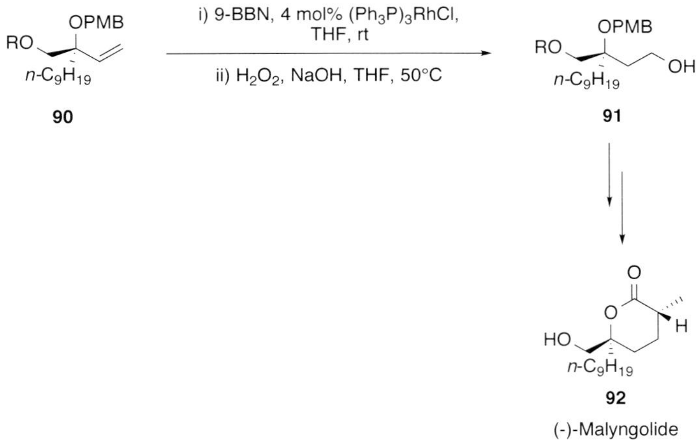

Scheme 3.12 Synthesis of (−)-malyngolide with hydroboration as a key step.

3.5
Conclusion

In the two decades since Männig and Nöth paved the way for the development of a rhodium-catalyzed olefin hydroboration process, the subsequent enantioselective variant has become a valuable tool in catalytic asymmetric synthesis and a plethora of chiral ligands have been employed. Much of the early groundwork relied on the application of homobidentate P,P ligands. However, it has been the advent of heterobidentate P,N ligands in particular that has advanced the scope of this reaction in terms of asymmetric induction. The development of new methods for transforming the carbon–boron bond into other functionalities will make a serious impact on the use of the catalytic asymmetric hydroboration in synthesis. Nevertheless, the future of enantioselective rhodium-catalyzed hydroboration as an important synthetic transformation seems assured due to the high reactivity and impressive enantioselectivities obtained thus far.

Acknowledgements

Dr. Anthony Coyne thanks Science Foundation Ireland (SFI) for financial support of his postdoctoral fellowship (Grant No. 04/BR/C0702).

References

1 Brown, H.C. and Subba Rao, B.C. (1956) *Journal of the American Chemical Society*, **78**, 2582.
2 Beletskaya, I. and Pelter, A. (1997) *Tetrahedron*, **53**, 4957.
3 Tucker, C.E., Davidson, J. and Knochel, P. (1992) *Journal of Organic Chemistry*, **57**, 3482.
4 Kono, H., Ito, K. and Nagai, Y. (1975) *Chemistry Letters*, **4**, 1095.
5 Westcott, S.A., Taylor, N.J., Marder, T.B., Baker, R.T., Jones, N.L. and Calabrese, J.C. (1991) *Chemical Communications*, 304.
6 Männig, D. and Nöth, H. (1985) *Angewandte Chemie (International Ed. in English)*, **24**, 878.
7 Fu, G.C. (1998) *Transition Metals for Organic Synthesis*, Vol. II, Wiley-VCH Verlag GmbH.
8 Hayashi, T. (1999) *Comprehensive Asymmetric Catalysis*, Vol. I, Springer, New York.
9 Hayashi, T., Matsumoto, Y. and Ito, Y. (1991) *Tetrahedron: Asymmetry*, **2**, 601.
10 Brown, J.M., Hulmes, D.I. and Layzell, T.P. (1993) *Chemical Communications*, 1673.
11 Chen, A., Ren, L. and Crudden, C.M. (1999) *Journal of Organic Chemistry*, **64**, 9704.
12 Chemler, S.R., Trauner, D. and Danishefsky, S.J. (2001) *Angewandte Chemie (International Ed. in English)*, **40**, 4544.
13 Rubina, R. and Rubin, M.G. (2003) *Journal of the American Chemical Society*, **125**, 7198.
14 O'Donnell, M.J., Cooper, J.T. and Mader, M.M. (2003) *Journal of the American Chemical Society*, **125**, 2370.
15 Fernandez, E., Maeda, K., Hooper, M.W. and Brown, J.M. (2000) *Chemistry – A European Journal*, **6**, 1840.
16 Brown, J.M. and Rhodium, M. (2004) *Catalyzed Organic Reactions*, Wiley-VCH Verlag GmbH, Weinheim.
17 Crudden, C.M. and Edwards, D. (2003) *European Journal of Organic Chemistry*, 4695.

18 Carroll, A-M., O'Sullivan, T.P. and Guiry, P.J. (2005) *Advanced Synthesis and Catalysis*, **347**, 609.
19 Gridnev, I.D., Miyaura, N. and Suzuki, A. (1993) *Organometallics*, **12**, 589.
20 Burgess, K. and Jaspars, M. (1993) *Organometallics*, **12**, 4197.
21 Evans, D.A., Fu, G.C. and Hoveyda, A.H. (1992) *Journal of the American Chemical Society*, **114**, 6671.
22 Evans, D.A., Muci, A.R. and Stürmer, R. (1993) *Journal of Organic Chemistry*, **58**, 5307.
23 He, X. and Hartwig, J.F. (1996) *Journal of the American Chemical Society*, **118**, 1696.
24 Pereira, S. and Srebnik, M. (1996) *Journal of the American Chemical Society*, **118**, 909.
25 Burgess, K. and Ohlmeyer, M.J. (1991) *Chemical Reviews*, **91**, 1179.
26 Widauer, C., Grützmacher, H. and Ziegler, T. (2000) *Organometallics*, **19**, 2097.
27 Rickard, C.E.F., Roper, W.R., Williamson, A. and Wright, L.J. (1999) *Angewandte Chemie (International Ed. in English)*, **38**, 1110.
28 Evans, D.A., Fu, G.C. and Anderson, B.A. (1992) *Journal of the American Chemical Society*, **114**, 6679.
29 Burgess, K., van der Donk, W.A., Westcott, S.A., Marder, T.B., Baker, R.T. and Calabrese, J.C. (1992) *Journal of the American Chemical Society*, **114**, 9350.
30 Huang, X. and Lin, Z.Y. (2002) *Computational Modelling of Homogeneous Catalysis*, Kluwer.
31 Dorigo, A.E. and von Ragué Schleyer, P. (1995) *Angewandte Chemie (International Ed. in English)*, **34**, 115.
32 Crudden, C.M., Hleba, Y.B. and Chen, A.C. (2004) *Journal of the American Chemical Society*, **126**, 9200.
33 Fazen, P.J. and Sneddon, L.G. (1994) *Organometallics*, **13**, 2867.
34 Smith, M.R., III (1999) *Progress in Inorganic Chemistry*, **48**, 505.
35 Ramachandran, P.V., Jennings, M.P. and Brown, H.C. (1999) *Organic Letters*, **1**, 1399.
36 Segarra, A.M., Claver, C. and Fernandez, E. (2004) *Chemical Communications*, 464.
37 Black, A., Brown, J.M. and Pichon, C. (2005) *Chemical Communications*, 5284.
38 Brown, J.M. and Lloyd-Jones, G.C. (1990) *Tetrahedron: Asymmetry*, **1**, 865.
39 Burgess, K. and Ohlmeyer, M.J. (1988) *Journal of Organic Chemistry*, **53**, 5178.
40 Sato, M., Miyaura, N. and Suzuki, A. (1990) *Tetrahedron Letters*, **31**, 231.
41 Hayashi, T., Matsumoto, Y. and Ito, Y. (1989) *Journal of the American Chemical Society*, **111**, 3426.
42 Kagan, H.B. and Dang, T.P. (1972) *Journal of the American Chemical Society*, **94**, 6429.
43 Takaya, H., Mashima, K., Koyano, K., Yagi, M., Kumobayashi, H., Taketomi, T., Akutagawa, S. and Noyori, R. (1986) *Journal of Organic Chemistry*, **51**, 629.
44 Burgess, K., van der Donk, W.A. and Ohlmeyer, M.J. (1991) *Tetrahedron: Asymmetry*, **2**, 613.
45 Vineyard, B.D., Knowles, W.S., Sabacky, M.J., Bachman, G.L. and Weinkauff, D.J. (1977) *Journal of the American Chemical Society*, **99**, 5946.
46 Ruiz, J.L., Flor, T. and Carles Bayón, J. (1999) *Inorganic Chemistry Communications*, **2**, 484.
47 Demay, S., Volant, F. and Knochel, P. (2001) *Angewandte Chemie (International Ed. in English)*, **40**, 1235.
48 Alcock, N.W., Brown, J.M. and Hulmes, D.I. (1993) *Tetrahedron: Asymmetry*, **4**, 743.
49 Alcock, N.W., Brown, J.M. and Hulmes, D.I. (1995) *Chemical Communications*, 395.
50 Lim, C.W., Tissot, O., Mattison, A., Hooper, M.W., Brown, J.M., Cowley, A.R., Hulmes, D.I. and Blacker, A.J. (2003) *Organic Process Research and Development*, **7**, 379.
51 Doucet, H., Fernandez, E., Layzell, T.P. and Brown, J.M. (1999) *Chemistry – A European Journal*, **5**, 1320.
52 Brunel, J.-M. and Buono, G. (1999) *Tetrahedron Letters*, **40**, 3561.
53 Brown, J.M., Hulmes, D.I. and Guiry, P.J. (1994) *Tetrahedron*, **50**, 4493.
54 Valk, J.-M., Claridge, T.D.W., Brown, J.M., Hibbs, D. and Hursthouse, M.B. (1995) *Tetrahedron: Asymmetry*, **6**, 2597.
55 Valk, J.-M., Whitlock, G.A., Layzell, T.P. and Brown, J.M. (1995) *Tetrahedron: Asymmetry*, **6**, 2593.

56 Korostylev, A., Gridnev, I. and Brown, J.M. (2003) *Journal of Organometallic Chemistry*, **680**, 329.
57 McCarthy, M. and Guiry, P.J. (2000) *Polyhedron*, **19**, 541.
58 McCarthy, M., Goddard, R. and Guiry, P.J. (1999) *Tetrahedron: Asymmetry*, **10**, 2797.
59 Lacey, P.M., McDonnell, C.M. and Guiry, P.J. (2000) *Tetrahedron Letters*, **41**, 2475.
60 Connolly, D.J., Lacey, P.M., McCarthy, M., Saunders, C.P., Carroll, A.-M., Goddard, R. and Guiry, P.J. (2004) *Journal of Organic Chemistry*, **69**, 6572.
61 McCarthy, M., Hooper, M.W. and Guiry, P.J. (2000) *Chemical Communications*, 1333.
62 Togni, A., Breutel, C., Schnyder, A., Spindler, F., Landert, H. and Tijani, A. (1994) *Journal of the American Chemical Society*, **116**, 4062.
63 Schnyder, A., Hintermann, L. and Togni, A. (1995) *Angewandte Chemie (International Ed. in English)*, **34**, 931.
64 Schnyder, A., Togni, A. and Wiesli, U. (1997) *Organometallics*, **16**, 255.
65 Kloetzing, R.J., Lotz, M. and Knochel, P. (2003) *Tetrahedron: Asymmetry*, **14**, 255.
66 Chen, A., Ren, L. and Crudden, C.M. (1999) *Chemical Communications*, 611.
67 Maeda, K. and Brown, J.M. (2002) *Chemical Communications*, 310.
68 May, S.A. and Grieco, P.A. (1998) *Chemical Communications*, 1597.
69 Scheidt, K.D., Bannister, A., Tasaka, T.D., Wendt, M.D. and Roush, W.R. (1999) *Angewandte Chemie (International Ed. in English)*, **38** (*11*), 1652.
70 Trost, B.M., Tang, W. and Schulte, J.L. (2000) *Organic Letters*, **2** (*25*), 4013.

4
Alkene Reduction: Hydrosilylation
Penelope A. Mayes and Patrick Perlmutter

4.1
Introduction

Catalytic, asymmetric hydrosilylations of alkenes were first reported in 1971 when the platinum complex *cis*-PtCl$_2$(C$_2$H$_4$) was used to catalyze the hydrosilylation of 1-methylstyrene in 5% ee [1]. Through the use of nickel this was raised to 18% ee [2]. A palladium complex with a ferrocene-based ligand successfully catalyzed the hydrosilylation of norbornene and styrene giving moderate ee (52–53%) [3]. The breakthrough came with the development of Hayashi's monophosphine (MOP) ligands in 1991 (see below), used in the palladium-catalyzed hydrosilylation of aliphatic alkenes with good regioselectivities and ee values greater than 90% being obtained. Early research into catalytic, asymmetric hydrosilylations of alkenes has been covered in reviews by Nishiyama and Hayashi [4], this review therefore focuses on research published after 1998.

4.2
Isolated Alkenes

4.2.1
Palladium

4.2.1.1 Aromatic Alkenes

Much of the recent research into catalytic, asymmetric hydrosilylation of alkenes has focused on developing catalysts for the hydrosilyation of aromatic alkenes, with many groups employing styrene **(1)** as the test substrate for their catalysts. These reactions generally require low catalyst loadings (0.1 to 0.25 mol% palladium) and often proceed at or below room temperature. The reactions are usually highly regioselective and often yield benzylsilane **(2)** as the only product (Scheme 4.1). Oxidation, generally employing the Tamao procedure, then yields alcohol **(3)** [5].

Modern Reduction Methods. Edited by Pher G. Andersson and Ian J. Munslow.
Copyright © 2008 WILEY-VCH Verlag GmbH & Co. KGaA, Weinheim
ISBN: 978-3-527-31862-9

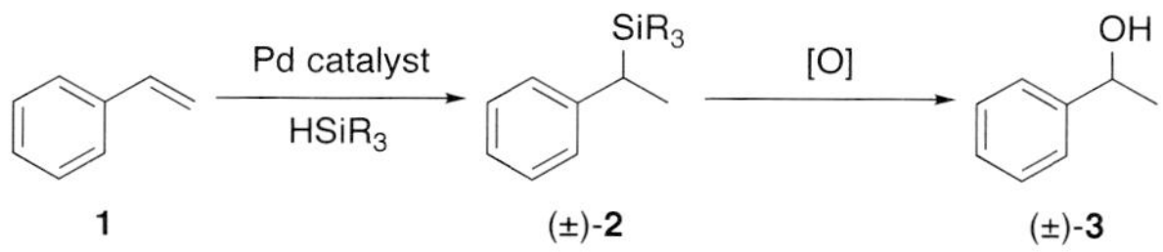

Scheme 4.1 Hydrosilylation of styrene.

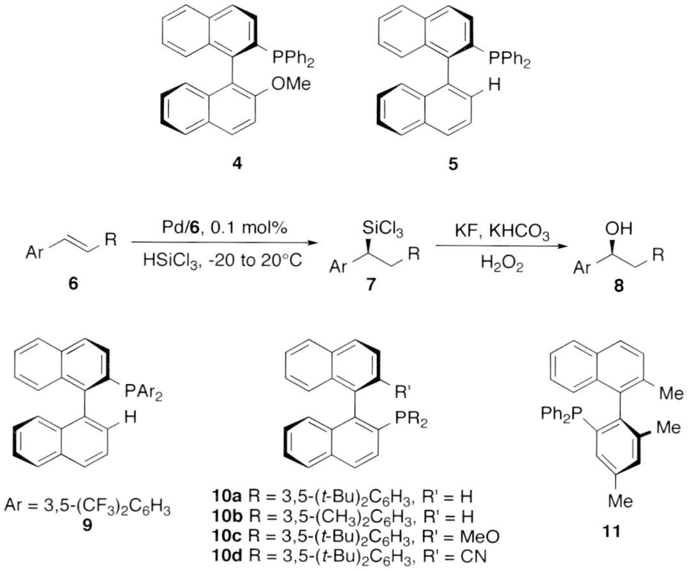

Scheme 4.2 Reduction of substituted styrenes.

Hayashi's original MeO-MOP ligand (**4**) gave only 71% ee for the palladium-catalyzed hydrosilylation of styrene; the simple H-MOP ligand (**5**) was found to perform much better (94% ee at −10 °C) [6, 7]. This ligand has since been fine-tuned through substituting the phenyl groups on the phosphine, and the best of these ligands (**9**) gave an ee of 97% for the hydrosilylation of styrene [8, 9]. Ligand (**9**) also performed well in experiments using substituted styrenes, with ee values generally ≥95% for substrates containing electron-donating or electron-withdrawing substituents (Scheme 4.2, Table 4.1). Yields of the silane were generally high. Reactions were usually complete within 24 hours, with notable exceptions being 3-Br and 3-NO_2 substrates (entries 8 and 11), both of which took 6 days. Substrates with substituents at R were also slower (up to 48 hours). 3,5-Dialkyl substitutions of the MOP phenyl groups (ligands (**10a–d**)) have also been investigated [10]. While t-butyl substitution did improve the performance of ligands (**10c**) and (**10d**), ee values remained moderate (57–86% ee). Dimethyl substitution did enable ligand (**10b**) to perform better than H-MOP (**5**) for the hydrosilylation of 4-methoxystyrene, but the enantiomeric excess was still only 80% and none of the ligands out-

Table 4.1 Hydrosilylations with MOP ligand (9).

Entry	Ar	R	Temperature (°C)	Time (h)	Yield silane	% ee alcohol
1[a]	C_6H_5	H	−20	24	85	98
2	2-$(CH_3)C_6H_4$	H	0	10	83	97
3	3-$(CH_3)C_6H_4$	H	0	4	91	97
4	4-$(CH_3)C_6H_4$	H	0	0.5	90	95
5	2-ClC_6H_4	H	0	15	88	91
6	3-ClC_6H_4	H	0	4	93	96
7	4-ClC_6H_4	H	0	4	92	98
8[a]	3-BrC_6H_4	H	0	6 days	90	94
9	4-BrC_6H_4	H	−10	15	95	97
10	4-$(OCH_3)C_6H_4$	H	−10	20	90	97
11	3-$(NO_2)C_6H_4$	H	0	6 days	89	98
12	C_6H_5	CH_3	0	48	81	98
13[a]	C_6H_5	CH_2OCH_3	0	30	85	97
14[a]	C_6H_5	CH_2OCH_2Ph	20	16	87	97

Reactions carried out in toluene (1.0 M) unless otherwise indicated.
Substrate/$HSiCl_3$/[$ClPd(C_3H_5)$]$_2$/ligand 1 : 1.2 : 0.0005 : 0.002.
a Reaction carried out solvent-free.

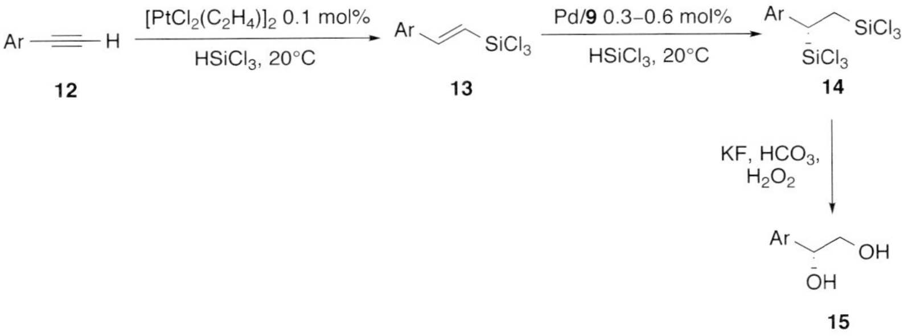

Scheme 4.3 Diol formation via double hydrosilylation.

performed the fine-tuned MOP ligand (9). The monophosphine biaryl ligand (11) has also been reported for the hydrosilylation of a small range of styrenes, which proceeded with only low selectivity (28–50% ee) [11].

The optimized MOP ligand (9) has also been successfully employed as part of a double-hydrosilylation of acetylenes to generate chiral 1,2-diols (Scheme 4.3) [12]. An achiral platinum complex initially catalyzed the hydrosilylation of alkyne (12) to the silane (13). Upon completion of this reaction, the palladium-MOP complex was added, giving the disilane (14) in 96% overall yield. Oxidation then yielded diol (15) in 95% ee. Attempts to use the palladium-MOP complex to catalyze both steps in one pot met with limited success due to dimerization during the first step.

A range of aromatic diols were synthesized in high enantiomeric excess and moderate to good yield (Table 4.2). Reaction times tended to be quite long (total time 42–192 hours) and reactions with alkyl-substituted alkynes were unsuccessful because of the very slow second step.

Chiral phosphoramidite ligands have also proved effective in the asymmetric hydrosilylation of styrenes; an ee value of 99% for (R)-**(3)** was obtained using ligand (S_A,R_C,R_C)-**(16)** (Scheme 4.4) [13], while ligand (R_A,R_C,R_C)-**(16)** gave (S)-**(3)** in 60% ee. The spirophosphoramidite ligands have also been effective, with ligand (R,R,R)-**(9)** giving (R)-**(3)** in 97% ee [14] and ligand (S,R,R)-**(17)** giving (S)-**(3)** in 55% ee. Both ligands performed well in hydrosilylations of substituted styrenes with ee values generally ≥95% for a range of substrates (Tables 4.3 and 4.4). Yields of the silanes were usually greater than 85%. Reaction times for ligand **(16)** tended to be quite long (typically 40–60 hours), and ligand **(17)** did not perform as well with electron-donating substituents (ligand **(16)** was not tested), with the enantiomeric excess dropping to 82% for 4-methoxystyrene (see Table 4.4, entry 9).

Chiral metallocene-based ligands are another class of ligands used in the asymmetric hydrosilylation of styrenes (Scheme 4.5). Ferrocene ligand **(18a)** gave (S)-**(3)** in only 70% ee, but subsequent substitution of the phenyl ring improved the

Table 4.2 Double hydrosilylations.

Entry	Ar	Time (h)		Pd cat. loading (mol%)	Yield diol	% ee diol
		Pt	Pd			
1	C_6H_5	24	48	0.3	75	95
2	4-$(CH_3)C_6H_4$	18	24	0.3	83	95
3	4-ClC_6H_4	24	72	0.3	75	94
4	4-$(CF_3)C_6H_4$	24	168	0.6	50	96
5	3-$(NO_2)C_6H_4$	20	72	0.6	67	98

Reactions carried out at 20°C in one pot. Pt and Pd catalysts were added sequentially. Substrate/$HSiCl_3$/Pt 1:4.5:0.0001.

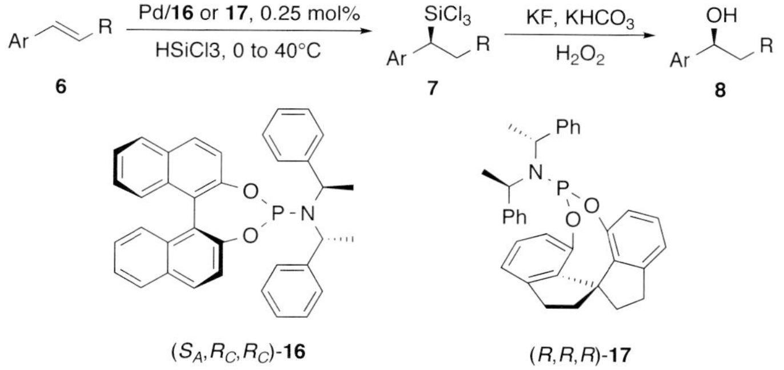

Scheme 4.4 Hydrosilylations with chiral phosphoramidite ligands.

4.2 Isolated Alkenes | 91

Table 4.3 Hydrosilylations with phosphoramidite ligand (S_A, R_C, R_C)-**(16)**.

Entry	Ar	R	Temperature (°C)	Time (h)	Yield silane	% ee alcohol
1	C₆H₅	H	20	16	87	99
2	2-(CH₃)C₆H₄	H	20	40	75	97
3	4-(CH₃)C₆H₄	H	20	40	95	86
4	2,4-(CH₃)C₆H₄	H	20	40	80	96
5	2-ClC₆H₄	H	20	40	89	96
6	3-ClC₆H₄	H	20	60	91	98
7	2-(CF₃)C₆H₄	H	20	40	74	95
8	3-(CF₃)C₆H₄	H	40	60	88	98
9	3-(NO₂)C₆H₄	H	40	144	94	95
10	C₆H₅	CH₃	40	40	91	98

Reactions carried out solvent-free.
Substrate/HSiCl₃/[ClPd(C₃H₅)]₂/ligand 1 : 1.2 : 0.001 25 : 0.005.

Table 4.4 Hydrosilylation with spirophosphoramidite ligand (R,R,R)-**(17)**.

Entry	Ar	R	Temperature (°C)	Time (h)	Yield silane	% ee alcohol
1	C₆H₅	H	rt	2	99	97
2	3-(CH₃)C₆H₄	H	rt	3	90	97
3	4-(CH₃)C₆H₄	H	rt	4	91	98
4	2-ClC₆H₄	H	rt	7	88	99
5	3-ClC₆H₄	H	rt	9	86	95
6	4-ClC₆H₄	H	rt	3	90	96
7	3-BrC₆H₄	H	rt	45	75	97
8	4-BrC₆H₄	H	rt	7	93	95
9	4-(OCH₃)C₆H₄	H	0	3	84	82
10	4-(CF₃)C₆H₄	H	rt	4	94	96
11	C₆H₅	CH₃	rt	36	86	95

Reactions carried out solvent-free.
Substrate/HSiCl₃/[ClPd(C₃H₅)]₂/ligand 1 : 2 : 0.001 25 : 0.005.

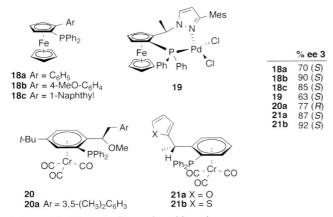

18a Ar = C₆H₅
18b Ar = 4-MeO-C₆H₄
18c Ar = 1-Naphthyl

19

	% ee 3
18a	70 (S)
18b	90 (S)
18c	85 (S)
19	63 (S)
20a	77 (R)
21a	87 (S)
21b	92 (S)

20
20a Ar = 3,5-(CH₃)₂C₆H₃

21a X = O
21b X = S

Scheme 4.5 Chiral metallocene-based ligands.

enantioselectivity [15, 16]. The most effective ligand was (**18b**), which raised the enantiomeric excess to 90%. The ratio of palladium to ligand was found to have an interesting effect on reaction times. When the standard 1:2 Pd:ligand ratio was employed, the reaction was complete in 5.5 hours. However, when the ratio was reduced to 1:1.4 the reaction proceeded in only 15 minutes, albeit with decreased selectivity (79% ee). Using ligand (**18c**) in the same ratio gave complete reaction in only 20 seconds; however, the enantiomeric excess was only 69% (85% ee with a 1:2 Pd:ligand ratio). The authors have proposed an alternate reaction pathway to account for this dramatic increase in rate [16]. Pyrazole-substituted ferrocene ligands have been developed; catalyst (**19**) gave only moderate yields and ee values (47–67%, 4-OMe gave 6% ee) for a small range of styrenes [17]. Interestingly, the presence of electron-donating substituents on the styrene reversed the enantioselectivity of the reaction. Arene complexes of chromium have also been investigated. A number of ligands based on structure (**20**) were synthesized and the best of these, ligand (**20a**), gave (*R*)-(**3**) in 95% yield and 77% ee after 48 hours [18]. Removal of the chromium from the ligand reduced the ee value to 28%. Ligands containing heterocyclic substituents, (**21a**) and (**21b**), performed better, giving (*S*)-(**3**) in 87% and 92% ee respectively, although the selectivity decreased when 4-substituted styrenes were used (71–83% ee) [19].

Although it has been shown that chelating bisphosphine ligands, such as BINAP, do not work in the catalytic hydrosilylation of 1-alkenes, the reaction has been successfully performed with BINAP-stabilized palladium nanoparticles [20]. (*S*)-BINAP-stabilized palladium catalyzed the hydrosilylation of styrene to give, after oxidation, (*S*)-(**3**) in 95% ee and 71% overall yield. The use of other BINAP derivatives as ligands in hydrosilylation has provided mixed results. Replacement of one of the phosphorus groups with antimony, ligand (**23a**), gave (*R*)-(**3**) in 78% yield and 95% ee (Scheme 4.6) [21]. Replacement of both phosphorus atoms, (**23b**), resulted in only 10% yield and 12% ee. Mono-oxidation of BINAP yielded ligand (**24**), which gave (*S*)-(**3**) in 72% ee [22]. Replacement of one of the phosphines with

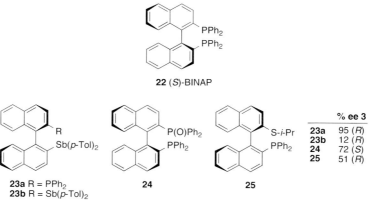

Scheme 4.6 BINAP derivatives.

sulfur, ligand **(25)**, gave (*R*)-**(3)** in only 51% ee and also gave rise to regioselectivity issues [23].

4.2.1.2 Nonaromatic Alkenes

Comparatively few reports of novel chiral catalysts for hydrosilylation of nonaromatic alkenes have appeared over recent years. The ferrocene-pyrazole ligand **(26)** has been successfully employed in the hydrosilylation of norbornene **(27)**, giving (1*R*,2*R*,4*S*)-**(28)** in 59% yield and >99% ee (Scheme 4.7) [17]. The spirophosphoramidite ligand (*R*,*R*,*R*)-**(17)**, developed for the hydrosilylation of styrenes (see above), was also used in the hydrosilylation of 1-hexene **(29)** (Scheme 4.7) [14]. However, despite the higher catalyst loading (1 mol% Pd) and longer reaction time (72 hours), alcohol **(30)** was synthesized in only 35% yield and 68% ee. The best ligand for palladium-catalyzed hydrosilylation of aliphatic alkenes remains Hayashi's MeO-MOP ligand **(4)** [3, 6].

Cationic palladium complexes have been successfully employed in one-pot cyclization/hydrosilylation of 1,6-dienes to give carbocycles (Scheme 4.8). Initial investigations employed a cationic complex generated *in situ* from **(33)** and NaBAr$_4$ to catalyze the hydrosilylation of **(31a)** with triethylsilane, giving **(32a)** in 82% yield, >95% de and 87% ee [24]. The reactions require a higher catalyst loading (5 mol%) than standard hydrosilylations. A range of diesters and protected diol substrates were tested and all proceeded in good yield with ee values generally greater than 85%. Reactions with internal alkenes such as **(31b)** were also successful. However, the use of triethylsilane prevented subsequent oxidation to the alcohol. Alternative silanes, including HSiMe$_2$OTMS, HSiMe$_2$OTBS and HSiMe$_2$CHPh$_2$ were investigated, with HSiMe$_2$CHPh$_2$ providing the best yields for the alcohols (71–98% from the dienes) and highest ee values (86–95%) over a range of substrates [25–28]. This methodology has been extended to the ring-opening cyclization/hydrosilylation of cyclopropyl-alkene **(31d)**, giving silane **(32d)** in 69% yield but only 73% ee [29]. A detailed mechanistic study for the cyclization/hydrosilylation reaction has recently been reported [30].

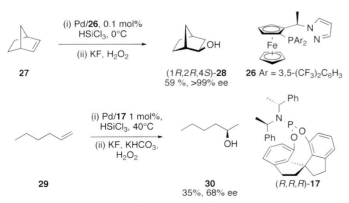

Scheme 4.7 Hydrosilylation of nonaromatic alkenes.

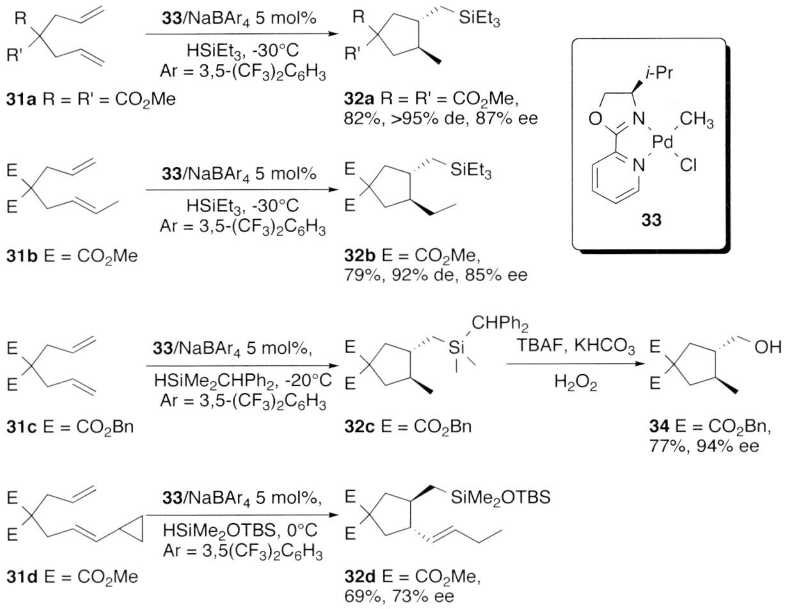

Scheme 4.8 One-pot cyclization/hydrosilylation of 1,6-dienes.

4.2.2
Metals Other Than Palladium

Only a handful of recent reports on non-palladium catalysts for the 1,2-hydrosilylation of alkenes have been published. The chiral yttrium complex (35) was used to catalyze the hydrosilylation of norbornene (27) giving silane (36) in 99% conversion and 90% ee (Scheme 4.9) [31]. Reactions with styrene and 1-hexene were also reported, but the enantioselectivities were not reported and regioselectivity for the styrene reaction was only moderate. The rhodium bisoxazoline complex (37) has been used to catalyze the hydrosilylation of a range of aromatic alkenes with $(EtO)_2MeSiH$ (Scheme 4.9) [32]. While ee values were often high (88–95%, 2-Cl gave 37% ee), the regioselectivities of the reactions were poor to moderate. The cyclization/hydrosilylation of α,ω-dienes with $PhSiH_3$ and $PhSiH_2Me$ using yttrocene (43) has also been performed (the addition of H_2 was required for successful reaction with $PhSiH_2Me$) [33]. While reactions of a small range of 1,5- and 1,6-dienes proceeded in good yield within 10 hours, ee values of only 10–50% were obtained (Scheme 4.9).

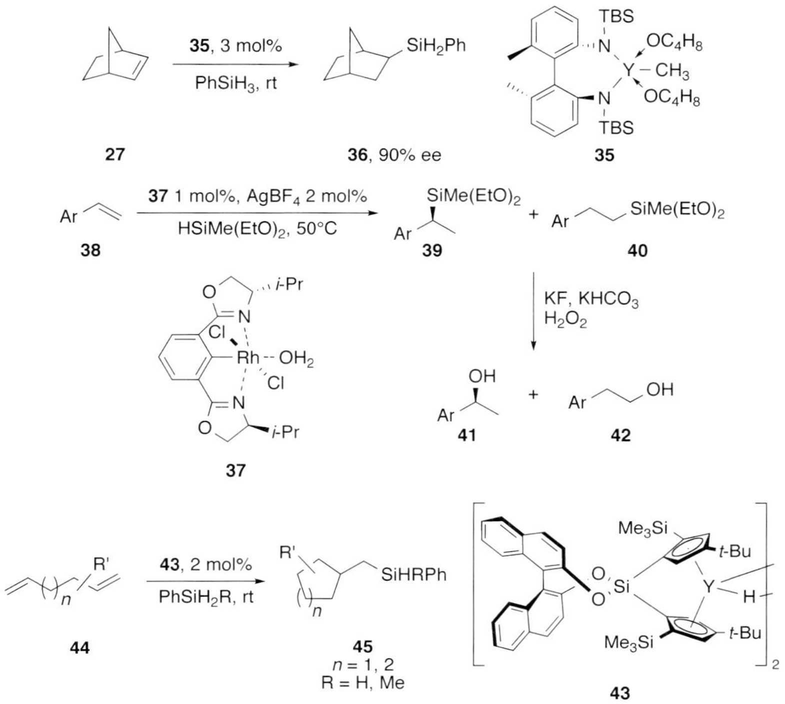

Scheme 4.9 Non-palladium catalysts for 1,2-hydrosilylations.

4.3
Conjugated Alkenes

4.3.1
Acyclic 1,3-Dienes

The transformation of allylic silanes into aldol-like structures presents an excellent opportunity for the development of asymmetric hydrosilylation of 1,3-dienes, summarized in Scheme 4.10 [34].

Many of the catalysts described earlier in this chapter have also been investigated for their efficiency with 1,3-dienes. However, unlike monoalkenes, 1,3-dienes have proved to be problematic to reduce with high enantioselectivities via hydrosilylation. In addition there is a regioselectivity issue that, in theory, could lead to a mixture of four products (ignoring absolute stereochemistry). In practice only three (Scheme 4.11) are usually observed [34]. The desired product, **(46)**, is always the major product, with the other two present in the range of 7–18%, depending on reaction conditions.

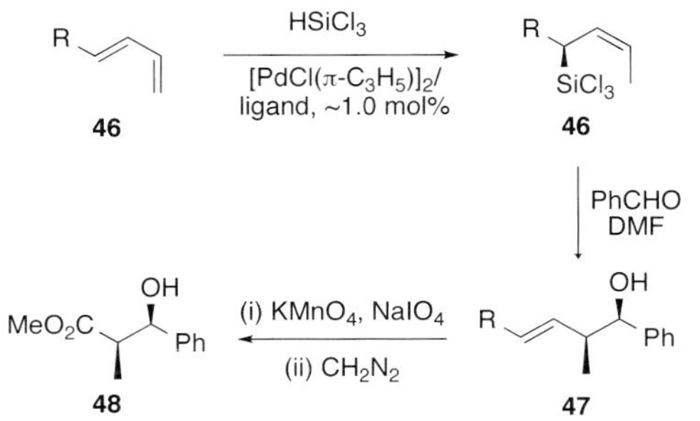

Scheme 4.10 Asymmetric hydrosilylation of 1,3-dienes.

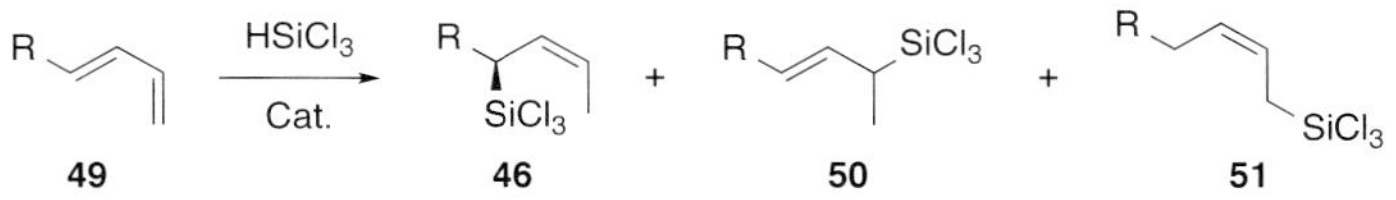

Scheme 4.11 Regioselectivity outcome.

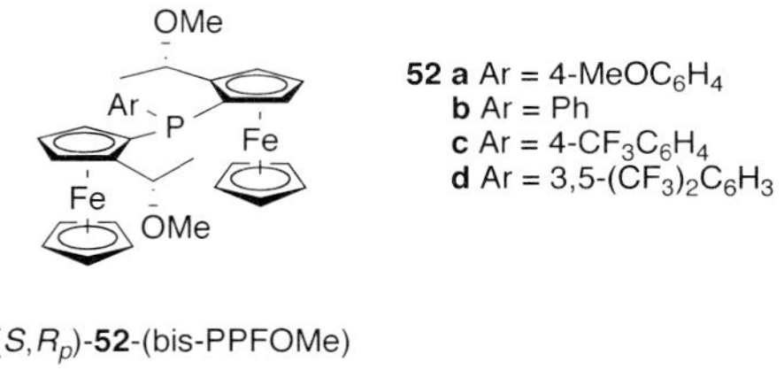

52 a Ar = 4-MeOC$_6$H$_4$
b Ar = Ph
c Ar = 4-CF$_3$C$_6$H$_4$
d Ar = 3,5-(CF$_3$)$_2$C$_6$H$_3$

(S,R_p)-**52**-(bis-PPFOMe)

Figure 4.1 Bisferrocenyl ligands **(52a–d)**.

Over the course of more than a decade, Hayashi's group has investigated the use of many of the MOP-related and ferrocenyl ligands. Only relatively recently have reproducibly high selectivities emerged from this work. In 2002 Hayashi reported that the bisferrocenyl ligands **(52a–d)** (shown in Figure 4.1, similar to those initially reported by Hayashi in 1980) gave good enantioselectivities with alkyl-substituted 1,3-dienes [34].

Several such ligands were prepared with variations of the phosphinyl aryl group. Selected results with these ligands are given in Table 4.5. For both dienes tested ligand **(52d)** worked best. The use of low temperatures gave slightly improved enantioselection, but at a price – the reactions took a week to achieve good conversions (entries 5 and 7).

In seemingly parallel studies the same research group has extended the basic MOP design to a new series of "Ar-MOPs" (originally developed for the catalytic, asymmetric arylation of activated imines) where the OMe group at C2' is replaced by an aryl moiety (Figure 4.2) [35]. The best of these gives around 90% ee in the catalytic, asymmetric hydrosilylation of acyclic 1,3-dienes.

Table 4.5 Hydrosilylation of acyclic 1,3-dienes with bisferrocenyl ligands **(52a–d)**.

Entry	Diene R (Scheme 4.11)	Ligand (52)	Temperature (°C)	Time (h)	Yield (silane)	(46):(50):(51)	% ee (46)
1	Me(CH$_2$)$_5$	a	20	96	85	82:3:15	68 (S)
2	Me(CH$_2$)$_5$	b	20	29	83	90:9:1	76 (S)
3	Me(CH$_2$)$_5$	c	20	8	91	90:8:2	78 (S)
4	Me(CH$_2$)$_5$	d	20	25	90	87:10:3	87 (S)
5	Me(CH$_2$)$_5$	d	−5	168	88	89:9:2	93 (S)
6	Cyclohexyl	d	20	26	93	92:6:2	88 (S)
7	Cyclohexyl	d	−5	168	92	93:5:2	90 (S)

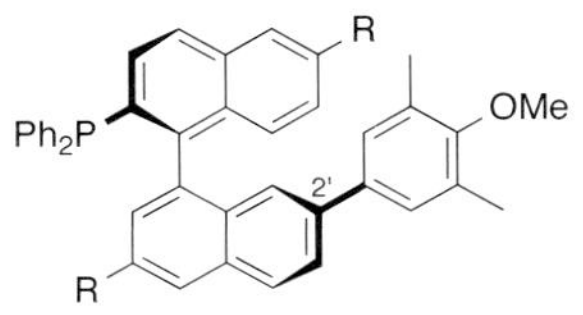

53a R = H Ar-MOP
53b R = n-Octyl

Figure 4.2 Hayashi's "Ar-MOPs" ligands.

Table 4.6 Hydrosilylation of acyclic 1,3-dienes with Ar-MOP ligands **(53a)** and **(53b)**.

Entry	Diene R (Scheme 4.11)	Ligand (53)	Temperature (°C)	Time (h)	Yield (silane)	% ee (46)
1	Me(CH$_2$)$_5$	a	20	24	81	69
2	Me(CH$_2$)$_5$	b	20	4	91	68
3	Me(CH$_2$)$_5$	a	−10	168	9	77
4	Me(CH$_2$)$_5$	b	−10	168	76	77
5	Ph	a	20	22	85	62
6	Ph	b	20	22	95	63
7	Ph	a	0	168	14	71
8	Ph	b	0	168	52	72
9	Ph	b	0	168	53	79

In order to gain access to lower reaction temperatures, the dioctyl derivative **(53b)** was prepared and shown to be effective at temperatures as low as −30 °C [36]. A significant rate enhancement was observed with the use of the more soluble ligand. This constitutes an early example of the use of solubilizing groups with ligands in catalytic, asymmetric processes. The ee values are consistently lower than for the bisferrocenyl ligands mentioned above (Table 4.6; however, results are much improved with cyclic dienes, see next section).

4.3.2
Cyclic 1,3-Dienes

A number of ligands, including some already discussed above, have been employed in the catalytic, asymmetric hydrosilylation of *cyclic* 1,3-dienes (Scheme 4.12) [4].

Until recently the ee values were somewhat lower than in similar studies with acyclic 1,3-dienes. It has now been shown that ee values between 80% and 90% are attainable with 1,3-cyclopentadiene and 1,3-cyclohexadiene using **(53a)** and **(53b)** [35]. An alternative class of (amino acid-derived) ligands, based on β-*N*-sulfonylaminoalkylphosphines such as **(56)** and **(57)** (Figure 4.3), have been reported by Frejd's group [37]. These ligands, similar to those originally reported by Achiwa in 1990 [38], are less effective in stereoinduction than the Ar-MOP ligands (84% ee with **(54b)** and 37% ee with **(54a)**). However, there is obvious opportunity for modification. Interestingly trichlorosilane, while adding readily to 1,3-cyclohexadiene (>80% yield), gave no enantioselectivity at all when employing **(56)** as ligand. The corresponding arsenium-based ligand, **(57)**, performed poorly compared to the phosphine.

4.3.3
Enynes

Hayashi's group has explored the catalytic, asymmetric hydrosilylation of 1,3-enynes. The products are axially chiral allenyl silanes that, while of intrinsic interest themselves, undergo a similar reaction with aldehydes to that of allyl silanes (see Scheme 4.7) giving homopropargyl alcohols [39]. Initial studies with MOP and some of its analogues gave very poor enantioselectivity (18–27%) although in conversion. Switching to the ferrocenyl-based phosphines **(61)** and **(62)** gave excellent ee values with good conversions (Scheme 4.13) [40].

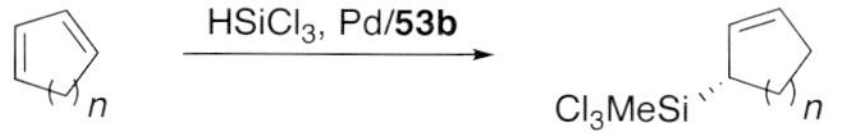

54a n = 1
 b n = 2

55a n = 1 (91% ee)
 b n = 2 (83% ee)

Scheme 4.12 Cyclic 1,3-dienes.

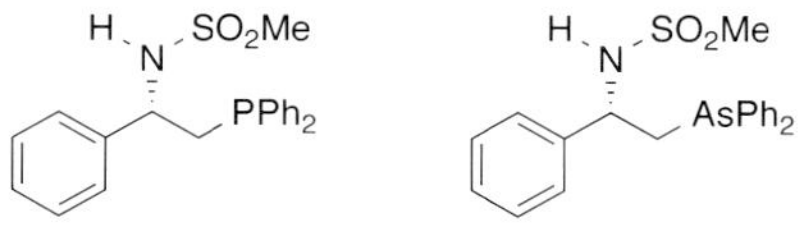

56 57

Figure 4.3 Amino acid-derived ligands **(56)** and **(57)**.

62a R = H, M = Fe (57%, 41% ee)
b R = Me, M = Fe (82%, 92% ee)
c R = Me, M = Ru (59%, 75% ee)

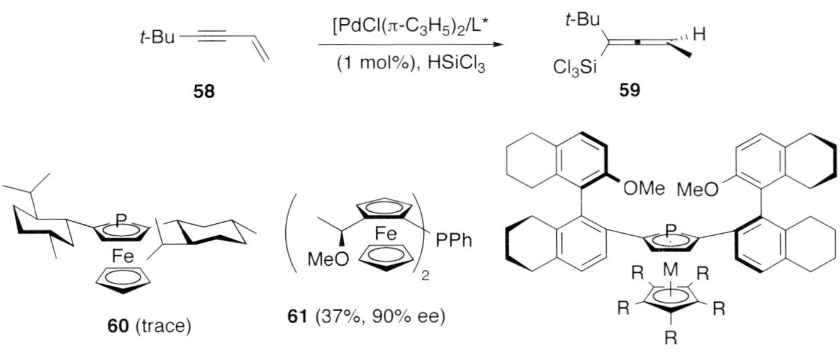

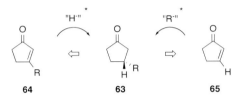

Scheme 4.13 Asymmetric hydrosilylation of 1,3-enynes.

Figure 4.4 Asymmetric conjugate additions of carbon nucleophiles.

4.4 α,β-Unsaturated Systems

4.4.1 Copper

Remarkable advances in the asymmetric, catalytic hydrosilylation of α,β-unsaturated systems have occurred over the last few years, largely due to the development of new copper-based catalysts [41]. Given the rich history of copper chemistry in conjugate addition reactions this is, perhaps, not surprising. However, these new catalysts should find wide application in synthetic chemistry due to their relative ease of use and constitute an attractive alternative to asymmetric conjugate additions of carbon nucleophiles (see, for example, Figure 4.4).

This area can be traced back to a seminal contribution by Brunner and Miehling in 1984 [42]. In that paper they reported the first asymmetric 1,2-hydrosilylation of ketones using copper(I) catalysis. Their conditions, which involved CuO-*t*-Bu in combination with a chiral diphosphine ligand [(−)-DIOP] and diphenylsilane, have stood the test of time and, with minor variations, are still employed today.

The first enantioselective catalytic, asymmetric hydrosilylation of α,β-unsaturated esters was published by Buchwald's group in 1999 [43]. The catalyst was generated *in situ* (from CuCl and NaO-*t*-Bu with (*S*)-Tol-BINAP (**68**) as the chiral ligand). The use of PMHS, {TMSO-[Si(H)Me]$_n$-OTMS}, was also introduced in that paper. Consistently good to excellent ee values (80–92%) were obtained over a range of substrates (see, for example, Scheme 4.14). A linear correlation between the enantiomeric excess of the ligand and the product was observed, indicating a 1:1 ligand-to-metal ratio (thus excluding nonlinear effects).

This process has since been applied by Buchwald's group to the reduction of a variety of other systems including α,β-unsaturated lactones and lactams [44], cycloalkenones [45] and β-enamino esters [46], the last providing a convenient route to β-amino esters. In addition the same group has shown that is possible to use their conditions to carry out dynamic kinetic resolution of 2,4-dialkylcyclopentenones and α,β-unsaturated lactones [47]. Unlike simple kinetic resolution, *dynamic* kinetic resolution allows, in principle, for the complete conversion of a racemic material into a single enantiomeric product. This was applied to a total synthesis of eupomatilone-3, one of a family of lignans found in the Australian shrub *Eupomatia bennettii* [48]. Thus asymmetric hydrosilylation of the advanced intermediate (**69**) gave (**70**), the immediate precursor to the natural product. In the example below in Scheme 4.15, (*R*)-MeO-BIPHEP (**71**) was used as it gave somewhat improved ee values compared with (**68**).

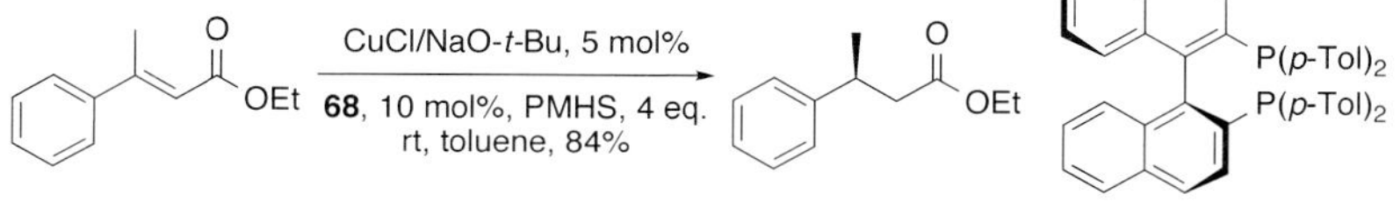

Scheme 4.14 The use of PMHS.

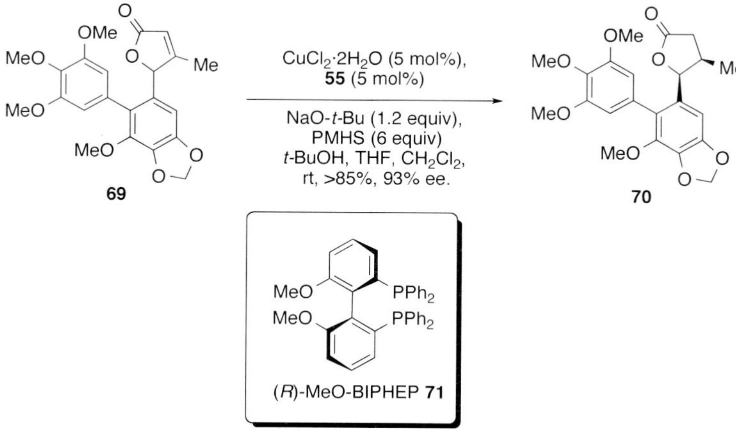

Scheme 4.15 Synthesis toward *Eupomatia bennettii*.

Carreira's group has developed a copper-catalyzed asymmetric hydrosilylation of nitroalkenes (Scheme 4.16) [49]. They found that the use of Josiphos **(74)** gave improved enantioselectivity compared with the use of **(68)**. They also found that the reaction worked best with preformed salt-free CuO-*t*-Bu. The presence of salts such as NaCl, KCN or LiCl leads to significant rate reductions. Over-reduction to the oxime was also a problem, which could be completely circumvented by the addition of 1.2 equivalents of water.

After extensive screening of potential chiral ligands, Lipshutz's group demonstrated the efficacy of Segphos and PPF-P(*t*-Bu)$_2$ in a series of papers involving "ligated" CuH. Outstanding results have so far been obtained in the 1,4-hydrosilylation of cycloalkenones [50], α,β-unsaturated esters [51] and β-silyl-α,β-unsaturated esters [52]. The choice of ligand is critical to the success, with each class of starting material having different requirements. Thus catalytic, asymmetric 1,4-hydrosilylation of acyclic enones [53] give excellent ee values, typically in the range 90–99%, with **(74)** or **(77)** [PPF-P(*t*-Bu)$_2$] but not with closely related ligands such as Segphos and Walphos (Scheme 4.17).

β-Substituted cyclic enones, on the other hand, give the highest enantioselectivities, again typically 90–99% ee, with Segphos **(80)** (Scheme 4.18) [51]. In this work

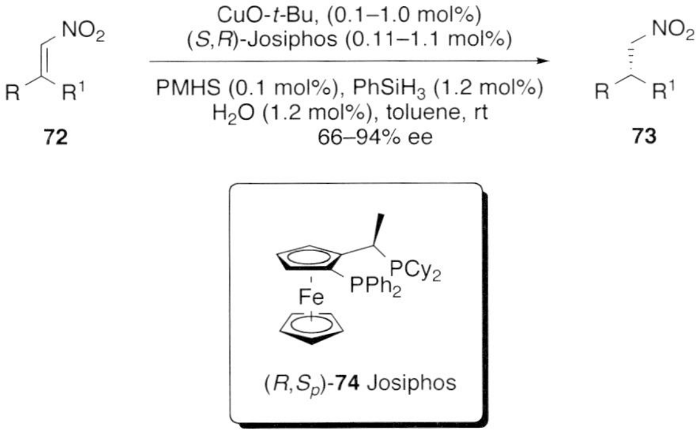

Scheme 4.16 Asymmetric hydrosilylation of nitroalkenes.

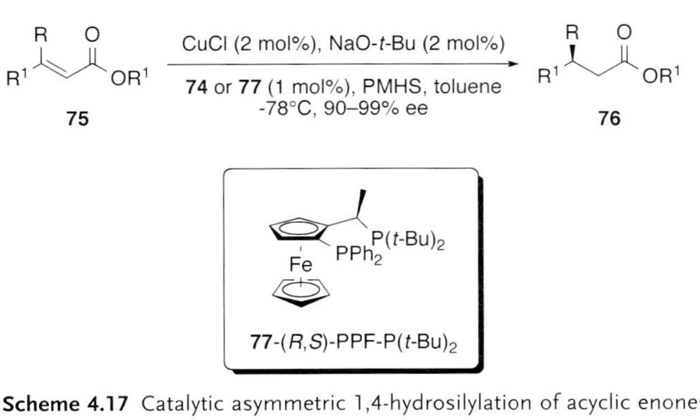

Scheme 4.17 Catalytic asymmetric 1,4-hydrosilylation of acyclic enones.

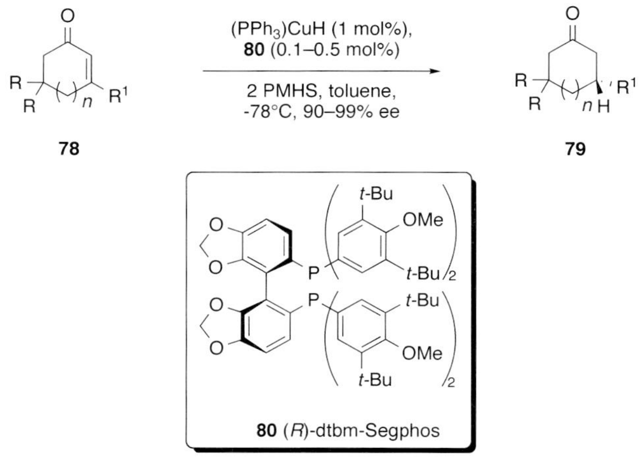

Scheme 4.18 Hydrosilylation of β-substituted cyclic enones.

Stryker's preformed [(PPh$_3$)CuH]$_6$ was predominantly employed. However, "standard" conditions, that is CuCl/NaO-t-Bu, worked equally well. The authors also found that NaO-t-Bu could be replaced with NaOMe. Remarkably, they also demonstrated that substrate/ligand ratios as high as 275 000 : 1 could be employed. For example, reduction of 64.76 g of isophorone (**(78)**, R = R^1 = Me) required only 2 mg of **(80)** for excellent conversion (88% isolated yield) and enantiomeric excess (98.5%) [54, 55].

A very recent report, again from Lipshutz's group, disclosed the first heterogeneous, copper-catalyzed asymmetric hydrosilylation of, *inter alia*, enones and enoates. In essence their new catalyst consists of "Cu-H in charcoal" [56]. Initially Cu(NO$_3$)$_2$ is mixed with charcoal in water; unlike in previous methods, ultrasonication is employed, which apparently improves distribution of Cu(II) into the solid. The water is then removed and the resultant solid is dried. A chiral ligand and excess of PMHS complete the list of ingredients. However, it was found that a catalytic amount of a sodium alkoxide or aryloxide (t-butoxide or phenoxide) was essential for activity, with the latter affording greater acceleration. Without this additive, no reduction took place. Several chiral ligands were screened and **(80)** was chosen. The yields and selectivities are very close to those of similar solution-phase reactions.

4.4.2
Tandem Processes

Several examples of tandem processes involving catalytic, asymmetric 1,4-hydrosilylation have appeared. Riant and Shibasaki have both recently reported a tandem reductive aldol reaction. Riant screened many candidate ligands and silanes and settled on Taniaphos-based ligands in combination with PhSiH$_3$. Although poor

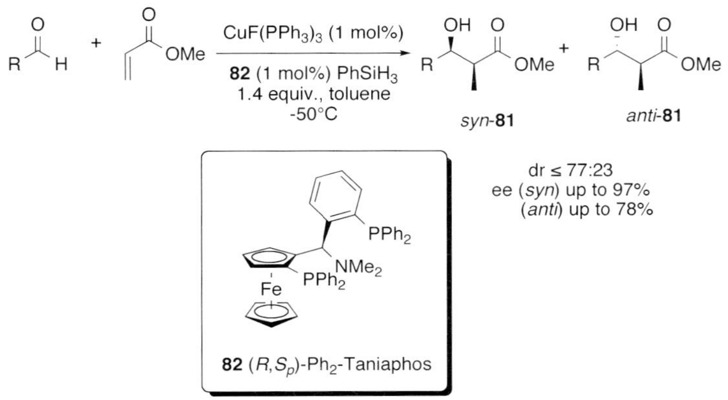

Scheme 4.19 Tandem reductive aldol reaction.

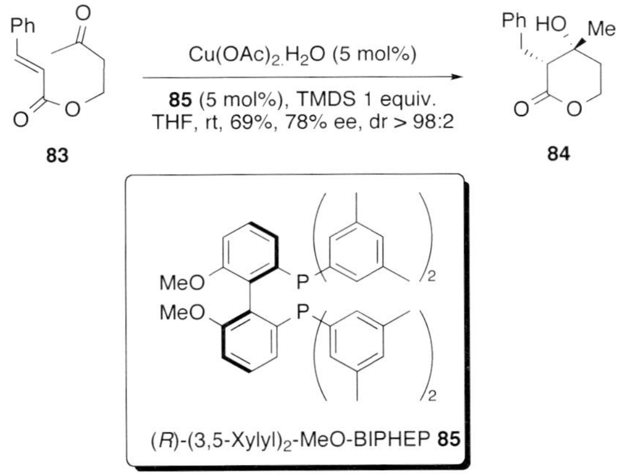

Scheme 4.20 Intramolecular tandem reductive reaction.

diastereoselectivity was obtained, the ee values of some of the individual diastereomers (especially the *anti*-isomers) were very high (Scheme 4.19).

Intramolecular tandem reductive processes are beginning to receive some attention. For example, lactone (**84**) is produced with reasonable enantiomeric excess by catalytic, asymmetric 1,4-hydrosilylation of unsaturated ester/ketone (**83**) (Scheme 4.20).

4.5
Conclusions

From the discussions in this chapter it is clear that catalytic, asymmetric 1,2- and 1,4-hydrosilylations hold enormous potential in synthesis. The next few years should see an expansion in the number of syntheses employing such asymmetric processes as key steps.

References

1 Yamamoto, K., Hayashi, T. and Kumada, M. (1971) *Journal of the American Chemical Society*, **93**, 5301.
2 Yamamoto, K., Uramoto, Y. and Kumada, M. (1971) *Journal of Organometallic Chemistry*, **31**, C9.
3 Hayashi, T., Tamao, K., Katsuro, Y., Nakae, I. and Kumada, M. (1980) *Tetrahedron Letters*, **21**, 1871.
4 For excellent discussions of the mechanism of transition-metal catalysed hydrosilylations see (a) Nishiyama, H. and Itoh, K. (2000) *Catalytic Asymmetric Synthesis*, 2nd edn, pp. 111–43. (b) Hayashi, T. (2000) *Accounts of Chemical Research*, **33**, 354–62. (c) Hayashi, T. (1999) *Comprehensive Asymmetric Catalysis I–III*, Vol. **I**, (eds E.N. Jacobsen, A. Pfaltz and H. Yamamoto), Springer-Verlag, Berlin, Heidelberg, pp 319.
5 (a) Tamao, K., Kakui, T. and Kumada, M. (1978) *Journal of the American Chemical Society*, **100**, 2268. (b) Tamao, K., Ishida, N., Tanaka, T. and Kumada, M. (1983) *Organometallics*, **2**, 1694.
6 Uozumi, Y. and Hayashi, T. (1991) *Journal of the American Chemical Society*, **113**, 9887.
7 Kitayama, K., Uozumi, Y. and Hayashi, T. (1995) *Journal of the Chemical Society, Chemical Communications*, 1533.
8 Hayashi, T., Hirate, S., Kitayama, K., Tsuji, H., Torii, A. and Uozumi, Y. (2000) *Chemistry Letters*, 1272.
9 Hayashi, T., Hirate, S., Kitayama, K., Tsuji, H., Torii, A. and Uozumi, Y. (2001) *Journal of Organic Chemistry*, **66**, 1441.
10 Dotta, P., Kumar, P.G.A., Pregosin, P.S., Albinati, A. and Rizzato, S. (2004) *Organometallics*, **23**, 2295.
11 Bringmann, G., Wuzik, A., Breuning, M., Henschel, P., Peters, K. and Peters, E.-M. (1999) *Tetrahedron: Asymmetry*, **10**, 3025.
12 Shimada, T., Mukaide, K., Shinohara, A., Han, J.W. and Hayashi, T. (2002) *Journal of the American Chemical Society*, **124**, 1584.
13 Jensen, J.F., Svendsen, B.Y., La Cour, T.V., Pedersen, H.L. and Johannsen, M. (2002) *Journal of the American Chemical Society*, **124**, 4558.
14 Guo, X.-X., Xie, J.-H., Hou, G.-H., Shi, W.-J., Wang, L.-X. and Zhou, Q.-L. (2004) *Tetrahedron: Asymmetry*, **15**, 2231.
15 Pedersen, H.L. and Johannsen, M. (1999) *Chemical Communications*, 2517.
16 Pedersen, H.L. and Johannsen, M. (2002) *Journal of Organic Chemistry*, **67**, 7982.
17 Pioda, G. and Togni, A. (1998) *Tetrahedron: Asymmetry*, **9**, 3903.
18 Gibson, S.E., Rendell, J.T. and Rudd, M. (2006) *Synthesis*, 3631.
19 Webber, I. and Jones, G.B. (2001) *Tetrahedron Letters*, **42**, 6983.
20 Tamura, M. and Fujihara, H. (2003) *Journal of the American Chemical Society*, **125**, 15742.
21 Yasuike, S., Kawara, S.-I., Okajima, S., Seki, H., Yamaguchi, K. and Kurita, J. (2004) *Tetrahedron Letters*, **45**, 9135.
22 Gladiali, S., Pulacchini, S., Fabbri, D., Manassero, M. and Sansoni, M. (1998) *Tetrahedron: Asymmetry*, **9**, 391.
23 Gladiali, S., Medici, S., Pirri, G., Pulacchini, S. and Fabbri, D. (2001) *Canadian Journal of Chemistry*, **79**, 670.
24 Perch, N.S. and Widenhoefer, R.A. (1999) *Journal of the American Chemical Society*, **121**, 6960.
25 Perch, N.S., Pei, T. and Widenhoefer, R.A. (2000) *Journal of Organic Chemistry*, **65**, 3836.
26 Pei, T. and Widenhoefer, R.A. (2000) *Tetrahedron Letters*, **41**, 7597.
27 Pei, T. and Widenhoefer, R.A. (2000) *Organic Letters*, **2**, 1469.
28 Pei, T. and Widenhoefer, R.A. (2001) *Journal of Organic Chemistry*, **66**, 7639.

29. Wang, X., Stankovich, S.Z. and Widenhoefer, R.A. (2002) *Organometallics*, **21**, 901.
30. Perch, N.S. and Widenhoefer, R.A. (2004) *Journal of the American Chemical Society*, **126**, 6332.
31. Gountchev, T.I. and Tilley, T.D. (1999) *Organometallics*, **18**, 5661.
32. Tsuchiya, Y., Uchimura, H., Kobayashi, K. and Nishiyama, H. (2004) *Synlett*, 2099.
33. Muci, A.R. and Bercaw, J.E. (2000) *Tetrahedron Letters*, **41**, 7609.
34. Han, J.W., Tokunaga, N. and Hayashi, T. (2002) *Helvetica Chimica Acta*, **85**, 3848.
35. (a) Hayashi, T., Han, J.W., Takeda, A., Tang, J., Nohmi, K., Mukaide, K., Tsuji, H. and Uozumi, Y. (2001) *Advanced Synthesis and Catalysis*, **343**, 279. (b) Han, J.W. and Hayashi, T. (2001) *Chemistry Letters*, 977.
36. Han, J.W. and Hayashi, T. (2002) *Tetrahedron: Asymmetry*, **13**, 325.
37. Gustafsson, M., Bergqvist, K.-E. and Frejd, T. (2001) *Journal of the Chemical Society, Perkin Transactions I*, 1452.
38. Okada, T., Morimoto, T. and Achiwa, K. (1990) *Chemistry Letters*, 999.
39. Han, J.W., Tokunaga, N. and Hayashi, T. (2001) *Journal of the American Chemical Society*, **123**, 12916.
40. Ogasawara, M., Ito, A., Yoshida, K. and Hayashi, T. (2006) *Organometallics*, **25**, 2715.
41. For an excellent account of the development of the "Cu-H" catalysis with silanes as well as mechanistic considerations see Rendler, S. and Oestreich, M. (2006) *Angewandte Chemie (International Ed. in English)*, **45**, 2.
42. Brunner, H. and Miehling, W. (1984) *Journal of Organometallic Chemistry*, **275**, C17.
43. Appella, D.H., Moritani, Y., Shintani, R., Ferreira, E.M. and Buchwald, S.L. (1999) *Journal of the American Chemical Society*, **121**, 9473.
44. Hughes, G., Kimura, M. and Buchwald, S.L. (2003) *Journal of the American Chemical Society*, **125**, 11253.
45. Chae, J., Yun, J. and Buchwald, S.L. (2004) *Organic Letters*, **6**, 4809.
46. Rainka, M.P., Aye, Y. and Buchwald, S.L. (2004) *Proceedings of the National Academy of Sciences of the United States of America*, **101**, 5821.
47. Jurkauskas, V. and Buchwald, S.L. (2002) *Journal of the American Chemical Society*, **124**, 2892.
48. Rainka, M.P., Milne, J.E. and Buchwald, S.L. (2005) *Angewandte Chemie (International Ed. in English)*, **44**, 6177.
49. Czekelius, C. and Carreira, E.M. (2003) *Angewandte Chemie (International Ed. in English)*, **42**, 4793.
50. Lipshutz, B.H., Servesko, J.M., Petersen, T.B., Papa, P.P. and Lover, A.A. (2004) *Organic Letters*, **6**, 1273.
51. Lipshutz, B.H., Servesko, J.M. and Taft, B.R. (2004) *Journal of the American Chemical Society*, **126**, 8352.
52. Lipshutz, B.H., Tanaka, N., Taft, B.R. and Lee, C.-T. (2006) *Organic Letters*, **8**, 1963.
53. Lipshutz, B.H. and Servesko, J.M. (2003) *Angewandte Chemie (International Ed. in English)*, **42**, 4789.
54. See also Lipshutz, B.H. and Frieman, B.A. (2005) *Angewandte Chemie (International Ed. in English)*, **44**, 6345.
55. These reductions are accelerated by microwave or thermal heating. See Lipshutz, B.H., Frieman, B.A., Unger, J.B. and Nihan, D.M. (2005) *Canadian Journal of Chemistry*, **83**, 606.
56. Lipshutz, B.H., Frieman, B.A. and Tomaso, A.E., Jr. (2006) *Angewandte Chemie (International Ed. in English)*, **45**, 1259.

Part Two
Carbonyl Reactions

5
Carbonyl Hydrogenation

Christian Hedberg

5.1
Introduction

Of comparable practical interest to the enantioselective reduction of C=C bonds is the hydrogenation of C=O bonds. Carbonyl compounds are among the most abundant starting materials for the synthetic organic chemist. β-Keto esters are readily available via various condensation reactions and aryl alkyl ketones from Friedel–Craft type reactions. Two general methods have been developed for catalytic homogeneous enantioselective hydrogenation of carbonyl compounds. The most efficient catalysts for hydrogenation of activated ketones, β-keto esters and ketones bearing a coordinating group in an α-position, are based on Ru(II)X$_2$-diphosphine complexes (X = halide). These highly selective hydrogenation reactions proceed with excellent values of turnover number (TON) and high rates (turnover frequency (TOF)), providing a robust synthetic route to optically enriched alcohols carrying functionality. Further, chemo- and enantioselective reductions of nonactivated ketonic carbonyl compounds have been a long-standing problem. Until recently, these transformations have only been affected via the use of stoichiometric chiral metal hydrides or chiral boron reagents. However, since 1995, the method of choice for reducing unactivated ketones is Noyori's bifunctional Ru(II)-diphosphine/diamine catalyst system. The development of chiral ligands over recent years has provided a vide variety of diphosphines suitable for Ru(II)-catalyzed asymmetric hydrogenation of ketonic substrates.

5.2
Asymmetric Hydrogenation of Activated Ketones and β-Keto Esters

One of the most important asymmetric reductions in the field of homogeneous catalysis is the reduction of β-keto esters and similar activated ketonic substrates. The corresponding reduction products are highly valuable starting materials for a vide variety of chemical applications. During the 1980s the development of Ru(II)-

Modern Reduction Methods. Edited by Pher G. Andersson and Ian J. Munslow.
Copyright © 2008 WILEY-VCH Verlag GmbH & Co. KGaA, Weinheim
ISBN: 978-3-527-31862-9

diphosphine chemistry established proof of concept of the method and it can now be considered as mature in terms of industrial applications. A typical catalyst consists of a Ru(II) complex bearing two halides and a chelating diphosphine, like BINAP, MeO-BIPHEP, Synphos, P-Phos and others (Figure 5.1). Such ligands are considered as privileged for the reduction of activated ketonic substrates. A overwhelming amount of literature has been written on the subject, including several excellent book chapters [1, 2]. This, coupled with the commercial availability of several classes of diphosphines and catalyst precursors, offers an excellent starting point for applied catalysis. The Ru(II)-diphosphine reduction of activated ketones is operationally simple and can easily be applied, even by an experimentalist not familiar with asymmetric hydrogenation technology.

Since the first report on efficient asymmetric β-keto ester hydrogenation by Noyori and coworkers [3], a large number highly enantioselective examples have been documented. The first examples of β-keto ester hydrogenation employed BINAP-Ru(II)-dihalide complexes and proved to be highly efficient (Table 5.1, entries 1–15) for a number of different substitution patterns, enantioselectivities in the range of 98–99% are routinely achieved. However, by changing the diphosphine to xylBINAP, the enantiomeric excess (ee) is raised to 99.9% for simple methyl acetoacetate (entry 16) [4]. Interestingly, racemic xylBINAP could be used if 50 mol% (relative to diphosphine) of a chiral diamine such as DPEN was added to the reaction mixture [5]. The chiral diamine efficiently poisons one of the enantiomers of the Ru-complex by forming a nonreactive diastereomeric complex. Some substrates do not perform well under these conditions, such as R^1 = Ph

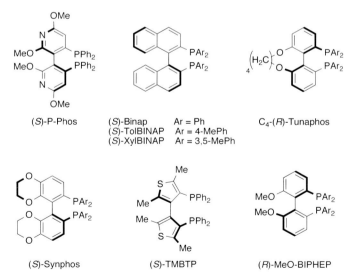

Figure 5.1 Diphosphine ligands commonly employed in Ru(II)-catalyzed asymmetric hydrogenation of activated ketonic substrates.

5.2 Asymmetric Hydrogenation of Activated Ketones and β-Keto Esters

Table 5.1 Asymmetric hydrogenation of β-keto esters.

$$R^1 \underset{O}{\overset{O}{\|}} \underset{O}{\overset{O}{\|}} OR^2 \xrightarrow{H_2, \text{ Chiral Catalyst}} R^1 \underset{*}{\overset{OH}{|}} \underset{O}{\overset{O}{\|}} OR^2$$

Entry	R^1	R^2	Catalyst	Temperature (°C)	H_2 (atm)	Yield (%)	ee (%)	S/C	Absolute configuration
1	Me	Me	(R)- a	25	100	99	99	2000	R
2	Me	Me	(R)- b	25	100	99	99	2100	R
3	Me	Me	c	25	100	99	99	1400	S
4	Me	Et	(R)- a	28	103	99	99	1000	R
5	Me	i-Pr	(R)- b	28	73	93	98	1100	R
6	Me	t-Bu	(R)- a	28	70	98	98	1000	R
7	Et	Me	(R)- b	28	98	99	100	1200	R
8	n-C_4H_9	Me	(S)- a	28	94	99	98	2000	S
9	i-Pr	Me	(R)- a	25	100	99	99	850	S
10	C_6H_5	Et	(R)- b	25	91	99.5	85	760	S
11	$C_6H_5CH_2OCH_2$	Et	(S)- b	28	100	100	78	560	R
12	$C_6H_5CH_2OCH_2$	Et	(S)- b	100	100	100	98	560	R
13	$(i\text{-Pr})_3\text{SiOCH}_2$	Et	(S)- b	27	100	100	95	290	R
14	CH_2Cl	Et	(S)- b	24	77	100	56	1080	R
15	CH_2Cl	Et	(S)- b	100	100	100	97	1300	R
16	Me	Me	d	25	100	100	99.9	1500	R
17	Me	Me	(R)- e	50	4	100	99		R
18	Me	Me	(R)- e	50	4	100	99		R
19	n-C_3H_7	Et	(R)- e	50	4	100	99		R
20	i-Pr	Et	(R)- e	50	4	100	99		R
21	C_6H_5	Et	(R)- e	80	10	100	97		S
22	CF_3	Et	(S)- e	100	10	100	49		R
23	CF_3CF_2	Et	(S)- e	100	10	100	63		R
24	Me	Me	f	70	5	100	98.5	400	S
25	Me	Et	f	70	5	100	98.6	400	S
26	Me	$CH_2C_6H_5$	f	80	5	100	96.6	2800	S
27	Et	Me	f	80	5	100	98	2800	S
28	CH_2Cl	Et	f	80	5	100	98	2800	S
29	C_6H_5	Et	f	90	5	100	95	2400	R
30	CF_3	Et	(S)- g	100	20	100	40	100	S
31	CF_3CF_2	Et	(S)- g	100	20	100	57	100	S
32	C_6H_5	Me	(R)- g	50	20	100	90	100	R
33	C_6H_5	Et	(R)- g	80	10	100	93	100	R
34	Me	Me	h	50	4	100	99	100	S
35	Me	Et	i	40	100	100	98	1000	S
36	Me	Me	j	25	52	100	99.1	100	R
37	Me	i-Pr	j	25	52	100	99.3	100	R
38	Me	t-Bu	j	25	52	100	99.2	100	R
39	Et	Et	j	25	52	100	99	100	R
40	C_6H_5	Et	j	25	52	100	82.3	100	R

a = RuCl$_2$[(R/S)-BINAP], b = RuBr$_2$[(R/S)-BINAP], c = RuI$_2$[(S)-BINAP], d = RuBr$_2$[(R)-xylBINAP], e = RuBr$_2$[(R/S)-Synphos], f = RuCl$_2$[(S)-P-Phos](DMF)$_n$, g = RuBr$_2$[(R/S)-MeO-BIPHEP], h = RuCl$_3$[(S)-MeO-BIPHEP], i = RuCl$_2$[TMBTP](C$_6$H$_6$), j = RuCl$_2$[(R)-C$_4$-Tunaphos](C$_6$H$_6$).

(entry 10) which gives only 85% ee when hydrogenated with the RuBr$_2$[(R)-BINAP]-complex. Genêt and coworkers employed the Synphos ligand together with Ru(II)(COD)(η^3-Metallyl)$_2$/HBr to hydrogenate a number of substituted keto esters with excellent results (entries 17–23) [6–8]. Chan and coworkers employed the P-Phos ligand as its RuCl$_2$[(S)-P-Phos](DMF)$_n$-complex to hydrogenate a series of β-keto esters with good results (entries 24–29) [9]. MeO-BIPHEP is an excellent ligand for the hydrogenation of β-keto esters, especially when the 3,5-xylyl and 3,5-t-Bu-4-MeO-derivatives are used. However, MeO-BIPHEP has mainly been employed in industry (Hoffman La Roche) and little has been published (entries 30–34). Sannicolo and coworkers employed their chiral bisthiophene diphosphine ligand (TMBTP) as its RuCl$_2$[TMBTP](C$_6$H$_6$) complex and hydrogenated a series of β-keto esters with good results (entry 35) [10]. Zhang and coworkers designed a series of ligands named "Tunaphos" based on the MeO-BIPHEP scaffold. After extensive optimization of the bite angle of the ligand via the cyclic backbone, they concluded that a C$_4$-spacer gave superior results for most substrates. The C$_4$-Tunaphos as its RuCl$_2$[(R)-C$_4$-Tunaphos](C$_6$H$_6$) complex hydrogenates β-keto esters with high enantioselectivity (entries 36–40) [11].

5.2.1
α-Keto Ester Hydrogenation

Asymmetric hydrogenations of α-keto esters are of comparable synthetic interest to the reduction of β-keto esters. Ru(II)-diphosphine-based hydrogenation catalysts perform well over a wide range of α-keto ester substrates. The mechanism of the reaction is similar to the reduction of β-keto esters, but requires higher catalyst loading when hydrogenated with the same catalyst. Mashima and coworkers hydrogenated a series of α-keto esters, employing RuCl$_2$[(S)-BINAP](C$_6$H$_6$)-complex as catalyst (Table 5.2) [4]. Interestingly, it was found that the addition of a small amount of strong acid (Table 5.2, HBF$_4$, even entries) increased both the enantioselectivity and rate of hydrogenation compared to the reactions performed in the absence of HBF$_4$ (odd entries).

5.2.2
1,3-Diketones

The double hydrogenation of 1,3-diketones can be carried out in the presence of a Ru(II)-diphosphine catalyst. Such reduction products offer valuable intermediates for the synthesis of statines and other biologically relevant molecules. The asymmetric hydrogenation of 2,4-pentanedione with the RuCl$_2$[(S)-BINAP]-complex resulted in the formation of enantiopure (R,R)-2,4-pentanediol (Table 5.3, entry 1) [12]. The same reaction is also valid for the corresponding 1,5-dichloro-2,4-pentanedione, which is hydrogenated to its (R,R)-diol-derivative in 92% ee [13]. The corresponding 1,3-dipheny-1,3-propanedione reacted sluggishly when hydrogenated in the presence of RuCl$_2$[(R)-BIPHEMP] complex as catalyst, with only

Table 5.2 Asymmetric hydrogenation of α-keto esters.

$$R^1\text{-CO-CO-OR}^2 \xrightarrow{H_2,\ \text{Chiral Catalyst}} R^1\text{-CH(OH)-CO-OR}^2$$

Entry	R^1	R^2	Catalyst	Temperature (°C)	H_2 (atm)	Yield (%)	ee (%)	S/C	Absolute configuration
1	Me	Me	a	30	100	100	88	580	S
2	Me	Me	b	30	100	100	81	590	S
3	i-Pr	Me	a	30	100	100	65	150	S
4	i-Pr	Me	b	30	100	100	91	150	S
5	c-C_6H_{11}	Me	a	30	100	100	41	150	S
6	c-C_6H_{11}	Me	b	30	100	100	90	150	S
7	C_6H_5	Me	a	30	100	100	79	560	S
8	C_6H_5	Me	b	30	100	100	89	540	S
9	4-MeO-C_6H_4	Me	a	30	100	44	69	150	S
10	4-MeO-C_6H_4	Me	b	30	100	82	86	150	S
11	4-Me-C_6H_4	Me	a	30	100	100	83	150	S
12	4-Me-C_6H_4	Me	b	30	100	100	93	150	S
13	4-Cl-C_6H_4	Me	a	30	100	100	80	150	S
14	4-Cl-C_6H_4	Me	b	30	100	100	93	150	S
15	4-NO_2-C_6H_4	Me	a	30	100	100	73	150	S
16	4-NO_2-C_6H_4	Me	b	30	100	100	88	150	S

a = $RuCl_2[(S)$-BINAP$](C_6H_6)$, b = $RuCl_2[(S)$-BINAP$](C_6H_6)$ + HBF_4.

Table 5.3 Asymmetric hydrogenation of 1,3-diketones.

$$R^1\text{-CO-CH}_2\text{-CO-}R^2 \xrightarrow{H_2,\ \text{Chiral Catalyst}} R^1\text{-CH(OH)-CH}_2\text{-CH(OH)-}R^2\ (anti) + R^1\text{-CH(OH)-CH}_2\text{-CH(OH)-}R^2\ (syn)$$

Entry	R^1	R^2	Catalyst	Temperature (°C)	H_2 (atm)	Yield (%)	ee (%)	dr	S/C
1	Me	Me	a	30	72	100	100	99:1	2000
2	Me	C_6H_5	b	26	83	98	94	94:6	360
3	Me	EtOCO	c	80	100	100	98	84:16	200
4	C_6H_5	C_6H_5	d	40	100	70	87	96:4	170

a = $RuCl_2[(R)$-BINAP$]$, b = $RuBr_2[(R)$-BINAP$]$, c = $RuBr_2[(S)$-MeO-BIPHEP$]$,
d = $RuCl_2[(R)$-BIPHEMP$]$.

70% yield (entry 4). In the case of nonsymmetrical 1,3-diketones (entries 2 and 3), good results can be achieved in terms of enantio- and diastereoselectivity.

5.2.3
Hydrogenation of β-Keto Ester Analogues

Ketones bearing electron-withdrawing groups, other than ester groups, in an α-position can be considered as structurally similar to β-keto esters, in both steric and electronic senses. Such activating moieties can be phosphonates, sulfonic acids, sulfoxides or sulfones. For example, $RuCl_2[(R)$-BINAP$](DMF)_n$-complex hydrogenates various β-keto sulfonates in the presence of HCl with excellent enantioselectivity, though low S/C-ratio (Table 5.4, entries 1–4) [14]. β-Keto

Table 5.4 Hydrogenation of electron-withdrawing group (EWG)-functionalized ketones.

EWG = Electron withdrawing group

Entry	R	EWG	Catalyst	Temperature (°C)	H_2 (atm)	Yield (%)	ee:(de) (%)	S/C	Absolute configuration
1	Me	SO_3Na	a	50	1	100	97	200	R
2	n-$C_{15}H_{31}$	SO_3Na	a	50	1	100	96	200	R
3	i-Pr	SO_3Na	a	50	1	100	97	200	R
4	C_6H_5	SO_3Na	a	50	1	100	96	200	R
5	Me	$SO_2C_6H_5$	(R)-b	65	1	100	>95	100	R
6	n-C_5H_{11}	$SO_2C_6H_5$	(R)-b	65	1	100	>95	100	R
7	c-C_6H_{11}	$SO_2C_6H_5$	(R)-b	65	1	100	>95	100	R
8	C_6H_5	$SO_2C_6H_5$	(S)-b	40	75	100	>95	100	S
9	n-C_6H_{13}	(R)-S(O)-C_6H_4Me	(S)-b	25	50	82	(99:1)	50	S,R
10	n-C_6H_{13}	(R)-S(O)-$C_6H_4CH_3$	(R)-b	25	50	74	(6:94)	50	R,R
11	C_6H_5	(R)-S(O)-$C_6H_4CH_3$	(S)-b	25	50	70	(99:1)	50	S,R
12	C_6H_5	(R)-S(O)-$C_6H_4CH_3$	(R)-b	25	50	95	(10:90)	50	R,R
13	CH_3	$P(O)(OCH_3)_2$	(R)-c	25	4	100	98	1200	R
14	Me	$P(O)(OC_2H_5)_2$	(S)-c	50	1	100	99	50	S
15	n-C_5H_{11}	$P(O)(OCH_3)_2$	(S)-c	25	100	100	98	100	S
16	i-Pr	$P(O)(OCH_3)_2$	(S)-c	80	4	100	96	530	S
17	C_6H_5	$P(O)(OCH_3)_2$	(R)-c	60	4	100	95	530	R
18	n-C_5H_{11}	$P(S)(OCH_3)_2$	d	25	100	100	94	100	S
19	i-Pr	$P(S)(OCH_3)_2$	d	25	10	100	93	100	S

a = $RuCl_2[(R)$-BINAP$](DMF)_n$ + HCl, b = $RuBr_2[(R/S)$-MeO-BIPHEP], c = $RuCl_2[(R/S)$-BINAP$](DMF)_n$, d = $RuCl_2[(S)$-MeO-BIPHEP].

sulfones can be hydrogenated in a similar manner; RuBr$_2$[(R)-MeO-BIPHEP]-catalyzed hydrogenation provided the corresponding chiral alcohols in high enantiomeric excess (entries 5–8) [15]. If the electron-withdrawing group is replaced by an enantiopure sulfoxide (entries 9–12), the corresponding cases (entries 9 and 11) provide the (R,R)-diastereomers with excellent selectivity (dr 99:1) [16]. Hydrogenation of substituted β-keto phosphonoates catalyzed by RuCl$_2$[(S)-BINAP](DMF)$_n$ complex proceeds with excellent enantioinduction (entries 13–17) [17]. The corresponding thionophosphonates can be hydrogenated efficiently by RuCl$_2$[(S)-MeO-BIPHEP]-complex, providing the enantioenriched alcohols in high enantiomeric excesses (entries 18 and 19) [18].

5.2.4
Mechanism

The catalytic cycle of β-keto ester hydrogenation starts with the hydrogenation of the precatalyst (1) (Scheme 5.1), which loses HCl and forms the mixed hydride–

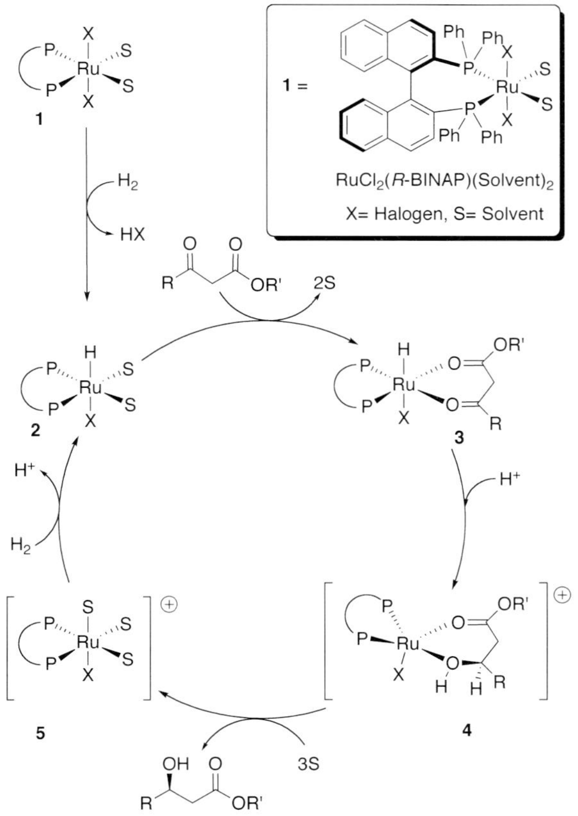

Scheme 5.1 Catalytic cycle of Ru(II)-diphosphine β-keto ester hydrogenation.

halide complex **(2)** containing two solvent molecules [19]. The substrate enters the catalytic cycle and the two coordinated solvent molecules are exchanged for the β-keto ester forming the σ-coordination complex **(3)**. Complex **(3)** is then protonated at the coordinated carbonyl oxygen, creating a strong polarization effect on the carbonyl carbon. Upon this event, the coordination mode of the carbonyl function changes from σ-coordination to π-coordination, subsequently acting as a hydride acceptor. At this stage of the catalytic cycle, the protonation of the carbonyl carbon is crucial for the catalytic activity and demands the forming of a strong mineral acid during the hydrogenolysis of the Ru–halide bond. This explains why the reaction is strongly promoted by the addition of small amounts of strong acid. After hydride transfer, two solvent molecules displace the β-hydroxy ester in **(4)**, yielding complex **(5)**. Upon heterolytic split of coordinated dihydrogen by complex **(5)**, starting complex **(2)** is regenerated and this event closes the catalytic cycle (Scheme 5.1).

The enantio-determining step is the irreversible hydride transfer from the Ru-center to the protonated π-coordinated β-keto ester ligand. The high degree of enantioselectivity originates from the unfavorable interaction between the alkyl substituent of the coordinated β-keto ester and the axial Ar-group of the phosphine ligand in the *Si*-coordination mode (Scheme 5.2, left path). However, in the *Re*-

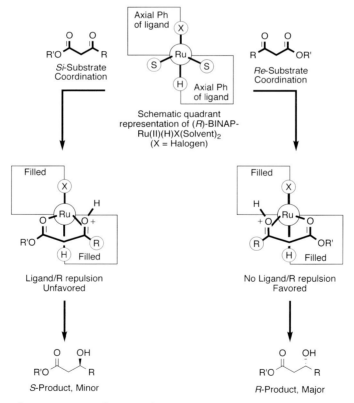

Scheme 5.2 Origin of enantioselectivity in asymmetric hydrogenation of β-keto esters.

coordination mode, this unfavorable interaction is not present and the reaction proceeds with almost complete enantiodiscrimination (Scheme 5.2, right path). This also explains why, 3,5-xylyl-substituted ligands perform better than unsubstituted phenyl ligands in the hydrogenation of keto esters bearing less bulky alkyl substituents. This mechanistic explanation is also valid for ketones bearing other activating groups than esters in an α-position, and the results are comparable in terms of enantioselectivity. However, if two coordinating groups are present in the substrate at the α- and α′-positions, this will result in competing coordination modes and lower enantioselectivity is observed. This is not only due to the coordination abilities, but also a function of steric hindrance. This is the case when hydrogenating keto esters within the statine series, for example.

5.2.5
Catalyst Preparation

Typical preparation of the Ru(II)-diphosphine catalyst involves the reduction of Ru(III)Cl$_3$ hydrate to a Ru(II) complex, often by refluxing in 1,4-cyclohexadiene, leading to [Ru(η^6-C$_6$H$_6$)Cl$_2$]$_2$ which is subsequently reacted with a diphosphine forming a RuCl[diphosphine]Cl(C$_6$H$_6$) complex [4]. This method employs forcing reaction conditions, which may damage (partially oxidize or epimerize, for example) sensitive diphosphines. Recently, Genet and coworkers reported the facile preparation of RuX$_2$[diphosphine] complexes (X = Cl or Br) by ligand displacement in Ru(II)(COD)(η^3-Metallyl)$_2$ with diphosphine, creating a Ru(II)[diphosphine](η^3-methallyl)$_2$-complex, followed by removal of the two η^3-methallyl ligands by treatment with anhydrous mineral acid (Scheme 5.3) [20]. Ru(II)(COD)(η^3-Metallyl)$_2$ shows excellent stability and can be considered as a first choice as it is both commercially available and straightforward to prepare. The preparation of the active catalysts is best done just prior to use. Typically, such preparation involves the dissolution of the metal precursor and diphosphine in a small amount of solvent, followed by formation of the active catalyst and its subsequent addition to the reaction mixture (Scheme 5.3).

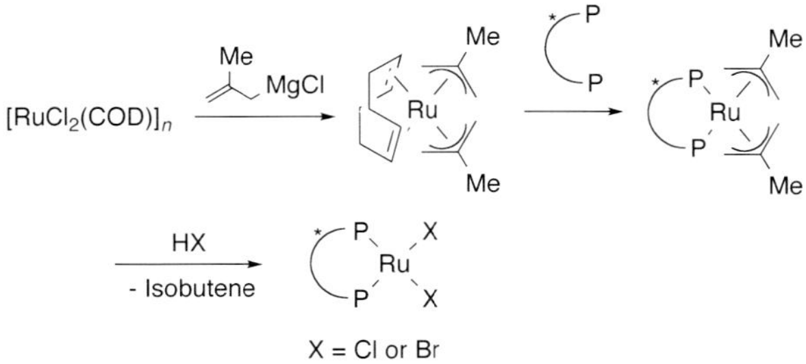

Scheme 5.3 Preparation of RuX$_2$[diphosphine] complexes from Ru(II)(COD)(η^3-Metallyl)$_2$ precursor.

5.2.6
Dynamic Kinetic Resolution (DKR) of β-Keto Esters

Hydrogenation of racemic α-substituted β-keto esters can give rise to four different stereoisomers. However, for most substrates, either the *syn-* or *anti-* product is produced with good diastereoselectivity and high enantiomeric excess. In contrast to kinetic resolution, where the maximum theoretical yield is only 50%, a dynamic kinetic resolution can reach 100%. In this process, the slower-reacting enantiomer is converted into the faster-reacting one (Scheme 5.4).

This is valid only if the racemization process is considerably faster than the reduction step. The outcome is strongly dependent on the substitution pattern of the substrate and the diphosphine ligand chosen. A good example of this reaction is the hydrogenation of methyl-2-acetamino-3-oxobutyrate into *N*-acyl-(L)-threonine by $RuCl_2[(S)$-BINAP] (Scheme 5.5) This method is of general applicability and tolerates various functional groups.

This method is suited for the hydrogenation of various β-keto keto esters bearing functional groups in an α-position such as acylamino, alkyl or halogens. For example, the asymmetric hydrogenation of methyl-2-benzamido methylene-3-oxobutyrate catalyzed by $RuX_2[(R/S)$-BINAP](C_6H_6) proceeds with excellent enantioselectivity (Table 5.5, entries 1–3) [4]. Cyclic β-keto esters can be hydrogenated with high selectivity, providing the corresponding alcohols with excellent enantioselectivity and diastereomeric ratio (entries 4–8) [21]. The asymmetric hydrogenation of a nonacylated 2-amino derivative as the HCl salt directly yields the corresponding amino alcohol with excellent ee value but poor de (entry 10). However, the hydrogenation of α-alkyl-β-keto esters gives poor diastereoselectivity, but the enantioselectivity for the major diastereomer remains high (entry 11) [22]. BINAP and tolBINAP dihalide complexes can be regarded as the prime choice for the DKR-hydrogenation of α-substituted β-keto esters. Excellent S/C-ratios can be achieved if the substrate is not too strongly coordinating. Further, if basic functional groups are present in the substrate, the addition of a strong acid to the reaction may be necessary.

An excellent example of applied dynamic kinetic resolution is Takasagos hydrogenation of methyl-2-benzamidomethylene-3-oxobutyrate (Scheme 5.6). The

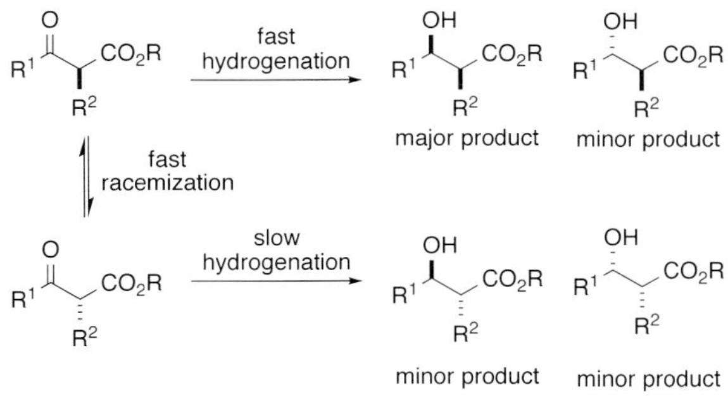

Scheme 5.4 General principle of dynamic kinetic resolution.

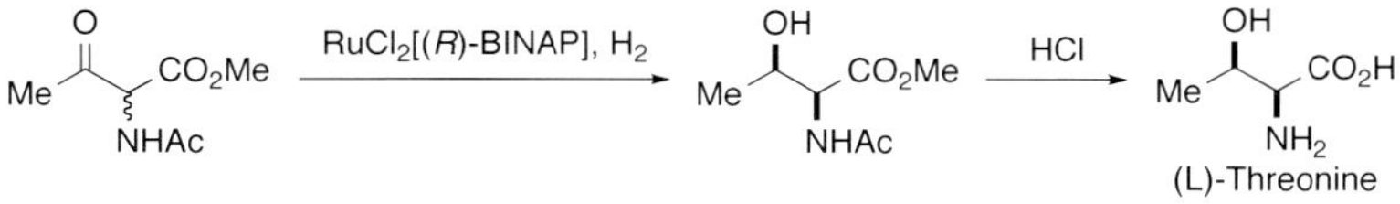

Scheme 5.5 Synthesis of (L)-threonine via dynamic kinetic resolution.

Table 5.5 Hydrogenation of acyclic and cyclic α-substituted β-keto esters via dynamic kinetic resolution.

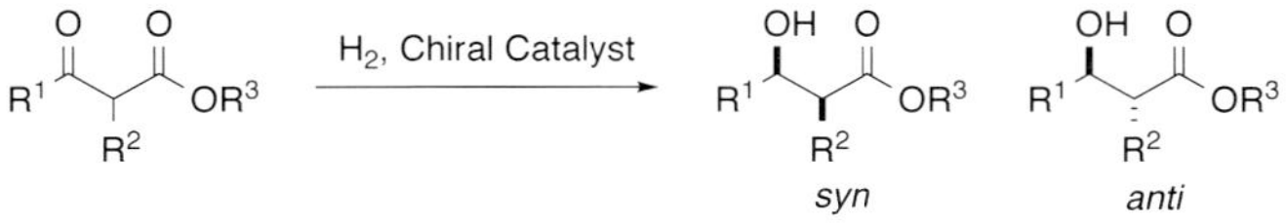

Entry	R^1	R^2	R^3	Catalyst	Yield (%)	ee (%)	de (dr) (%)	Absolute configuration
1	Me	$CH_2NHCOC_6H_5$	Me	a	74	96	90	2S,3R
2	Me	$CH_2NHCOC_6H_5$	Me	b	91	98	79	2S,3R
3	Me	$CH_2NHCOC_6H_5$	Me	c	100	98	87	2R,3S
4	$-CH_2-$	$-CH_2-$	Me	a	>95	92	(99:1)	1R,2R
5	$-(CH_2)_2-$	$-CH_2-$	Me	a	>95	90	(95:5)	1R,2R
6	$-(CH_2)_3-$	$-CH_2-$	Me	a	>95	93	(93:7)	1R,2R
7	Me	$-CH_2-$	$-CH_2-$	a	100	94	(98:2)	3R,6S
8	Me	$-CH_2-$	$-CH_2-$	c	100	97	(99:1)	3R,6S
9	Me	$NHCOCH_3$	Me	d	>90	98	(99:1)	2S,3R
10	i-Pr	NH_3Cl	Me	e	>90	98	(1:99)	2S,3S
11	Me	Me	Me	a	>90	94	(32:68)	2R,3R

a = $RuCl_2[(R)$-BINAP$](C_6H_6)$, b = $RuBr_2[(R)$-tolBINAP$](C_6H_6)$, c = $RuI_2[(S)$-BINAP$](p$-cymene),
d = $RuBr_2[(R)$-BINAP$]$, e = $RuCl_2[(S)$-BINAP$](DMF)_n$.

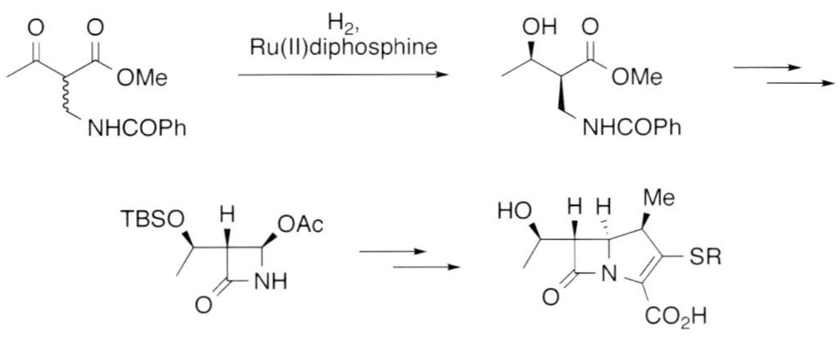

Scheme 5.6 Asymmetric hydrogenation via dynamic kinetic resolution of an antibiotic intermediate (Takasago Co.).

product is an intermediate in the production of carbapenem antibiotics, a process carried out on 150 tonnes per year.

5.3
Ketone Hydrogenation

Until recently, direct hydrogenation of simple unactivated ketones has proved to be problematic. Due to the importance of chiral secondary alcohols as chiral building blocks, this issue has stood out as a long-term unsolved problem. This changed in 1995 when Noyori and coworkers published the first enantioselective hydrogenation of ketonic substrates catalyzed by Ru(II)Cl$_2$-(BINAP)(diamine) complexes (Figure 5.2) [19, 23]. This hydrogenation protocol makes use of an alcoholic solvent like i-PrOH and is dependent on the addition of an alkoxide base. The hydrogenation proceeds at slightly elevated temperature under a reasonable H$_2$ pressure. It was also found that the use of 3,5-xylyl-substituted BINAP gave almost complete enantioselectivity.

To-date this hydrogenation protocol has been evaluated with a large number of different diphosphine ligands. The most successful ones, except xylBINAP, are xylPhanephos and xylSPD. Further, this hydrogenation protocol has proved to be completely chemoselective for ketones, tolerating functionalities such as halides, amino groups, nitro substituents and double bonds, without affecting them. The best reported example of the hydrogenation of acetophenone had a TON of 2 400 000. During the 1990s, examples of Rh-catalyzed ketone hydrogenation appeared in the literature, Zhang and coworkers hydrogenated various acetophenones employing Pennphos-Rh(I)/KBr system together with lutidine as additive [24]. The next major improvement came when Noyori and coworkers published the use of trans-RuH(η^1-BH$_4$)[(S)-xylBINAP][(S,S)-DPEN] complex in asymmetric ketone hydrogenation, which is not dependent on the addition of alkoxide base to the reaction mixture [25]. The base-free procedure can be applied to ketonic substrates containing substituents that may be affected by strong base, such as esters and β-halo-ketonic functionalities. The base-free catalyst is even more potent than the original catalyst leading to superior TONs and TOFs.

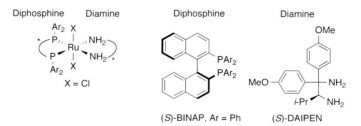

Figure 5.2 General structure of trans-Ru(II)X$_2$-(diphosphine)(diamine) complexes employed as hydrogenation catalysts for unfunctionalized ketones.

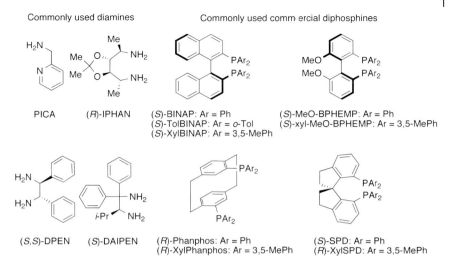

Figure 5.3 Commonly used diamine-ligands (left) and diphosphine ligands (right) for the construction of Ru(II)-diamine-diphosphine complexes for asymmetric hydrogenation of ketones.

Preparation of the diamine-diphosphine Ru(II) complexes is straightforward from shelf-stable precursors such as a diamine, diphosphine and a metal precursor. Simply heating $RuCl_2(DMF)_n$ together with the diphosphine of choice, followed by addition of a diamine, gives complexes in high yield. Today, a number of suitable diamines and diphosphines are commercially available from various sources. The most commonly employed diphosphine and diamine ligands for ketone hydrogenation are shown below (Figure 5.3).

5.3.1
Mechanism

The extraordinary performance of the Ru(II)-diamine-diphosphine catalysts in the hydrogenation of unfunctionalized ketones has several explanations. First, the catalytically active complex does not form any covalent bonds to the substrate, compared to mechanisms for C—C double bond hydrogenation with Ru, Rh and Ir. Further, the prime function of the Ru-center is to activate dihydrogen during the catalytic cycle and to support the hydride prior to transfer. The Ru-center does not change oxidation state during the reaction. Treatment of Ru(II)-diamine-diphosphine dihalide precursor (**6**) with strong base (step **A** in Scheme 5.7) eliminates HX from the precatalyst, giving the corresponding zwitterionic complex (**7**). In step **B**, dihydrogen is coordinated and the anionic nitrogen is protonated by an alcohol solvent molecule, yielding complex (**8**). The resulting transient species (**8**) eliminates the alcohol molecule and forms a mixed hydride–halide complex (**9**).

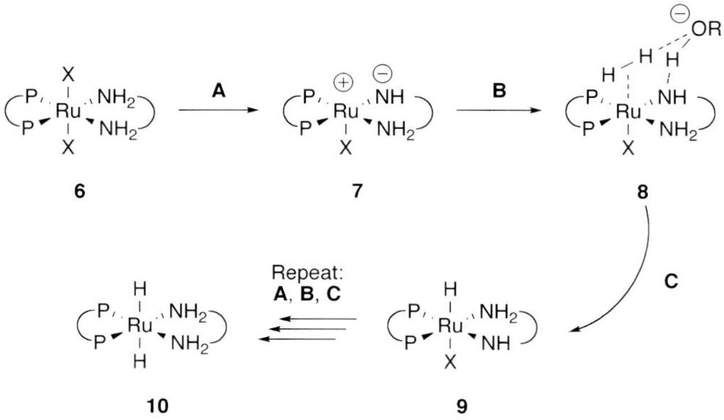

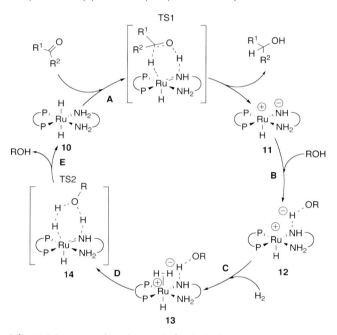

Scheme 5.7 Proposed mechanism for the formation of active catalyst from Ru(II)-diamine-diphosphine dihalide precursor.

Scheme 5.8 Proposed catalytic cycle for the hydrogenation of nonfunctionalized ketones by Ru(II)-diamine-diphosphine catalyst.

Steps **A**, **B** and **C** are then repeated and the second halide ion is eliminated, yielding the final active catalyst **(10)** [26].

It has long been postulated, and nowadays experimentally supported, that the mechanism of ketone hydrogenation by Ru-catalysts containing at least one sp^3-NH donor atom proceeds via a concerted proton-hydride transfer from the complex to the substrate. Starting out from the active catalyst **(10)**, the ketone substrate forms a six-membered pericyclic transition state (TS1) (Scheme 5.8) resulting in a low energy barrier in the reduction step. The strongly dipolar RuH-NH-complex interacts in a favorable way with the carbonyl function and the Ru-hydride is transferred to the carbonyl carbon simultaneously as the carbonyl oxygen is protonated by the NH-functionality of the complex. The resulting zwitterionic Ru-complex **(11)** is then protonated by a solvent molecule, which increases the electropositive character of the Ru-center, leading to an increased affinity for dihydrogen (step **B**). Complex **(12)** then coordinates one molecule of dihydrogen (step **C**), leading to complex **(13)**. This complex is labile and a strong interaction between the alkoxide and the coordinated dihydrogen molecule develops. This leads to a six-membered cyclic transition state (TS2) where dihydrogen is heterolytically cleaved by the alkoxide base, regenerating the starting dihydride complex **(10)**, which closes the catalytic cycle (Scheme 5.8) [26].

This is the commonly accepted mechanism concerning the reduction step, but very little is known about how the dihydrogen is heterolytically activated into a metal-bound hydride and nitrogen bound proton. This part of the reaction is of considerable interest as its calculated barrier is estimated to be high (ca. 20 kcal mol^{-1}) and possibly rate determining (Scheme 5.9, path B) [27]. In conflict with the calculated barrier, Morris and coworkers reported an experimental activation enthalpy of only 7.6–8.6 kcal mol^{-1} [28]. The difference between the calculated and experimental results could be a consequence of the active participation of an alcohol molecule in the coordination and cleavage of dihydrogen (path A). Such a

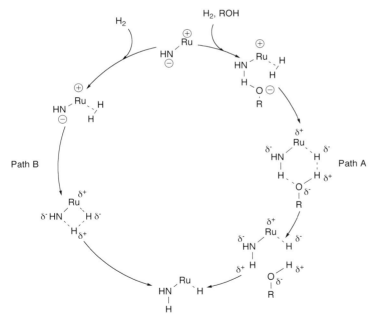

Scheme 5.9 Activation through heterolytic cleavage of dihydrogen by zwitterionic 16e Ru(II) complex (left), by either alcohol mediated reaction (path A) or direct activation (path B), leading to active catalyst RuH-NH$_2$ (right).

mechanism has been proposed and experimentally supported by deuterium labeling experiments by Ikariya and coworkers in 2001 [29]. Further, in a recent computational study of the dihydrogen activation by Andersson and coworkers, the alcohol-assisted pathway has found strong support [27].

The enantiodetermining step in ketone hydrogenation takes place when the ketonic substrate is coordinated to the RuH-NH bifunctional catalyst by initial hydrogen bonding prior to the concerted hydride/proton transfer. As both the carbonyl oxygen and carbon need to access their reaction partners simultaneously, only the axially oriented NH-protons can be transferred to the carbonyl oxygen. The reason for this is that the dihedral angle would otherwise be too small. Comparing *Si*- and *Re*-face coordination of acetophenone with a schematically illustrated catalyst complex (**15**), one can clearly see that the coordination mode leading to the *S*-enantiomer (**16**) will suffer from a steric repulsion with the axially oriented aryl group on the phosphine moiety. This leads to the formation of the *R*-enantiomer exclusively. This is also the reason for the high selectivity observed when 3,5-xylyl biaryl diphosphines like xylBINAP and xylPhanephos are employed in this reaction. The methyl groups on the aromatic rings increase the steric bulk and thus increase the relative energies between the enantiotopic transition states. This leads to a major improvement in enantioselectivity (Scheme 5.10) [26, 30].

5.3.2
Aryl Alkyl Ketones

Noyori and coworkers hydrogenated various substituted acetophenones and other aryl alkyl ketones with excellent results, employing *trans*-RuCl$_2$[(*S*)-xylBINAP][(*S*,*S*)-DPEN] (Table 5.6, entries 1–8) and *trans*-RuCl$_2$[(*S*)-xylBINAP][(*S*)-daipen] (entries 9–19) [23, 31, 32]. A further improvement on the hydrogenation of sterically hindered aryl alkyl ketones was reported, illustrated in the hydrogenation of *t*-Bu-phenylketone by the RuCl$_2$[(*S*)-xylBINAP][Pica] catalyst (entry 20) [33]. Zhu and coworkers employed the *trans*-RuCl$_2$[(*S*)-xylSPD][(*R*,*R*)-DPEN] catalyst to hydrogenate a number of aryl ketones with good results (entries 40–16) [34]. Burk and coworkers employed the *trans*-RuCl$_2$[(*S*)-xylPhanephos][(*S*,*S*)-DPEN] catalyst to hydrogenate a series of acetophenones with excellent results (entries 25–39) [35]. All previous mentioned hydrogenations are dependent on the addition of a strong alkoxide base in order to display activity, according to the previous discussion on the generation of the active 16e catalytic complex. The usual loading for the base additive is between 1–5 mol% in relation to the substrate. However, base-sensitive substrates can be damaged from these conditions, and further it complicates the work-up procedure of the hydrogenation mixture, calling for an extractive aqueous work-up in order to neutralize the base. Noyori and coworkers reported the base-free hydrogenation of unfunctionalized ketones employing the *trans*-RuH(η^1-BH$_4$)[(*S*)-xylBINAP][(*S*,*S*)-DPEN] complex as a catalyst (entries 21–24) [25]. This method offers excellent performance with various base-sensitive substrates. The method allows for the hydrogenation of substrates carrying the sensitive epoxide moiety (entry 24) and also for ester functionality that would otherwise be

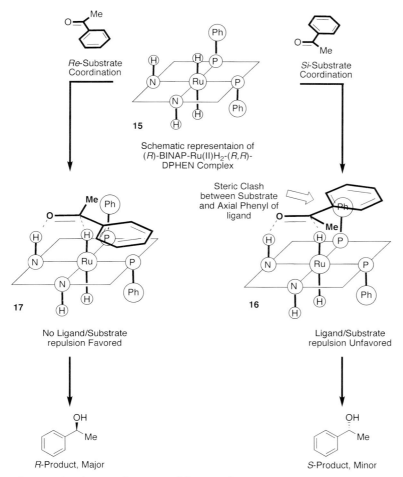

Scheme 5.10 Schematic illustration of the RuH$_2$-diamine-diphosphine catalyst in the enantiodetermining transition states. *Re*-side coordination (left) leads to *R*-product due to lack of steric repulsions in the transition state. *Si*-side coordination suffers from steric repulsions between the axially oriented aromatic substituent of the phosphine ligand (right).

trans-esterified by the alcoholic reaction solvent (entry 22). Interestingly, it was found that the addition of a small amount of base (*t*-BuOK) increased the rate by an order of magnitude. Although the asymmetric hydrogenation of aryl alkyl ketones is among one of the most studied reactions, the substrate scope is still relative narrow, only allowing a few types of substrates such as various acetophenones and similar. However, with modification of the diamine ligand in the Ru-complexes, Noyori and coworkers were able to hydrogenate various indanone and tetralones with excellent enantiomeric excess by employing the *trans*-RuCl$_2$[(*S*)-

5 Carbonyl Hydrogenation

Table 5.6 Hydrogenation of aryl alkyl ketones with bifunctional Ru(II)-diphosphine-diamine catalysts.

$$Ar\underset{Alkyl}{\overset{O}{\|}} \xrightarrow{H_2,\ Chiral\ Catalyst} Ar\underset{*}{\overset{OH}{|}}Alkyl$$

Entry	Ar	Alkyl	Temperature (°C)	H_2 (atm)	Yield (%)	ee (%)	S/C	Absolute configuration
1	C_6H_5	Me	45	8	100	99	100 000	S
2	4-Me-C_6H_4	Me	28	4	100	98	2 000	S
3	4-n-Bu-C_6H_4	Me	28	4	100	98	2 000	S
4	3-F-C_6H_4	Me	28	4	99	98	2 000	S
5	2-Cl-C_6H_5	Me	28	4	99.5	98	2 000	S
6	1-Naphthyl	Me	28	4	99	99	2 000	S
7	2-Naphthyl	Me	28	4	99	98	2 000	S
8	C_6H_5	Et	28	4	99	99	2 000	S
9	4-F-C_6H_4	Me	28	4	99	99	2 000	S
10	3-Me-C_6H_4	Me	28	10	98	100	10 000	R
11	2-F-C_6H_4	Me	28	8	100	97	2 000	R
12	4-Br-C_6H_4	Me	28	8	99.9	99.6	20 000	R
13	4-I-C_6H_4	Me	28	8	99.7	99	2 000	R
14	4-CF_3-C_6H_4	Me	28	10	100	99.6	10 000	R
15	4-MeO-C_6H_4	Me	28	10	100	100	2 000	R
16	4-NO_2-C_6H_4	Me	28	8	100	99.8	2 000	R
17	4-NH_2-C_6H_4	Me	28	8	100	99	2 000	R
18	4-Cl-C_6H_4	Et	28	8	99.9	99	20 000	R
19	C_6H_5	cyclo-C_3H_5	28	8	99.7	96	2 000	R
20	C_6H_5	t-Bu	25	5	100	97	2 000	R
21	C_6H_5	Me	25	8	100	99	100 000	R
22	4-$EtCO_2$-C_6H_5	Me	25	8	100	99	4 000	R
23	C_6H_5	$(CH_2)_2N(CH_3)_2$	25	8	100	97	4 000	R
24	(R-glycidyl)-3-O-C_6H_4	Me	25	8	99	99	4 000	R,R
25	C_6H_5	Me	20	8	100	99	20 000	R
26	4-CF_3-C_6H_4	Me	20	8	100	97	3 000	R
27	4-Br-C_6H_4	Me	20	8	100	99	3 000	R
28	4-MeO-C_6H_4	Me	20	8	100	97	3 000	R
29	3-CF_3-C_6H_4	Me	20	8	100	99	3 000	R
30	2-CF_3-C_6H_4	Me	20	8	100	99	3 000	R
31	2-CH_3-C_6H_4	Me	20	8	100	97	3 000	R
32	2-Br-C_6H_4	Me	20	8	100	99	3 000	R
33	2-MeO-C_6H_4	Me	20	8	100	94	500	R
34	1-Naphthyl	Me	20	8	100	98	3 000	R
35	2-Napthyl	Me	20	8	100	99	3 000	R
36	C_6H_5	Et	20	8	100	98	3 000	R
37	C_6H_5	n-Bu	20	8	100	96	3 000	R
38	C_6H_5	$CH_2C_6H_5$	20	8	100	98	3 000	R
39	C_6H_5	i-Pr	20	8	100	71	3 000	R
40	C_6H_5	Me	25	50	100	99	5 000	S
41	2-Cl-C_6H_4	Me	25	50	99	98	5 000	S
42	2-Br-C_6H_4	Me	25	50	100	99	5 000	S
43	3-Br-C_6H_4	Me	25	50	99	99	5 000	S

Table 5.6 *Continued*

$$\underset{Ar}{\overset{O}{\underset{\|}{\bigvee}}}\underset{Alkyl}{\overset{}{\longrightarrow}} \xrightarrow{H_2,\ Chiral\ Catalyst} \underset{Ar}{\overset{OH}{\underset{*}{\bigvee}}}\underset{Alkyl}{\overset{}{}}$$

Entry	Ar	Alkyl	Temperature (°C)	H$_2$ (atm)	Yield (%)	ee (%)	S/C	Absolute configuration
44	3-CF$_3$-C$_6$H$_4$	Me	25	50	99	99	5 000	S
45	4-CH$_3$-C$_6$H$_4$	Me	25	50	100	99	5 000	S
46	4-Cl-C$_6$H$_4$	Me	25	50	100	99	5 000	S
47	4-Br-C$_6$H$_4$	Me	25	50	100	99	5 000	S
48	C$_6$H$_5$	Et	25	50	99	99.5	5 000	S
49	C$_6$H$_5$	CH$_2$C$_6$H$_5$	25	50	100	98	5 000	S
50	2-Naphthyl	Me	25	50	98	99.2	5 000	S

Catalysts used:
Entries 1–8:   *trans*-RuCl$_2$[(S)-xylBINAP][(S,S)-DPEN] + *t*-BuOK.
Entries 9–19:   *trans*-RuCl$_2$[(S)-xylBINAP][(S)-DAIPEN] + *t*-BuOK.
Entry 20:   RuCl$_2$[(S)-xylBINAP][Pica] + *t*-BuOK.
Entries 21–24:   *trans*-RuH(η^1-BH$_4$)[(S)-xylBINAP][(S,S)-DPEN].
Entries 25–39:   *trans*-RuCl$_2$[(S)-xylPhanephos][(S,S)-DPEN] + *t*-BuOK.
Entries 40–50:   *trans*-RuCl$_2$[(S)-xyl-SPD][(R,R)-DPEN] + *t*-BuOK.

xylBINAP][(R)-IBAN] complex as a catalyst [36]. However, indanone and tetralone substrates require harsher conditions compared to other noncyclic aryl alkyl ketones and cannot compete in terms of enantioselectivity.

5.3.3
Dialkyl Ketones

Dialkyl ketones have proved to be some of the most demanding substrates so far to hydrogenate. In order to gain enantioselectivity, one side needs to be sterically differentiated from the other. Such cases are linear alkyl tertiary alkyl ketones and similar. Pinacolone, for example, is hydrogenated by *trans*-RuH(η^1-BH$_4$) [(S)-tolBINAP][Pica] complex, under base-free conditions, in 94% ee [33]. Other successful hydrogenations by *trans*-RuCl$_2$[(S)-xylBINAP][(S)-daipen] have involved cyclopropylmethyl ketone and cyclohexylmethyl ketone in 95% and 85% ee, respectively [32]. The most interesting substrates for the fragrance industry, such as long-chain aliphatic methyl ketones, have until recently not been successfully hydrogenated by Ru-diamine-diphosphine catalysts.

5.3.4
Diaryl and Aryl Heteroaryl Ketones

Asymmetric hydrogenation of diaryl ketones offers a straightforward route to enantioenriched diaryl methanols. Metal- or borohydride reductions of diaryl ketones often yield over-reduction, resulting in diaryl methanes. However, under basic hydrogenation conditions, no such side-products are detectable. Further, very few examples of kinetic resolution of racemic diaryl methyl alcohols by enzymatic methods have been reported, making the hydrogenation route to enantio-enriched material even more attractive. The enantioselective hydrogenation of diaryl ketones, $Ar^1 \neq Ar^2$, requires a steric or electronic differentiation between the two Ar-substituents, typically a substituent in the 2-position is essential for high enantioselectivity. Substrates carrying only substituents in the 3- or 4-position are generally hydrogenated with only modest enantioselectivity. A representative series of 2- and 4-substituted diaryl methanols, including a single example of benzoylferrocene, were hydrogenated by Noyori and coworkers, employing trans-RuCl$_2$[(S)-xylBINAP][(S)-DAIPEN] complex (Table 5.7, entries 1–10) with excellent results [37]. Few examples of asymmetric hydrogenation of aryl heteroaryl ketones have been reported. Chen and coworkers hydrogenated various aryl heteroaryl ketones, employing trans-RuCl$_2$[(R)-xylBINAP][(R)-DAIPEN] complex in combination with an inorganic base (K$_2$CO$_3$) with good results (entries 11–21) [38]. 2,8-Bis(trifluoromethyl)-quinolin-4-yl-2-pyridinyl ketone, a precursor to the antimalarial drug mefloquine, had previously been hydrogenated by Schmid and coworkers at Roche, employing a (S)-MeO-BIPHEP-Rh(I) complex (entry 21) [39]. Chan and coworkers hydrogenated aryl and heteroaryl ferrocenyl ketones with high selectivity, employing trans-RuCl$_2$[(R)-tol-P-Phos][(R,R)-DPEN]-complex [40]. Chiral aryl ferrocenyl methanols are useful starting materials for the preparation of chiral phosphine ligands within the Josiphos series (entries 22–26).

Heteroaryl alkyl ketones are readily reduced under normal hydrogenation conditions, typically yielding complete conversion and high optical purities. Noyori and coworkers employed the trans-RuCl$_2$[(S)-xylBINAP][(S)-DAIPEN] system with excellent result on a series of heteroaryl alkyl ketones (Table 5.8, entries 1–11) [41]. Catalyst inhibition from highly chelating products (entries 6–7) was eliminated in a very elegant way by employing B(O-i-Pr)$_3$ as an additive. Further, the sterically hindered t-butyl ketones (entries 12–13) were hydrogenated with high enantioselectivities employing the corresponding Pica-complex [33]. Zhu and coworkers employed the trans-RuCl$_2$[(S)-xylSPD][(R,R)-DPEN] catalyst to hydrogenate a number of heteroaromatic ketones with good results (entries 14–16) [34]. The hydrogenation of heteroaryl alkyl ketones can be considered a method of general applicability, employing a diphosphine-diamine Ru(II) complex.

5.3.5
Phosphine-Free Hydrogenation of Alkyl Aryl Ketones

During the last years, several phosphine-free hydrogenations of simple ketones have been reported. Ikariya and coworkers employed (S)-(1-ethylpyrrolidin-2-yl)methanamine **(17)** as a ligand, together with Cp*Ru(II)(COD)Cl in i-PrOH for the hydrogenation of a series of acetophenones [29], the best result in this study was 90% ee. Another recent study by Andersson and coworkers employed [Cp*Ru(II)Cl]$_4$ as metal precursor and the highly basic QCD-amine **(18)** under

5.3 Ketone Hydrogenation

Table 5.7 Hydrogenation of aryl aryl, heteroaryl aryl, heteroaryl heteroaryl and ferrocenyl aryl ketones.

$$Ar^1\text{-CO-}Ar^2 \xrightarrow[\text{Base, additive}]{H_2, \text{ Chiral Catalyst}} Ar^1\text{-CH(OH)-}Ar^2$$

Ar = Aromatic, Heteroaromatic, Ferrocenyl

Entry	Ar^1	Ar^2	Temperature (°C)	H$_2$ (atm)	Yield (%)	ee (%)	S/C	Absolute configuration
1	2-Me-C$_6$H$_4$	C$_6$H$_5$	30	8	99	93	2000	S
2	2-MeO-C$_6$H$_4$	C$_6$H$_5$	30	8	100	99	2000	S
3	2-F-C$_6$H$_4$	C$_6$H$_5$	30	8	99	97	2000	S
4	2-Cl-C$_6$H$_4$	C$_6$H$_5$	30	8	99	97	2000	S
5	2-Br-C$_6$H$_4$	C$_6$H$_5$	30	8	99	96	20000	S
6	2-Br-C$_6$H$_4$	4-Me-C$_6$H$_4$	30	8	99	98	2000	S
7	4-Me-C$_6$H$_4$	C$_6$H$_5$	30	8	95	35	2000	R
8	4-CF$_3$-C$_6$H$_4$	C$_6$H$_5$	30	8	99	47	2000	S
9	2-Me-O-C$_6$H$_4$	4-CF$_3$-C$_6$H$_4$	30	8	97	61	2000	–
10	Ferrocenyl	C$_6$H$_5$	30	8	100	95	2000	S
11	C$_6$H$_5$	2-Pyridinyl	25	50	87	75	100	S
12	C$_6$H$_5$	3-Pyridinyl	25	50	94	73	100	R
13	C$_6$H$_5$	4-Pyridinyl	25	50	100	57	100	R
14	4-Cl-C$_6$H$_4$	2-Pyridinyl	25	50	98	61	100	S
15	4-Me-C$_6$H$_4$	2-Pyridinyl	25	50	100	90	100	S
16	2-Me-C$_6$H$_4$	2-Pyridinyl	25	50	93	99	100	S
17	C$_6$H$_5$	1-Isoquinolinyl	25	50	87	91	100	R
18	2-MeO-C$_6$H$_4$	N-1-phenyl-pyrazol-4-yl	25	50	65	94	100	R
19	C$_6$H$_5$	2-methylthiazol-4-yl	25	50	89	83	100	S
20	2-Pyridinyl	2,8-(CF$_3$)-quinolin-4-yl	25	50	92	88	100	R
21	2-Pyridinyl	2,8-(CF$_3$)-quinolin-4-yl	–	–	95	91	500	R
22	Ferrocenyl	C$_6$H$_5$	25	50	99	89	1000	(R)/(-)
23	Ferrocenyl	2-MeO-C$_6$H$_4$	25	50	94	94	1000	(-)
24	Ferrocenyl	4-Br-C$_6$H$_4$	25	50	37	73	1000	(-)
25	Ferrocenyl	1-Naphthyl	25	50	90	98	1000	(-)
26	Ferrocenyl	2-Naphthyl	25	50	94	83	1000	(-)

Catalysts used:
Entry 1–10: trans-RuCl$_2$[(S)-xylBINAP][(S)-DAIPEN].
Entry 11–20: trans-RuCl$_2$[(R)-xylBINAP][(R)-DAIPEN]-K$_2$CO$_3$ (25 mol%).
Entry 21: (S)-MeO-BIPHEP-[Rh(COD)Cl]$_2$.
Entry 22–26: trans-RuCl$_2$[(R)-tol-P-Phos][(R,R)-DPEN].

similar conditions [27]. In this case, the best results were also in the range of 90% ee, but the catalytic system employing the more basic **(18)** was considerably faster than ligand **(17)**. In both cases, the S/C ratios were a modest 200:1. The first breakthrough in the field of phosphine-free ketone hydrogenation came in 2006 when Kitamura and coworkers published their work on asymmetric hydrogenation with Goodwins-Lion type hybrid ligands **(19a–c)** [42]. These complexes gave good

Table 5.8 Hydrogenation of heteroaryl alkyl ketones.

$$\text{Het}-\overset{O}{\underset{}{C}}-\text{Alkyl} \xrightarrow[\text{Base, additive}]{\text{H}_2, \text{Chiral Catalyst}} \text{Het}-\overset{OH}{\underset{}{C}}-\text{Alkyl}$$

Het = Heteroaromatic

Entry	Het	Alkyl	Temperature (°C)	H_2 (atm)	Yield (%)	ee (%)	S/C	Absolute configuration
1	2-Furyl	Me	25	50	96	99	40 000	S
2	2-Furyl	n-C_5H_{11}	25	8	100	98	2 000	S
3	2-Thienyl	Me	25	8	100	99	5 000	S
4	3-Thienyl	Me	25	8	100	<99	5 000	S
5	2-(N-Tosyl)-pyrrole	Me	25	8	93	98	1 000	S
6	2-Thiazolyl	Me	25	8	100	96	2 000	S
7	2-Pyridinyl	Me	25	8	100	96	2 000	S
8	3-Pyridinyl	Me	25	8	100	99	5 000	S
9	4-Pyridinyl	Me	25	8	100	99	2 000	S
10	2-Pyridinyl	i-Pr	25	8	100	94	2 000	S
11	2,6-Pyridinyl	Me	25	8	100	99	10 000	S,S
12	2-Furyl	t-Bu	25	8	99	97	2 400	R
13	2-Thienyl	t-Bu	25	8	100	98	2 100	R
14	1-Ferrocenyl	Me	25	50	100	98	5 000	S
15	2-Thienyl	Me	25	50	98	98	5 000	S
16	2-Furyl	Me	25	50	99	98	5 000	S
17	2-Thienyl	Me	20	8	>99	92	5 000	S
18	2-Furyl	Me	20	8	>99	96	5 000	S
19	3-Thienyl	Me	20	8	>99	96	5 000	S
20	2-Pyridinyl	Me	20	8	>99	98	1 500	S
21	3-Pyridinyl	Me	20	8	>99	78	1 500	S
22	4-Pyridinyl	Me	20	8	>99	99	1 500	S
23	2-Benzofuryl	Me	20	8	>99	96	1 000	S
24	1-Ferrocenyl	Me	20	8	>99	97	5 000	S

Catalysts used:
Entries 1–11: trans-RuCl$_2$[(S)-xylBINAP][(S)-Daipen] + t-BuOK.
Entries 12–13: RuCl$_2$[(S)-xylBINAP][Pica] + t-BuOK.
Entries 14–16: trans-RuCl$_2$[(S)-xylSPD][(R,R)-DPEN] + t-BuOK.
Entries 17–24: trans-RuCl$_2$[(S)-xylPhanephos][(S,S)-DPEN] + t-BuOK.
Entry 10: DMF used as co-solvent (10 %).
Entries 12–13: B(O-i-Pr)$_3$ used as base.

to excellent enantioselectivity for a variety of acetophenone derivatives. The best results were achieved with Ru(II)(COD)(η^3-methallyl)$_2$ as Ru-precursor, and excellent S/C ratios were achieved, up to 10 000 (Figure 5.4).

A recent example of phosphine free hydrogenation is reported by Noyori and coworkers, who employed η^6-arene-N-tosylethylenediamine-(Ru)(II)(TfO) catalyst (20) to hydrogenate 4-chromanones and α-halo ketones (Table 5.9, entries 1–10)

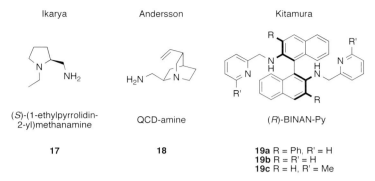

Figure 5.4 Chiral diamines employed in phosphine-free Ru(II)-catalyzed asymmetric ketone hydrogenation.

Table 5.9 Asymmetric hydrogenation of phenacyl chlorides by η'-arene-N-tosylethylenediamine-Ru(II)(TfO) catalyst **(20)**.

Entry	Ar	R	Temperature (°C)	H₂ (atm)	Yield (%)	ee (%)	S/C	Absolute configuration
1	C₆H₅	CH₂Cl	30	10	>99	96	1000	R
2	3-Me-C₆H₄	CH₂Cl	30	10	98	96	1000	R
3	4-Me-C₆H₄	CH₂Cl	30	10	>99	95	1000	R
4	4-MeO-C₆H₄	CH₂Cl	30	10	>99	95	1000	R
5	3-HO-C₆H₄	CH₂Cl	30	10	>99	96	1000	R
6	2-ClC₆H₄	CH₂Cl		10	>99	95	500	R
7	3-ClC₆H₄	CH₂Cl	30	10	>99	94	1000	R
8	4-ClC₆H₄	CH₂Cl	30	10	>99	93	1000	R
9	3-CF₃-C₆H₄	CH₂Cl	30	10	>99	93	500	R
10	4-(CO₂Me)-C₆H₄	CH₂Cl	30	10	>99	94	1000	R

[43–45]. This reaction operates under acidic/neutral conditions. Several different phenacyl chlorides were hydrogenated with excellent enantioselectivity. However, it seems that the method is not general, as no example of simple ketones except 4-chromanones were presented. This method offers several advantages over traditional methods (CBS-reduction or enzymatic epoxide resolution) for the prepara-

5.3.6
α,β-Unsaturated Ketones

Asymmetric hydrogenation of α,β-unsaturated ketones can be considered a very valuable tool for the synthetic organic chemist, due to the rich plethora of useful reactions of chiral secondary allylic alcohols. Ru(II)-(diamine)(diphosphine)-complexes perform excellent in these reductions with very little olefinic over-reduction. An illustrative example is the asymmetric hydrogenation of 6-methyl-2-heptene-2-one, which is the starting material for the synthesis of the α-tocopherol side chain of vitamin E. This hydrogenation proceeds with high enantioselectivity (Table 5.10, entry 5) [32]. Most α,β-unsaturated ketones can be considered as base sensitive substrates and thereby easily affected by the highly basic conditions (i-PrOK in i-PrOH) generally employed. Potassium carbonate has been used as a base additive in order to avoid this issue (entries 1–6) [32]. However, weak inorganic bases lead to relatively low TOF numbers. A good alternative is to employ the *trans*-RuH(η^1-BH$_4$)[(S)-xylBINAP][(S,S)-DPEN] catalyst, which is compatible with base-sensitive substrates [25]. Burk and coworkers hydrogenated a series of unsaturated ketones with good results, employing their RuCl$_2$[(S)-xylPhanephos][(S,S)-DPEN] complex

Table 5.10 Hydrogenation of α,β-unsaturated ketones with Ru(II)-diamine-diphosphine complexes.

Entry	R^1	R^2	R^3	R^4	H_2 (atm)	Yield (%)	ee (%)	S/C	Absolute configuration
1	Me	H	C$_6$H$_5$	H	80	100	97	100000	R
2	Me	H	C$_6$H$_5$	H	10	100	96	10000	R
3	Me	H	2-Thienyl	H	8	100	86	2000	R
4	i-Pr	H	C$_6$H$_5$	H	8	100	91	5000	S
5	C(CH$_3$)$_2$CHCH$_2$	H	Me	H	10	100	90	2000	S
6	Me	H	Me	Me	8	100	93	10000	R
7	Me	H	C$_6$H$_5$	H	8	100	97	3000	R
8	Me	R^3	–[(CH$_2$)$_4$]–	H	8	100	94	3000	R
9	Me	H	C$_6$H$_5$	H	50	100	96	5000	S

Catalysts used:
Entries 1–5: *trans*-RuCl$_2$[(S)-xylBINAP][(S)-DAIPEN] + K$_2$CO$_3$.
Entry 6: *trans*-RuCl$_2$[(S)-xylBINAP][(S,S)-DPEN] + K$_2$CO$_3$.
Entries 7–8: *trans*-RuCl$_2$[(S)-xylPhanephos][(S,S)-DPEN] + t-BuOK.
Entry 9: *trans*-RuCl$_2$[(S)-xylSPD][(R,R)-DPEN] + t-BuOK.

(entries 7–8) [35]. Xyl-SPD also performs well as its *trans*-RuCl$_2$[(*S*)-xylSPD][(*R,R*)-DPEN]-complex in the hydrogenation of an unsaturated ketone (entry 9) [34].

References

1 Ohkuma, T. and Noyori, R. (2004) *Comprehensive Asymmetric Catalysis, Supplement*, **1**, 1.
2 Ohkuma, T. and Noyori, R. (2004) *Transition Metals for Organic Synthesis*, 2nd edn, Vol. **2**, 29.
3 Noyori, R., Ohkuma, T., Kitamura, M., Takaya, H., Sayo, N., Kumobayashi, H. and Akutagawa, S. (1987) *Journal of the American Chemical Society*, **109**, 5856.
4 Mashima, K., Sato, K.-h., Kusano, N., Nozaki, Y.-i., Matsumura, K., Kumobayashi, H., Sayo, N., Hori, Y., Ishizaki, T., Akutagawa, S. and Takaya, H. (1994) *Journal of Organic Chemistry*, **59**, 3064.
5 Mikami, K., Korenaga, T., Yusa, Y. and Yamanaka, M. (2003) *Advanced Synthesis and Catalysis*, **345**, 246.
6 Duprat De Paule, S., Jeulin, S., Ratovelomanana-Vidal, V., Genêt, J.-P., Champion, N. and Dellis, P. (2003) *European Journal of Organic Chemistry*, 2003, **10**, 1931.
7 Duprat De Paule, S., Jeulin, S., Ratovelomanana-Vidal, V., Genêt, J.-P., Champion, N. and Dellis, P. (2003) *Tetrahedron Letters*, **44**, 823.
8 Duprat De Paule, S., Jeulin, S., Ratovelomanana-Vidal, V., Genêt, J.-P., Champion, N., Deschaux, G. and Dellis, P. (2003) *Organic Process Research and Development*, **7**, 399.
9 Pai, C.-C., Lin, C.-W., Lin, C.-C., Chen, C.-C., Chan, A.S.C. and Wong, W.T. (2000) *Journal of the American Chemical Society*, **122**, 11513.
10 Benincori, T., Cesarotti, E., Piccolo, O. and Sannicolo, F. (2000) *Journal of Organic Chemistry*, **65**, 2043.
11 Zhang, Z., Qian, H., Longmire, J. and Zhang, X. (2000) *Journal of Organic Chemistry*, **65**, 6223.
12 Kitamura, M., Ohkuma, T., Inoue, S., Sayo, N., Kumobayashi, H., Akutagawa, S., Ohta, T., Takaya, H. and Noyori, R. (1988) *Journal of the American Chemical Society*, **110**, 629.
13 Rychnovsky, S.D., Griesgraber, G., Zeller, S. and Skalitzky, D.J. (1991) *Journal of Organic Chemistry*, **56**, 5161.
14 Kitamura, M., Yoshimura, M., Kanda, N. and Noyori, R. (1999) *Tetrahedron*, **55**, 8769.
15 Bertus, P., Phansavath, P., Ratovelomanana-Vidal, V., Genet, J.P., Touati, A.R., Homri, T. and Hassine, B.B. (1999) *Tetrahedron: Asymmetry*, **10**, 1369.
16 Duprat De Paule, S., Piombo, L., Ratovelomanana-Vidal, V., Greck, C. and Genêt, J.-P. (2000) *European Journal of Organic Chemistry*, 2000, **8**, 1535.
17 Kitamura, M., Tokunaga, M. and Noyori, R. (1995) *Journal of the American Chemical Society*, **117**, 2931.
18 Gautier, I., Ratovelomanana-Vidal, V., Savignac, P. and Genet, J.-P. (1996) *Tetrahedron Letters*, **37**, 7721.
19 Noyori, R. (2002) *Angewandte Chemie (International Ed. in English)*, **41**, 2008.
20 Genêt, J.-P., Pinel, C., Mallart, S., Juge, S., Cailhol, N. and Laffitte, J.A. (1992) *Tetrahedron Letters*, **33**, 5343.
21 Noyori, R., Ikeda, T., Ohkuma, T., Widhalm, M., Kitamura, M., Takaya, H., Akutagawa, S., Sayo, N., Saito, T., *et al.* (1989) *Journal of the American Chemical Society*, **111**, 9134.
22 Kitamura, M., Ohkuma, T., Tokunaga, M. and Noyori, R. (1990) *Tetrahedron: Asymmetry*, **1**, 1.
23 Ohkuma, T., Ooka, H., Ikariya, T. and Noyori, R. (1995) *Journal of the American Chemical Society*, **117**, 10417.
24 Jiang, Q., Jiang, Y., Xiao, D., Cao, P. and Zhang, X. (1998) *Angewandte Chemie (International Ed. in English)*, **37**, 1100.
25 Ohkuma, T., Koizumi, M., Muniz, K., Hilt, G., Kabuto, C. and Noyori, R. (2002) *Journal of the American Chemical Society*, **124**, 6508.

26 Sandoval, C.A., Ohkuma, T., Muniz, K. and Noyori, R. (2003) *Journal of the American Chemical Society*, **125**, 13490.
27 Hedberg, C., Källström, K., Arvidsson, P.I., Brandt, P. and Andersson, P.G. (2005) *Journal of the American Chemical Society*, **127**, 15083.
28 Abdur-Rashid, K., Faatz, M., Lough, A.J. and Morris, R.H. (2001) *Journal of the American Chemical Society*, **123**, 7473.
29 Ito, M., Hirakawa, M., Murata, K. and Ikariya, T. (2001) *Organometallics*, **20**, 379.
30 Noyori, R., Yamakawa, M. and Hashiguchi, S. (2001) *Journal of Organic Chemistry*, **66**, 7931.
31 Doucet, H., Ohkuma, T., Murata, K., Yokozawa, T., Kozawa, M., Katayama, E., England, A.F., Ikariya, T. and Noyori, R. (1998) *Angewandte Chemie (International Ed. in English)*, **37**, 1703.
32 Ohkuma, T., Koizumi, M., Doucet, H., Pham, T., Kozawa, M., Murata, K., Katayama, E., Yokozawa, T., Ikariya, T. and Noyori, R. (1998) *Journal of the American Chemical Society*, **120**, 13529.
33 Ohkuma, T., Sandoval, C.A., Srinivasan, R., Lin, Q., Wei, Y., Muniz, K. and Noyori, R. (2005) *Journal of the American Chemical Society*, **127**, 8288.
34 Xie, J.-H., Wang, L.-X., Fu, Y., Zhu, S.-F., Fan, B.-M., Duan, H.-F. and Zhou, Q.-L. (2003) *Journal of the American Chemical Society*, **125**, 4404.
35 Burk, M.J., Hems, W., Herzberg, D., Malan, C. and Zanotti-Gerosa, A. (2000) *Organic Letters*, **2**, 4173.
36 Ohkuma, T., Hattori, T., Ooka, H., Inoue, T. and Noyori, R. (2004) *Organic Letters*, **6**, 2681.
37 Ohkuma, T., Koizumi, M., Ikehira, H., Yokozawa, T. and Noyori, R. (2000) *Organic Letters*, **2**, 659.
38 Reamer, C.-y. Chen, R.A., Chilenski, J.R. and McWilliams, C.J. (2003) *Organic Letters*, **5**, 5039.
39 Schmid, R., Broger, E.A., Cereghetti, M., Crameri, Y., Foricher, J., Lalonde, M., Muller, R.K., Scalone, M., Schoettel, G. and Zutter, U. (1996) *Pure and Applied Chemistry. Chimie Pure et Appliquée*, **68**, 131.
40 Lam, W.-S., Kok, S.H.L., Au-Yeung, T.T.L., Wu, J., Cheung, H.-Y., Lam, F.-L., Yeung, C.-H. and Chan, A.S.C. (2006) *Advanced Synthesis and Catalysis*, **348**, 370.
41 Ohkuma, T., Koizumi, M., Yoshida, M. and Noyori, R. (2000) *Organic Letters*, **2**, 1749.
42 Huang, H., Okuno, T., Tsuda, K., Yoshimura, M. and Kitamura, M. (2006) *Journal of the American Chemical Society*, **128**, 8716.
43 Ohkuma, T., Tsutsumi, K., Utsumi, N., Arai, N., Noyori, R. and Murata, K. (2007) *Organic Letters*, **9**, 255.
44 Sandoval, C.A., Ohkuma, T., Utsumi, N., Tsutsumi, K., Murata, K. and Noyori, R. (2006) *Chemistry – An Asian Journal*, **1**, 102.
45 Ohkuma, T., Utsumi, N., Tsutsumi, K., Murata, K., Sandoval, C. and Noyori, R. (2006) *Journal of the American Chemical Society*, **128**, 8724.

6
Reduction of Carbonyl Compounds by Hydrogen Transfer

Serafino Gladiali and Rossana Taras

6.1
Introduction

Among the synthetic methodologies amenable to the reduction of carbonyl compounds, the addition of hydrogen from a suitable donor to a C—O double bond (hydrogen transfer; H-transfer) has received increasing attention since the mid of the nineties and has become one of the procedures of choice for the preparation of alcohols from ketones and, to a lesser extent, aldehydes. The increasing popularity of this technique is evident from the number of papers published in this field, over 250 reports in 2006. H-transfer has many positives compared to alternative techniques available for the same transformation: operational simplicity; reduction of safety constraints associated with the use of gaseous hydrogen or of metal hydrides in stoichiometric amounts; safe, benign and low-cost reducing agents; and a large variety of metal catalysts of high activity and selectivity for the process.

As the most recent surveys of this subject have covered the pertinent literature up to the end of 2004 [1], in this contribution particular attention is paid to the most recent advances in the field. This is preceded by a brief historical overview highlighting the milestones in the development of the methodology. Illustrations are included of the basic concepts of the chemistry underlying the process.

6.2
Historical Overview

The first examples of the H-transfer reduction of unsaturated organic compounds were reported in the first decade of the last century [2]. However, the utility of such a process for the reduction of carbonyl compounds only became apparent to chemists several years later, when the ability of aluminum alkoxides in promoting hydrogen transfer reactions was reported and established in the form of a synthetic protocol useful for the reduction of ketones to their corresponding alcohols

Modern Reduction Methods. Edited by Pher G. Andersson and Ian J. Munslow.
Copyright © 2008 WILEY-VCH Verlag GmbH & Co. KGaA, Weinheim
ISBN: 978-3-527-31862-9

(Meerwein–Ponndorf–Verley (MPV) reduction) [3]. Several years later the reversibility of this reaction, aluminum alkoxides with ketones, was exploited. This backward reaction allowed the useful synthetic procedure for the oxidation of secondary alcohols to ketones (Oppenauer oxidation) [4]. The next milestone was the discovery that some transition metal complexes were able to catalyze the H-transfer reduction of ketones homogeneously. The first reports illustrating this property, published in 1967 [5], were preceded by preliminary work reporting the Ir-catalyzed reduction of cyclohexanones to alcohols with 2-propanol, which has been almost completely overlooked [6]. However, it was only in the second half of the 1970s with impressive results obtained in catalytic homogeneous hydrogenation that this topic began to attract significant interest. A seminal contribution to the development of this area was provided by the pioneering work of Sasson and Blum, who demonstrated that Ru-triphenylphosphine complexes had good catalytic activity in the transfer hydrogenation of acetophenone with 2-propanol [7]. Their contributions were shortly followed by the first reports on the Ru-catalyzed asymmetric transfer hydrogenation of ketones [8]. Since then, most research efforts have concerned the use of chiral transition metal catalysts for asymmetric transfer hydrogenation; the first significant results were obtained at the end of the 1980s [9]. In the last two decades the importance of this process as a synthetic tool for the reduction of ketones has grown steadily and it can now be considered a methodology of choice for accomplishing this transformation.

6.3
General Background

Transfer hydrogenation of carbonyl compounds implies the use of a hydrogen donor, DH_2, as the reducing agent. The hydrogen donor is characterized by the presence of two hydrogens that, under the influence of a suitable promoter, can be mobilized in such a way as to add to an unsaturated functional group in the substrate. At the same time the hydrogen donor is converted to its dehydrogenated counterpart, D (Scheme 6.1). In the large majority of cases the two hydrogens of the donor are nonequivalent and they are transferred sequentially, one to the carbonyl carbon as a formal hydride and the other to the oxygen as a formal proton.

In the reduction of carbonyl compounds, H-transfer offers certain advantages over alternative methodologies. Compared with the use of molecular hydrogen, positive aspects are the reduction of risks and safety restrictions associated with the use of pressure vessels as well as the added value of exploiting molecular diversity in the reducing agent. Advantages compared with other conventional

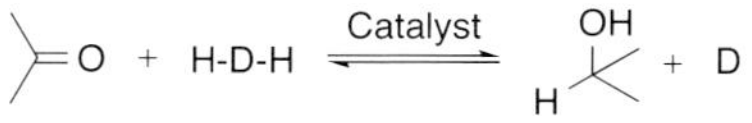

$$\ce{>C=O + H-D-H <=>[\text{Catalyst}] >CH-OH + D}$$

Scheme 6.1 Carbonyl H-transfer reduction.

methodologies are cheaper reagents as well ones that are less hazardous than boron, aluminum or silicon hydrides. Furthermore, in the case of large-scale applications, the downstream operations are much easier in transfer hydrogenation as the accompanying co-products can easily be disposed of and are not associated with any significant environmental issues.

Formation of a co-product, D, is the main drawback of H-transfer processes when compared with processes such as catalytic hydrogenation. The presence of this unwanted compound in the reaction mixture can lead to undesired complications in isolation of the required product as well as separation of the catalyst. Furthermore, upon losing two hydrogen atoms the hydrogen donor (DH_2) itself becomes a hydrogen acceptor that may be in competition with the substrate until equilibrium is reached, such as where an alcohol is used as the hydrogen source in the transfer hydrogenation of ketones. In these cases if the use of H-donors such as formic acid or formates, which are irreversibly dehydrogenated to CO_2, is not possible, an alcohol of lower oxidation potential is the hydrogen donor of choice.

6.4
Hydrogen Donors

In principle any chemical compound whose hydrogens can be mobilized under appropriate conditions could act as the hydrogen source; however, hydrogen donors of practical use in the transition metal-catalyzed reduction of carbonyl compounds are just two: 2-propanol and formic acid and its salts.

Where 2-propanol is the reducing agent, the reaction is under thermodynamic control and a large excess of the hydrogen donor is required to shift the equilibrium. For this reason, 2-propanol is the most common solvent for these reactions. Additionally, 2-propanol is cheap, is a good solvent for most substrates, can be easily disposed of, is readily recycled and, like the relevant dehydrogenation product acetone, is environmentally benign. Furthermore, in this solvent many catalysts have a long enough lifetime for high conversions to be obtained, even at reflux.

When using 2-propanol as a reductant, a base such as an alkali metal hydroxide or alkoxide is usually necessary to enable the extraction of hydrogen from the alcohol. The amount of this promoter can vary over a wide range depending on the nature of the catalyst. Typical base-to-metal ratios are between 5:1 and 20:1, but some catalysts can operate with lower amounts or even without any base, while others require a ratio as high as 200:1. The presence of base may also cause undesired side-reactions and reduce the enantiomeric purity of the alcohol.

Formic acid and its salts are the hydrogen donors of choice for the reduction of carbonyl compounds as they are dehydrogenated to CO_2, making the reaction irreversible and kinetically controlled. Moreover, these derivatives can be activated toward hydrogen extraction by bases as weak as triethylamine and, under appropriate conditions, can sustain hydrogen transfer reductions in aqueous solution or

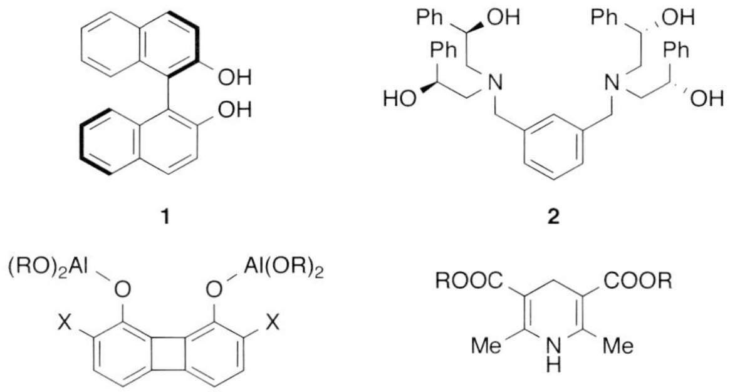

Figure 6.1 Example ligands and the Hantzsch ester (**4**).

in water–organic biphasic conditions. The inherent acidity of formic acid, however, constitutes a significant drawback to its general use since it may favor a stronger interaction of the hydrogen donor with the catalytic system, eventually resulting in complete inhibition or even decomposition of the catalytic species. This renders formic acid incompatible with a large number of the most active hydrogen transfer catalysts and limits its scope as a hydrogen source in the reduction of carbonyl compounds.

The ability of Hantzsch esters like (**4**) (Figure 6.1) to act as hydrogen donors in a process mimicking the action of hydride-based reducing biocofactors such as nicotinamide-adenine dinucleotide (NADH) have long been recognized. Their use in metal-catalyzed hydrogen transfer reactions, however, has very few precedents in the literature due to their very unfavorable atom-efficiency (1 kg of Hantzsch ester provides less than four equivalents of hydride compared to 22 equivalents for formic acid and 17 equivalents for 2-propanol) and to the severe interference generated by the dehydrogenated counterpart in the downstream separation/purification processes. Nevertheless, the recent flourishing of organocatalytic hydrogen transfer chemistry has contributed to their use becoming more frequent.

6.5
Catalysts

Ruthenium-based catalysts have become the catalysts of choice for the transfer hydrogenation of ketones. In the last two or three years the number of papers dealing with Ru-catalyzed H-transfer processes has far exceeded the sum of papers detailing other transition metals. This is even more pronounced in the case of asymmetric transfer hydrogenation reactions, where the number of papers on ruthenium is twice that of all other metals. Second and third, behind ruthenium, are iridium and rhodium catalysts, respectively. Recently, some lanthanides, La, Sm and Gd, in combination with chiral polydentate ligands such as binaphthol (**1**) and the amino alcohol (**2**) have shown high activities and an excellent degree of stereoselectivity (over 99% ee) in the reduction of aryl alkyl ketones [10]. The MPV reduction and Oppenauer oxidation have received new interest thanks to the discovery that low-aggregation aluminum alkoxides are able to induce the hydrogen transfer process in both directions even when used in catalytic amounts [11a]. These catalysts can be prepared *in situ* from trialkylaluminum and the alcohol used as H-donor or, alternatively, they can be preformed by reaction with binaphthol or with other diols of appropriate design that react with alkylaluminum giving bidentate Al-alkoxides such as (3). Notably, significant enantioselectivities can be obtained when a chiral alcohol is utilized as the H-donor [11b] or a chiral Al-alkoxide is used as the catalyst in conjunction with an achiral H-donor [11c].

Iron- [12] and copper- [13] based catalysts for H-transfer have recently been reported. While some results obtained in the asymmetric reductions with Fe-based complexes with the tetradentate ligand (26) have not easily been reproduced [12b], reliable results and good catalytic activity have been obtained in the reduction of aryl alkyl and dialkyl ketones with Fe-porphyrin complexes [12a]. Cu-complexes with chiral bisoxazolines act as mimics of alcohol dehydrogenase and in the presence of Hantzsch esters like (4) are able to reduce α-keto esters to α-hydroxy esters in over 99% ee [13].

Hantzsch esters themselves are routinely used as hydrogen donors in organocatalytic H-transfer processes. The first paper on the application of this technique to asymmetric H-transfer reductions appeared in 2005 [14].

6.6
Mechanisms

The transfer of hydrogen from a donor to an acceptor can proceed in different manners depending on the presence or not of a metal catalyst and on the nature of any catalytic species. One possibility is that the migration of covalently bonded hydrogen from the hydroxy-substituted sp^3-C of an alcohol to the sp^2-C of a carbonyl group takes place under purely thermal conditions in the absence of any metal, base or acid catalyst. It has been reported that this process is feasible and that aryl and alkyl aldehydes and, to a lesser extent, ketones can be reduced to the relevant alcohols upon simple heating at high temperature with ethanol or 1- or 2-propanol (but not methanol) [15]. No hypothesis has been formulated concerning the reaction path, but it seems conceivable that this reaction proceeds through a mechanism where the two hydrogens are transferred simultaneously to the carbonyl group in a concerted process involving a planar six-membered transition state (Scheme 6.2). Such a mechanism was suggested for the reduction of C—C double bonds by diimide [16a], which has recently been confirmed by deuterium labeling experiments and kinetic isotope effects [16b].

From experimental evidence, two basic reaction pathways have been proposed for the metal-catalyzed process, the "direct hydrogen transfer" and the "hydridic route." The "direct hydrogen transfer" mechanism requires that the substrate and hydrogen donor interact simultaneously with the catalyst to form an intermediate

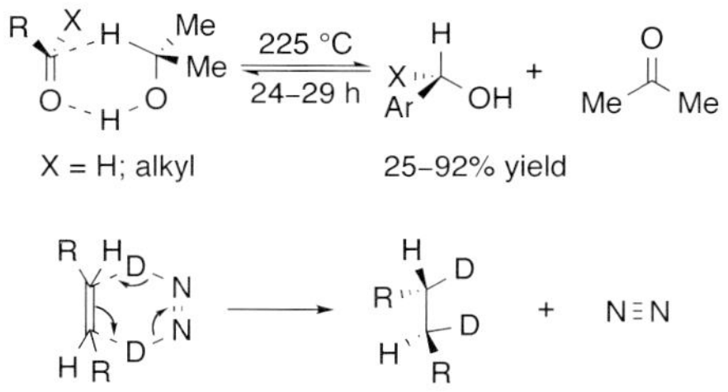

Scheme 6.2 Uncatalyzed concerted H-transfer process.

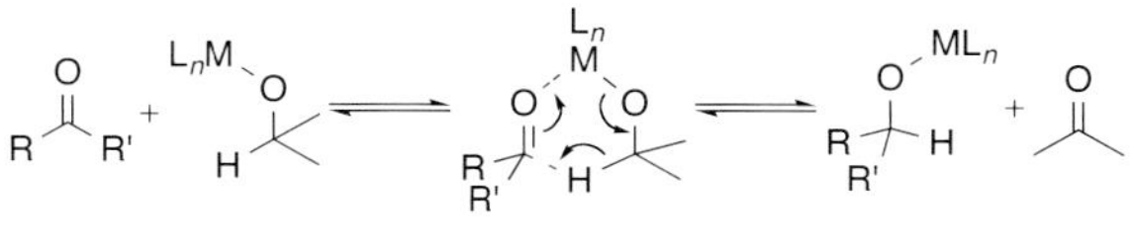

$M = Al; La; Sm; \ldots$

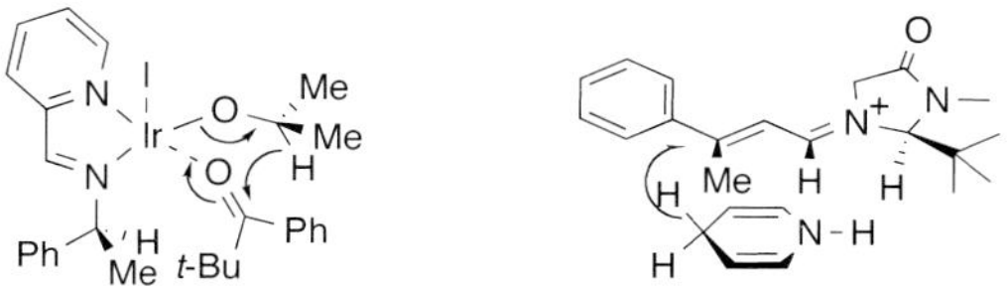

Scheme 6.3 Direct hydrogen transfer pathway.

where the hydrogen is delivered as a formal hydride from the donor to the acceptor in a concerted process (Scheme 6.3). This reorganization of the chemical bonds within the adduct may involve a six-membered transition state such as the one originally proposed for the MPV reduction. Distinctive features of this mechanism are the direct interaction between the H-donor and H-acceptor in the transition state and the absence of any involvement of metal hydride intermediates. The catalyst here has a double role: first in enhancing the electrophilic character of the carbonyl group so as to become more receptive to the hydride attack; and second in providing a highly organized transition state where the correct orientations for a concerted hydride shift are met. While this mechanism is typically observed with electropositive metals, such as Al or lanthanides, it is fairly rare with other transition metal derivatives. Experimental evidence in its support has been reported in the case of the Ir-catalyst (**5**) (Scheme 6.3) [17]. This finding has been corroborated by more recent theoretical calculations [18].

A more sophisticated model of this mechanism has recently been adapted to the organocatalytic H-transfer reductions of open chain α,β-unsaturated aldehydes, such as in (**6**), and of cyclic enones [19].

When direct hydrogen transfer is operating, the use of enantiopure chiral H-donors can result in a stereoselective reduction. It has been demonstrated that

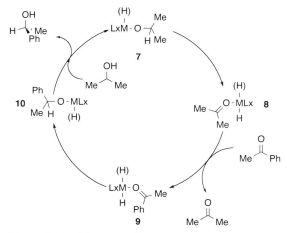

Scheme 6.4 Hydric route pathway.

under these conditions this reaction can deliver good to high enantioselectivities when Al-alkoxides are used as catalysts in the presence of chiral alcohols [11b]. An optically enriched product, however, is just one piece of evidence and on its own is not sufficient to show that the reaction proceeds via this reaction path. A chiral hydrogen donor can indirectly influence the stereoselectivity of a reaction, even if it proceeds through the hydridic route.

H-transfer processes that follow the hydridic route are typical where transition metal derivatives are involved as catalysts. This reaction pathway can be further categorized as the mono- or dihydridic route depending on the number of hydrogens, one or two, that the intermediate hydridic species contains. Common features shared by the hydridic mechanisms are the intermediacy of a metal hydride as the catalytically active species and the fact that H-donor and H-acceptor interact separately with the metal in different steps of the catalytic cycle. In most cases the metal hydride catalyst is not introduced as such in the reaction, but is formed *in situ* by reaction of a suitable precursor with the hydrogen donor in the initial stages of the reaction (pre-activation). In this case an induction period before the reaction reaches a steady state is usually observed. As this may negatively affect the initial reaction rate and conversion, it is advisable to feed the substrate into the reactor only when a constant concentration of the active catalyst has built up in solution.

Whichever the nature of the hydride, the reaction proceeds through the same sequence of elementary steps as shown in Scheme 6.4 for the reduction of acetophenone with 2-propanol as the hydrogen source. Activation of the H-donor occurs upon coordination to the metal of a molecule of 2-propanol, providing the alkoxy complex (**7**), which subsequently undergoes an intramolecular β-hydrogen extraction to produce the metal-hydride (**8**) and acetone. The hydride is then delivered to the carbonyl group according to two different limiting mechanisms. The most frequently encountered implies displacement of the coordinated acetone by acetophenone giving **(9)**, followed by migratory insertion of the ketone into the metal–hydrogen bond to provide the new alkoxy derivative **(10)** (inner sphere mechanism). Displacement of the alkoxy ligand from the coordination sphere of the metal by a new molecule of 2-propanol delivers the reduced product and restarts the catalytic cycle.

An alternative pathway has been proposed by Noyori for Ru-complexes, where a chelating ligand with a primary amino group on a donor arm is coordinated to the metal is such a way as to achieve *syn*-periplanar geometry of the two hydrogens on the vicinal Ru- and N-centers in an intermediate Ru-hydride **(11)** [20]. In this case both the hydrogens can be delivered simultaneously to the carbonyl group in a pericyclic transition state such as **(13)**, which does not require either the substrate or the hydrogen donor to enter the coordination sphere of the metal at any time (outer sphere mechanism). Hydrogen bond and dipolar interactions between the carbonyl group and the hydrogens on Ru and on N assist in the periplanarity of the transition state and the appropriate docking of the substrate to the catalyst. Upon delivering the hydrogens to the substrate, the metal hydride **(11)** gives the coordinatively unsaturated 16-electron amido complex **(12)**, which promptly dehydrogenates 2-propanol to regenerate **(11)** (Scheme 6.5). While for this reaction path to be enabled a vicinal N—H group is mandatory, recently it has been pointed out [21] that alternative pathways can operate in the reduction of ketones, including the inner sphere process [22], even in the presence of this structural feature.

Bäckvall has devised a procedure that demonstrates whether the hydrogen transfer reaction operates through a monohydride or a dihydride mechanism [23]. In the first case the hydrogens of the donor keep their identity and migrate selectively to the acceptor as described in path A (Scheme 6.6), while in the second case they are randomized due to the coexistence of both paths A and B. A protocol reliant on the use of enantiopure α-deuterated α-phenylethyl carbinol has been developed for the procedure. Using this tool it has been shown that both the Rh- and the Ir-catalyzed reactions thus far examined proceed through monohydride intermediates.

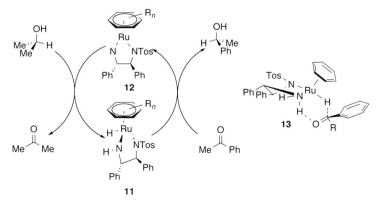

Scheme 6.5 Metal–ligand bifunctional catalysis. Outer sphere mechanism.

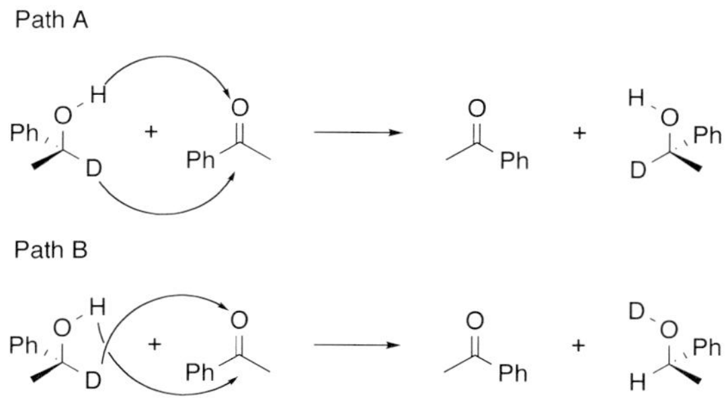

Scheme 6.6 Hydridic route, tracing the hydrogens.

With ruthenium the outcome is less clear-cut: with dichloride complexes a dihydride mechanism is observed, while a monohydride pathway is in operation when Ru-arene or Ru-cyclopentadienyl monochloro complexes are utilized as precatalysts [24].

6.7
Ligands

A number of ligands of diverse structure have been developed for the synthesis of Ru, Rh and Ir complexes in d^6 and d^8 electronic configurations, which have been used in hydrogen transfer reductions. The ligands differ in the number and identity of the donor atom(s), in hapticity and in symmetry properties and, most importantly, they can be neutral or anionic. This last property is crucial for enabling an outer or an inner sphere mechanism in H-transfer. A selection of neutral achiral and chiral ligands recently reported in the Ru-catalyzed transfer hydrogen of ketones is shown in Figure 6.2. For chiral derivatives, the ee values obtained in the asymmetric reduction of acetophenone are quoted.

While the stereoselectivity ranges from good to excellent with all the chiral ligands reported above, the catalytic activity of the Ru-catalysts derived from these ligands is spread over a wide range of magnitudes. While the best turnover frequency obtained with ligand **(15)** is impressive (TOF = 1 200 000 h^{-1}), the activity of the other ligands is much lower and among these only **(21)** approaches 10 000 h^{-1}. In contrast to asymmetric hydrogenation, bidentate P-donors are poorly represented among the ligands of choice in hydrogen transfer reduction of ketones. The recent report of the diphosphonite **(25)**, possessing a large-bite angle, is an exception and may open new research in this area. This is the first Ru-based catalyst utilizing P-based ligands capable of high stereoselectivity in the H-transfer reduction of ketones. Importantly, it has outperformed the majority of the other chiral ligands in the asymmetric reduction of dialkyl ketones that are notoriously difficult to reduce in high stereoselectivity. The high performance of the potentially tetradentate derivative **(26)** deserves particular attention, the structure of **(26)** can be considered as two chelating P,N-heterodonor ligands or as one P,P- and one N,N-chelating ligand.

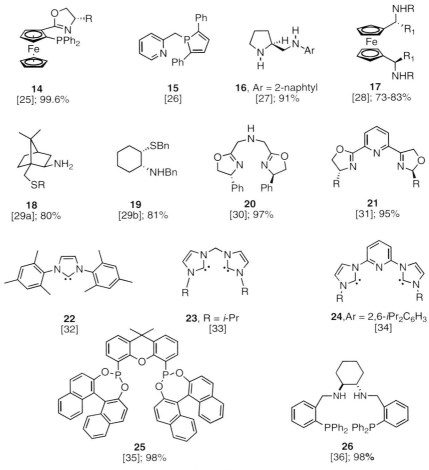

Figure 6.2 Neutral ligands for Ru-catalyzed transfer hydrogenation of acetophenone [25–36]; ee %.

With this in mind, a number of Ru-catalysts such as **(27)** (Figure 6.3) have been reported, containing one chiral C_2-symmetric diphosphine, such as BINAP **(28)**, and one chiral C_2-symmetric chelating diamine such as DPEN **(29)** [37]. While most of these complexes have shown impressive activity and stereoselectivity in the hydrogenation of ketones, they are less efficient in H-transfer reductions [38]. Recent work from Baratta [39] and Morris [40], however, has succeeded in discovering matched pairs of cooperative P- and N-donors which provide H-transfer catalysts of high activity (over 200 000 h^{-1} **(31)** and **(32)**) and over 90% ee in the reduction of acetophenone (both **(31)**/**(32)**, and **(29)**/**(30)**) (Figure 6.3).

The most popular anionic ligand for asymmetric transfer hydrogenation is the monotosyl diamine **(33)** (TsDPEN) reported by Noyori and coworkers in a series

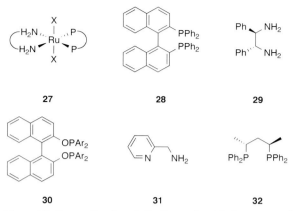

Figure 6.3 Cooperative ligands for Ru-catalyzed hydrogen transfer reduction of ketones.

of papers initiated in 1995 [41]. Since then the library of monosulfonyl diamines for the H-transfer reduction of carbonyl compounds has increased steadily and this is now the most populous class of chiral anionic ligands for this process. A selection of these ligands is shown in Figure 6.4 **(33–38)**. While the activity of the catalysts obtained from these ligands is fair to good, their ability in stereodifferentiation, in general, is very good. Complexes of these ligands even maintain their catalytic activity in formic acid. Notably, changing the hydrogen donor only marginally effects the selectivity of the reaction. The stereochemistry of the products is predictable and depends on the configuration of the carbon bound to the tosyl-amino group [42]. This is readily apparent from a comparative examination of the entries **(33)–(36)** of Figure 6.4. These ligands have also been used in the preparation of Rh(III) and Ir(III) cyclopentadienyl complexes that have resulted in the successful H-transfer reduction of carbonyl compounds under nonconventional conditions (see below).

The second most abundant class of anionic ligands is that of β-amino alcohols, introduced by Noyori, almost complementary to tosyl diamines [53]. As a general rule, β-amino alcohols produce catalysts of higher activity than tosyl diamine derivatives and, for instance, ligand **(40)** achieves 96% ee with high catalytic activity ($8500\,h^{-1}$) [45].

Several chiral β-amino alcohols are commercially available and monosubstitution at nitrogen exerts a positive effect on stereoselectivity. The main limit of these ligands is that the relevant catalysts do not tolerate the use of formic acid as the hydrogen donor. The presence of an ancillary arene ligand is a constant feature in the Ru-complexes of these anionic ligands. This η^6-arene fragment, preserved in the catalytic species, does not act as an passive ligand but exerts a marked effect on the performance of the catalysts and provides a non-negligible contribution to the stereoselectivity.

Wills has connected this arene fragment to the backbone of the chiral ligand, building a new range of tethered ligands. The underlying intention was to provide

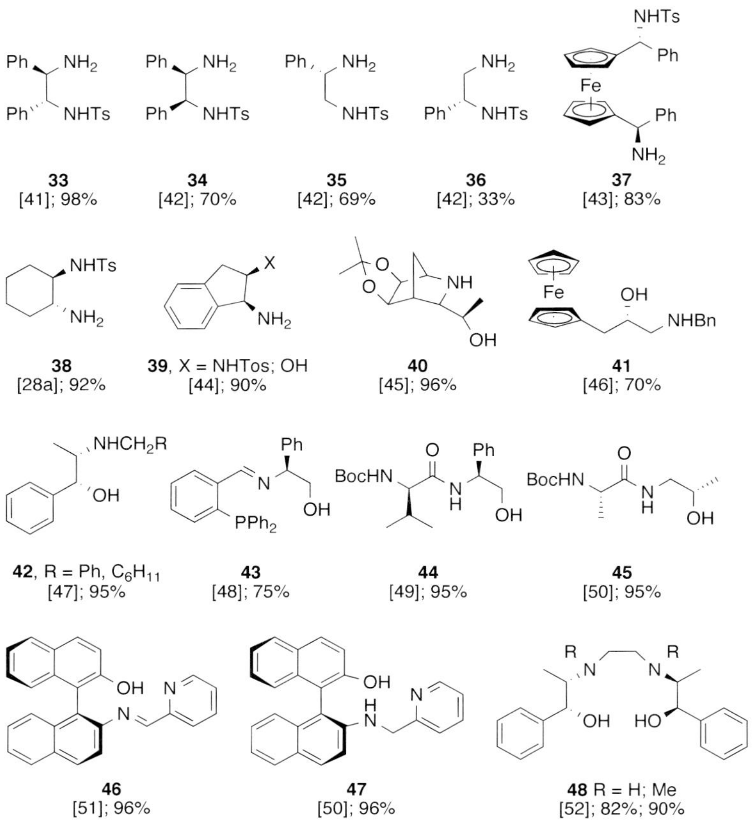

Figure 6.4 Anionic ligands for Ru-catalyzed transfer reduction of ketones [41–52]; ee %.

a new ligand framework with the appropriate conditions to allow a transition state of high organization. This goal has effectively been achieved due to the locking of the chiral elements by the three coordination sites of the ligand. As an additional benefit, when compared with the untethered counterparts, the lifetime of the catalyst was improved significantly. The structures of the tethered ligands and of the relevant arene-Ru or cyclopentadienyl-Rh complexes are illustrated in Figure 6.5. While the first catalysts with this class of ligand, Ru-**(49)** and Ru-**(51)**, showed no improvement in catalytic activity compared with the parent Noyori catalysts, the "reverse-tethered" Ru-complex derived from ligand **(50)** was 1–2 orders of magnitude more active while preserving high levels of enantioselectivity.

The utility of cyclometalated Ru-complexes in H-transfer reductions was first reported in 2000, when the Ru-pincer complex obtained from **(53a)** was shown to display a high catalytic activity, with turnover frequency (TOF) of $27\,000$–$33\,000\,h^{-1}$ for cyclohexanone (Figure 6.6) [58a]. Increasing the electron density of the P-donors as in **(54)** and **(55)** had a negative affect [59]. Further, the introduction of

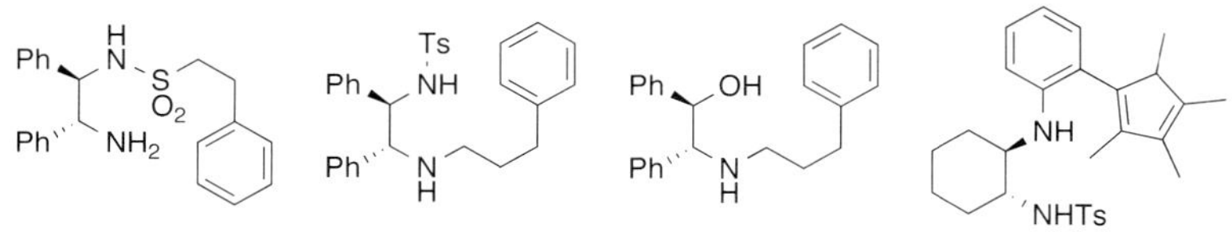

49 [54]; 96% | 50 [55]; 96% | 51 [56]; 67% | 52 [57]; 96%

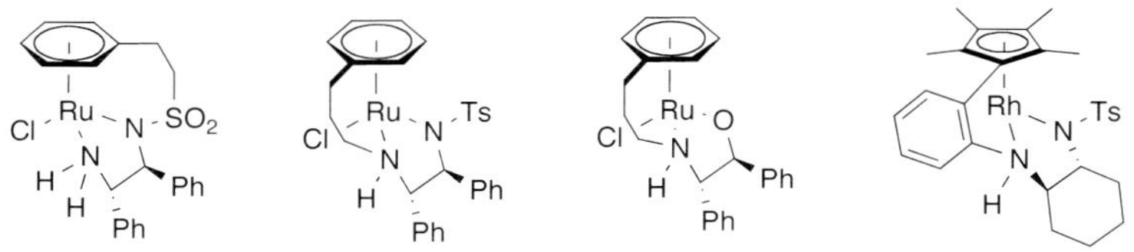

Ru-**49** | Ru-**50** | Ru-**51** | Rh-**52**

Figure 6.5 Chiral tethered catalysts for transfer reduction of ketones [54–57]; ee %.

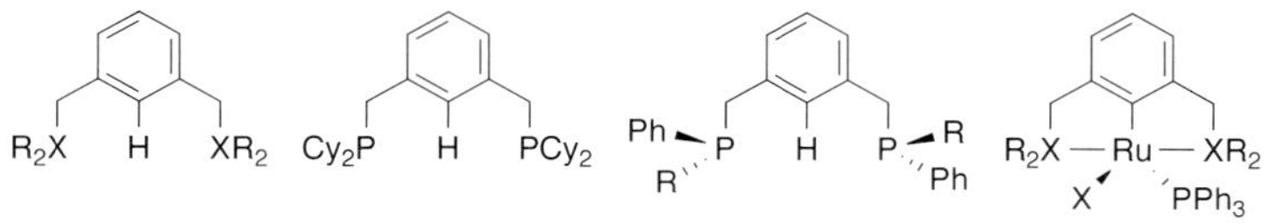

53a XR_2 = PPh_2
53b XR_2 = NMe_2
[58]

54
[59]

55 R = Me; *i*-Pr
[60]; 18%

Ru-pincer

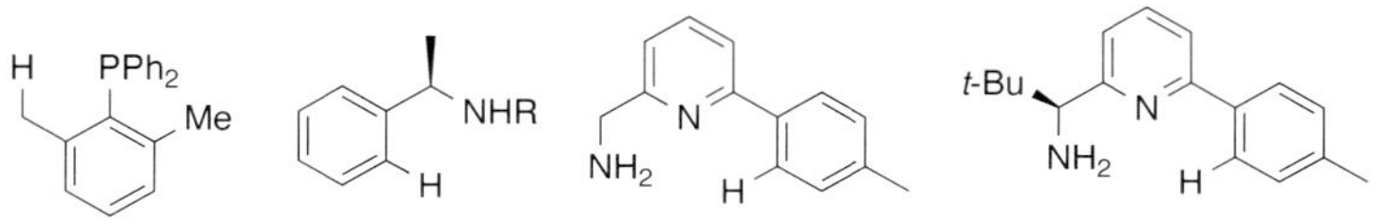

56 [61] | **57** [62]; 10–90% | **58** [22b] | **59** [63]; 90–95%

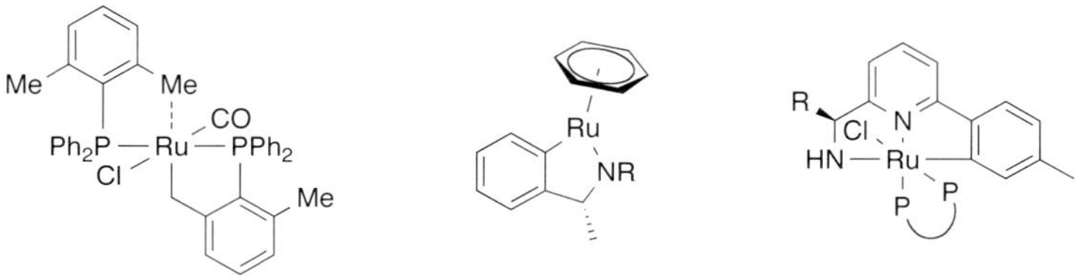

Ru-**56** | Ru-**57** | Ru-**58**

Figure 6.6 Cyclometalated Ru-catalysts for transfer reduction of ketones [58–63]; ee %.

stereogenic P-centers did not result in high ee values [60]. A later paper [58b] has shown that the catalytic activity of PCP-pincer Ru-complexes **(53a)** can be increased by the incorporation of electron-withdrawing substituents, such as CF_3, in the *para*-position of the P-phenyl rings. Ruthenacycle complexes obtained from monodentate ligands such as **(56)** and **(57)** possess high catalytic activity, TOF 63 000 h^{-1}, when the complex Ru-**(56)** is used in conjunction with 2-(aminomethyl)pyridine **(31)** [61]. This figure has been improved substantially and now impressive TOFs 500 000–2 500 000 h^{-1} can be obtained, depending on the substrate, with complex Ru-**(58)** (R = H) where the aminomethylpyridine fragment is internally cyclometalated to a pincer-like coordination and matched with a chelating diphosphine such as **(32)** [22b]. These are the highest catalytic activities thus far reported for the H-transfer reduction of ketones. Notably these catalysts are also very stereoselective, 90–95% ee when chiral diphosphines such as **(32)** are used in conjunction with achiral or chiral aminopyridines such as **(58)** and **(59)**, respectively [63]. Lower and variable stereoselectivities (10–90% ee) are observed with the cycloruthenated complexes Ru-**(57)**. These catalysts are active enough to allow the reduction process to be run even at room temperature with a sizable TOF of 190 h^{-1}.

6.8
Substrates

6.8.1
Ketones

The transfer hydrogenation of aryl alkyl ketones is frequently employed and the method is comparable to the alternative synthetic procedures for their reductions. Acetophenone is generally used as the benchmark substrate for these reductions. Structural diversity has been introduced on the aryl ring, on the alkyl chain and on benzocondensed cycloalkanones (Figure 6.7).

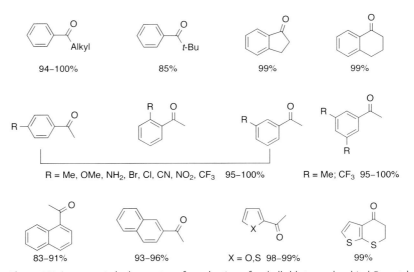

Figure 6.7 Asymmetric hydrogen transfer reduction of aryl alkyl ketones by chiral Ru-catalysts.

6.8 Substrates

A number of transition metal complexes have been screened for their activity and stereoselectivity in this reaction and information on these catalysts can be extracted from the data reported in previous sections.

Since most catalysts have been developed for the enantioselective reduction of ketones, enantiomeric excess is important in assessing their performances. Due to their broad scope, Noyori's catalysts, derived both from chiral amino alcohols and monosulfonyl diamines (Figure 6.4), and Wills' tethered ligands (Figure 6.5) are the most favored for this transformation among the Ru-complexes supported by anionic ligands. The Ru-catalysts supported by neutral ligands such as the phosphino oxazoline **(14)** and the Reetz diphosphonite **(25)** are also able to obtain high enantiomeric excesses over a wide substrate scope. As is apparent from the selected data quoted in Figure 6.7, secondary aryl alkyl alcohols can be obtained in excellent enantiomeric excess via Ru-catalyzed H-transfer reduction with these ligands, with either 2-propanol or formic acid as the H-donor.

The highly enantioselective reduction of dialkyl ketones is more challenging and successful catalysts have only recently been reported (Figure 6.8). The first example was a Noyori-type complex linked to β-cyclodextrin enabling, in an aqueous solution of sodium formate, the asymmetric transfer hydrogenation of a range of aliphatic ketones in 42–95% ee [64]. The second example is the above cited Ru-catalyst of Reetz which, with 2-propanol, reduces a range of aliphatic ketones in 76–99% ee.

Highly stereoselective Ru-complexes are often not the most active, usually obtaining TOF numbers in the range 100–$1000\,h^{-1}$, although exceptionally they have achieved $10\,000\,h^{-1}$, as in the case of the bicyclic amino alcohol **(40)** [45]. Ru-catalysts of high activity have been reported, but most of them are characterized by the presence of a Ru-metallacycle unit either in a chelate bidentate fashion [61, 62] or with a tridentate pincer ligand [22b, 58, 63]. Further, there are several examples of catalysts of high activity based on neutral P,N- [26] or P,P/N,N-bischelating ligands [39]. If both catalyst activity and stereoselectivity are considered, then the catalysts containing the aminomethylpyridine fragment developed by Baratta, particularly the complex Ru-**(58)** (Figure 6.6), are the most efficient, exhibiting high ee values (90–95%) and activity (TOF 10^5–$10^6\,h^{-1}$).

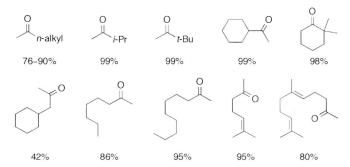

Figure 6.8 Asymmetric hydrogen transfer reduction of alkyl ketones by chiral Ru-catalysts.

While Rh and Ir catalysts have received less coverage than those using Ru, excellent results have been published. In the asymmetric reduction of ketones, where stereoselectivity is basically dictated by the chiral structure of the ligand, the differences in ee values among the three metals are only small [1a]. However, the use of Rh(III) or Ir(III) complexes is occasionally advantageous over the isoelectronic Ru(II) derivatives, especially in areas such as reaction conditions or downstream processing. For instance, with 2-propanol as the solvent, Ir and Rh complexes with the chiral tetradentate ligand **(26)** exhibit high catalytic activity and excellent stereoselectivity in the hydrogenation of ketones even in the absence of base [65]. In many cases it is not necessary to operate under strictly anhydrous or inert conditions with these metals; in fact the reactions can be performed under aqueous conditions and in an open vessel without observing any catalyst deactivation or reduction in the observed enatioselectivity. Complexes derived from [Cp*MCl$_2$]$_2$ (M = Rh or Ir) and monosulfonylated diamines such **(33)** or **(38)** use an aqueous solution of HCOONa as the hydrogen source in the reduction of ketones in high enantiomeric excess [57, 66].

A number of the ligands shown in the schemes above, as well as others not reported here, have been utilized in the synthesis of Rh and Ir complexes and screened for their activity in the transfer hydrogenation of ketones.

Results obtained with the same ligands complexed to different metals, even where the relevant complexes are isoelectronic, are hard to compare because of the differences in the reaction conditions and, sometimes, in reaction pathways. Where comparisons are possible [1a] some similarities can be observed with regard to stereoselectivity, which appears to be dictated by the chiral structure of the ligand. However, no trend in catalytic activity can be observed.

6.8.2
Functionalized Carbonyl Compounds

Changing the metal center can result, apart from different reaction conditions, in a variation of the chemical behavior of the catalyst, often leading to new reactivity or selectivity. This can be particularly significant with bifunctional substrates where the presence of two polar groups can greatly influence the manner in which the substrate interacts with the catalyst. Reduction of these substrates usually proceeds in high yield, although sometimes they are affected by catalyst inhibition either by the reduction product [67] or by the substrate [68]. Ru-catalysts appear to be sensitive to these effects and are often outperformed by Ir- or Rh-centered catalysts in the transfer hydrogenation of bifunctional substrates. Major efforts have been made in the enantioselective reduction of these substrates, as the relevant alcohols are valuable intermediates in the fine chemicals industry (Figure 6.9).

The reduction of α-substituted acetophenones can differ sharply depending on the catalyst metal center. 2-Chloro-1-phenylethanol can be obtained in high yield and enantioselectivity by the transfer hydrogenation of α-chloroacetophenone with the complex Cp*RhCl[(R,R)-**(33)**] in formic acid, [69] while the isoelectronic Ru(II)-complex gives only poor results. Poor yields and ee values are also obtained with

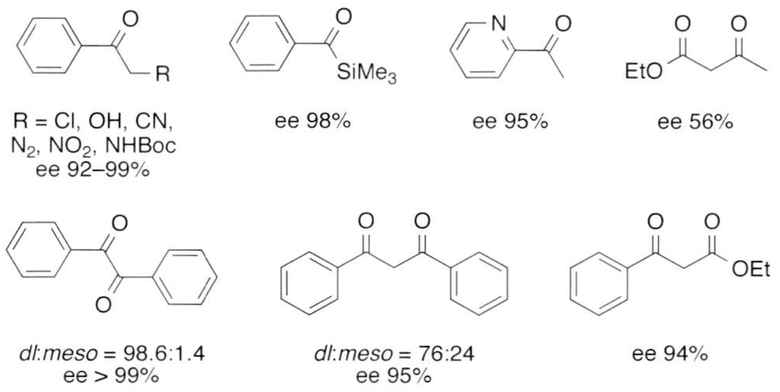

Figure 6.9 Asymmetric hydrogen transfer reduction of functionalized ketones by chiral catalysts.

- R = Cl, OH, CN, N$_3$, NO$_2$, NHBoc; ee 92–99%
- PhCOCH$_2$SiMe$_3$: ee 98%
- 2-acetylpyridine: ee 95%
- ethyl acetoacetate: ee 56%
- benzil: dl:meso = 98.6:1.4, ee > 99%
- 1,3-diphenyl-1,3-propanedione: dl:meso = 76:24, ee 95%
- ethyl benzoylacetate: ee 94%

catalyst Rh-**(52)** (Figure 6.5), derived from a tethered ligand [70]. In contrast, Noyori's catalysts **(11)** is particularly well suited for the highly enantioselective reduction of 2-cyano-, 2-azido-, 2-nitro and 2-trimethylsilylacetophenone [71].

The transfer hydrogenation of benzil results in the almost quantitative production of hydrobenzoin with excellent diastereomeric (*dl* : *meso* = 98.6 : 1.4) and enantiomeric purities (>99% ee). The reaction proceeds in stepwise fashion through benzoin, which, having a labile stereogenic center, is converted to the major stereoisomer via dynamic kinetic resolution [72a]. At low temperatures, unsymmetrically substituted aryl alkyl 1,2-diketones lead to optically active α-hydroxy ketones by preferential reduction of the less hindered carbonyl group adjacent to the alkyl substituent in up to 99% ee and 89% yield [72b]. The reduction of symmetric 1,3-diaryl-1,3-diketones affords diols in comparably higher de and ee (up to 90%) than unsymmetrically substituted 1,3-diketones [73].

6.8.3
Aldehydes

The transfer hydrogenation of aldehydes is a reaction of synthetic utility for the preparation of primary alcohols, in particular benzylic alcohols. Aluminum alkoxides, Ru(II), Rh(III) and Ir(III) complexes are all effective catalysts for this process. Among the transition metals, iridium is the most frequently used with these substrates as it provides the most active catalysts. The H-transfer reduction of aldehydes is favored on thermodynamic grounds even when using 2-propanol as the H-donor, but under these conditions it is frequently affected by several side-reactions caused by the base used as promoter (self-condensation; disproportionation; catalyst inhibition). Major advances in this transformation have been achieved in the last few years that have rendered the process synthetically useful. It is now known that the reaction can be performed even in refluxing 2-propanol if a base of low basicity, such as potassium carbonate, is used in conjunction with an Ir(III) catalyst supported by strongly electron-donating ligands such as **(23)**. This catalytic system allows the hydrogenation of a range of aldehydes in high yield and selectivity [74]. Notably, the reduction takes place to some extent in the absence of the Ir-catalyst, presumably according to a MPV mechanism.

An alternative possibility for the H-transfer reduction of aldehydes is the use of formic acid or its salts as the hydrogen source. This approach has recently been developed into a synthetic methodology of practical use for the reduction of aldehydes with sodium formate in neat water and in open vessels with Ir-catalysts and monosulfonylated diamine ligands. Under these conditions the reduction of the formyl group proceeds with complete selectivity for aryl, alkyl and α,β-unsaturated aldehydes, with the reaction providing high yields even with substrates that have a high tendency to undergo aldol condensations [75].

The asymmetric synthesis of stereoselectively labeled deuterobenzylic alcohols can be accomplished in up to 98% ee via enantioselective deuterohydrogenation of benzaldehydes using the Noyori catalyst (11) and deuteroformic acid as the deuterium donor [76]. The same reaction can be performed in modest to good enantiomeric excess using *cis*-1-amino-2-indanol (39) as the chiral ligand with several different transition metal catalysts [77]. However, the results are less satisfactory with conjugated aldehydes and poor with aliphatic substrates.

6.8.4
Conjugated Carbonyl Compounds

The H-transfer reduction of activated C—C double bonds is a thermodynamically favored reaction even when alcohols are used as H-donors and conjugated acid derivatives smoothly add hydrogen at the C—C double bond under mild conditions. Under certain circumstances in the H-transfer reduction of α,β-unsaturated carbonyl derivatives a competition may arise between vinyl and carbonyl group reductions. The general trend is that α,β-unsaturated aldehydes are preferentially hydrogenated at the formyl group to the corresponding allylic alcohols.

While selective carbonyl reduction can be achieved without any catalyst by reaction at high temperature in supercritical 2-propanol [16b], a catalyst is normally required for the reaction to proceed under less forcing conditions. Iridium complexes are the catalysts of choice for this transformation, although several Ru- and Rh-complexes are also efficient. With Ir-monosulfonyldiamine catalysts and sodium formate in water, the reduction is highly chemoselective and the conjugate formyl group is preferentially hydrogenated with respect to a ketonic functionality when two different carbonyl groups are present in the same substrate, as in the case of (60) (Scheme 6.7) [75].

The reduction of conjugated ketones gives different products with regard to the number and the nature of the substituents on C—C double bond. Thus, the presence on the double bond of an additional electron-withdrawing group, or even a phenyl substituent, may cause a switch in the chemoselectivity of the Ru-catalyzed reduction of conjugated ketones from C—O to C—C bond hydrogenation (Scheme 6.8) [78].

The chemoselectivity of the reaction can sometimes be addressed by selecting an appropriate metal–ligand combination. While Ir(I) complexes with heterodonor

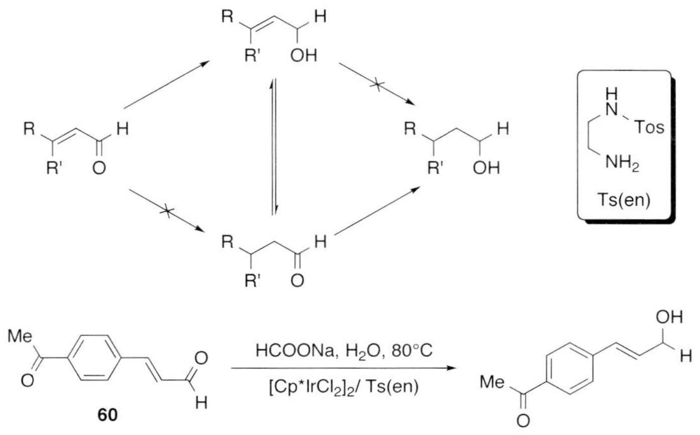

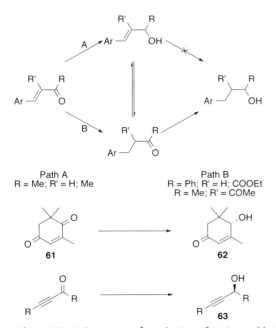

Scheme 6.7 Hydrogen transfer reduction of conjugated aldehydes.

Scheme 6.8 Hydrogen transfer reduction of conjugated ketones.

P,N ligands favor the reduction of carbonyl groups affording the corresponding allylic alcohols [79], ligands with chelating nitrogen atoms form complexes that lead to the preferential formation of saturated ketones [80]. These are the anticipated primary products of the conjugate hydride addition, but they can only be isolated under specific circumstances when the H-transfer catalyst has a low tendency to reduce carbonyl groups [81]. Otherwise, their fate is further reduction by the same catalyst to their respective saturated alcohols. The origin of saturated ketones in this reaction, however, may be not unequivocal proof of the conjugate addition of hydride to the C—C bond since allylic alcohols can be isomerized to carbonyl compounds by H-transfer catalysts (Scheme 6.7) [82]. Whatever the reaction pathway and the metal catalyst, conjugated ketones are more prone to providing the corresponding saturated derivatives and therefore the relevant saturated alcohols than the corresponding allylic alcohols even when the reaction is run in neat water [83]. In conclusion, the chemoselectivity of the H-transfer reduction of α,β-unsaturated ketones is critically dependent on the ability of the catalyst to prevent further reactions of the initially formed products.

A notable example of this behavior is the regioselective reduction of an oxo group in the diketone **(61)** (Scheme 6.8). In the presence of a Ru catalyst, with chiral amino alcohol ligands, this reduction proceeds with high selectivity at the most hindered carbonyl group to give the isophorone derivative **(62)** in over 95% ee [84].

The lack of reactivity of the C—C triple bond can be exploited for the preparation of chiral propargylic alcohols such as **(63)**, (Scheme 6.8) which is accessible in over 95% ee from the corresponding ketone with Noyori's catalyst, 2-propanol [85].

The organocatalytic H-transfer reduction of open-chain α,β-unsaturated aldehydes and cyclic enones by Hantzsch esters proceeds consistently through conjugate hydride addition to the C—C bond as depicted in Scheme 6.3, and always leads to saturated carbonyl compounds.

References

1 (a) Gladiali, S. and Alberico, E. (2004) *Transition Metals for Organic Synthesis*, 2nd edn, Vol. **2** (eds M. Beller and C. Bolm), Wiley-VCH Verlag GmbH, Weinheim, p. 145. (b) Gladiali, S. and Alberico, E. (2006) *Chemical Society Reviews*, **35**, 226. (c) Samec, J.S.M., Bäckvall, J.-E., Andersson, P. and Brandt, P. (2006) *Chemical Society Reviews*, **35**, 237.

2 (a) Knoevenagel, E. and Bergdolt. B. (1903) *Chemische Berichte*, **36**, 2857. (b) Wieland, H. (1912) *Chemische Berichte*, **45**, 484.

3 Meerwein, H. and Schmidt, R. (1925) *Liebigs Annalen der Chemie*, **444**, 221.

4 Oppenauer, R.V. (1937) *Recueil des Travaux Chimiques des Pays-Bas*, **56**, 137.

5 (a) Bailar, J.C. and Hatani, H. (1967) *Journal of the American Chemical Society*, **89**, 1592. (b) Trocha-Grimshaw, J. and Henbest, H.B. (1967) *Chemical Communications*, 544.

6 Haddad, Y.M.Y., Henbest, H.B., Husbands, J. and Mitchell, T.R.B. (1964) *Proceedings. Chemical Society (Great Britain)*, 361.

7 (a) Sasson, Y. and Blum, J. (1971) *Tetrahedron Letters*, **12**, 2167. (b) Sasson, Y. and Blum, J. (1975) *Journal of Organic Chemistry*, **40**, 1887.

8 (a) Bianchi, M., Matteoli, U., Menchi, G., Frediani, P., Pratesi, U., Piacenti, F. and Botteghi, C. (1980) *Journal of Organometallic Chemistry*, **198**, 73. (b) Matteoli, U., Frediani, P., Bianchi, M., Botteghi, C. and Gladiali, S. (1981) *Journal of Molecular Catalysis*, **12**, 265.

9 Zassinovich, G., Mestroni, G. and Gladiali, S. (1992) *Chemical Reviews*, **92**, 1051.

10 (a) Ohno, K., Kataoka, Y. and Mashima, K. (2004) *Organic Letters*, **6**, 4695. (b) Yan, P., Nie, C., Li, G., Hou, G., Sun, W. and Gao, J. (2006) *Applied Organometallic Chemistry*, **20**, 338.

11 (a) Ooi, T., Miura, T. and Maruoka, K. (1998) *Angewandte Chemie (International Ed. in English)*, **37**, 2347. (b) Campbell, E.J., Zhou, H. and Nguyen, S.T. (2001) *Organic Letters*, **3**, 2391. (c) Campbell, E.J., Zhou, H. and Nguyen, S.T. (2002) *Angewandte Chemie (International Ed. in English)*, **41**, 1020.

12 (a) Enthaler, S. Erre, G. Tse, M.K. and Beller, M. (2006) *Tetrahedron Letters*, **47**, 8095. (b) Chen, J.-S., Chen, L.-L., Xing, Y., Chen, G., Shen, W.-Y., Dong, Z.-R., Li, Y.-Y. and Gao, J.-X. (2004) *Huaxue Xuebao*, **62**, 1745.

13 Yang, J.W. and List, B. (2006) *Organic Letters*, **8**, 5653.

14 Yang, J.W., Hechevarria Fonseca, M.T., Vignola, N. and List, B. (2005) *Angewandte Chemie (International Ed. in English)*, **44**, 108.

15 Bagnell, R. and Strauss, C.R. (1999) *Chemical Communications*, 287.

16 (a) Corey, E.J., Pasto, D.J. and Mock, W.L. (1961) *Journal of the American Chemical Society*, **83**, 2957. (b) Kamitanaka, T., Matsuda, T. and Harada, T. (2007) *Tetrahedron*, **63**, 1429.

17 Zassinovich, G., Bettella, R., Mestroni, G., Bresciani-Pahor, N., Geremia, S. and Randaccio, L. (1989) *Journal of Organometallic Chemistry*, **370**, 187.

18 (a) Petra, D.G.I., Kamer, P.C.J., Speck, A.L., Schoemaker, H.E. and van Leeuwen, P.W.N.M. (2000) *Journal of Organic Chemistry*, **65**, 3010. (b) Handgraaf, J.-W., Reek, J.N.H. and Meijer. E.J. (2003) *Organometallics*, **22**, 3150.

19 (a) Tuttle, J.B., Ouellet, S.G. and MacMillan, D.W.C. (2005) *Journal of the American Chemical Society*, **127**, 32. (b) Tuttle, J.B., Ouellet, S.G. and MacMillan, D.W.C. (2006) *Journal of the American Chemical Society*, **128**, 12662.

20 Noyori, R. and Hashiguchi, S. (1997) *Accounts of Chemical Research*, **30**, 97.

21 (a) Handgraaf, J.-W. and Meijer, E.J. (2007) *Journal of the American Chemical Society*, **129**, 3099. (b) For a review on the hydrogenation mechanisms promoted by Ru-hydrides see: Clapham, S.E., Hadzovic, A. and Morris, R.H. (2004) *Coordination Chemistry Reviews*, **248**, 2201.

22 (a) Yi, C.S. and He, Z. (2001) *Organometallics*, **22**, 3641. (b) Baratta, W., Chelucci, G., Gladiali, S., Siega, K., Toniutti, M., Zanette, M., Zangrando, E. and Rigo, P. (2005) *Angewandte Chemie (International Ed. in English)*, **44**, 6214.

23 Laxmi, Y.R.S. and Bäckvall, J.-E. (2000) *Chemical Communications*, 611.

24 Pàmies, O. and Bäckvall, J.-E. (2001) *Chemistry – A European Journal*, **7**, 5052.

25 (a) Sammakia, T. and Stangeland, E.L. (1997) *Journal of Organic Chemistry*, **62**, 6104. (b) Nishibayashi, Y., Takei, I., Uemura, S. and Hidai, M. (1999) *Organometallics*, **18**, 2291. (c) Arikawa, Y., Ueoka, M., Matoba, K., Nishibayashi, Y., Hidai, M. and Uemura, S. (1999) *Journal of Organometallic Chemistry*, **572**, 163.

26 Thoumazet, C., Melaimi, M., Ricard, L., Mathey, F. and Le Floch, P. (2003) *Organometallics*, **22**, 1580.

27 Aitali, M., Allaoud, S., Karim, A., Meliet, C. and Mortreux, A. (2000) *Tetrahedron: Asymmetry*, **11**, 1367.

28 (a) Püntener, K., Schwink, L. and Knochel, P. (1996) *Tetrahedron Letters*, **37**, 8165. (b) Fukuzawa, S. and Suzuky, T. (2006) *European Journal of Organic Chemistry*, 1012.

29 (a) Gayet, A., Bolea, C. and Andersson, P. (2004) *Organic and Biomolecular Chemistry*, **2**, 1887. (b) Schiffers, I., Rantanen, T., Schmidt, F., Bergmans, W., Zani, L. and Bolm, C. (2006) *Journal of Organic Chemistry*, **71**, 2320.

30 Jiang, Y., Jiang, Q. and Zhang, X. (1998) *Journal of the American Chemical Society*, **120**, 3817.

31 Cuervo, D., Pilar Gamasa, M. and Gimeno, J. (2004) *Chemistry – A European Journal*, **10**, 425.

32 Burling, S., Whittlesey, M.K. and Williams, J.M.J. (2005) *Advanced Synthesis and Catalysis*, **347**, 591.

33 Albrecht, M., Crabtree, R.H., Mata, J. and Peris, E. (2002) *Chemical Communications*, 32.

34 Danopoulos, A.A., Winston, S. and Motherwell, W.B. (2002) *Chemical Communications*, 1376.

35 Reetz, M.T. and Li, X. (2006) *Journal of the American Chemical Society*, **128**, 1044.

36 Gao, J.-X., Ikariya, T. and Noyori, R. (1996) *Organometallics*, **15**, 1087.

37 Review: Noyori, R. and Ohkuma, T. (2001) *Angewandte Chemie (International Ed. in English)*, **40**, 40.
38 Rautenstrauch, V., Hoang-Cong, X., Churlaud, R., Abdur-Rashid, K. and Morris, R.H. (2003) *Chemistry A European Journal*, **9**, 4945.
39 (a) Baratta, W., Da Ros, P., Del Zotto, A., Sechi, A., Zangrando, E. and Rigo, P. (2004) *Angewandte Chemie (International Ed. in English)*, **43**, 3584. (b) Baratta, W., Herdtweck, E., Siega, K., Toniutti, M. and Rigo, P. (2005) *Organometallics*, **24**, 1660.
40 Guo, R., Elpelt, C., Chen, X., Song, D. and Morris, R.H. (2005) *Chemical Communications*, 3050.
41 Hashiguchi, S., Fujii, A., Takehara, J., Ikariya, T. and Noyori, R. (1995) *Journal of the American Chemical Society*, **117**, 7562.
42 Hayes, A., Clarckson, G. and Wills, M. (2004) *Tetrahedron: Asymmetry*, **15**, 2079.
43 Schwink, L., Ireland, T., Püntener, K. and Knochel, P. (1998) *Tetrahedron: Asymmetry*, **9**, 1143.
44 Palmer, M., Kenny, J., Walsgrove, T., Kawamoto, A.M. and Wills, M. (2002) *Journal of the Chemical Society, Perkin Transactions I*, 416.
45 Nordin, S.J.M., Roth, P., Tarnai, T., Alonso, D.A., Brandt, P. and Andersson, P.G. (2001) *Chemistry—A European Journal*, **7**, 1431.
46 Patti, A. and Pedotti, S. (2003) *Tetrahedron: Asymmetry*, **14**, 597.
47 (a) Everaere, K., Mortreux, A. and Carpentier, J.-F. (2003) *Advanced Synthesis and Catalysis*, **345**, 67. (b) Frost, C.G. and Mendonça, P. (2000) *Tetrahedron: Asymmetry*, **11**, 1845.
48 Kwong, H.-L., Lee, W.-S., Lai, T.-S. and Wong, W.-T. (1999) *Inorganic Chemistry Communications*, **2**, 66.
49 Pastor, I.M., Västilä, P. and Adolfsson, H. (2002) *Chemical Communications*, 2046.
50 (a) Bøgevig, A., Pastor, I.M. and Adolfsson, H. (2004) *Chemistry—A European Journal*, **10**, 294. (b) Wettergren, J., Bøgevig, A., Portier, M. and Adolfsson, H. (2006) *Advanced Synthesis and Catalysis*, **348**, 1277.
51 Brunner, H., Henning, F. and Weber, M. (2002) *Tetrahedron: Asymmetry*, **13**, 37.
52 Petra, D.G.I., Kamer, P.C.J., van Leeuwen, P.W.N.M., Goubitz, K., van Loon, A.M., de Vries, J.G. and Schoemaker, H.E. (1999) *European Journal of Inorganic Chemistry*, 2335.
53 Takehara, J., Hashiguchi, S., Fujii, A., Inoue, S.-I., Ikariya, T. and Noyori, R. (1996) *Chemical Communications*, 233.
54 Cheung, F.K., Hayes, A.M., Hannedouche, J., Yim, A.S.Y. and Wills, M. (2005) *Journal of Organic Chemistry*, **70**, 3188.
55 Hayes, A.M., Morris, D.J., Clarkson, G.J. and Wills, M. (2005) *Journal of the American Chemical Society*, **127**, 7318.
56 Hannedouche, J., Clarkson, G.J. and Wills, M. (2004) *Journal of the American Chemical Society*, **126**, 986.
57 Matharu, D.S., Morris, D.J., Clarkson, G.J. and Wills, M. (2006) *Chemical Communications*, 3232.
58 (a) Dani, P., Karlen, T., Gossage, R.A., Gladiali, S. and van Koten, G. (2000) *Angewandte Chemie (International Ed. in English)*, **39**, 743. (b) Gagliardo, M., Chase, P.A., Brouwer, S., van Klink, G.P.M. and van Koten, G. (2007) *Organometallics*, **26**, 2219.
59 Amoroso, D., Jabri, A., Yap, G.P.A., Gusev, D.G., dos Santos, E.N. and Fogg, D.E. (2004) *Organometallics*, **23**, 4047.
60 Medici, S., Gagliardo, M., Scott Williams, B., Chase, P.A., Gladiali, S., Lutz, M., Spek, A.L., van Klink, G.P.M. and van Koten, G. (2005) *Helvetica Chimica Acta*, **88**, 694.
61 Baratta, W., Del Zotto, A., Esposito, G., Sechi, A., Toniutti, M., Zangrando, E. and Rigo, P. (2004) *Organometallics*, **23**, 6264.
62 (a) Sortais, J.-B., Ritleng, V., Voelklin, A., Holuigue, A., Smail, H., Barloy, L., Sirlin, C., Verzijl, G.K.M., Boogers, J.A.F., de Vries, A.H.M., de Vries, J.G. and Pfeffer, M. (2005) *Organic Letters*, **7**, 1247. (b) Sortais, J.-B., Barloy, L., Sirlin, C., de Vries, A.H.M., de Vries, J.G. and Pfeffer, M. (2006) *Pure and Applied Chemistry. Chimie Pure et Appliquée*, **78**, 457.
63 Baratta, W., Bosco, M., Chelucci, G., Del Zotto, A., Siega, K., Toniutti, M., Zangrando, E. and Rigo, P. (2006) *Organometallics*, **25**, 4611.

64 Schlatter, A., Kundu, M.K., Woggon, W.-D. (2004) *Angewandte Chemie (International Ed. in English)*, **43**, 6731.

65 Dong, Z.-R., Li, Y.-Y., Chen, J.-S., Li, B.-Z., Xing, Y. and Gao, J.-X. (2005) *Organic Letters*, **7**, 1043.

66 (a) Wu, X., Li, X., King, F. and Xiao, J. (2005) *Angewandte Chemie (International Ed. in English)*, **44**, 3407. (b) Wu, X., Vinci, D., Ikarya, T. and Xiao, J. (2005) *Chemical Communications*, 4447.

67 Kenny, J.A., Palmer, M.J., Smith, A.R.C., Walsgrove, T. and Wills, M. (1999) *Synlett*, 1615.

68 Everaere, K., Mortreux, A., Bulliard, M., Brussee, J., van der Gen, A., Nowogrocki, G. and Carpentier, J.-F. (2001) *European Journal of Organic Chemistry*, 275.

69 Hamada, T., Torii, T., Izawa, K. and Ikariya, T. (2004) *Tetrahedron*, **60**, 7411.

70 Matharu, D.S., Morris, D.J., Kawamoto, A.M., Clarkson, G.J. and Wills, M. (2005) *Organic Letters*, **7**, 5489.

71 (a) Watanabe, M., Murata, K. and Ikariya, T. (2002) *Journal of Organic Chemistry*, **67**, 1712. (b) Cossrow, J. and Rychnovsky. S.D. (2002) *Organic Letters*, **4**, 147.

72 (a) Murata, K., Okano, K., Miyagi, M., Iwane, H., Noyori, R. and Ikariya, T. (1999) *Organic Letters*, **1**, 1119. (b) Koike, T., Murata, K. and Ikariya, T. (2000) *Organic Letters*, **2**, 3833.

73 Cossy, J., Eustache, F. and Dalko, P.I. (2001) *Tetrahedron Letters*, **42**, 5005.

74 Miecznikowski, J.R. and Crabtree, R.H. (2004) *Organometallics*, **23**, 629.

75 Wu, X., Liu, J., Li, X., Zanotti-Gerosa, A., Hancock, F., Vinci, D., Ruan, J. and Xiao, J. (2006) *Angewandte Chemie (International Ed. in English)*, **45**, 6718.

76 Yamada, I. and Noyori, R. (2000) *Organic Letters*, **2**, 3425.

77 Faller, J.W. and Lavoie, A.R. (2002) *Organometallics*, **21**, 3493.

78 Xue, D., Chen, Y.-C., Cui, X., Wang, Q.-W., Zhu, J. and Deng, J.-G. (2005) *Journal of Organic Chemistry*, **70**, 3584.

79 (a) Farnetti, E., Nardin, G. and Graziani, M. (1989) *Chemical Communications*, 1264. (b) Bianchini, C., Farnetti, E., Graziani, M., Nardin, G., Vacca, A. and Zanobini, F. (1990) *Journal of the American Chemical Society*, **112**, 9190.

80 Farnetti, E., Vinzi, F. and Mestroni, G. (1984) *Journal of Molecular Catalysis*, **24**, 147.

81 Doi, T., Fukuyama, T., Horiguchi, J., Okamura, T. and Ryu, I. (2006) *Synlett*, 721.

82 (a) Bernard, M., Delbecq, F., Sautet, P., Fache, F. and Lemaire, M. (2000) *Organometallics*, **19**, 5715. (b) Crochet, P., Fernandez-Zumel, M.A., Gimeno, J. and Scheele, M. (2006) *Organometallics*, **25**, 4846. (c) Bianchini, C., Farnetti, E., Graziani, M., Peruzzini, M. and Polo, A. (1993) *Organometallics*, **12**, 3753. (d) Bäckvall, J.-E. and Andreasson, U. (1993) *Tetrahedron Letters*, **34**, 5459.

83 Li, X., Blacker, J., Houston, I., Xiao, X. and Wu, J. (2006) *Synlett*, 1155.

84 Henning, M., Püntener, K. and Scalone, M. (2000) *Tetrahedron: Asymmetry*, **11**, 1849.

85 Matsumura, K., Hashiguchi, S., Ikariya, T., Noyori, R. (1997) *Journal of the American Chemical Society*, **119**, 8738.

7
Carbonyl Hydroboration

Noriyoshi Arai and Takeshi Ohkuma

7.1
Introduction

The hydride reduction of carbonyl compounds is one of the most important synthetic methods for the preparation of secondary alcohols. The great contribution of Brown to this research area should be noted at the outset [1]. Since the first report by Kagan in 1969, chiral versions of this transformation have been studied extensively [2]. So important is it in synthetic organic chemistry that a large number of excellent reviews have already been published [3]. Current research interest in this field is focused mainly on the development of highly stereoselective reagents applicable to the syntheses of complicated organic molecules. In order to avoid overlap with previous reviews, the present chapter describes recent advances in diastereoselective and enantioselective hydroboration of carbonyl compounds, mainly since the late 1990s.

7.2
Recent Topics in Diastereoselective Reduction

The 1,2-asymmetric induction of α-chiral ketones has been well studied, with the stereochemical outcome being predicted by the Felkin–Anh model (nucleophilic) [4] or Houk's model (electrophilic) (Scheme 7.1) [5]. A hydride can attack the prochiral carbonyl either on the *Re* or on the *Si* face, leading to a pair of diastereomers, **(1B)** and **(1D)**. According to the Felkin–Anh model, the nucleophilic attack of the hydride takes place *anti* to the most bulky group L (or polar group), which is located perpendicular to the carbonyl plane. To minimize steric interactions, the attack preferentially occurs through the conformer **(1A)** rather than through **(1C)**. When a stereoelectronic interaction between the Lewis acidic reductant (such as borane) and the carbonyl group exists, different repulsive interactions become predominant. In this case, conformation **(1E)** is favored, especially when the substituents on the boron (R′) are bulky, affording the product **(1D)**. A representative example

Modern Reduction Methods. Edited by Pher G. Andersson and Ian J. Munslow
Copyright © 2008 WILEY-VCH Verlag GmbH & Co. KGaA, Weinheim
ISBN: 978-3-527-31862-9

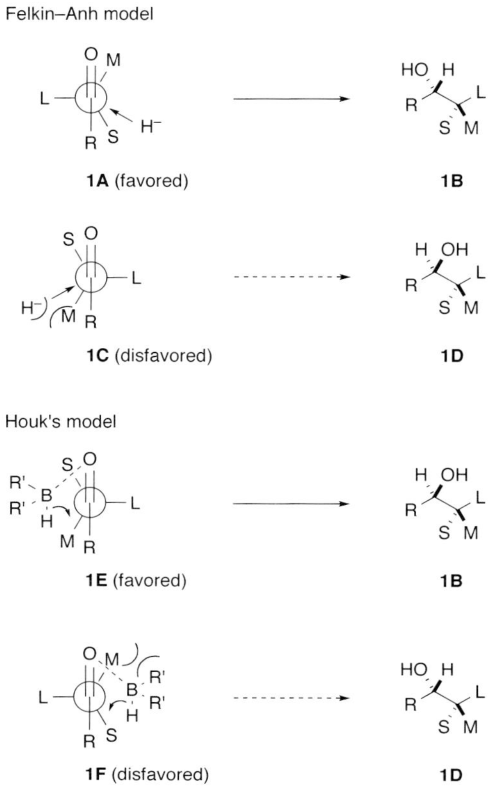

Felkin–Anh model

1A (favored) → **1B**

1C (disfavored) ⇢ **1D**

Houk's model

1E (favored) → **1B**

1F (disfavored) ⇢ **1D**

Scheme 7.1 Typical models to rationalize the stereoselectivity.

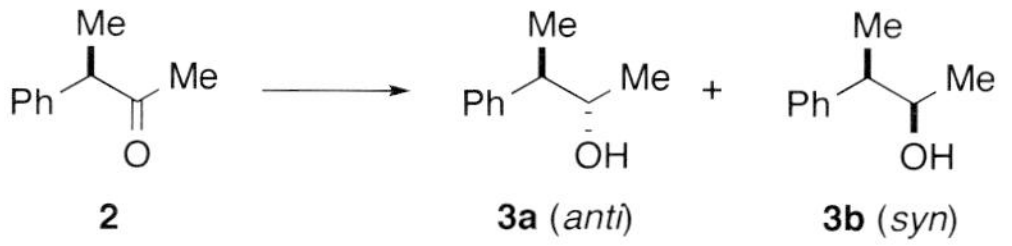

2 → **3a** (anti) + **3b** (syn)

Product distribution

Li(s-Bu)$_3$BH / THF	96 : 4
Sia$_2$BH[a] / THF	20 : 80

[a]Sia = (CH$_3$)$_2$CHCH(CH$_3$)-

Scheme 7.2 1,2-Asymmetric induction of an α-chiral ketone.

is shown in Scheme 7.2. Thus, reduction of 3-phenyl-2-butanone (**2**) by lithium tri-sec-butylborohydride (L-Selectride), a typical nucleophilic borohydride reagent, selectively gives *anti*-alcohol (**3a**), while the Lewis acid disiamylborane preferably affords the *syn*-product (**3b**) [6]. As shown in Scheme 7.3, these models for stereoselection can be applied to reductions of complicated compounds, such as (**4**) [6].

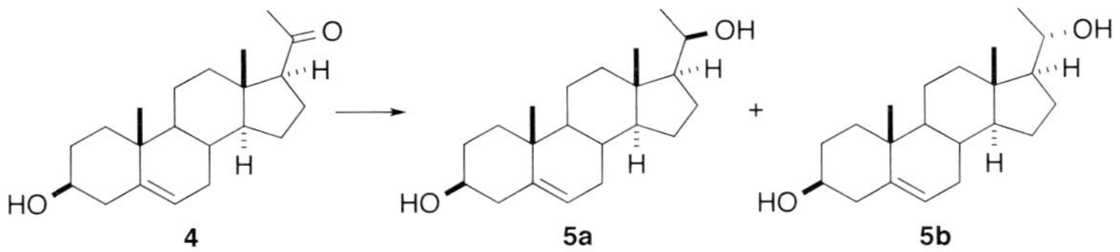

7.2 Recent Topics in Diastereoselective Reduction | 161

	Product distribution		
Li(s-Bu)$_3$BH / THF	99	:	1
Sia$_2$BH[a] / THF	10	:	90

[a]Sia = (CH$_3$)$_2$CHCH(CH$_3$)-

Scheme 7.3 1,2-Asymmetric induction of a steroidal ketone.

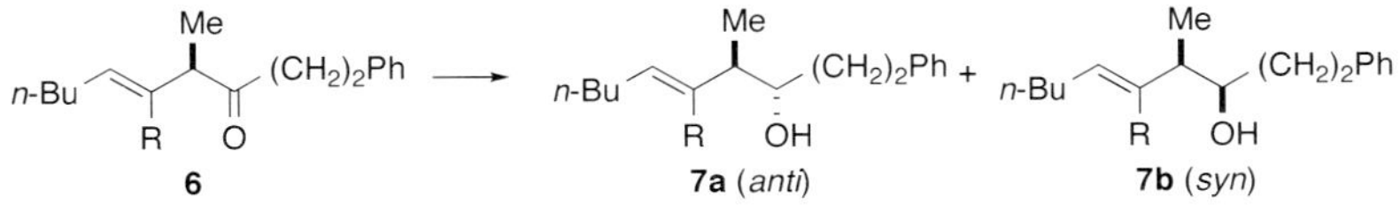

		Product distribution		
R = H	Li(s-Bu)$_3$BH / THF	93	:	7
	DIBAL / toluene	52	:	48
R = TMS[a]	Li(s-Bu)$_3$BH / THF	>99	:	1
	DIBAL / toluene	94	:	6

[a]TMS = Me$_3$Si

Scheme 7.4 Steric effect in 1,2-asymmetric induction.

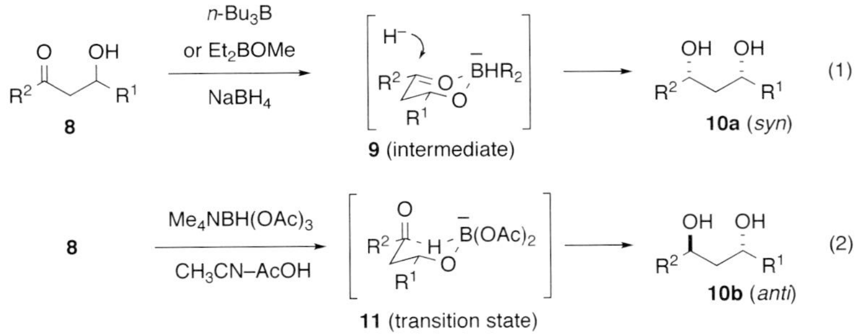

Scheme 7.5 1,3-Asymmetric induction of β-hydroxy ketones.

In general, the steric environment around the reaction site has a large effect on the stereoselectivity. For example, the degree of *anti*-selectivity in the reduction of α-methyl β,γ-unsaturated ketone **(6)** (R = H) is markedly increased by introduction of the bulky trimethylsilyl group at the β-position (R = TMS) shown in Scheme 7.4 [7].

1,3-Asymmetric induction is a known reliable method for constructing a chiral center, as is the 1,2-relationship. As shown in Scheme 7.5, either 1,3-*syn* or 1,3-*anti* products can be obtained by use of the appropriate system. Thus, the carbonyl reduction of a β-hydroxy ketone (**8**) with the combination of NaBH$_4$ and a chelating borane reagent gives the 1,3-*syn* diol (**10a**) in excellent diastereoselectivity, through intermolecular hydride attack of the six-membered boron-chelate intermediate (**9**) (Scheme 7.5, equation 1) [8]. Conversely, reaction with Me$_4$NBH(OAc)$_3$ in an acetonitrile–acetic acid mixture predominantly affords the 1,3-*anti* diol (**10**)b [Scheme 7.5, equation 2] [9]. The intramolecular hydride transfer in the borate chair-like transition state (**11**) explains this stereoselectivity. Formation of the six-membered boron-chelate intermediate is also effective in the 1,2-asymmetric induction in the reduction of α-substituted β-hydroxy ketones as shown in Scheme 7.6 [10]. NaBH$_4$ reduction of the TBS-protected alkoxy ketone (**12**) (R = TBS) exhibits moderate *syn* selectivity, while BH$_3$ reduction of the nonprotected compound (**12**) (R = H) shows very high diastereoselectivity through the chelation-control mechanism.

The stereoselective reduction of cyclic ketones is an important process for studies of stereochemistry as well as synthetic organic chemistry. Many excellent reviews have been published, and the rationale of the selectivity has been discussed in detail [11, 12]. A typical example is shown in Table 7.1. Generally, stereoelectronic control favors axial attack on the rigid cyclohexanone due to orbital interac-

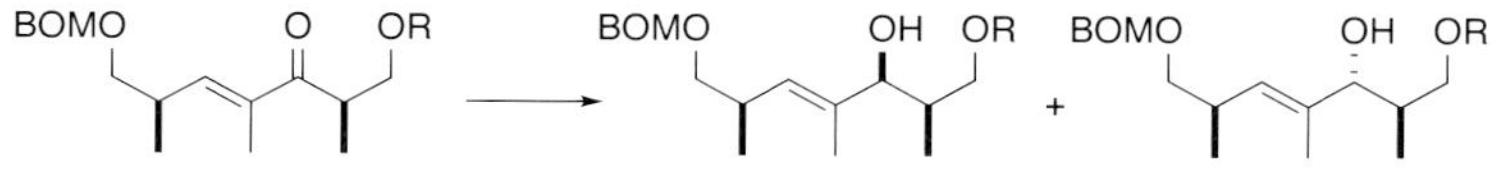

BOM = PhCH$_2$OCH$_2$ NaBH$_4$ / THF   R = TBS   83%   1.5:1
                       BH$_3$ / THF   R = H   94%   >99:1

Scheme 7.6 Asymmetric induction through nonchelation- and chelation-control.

Table 7.1 Stereoselectivity in the reduction of a cyclic ketone.

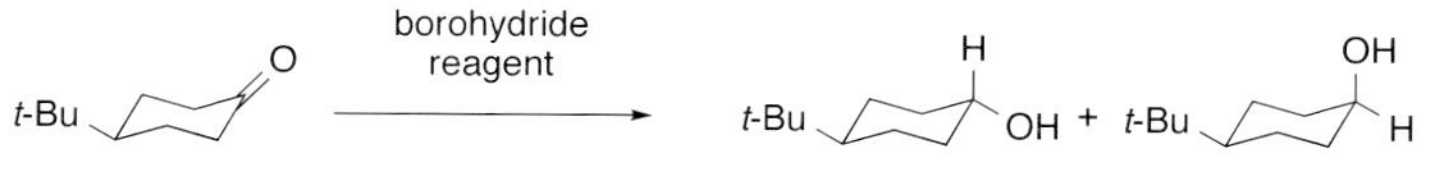

Reagent	Solvent	Temperature (°C)	Trans:cis	Reference
NaBH$_4$	CH$_3$OH	0	86:14	[13]
Li(n-C$_4$H$_9$)BH$_3$	THF-hexane	−78	98:2	[14]
Li(n-C$_3$H$_7$)$_2$NBH$_3$	THF	0	99:1	[15]
Cp$_2$TiBH$_4$[a]	THF	25	97:3	[16]
(i-C$_3$H$_7$O)$_2$TiBH$_4$	CH$_2$Cl$_2$	−78 to −20	97:3	[17]
LiSia$_3$BH[b]	THF	−78	0.6:99.4	[18]

a Cp = η^5-cyclopentadienyl.
b Sia = (CH$_3$)$_2$CHCH(CH$_3$)–.

tions between the newly forming C—H σ and C—H σ* adjacent to the carbonyl group, whereas steric interactions work against this pathway, leading to equatorial attack [12]. Thus, a relatively small hydride donor (BH_4^-, BRH_3^-) gives *trans*-alcohols (axial attack product), while a bulky reagent yields *cis*-alcohols (equatorial attack product) [13–18].

7.3 Enantioselective Reduction

7.3.1 Reagents (Introduction)

Stoichiometric reduction with a chiral alkylborane reagent, developed mainly by Midland and Brown, is a reliable method for the preparation of optically active secondary alcohols. Representative reagents are shown in Figure 7.1.

Alpine-borane [(**16**); *B*-(3-pinanyl)-9-borabicyclo[3.3.1]nonane], reported by Midland and coworkers, is readily prepared by the hydroboration of commercially available (+)-α-pinene by 9-BBN (9-borabicyclo[3.3.1]nonane) [19]. This reagent can reduce various prochiral ketones in moderate to high enantioselectivity. As depicted in Scheme 7.7 the stereochemical outcome of the reaction is determined when hydrogen is transferred via a six-membered transition state [19b, 20]. Due to the rigid structure of pinene ring, the C2-methyl effectively prevents the large group (R^L) of the ketone from approaching. Although alpine-borane (**16**) is a very effective

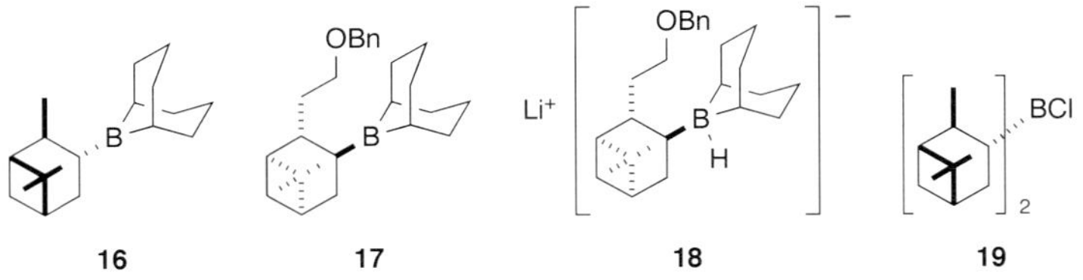

Figure 7.1 Structures of selected chiral alkylborane reagents.

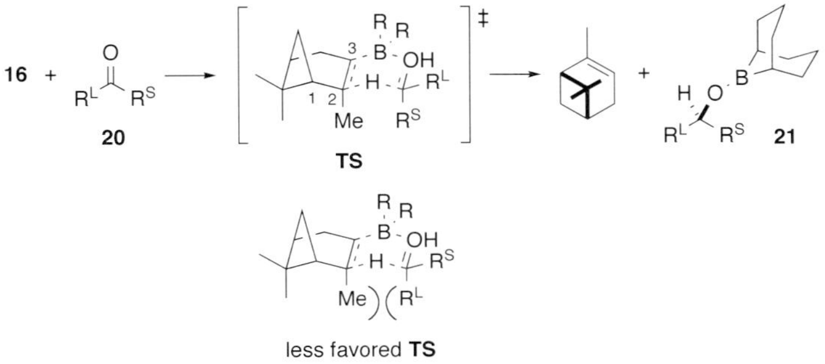

Scheme 7.7 Proposed "boat-like" transition state in the reduction with (**16**).

asymmetric reducing agent for 1-deuterio aldehydes and less-hindered α,β-acetylenic ketones, prochiral ketones of moderate steric bulk are only sluggishly reduced [19a, 21]. Under the relatively dilute reaction conditions of the original procedure, unfavorable achiral reductions caused by the reversible dissociation of the reagent into α-pinene and 9-BBN were noticed. Later this drawback was dramatically improved by conducting the reaction under high-pressure or high concentration conditions [20, 21]. Midland *et al.* also developed NB-Enantrane **(17)**, which is derived from the low-cost nopol {6,6-dimethylbicyclo[3.1.1]hept-2-ene-2-ethano1} structurally related to (−)-α-pinene [22]. Midland *et al.* also detailed the borohydride derivative of NB-Enantrane (NB-Enantride, **(18)** [23].

(−)-Ipc₂BCl (**(19)**, *B*-chlorodiisopinocamphenylborane) reported by Brown *et al.* is readily prepared from commercially available (+)-α-pinene (92% ee) in high chemical and optical purities (99% ee) by hydroboration followed by treatment with dry hydrogen chloride in diethyl ether [24]. Reagent **(19)** shows higher reactivity than alpine-borane **(16)** due to increased Lewis acidity on the chloro-substituted boron center [24b]. A chiral reducing agent derived from D-glucose (**(22)**, K-glucoride) affords alcohols in high enantiomeric excess in the reduction of acetophenone derivatives and β-keto esters (Figure 7.2) [25]. These reagents exhibit unique stereoselectivity features in some cases, discussed in the following sections.

Since the pioneering work by Itsuno and Corey in oxazaborolidine chemistry, a number of related chiral catalysts for enantioselective borane reductions have been developed [26, 27]. Representative stoichiometric or catalytic reaction systems and some recent examples are shown in Figures 7.3 and 7.4. In 1981, Itsuno *et al.* reported the asymmetric reduction of aromatic ketones using stoichiometric amounts of amino alcohol–borane reagents (e.g. **(23)**-BH₃, **(24)**-BH₃), which are simply prepared by mixing the two compounds in THF [26]. The reduction with **(23)**-BH₃ gave the corresponding secondary alcohols in only 7–46% ee, while with

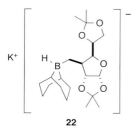

Figure 7.2 Structure of K-glucoride.

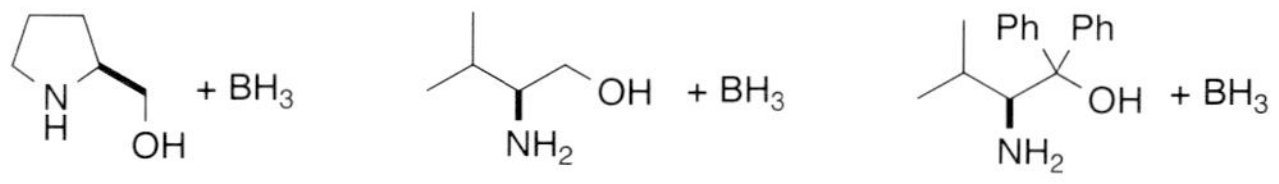

Figure 7.3 Amino alcohol–borane reagents for enantioselective reduction.

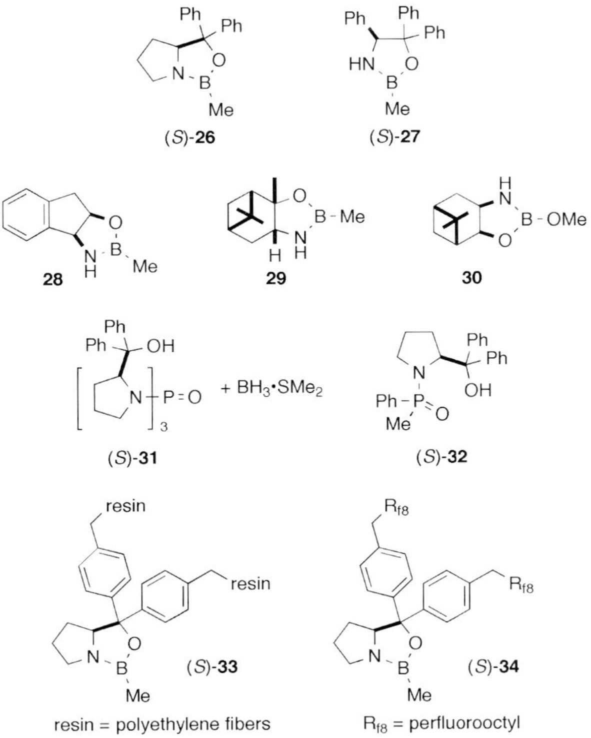

Figure 7.4 Selected oxazaborolidine catalysts for enantioselective borane reduction.

reagent **(24)**-BH$_3$ higher ee values were obtained, up to 60%. This enantioselectivity was subsequently improved to 94–99% by the use of the modified reagent **(25)**-BH$_3$, also reported by Itsuno et al. [28, 29]. Corey's group investigated the mechanism of this type of reaction, which led to a powerful catalytic version of the original stoichiometric reaction (so-called CBS reduction) [3a, 27]. This reduction system is applicable to a wide range of ketonic substrates to afford the chiral alcohols in excellent enantiomeric excess as described in the following sections. The enantioselectivity in this reaction using oxazaborolidine, (S)-**(26)**, can be explained by the mechanism depicted in Scheme 7.8 [27, 30, 31]. The initial step is the coordination of BH$_3$ to the Lewis basic nitrogen atom of oxazaborolidine **(26)**, forming the active complex **(35)**. This complex has both a Lewis acidic site (boron in the borolidine ring) and a nucleophilic R$_3$NBH$_3$ moiety. Thus, the complex binds to a ketone at the less hindered carbonyl lone-pair, then face-selective hydrogen transfer via a six-membered transition state ((**TS-1**) in Scheme 7.8) gives the R-configured product. Other oxazaborolidine-catalyzed reductions are believed to proceed by way of a similar mechanism. Chirally modified titanium reagents, **(37)** and **(38)**, can also reduce aromatic ketones in high enantioselectivity (Figure 7.5) [32, 44].

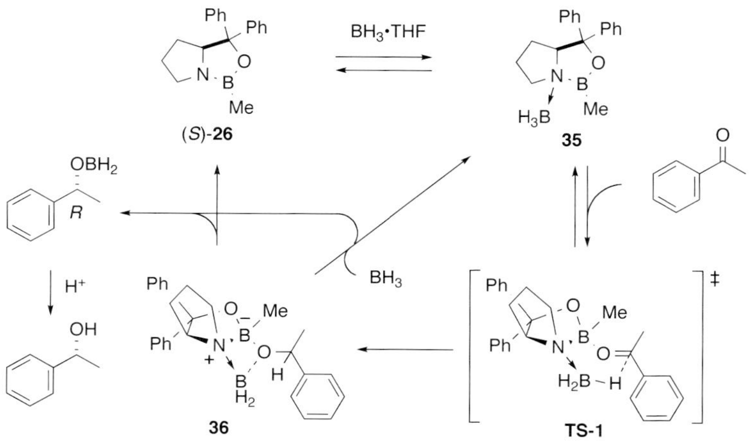

Scheme 7.8 Proposed mechanism for the catalytic enantioselective reduction with (S)-(26).

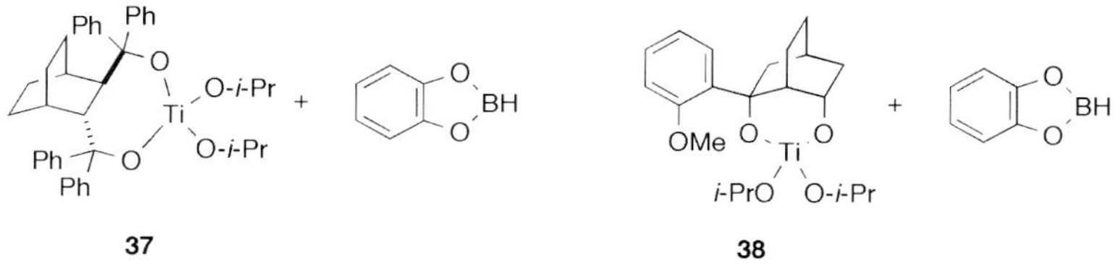

Figure 7.5 Titanium–catecholborane catalytic systems.

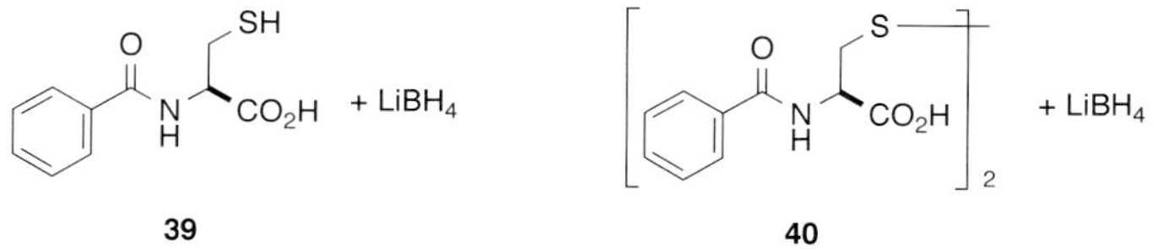

Figure 7.6 L-Cystine-derived reagents.

Studies on enantioselective hydride reductions by the use of metal borohydride reagents have also been carried out extensively as well as the borane reduction. LiBH$_4$ combined with benzoylcystine derivatives, **(39)** and **(40)**, are effective stoichiometric reagents in the reduction of aromatic ketones (Figure 7.6) [33, 34]. Recently, Yamada and Mukaiyama reported a series of NaBH$_4$ reductions catalyzed by the chiral cobalt-ketoiminato complex **(41)** (Figure 7.7), high levels of enantioselectivity were accomplished with a wide range of ketonic substrates [35, 36].

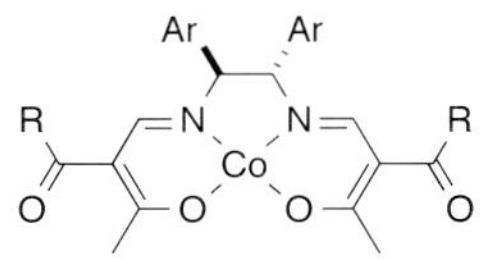

7.3 Enantioselective Reduction

Figure 7.7 Cobalt–ketoiminato complexes.

Ar = R = 2,4,6-Me$_3$C$_6$H$_2$ (S,S)-**41a**
Ar = 2,4,6-Me$_3$C$_6$H$_2$, R = Me   (S,S)-**41b**

Table 7.2 Asymmetric hydroboration of aromatic ketones.

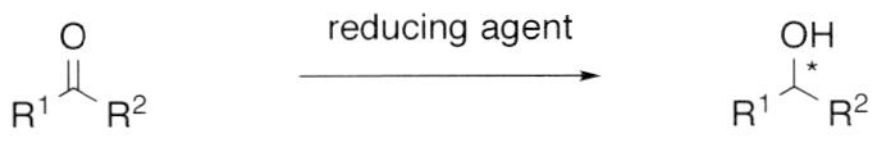

R^1	R^2	Reducing agent	% Yield	% ee	Configuration	Reference
Ph	Me	(S)-(**26**)/BH$_3$·THF	>99.7	97	R	[27]
Ph	Me	(S)-(**26**)/BH$_3$·DEAN^a	b	99	R	[37]
Ph	Me	(S)-(**25**)/BH$_3$·THF	90	94	R	[29b]
Ph	Me	(**30**)/B(OMe)$_3$, BH$_3$·Me$_2$S	98	98	S	[38]
Ph	Me	(S)-(**31**)/BH$_3$·Me$_2$S	94	95	R	[39]
Ph	Me	(S)-(**32**)/BH$_3$·Me$_2$S	84	55	R	[40]
Ph	Me	(S)-(**33**)/BH$_3$·Me$_2$S	90	95	R	[41]
Ph	Me	(S)-(**34**)/BH$_3$·THF	99	97	S	[42]
Ph	Me	(**16**)	68	87	S	[43c]
Ph	Me	(**18**)	70–80	70	S	[23]
Ph	Me	(**19**)	72	98	S	[24c]
Ph	Me	(**22**)	95	78	R	[25a]
Ph	Me	(**39**)/LiBH$_4$	66	87	R	[34]
Ph	Me	(**40**)/LiBH$_4$	68	87	S	[33]
Ph	Me	(**37**)/catecholborane	90	84	R	[32]
Ph	Me	(**38**)/catecholborane	>97	97	R	[44]
Ph	n-Pr	(S,S)-(**41a**)/NaBH$_4$, THFA^c	100	97	S	[36a]
Ph	n-Pr	(S)-(**25**)/BH$_3$·THF	b	96	R	[29b]
2-Naphthyl	Me	(S)-(**26**)/BH$_3$·THF	b	96	R	[45]
2-Thienyl	Me	(S)-(**26**)/BH$_3$·Me$_2$S	91	94	S	[46]

a   DEAN = N,N-diethylaniline.
b   Not shown in the literature.
c   THFA = tetrahydrofurfuryl alcohol.

7.3.2 Simple Ketones

7.3.2.1 Aromatic Ketones

Acetophenone and its derivatives have been the most popular substrates for studies on enantioselective reductions since the first successful example using (S)-diphenylvalinol-BH$_3$ [26, 28, 29]. There are extensive reports concerning this type of reduction and some representative examples and recently reported results are described below. In most cases, optically active secondary alcohols are obtained in high enantiomeric excess and in nearly quantitative yield as shown in Table 7.2.

Since the first successful results, various types of modified reagents have been devised [3, 47]. A small number of representative examples are shown in Table 7.2. The unique C_3-symmetric chiral phosphoric triamide ligand (**31**) showed a good level of selectivity [39], whereas the monoamide ligand (**32**) afforded only 55% optical yield [40]. The polymer-supported and polyfluorinated derivatives of oxazaborolidine, (**33**) and (**34**), devised to decrease the work of product separation and purification, also gave good enantioselectivities [41, 42]. Catalyst (**34**) proved to be sufficiently robust to stand being recycled three times without any deleterious effect on enantioselectivity [42]. The metal complex-catalyzed hydroboration was also effective in enantioselective reduction of ketones to alcohols. Examples using titanium catalysts, (**37**) and (**38**), are also shown in the table [32, 44]. The cobalt-ketoiminato complex (**41a**) reduces ketones with $NaBH_4$ to afford the corresponding secondary alcohols in excellent enantiomeric excess [36].

7.3.2.2 Aliphatic Ketones

Enantioselective reduction of aliphatic ketones is more difficult than that of aromatic ketones, since the steric and electronic character of substituents on the carbonyl group (R^1, R^2) are similar to each other (Table 7.3). Selected results are summarized in Table 7.3. Corey's oxazaborolidine-BH_3 catalytic system gives alcohols in high enantioselectivity in the case of *t*-butyl methyl or cyclohexyl methyl ketone [37]. However, this reagent does not work well in the reduction of a *n*-alkyl

Table 7.3 Asymmetric hydroboration of aliphatic ketones.

R^1	R^2	Reducing agent	% Yield	% ee	Configuration	References
Me	*t*-Bu	(S)-(**26**)/$BH_3 \cdot DEAN^a$	b	97.4	R	[37]
Me	*t*-Bu	(S)-(**25**)/$BH_3 \cdot THF$	b	73	R	[29b]
Me	*t*-Bu	(**16**)	b	0.6	S	[43c]
Me	*t*-Bu	(R,R)-(**42**)	72	99.3	R	[48]
Me	*t*-Bu	(**43**)	80	94	R	[49]
Me	*t*-Bu	(**19**)	50	95	S	[24a]
Me	*c*-Hex	(S)-(**26**)/$BH_3 \cdot DMA$	b	>99	R	[37]
Me	*c*-Hex	(R,R)-(**42**)	83	99.5	R	[48]
Me	*c*-Hex	(**43**)	82	82	R	[49]
Me	*n*-Hex	(**19**)	b	7	R	[23]
Me	*n*-Hex	(R,R)-(**42**)	68	81.3	R	[48]
Me	*n*-Hex	(**43**)	83	60	R	[49]
Me	*n*-Hex	(**18**)	76	79	S	[23]
n-Bu	i-Bu	(R,R)-(**42**)	72	96.8	R	[48]

a DEAN = N,N-diethylaniline.
b Not shown in the literature.

methyl ketone, such as 2-octanone [3a]. It should be noted that the reagent 2,5-dimethylborolidine, **(42)** (Figure 7.8), developed by Masamune *et al.* shows high stereoselectivity in the reduction of 2-octanone [48]. This is recognized as one of the most general enantioselective reagents for reduction of primary alkyl ketones. The recently reported borate reagent **(43)** (TarB-NO$_2$), prepared from tartaric acid, gives moderate selectivity in the reduction of 2-octanone [49].

7.3.3
α,β-Unsaturated Ketones

Enantiomerically pure allylic alcohols are versatile intermediates participating in many useful synthetic transformations, hence asymmetric reductions of α,β-unsaturated ketones are of great interest. In the enantioselective reduction of these conjugated enones, boranes are usually employed as reductants rather than metal tetrahydroborates, which often give rise to 1,4-reductions in the absence of a Lewis acid, such as CeCl$_3$ [50]. Selected examples are listed in Table 7.4 and Figure 7.9.

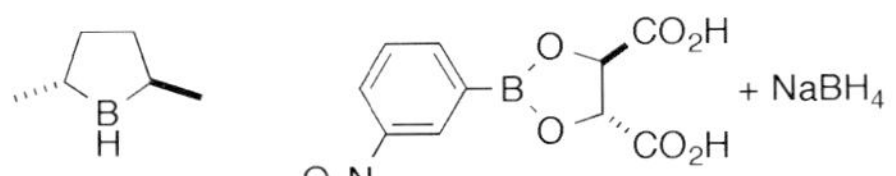

Figure 7.8 Structures of reducing agents for aliphatic ketones.

Table 7.4 Asymmetric hydroboration of α,β-unsaturated ketones.

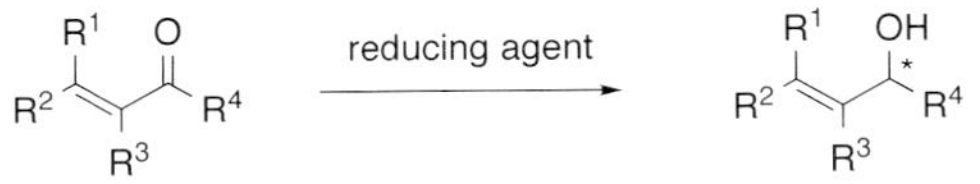

R^1	R^2	R^3	R^4	Reducing agent	% Yield	% ee	Configuration	References
H	Ph	H	Me	(S)-**(45)**/CB^a	90	97	R	[51]
H	4-MeOC$_6$H$_4$	H	Me	(S)-**(45)**/CB^a	88	95	R	[51]
H	4-O$_2$NC$_6$H$_4$	H	Me	(S)-**(45)**/CB^a	88	72	R	[51]
H	Me	Me	CH$_2$SiMe$_2$Ph	(R)-**(26)**/BH$_3$·THF	84	95	S	[52]
Me	Me	H	CH$_2$S(4-MeC$_6$H$_4$)	(S)-**(26)**/BH$_3$·EIPAN^b	93	90	S	[53]
H	Ph	H	CH$_2$S(4-MeC$_6$H$_4$)	(S)-**(26)**/BH$_3$·EIPAN^b	97	98	S	[53]
H	-(CH$_2$)$_4$-		CH$_2$S(4-MeC$_6$H$_4$)	(S)-**(26)**/BH$_3$·EIPAN^b	95	95	S	[53]
H	2-Me-4-thiazolyl	Me	(CH$_2$)$_2$OTBS	(R)-**(26)**/BH$_3$·Me$_2$S	98	95	S	[54]
H	Ph	SPh	Me	(S)-**(44)**/BH$_3$·Me$_2$S	93	95	R	[55]
H	Ph	H	Me	**(16)**	>97	71	S	[20]
H	Ph	H	Me	**(19)**	65	81	S	[24b]

a CB = catecholborane.
b EIPAN = *N*-ethyl-*N*-isopropylaniline.

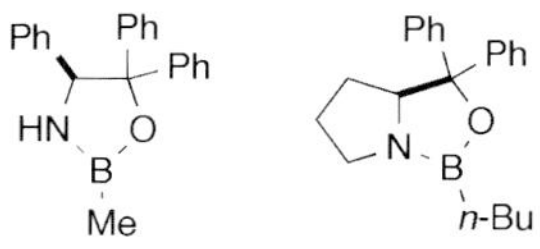

Figure 7.9 Structures of reagents for Table 7.4.

Table 7.5 Asymmetric hydroboration of acetylenic ketones.

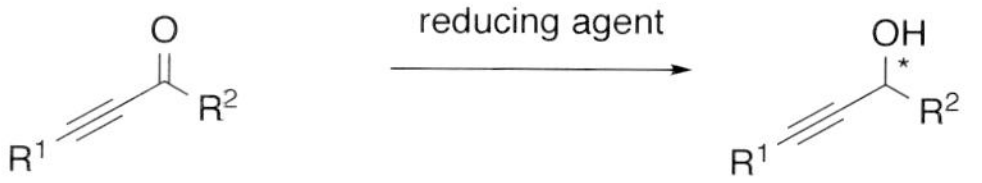

R^1	R^2	Reducing agent	% Yield	% ee	Configuration	References
n-Bu	Ph	(16)	72	89	R	[21]
Ph	Me	(16)	98	78	R	[21]
n-Hex	n-Pr	(16)	68	77	R	[21]
H	n-Pent	(16)	65	92	R	[21]
H	$(CH_2)_7OBn$	(16)	60	97	S	[56]
H	$(CH_2)_7OBn$	(19)	76	21	R	[24b]
H	Me	(19)	83	17	R	[57]
Ph	Me	(19)	92	21	R	[57]
Ph	i-Pr	(19)	85	53	R	[57]
Ph	t-Bu	(19)	80	>99	R	[57]
n-Oct	t-Bu	(19)	76	>99	R	[57]
H	$(CH_2)_7OBn$	(R)-(26)/$BH_3 \cdot Me_2S$	56	84	S	[56]
Ph	Me	(S)-(26)/$BH_3 \cdot Me_2S$	80	71	S	[58]
Ph	$CH(CH_3)_2$	(S)-(26)/$BH_3 \cdot Me_2S$	85	94	S	[58]
H	n-Hept	(S)-(26)/$BH_3 \cdot Me_2S$	54	95	S	[58]
$CH_3(CH_2)_5$	Ph	(S)-(26)/$BH_3 \cdot Me_2S$	93	99	R	[59]
$(CH_3)_2CH$	(E)-CH=CHCH$(CH_3)_2$	(S)-(26)/$BH_3 \cdot Me_2S$	85	86	S	[60]
H	2-Bn-2-dithianyl	(46)	58	98	S	[61]
i-Pr_3Si	n-Pent	(S)-(47)/CB^a	98	97	R	[62]

a CB = catecholborane.

CBS-type oxazaborolidine catalytic systems ((44)-BH$_3$ or (45)-BH$_3$) give superior results to other organoborane reagents, generally affording the corresponding allylic alcohol in greater than 90% ee [51–55]. In addition to simple α,β-unsaturated ketones, recent advance include the highly enantioselective reduction of α-heteroatom-substituted (O, S, Si) enones. It is interesting to note that in the case of acyclic enones the olefinic moiety generally behaves as the large group. The stereoselective reduction of enones has been applied in the syntheses of natural products (see, Section 7.4). The reduction of α,β-unsaturated ketones with alpine-borane (16) or (−)-Ipc$_2$BCl (19) results in relatively low ee values [20, 24b].

Enantioselective reductions of α,β-ynones are exemplified in Table 7.5 and Figure 7.10. Alpine-borane (16) and (−)-Ipc$_2$BCl (19) show moderate to good ste-

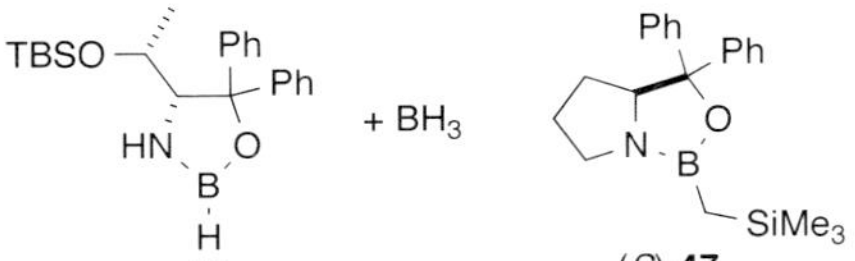

Figure 7.10 Structures of reagents for Table 7.5.

reoselectivity [21, 24b, 56, 57], while in the case of the terminal ynones **(19)** tends to give propargylic alcohols with inferior ee values compared to **(16)** [56]. Alkynyl *t*-alkyl ketones are reduced with exceptionally high enantioselectivity by this reagent. The oxazaborolidine-catalyzed reduction of α,β-ynones to propargylic alcohols with high optical purity has recently been accomplished by two different protocols. The use of *B*-Me catalyst (*S*)-**(26)** with $BH_3 \cdot Me_2S$ in THF leads to the (*S*)-alcohols, recognizing ethynyl as the small group [58]. In contrast, *B*-CH_2SiMe_3 catalyst (*S*)-**(47)** with catecholborane in dichloromethane recognizes the alkynyl moiety as the large group in the reduction of triisopropylsilyl- or trimethylsilylacetylenic ketones, furnishing (*R*)-alcohols [62]. The long-range interaction between *B*-CH_2SiMe_3 and the silyl group of the ynone causes reverse-sense enantioselection. Modified oxazaborolidine reductant **(46)** exhibits good selectivity in the reduction of dithianyl ketones [61].

7.3.4
α-Hetero Substituted Ketones

CBS-type catalysts perform well in the enantioselective reduction of α-halo ketones but newly reported reductants also show good selectivity (Table 7.6 and Figure 7.11). BH_3 with a proline-derived diamine **(48)** reduces phenacyl bromide in 91% optical yield [64]. The phosphoramide-type reducing agents, **(49)**–**(51)**, also produce the halohydrin with high enantioselectivity [40, 65, 66]. α-Thio ketone is reduced in moderate selectivity by the CBS system [67].

7.3.5
Keto Esters

Generally, the keto-carbonyl group in a keto ester can be selectively reduced to an alcohol, leaving the ester moiety intact by using borohydride reagents, though a few examples of borane reductions have been reported. Enantioselective hydroboration of α- or β-keto esters is rare. The reduction is usually performed via asymmetric catalytic hydrogenation or biochemical transformations [63, 68]. Keto esters behave like simple ketones when the keto-carbonyl is separated from the ester group by more than two methylene groups. The resulting hydroxy esters are obtained in high enantiomeric excess and moderate to good yield. Selected results are summarized in Table 7.7 and Figure 7.12. An alkynyl β-keto ester was converted to its corresponding hydroxy ester in 91% ee, though in poor yield [69]. A

Table 7.6 Asymmetric hydroboration of α-hetero substituted ketones.

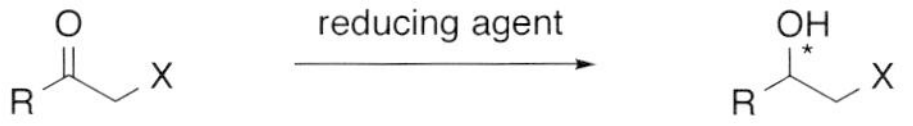

R	X	Reducing agent	% Yield	% ee	Configuration	References
Ph	Cl	(S)-(26)/BH$_3$·DEAN[a]	b	98	S	[3a]
Ph	Cl	(S)-(49)/BH$_3$·Me$_2$S	85	87	S	[58]
Ph	Cl	(S)-(50)/BH$_3$·Me$_2$S	91	94	R	[33]
Ph	Cl	(S)-(51)/BH$_3$·Me$_2$S	91	94	R	[63]
Ph	Br	(S)-(48)/BH$_3$·Me$_2$S	82	91	S	[60]
Ph	Br	(S)-(49)/BH$_3$·Me$_2$S	86	89	S	[58]
2,4-F$_2$C$_6$H$_3$	Cl	(S)-(26)/BH$_3$·DEAN[a]	99	99	S	[61]
n-C$_{10}$H$_{21}$	S(4-MeC$_6$H$_4$)	(S)-(26)/BH$_3$·EIPAN[c]	99	79	S	[62]

a DEAN = N,N-diethylaniline.
b Not shown in the literature.
c EIPAN = N-ethyl-N-isopropylaniline.

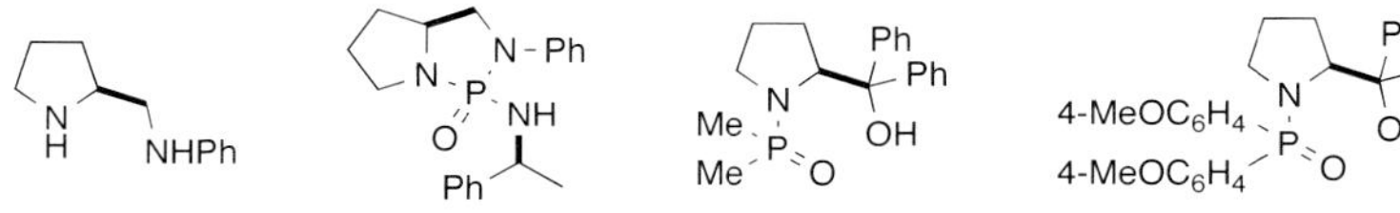

Figure 7.11 Chiral modifiers for the asymmetric hydroboration of α-hetero substituted ketones.

Table 7.7 Asymmetric hydroboration of keto esters.

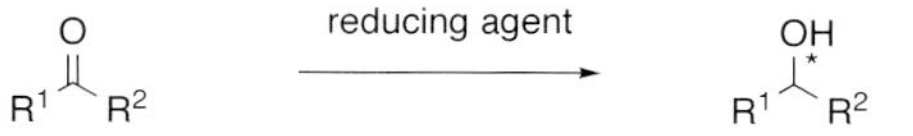

R^1	R^2	Reducing agent	% Yield	% ee	Configuration	References
TBS≡—[a]	CH$_2$CO$_2$-t-Bu	(16)	21	91	R	[69]
CH$_3$(CH$_2$)$_7$	R^{3b}	(S)-(26)/BH$_3$·THF	89	87	R	[70]
Me	(CH$_2$)$_{11}$CO$_2$Me	(52)	57	94	S	[71]
CH$_2$Cl	CH$_2$P(O)(OEt)$_2$	(R,R)-(53)	82	80	S	[72]
Ph	CHMeCO$_2$Et	(S,S)-(41b)	97[c]	99	R	[73]

a TBS = t-BuMe$_2$Si.
b R^3 = 
c 99% anti-selective.

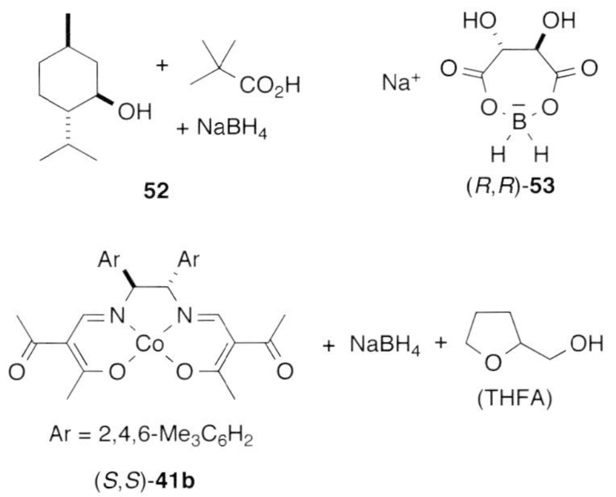

Figure 7.12 Structures of reducing agents for Table 7.7.

keto ester with a long aliphatic chain was selectively reduced with a menthol–pivalic acid–NaBH$_4$ system **(52)** [71]. Reagent **(53)**, a borohydride modified by tartaric acid, reduces ketophosphonates in moderate selectivity [72]. In the reduction of α-substituted β-keto esters the cobalt-ketoiminato complex **(41b)**–NaBH$_4$ system is highly effective, affording the corresponding hydroxy ester with excellent de and ee values via dynamic kinetic resolution [36a, 73].

7.3.6
Diketones

Diketones are reduced to diols with borane reagents or borohydride reagents. Representative results are shown in Table 7.8 and Figure 7.13. CBS-type catalytic systems give the corresponding diols in excellent enantioselectivity. The typical substrate 1,4-diphenyl-1,4-butanedione is converted to the chiral *dl*-diol with oxazaborolidine **(54)** or cobalt complex **(41)** in excellent enantioselectivity [76–78]. The minor *meso*-diols are formed by intramolecular stereoselection. It is worth noting that only the aromatic carbonyl group was reduced by the chiral cobalt complex **(41b)** system in the case of 1-aryl-1,3-alkanedione **(59)**, affording the β-hydroxy ketone **(60)** in high diastereomeric and enantiomeric excess (Scheme 7.9) [83].

7.4
Synthetic Applications

7.4.1
Reduction of Chiral Ketones with Chiral Reducing Agents

When using chiral reducing agents in reductions of chiral ketones, the diastereoselectivity is determined by the combined efficiency of the reagent's stereoregulation ability and the substrate's stereoinductive effect [84]. Usually, the stereochemical

Table 7.8 Asymmetric hydroboration of diketones.

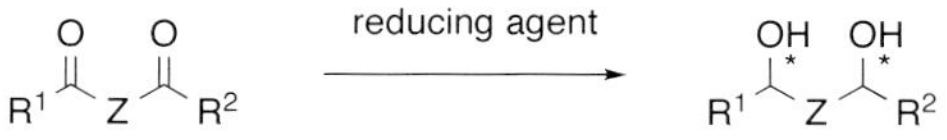

R^1	R^2	Z	Reducing agent	% Yield	dl : meso	% ee	Configuration	References
Ph	Ph	None	(R)-(57)/BH$_3$·Me$_2$S	85	88:12	>99	S,S	[74]
—(CH$_2$)$_2$—		CMeCH$_2$Ph	(S)-(58)/BH$_3$·THF	83	>99:1	>99	R,R	[75]
Ph	Ph	(CH$_2$)$_2$	(S)-(54)/B(OMe)$_3$/NaBH$_4$/TMSCl[a]	70	93:7	93	R,R	[76]
Ph	Ph	(CH$_2$)$_2$	(S)-(54)/B(OMe)$_3$/BH$_3$·THF	85	88:12	97	R,R	[76]
Ph	Ph	(CH$_2$)$_2$	(R)-(54)/B(OMe)$_3$/BH$_3$·Me$_2$S	96	>95:5	99	S,S	[77]
Ph	Ph	(CH$_2$)$_2$	(R,R)-(41b)	100	87:13	99	R,R	[78]
Ph	Ph	(CH$_2$)$_2$	(1S, 2R)-(55)/BH$_3$·THF	[b]	84:16	99	S,S	[79]
Me	Me	(CH$_2$)$_2$	(R)-(56)/BH$_3$·THF	92	68:32	92	S,S	[80]
Me	Me	—Ar—[c]	(R)-(26)/BH$_3$·Me$_2$S	85	94:6	>99	S,S	[81]
Ph	Ph	CH$_2$	(S,S)-(41b)	100	84:16	98	S,S	[82]

a TMS = Me$_3$Si.
b Not shown in the literature.
c —Ar— = (2,2'-biphenyl with OMe, MeO substituents)

(S)-54, (1S,2R)-55, (R)-56

(R)-57, (S)-58

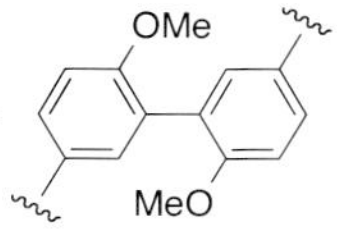

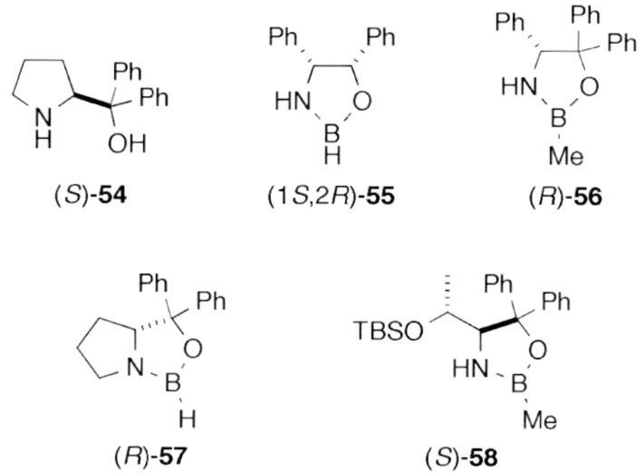

Figure 7.13 Structures of reagents for Table 7.8.

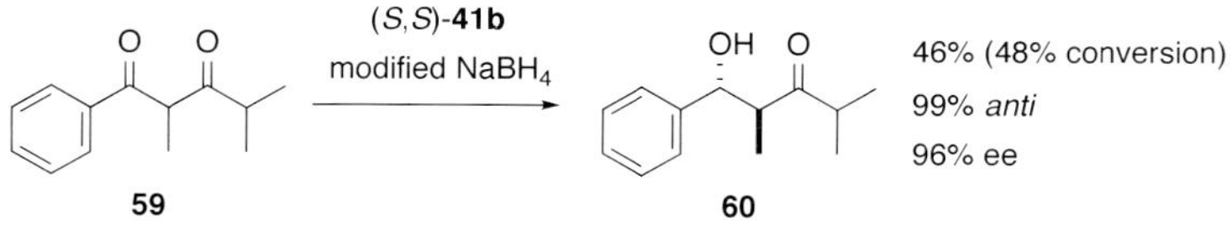

59 → **60**, (S,S)-41b, modified NaBH$_4$

46% (48% conversion)
99% anti
96% ee

Scheme 7.9 Chemoselective reduction of a diketone with cobalt-ketoiminato complex.

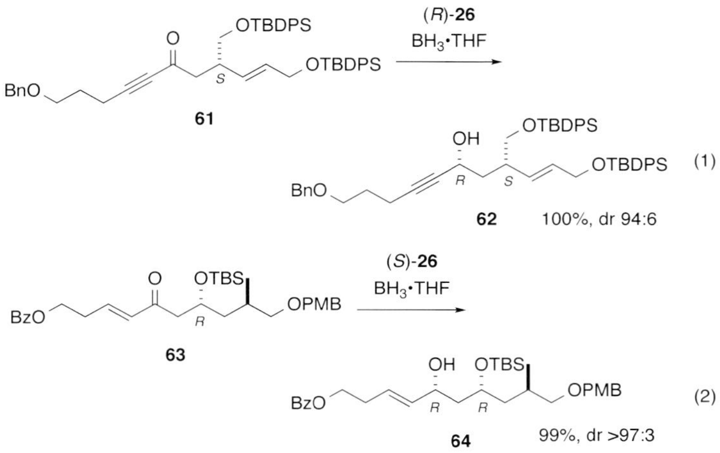

Scheme 7.10 Reduction of chiral ketones with chiral reducing agents.

Bn = PhCH$_2$, TBDPS = t-BuPh$_2$Si, Bz = PhCO, TBS = t-BuMe$_2$Si, PMB = 4-MeOC$_6$H$_4$.

outcome is markedly affected by the structure of ketonic substrates, although the CBS-catalytic system is relatively insensitive to variety in the structure. For example, borane reduction of the (S)-β-siloxymethyl ketone (61) catalyzed by the (R)-Me-CBS catalyst, (R)-(26), selectively gives the R,S alcohol (62) (Scheme 7.10, equation 1) [85]. The R stereochemistry is mainly controlled by the chirality of the catalyst, recognizing the alkynyl moiety as a small group according to the selectivity rule. In contrast, reaction of the (R)-β-siloxy ketone (63) with the (S)-(26) (not R) catalyst results in the R,R alcohol (64) predominantly (Scheme 7.10, equation 2) [86], suggesting that the alkenyl group behaves as the large substituent.

7.4.2
Application to Natural Product Synthesis

Recently, the stereoselective reduction of carbonyl compounds has become an indispensable tool in organic synthesis. Numerous examples are known and some selected recent examples are shown in this section. The CBS reduction of acetylenic ketone (65) was utilized in the total synthesis of 10-hydroxyasimicin to construct the stereocenter that is located in the middle of the alkyl chain tether (Scheme 7.11) [87]. The propargylic alcohol (66) is obtained in excellent diastereoselectivity in the reaction of this long chain-like molecule. The dienone compound (68), an intermediate in the total synthesis of (+)-spongistatin, is converted to alcohol (69) with more than 90% selectivity (Scheme 7.12) [88]. It is notable that the reduction system can be applied to the reduction of such highly functionalized molecules. As described in Section 7.3.3, olefinic moieties generally behave as the

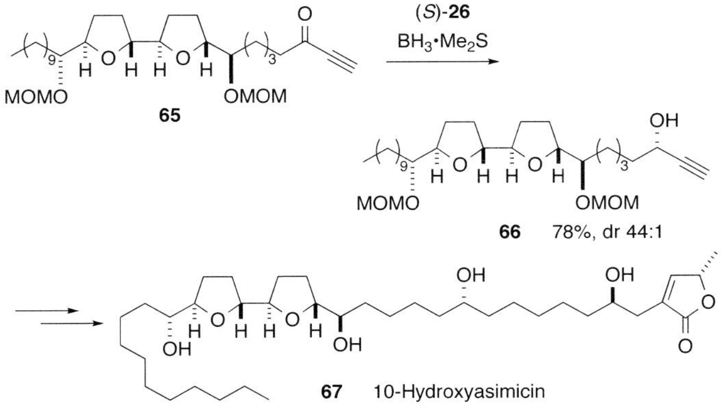

MOM = $MeOCH_2$

Scheme 7.11 Synthetic application to the total synthesis of 10-hydroxyasimicin.

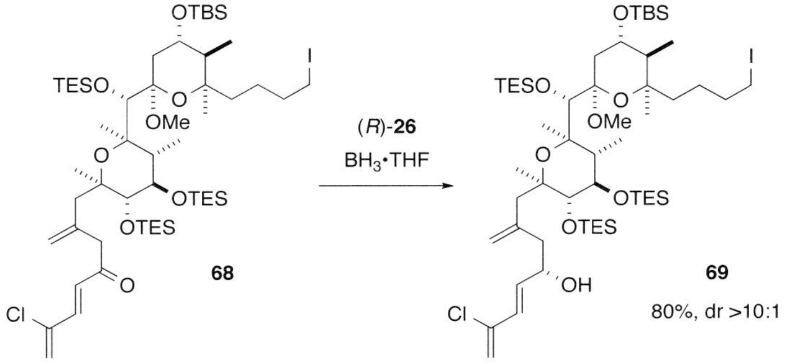

TES = Et_3Si, TBS = t-$BuMe_2Si$.

Scheme 7.12 Application to the total synthesis of (+)-spongistatin.

large group in CBS reductions. However, the reduction of enone **(70)** in the presence of (*S*)-Me-CBS catalyst, (*S*)-**(26)**, results in the unusual *S*-configured chiral center, suggesting that intramolecular asymmetric induction of the β-methoxyl group in **(70)** mainly controls the stereochemistry (Scheme 7.13) [89]. Diastereoselective reductions with achiral boron reagents are also useful procedures for the total synthesis of natural compounds through intramolecular asymmetric induction, utilizing chirality close to the reaction site, as in the reduction of β-hydroxy ketone **(73)** (Scheme 7.14) where 1,3-asymmetric induction is employed [90].

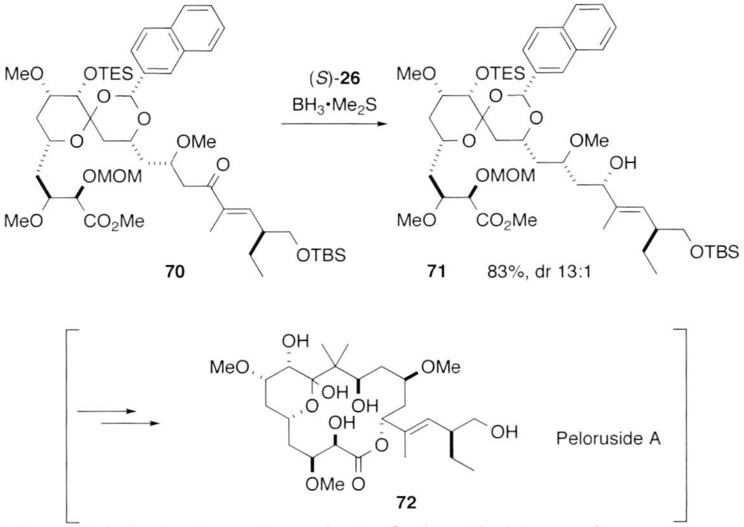

7.4 Synthetic Applications | 177

Scheme 7.13 Application to the synthesis of peloruside A intermediate.

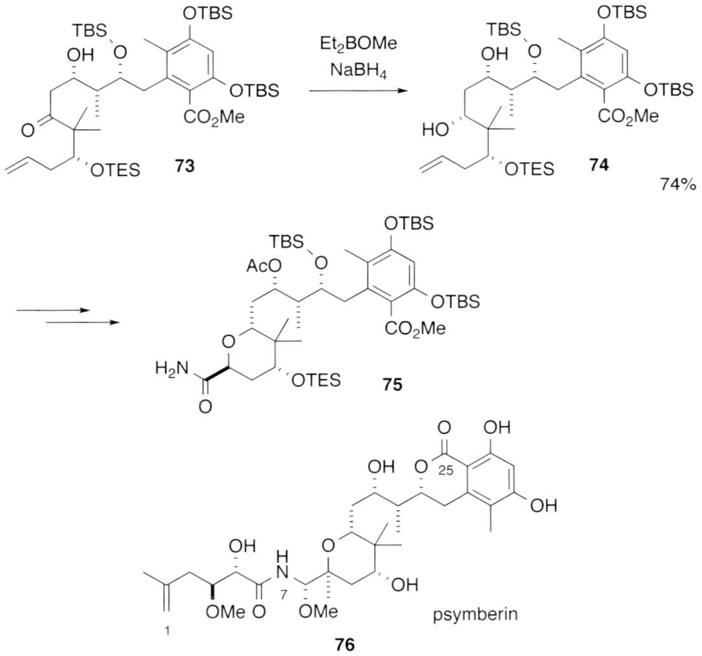

Scheme 7.14 Application of the diastereoselective reduction to synthesis of the N_7–C_{25} fragment of psymberin.

References

1. (a) Brown, H.C. and Krishnamurthy, S. (1979) *Tetrahedron*, **35**, 567. (b) Brown, H.C. and Ramachandran, P.V. (1996) *Sixty years of Hydride Reduction*, in *Reductions in Organic Synthesis* (ed. A.F. Abdel-Magid), American Chemical Society, Washington, DC.
2. Fiaud, J.C. and Kagan, H.B. (1969) *Bulletin de la Societe Chimique de France*, 2742.
3. (a) Corey, E.J. and Heral, C.J. (1998) *Angewandte Chemie (International Ed. in English)*, **37**, 1986. (b) Itsuno, S. (1998) *Organic Reactions*, **52**, 395. (c) Carboni, B. and Monnier, L. (1999) *Tetrahedron*, **55**, 1197. (d) Itsuno, S. (1999) *Hydroboration of Carbonyl Groups*, in *Comprehensive Asymmetric Catalysis*, Vol. 1 (eds E.N. Jacobsen, A. Pfaltz and H. Yamamoto), Springer, Berlin. (e) Seyden-Penne, J. (1991) *Reductions by the Alumino- and Borohydrides in Organic Synthesis*, Wiley-VCH Verlag GmbH, New York. (f) Greeves, N. (1991) *Reduction of C=O to CHOH by Metal Hydrides*, in *Comprehensive Organic Synthesis*, Vol. 8 (eds B.M. Trost and I. Fleming), Pergamon Press, Oxford. (g) Cho, B.T. (2002) *Aldrichimica Acta*, **35**, 3.
4. (a) Cram, D.J. and Greene, F.D. (1953) *Journal of the American Chemical Society*, **75**, 6005. (b) Chérest, M., Ferkin, H. and Prudent, N. (1968) *Tetrahedron Letters*, **9**, 2199. (c) Anh, N.T. and Eisenstein, O. (1976) *Tetrahedron Letters*, **17**, 155. (d) Mengel, A. and Reiser, O. (1999) *Chemical Reviews*, **99**, 1191. (e) Review: Davis, A.P. (1995) *Formation of C—H Bonds by the Reduction of Carbonyl Groups with Metal Hydrides*, in *Methods of Organic Chemistry*, Vol. E21d (eds G. Helmchen, R.W. Hoffmann, J. Mulzer and E. Schaumann), Thieme, Stuttgart.
5. Paddon-Row, M.N., Rondan, N.G. and Houk, K.N. (1982) *Journal of the American Chemical Society*, **104**, 7162.
6. Midland, M.M. and Kwon, Y.C. (1983) *Journal of the American Chemical Society*, **105**, 3725.
7. Suzuki, K., Katayama, E. and Tsuchihashi, G.-I. (1984) *Tetrahedron Letters*, **25**, 2479.
8. (a) Narasaka, K. and Pai, F.-C. (1984) *Tetrahedron*, **40**, 2233. (b) Chen, K.-M., Hardtman, G.E., Prasad, K., Repic, O. and Shapiro, M.J. (1987) *Tetrahedron Letters*, **28**, 155.
9. Evans, D.A., Chapman, K.T. and Carreira, E.M. (1988) *Journal of the American Chemical Society*, **110**, 3560.
10. Mulzer, J., Sieg, A., Brücher, C., Müller, D. and Martin, H.J. (2005) *Synlett*, 685.
11. (a) Tomoda, S. (1999) *Chemical Reviews*, **99**, 1243. (b) Senda, Y. (2002) *Chirality*, **14**, 110. (c) Gung, B.W. (1996) *Tetrahedron*, **52**, 5263. (d) Senda, E.L. and Eliel, Y. (1970) *Tetrahedron*, **26**, 2411.
12. (a) Cieplak, A.S. (1981) *Journal of the American Chemical Society*, **103**, 4540. (b) Cieplak, A.S. (1999) *Chemical Reviews*, **99**, 1265.
13. Lansbury, P.T. and MacLeay, R.E. (1963) *Journal of Organic Chemistry*, **28**, 1940.
14. Kim, S., Moon, Y.C. and Ahn, K.H. (1982) *Journal of Organic Chemistry*, **47**, 3311.
15. Fisher, G.B., Fuller, J.C., Harrison, J., Alvarez, S.G., Burkhardt, E.R., Goralski, C.T. and Singaram, B. (1994) *Journal of Organic Chemistry*, **59**, 6378.
16. Barden, M.C. and Schwartz, J. (1995) *Journal of Organic Chemistry*, **60**, 5963.
17. Ravikumar, K.S. and Chandrasekaran, S. (1996) *Journal of Organic Chemistry*, **61**, 826.
18. Krishnamurthy, S. and Brown, H.C. (1976) *Journal of the American Chemical Society*, **98**, 3383.
19. (a) Midland, M.M., Tramontano, A. and Zderic, S.A. (1977) *Journal of the American Chemical Society*, **99**, 5211. (b) Midland, M.M., Greer, S., Tramontano, A. and Zderic, S.A. (1979) *Journal of the American Chemical Society*, **101**, 2352. (c) Review: Midland, M.M. (1989) *Chemical Reviews*, **89**, 1553.
20. Midland, M.M. and McLoughlin, J.I. (1984) *Journal of Organic Chemistry*, **49**, 1316.
21. Midland, M.M., Mcdowell, D.C., Hatch, R.L. and Tramontano, A. (1980) *Journal of the American Chemical Society*, **102**, 867.
22. Midland, M.M. and Kazubski, A. (1982) *Journal of Organic Chemistry*, **47**, 2814.

23 Midland, M.M. and Kazubski, A. (1982) *Journal of Organic Chemistry*, **47**, 2495.
24 (a) Chandrasekharan, J., Ramachandran, P.V. and Brown, H.C. (1985) *Journal of Organic Chemistry*, **50**, 5446. (b) Brown, H.C. and Chandrasekharan, J. (1988) *Journal of the American Chemical Society*, **110**, 1539. (c) Srebnik, M., Ramachandran, P.V. and Brown, H.C. (1988) *Journal of Organic Chemistry*, **53**, 2916. (d) Review: Dhar, R.K. (1994) *Aldrichimica Acta*, **27**, 43.
25 (a) Brown, H.C., Park, W.S. and Cho, B.T. (1986) *Journal of Organic Chemistry*, **51**, 1934. (b) Brown, H.C., Cho, B.T. and Park, W.S. (1988) *Journal of Organic Chemistry*, **53**, 1231.
26 Hirao, A., Itsuno, S., Nakahama, S. and Yamazaki, N. (1981) *Journal of the Chemical Society, Chemical Communications*, 315.
27 (a) Corey, E.J., Bakshi, R.K. and Shibata, S. (1987) *Journal of the American Chemical Society*, **109**, 5551. (b) Review: Wallbaum, S. and Martens, J. (1992) *Tetrahedron: Asymmetry*, **3**, 1475.
28 Itsuno, S., Hirao, A., Nakahama, S. and Yamazaki, N. (1983) *Journal of the Chemical Society, Perkin Transactions*, **1**, 1673.
29 (a) Itsuno, S., Ito, K., Hirao, A. and Nakahama, S. (1983) *Journal of the Chemical Society, Chemical Communications*, 469. (b) Itsuno, S., Nakano, M., Miyazaki, K., Masuda, H., Ito, K., Hirao, A. and Nakahama, S. (1985) *Journal of the Chemical Society, Perkin Transactions I*, 2039.
30 Corey, E.J., Bakshi, R.K., Shibata, S., Chen, C.-P. and Singh, V.K. (1987) *Journal of the American Chemical Society*, **109**, 7925.
31 (a) Corey, E.J., Azimioara, M. and Sarshar, S. (1992) *Tetrahedron Letters*, **33**, 3429. (b) Mathre, D.J., Thompson, A.S., Douglas, A.W., Hoogsteen, K., Carroll, J.D., Corley, E.G. and Grabowski, E.J.J. (1993) *Journal of Organic Chemistry*, **58**, 2880.
32 Giffels, G., Dreisbach, C., Kragl, U., Weigerding, M., Waldmann, H., Wandrey, C. (1995) *Angewandte Chemie (International Ed. in English)*, **34**, 2005.
33 (a) Soai, K., Oyamada, H. and Yamanoi, T. (1984) *Journal of the Chemical Society, Chemical Communications*, 413. (b) Soai, K., Yamanoi, T., Hikima, H. and Oyamada, H. (1985) *Journal of the Chemical Society, Chemical Communications*, 138.
34 Soai, K., Yamanoi, T. and Oyamada, H. (1984) *Chemistry Letters*, **13**, 251.
35 Nagata, T., Yorozu, K., Yamada, T. and Muakaiyama, T. (1995) *Angewandte Chemie (International Ed. in English)*, **34**, 2145.
36 (a) Yamada, T., Nagata, T., Sugi, K.D., Yorozu, K., Ikeno, T., Ohtsuka, Y., Miyazaki, D. and Muakaiyama, T. (2003) *Chemistry – A European Journal*, **9**, 4485. (b) Iwakura, I., Hatanaka, M., Kokura, A., Teraoka, H., Ikeno, T., Nagata, T. and Yamada, T. (2006) *Chemistry – An Asian Journal*, **1**, 656.
37 Salunkhe, A.M. and Burkhardt, E.R. (1997) *Tetrahedron Letters*, **38**, 1523.
38 Krzeminski, M.P. and Wojtczak, A. (2005) *Tetrahedron Letters*, **46**, 8299.
39 Du, D.-M., Fang, T., Zhang, J. and Xu, S.-W. (2006) *Organic Letters*, **8**, 1327.
40 Gamble, M.P., Studley, J.R. and Wills, M. (1996) *Tetrahedron Letters*, **37**, 2853.
41 Degni, S., Wilén, C.-E. and Rosling, A. (2004) *Tetrahedron: Asymmetry*, **15**, 1495.
42 Dalicsek, Z., Pollreisz, F., Gömöry, Á. and Soós, T. (2005) *Organic Letters*, **7**, 3243.
43 (a) Brown, H.C. and Pai, G.G. (1982) *Journal of Organic Chemistry*, **47**, 1606. (b) Brown, H.C. and Pai, G.G. (1983) *Journal of Organic Chemistry*, **48**, 1784. (c) Brown, H.C. and Pai, G.G. (1985) *Journal of Organic Chemistry*, **50**, 1384.
44 (a) Almqvist, F., Torstensson, L., Gudmundsson, A. and Frejd, T. (1997) *Angewandte Chemie (International Ed. in English)*, **36**, 376. (b) Sarvary, I., Almqvist, F. and Frejd, T. (2001) *Chemistry – A European Journal*, **7**, 2158.
45 Chen, C.-P., Prasad, K. and Repic, O. (1991) *Tetrahedron Letters*, **32**, 7175.
46 Quallich, G.J. and Woodall, T.M. (1993) *Tetrahedron Letters*, **34**, 785.
47 Singh, V.K. (1992) *Synthesis*, 605.
48 Imai, T., Tamura, T., Yamamoto, A., Sato, T., Wollmann, T.A., Kennedy, R.M. and Masamune, S. (1986) *Journal of the American Chemical Society*, **108**, 7402.

49 Kim, J. and Singaram, B. (2006) *Tetrahedron Letters*, **47**, 3901.

50 (a) Luche, J.-L. (1978) *Journal of the American Chemical Society*, **100**, 2226. (b) Gemal, A.L. and Luche, J.-L. (1981) *Journal of the American Chemical Society*, **103**, 5454.

51 Corey, E.J. and Helal, C.J. (1995) *Tetrahedron Letters*, **36**, 9153.

52 Rodgen, S.A. and Schaus, S.E. (2005) *Angewandte Chemie (International Ed. in English)*, **45**, 4929.

53 Cho, B.T. and Shin, S.H. (2005) *Tetrahedron*, **61**, 6959.

54 Reiff, E.A., Nair, S.K., Reddy, B.S.N., Inagaki, J., Henri, J.T., Greiner, J.F. and Georg, G.I. (2004) *Tetrahedron Letters*, **45**, 5845.

55 Berenguer, R., Cavero, M., Garcia, J. and Munoz, M. (1998) *Tetrahedron Letters*, **39**, 2183.

56 Tominaga, H., Maezaki, N., Yanai, M., Kojima, N., Urabe, D., Ueki, R. and Tanaka, T. (2006) *European Journal of Organic Chemistry*, 1422.

57 Ramachandran, P.V., Teodorovic, A.V., Rangaishenvi, M.V. and Brown, H.C. (1992) *Journal of Organic Chemistry*, **57**, 2379.

58 Parker, K.A. and Ledeboer, M.W. (1996) *Journal of Organic Chemistry*, **61**, 3214.

59 Morrill, C. and Grubbs, R.H. (2005) *Journal of the American Chemical Society*, **127**, 2842.

60 Parker, K.A. and Katsoulis, I.A. (2004) *Organic Letters*, **6**, 1413.

61 Shimizu, M., Ikari, Y. and Wakabayashi, A. (2001) *Journal of the Chemical Society, Perkin Transactions I*, 2519.

62 Helal, C.J., Magriotis, P.A. and Corey, E.J. (1996) *Journal of the American Chemical Society*, **118**, 10938.

63 (a) Wipf, B., Kupfer, E., Bertazzi, R. and Leuenberger, H.G.W. (1983) *Helvetica Chimica Acta*, **66**, 485. (b) Mori, K., Mori, H. and Sugai, T. (1985) *Tetrahedron*, **41**, 919. (c) Zhou, B., Gopalan, A.S., van Middesworth, F., Sieh, W.R. and Sih, C.J. (1983) *Journal of the American Chemical Society*, **105**, 5925.

64 Basavaiah, D., Rao, K.V. and Reddy, B.S. (2006) *Tetrahedron: Asymmetry*, **17**, 1041.

65 Basavaiah, D., Reddy, B.S. and Rao, K.V. (2004) *Tetrahedron: Asymmetry*, **15**, 1881.

66 Gamble, M.P., Smith, A.R.C. and Wills, M. (1998) *Journal of Organic Chemistry*, **63**, 6068.

67 Cho, B.T. and Kim, D.J. (2003) *Tetrahedron*, **59**, 2457.

68 (a) Noyori, R., Ohkuma, T., Kitamura, M., Takaya, H., Sayo, N., Kumobayashi, H. and Akutagawa, S. (1987) *Journal of the American Chemical Society*, **109**, 5856. (b) Kitamura, M., Ohkuma, T., Takaya, H. and Noyori, R. (1988) *Tetrahedron Letters*, **29**, 1555. (c) Nishi, T., Kitamura, M., Ohkuma, T. and Noyori, R. (1988) *Tetrahedron Letters*, **29**, 6327. (d) Kitamura, M., Tokunaga, M., Ohkuma, T. and Noyori, R. (1991) *Tetrahedron Letters*, **32**, 4163.

69 Imagawa, H., Fujikawa, Y., Tsuchihiro, A., Kinoshita, A., Yoshinaga, T., Takao, H. and Nishizawa, M. (2006) *Synlett*, 639.

70 Manthati, V.L., Grée, D. and Grée, R. (2005) *European Journal of Organic Chemistry*, 3825.

71 Hasdemir, B. and Yusufoğlu, A. (2004) *Tetrahedron: Asymmetry*, **15**, 65.

72 Nesterov, V.V. and Kolodiazhnyi, O.I. (2006) *Tetrahedron: Asymmetry*, **17**, 1023.

73 Ohtsuka, Y., Miyazaki, D., Ikeno, T. and Yamada, T. (2002) *Chemistry Letters*, **31**, 24.

74 Prasad, K.R.K. and Joshi, N.N. (1996) *Journal of Organic Chemistry*, **61**, 3888.

75 Shimizu, M., Yamada, S., Fujita, Y. and Kobayashi, F. (2000) *Tetrahedron: Asymmetry*, **11**, 3883.

76 Periasamy, M., Seenivasaperumal, M. and Rao, V.D. (2003) *Synthesis*, 2507.

77 Aldous, D.J., Dutton, W.M. and Steel, P.G. (2000) *Tetrahedron: Asymmetry*, **11**, 2455.

78 Sato, M., Gunji, Y., Ikeno, T. and Yamada, T. (2004) *Synthesis*, 1434.

79 Quallich, G.J., Keavey, K.N. and Woodall, T.M. (1995), *Tetrahedron Letters*, **36**, 4729.

80 Bach, J., Berenguer, R., Garcia, J., López, M., Manzanal, J. and Vilarrasa, J. (1998) *Tetrahedron*, **54**, 14947.

81 Delogu, G., Dettori, M.A., Patti, A., Pedotti, S., Forni, A. and Casalone, G. (2003) *Tetrahedron: Asymmetry*, **14**, 2467.

82 Ohtsuka, Y., Kubota, T., Ikeno, T., Nagata, T. and Yamada, T. (2000) *Synlett*, 535.

83 Ohtsuka, Y., Koyasu, K., Miyazaki, D., Ikeno, T. and Yamada, T. (2001) *Organic Letters*, **3**, 3421.
84 Masamune, S., Choy, W., Petersen, J.S. and Sita, L.R. (1985) *Angewandte Chemie (International Ed. in English)*, **24**, 1.
85 Pichlmair, S., Ruiz, M.L., Basu, K. and Paquette, L.A. (2006) *Tetrahedron*, **62**, 5178.
86 Uenishi, J. and Ohmi, M. (2005) *Angewandte Chemie (International Ed. in English)*, **44**, 2756.
87 Nattrass, G.L., Díez, E., McLachlan, M.M., Dixon, D.J. and Ley, S.V. (2005) *Angewandte Chemie (International Ed. in English)*, **44**, 580.
88 Smith, A.B., III, Sfouggatakis, C., Gotchev, D.B., Shirakami, S., Bauer, D., Zhu, W. and Doughty, V.A. (2004) *Organic Letters*, **6**, 3637.
89 Liao, X., Wu, Y. and De Brabander, J.K. (2003) *Angewandte Chemie (International Ed. in English)*, **42**, 1648.
90 Rech, J.C. and Floreancig, P.E. (2005) *Organic Letters*, **7**, 5175.

8
Diverse Modes of Silane Activation for the Hydrosilylation of Carbonyl Compounds

Sebastian Rendler and Martin Oestreich

8.1
Introduction

The utilization of silanes in the reduction of carbonyl compounds, that is C—H accompanied by Si—O bond formation, is of major importance in modern organic synthesis [1–6]. Spanning a period of three decades, numerous efforts have led to a plethora of well-understood procedures covering almost all areas of organic chemistry. This research topic is covered by several comprehensive reviews that have appeared in recent years [1–6]. This chapter aims at a survey of the current state-of-the-art of hydrosilylative C=O reductions arranged by their individual mechanisms.

The attractiveness of hydrosilylation chemistry derives from the ease with which mildly hydridic silanes, chemically stable, neat and easy-to-handle and usually not requiring particular precautions, serve as a hydride source upon activation by a catalyst or an additive. The primary product of the addition of a Si—H bond across a C=O double bond is a silyl ether, which is normally prone to facile hydrolytic cleavage, thus liberating the desired alcohol.

It appeared expedient to structure this chapter according to the mechanisms, that is the different modes of Si—H bond activation (Scheme 8.1). The chapter is divided into sections summarizing transition metal-catalyzed methods (Section 8.2) and transition metal-free or even metal-free (organocatalytic) procedures (Section 8.3).

In all cases involving a transition metal, the Si—H bond of **A** (R = aryl, alkyl, alkoxy or halo) is activated by interaction with the transition metal center (Scheme 8.1). Most commonly, the oxidative addition of a group 8–10 transition metal complex into the Si—H bond leads to intermediate **B** (**A**→**B**) which subsequently transfers both the silicon fragment and the hydrogen in a sequence of migratory insertion and reductive elimination. This resembles a mechanism known for the related hydrogenation and, thus, silane **A** might well be considered as a dihydrogen surrogate in these transformations. The discussion in Section 8.2.1 focuses on asymmetric variants.

Modern Reduction Methods. Edited by Pher G. Andersson and Ian J. Munslow.
Copyright © 2008 WILEY-VCH Verlag GmbH & Co. KGaA, Weinheim
ISBN: 978-3-527-31862-9

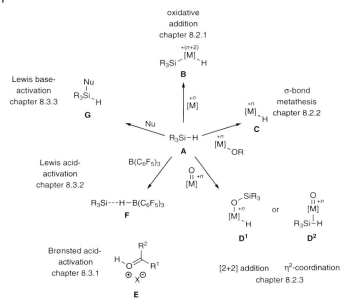

Scheme 8.1 Modes of silane activation.

Early transition metals of group 4 and those of groups 11 and 12 represent another class of hydrosilylation catalysts. Silane activation by σ-bond metathesis is here the pivotal step leading to metal hydride **C**, which is suited for reduction of the carbonyl compound by hydrometallation (**A→C**). Section 8.2.2 provides a summary of this reaction type, again with major emphasis on catalytic asymmetric processes. Yet another conceptually different approach is catalysis by high-valent oxo complexes (**A→D**). Within this new development, silane activation by [2+2] addition (**A→D¹**) or by η²-coordination (**A→D²**) will be discussed as conceivable pathways (Section 8.2.3).

Transition metal-free hydrosilylations either involve Brønsted or Lewis acids as promoters or Lewis basic activators (Section 8.3). The Brønsted acid-promoted reaction is an exception since the carbonyl compound (**E**, Scheme 8.1) and not the silane **A** is activated by protonation (Section 8.3.1). In turn, a few Lewis acids are capable of activating the silane by interaction with the hydride; as a result of the polarization and the weakening of the Si—H bond, hydride transfer from and increase of the Lewis acidic properties of the silicon center promotes the formation of a silylium ion equivalent (**A→F**). Remarkably, catalysis is possible (Section 8.3.2). The application of nucleophilic promoters to hydrosilylation complements the transition metal-free repertory (Section 8.3.3). This activation mode is characterized by intermediates **G** (**A→G**). These hypervalent species facilitate hydride transfer from silicon to the carbonyl carbon; moreover, the Lewis acidity of silicon is increased. When organic nucleophiles are used in catalytic quantities, this is referred to as an organocatalytic protocol.

8.2
Metal-Catalyzed Hydrosilylations

After several decades of intense activity in this field, the transition metal-catalyzed hydrosilylation of carbonyl compounds is highly developed [1–6]. In particular, catalytic asymmetric hydrosilylation has evolved as a competitive strategy for the preparation of chiral nonracemic secondary alcohols [2–6].

8.2.1
Silane Activation by Oxidative Addition

The common strategy for carbonyl hydrosilylation is based on low-valent group 8–10 transition metal catalysts being able to undergo oxidative addition into the Si–H bond (cf. **A**→**B**, Scheme 8.1). Despite occasional reports using nickel, platinum, ruthenium, iridium as well as tungsten [1–6], rhodium-based catalysts are more prevalent.

The generally accepted mechanism is shown in Scheme 8.2. A solvated cationic Rh(I) complex **1**, coordinated by a bidentate ligand, oxidatively adds into the Si–H bond of silane **2** (**1**→**3**). The resulting Rh(III) intermediate **3** then captures the carbonyl donor **4** (**3**→**5**) and subsequently undergoes *silyl*metallation of the C=O double bond (**5**→**6**). In the final reductive elimination step, the C–H bond is formed and silyl ether **7** is liberated (**6**→**1**).

Ojima and coworkers not only postulated this mechanism as early as 1972 [7a], but also validated it for rhodium-catalyzed asymmetric hydrosilylations in 1976 [7b]. An alternative fate of intermediate **5** is conceivable: *hydro*metallation of the carbonyl group producing a Rh(III) alkoxide (**5**→**6'**) which might suffer reductive

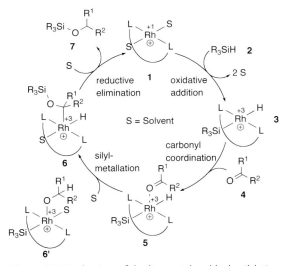

Scheme 8.2 Mechanism of rhodium-catalyzed hydrosilylation.

elimination, thereby forming the Si—O linkage. This pathway was precluded on the basis of the experimental observation that the substitution pattern at silicon had a marked impact on the enantioselectivity of the overall process, which was, in turn, interpreted as silicon being directly involved in the enantiodiscriminating step [7b]. This rather vague theory was later substantiated in spin-trapping experiments by Peyronel and Kagan [8] as well as kinetic analysis by Kolb and Hetfleijs [9]. Prock and Giering, more recently, corroborated these results by kinetic measurements supported by kinetic modeling data [10].

The catalytic cycle (Scheme 8.2) is shown for a rhodium complex with a *trans*-chelating bidentate ligand. Indeed, this unusual binding motif emerged as particularly effective in the asymmetric hydrosilylation of carbonyl compounds [11–13].

In 1999, Ito and coworkers introduced a bisferrocene-based diphosphine **8** ("TRAP") and demonstrated the utility of in situ formed Rh(I)-complexes for the asymmetric hydrosilylation of a variety of prochiral ketones (**4→9**, Scheme 8.3) [11]. The enantiomeric excesses typically ranged between 80% and 90%. In comparison with ligand **8**, combining central and planar chirality, related ligands lacking the former chiral element gave superior results for dialkyl ketones **4** [11].

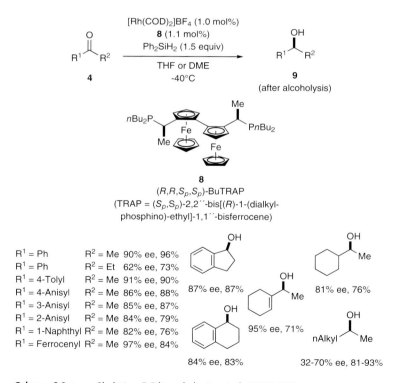

Scheme 8.3 *trans*-Chelating P,P-ligands by Ito *et al.* (1999) [11].

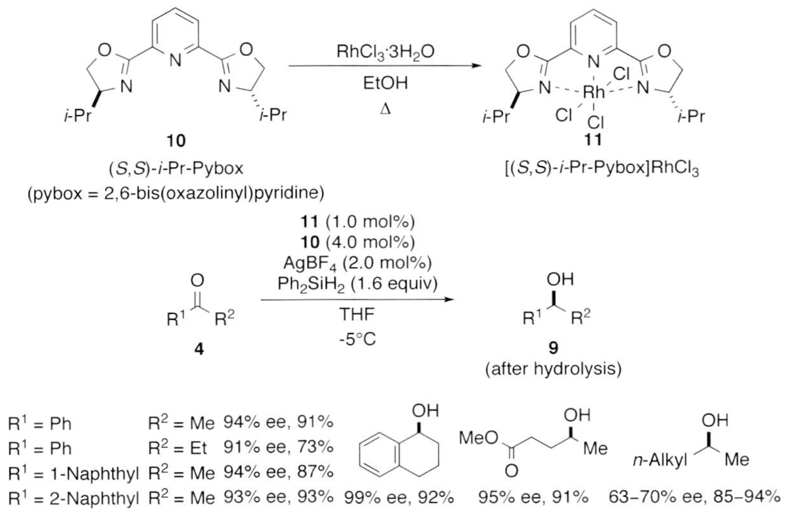

8.2 Metal-Catalyzed Hydrosilylations

Scheme 8.4 Tridentate N,N,N-ligands by Nishiyama et al. (1989) [12].

Reaction conditions (Scheme 8.4):
- **10** (S,S)-i-Pr-Pybox (pybox = 2,6-bis(oxazolinyl)pyridine) + RhCl$_3$·3H$_2$O, EtOH, Δ → **11** [(S,S)-i-Pr-Pybox]RhCl$_3$
- Ketone **4** (R^{1}COR^2): **11** (1.0 mol%), **10** (4.0 mol%), AgBF$_4$ (2.0 mol%), Ph$_2$SiH$_2$ (1.6 equiv), THF, −5 °C → alcohol **9** (after hydrolysis)

Results:
- R^1 = Ph, R^2 = Me: 94% ee, 91%
- R^1 = Ph, R^2 = Et: 91% ee, 73%
- R^1 = 1-Naphthyl, R^2 = Me: 94% ee, 87%
- R^1 = 2-Naphthyl, R^2 = Me: 93% ee, 93%
- 1-tetralol type: 99% ee, 92%
- MeO-CH$_2$-C(O)-CH(OH)-Me: 95% ee, 91%
- n-Alkyl, Me: 63–70% ee, 85–94%

As exemplified by tridentate pyridinebisoxazoline ligand **10** (Scheme 8.4) [12], ligands with three [12] or even four [13] donor atoms form powerful rhodium complexes for asymmetric ketone hydrosilylation. The successful application of this ligand class for asymmetric ketone reduction by way of hydrosilylation was reported by Nishiyama and coworkers in 1989. The catalytic system consisting of the preformed Rh(III)-complex **11**, AgBF$_4$ as a halide scavenger and excess **10** enabled the reduction of aryl alkyl ketones in good to excellent optical purities; dialkyl ketones were prepared in modest to good enantiomeric excesses (**4**→**9**, Scheme 8.4) [12].

Apart from these contributions, several cognate N,N-ligands (pyridineoxazolines [14a] or bisoxazolines [14b]) and P,N-ligands (phosphinoimines [15a, b], phosphinooxazolines [15c–f] and bis(phosphinooxazolines) [15g, h]) were assessed according to catalyst reactivity and asymmetric induction in carbonyl hydrosilylation.

A privileged class of P,N-ligands **12** was introduced by Tao and Fu in 2002 (Scheme 8.5) [16]. A rhodium complex of the planar-chiral pyridinophosphine ligand **12** enabled the preparation of alcohols **9** in remarkably high enantioselectivities (**4**→**9**). Not only did aryl alkyl ketones give excellent results, but also selected dialkyl ketones. Within this investigation an often ignored, but significant, feature was systematically studied. Seminal observations by Ojima had already indicated that the substituents of the dihydrosilane used noticeably affected the level of enantioselection [7b]. Fu and Tao were able to demonstrate that optical purities from 1% to 98% ee could be obtained by simply altering the substitution pattern at silicon (Scheme 8.5). Sterically hindered MesPhSiH$_2$ or (2-tolyl)$_2$SiH$_2$ performed best [16]. Notably, this argument holds also true for silicon-stereogenic silanes in C—C double hydrosilylation [17].

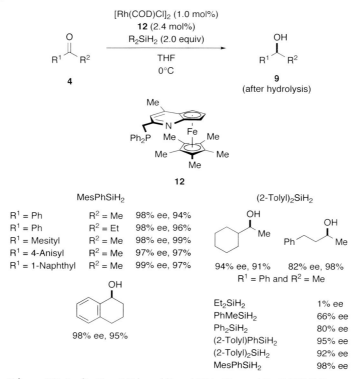

Scheme 8.5 Application P,N-based ligand **12** by Tao and Fu (2002) [16].

A selection of rhodium complexes worth mentioning is shown in Figure 8.1. Their skeletons are characterized by completely different structural motifs compared with those presented so far. Imamoto and coworkers used a bidentate phosphorus-stereogenic ligand to prepare corresponding **13**, which proved to be a potent catalyst [18]. Based on earlier work on TADDOL-derived phosphonites [19a], Seebach *et al.* developed ketone hydrosilylations with the rhodium complex **14** derived from an oxazolinophosphite [19b, 20]. The Evans group successfully introduced P,S-ligands: **15** allowed for conducting highly enantioselective ketone hydrosilylations [21]. Gade and Bellemin-Laponnaz prepared phosphine-free **16** incorporating a N-heterocyclic carbene unit, which is a potent catalyst for hydrosilylations [22].

8.2.2
Silane Activation by σ-Bond Metathesis

In contrast to group 8–10 transition metals, early transition metals (e.g. titanium or zirconium) and group 11 or 12 metals (e.g. copper and zinc) will normally not undergo oxidative addition into Si—H bonds (cf. **A→B**, Scheme 8.1). This is rationalized by the electronic structure of the metal center: d^1 configuration for Ti(III)

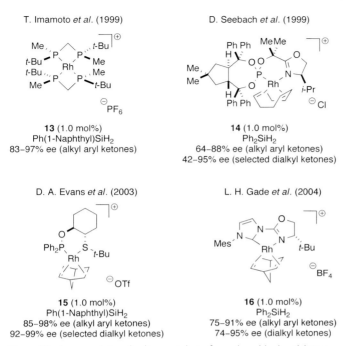

Figure 8.1 Representative rhodium catalysts for carbonyl hydrosilylation.

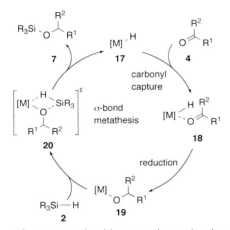

Scheme 8.6 Hydrosilylation involving σ-bond metathesis.

and d^{10} configuration for Cu(I) and Zn(II). The former metal is devoid of any d electrons available for an oxidative addition pathway, whereas the latter favors retaining an energetically preferred closed-shell electronic configuration. Instead, σ-bond metathesis becomes the predominant pathway (cf. **A→C**, Scheme 8.1); Scheme 8.6 illustrates the currently accepted catalytic cycle.

The catalytically active species is the metal hydride **17**, which is preformed (Ti) or generated by a single catalytic turnover (Cu and Zn, **19**→**17**). Upon coordination of prochiral ketone **4** (**17**→**18**), hydride transfer from metal to carbon occurs (**18**→**19**). Silane **2** then coordinates to the metal alkoxide **19** followed by irreversible σ-bond metathesis via a four-centered transition state **20**, expelling silyl ether **7** accompanied by regeneration of the metal hydride **17**. In the case of copper, a four-centered transition state **20** is also supported by the related dehydrogenative Si–O coupling of alcohols and silanes; these are known to proceed with retention of configuration at silicon when using silicon-stereogenic silanes [23].

In an early contribution, Nagano and Nagai reported the catalytic activity of Cp_2TiPh_2 in ketone hydrosilylations [24]. Prior to this discovery, Samuel and Harrod had already shown that titanium hydrides were accessible by the reaction of silanes with Cp_2TiMe_2 [25a]; these were tested in hydrosilylations of C=C double bonds [25b]. A few years later, the Halterman group [26] and Buchwald and coworkers [27] independently examined catalytic asymmetric protocols, both using C_2-symmetric titanocene complexes. Whereas the enantiomeric excesses for a binaphthyl-derived catalyst remained low [26], the ethylene-bridged bis(tetrahydroindenyl) titanium catalyst **22** provided excellent enantiofacial selectivity (Scheme 8.7) [27].

In a modified Buchwald procedure, precatalyst **21** was first reacted with pyrrolidine, phenylsilane and methanol, thereby generating the catalytically active Ti(III) complex **22** (Scheme 8.7) [27b]. Prochiral ketone **4** and polymeric hydrosiloxane PMHS were subsequently added. Moreover, the slow addition of methanol was found to be beneficial to both the yield and the optical purity of **9**. The observed enhanced reaction rate was explained by alkoxide exchange of a secondary titanium alkoxide **19** via **23**, leading to the primary alkoxide **19′** (Scheme 8.7). Sterically less encumbered **19** then undergoes σ-bond metathesis with silane **2** more rapidly (cf. **20**, Scheme 8.6). This procedure allowed for the high-yielding preparation of allylic and benzylic alcohols **9** in excellent enantiomeric ratios. Conversely, nonconjugated dialkyl ketones **4** gave poor results [27], which were also observed in an independent investigation [25c]. In addition to $(ebthi)_2TiF_2$ (**21**), other titanocene-based catalysts [26, 28], BINOL-Ti(IV) [29] as well as bisoxazoline-Ti(IV) [30] precatalysts turned out to be less effective.

A related concept was developed using copper catalysts [31, 32]. In spite of a pioneering report by Brunner and Miehling on the catalytic asymmetric hydrosilylation of prochiral ketones using silane/copper/diphosphine combinations as early as 1984 [33] and even with Stryker's synthetic chemistry of "Cu–H" at hand [34], the full potential of chirally modified copper hydrides in catalytic asymmetric carbonyl reductions was not realized until recently [32]. Shortly after Lipshutz and coworkers had discovered a significant ligand-acceleration by bidentate phosphines for ketone reductions employing Stryker's reagent [{CuH(PPh$_3$)}$_6$] [35a], a highly enantioselective procedure for ketone hydrosilylation was elaborated by the same group [35b]. In a series of publications, Lipshutz *et al.* established the use of axially chiral bidentate phosphines such as MeO-BIPHEP derivative **24** or SEGPHOS ligand **26** in combination with copper salts, an alkoxide base and stoichiometric

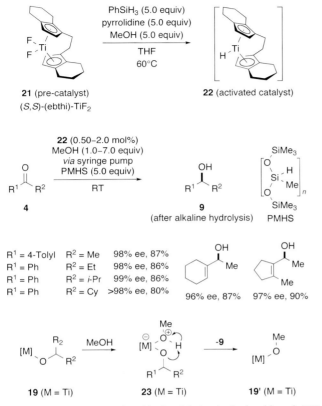

Scheme 8.7 Titanium-catalyzed hydrosilylation by Buchwald *et al.* (1994) [27].

amounts of silanes as an ideal system for the reduction of prochiral ketones (Scheme 8.8) [35b–g]. This method proved to be generally applicable to a variety of (functionalized) aryl alkyl and heteroaryl alkyl ketones, respectively (**4→9**, Scheme 8.8) [35]. It was also shown that these copper hydride complexes **17** (M = Cu(I)) were sufficiently stable to be stored in solution for a prolonged period [35g]; a heterogeneous variant using copper in charcoal (Cu/C) was also reported [35h].

The use of BINAP (**25**) as a ligand was independently reported by Riant [36a, b] and Yun [36c]; the optical purities obtained were recently improved by changing the silane source from PMHS to PhMeSiH$_2$ (Scheme 8.8) [36d]. Several other ligands such as bipyridine-derived phosphines [37] and *N*-heterocyclic carbenes have been used with considerable success [38]. The option of extremely low catalyst loadings must be emphasized as one of the striking advantages of copper-catalyzed hydrosilylations. Lipshutz *et al.* reported that conversion and enantiomeric excess are still unaffected by substrate-to-ligand ratios as low as 100 000 : 1 [35d]. Moreover, ligand-to-copper ratios less than 1 are not detrimental to enantioselection

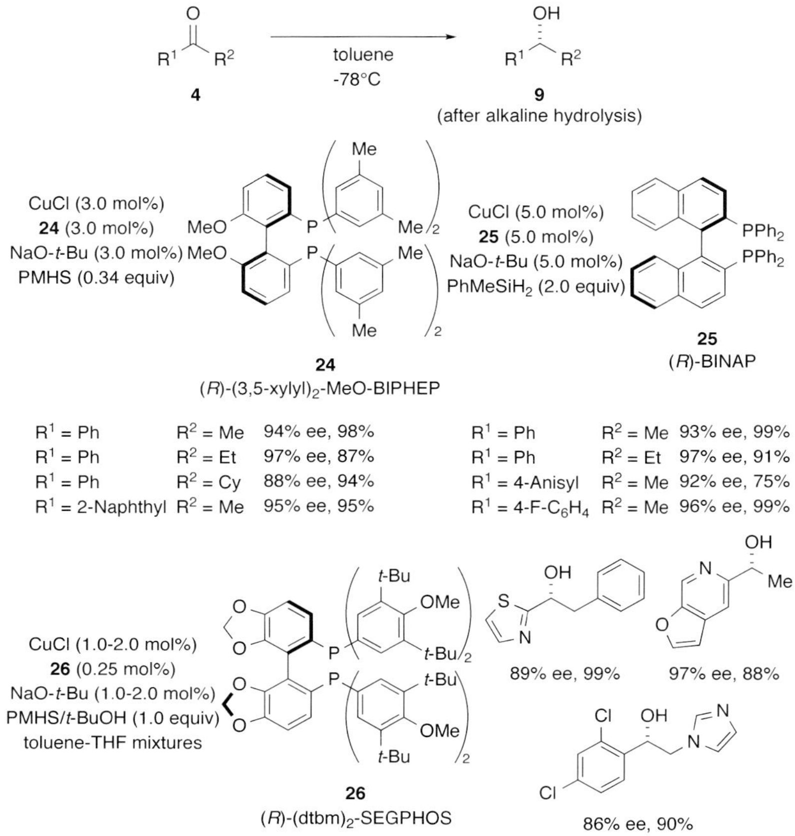

Scheme 8.8 Chirally modified copper hydrides for ketone hydrosilylation.

due to the latent instability of nonstabilized, that is phosphine-ligated, copper hydride complexes. Coupled with the straightforward experimental procedure, this methodology is certainly an inexpensive alternative in asymmetric reduction chemistry, not only on laboratory scale.

A third class of catalysts, which is also believed to involve a σ-bond metathesis as a crucial step, is based on zinc complexes [39]. Seminal contributions by Caubère *et al.* demonstrated that zinc hydrides complexed by a chiral amino alcohol catalyze asymmetric ketone reductions [40a]; Noyori and coworkers later used zinc salts and a mixture of Me_3SiCl/LiH [40b]. With these preliminary insights as a starting point, Mimoun *et al.* succeeded in the elaboration of a general protocol for the reduction of aryl methyl (ethyl) ketones using PMHS as the stoichiometric reducing agent and chiral diamines as the source of chirality (Scheme 8.9) [41].

Preparation of the catalyst precursors is facile (Scheme 8.9), by reaction of either zinc carboxylates (**27**→**28**) or diethyl zinc (**27**→**29**) with the diamine **27**. Whereas

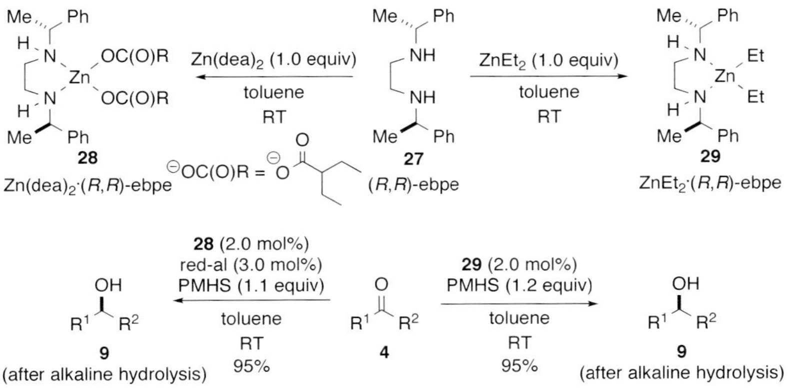

Scheme 8.9 Zinc-catalyzed hydrosilylation.

4 → 9 via **28**	R¹, R²	4 → 9 via **29**
74% ee	R¹ = Ph, R² = Me	76% ee
81% ee	R¹ = Ph, R² = Et	77% ee
66% ee	R¹ = 4-Tolyl, R² = Me	71% ee
73% ee	R¹ = 3-Anisyl, R² = Me	76% ee
72% ee	R¹ = 4-Anisyl, R² = Me	76% ee
72% ee	R¹ = 1-Naphthyl, R² = Me	75% ee
76% ee	R¹ = 2-Naphthyl, R² = Me	80% ee
68% ee	R¹ = 2-Furyl, R² = Me	66% ee

28 requires further activation with aluminum hydrides, **29** is ready to use [41b]. Under both reaction conditions, moderate to good enantiomeric excesses were obtained, even on large scales [41]. Although the precise nature of the catalytically active species is unclear at present, the involvement of zinc hydrides originating from a σ-bond metathesis is almost certainly out of the question [41]. Related investigations appeared in subsequent years, but no substantial improvement was achieved [42].

8.2.3
Silane Activation by High-valent Oxo Complexes

In recent times, a truly different type of potent transition metal catalysts for the hydrosilylation of carbonyl compounds has extended the spectrum. The traditional hydrosilylation catalysts (group 8–10 metals in low oxidation states, for example rhodium in d^8 configuration, Section 8.2.1; and group 4 metals such as titanium in d^1 configuration, Section 8.2.2; as well as group 11/12 metals such as copper or zinc in d^{10} configuration) were complemented by metals in high oxidation states, namely rhenium and molybdenum oxo complexes [43, 44b, 45]. Silane activation by these complexes proceeds through unprecedented mechanisms of action (Schemes 8.11 and 8.13).

The reactivity of air- and moisture-stable rhenium(V) dioxo complex **30** towards silanes and their catalytic activity for carbonyl hydrosilylation was first demonstrated by Toste and coworkers (4→7, Scheme 8.10) [43]. A selection of ketones

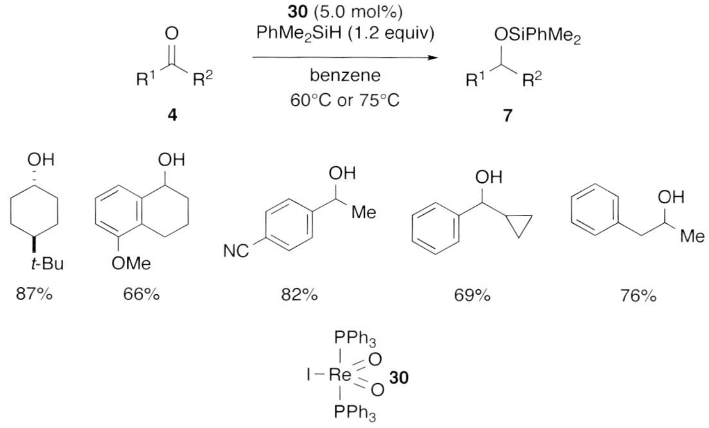

Scheme 8.10 Rhenium-catalyzed hydrosilylation of ketones.

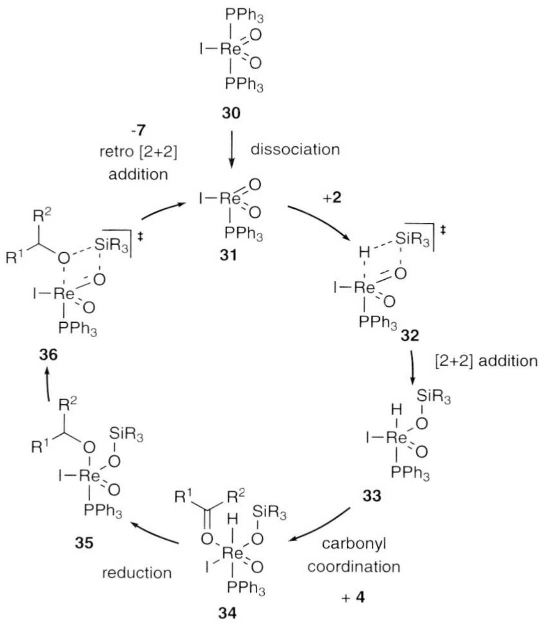

Scheme 8.11 Mechanism of the Re(V)-catalyzed hydrosilylation.

4 and aldehydes (not shown) was transformed chemoselectively into their corresponding ethers 7 using catalytic amounts of rhenium complex 30 at elevated temperatures.

Apart from the synthetic attractiveness of this process, its presumed mechanism is even more intriguing. In fact, the current mechanistic rationale is strikingly different from all known catalyst systems (Scheme 8.11) [43, 46]. The original proposal was based on the assumption that silane activation might occur by a concerted [2+2] addition of the Re=O double bond and the Si—H bond, which is later followed by intramolecular hydride transfer onto a coordinated carbonyl unit [43]. This picture was supported by NMR analysis of observable intermediates [43] and, very recently, by a computational investigation (Scheme 8.11) [46]. The catalysis commences with the formation of the catalytically active **31** from **30** by dissociation of a phosphine ligand (**30**→**31**). The above-mentioned addition step would then lead to a hydrido-rhenium intermediate **33** bearing a siloxy substituent (**31**→**33**). In principle, transition state **32** resembles silane activation by metathesis (cf. Section 8.2.2); it must be noted that a M=O double bond reacts instead (M=O ↔ M$^+$—O$^-$). Coordination of substrate **4** (**33**→**34**) and subsequent hydride transfer through a four-centered transition state affords rhenium alkoxide **35** (**34**→**35**). In the final step, the Si—O bond is formed in a retro [2+2] addition, that is an *O*-to-*O* [1,3] silyl shift through **36**, thereby liberating **7** and regenerating catalytically active **31**. Throughout the catalytic cycle, rhenium does not change its oxidation state of +V. Further evidence is provided by reports on the ability of hydride transfer from rhenium oxo complexes onto acetaldehyde [47a] and the silylation of rhenium dioxo complexes with chlorosilanes [47b].

After this seminal publication, Royo *et al.* developed a similar protocol using molybdenum dioxo complex MoO_2Cl_2 as a catalyst as well as other simple catalyst systems based on rhenium [45].

The synthetic potential of rhenium oxo complexes as catalysts for Si—O bond formation was also grasped by Abu-Omar and coworkers. Initially targeting the catalytic hydrolysis of hydrosilanes with **37** [44a], these authors found that the same catalyst facilitates the hydrosilylation of carbonyl compounds (**4**→**7**, Scheme 8.12) [44b]. Cationic rhenium(V) complex **37** displayed high reactivity even at low catalyst loadings.

The mechanism proposed by Abu-Omar *et al.* was postulated by conscientious exclusion of several other reaction pathways on the basis of experimental data obtained from kinetic measurements and isotopic labeling (Scheme 8.13). The essence of this investigation is that **37** functions as a Lewis acid but neither by way of LUMO-lowering carbonyl activation nor by generation of a silylium ion (cf. Section 8.3.2). Instead, η^2-coordination of the silane **2** furnishes **38** (**37**→**38**), which establishes increased electrophilicity at silicon; this then allows for a concerted [2+2] addition (**39**, Scheme 8.13) directly liberating both product **7** and active catalyst **37**.

The set of experimental data for **37** and the unusual mechanism derived from it (Scheme 8.13) has put the initially proposed catalytic cycle into question (Scheme 8.11) [44b]. While these mechanisms were still under debate, a computational investigation by Lin and Wu further supported the Toste catalysis yet indicated that these seemingly identical catalysts **30** and **37** might not be comparable; data obtained for the individual systems are not mutually applicable [46]. To illustrate this, one might take a closer look at the coordination spheres of **30** and **37**. The former, **30**, is decorated with five ligands, of which the monodentate phosphines are prone to facile dissociation; moreover, hydrido-rhenium intermediate

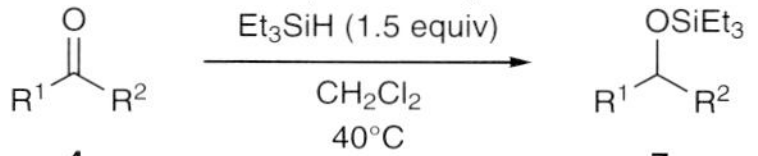

196 | *8 Diverse Modes of Silane Activation for the Hydrosilylation of Carbonyl Compounds*

Scheme 8.12 Cationic rhenium complex **37** for ketone hydrosilylation.

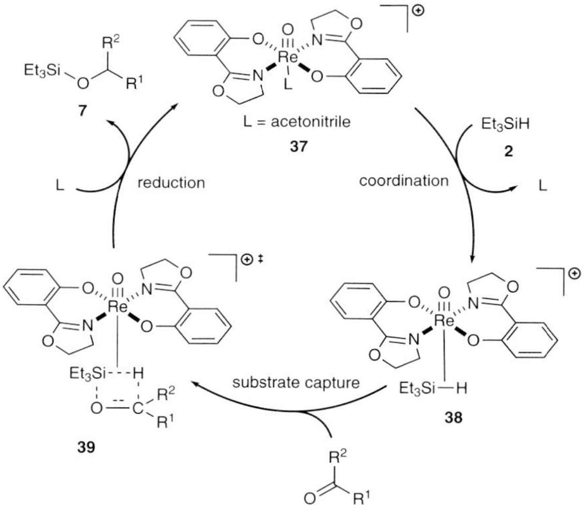

R¹ = Ph, R² = Me, 86%
R¹ = Me, R² = Me, 67%
R¹ = Me, R² = Et, 71%
R¹ = Me, R² = *i*Pr, 72%
R¹ = Et, R² = Et, 59%

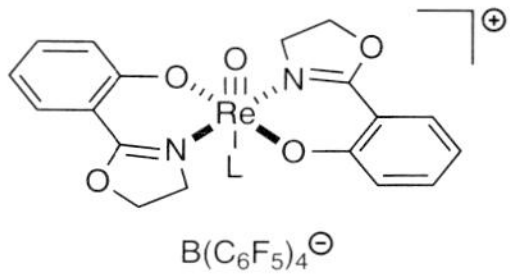

37
[Re(O)(hoz)$_2$L][TFPB]
(hoz = 2-(2′-hydroxyphenyl)-2-oxazoline,
TFPB = tetrakis(pentafluorophenyl)borate,
L = acetonitrile)

Scheme 8.13 Mechanism of the "cationic" Re(V)-catalyzed hydrosilylation.

33 is still coordinatively unsaturated, thus providing a vacant site for the carbonyl donor. In the latter, **37**, there are also five donor atoms arranged around the metal center, but two ligands are bidentate. By this, an addition of the Si—H bond across the Re—O multiple bond would lead to a coordinatively saturated rhenium hydride; without any further coordination site available for substrate coordination, this reaction channel would hence come to a dead end. The anionic dioxo complex Re(O)$_2$(hoz)$_2$ generated from **37** was shown to be inactive for the related dehydrogenative silane hydrolysis [44a]. However, this octahedral complex is again not comparable to key intermediate **31** (generated from **30**) as it is coordinatively unsaturated and has equatorial oxo substituents. Thus, the elucidation of these mechanisms remains an interesting task [43b, 44d, 45c].

8.3
Transition-metal-free Hydrosilylations

All of the previously discussed catalytic systems are based on transition metals; the manifold mechanisms of action were each determined by the inherent properties of the metal. Conversely, there are several hydrosilylations reported in the literature in which catalytic turnover is not dependent upon a (transition) metal. The following is devoted to transition metal-free or even completely metal-free carbonyl reductions by hydrosilylation.

8.3.1
Brønsted Acid-promoted Hydrosilylations

Carbonyl reductions by means of addition of silanes in a Brønsted-acidic media is a well-established methodology [48]. These mixtures are traditionally used for the deoxygenation of ketones and aldehydes or for the preparation of ethers. Both transformations work particularly well if stabilized carbenium ions are plausible intermediates [48]. When using trifluoroacetic acid, partial reduction to the (protected) alcohol is possible [49]. With regard to the mechanism, carbonyl activation by the Brønsted acid by way of protonation of the carbonyl oxygen is evident (cf. **E**, Scheme 8.1) [49]. Conversion is only secured by activation of the acceptor; silane activation in almost non-nucleophilic media (trifluoroacetic acid is often used as the solvent) is, if at all, of minor relevance.

Hiyama and coworkers established a highly diastereoselective carbonyl reduction of α-amino carbonyl compounds and β-keto amides **40** (**40**→**41**, Scheme 8.14) [50]. Chelation-controlled formation of protonated conformer **42** is believed

R^1 = Ph	R^2 = NH(O)OEt	87%	>99:1
R^1 = Ph	R^2 = OC(O)Ph	72%	93:7
R^1 = Ph	R^2 = C(O)NEt$_2$	98%	>99:1
R^1 = Me	R^2 = C(O)NEt$_2$	94%	98:2
R^1 = *i*-Pr	R^2 = C(O)NEt$_2$	89%	99:1

Scheme 8.14 *syn*-Selective reduction in Brønsted-acidic media.

to account for the observed *syn*-selectivity; subsequent reduction of **42** proceeds in diastereoselective fashion according to Cram's rule. Notably, no epimerization was observed at the adjacent carbon when enantiopure precursors **40** were used. A complementary Lewis base-catalyzed *anti*-selective reduction of the same substrate class was reported simultaneously (see Section 8.3.3) [50a, b].

8.3.2
Lewis Acid-catalyzed Hydrosilylations

In order to overcome the issue of over-reduction, Lewis acid-catalyzed hydrosilylations became the next logical step [51]. Several Lewis acids ($ZnCl_2$, $AlCl_3$, $SnCl_2$ and $BF_3 \cdot OEt_2$) were found to effect the desired reduction. In particular, the reaction of $BF_3 \cdot OEt_2$ was investigated in somewhat greater detail, strongly suggesting a carbonyl activation pathway by the Lewis acid (cf. **E**, Scheme 8.1, BF_3 instead of H$^+$). Stoichiometric amounts were needed since silyl fluoride was formed quantitatively as a by-product.

Truly catalytic carbonyl hydrosilylation using a Lewis acid was accomplished two decades later by Parks and Piers [52]. Replacing $BF_3 \cdot OEt_2$ by $B(C_6F_5)_3$ (**43**) allowed ketone hydrosilylations with catalyst loadings as low as 2.0 mol%. The superior catalytic activity of **43** is mainly attributable to its chemical stability compared to the lability of B—F bond in $BF_3 \cdot OEt_2$. Detailed mechanistic investigations by kinetic analysis and NMR studies revealed a mechanism including both silane and carbonyl activation (Scheme 8.15) [52b].

The strong Lewis acidity of **43** catalyzes a hydride shuffle from silicon to boron in two steps (**43→45**, Scheme 8.15). The carbonyl oxygen attacks the initially formed hydride-bridged intermediate **44**, presumably by a S_N2-Si mechanism at silicon [53] (**44→45**); thus, **44** corresponds to a silylium ion equivalent. Rapid

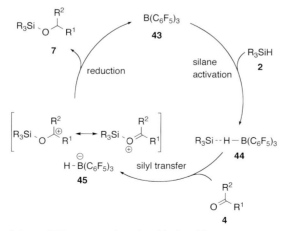

Scheme 8.15 Lewis acid-catalyzed hydrosilylation.

exchange of H and D between adequately labeled silanes by **43** substantiated the silylium borohydride intermediacy under these reaction conditions. The newly formed silylated carbonyl compound is then reduced by boron hydride regenerating the catalyst **43** and liberating **7**. An alternative pathway catalyzed by the silylium ion R_3Si^+ with the free silane **2** as the hydride source was examined in a control experiment. Preformed $R_3Si^+[B(C_6F_5)_4]^-$ together with **2** predominantly produced the deoxygenated product thereby indicating that, in the $B(C_6F_5)_3$-catalyzed process, borohydride **45** and not silane **2** delivers the hydride [52b].

While applying catalyst **43** to the hydrosilylation of α-stereogenic ketones **46** incorporating a C–C triple bond (**46**→**47**, Scheme 8.16), Yamamoto *et al.* discovered the previously unprecedented mode of σ-π chelation [54]. In this remarkable chelation-controlled reaction, the silylium ion functions as a bidentate Lewis acid coordinating not only to the Lewis basic carbonyl oxygen but also to the π system of the proximate triple bond. The reduction product **47** was isolated in good to excellent *syn*-selectivity, which is rationalized by minimized steric repulsion (^{1,2}A interaction) in one of two conceivable transition states (**48** and **49**, Scheme 8.16). Hydride attack from the least hindered face of the σ-π chelate accounts for the predominant relative configuration.

The same model was also valid for the *anti*-selective reduction of β-stereogenic substrates **50** (**50**→**51**, Scheme 8.17) [54]. Preferential equatorial orientation of the β-methyl group in the transition state **52** explains the formation of *anti*-**51**.

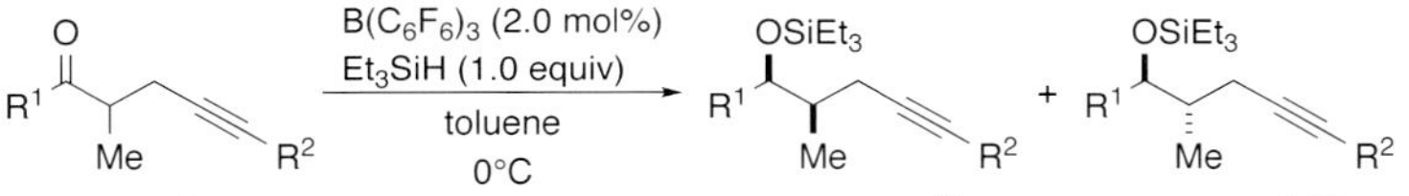

46 *syn*-**47** *anti*-**47**

R^1 = Ph	R^2 = H	90%	88:12
R^1 = Ph	R^2 = Me	99%	83:17
R^1 = Ph	R^2 = Ph	99%	75:25
R^1 = Ph	R^2 = $SiMe_3$	99%	89:11
R^1 = 2-Tolyl	R^2 = H	94%	94:6
R^1 = *t*-Bu	R^2 = H	99%	97:3

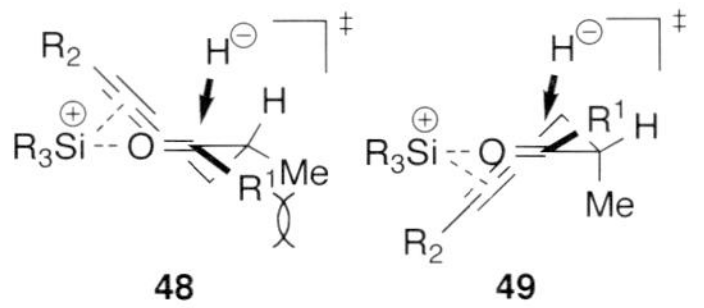

48 **49**

Scheme 8.16 σ-π Chelation controlled hydrosilylation.

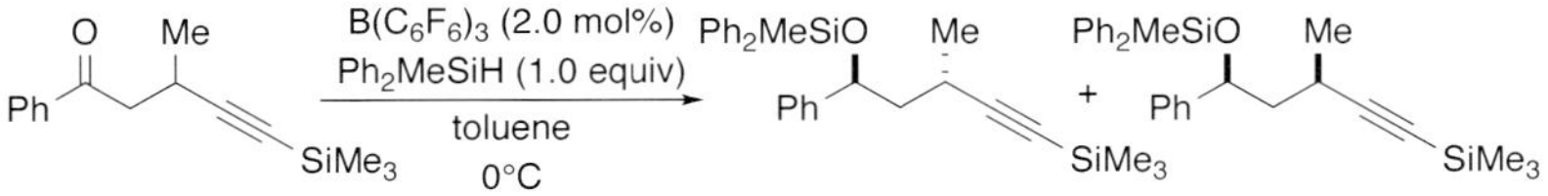

50 99% *anti*-**51** *syn*-**51**

87:13

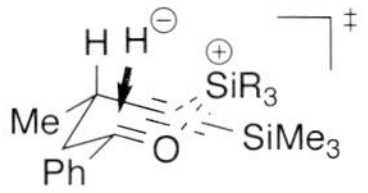

52

Scheme 8.17 *anti*-Selective Lewis acid-catalyzed hydrosilylation.

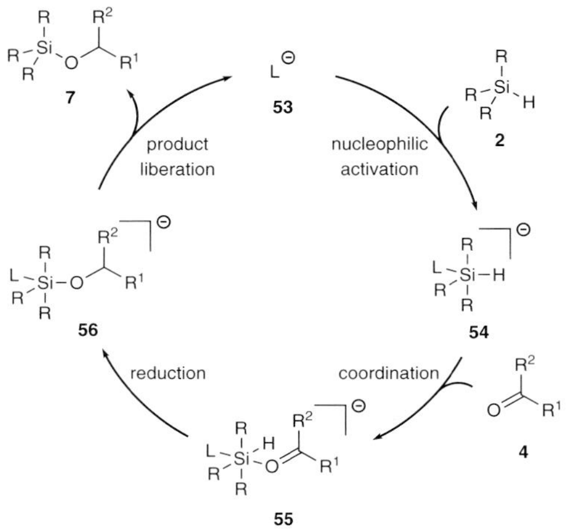

Scheme 8.18 Lewis base catalysis.

8.3.3
Lewis Base-catalyzed Hydrosilylations

The activation of silanes by Lewis bases is a conceptually unique strategy. Hydride transfer from silane to the carbonyl carbon is promoted by nucleophilic addition of a Lewis base to weakly Lewis acidic silicon centers of **A** (**A**→**G**, Scheme 8.1). Hypervalent silicon intermediates **G** display increased Lewis acidity and are potent hydride transfer reagents [55]. Nowadays, this counterintuitive reactivity is generally referred to as "Lewis base activation of Lewis acids" [56]. By exploitation of this strategy, many organocatalytic C—H and C—C bond-forming processes have been elaborated [55].

The generally accepted mechanism of carbonyl reduction by hypervalent silicates starts with the nucleophilic activation of silane **2** (R = alkoxy, halo, aryl, or alkyl) by a nucleophilic promoter L (**53**→**54**, Scheme 8.18). It must be noted that, apart from charged (anionic) activators **53**, neutral donors are also useful Lewis bases [55]. Pentacoordinate intermediate **54**, which is more Lewis acidic than **2**, readily expands its coordination sphere by coordination of a second Lewis base, the carbonyl oxygen of **4** (**54**→**55**). The Si—H bond in **55** is sufficiently weakened to enable the reduction step (**55**→**56**). Final liberation of product **7** regenerates the nucleophilic organocatalyst **53**.

Such Lewis base-mediated carbonyl reductions through hypervalent silicon intermediates were initially pioneered by the Corriu group [57]. Their original procedures included the use of alkaline metal fluorides as nucleophilic activators. This methodology was later adapted by Hiyama and coworkers using quaternary ammonium salts such as TBAF ($Bu_4N^+F^-$) or sulfonium salts such as TASF

Scheme 8.19 *anti*-Selective reduction using fluoride ion catalysis.

[(Et$_2$N)$_3$S$^+$SiMe$_3$F$_2^-$] in polar aprotic solvents (Scheme 8.19) [50a, b, 58a]. Under these conditions, α-oxygenated or α-amino carbonyl compounds as well as β-ketoamides **40** were reduced with good to excellent diastereoselectivities without racemization of the α-stereogenic carbon atom (**40**→**41**). The *anti*-product was formed predominantly; thus, the Lewis base-catalyzed protocol is complementary to the Brønsted acid-mediated variant (cf. Scheme 8.14) [50]. The diastereofacial selectivity was explained by the Felkin–Anh transition state **57**, which is governed by electronic factors. In contrast to the general catalytic cycle (Scheme 8.18), substrate coordination (cf. **55**) was thwarted in the presence of excess HMPA or DMPU. Kinetic analysis indicated that the strong Lewis bases occupy the vacant Lewis acidic sites in **54**, which, in turn, ensured exclusive *inter*molecular hydride attack from hexacoordinate [R$_3$SiF(HMPA)H]$^-$ (**58**) (Scheme 8.19) [58b].

In the late 1980s, several groups described nucleophilic activators for carbonyl reduction such as alkaline alkoxides [59a, d] and alkaline hydrides [59e] as well as the use of preformed silicates [59b, c]. It, therefore, suggests itself that a survey of chiral alkoxides might render this reaction asymmetrically. The Hosomi group was first to develop such an organocatalytic reduction (Scheme 8.20) [60]. Tartrate-derived lithium diolates as well as several dianions of amino alcohols were tested; the dilithium salt of prolinol **59** was a particularly valuable chiral modifier. With (MeO)$_3$SiH as the reducing agent, moderate optical purities were seen in the reduction of acetophenone, even in the presence of catalytic amounts of **59** (**4**→**9**, Scheme 8.20). Of course, stoichiometric quantities of **59** gave higher enantiomeric excesses [60]. Autocatalytic turnover as a competing reaction pathway was recognized by Schiffers and Kagan; the newly formed chiral alkoxide, if expelled from intermediate **56** instead of catalyst **53** (Scheme 8.18), acts itself as a chiral activator [61]. In an attempt to eliminate this issue, less basic phenoxides as cata-

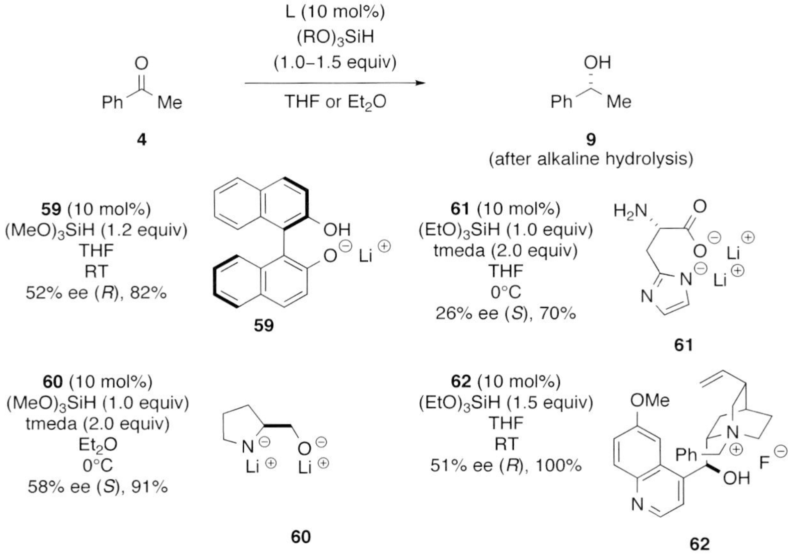

Scheme 8.20 Organocatalytic ketone hydrosilylation.

lysts were considered as an alternative to alkoxides. However, the results obtained with binolate **60** showed only very minor improvement (**4→9**, Scheme 8.20). Later, Brook et al. used the dianion of the imidazole-containing amino acid histidine **61** as a bidentate optically pure catalyst for ketone hydrosilylation [62]. Again, enantiomeric excesses were not satisfying.

Lawrence and coworkers introduced a progressive approach: chiral counterion catalysis [63]. Using the optically pure quaternary cinchona alkaloid-derived ammonium fluoride **62** as a catalyst, the stereochemical information is not located in the Lewis base and, hence, not in a chirally modified silicon hydride. With fluoride as the activator, any stereoinduction stems from the chiral counterion! Remarkable enantioselection of 51% ee was achieved.

8.4
Closing Remarks

The present summary focuses on the current state of the art of each individual method for carbonyl hydrosilylation thereby reflecting their different levels of sophistication.

From a preparative standpoint, conventional rhodium-catalyzed as well as modern "Cu–H"-catalyzed hydrosilylations are certainly highly developed, reliably providing synthetically useful yields and excellent enantioselectivities. A comparison of the efficiency of these catalyses is informative, substrate-to-catalyst as well as substrate-to-ligand ratios differ significantly. Aside from a few important reports in the racemic series [64], S : C and S : L ratios greater 1000 : 1 are scarce in rhodium catalysis whereas copper-based systems seem to not be afflicted by this shortcoming. Optimization is likely to generally result in S : L ratios of 100 000 : 1 and even more [35]. The inexpensive metal source as well as the chemical stability of catalysts [35f] and the availability of heterogeneous protocols with negligible catalyst-leaching [35h] qualify these catalysts as the method of choice [32]. This might foreshadow a change of paradigm in asymmetric hydrosilylation chemistry.

From an academic standpoint, the fascinating rhenium-catalyzed processes stand out, simply because new mechanisms are inherently attractive and asymmetric variants are to be discovered. The transition metal-free hydrosilylations have so far occupied interesting niches. Complementary Brønsted acid-promoted (*syn*-selective) and Lewis base-catalyzed (*anti*-selective) reduction of chiral substrates have gained some synthetic relevance. Enantioselective versions of the latter are still not competitive; it is interesting to note that the related imine reduction is [65]. After submission of this manuscript in December 2006, a notable catalytic enantioselective reduction of prochiral ketones using a chiral Lewis base was reported: Zhou, L., Wang, Z., Wei, S. and Sun, J. (2007) *Chemical Communications*, 2977.

The true value of a method is, however, determined by its comparison with competing techniques: asymmetric ketone hydrogenation and transfer hydrogenation [66, 67]. Catalytic hydrosilylation is now at a similar level to ruthenium-based hydrogenation catalysts [68]. In fact, substrate-to-catalyst ratios of 100 000 : 1 are matching the benchmark set by ruthenium catalysis, which is "top notch" from the industrial standpoint!

References

1 (a) Ojima, I. (1989) The hydrosilylation reaction, in *The Chemistry of Organic Silicon Compounds*, Vol. 1 (eds S. Patai and Y. Apeloig), John Wiley & Sons, Ltd., Chichester, pp. 1479–526. (b) Ojima, I., Li, Z. and Zhu, J. (1998) Recent advances in hydrosilylation and related reactions, in *The Chemistry of Organic Silicon Compounds*, Vol. 2 (eds Z. Rappoport and Y. Apeloig), John Wiley & Sons, Ltd., Chichester, pp. 1687–792.

2 (a) Nishiyama, H. (1999) Hydrosilylation of carbonyl and imino groups, in *Comprehensive Asymmetric Catalysis*, Vol. 1 (eds E.N. Jacobsen, A. Pfaltz and H. Yamamoto), Springer-Verlag, Berlin, pp. 267–88. (b) Ohkuma, T. and Noyori, R. (2004) Hydrosilylation of carbonyl and imino groups, in *Comprehensive Asymmetric Catalysis – Supplement 1* (eds E. N. Jacobsen, A. Pfaltz and H. Yamamoto), Springer-Verlag, Berlin, pp. 55–71.

3 (a) Brunner, H., Nishiyama, H. and Itoh, K. (1993) Asymmetric hydrosilylation and related reactions, in *Catalytic Asymmetric Synthesis* (ed. I. Ojima), Wiley-VCH Verlag GmbH, New York, pp. 303–22. (b) Nishiyama, H. and Itoh, K. (2000) Asymmetric hydrosilylation and related reactions, in *Catalytic Asymmetric Synthesis* (ed. I. Ojima), Wiley-VCH Verlag GmbH, New York, pp. 111–43.

4 Ojima, I. and Hirai, K. (1985) Asymmetric hydrosilylation and hydrocarbonylation, in *Asymmetric Synthesis*, Vol. 5 (ed. J.D. Morrison), Academic Press, Orlando, pp. 104–46.

5 (a) Brunner, H. (1998) Hydrosilylations of carbonyl compounds, in *Transition Metals for Organic Synthesis*, Vol. 2 (eds M. Beller and C. Bolm), Wiley-VCH Verlag GmbH, Weinheim, pp. 131–40. (b) Nishiyama, H. (2004) Hydrosilylations of carbonyl and imine compounds, in *Transition Metals for Organic Synthesis*, Vol. 2 (eds M. Beller and C. Bolm), Wiley-VCH Verlag GmbH, Weinheim, pp. 182–94.

6 Riant, O., Mostefai, N. and Courmarcel, J. (2004) *Synthesis*, 2943.

7 (a) Ojima, I., Nihonyanagi, N. and Nagai, Y. (1972) *Bulletin of the Chemical Society of Japan*, **45**, 3722. (b) Ojima, I., Kogure, T., Kumagai, M., Horiuchi, S. and Sato, T. (1976) *Journal of Organometallic Chemistry*, **122**, 83.

8 Peyronel, J.F. and Kagan, H.B. (1978) *Nouveau Journal de Chimie*, **2**, 211.

9 (a) Kolb, I. and Hetflejs, J. (1980) *Collection of Czechoslovak Chemical Communications*, **45**, 2224. (b) Kolb, I. and Hetflejs, J. (1980) *Collection of Czechoslovak Chemical Communications*, **45**, 2808.

10 (a) Reyes, C., Prock, A. and Giering, W.P. (2002) *Organometallics*, **21**, 546. (b) Reyes, C., Prock, A. and Giering, W.P. (2003) *Journal of Organometallic Chemistry*, **671**, 13.

11 (a) Kuwano, R., Uemura, T., Saitoh, M. and Ito, Y. (1999) *Tetrahedron Letters*, **40**, 1327. (b) Kuwano, R., Sawamura, M., Shirai, J., Takahashi, M. and Ito, Y. (2000) *Bulletin of the Chemical Society of Japan*, **73**, 485. (c) Kuwano, R., Uemura, T., Saitoh, M. and Ito, Y. (2004) *Tetrahedron: Asymmetry*, **15**, 2263.

12 (a) Nishiyama, H., Sakaguchi, H., Nakamura, T., Horihata, M., Kondo, M. and Itoh, K. (1989) *Organometallics*, **8**, 846. (b) Nishiyama, H., Kondo, M., Nakamura, T. and Itoh, K. (1991) *Organometallics*, **10**, 500. (c) Nishiyama, H., Yamaguchi, S., Kondo, M. and Itoh, K. (1992) *Journal of Organic Chemistry*, **57**, 4306.

13 (a) Nishibayashi, Y., Singh, J.D., Segawa, K., Fukuzawa, S. and Uemura, S. (1994) *Journal of the Chemical Society, Chemical Communications*, 1375. (b) Nishibayashi, Y., Segawa, K., Singh, J.D., Fukuzawa, S., Ohe, K. and Uemura, S. (1996) *Organometallics*, **15**, 370.

14 (a) Brunner, H. and Obermann, U. (1989) *Chemische Berichte*, **122**, 499. (b) Helmchen, G., Krotz, A., Ganz, K.-T. and Hansen, D. (1991) *Synlett*, 257.

15 (a) Brunner, H. and Rahman, A.F.M. (1984) *Chemische Berichte*, **117**, 710. (b) Hayashi, T., Hayashi, C. and Uozumi, Y. (1995) *Tetrahedron: Asymmetry*, **6**, 2503. (c) Nishibayashi, Y., Segawa, K., Ohe, K. and Uemura, S. (1995) *Organometallics*, **14**, 5486. (d) Newman, L.M., Williams, J.M.J., McCague, R. and Potter, G.A. (1996) *Tetrahedron: Asymmetry*, **7**, 1597. (e) Langer, T., Janssen, J. and Helmchen, G. (1996) *Tetrahedron: Asymmetry*, **7**, 1599. (f) Sudo, A., Yoshida, H. and Saigo, K. (1997) *Tetrahedron: Asymmetry*, **8**, 3205. (g) Lee, S., Lim, C.W., Song, C.E. and Kim, I.O. (1997) *Tetrahedron: Asymmetry*, **8**, 4027. (h) Lee, S. and Lim, C.W. (2001) *Bulletin of the Korean Chemical Society*, **22**, 231.

16 Tao, B. and Fu, G.C. (2002) *Angewandte Chemie (International Ed. in English)*, **41**, 3892.

17 (a) Oestreich, M. and Rendler, S. (2005) *Angewandte Chemie (International Ed. in English)*, **44**, 1661. (b) Oestreich, M. (2006) *Chemistry – A European Journal*, **12**, 30.

18 Yamanoi, Y. and Imamoto, T. (1999) *Journal of Organic Chemistry*, **64**, 2988.

19 (a) Sakai, J., Schweizer, W.B. and Seebach, D. (1993) *Helvetica Chimica Acta*, **76**, 2654. (b) Heldmann, D.K. and Seebach, D. (1999) *Helvetica Chimica Acta*, **82**, 1096.

20 Yao, S., Meng, J.-C. and Siuzdak, G. (2003) *Journal of Organic Chemistry*, **68**, 2540.

21 Evans, D.A., Michael, F.E., Tedrow, J.S. and Campos, K.R. (2003) *Journal of the American Chemical Society*, **125**, 3534.

22 (a) Gade, L.H., César, V. and Bellemin-Laponnaz, S. (2004) *Angewandte Chemie (International Ed. in English)*, **43**, 1014. (b) César, V., Bellemin-Laponnaz, S., Wadepohl, H. and Gade, L.H. (2005) *Chemistry – A European Journal*, **11**, 2862.

23 (a) Ito, H., Ishizuka, T., Okumura, T., Yamanaka, H., Tateiwa, J., Sonoda, M. and Hosomi, A. (1999) *Journal of Organometallic Chemistry*, **574**, 102. (b) Rendler, S., Auer, G. and Oestreich, M. (2005) *Angewandte Chemie (International Ed. in English)*, **44**, 7620.

24 Nakano, T. and Nagai, Y. (1988) *Chemistry Letters*, 481.
25 (a) Samuel, E. and Harrod, J.F. (1984) *Journal of the American Chemical Society*, **106**, 1859. (b) Harrod, J.F. and Yun, S.S. (1987) *Organometallics*, **6**, 1381. (c) Xin, S. and Harrod, J.F. (1995) *Canadian Journal of Chemistry*, **73**, 999.
26 Halterman, R.L. and Ramsey, T.M. (1994) *Journal of Organic Chemistry*, **59**, 2642.
27 (a) Carter, M.B., Schiøtt, B., Gutiérrez, A. and Buchwald, S.L. (1994) *Journal of the American Chemical Society*, **116**, 11667. (b) Yun, J. and Buchwald, S.L. (1999) *Journal of the American Chemical Society*, **121**, 5640.
28 (a) Beagley, P., Davies, P., Adams, H. and White, C. (2001) *Canadian Journal of Chemistry*, **79**, 731. (b) Beagley, P., Davies, P.J., Blacker, A.J. and White, C. (2002) *Organometallics*, **21**, 5852.
29 Imma, H., Mori, M. and Nakai, T. (1996) *Synlett*, 1229.
30 (a) Bandini, M., Cozzi, P.G., Negro, L. and Umani-Ronchi, A. (1999) *Chemical Communications*, 39. (b) Bandini, M., Bernardi, F., Bottomi, A., Cozzi, P.G., Miscione, G.P. and Umani-Ronchi, A. (2003) *European Journal of Organic Chemistry*, 2972.
31 Lipshutz, B.H. (2002) Copper(I)-mediated 1,2- and 1,4-reductions, in *Modern Organocopper Chemistry* (ed. N. Krause), Wiley-VCH Verlag GmbH, Weinheim, pp. 167–87.
32 Rendler, S. and Oestreich, M. (2007) *Angewandte Chemie (International Ed. in English)*, **46**, 498.
33 Brunner, H. and Miehling, W. (1984) *Journal of Organometallic Chemistry*, **275**, C17.
34 Stryker, J.M., Mahoney, W.S., Daeuble, J.F. and Bestrensky, D.M. (1992) Hydride-mediated homogeneous catalysis: chemoselective catalytic hydride reductions via heterolytic hydrogen activation, in *Catalysis of Organic Reactions* (ed. W.E. Pascoe), Marcel Dekker, New York, pp. 29–44.
35 (a) Lipshutz, B.H., Chrisman, W. and Noson, K. (2001) *Journal of Organometallic Chemistry*, **624**, 367. (b) Lipshutz, B.H., Noson, K. and Chrisman, W. (2001) *Journal of the American Chemical Society*, **123**, 12917. (c) Lipshutz, B.H., Lower, A. and Noson, K. (2002) *Organic Letters*, **4**, 4045. (d) Lipshutz, B.H., Noson, K., Chrisman, W. and Lower, A. (2003) *Journal of the American Chemical Society*, **125**, 8779. (e) Lipshutz, B.H., Caires, C.C., Kuipers, P. and Chrisman, W. (2003) *Organic Letters*, **5**, 3085. (f) Lipshutz, B.H. and Frieman, B.A. (2005) *Angewandte Chemie (International Ed. in English)*, **44**, 6345. (g) Lipshutz, B.H., Lower, A., Kucejko, R.J. and Noson, K. (2006) *Organic Letters*, **8**, 2969. (h) Lipshutz, B.H., Frieman, B.A. and Tomaso, A.E., Jr. (2006) *Angewandte Chemie (International Ed. in English)*, **45**, 1259.
36 (a) Sirol, S., Courmarcel, J., Mostefai, N. and Riant, O. (2001) *Organic Letters*, **3**, 4111. (b) Courmarcel, J., Mostefai, N., Sirol, S., Choppin, S. and Riant, O. (2001) *Israel Journal of Chemistry*, **41**, 231. (c) Lee, D. and Yun, J. (2004) *Tetrahedron Letters*, **45**, 5415. (d) Issenhuth, J.T., Dagorne, S. and Bellemin-Laponnaz, S. (2006) *Advanced Synthesis and Catalysis*, **348**, 1991.
37 Wu, J., Ji, J.-X. and Chan, A.S.C. (2005) *Proceedings of the National Academy of Sciences of the United States of America*, **102**, 3570.
38 (a) Kauer, H., Kauer Zinn, F., Stevens, E.D. and Nolan, S.P. (2004) *Organometallics*, **23**, 1157. (b) Diez-Gonzalez, S., Kauer, H., Kauer Zinn, F., Stevens, E.D. and Nolan, S.P. (2005) *Journal of Organic Chemistry*, **70**, 4784. (c) Diez-Gonzalez, S., Scott, N.M. and Nolan, S.P. (2006) *Organometallics*, **25**, 2355.
39 Carpentier, J.-F. and Bette, V. (2002) *Current Organic Chemistry*, **6**, 913.
40 (a) Feghouli, A., Vanderesse, R., Fort, Y. and Caubère, P. (1989) *Journal of the Chemical Society, Chemical Communications*, 224. (b) Ohkuma, T., Hashiguchi, S. and Noyori, R. (1994) *Journal of Organic Chemistry*, **59**, 217.
41 (a) Mimoun, H. (1999) *Journal of Organic Chemistry*, **64**, 2582. (b) Mimoun, H., de Saint Laumer, J.Y., Giannini, L., Scopelliti, R. and Floriani, C. (1999) *Journal of the American Chemical Society*, **121**, 6158.
42 (a) Mastranzo, V.M., Quintero, L., de Parrodi, C.A., Juaristi, E. and Walsh, P.J. (2004) *Tetrahedron*, **60**, 1781. (b) Bette, V.,

Montreux, A., Savoia, D. and Carpentier, J.-F. (2004) *Tetrahedron*, **60**, 2837. (c) Gérard, S., Pressel, Y. and Riant, O. (2005) *Tetrahedron: Asymmetry*, **16**, 1889.

43 (a) Kennedy-Smith, J.J., Nolin, K.A., Gunterman, H.P. and Toste, F.D. (2003) *Journal of the American Chemical Society*, **125**, 4056. (b) Nolin, K.A., Krumper, J.R., Pluth, M.D., Bergman, R.G. and Toste, F.D. (2007) *Journal of the American Chemical Society*, **129**, 14684.

44 (a) Ison, E.A., Corbin, R.A. and Abu-Omar, M.M. (2005) *Journal of the American Chemical Society*, **127**, 11938. (b) Ison, E.A., Trivedi, E.R., Corbin, R.A. and Abu-Omar, M.M. (2005) *Journal of the American Chemical Society*, **127**, 15374. (c) Du, G. and Abu-Omar, M.M. (2006) *Organometallics*, **25**, 4920. (d) Du, G., Fanwick, P.E. and Abu-Omar, M.M. (2007) *Journal of the American Chemical Society*, **129**, 5180.

45 (a) Fernandes, A.C., Fernandes, R., Romão, C.C. and Royo, B. (2005) *Chemical Communications*, 213. (b) Royo, B. and Romão, C.C. (2005) *Journal of Molecular Catalysis A: Chemical*, **236**, 107. (c) Costa, P.J., Romão, C.C., Fernandes, A.C., Royo, B., Reis, P.M. and Calhorda, M.J. (2007) *Chemistry – A European Journal*, **13**, 3934.

46 Chung, L.W., Lee, H.G., Lin, Z. and Wu, Y.-D. (2006) *Journal of Organic Chemistry*, **71**, 6000.

47 (a) Matano, Y., Brown, S.N., Northcutt, T.O. and Mayer, J.M. (1998) *Organometallics*, **17**, 2939. (b) Paulo, A., Domingos, Â., Garcia, R. and Santos, I. (2000) *Inorganic Chemistry*, **39**, 5669.

48 Kursanov, D.M., Parnes, Z.N. and Loim, N.N. (1974) *Synthesis*, 633.

49 Doyle, M.P. and West, C.T. (1975) *Journal of Organic Chemistry*, **40**, 3835.

50 (a) Fujita, M. and Hiyama, T. (1984) *Journal of the American Chemical Society*, **106**, 4629. (b) Fujita, M. and Hiyama, T. (1985) *Journal of the American Chemical Society*, **107**, 8294. (c) Fujita, M. and Hiyama, T. (1988) *Journal of Organic Chemistry*, **53**, 5415.

51 (a) Doyle, M.P., West, C.T., Donnelly, S.J. and McOsker, C.C. (1976) *Journal of Organometallic Chemistry*, **117**, 129. (b) Fry, J.L., Orfanopulo, M., Adlington, M.G., Dittman, W. and Silverman, S.B. (1978) *Journal of Organic Chemistry*, **43**, 374.

52 (a) Parks, D.J. and Piers, W.E. (1996) *Journal of the American Chemical Society*, **118**, 9440. (b) Parks, D.J., Blackwell, J.M. and Piers, W.E. (2000) *Journal of Organic Chemistry*, **65**, 3090.

53 (a) Sommer, L.H. (1965) *Stereochemistry, Mechanism and Silicon*, McGraw-Hill, New York. (b) Sommer, L.H. (1973) *Intra-Science Chemistry Reports*, **7**, 1.

54 (a) Asao, N., Ohishi, T., Sato, K. and Yamamoto, Y. (2001) *Journal of the American Chemical Society*, **123**, 6931. (b) Asao, N., Ohishi, T., Sato, K. and Yamamoto, Y. (2002) *Tetrahedron*, **58**, 8195.

55 (a) Chuit, C., Corriu, R.J.P., Reye, C. and Young, J.C. (1993) *Chemical Reviews*, **93**, 1371. (b) Kira, M. and Zhang, L.C. (1999) Hypercoordinate silicon species in organic syntheses, in *Chemistry of Hypervalent Compounds* (ed. K. Akiba), Wiley-VCH Verlag GmbH, New York, pp. 147–69. (c) Rendler, S. and Oestreich, M. (2005) *Synthesis*, 1727.

56 Denmark, S.E., Coe, D.M., Pratt, N.E. and Griedel, B.D. (1994) *Journal of Organic Chemistry*, **59**, 6161.

57 (a) Boyer, J., Corriu, R.J.P., Perz, R. and Reye, C. (1979) *Journal of Organometallic Chemistry*, **172**, 143. (b) Boyer, J., Corriu, R.J.P., Perz, R. and Reye, C. (1981) *Tetrahedron*, **37**, 2165. (c) Corriu, R.J.P., Perz, R. and Reye, C. (1983) *Tetrahedron*, **39**, 999.

58 (a) Fujita, M. and Hiyama, T. (1988) *Journal of Organic Chemistry*, **53**, 5405. (b) Fujita, M. and Hiyama, T. (1987) *Tetrahedron Letters*, **28**, 2263.

59 (a) Hosomi, A., Hayashida, H., Kohra, S. and Tominaga, Y. (1986) *Journal of the Chemical Society, Chemical Communications*, 1411. (b) Kira, M., Sato, K. and Sakurai, H. (1987) *Journal of Organic Chemistry*, **52**, 948. (c) Kira, M., Sato, K. and Sakurai, H. (1987) *Chemistry Letters*, **16**, 2243. (d) Corriu, R., Guérin, C., Henner, B. and Wang, Q. (1989) *Journal of Organometallic Chemistry*, **365**, C7. (e) Becker, B., Corriu, R.J.P., Guérin, C., Henner, B. and Wang, Q. (1989) *Journal of Organometallic Chemistry*, **368**, C25.

60 Kohra, S., Hayashida, H., Tominaga, Y. and Hosomi, A. (1988) *Tetrahedron Letters*, **29**, 89.
61 Schiffers, R. and Kagan, H.B. (1997) *Synlett*, 1175.
62 (a) LaRonde, F.J. and Brook, M.A. (1999) *Tetrahedron Letters*, **40**, 3507. (b) LaRonde, F.J. and Brook, M.A. (1999) *Inorganica Chimica Acta*, **296**, 208.
63 Drew, M.D., Lawrence, N.J., Watson, W. and Bowles, S.A. (1997) *Tetrahedron Letters*, **38**, 5857.
64 (a) Niyomura, O., Tokunaga, M., Obora, Y., Iwasawa, T. and Tsuji, Y. (2003) *Angewandte Chemie (International Ed. in English)*, **42**, 1287. (b) Niyomura, O., Iwasawa, T., Sawada, N., Tokunaga, M., Obora, Y. and Tsuji, Y. (2005) *Organometallics*, **24**, 3468. (c) Ito, H., Kato, T. and Sawamura, M. (2006) *Chemistry Letters*, **35**, 1038.
65 Claver, C. and Fernández, E. (2008) Imine hydrogenation, in *Modern Reduction Methods* (eds P. Andersson and I. Munslow), Wiley-VCH Verlag GmbH, Weinheim.
66 (a) Ohkuma, T. and Noyori, R. (1999) Hydrogenation of carbonyl groups, in *Comprehensive Asymmetric Catalysis I-III* (eds E.N. Jacobsen, A. Pfaltz and H. Yamamoto), Springer-Verlag, Berlin, pp. 199–246. (b) Ohkuma, T. and Noyori, R. (2004) Hydrogenation of carbonyl groups, in *Comprehensive Asymmetric Catalysis – Supplement 1* (eds E.N. Jacobsen, A. Pfaltz and H. Yamamoto), Springer-Verlag, Berlin, pp. 1–41.
67 Ohkuma, T., Kitamura, M. and Noyori, R. (2000) Asymmetric hydrogenation, in *Catalytic Asymmetric Synthesis* (ed. I. Ojima), Wiley-VCH Verlag GmbH, New York, pp. 1–110.
68 Hedberg, C. (2008) Carbonyl hydrogenation, in *Modern Reduction Methods* (eds P. Andersson and I. Munslow), Wiley-VCH Verlag GmbH, Weinheim.

9
Enzyme-catalyzed Reduction of Carbonyl Compounds

Kaoru Nakamura and Tomoko Matsuda

9.1
Introduction

Catalysts for asymmetric reductions can be classified into two categories: chemical and biological. Both have their own peculiarities, and development of both to enable the appropriate selection of a catalyst for particular purpose is necessary to promote green chemistry. For example, biocatalysts have the advantages of being natural, having high chemo-, regio- and enantioselectivity, and being active under mild reaction conditions. Here, fundamentals and new methodology for improving reactivity and selectivity of enzymatic reductions will be explained. Synthetic applications of enzymatic reductions are also discussed [1].

9.1.1
Differences between Chemical and Biological Reductions

Biocatalysts have unique characteristics when compared with chemical (homogeneous and heterogeneous) catalysts. Some features that distinguish biocatalysts from chemical catalysts are listed below.

9.1.1.1 Selectivity

Very high enantio-, regio- and chemo-selectivities can be achieved due to the strict recognition of the substrate by the enzyme. For example, the reduction of 2-octanone by a biocatalyst gave >99% ee at room temperature [2a] while chemical methods resulted in less than 87% ee even at very low temperatures (−100 °C to −20 °C [2b, c]. Even in the reduction of ethyl propyl ketone, biocatalysts have achieved high enantioselectivities (98% ee) [2d]. In contrast, chemical catalysts can perform highly enantioselective reductions usually when two adjacent groups of the carbonyl carbon of the ketones are significantly different (Figure 9.1). However, in biocatalytic reductions, the synthesis of both enantiomers is rather difficult compared with chemical reductions. In the latter case, a change of ligand chirality easily affords the other enantiomer of the product.

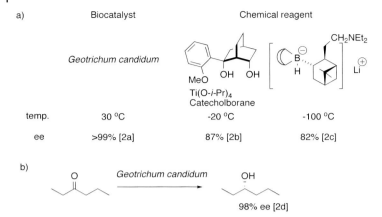

Figure 9.1 Chemical versus biological reduction:
(a) Reduction of 2-octanone; (b) Reduction of 3-hexanone.

9.1.1.2 Safety of the Reaction

Biocatalytic reductions are generally safe, with mild reaction conditions; the solvent is usually water and dangerous reagents are not necessary. For example, ethanol, glucose and so forth are used as hydrogen sources instead of explosive hydrogen gas.

9.1.1.3 Natural Catalysts

The biocatalysts, that is microorganisms, plants and animals, or their isolated enzymes, are reproducible and decompose easily in the environment after use.

9.1.1.4 Catalyst Preparation

Some of the biocatalysts in reductions–isolated enzymes and whole cells–are commercially available and ready to use like chemical catalysts or hydrolytic enzymes. Commercially available biocatalysts include baker's yeast, alcohol dehydrogenase from baker's yeast, *Thermoanaerobium brockii* (TBADH), horse liver and the hydroxysteroid dehydrogenase from *Pseudomonas testosteroni* and *Bacillus spherisus*. However, to obtain other biocatalysts, it is necessary to culture cells from seed cultures that may be commercially available.

9.1.1.5 Large-scale Synthesis and Space–Time Yield

One of the disadvantages of using biocatalysts is the difficulties encountered in large-scale synthesis: (i) work-up procedures may be complicated; (ii) large spaces for the cultivation of the cell may be necessary; or (iii) the space–time yields are not high due to the low substrate concentrations and long reaction times. However, these disadvantages have been surmounted by improving the biocatalysts using genetic methods and by investigating the reaction conditions.

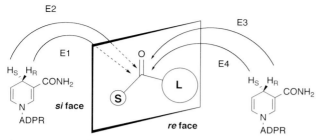

Figure 9.2 Stereochemistry of the hydride transfer from NAD(P)H to the carbonyl carbon on the substrate (S is a small group and L is a large group).

9.1.2
Reaction Mechanism

Dehydrogenases and reductases, classified under EC 1.1.1., are enzymes that catalyze the reduction of carbonyl groups [3]. The natural substrates of the enzymes are alcohols such as ethanol, lactate, glycerol and so forth and the corresponding carbonyl compounds; however, non-natural ketones can also be reduced enantioselectively. To exhibit catalytic activities, the enzymes require a coenzyme such as nicotinamide–adenine dinucleotide (reduced form) (NADH) or the phosphate NADPH from which a hydride is transferred to the substrate carbonyl carbon.

There are four stereochemical patterns that enable the transfer of the hydride from the coenzyme, NAD(P)H, to the substrate, as shown in Figure 9.2. With E1 [4] and E2 [5] enzymes, the hydride attacks the *si*-face of the carbonyl group, whereas with E3 [6] and E4 enzymes, the hydride attacks the *re*-face, which results in the formation of (*R*)- and (*S*)- alcohols, respectively. On the other hand, E1 and E3 enzymes transfer the *pro-R* hydride of the coenzyme, and E2 and E4 enzymes use the *pro-S* hydride. Examples of the E1–E3 enzymes are as follows:

- E1: *Pseudomonas* spp. alcohol dehydrogenase [4a]
  Lactobacillus kefir alcohol dehydrogenase [4b]
- E2: *Geotrichum candidum* glycerol dehydrogenase [5a–c]
  Mucor javanicus dihydroxyacetone reductase [5d]
- E3: Yeast alcohol dehydrogenase [6a]
  Horse liver alcohol dehydrogenase [6b–e]
  Moraxella spp. alcohol dehydrogenase [6f]

9.2
Hydrogen Sources

Enzymes that perform the reduction of carbonyl groups usually require a coenzyme from which a hydride is transferred to the carbonyl carbon. Since reduction of the substrate is concomitant with oxidation of the coenzyme, and the coenzyme

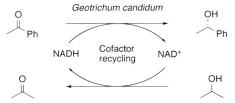

Scheme 9.1 NADH recycling using alcohol as a hydrogen source for reduction [2a, 7].

is too expensive to use as a throwaway reagent, it is necessary to recycle and reuse the oxidized form of the coenzyme [1]. The oxidized forms of the coenzymes have to be transformed back to their reduced form for the next cycle of the reduction. Hydrogen sources are necessary to perform this reduction reaction. In biocatalytic reductions, alcohols such as ethanol and 2-propanol, sugars such as glucose, glucose 6-phosphate and glucose 6-sulfate, formic acid and dihydrogen can be used. Some examples are shown in this section.

9.2.1
Alcohol as Hydrogen Sources for Reduction

Alcohols such as ethanol, 2-propanol and so forth have been widely used to recycle the coenzyme for the reduction catalyzed by alcohol dehydrogenase since the enzyme catalyzes both reduction and oxidation. Usually, an excess of the hydrogen source is used to push the equilibrium to the formation of the desired alcohols.

There is an interesting example of the use of secondary alcohols in which the enantioselectivity and reaction yield were improved by recycling the coenzyme using secondary alcohols for the reduction of ketones with *Geotrichum candidum* (Scheme 9.1) [2a, 7]. Although the enantioselectivity of the reduction of acetophenone with resting cell of the microbe was low, the use of the dried cells and the addition of a catalytic amount of NAD(P)$^+$ and an excess of secondary alcohols such as 2-propanol or cyclopentanol increased both the enantioselectivity and chemical yield of the reduction; the (*S*)-alcohol was obtained in >99% ee with high yield.

9.2.2
Sugars as Hydrogen Sources for Reduction

Glucose and glucose 6-phosphate have been widely used as a reducing source. A recent example has shown that the thermostable glucose-6-phosphate dehydrogenase (G6PDH), from *Bacillus stearothermophilus*, can be used for recycling NADPH at high temperatures (55 °C) for the reduction of 2-butanone by the thermostable alcohol dehydrogenase from *Thermoanaerobacter brockii*. Glucose-6-sulfate was used instead of glucose-6-phosphate as the sulfate is three times more effective than the phosphate, the natural substrate for G6PDH (Scheme 9.2) [8].

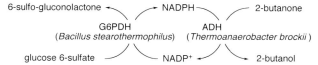

Scheme 9.2 Reduction of ketone with alcohol dehydrogenase from *Thermoanaerobacter brockii* using glucose 6-sulfate as a hydrogen source [8].

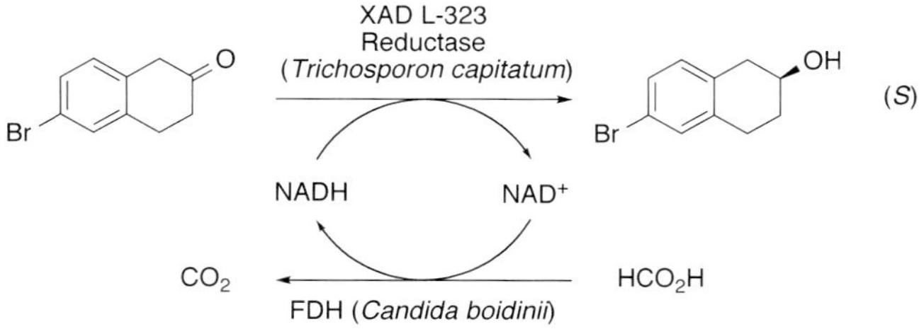

Scheme 9.3 Reduction of 6-bromotetralone with reductase from *Trichosporon capitatum* using formate as a hydrogen source [9].

9.2.3
Formate as a Hydrogen Source for Reduction

Formate is one of the most representative hydrogen sources for the biocatalytic reduction as CO_2, formed by the oxidation of formate, is released from the reaction system. For example, a formate dehydrogenase (FDH) system was applied in the reduction of 6-bromotetralone to (S)-6-bromotetralol, a potential pharmaceutical precursor, with the NADH-dependent ketone reductase from *Trichosporon capitatum* [9]. A resin (XAD L-323) was used to bind the product (Scheme 9.3).

9.2.4
Molecular Hydrogen as a Hydrogen Source for Reduction

Molecular hydrogen has been used for the recycling of coenzymes [10]. The soluble hydrogenase I (H_2: $NADP^+$ oxidoreductase, EC 1.18.99.1) from the marine hyperthermophilic strain of the archaeon *Pyrococcus furiosus* (PF H2ase I) has been used as a biocatalyst in the enzymatic production and regeneration of NADPH employing molecular hydrogen. Utilizing the thermophilic NADPH-dependent alcohol dehydrogenase from *Thermoanaerobium* sp. (ADH M) coupled to the PF H2ase I in situ NADPH-regenerating system, (2S)-hydroxy-1-phenyl-propanone was quantitatively reduced to the corresponding (1R,2S)-diol in >98% de, with total turnover numbers (ttn: mol product/mol consumed cofactor $NADP^+$) of 160 being obtained (Scheme 9.4) [10].

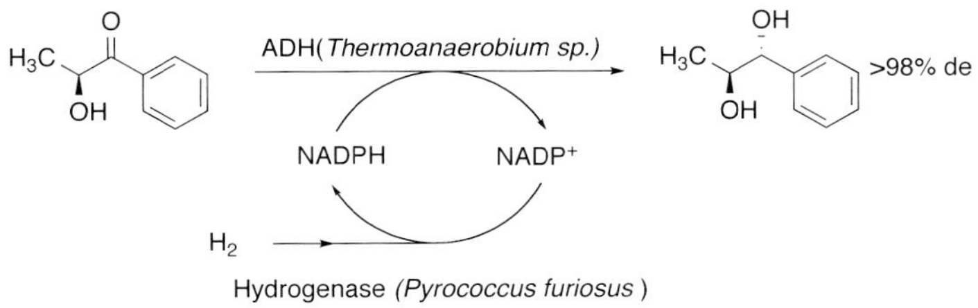

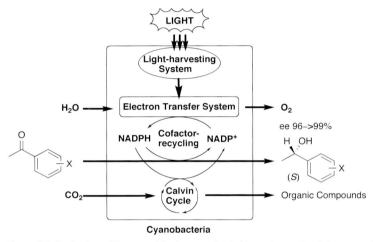

Scheme 9.4 Reduction of ketone with alcohol dehydrogenase from *Thermoanaerobium* sp. using molecular hydrogen as a hydrogen source [10].

Figure 9.3 Reduction of ketone with photosynthetic biocatalyst using light energy [11b, c].

9.2.5
Light Energy as a Hydrogen Source for Reduction

Photochemical methods [11] have been developed that provide an environmentally friendly system employing light energy to regenerate NAD(P)H, for example the use of a cyanobacterium, a photosynthetic biocatalyst. Using this biocatalysts, the reduction of acetophenone derivatives occurred more effectively under illumination than in the dark (Figure 9.3) [11b, c]. The light energy harvested by the cyanobacterium is converted into chemical energy in the form of NADPH through an electron-transfer system, and, subsequently, the chemical energy (NADPH) is used to reduce the substrate to the corresponding chiral alcohol (96→99% ee).

9.2.6
Electric Power as a Hydrogen Source for Reduction

The electrochemical regeneration of NAD(P)H has long been recognized as a potentially powerful technology from the viewpoint of green chemistry, since this

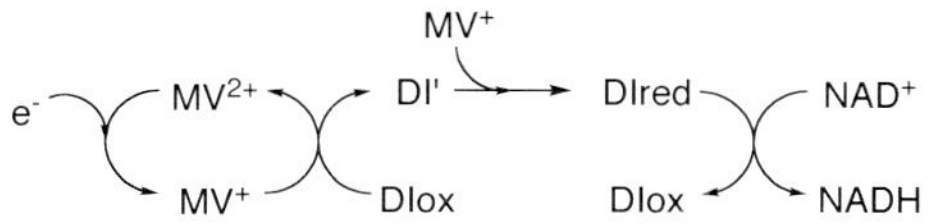

Scheme 9.5 Regeneration of NADH using diaphorase (DI) and electric power [12a].

does not required a second enzyme and co-substrate [1b]. However, the method is not totally effective due to the necessity for high overpotentials for direct reduction of the coenzyme, electrode fouling, dimerization of the coenzyme and the fact that only the enzyme in the immediate vicinity of the electrode is productive. Viologen-diaphorase (lipoamide dehydrogenase) was used for reduction of $NAD(P)^+$, where viologens were used as mediator of an electron from electrode to diaphorase (Scheme 9.5) [12a].

Organometallics such as rhodium complexes have also used for the electrochemical regeneration of NAD(P)H from electrodes [12b].

9.3
Methodology for Stereochemical Control

Since the stereoselectivities of biocatalytic reductions are not always satisfactory, modifications are often necessary for their practical use. This section explains how to find, prepare and modify suitable biocatalysts, how to recycle the coenzyme and how to improve the productivity and enantioselectivity of the reactions.

9.3.1
Screening of Biocatalysts

Screening for a novel enzyme is a classical method and still one of the most powerful tools for finding a biocatalytic reduction system [13]. It is possible to discover suitable biocatalysts by applying the latest screening and selection technologies, allowing rapid identification of enzyme activities from diverse sources [13a]. Enzyme sources used for screening can include soil samples, commercial enzymes, culture sources or a clone bank. Their origin can be microorganisms, animals or plants. From these sources, enzymes that are regularly expressed and those that are not expressed in the original host can be tested to determine whether they are suitable for the transformation of certain substrates [13a].

For example, a screening of 416 strains (71 bacterial strains, 45 actinomycetes, 59 yeast, 60 basidiomycetes, 33 marine fungi and 148 filamentous fungi) was performed to look for microorganisms that display reductase activity in the absence of oxidase activity [13b]. A new microorganism, *Diplogelasinospora grovesii* IMI 171018 was isolated and showed very high activity and stereoselectivity in the reduction of cyclic ketones (Scheme 9.6a).

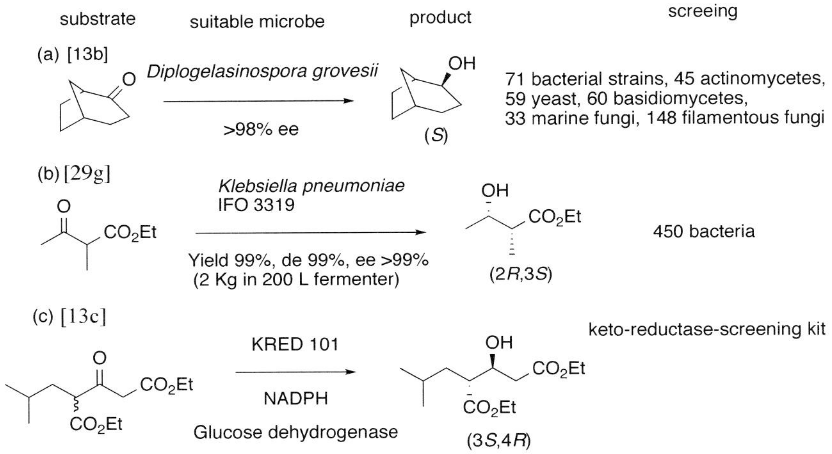

216 | 9 Enzyme-catalyzed Reduction of Carbonyl Compounds

Scheme 9.6 Screening for a biocatalysts [13].

In another example, *Klebsiella pneumoniae* IFO 3319, out of 450 bacterial strains, was found to give the corresponding (2R,3S)-hydroxy ester with 99% de and >99% ee (Scheme 9.6b) [14g]. An eto-reductase-screening kit is now available (10 enzymes for US$2295 from Biocatalytics Inc. at the time of writing) and this kit was used in the screening of the reduction of a keto diester (Scheme 9.6c), KRED101 afforded (3S,4R)-hydroxydiester in 90% de with 100% ee [13c].

9.3.2
Modification of Biocatalysts by Genetic Methods

9.3.2.1 Engineered Yeast
Recently, various genetic methods for screening, as well as classical methods, have been reported for finding effective biocatalytic reductions. One of the most interesting examples is the use of glutathione S-transferase, (GST) fusion proteins, to allow rapid identification of synthetically useful biocatalysts [15]. A set of fusion proteins consisting of GST linked to the N-terminus of putative dehydrogenases produced by baker's yeast *(Saccharomyces cerevisiae)* were screened for the reduction of various substrates. For example, ethyl 2-oxo-4-phenylbutyrate was reduced rapidly in the presence of NADH and NADPH by two dehydrogenases, Ypr1p and Gre2p, providing the (R)- and (S)-alcohols, respectively, with high stereoselectivities (Scheme 9.7a) [15a]. The same enzymes were overexpressed in their native forms in *Escherichia coli* and growing cells of the engineered strains could also be used to carry out the reductions without the need for exogenous cofactor.

A representative set of α- and β-keto esters was also tested as substrates (11 total) for each purified fusion protein (Scheme 9.7b) [15b]. The stereoselectivities of β-keto ester reductions depended on both the identity of the enzyme and the substrate structure, with some reductases yielding both L- and D-alcohols with

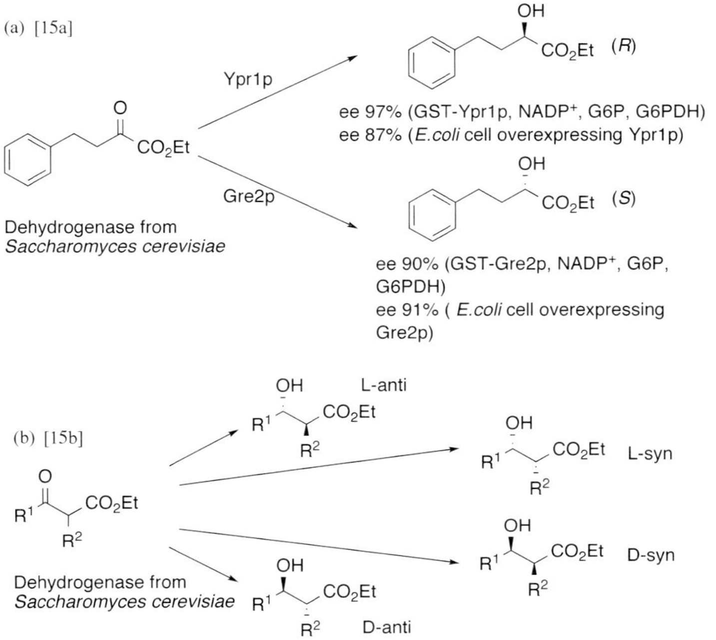

9.3 Methodology for Stereochemical Control | 217

Scheme 9.7 Identification of appropriate biocatalysts from fusion protein libraries from baker's yeast [15].

high stereoselectivities. The results demonstrate the power of genomic fusion protein libraries to identify appropriate biocatalysts rapidly and expedite process development.

9.3.2.2 Overexpression

Biocatalysts discovered by screening can been prepared in a large quantity by overexpression of the enzymes in transformed *Escherichia coli* [16]. For example, the gene encoding (6R)-2,2,6-trimethyl-1-4-cyclohexanedione (levodione) reductase was cloned from the genomic DNA of the soil-isolated bacterium *Corynebacterium aquaticum* M-13 (Scheme 9.8a) [16a]. The enzyme was sufficiently produced in recombinant *E. coli* cells and the enzyme purified from *E. coli* catalyzed stereo- and regio-selective reduction of levodione.

Enantioselective microbial reductions of methyl, ethyl and *t*-butyl 4-(2'-acetyl-5'-fluorophenyl)butanoates were demonstrated using the reductase from *Pichia methanolica* SC 13825. It was cloned and expressed in *E. coli* with recombinant cultures used for the enantioselective reduction of keto ester to the corresponding (S)-hydroxymethyl ester. On a preparative scale, a reaction yield of 98% and an ee of 99% were obtained (Scheme 9.8b).

The synthesis of ethyl (R)-4-chloro-3-hydroxy-butanoate((R)-ECHB) from ethyl 4-chloroacetoacetate (ECAA) was studied using whole recombinant cells of *E. coli*

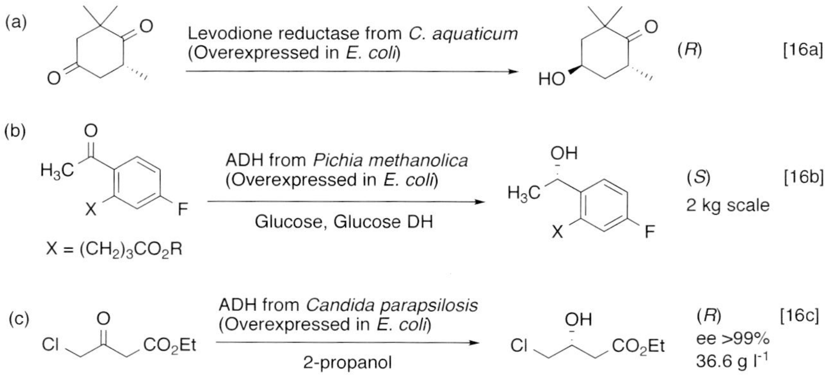

Scheme 9.8 Overexpressed reductase in *E. coli* [16].

expressing a secondary alcohol dehydrogenase of *Candida parapsilosis* [16c]. Using 2-propanol as an energy source to regenerate NADH, the yield of (*R*)-ECHB reached $36.6 \, gl^{-1}$ (more than 99% ee and 95.2% yield) without addition of NADH to the reaction mixture (Scheme 9.8c). On the other hand, a novel carbonyl reductase (KLCR1) that reduced ECAA to (*S*)-ECHB was purified from *Kluyveromyces lactis* [16d]. KLCR1 catalyzed the NADPH-dependent reduction of ECAA enantioselectively but not the oxidation of (*S*)-ECHB.

9.3.2.3 Modification of Biocatalysts: Directed Evolution

Directed evolution of enzymes has been used to improve the reducing function of the enzymes, and this method was used to eliminate the cofactor requirement of *Bacillus stearothermophilus* lactate dehydrogenase, which is activated in the presence of fructose 1,6-bisphosphate [17]. The activator is expensive and representative of the sort of cofactor complications that are undesirable in industrial processes. Three rounds of random mutagenesis and screening produced a mutant that is almost fully activated in the absence of fructose 1,6-bisphosphate. The K_m of the enzyme without the cofactor was improved from $5 \, mM$ to $0.07 \, mM$ for the pyruvate.

9.3.3
Modification of Substrates

The enantioselectivity of a biocatalytic reduction can be controlled by modifying the substrate as the enantioselectivity of the reduction reaction is profoundly affected by the substrate's structure [18]. In the reduction of 4-chloro-3-oxobutanoate by baker's yeast, the length of the ester moiety controls the stereochemical course of the reduction [18a–c]. When the ester moiety was smaller than a butyl group (*S*)-alcohols were obtained, and when it was larger than a pentyl group (*R*)-alcohols were obtained (Figure 9.4a). After reduction, the ester moiety

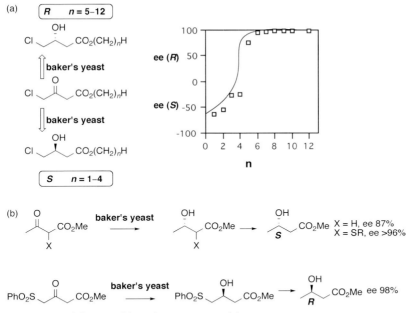

Figure 9.4 Modification of the substrate to control the enantioselectivities: (a) effect of the length of the ester moiety [18a–c]; (b) effect of introducing sulfur functionalities [18d, e].

can easily be exchanged without racemization, so both enantiomers of an equivalent synthetic building block could be obtained using the same reaction system.

Another example is the introduction of sulfur functionalities into the keto esters at the α- or α′-positions, which can be eliminated after reduction. Methylthio1 [18d] and phenylsulfonyl [18e] groups were used to improve the enantioselectivities (Figure 9.4b).

9.3.4
Modification of Reaction Conditions

9.3.4.1 Acetone Treatment of the Cell

A dried cell mass is often used as a biocatalyst for a reduction since it can be stored for a long time and can be used whenever needed without cultivation. One convenient method of drying the cell mass is acetone dehydration. Dried cells of G. candidum IFO 4597 are easily obtained by mixing the cells with cold acetone (−20 °C) followed by filtration and drying under reduced pressure.

Cell drying not only aids the preservation of the cell, but also contributes to the stereochemical control (Table 9.1). The reduction of acetophenone catalyzed by untreated wet whole cells of G. candidum IFO 4597 resulted in poor enantioselectivity (28% ee (R)), while the reduction by dried powdered cell with the aid of a catalytic amount of NAD(P)$^+$ and an excess of secondary

Table 9.1 Acetone treatment of *Geotrichum candidum* for the improvement of enantioselectivity [19].

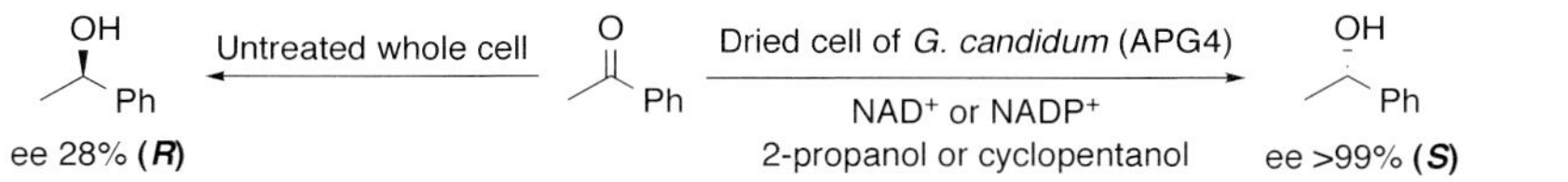

OH
| Ph ← Untreated whole cell ― O=Ph → Dried cell of *G. candidum* (APG4) / NAD⁺ or NADP⁺ / 2-propanol or cyclopentanol → OH | Ph
ee 28% (**R**)   ee >99% (**S**)

Catalyst	Coenzyme	Additive	Yield (%)	ee (%)
Untreated whole cell	None	None	52	28 (R)
Acetone-dried cell (APG4)	None	None	0	–
Acetone-dried cell (APG4)	NAD⁺	None	1	–
Acetone-dried cell (APG4)	NAD⁺	cyclopentanol	97	>99 (S)

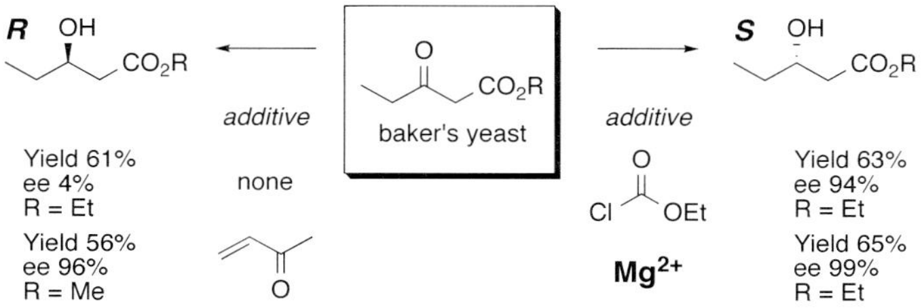

Scheme 9.9 Improvement of the enantioselectivity by using an inhibitor of undesired enzymes [20a–c].

alcohol, such as 2-propanol and cyclopentanol, gave the (*S*)-alcohol in excellent ee. The improvement in the enantioselectivity from 28% (*R*) to >99% (*S*) was due to the suppression of all enzyme that reduces the substrate, followed by the stimulation of an *S*-directing enzyme by the addition of the coenzyme and an excess of 2-propanol [19c].

9.3.4.2 Selective Inhibitors

In the case of poor enantioselectivities, due to the presence of two competing enzymes exhibiting different selectivities, one of the most straightforward methods to improve the enantioselectivity is to use an inhibitor of the unwanted enzyme(s). Ethyl chloroacetate, methyl vinyl ketone, allyl alcohol, allyl bromide, sulfur compounds, Mg^{2+} and Ca^{2+}, have all been reported as inhibitors of enzymes in baker's yeast [20]. For example, the low enantioselectivity in the yeast reduction of an α-keto ester was improved by addition of methyl vinyl ketone, Mg^{2+} or ethyl chloroacetate as described in Scheme 9.9 [20a–c]. From enzymatic studies, using purified enzymes from baker's yeast, the enzymes inhibited were identified and the inhibition mechanism was reported as being noncompetitive [20k].

9.4
Medium Engineering

Biocatalytic reductions have been performed in nonaqueous solvents to improve the efficiency of the reaction. This section explains the use of organic solvents, supercritical fluids and ionic liquids in enzymatic reductions.

9.4.1
Organic Solvent

9.4.1.1 Water-soluble Organic Solvent
Water-soluble organic solvents have often been used as a co-solvent to solubilize miscible organic substrates. Since organic compounds, including solvents, can be incorporated inside an enzyme, they may affect the stereoselectivity of enzymatic reactions. The addition of dimethyl sulfoxide (DMSO) (10%) has been shown to enhance not only chemical yields but also enantioselectivity of yeast reductions. Thus, in the reduction of 2-chlorofluoren-9-one with baker's yeast the poor yield (23%) and ee (80%) were increased to 65% and >99% respectively (Scheme 9.10) [21].

9.4.1.2 Aqueous–Organic Two-Phase Reaction
Aqueous–organic two-phase reactions have been widely performed [22]; one of the purposes of using two-phase reaction systems is to control substrate concentrations in the aqueous phase where the biocatalysts reside. Hydrophobic substrates and products dissolve easily in the organic phase, decreasing their concentration in the aqueous phase. The merits of controlling and decreasing the substrate concentration in the aqueous phase are as follows:
1. Substrate and product inhibition can be prevented.
2. When whole cells containing multiple enzymes with opposite selectivities and different K_m (Michaelis–Menten constant) values are used, problems of low selectivities occur. If the substrate concentration is decreased, one of the enzymes with a low K_m value catalyzes the reaction so that the selectivity can be improved.
3. The decomposition of unstable substrate/product by aqueous buffer can be prevented by dissolving the substrate and product in the organic phase.

Accordingly, organic solvents are widely used in biocatalytic reductions; an interesting example is the stereochemical control observed when using organic

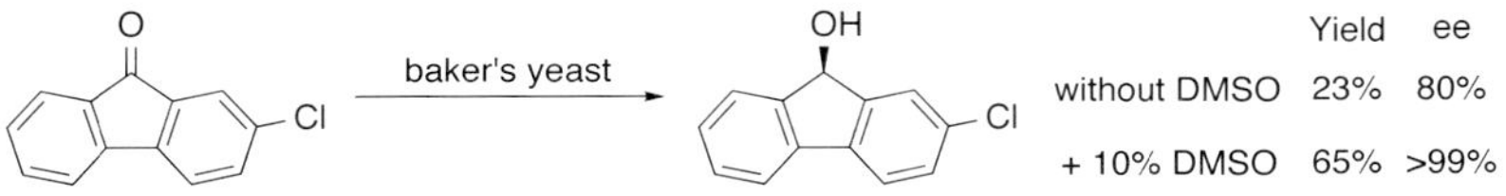

Scheme 9.10 Stereochemical control by soluble organic solvent [21].

Table 9.2 Reduction of acetophenone in hexane by the immobilized resting cell of *Geotrichum candidum* and 2-hexanol [22].

Substrate	Yield (%)	ee (%)
Acetophenone	52[a]	28 (R) [a]
Acetophenone	2[b]	–
Acetophenone	73	>99 (S)
o-Methylacetophenone	59	>99 (S)
m-Methylacetophenone	56	>99 (S)
p-Methylacetophenone	40	>99 (S)
1-Phenyl-2-propanone	97	>99 (S)
Benzylacetone	88	>99 (S)

a Reaction in water without immobilization and without 2-hexanol.
b Reaction without 2-hexanol.

solvents in reductions with *Geotrichum candidum* IFO 4597. The reduction of acetophenone with *G. candidum* (immobilized on a water-absorbing polymer) in hexane in the presence of 2-hexanol afforded (S)-phenylethanol in excellent enantiomeric excess, while the reaction in water alone gave low ee values (Table 9.2) [22]. Secondary alcohols such as 2-propanol and cyclopentanol can also be used as additives.

9.4.2
Use of Hydrophobic Resin

Instead of organic solvents a hydrophobic resin, Amberlite XAD, can be used to control the substrate concentration [23]. When XAD is added to the reaction mixture, substrate and products are adsorbed onto the hydrophobic resin, since the substrate and product are usually hydrophobic, and so the effective concentration of the substrate around the enzyme is decreased. In the presence of XAD, simple aliphatic and aromatic ketones were reduced to their corresponding (S)-alcohols in excellent enantioselectivity, while low enantioselectivities were observed in the absence of the polymer (Scheme 9.11a).

In the reduction of benzyloxypropanone, the hydrophobic polymer, XAD-7 was used to prevent product inhibition and to increase substrate concentration [23b]. Thus, the reduction proceeded in $70\,g\,l^{-1}$ substrate concentration and afforded 87%

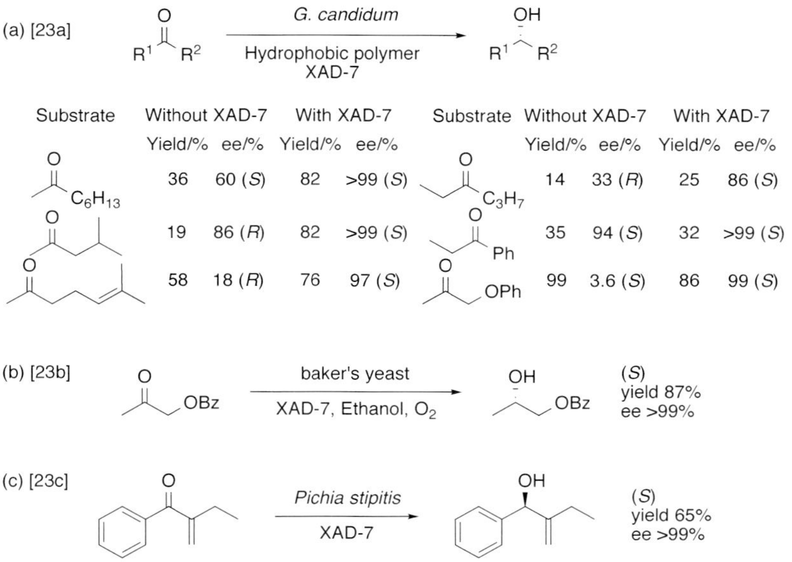

Scheme 9.11 Reduction of ketones in the presence of hydrophobic polymer XAD.
(a) Reduction of ketones with *Geotrichum candidum* [23a].
(b) Reduction of ketone with baker's yeast [23b].
(c) Reduction of ketone with *Pichia stipitis* [23c].

(12.4 g) of (S)-1-benzoyloxy-2-propanol in >99% ee (Scheme 9.11b). The acyclic enone, 2-ethyl-1-phenylprop-2-en-1-one, was reduced with the yeast *Pichia stipitis* CCT 2617 [23c]. The reduction proceeded chemo- and enantioselectively affording (S)-2-ethyl-1-phenylprop-2-en-1-ol (65% yield, >99% ee), XAD-7 was used to decrease and control the concentration of both substrate and product (Scheme 9.11c).

9.4.3
Supercritical Carbon Dioxide

The difficulties involved in conducting biocatalytic reductions in aqueous media, such as the extraction of products that dissolve in aqueous media at low concentrations, can be overcome by the use of supercritical carbon dioxide (scCO$_2$) since CO$_2$ transforms into a gas as the pressure decreases. The reduction of carbonyl groups using alcohol dehydrogenases has been conducted in scCO$_2$ using immobilized cell of *G. candidum* [24a, b]. Since whole resting cells were used, the addition of expensive coenzymes was avoided; moreover, the solubility of the coenzymes in scCO$_2$ did not need to be considered. In fact, the enzymatic reduction in scCO$_2$ proceeded for various ketones, such as acetophenone, acetophenone derivatives, benzyl acetone and cyclohexanone. The enantioselectivities obtained for this

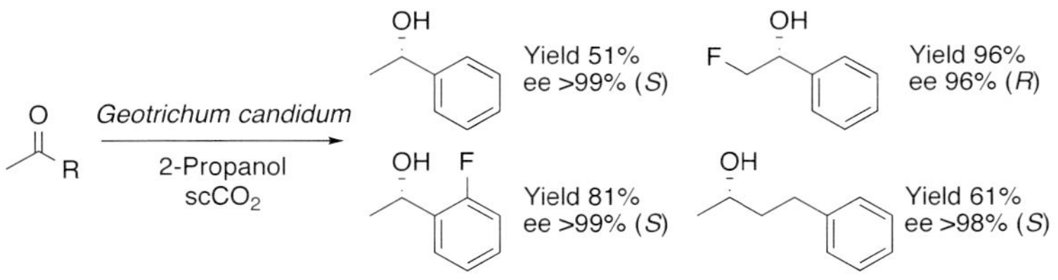

Scheme 9.12 Asymmetric reduction of ketones in supercritical CO_2 by *Geotrichum candidum* immobilized whole cell [24].

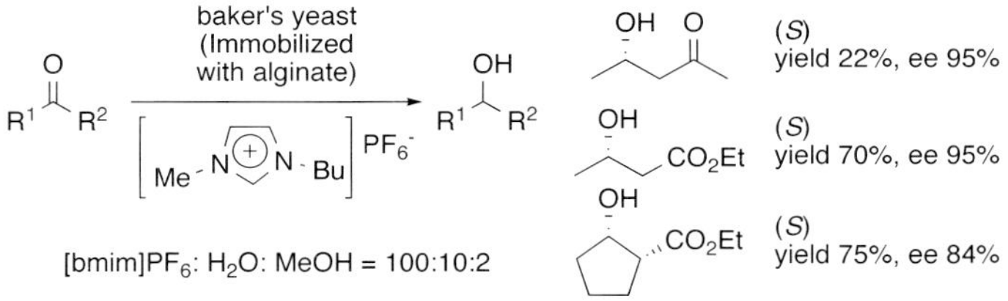

Scheme 9.13 Use of ionic liquids in yeast reduction [25a].

system are superior to or at least equal to those for most other biocatalytic and chemical systems (Scheme 9.12) [24a].

The immobilized resting-cell of *G. candidum* was also used as a catalyst for the reduction of *o*-fluoroacetophenone and cyclohexanone in a semi-continuous flow process using $scCO_2$ [24b]. With flow reactors, the addition of a substrate to the column with a catalyst yields the product and CO_2, whereas, with a batch reactor separation of the product from the biocatalyst is necessary after depressurization. Moreover, the size of flow reactor required to generate an amount of product comparable to that from a batch reactor is smaller. Reactions using a semi-continuous flow process also result in a higher space–time yield than the corresponding batch process.

9.4.4
Ionic Liquid

The ionic liquid [bmim]PF_6 has also been used as a solvent in yeast reductions [25a]. The reduction of ketones with immobilized baker's yeast (alginate) in a 100:10:2 [bmim]PF_6:H_2O:MeOH mix afforded chiral alcohols (Scheme 9.13). A two-phase system containing a buffer and the ionic liquid [bmim][$(CF_3SO_2)_2$N] has been used in the reduction of 2-octanone with an alcohol dehydrogenase from *Lactobacillus brevis*. Due to favorable partition coefficients, the reduction is faster than that in a biphasic system containing buffer and *t*-butyl ether [25b].

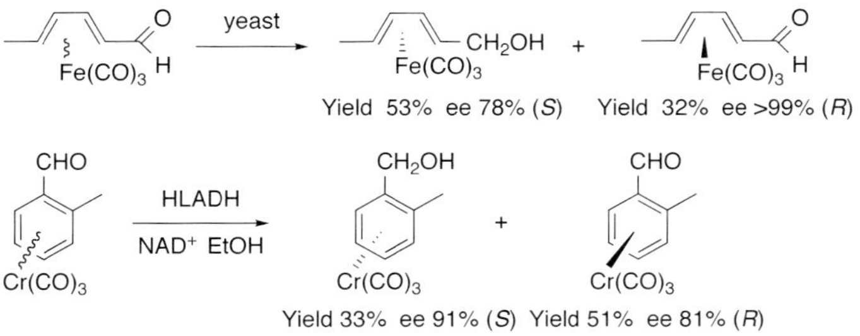

Scheme 9.14 Reduction of organometallic aldehydes to produce alcohols with planar chiralities [26c–e].

9.5 Synthetic Applications

9.5.1 Reduction of Aldehydes

Many aldehyde reductases transform both aldehydes and ketones, such as phenylacetaldehyde reductase from a styrene-assimilating *Corynebacterium* strain, ST-10, which reduces hexylaldehyde and phenylacetaldehyde [26a]. Other aldehyde reductases such as that from *Sporobolomyces salmonicolor* also reduce aldehydes as well as ketones [26b].

Organometallic aldehydes can be reduced enantioselectively with dehydrogenases; for example, optically active organometallic compounds having planar chiralities were obtained by biocatalytic reduction of racemic aldehydes with yeast [26c, d] or horse liver alcohol dehydrogenase (HLADH) [26e] as shown in Scheme 9.14.

9.5.2 Reduction of Ketones

Simple aliphatic ketones as well as aromatic ketones can be reduced with very high enantioselectivity using biocatalysts. Aliphatic ketones such as 2-pentanone, 2-butanone and 3-hexanone were reduced with excellent enantioselectivities to their corresponding (S)-alcohols using the dried cells of *Geotrichum candidum* (Scheme 9.15) [19]. The dried-cell *G. candidum* system can distinguish between two alkyl groups with a difference of a single methylene unit. The system can also be applied to the reduction of aromatic ketones. Trifluoromethyl ketones were reduced by the same system and afforded (S)-alcohols in excellent enantioselectivity [19h].

Resting cells of *G. candidum* as well as dried cells have been shown to be an effective catalyst for this asymmetric reduction. Both enantiomers of secondary alcohols were prepared by reduction of the corresponding ketones with a single microbe [27]. The reduction of aromatic ketones with *Geotrichum candidum* IFO 5767 afforded the corresponding (S)-alcohols in excellent enantioselectivity when

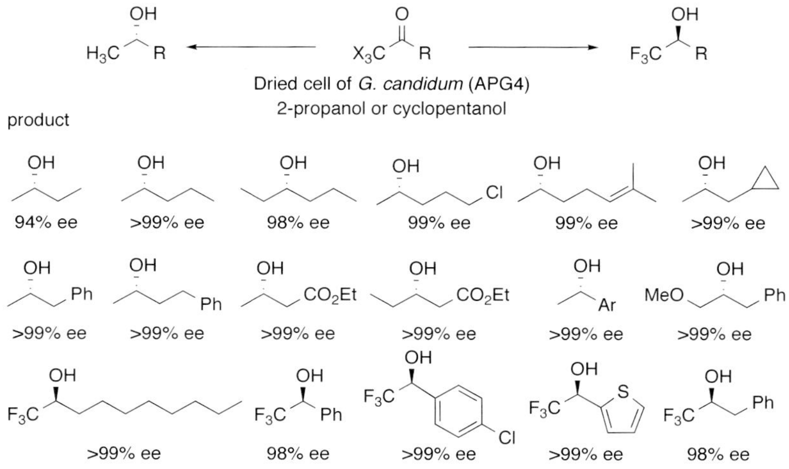

Scheme 9.15 Reduction of ketones with dried cells of G. candidum, NAD(P)⁺ and a secondary alcohol [19].

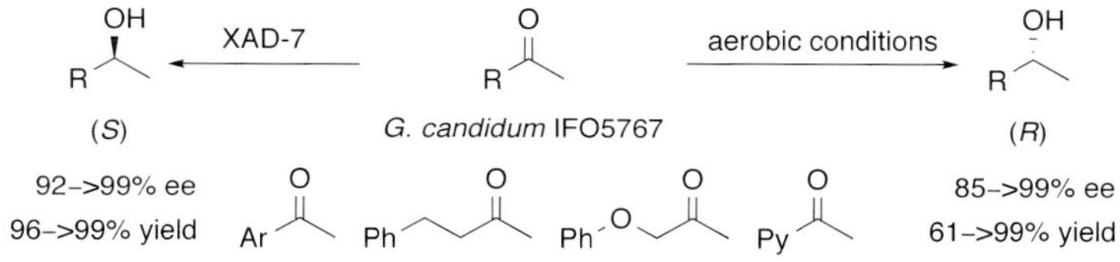

Scheme 9.16 Synthesis of both enantiomers of secondary alcohols with one kind of microorganism [27].

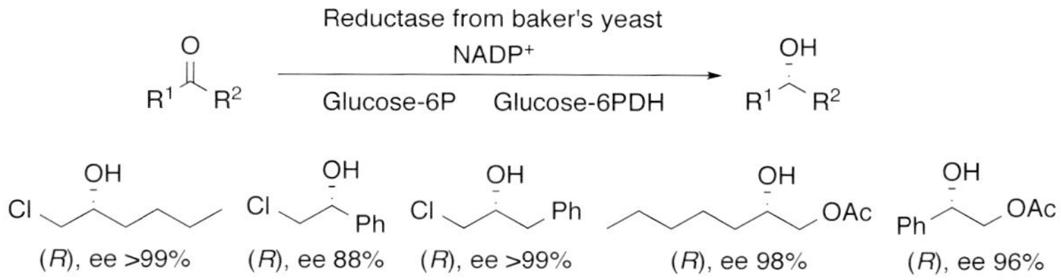

Scheme 9.17 Reduction of ketones with reductase from baker's yeast [28a].

Amberlite XAD-7, a hydrophobic polymer, was added to the reaction system. The same microbe also afforded (R)-alcohols in excellent enantioselectivity when the reaction was conducted under aerobic conditions (Scheme 9.16).

Baker's yeast has been widely used for the reduction of ketones. The substrate specificity and enantioselectivity of the carbonyl reductase from baker's yeast, which is known to catalyze the reduction of β-keto ester to L-hydroxyester (L2-enzyme) [20k], was investigated and was found to reduce chloro- and acetoxy ketones with high stereoselectivities (Scheme 9.17) [28a].

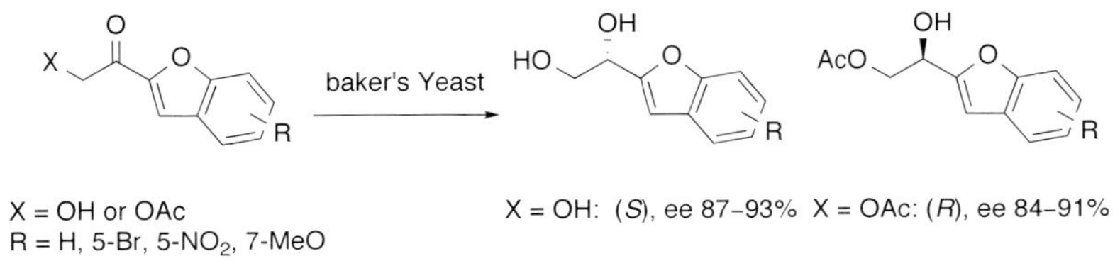

Scheme 9.18 Reduction of hydroxy and acetoxy ketones [28b].

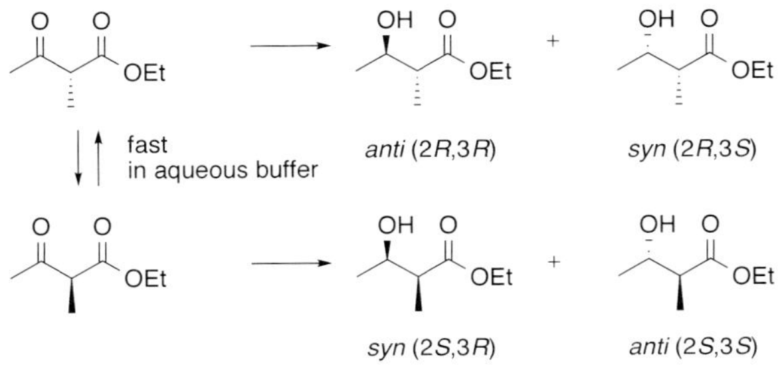

X = OH or OAc
R = H, 5-Br, 5-NO$_2$, 7-MeO

X = OH: (S), ee 87–93%   X = OAc: (R), ee 84–91%

Scheme 9.19 Possible products for the reduction of α-methyl β-keto ester.

The reduction of hydroxy or acetoxy ketones by baker's yeast shows an interesting stereoselectivity. For the reduction of acetylbenzofuran derivatives with baker's yeast, the methyl ketones afforded the corresponding (S)-alcohols in 20–68% ee; the hydroxyl derivatives afforded the (S)-diols in 87–93% ee and the acetoxy derivatives gave (R)-alcohols in 84–91% ee (Scheme 9.18) [28b].

In addition to baker's yeast and *Geotrichum candidum*, it has been shown that two microbes; *Corynebacterium* [29a] and *Rhodococcus ruber* [29b] have broad substrate specificities and show high enantioselectivities.

9.5.3
Dynamic Kinetic Resolution and Deracemization

9.5.3.1 Dynamic Kinetic Resolution

With dynamic kinetic resolution (DKR) it is possible to convert a racemic reactant with 100% completion as both (reactant) enantiomers form a chemical equilibrium and exchange under the applied experimental conditions. The faster-reacting enantiomer is replenished during the course of the reaction at the expense of the slower-reacting enantiomer.

The DKR of an α-alkyl β-keto ester has been performed through enzymatic reduction. Out of the four possible products for the unselective reduction (Scheme 9.19), one isomer can be selectively synthesized using biocatalyst, and by changing the biocatalyst or conditions all of the isomers can be selectively synthesized. For example, baker's yeast selectively gives the *syn*-(2R,3S)-product [14a] and the selectivity can be enhanced by using a selective inhibitor [14b] or heat treatment of the yeast [14c]. Organic solvents were used for stereochemical control when using *Geotrichum candidum* [14d]. Plant cell cultures were used for reduction of 2-methyl-3-oxobutanoate and afforded the *anti*-alcohol with *Marchantia* [14e, f] and the *syn*-isomer with *Glycine max* [14f].

Extensive screening was used to find a suitable microorganism and *Klebsiella pneumoniae* IFO 3319, out of 450 bacterial strains, was found to give the corresponding (2*R*,3*S*)-hydroxy ester with 99% de and >99% ee quantitatively on the kilogram scale (Scheme 9.6b) [14g].

The reduction of a series of 2-(4-chlorophenoxy)-3-oxoalkanoates was mediated by baker's yeast and *Kluyvromyces marxianus*. Yeast reduction of ethyl 2-(4-chlorophenoxy)-3-oxo-4-phenylpropanoate afforded only (2*R*,3*S*)-2-(4-chlorophenoxy)-3-hydroxy-4-phenylpropanoate out of the four possible stereoisomers in >99% de [14h]. Although the baker's yeast reduction of butanoate (R = CH_3) was not selective (92% de), the use of *Kluyvromyces marxianus* afforded the (2*R*,3*S*) isomer selectively [14i]. These products are intermediates for potential peroxisome proliferator-activated receptor isoform α-agonists (Scheme 9.20a).

It is also possible to reduce β-diketones in a diastereoselective manner; a recombinant alcohol dehydrogenase from *Lactobacillus brevis* (recLBADH), overexpressed in *E. coli*, was used for the reduction of *t*-butyl 4-methyl-3,5-dioxohexanoate. The

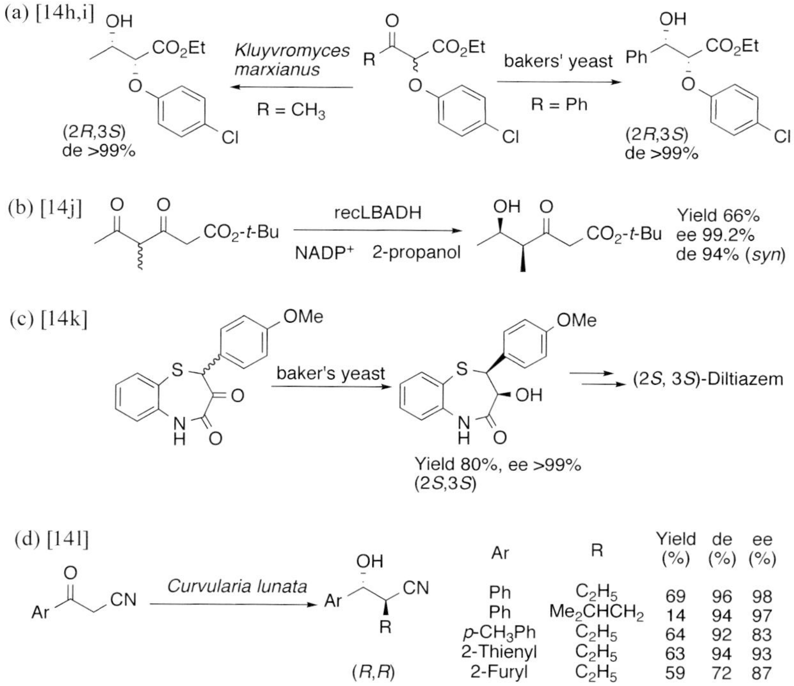

Ar	R	Yield (%)	de (%)	ee (%)
Ph	C_2H_5	69	96	98
Ph	Me_2CHCH_2	14	94	97
p-CH_3Ph	C_2H_5	64	92	83
2-Thienyl	C_2H_5	63	94	93
2-Furyl	C_2H_5	59	72	87

Scheme 9.20 Dynamic kinetic resolution.

syn-hydroxyester was obtained in 66% yield with 94% de and 99.2% ee (Scheme 9.20b) [14j].

Another example of DKR is the reduction of sulfur-substituted ketones such as (*R/S*)-2-(4-methoxyphenyl)-1,5-benzothiazepin-3,4(2*H*,5*H*)-dione. The yeast reduction gives the corresponding (2*S*,3*S*)-alcohol as the sole product, out of four possible isomers, as shown in Scheme 9.20c [14k]. Only the (*S*)-ketone is recognized by the enzyme as a substrate and the resulting product was subsequently used for the synthesis of (2*S*,3*S*)-diltiazem, a coronary vasodilator.

White-rot fungus has been used as a biocatalyst for reductions and alkylations. The reaction of aromatic β-keto nitriles with the white-rot fungus *Curvularia lunata*, CECT 2130, in the presence of alcohols affords the alkylation-reduction reaction [14l], alcohols such as ethanol, propanol, butanol and isobutanol were used (Scheme 9.20d).

The dynamic kinetic resolution of an aldehyde is shown in Table 9.3. Racemization of the starting aldehyde and enantioselective reduction of the carbonyl group by baker's yeast resulted in the formation of chiral alcohols. The enantiomeric excess of the product was improved from 19% to 90% by changing the ester moiety from an isopropyl group to a neopentyl group [30a]. Other biocatalysts were also used to perform this DKR; for example, *Thermoanaerobium brockii* reduced the aldehyde with only moderate enantioselectivity [30b, c], while *Candida humicola* was found, as a result of screening from 107 microorganisms, to give (*R*)-alcohols with up to 98.2% ee, when the ester group was a methyl [30d].

Table 9.3 Dynamic kinetic resolution of aldehydes.

Biocatalysts	R	Yield (%)	ee (%)	Reference
Baker's yeast	$-CH_2CH_3$	70–80	60–65	[30b]
Baker's yeast	$-CH(CH_3)_2$	49	19	[30a]
Baker's yeast	$-CH_2CH(CH_3)_2$	84	64	[30a]
Baker's yeast	$-CH_2C(CH_3)_3$	78	90	[30a]
Thermoanaerobium brockii	$-CH_2CH_3$	50–80	72	[30c]
Candida humicola	$-CH_3$	–	98.2	[30d]
Candida humicola	$-CH_2CH_3$	–	73.6	[30d]

9.5.3.2 Deracemization through Oxidation and Reduction

Many biocatalysts exist as mixtures, for example in microbial reductions; multiple enzymes may contribute to the reduction. A negative of having multiple enzymes that catalyze a reduction is the low enantioselectivity of the product due to the existence of enzymes possessing different stereoselectivities. However, these multiple enzymes show merit in being able to synthesize an enantiopure product from a racemate.

A deracemization reaction, which converts the racemic compounds into chiral products in one step in one pot without changing their chemical structures, can be performed using microorganisms containing several different stereochemical enzymes (Scheme 9.21) [31]. In the deracemization of 1,2-pentanediol (Scheme

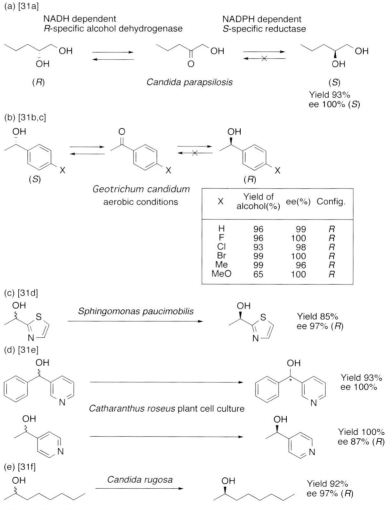

Scheme 9.21 Deracemization via oxidation-reduction.

9.21a), the (R)-specific NADH-enzyme in *Candida parapsilosis* is reversible; however, the (S)-specific NADPH-enzyme in the same microorganism is irreversible [31a]. As a result, whole-cell reaction of racemic 1,2-pentanediol gave the (S)-diol in 93% yield and 100% ee.

In the deracemization of phenylethanol derivatives using *Geotrichum candidum* under aerobic conditions (Scheme 9.21b), the (S)-specific enzyme was reversible and (R)-enzyme was irreversible, and so the (R)-alcohol accumulated when the cells and racemic alcohols were mixed [31b, c]. *para*-Substituted phenylethanol derivatives gave better results than *meta*-substituted derivatives. *Sphingomonas* was used in the deracemization of a thiazole derivative (Scheme 9.21c) [31d], while the deracemization of aromatic alcohols has been achieved with plant cell cultures (Scheme 9.21d) [31e]. The simple aliphatic alcohol 2-octanol was also deracemized to the (R)-alcohol in 92% yield with 97% ee (Scheme 9.21e) [31f].

9.6
Conclusions

Asymmetric reduction of carbonyl groups by biocatalysts is a useful method for the preparation of valuable compounds as shown in this chapter. Increasingly, bioengineering technology has been applied to alcohol dehydrogenases and reductases as well as hydrolytic enzymes to improve enzyme stability, reactivity and enantioselectivity and to extend substrate specificity. Novel enzymes for reductions created by this technique will be available in large quantities and varieties within several years. In the near future, many useful biocatalysts for reduction will be on the market, and an increasing number of chemists will be able to use enzymes for reductions more freely than at present, due to simplification of experimental procedures.

These biocatalysts will be even more important with the shift of raw materials from oil to biomass. Since biomass is a mixture of various multifunctional compounds, chemo-, regio- and enantioselective catalysts will be indispensable, and biocatalysts will play an important role. Moreover, biocatalysts typically perform better using natural substrates from biomass than with man-made substrates.

References

1 Reviews for enzymatic reductions:
(a) Nakamura, K., Yamanaka, R., Matsuda, T. and Harada, T. (2003) *Tetrahedron: Asymmetry*, **14**, 2659.
(b) van der Donk, W.A. and Zhao, H. (2003) *Current Opinion in Biotechnology*, **14**, 421. (c) Nakamura, K. and Matsuda, T. (2002) *Enzyme catalyzed reduction reaction*, in *Enzyme Catalysis in Organic Synthesis A Comprehensive Handbook* (eds K. Drauz and H. Waldmann), Wiley-VCH Verlag GmbH, Weinheim, pp. 991–1047.
(d) Santaniello, E., Ferraboschi, P. and Manzocchi, A. (2000) *Recent Advances on Bioreductions Mediated by Baker's yeast and other Microorganisms*, in *Enzymes in Action* (eds B. Zwanenburg, M. Mikolajczyk and P. Kielbasinski),

Kluwer Academic Publishers, Dordrecht, pp. 95–115.

2 (a) Nakamura, K. and Matsuda, T. (1998) *Journal of Organic Chemistry*, **63**, 8957. (b) Sarvary, I., Almqvist, F. and Frejd, T. (2001) *Chemistry – A European Journal*, **7**, 2158. (c) Weissman, A.S. and Ramachandran, P.V. (1996) *Tetrahedron Letters*, **37**, 3791. (d) Matsuda, T., Nakajima, Y., Harada, T. and Nakamura, K. (2002) *Tetrahedron: Asymmetry*, **13**, 971.

3 Schomburg, D. and Salzmann, M. (1990) *(GFB-Gesellschaft fur Biotechnologische Forschung), Enzyme Handbook*, Springer-Verlag, Berlin.

4 (a) Bradshaw, C.W., Fu, H., Shen, G.-J. and Wong, C.-H. (1992) *Journal of Organic Chemistry*, **57**, 1526. (b) Bradshaw, C.W., Hummel, W. and Wong, C.-H. (1992) *Journal of Organic Chemistry*, **57**, 1532.

5 (a) Nakamura, K., Shiraga, T., Miyai, T. and Ohno, A. (1990) *Bulletin of the Chemical Society of Japan*, **63**, 1735. (b) Nakamura, K., Takano, S., Terada, K. and Ohno, A. (1992) *Chemistry Letters*, **12**, 951. (c) Nakamura, K., Yoneda, T., Miyai, T., Ushio, K., Oka, S. and Ohno, A. (1988) *Tetrahedron Letters*, **29**, 2453. (d) Dutler, H., Van Der Baan, J.L., Hochuli, E., Kis, Z., Taylor, K.E. and Prelog, V. (1977) *European Journal of Biochemistry/FEBS*, **75**, 423.

6 (a) Prelog, V. (1964) *Pure and Applied Chemistry Chimie Pure et Appliquée*, **9**, 119. (b) Jones, J.B. (1986) *Tetrahedron*, **42**, 3351. (c) Jones, J.B. and Beck, J.F. (1976) *Asymmetric Synthesis and resolutions using enzymes*, in *Applications of Biochemical Systems in Organic Chemistry* (eds J.B. Jones, C.J. Sih and D. Perlman), John Wiley & Sons, New York, pp. 107–401. (d) Davies, J. and Jones, J.B. (1979) *Journal of the American Chemical Society*, **101**, 5405. (e) Lam, K.K.P., Gair, I.A. and Jones, J.B. (1988) *Journal of Organic Chemistry*, **53**, 1611. (f) Velonia, K., Tsigos, I., Bouriotis, V. and Smonou, I. (1999) *Bioorganic and Medicinal Chemistry Letters*, **9**, 65.

7 (a) Matsuda, T., Harada, T., Nakajima, N. and Nakamura, K. (2000) *Tetrahedron Letters*, **41**, 4135. (b) Matsuda, T., Harada, T., Nakajima, N., Itoh, T. and Nakamura, K. (2000) *Journal of Organic Chemistry*, **65**, 157.

8 Iyer, R.B. and Bachas, L.G. (2004) *Journal of Molecular Catalysis B: Enzymatic*, **28**, 1.

9 Shorrock, V.J., Chartrain, M. and Woodley, J.M. (2004) *Tetrahedron*, **60**, 781.

10 Mertens, R., Greiner, L., Haaker, E.C.D., van den Ban, H.B.C.M. and Liese, A. (2003) *Journal of Molecular Catalysis B: Enzymatic*, **39**, 24–5.

11 (a) Nakamura, K., Yamanaka, R., Tohi, K. and Hamada, H. (2000) *Tetrahedron Letters*, **41**, 6799. (b) Nakamura, K. and Yamanaka, R. (2002) *Chemical Communications*, 1782. (c) Nakamura, K. and Yamanaka, R. (2002) *Tetrahedron: Asymmetry*, **13**, 2529. (d) Utsukihara, T., Chai, W., Kato, N., Nakamura, K. and Horiuchi, C.A. (2004) *Journal of Molecular Catalysis B: Enzymatic*, **31**, 19. (e) Utsukihara, T., Misumi, O., Kato, N., Kuroiwa, T. and Horiuchi, C.A. (2006) *Tetrahedron: Asymmetry*, **17**, 1179. (f) Itoh, K., Sakamaki, H., Nakamura, K. and Horiuchi, C.A. (2005) *Tetrahedron: Asymmetry*, **16**, 1403.

12 (a) Kim, S., Yun, S.-E. and Kang, C. (1999) *Journal of Electroanalytical Chemistry*, **465**, 153. (b) Hollmann, F., Schmid, A. and Steckhan, E. (2001) *Angewandte Chemie (International Ed. in English)*, **40**, 169.

13 (a) Demirjian, D.C., Shah, P.C. and Moris-Varas, F. (2000) *Screening for Novel Enzymes*, in *Biocatalysis from Discovery to Application* (ed. W.-D. Fessner), Springer-Verlag, Berlin, pp. 1–29. (b) Carballeira, J.D., Campillo, E., Álvarez, M., Pardo, L. and Sinisterra, J.V. (2004) *Tetrahedron: Asymmetry*, **15**, 951. (c) Miya, H., Kawada, M. and Sugiyama, Y. (1996) *Bioscience, Biotechnology and Biochemistry*, **60**, 95. (d) Kambourakis, S. and Rozzell, J.D. (2004) *Tetrahedron*, **60**, 663. (e) Homann, M.J., Vail, R.B., Previte, E., Tamarez, M., Morgan, B., Dodds, D.R. and Zaks, A. (2004) *Tetrahedron*, **60**, 789.

14 (a) Nakamura, K., Kawai, Y., Miyai, T. and Ohno, A. (1990) *Tetrahedron Letters*, **31**, 3631. (b) Nakamura, K., Kawai, Y., Nakajima, N., Miyai, T., Honda, S. and Ohno, A. (1991) *Bulletin of the Chemical Society of Japan*, **64**, 1467. (c) Nakamura, K., Kawai, Y. and Ohno, A. (1991) *Tetrahedron Letters*, **32**, 2927. (d)

Nakamura, K., Takano, S. and Ohno, A. (1993) *Tetrahedron Letters*, **34**, 6087. (e) Speicher, A., Roeser, H. and Heisel, R. (2003) *Journal of Molecular Catalysis B: Enzymatic*, **22**, 71. (f) Nakamura, K., Miyoshi, H., Sugiyama, T. and Hamada, H. (1995) *Phytochemistry*, **40**, 1419. (g) Miya, H., Kawada, M. and Sugiyama, Y. (1996) *Bioscience, Biotechnology and Biochemistry*, **60**, 95. (h) Perrone, M.G., Santandrea, E., Scilimati, A., Tortorella, V., Capitelli, F. and Bertolasi, V. (2004) *Tetrahedron: Asymmetry*, **15**, 3501. (i) Perrone, M.G., Santandrea, E., Scilimati, A., Syldatk, C., Tortorella, V., Capitelli, F. and Bertolasi, V. (2004) *Tetrahedron: Asymmetry*, **15**, 3511. (j) Ji, A., Wolberg, M., Hummel, W., Wandrey, C. and Müller, M. (2001) *Chemical Communications*, 57. (k) Kometani, T., Sakai, Y., Matsumae, H., Shibatani, T. and Matsuno, R. (1997) *Journal of Fermentation and Bioengineering*, **84**, 195. (l) Dehli, J.R. and Gotor, V. (2001) *Tetrahedron: Asymmetry*, **12**, 1485.

15 (a) Kaluzna, I., Andrew, A.A., Bonilla, M., Martzen, M.R. and Stewart, J.D. (2002) *Journal of Molecular Catalysis B: Enzymatic*, **17**, 101. (b) Kaluzna, I.A., Matsuda, T., Sewell, A.K. and Stewart, J.D. (2004) *Journal of the American Chemical Society*, **126**, 12827.

16 (a) Yoshizumi, A., Wada, M., Takagi, H., Shimizu, S. and Nakamori, S. (2001) *Bioscience, Biotechnology, and Biochemistry*, **65**, 830. (b) Patel, R.N. *et al.* (2004) *Tetrahedron: Asymmetry*, **15**, 1247. (c) Yamamoto, H., Matsuyama, A. and Kobayashi, Y. (2002) *Bioscience, Biotechnology and Biochemistry*, **66**, 481. (d) Yamamoto, H., Kimoto, N., Matsuyama, A. and Kobayashi, Y. (2002) *Bioscience, Biotechnology and Biochemistry*, **66**, 1775.

17 Allen, S.J. and Holbrook, J.J. (2000) *Protein Engineering*, **13**, 5.

18 (a) Zhou, B.-N., Gopalan, A.S., VanMiddlesworth, F., Shieh, W.-R. and Sih, C.J. (1983) *Journal of the American Chemical Society*, **105**, 5925. (b) Chen, C.-S., Zhou, B.-N., Girdaukas, G., Shieh, W.-R., VanMiddlesworth, F., Gopalan, A.S. and Sih, C.J. (1984) *Bioorganic Chemistry*, **12**, 98. (c) Shieh, W.-R., Gopalan, A.S. and Sih, C.J. (1985) *Journal of the American Chemical Society*, **107**, 2993. (d) Fujisawa, T., Itoh, T. and Sato, T. (1984) *Tetrahedron Letters*, **25**, 5083. (e) Nakamura, K., Ushio, K., Oka, S., Ohno, A. and Yasui, S. (1984) *Tetrahedron Letters*, **25**, 3979.

19 (a) Nakamura, K., Kitano, K., Matsuda, T. and Ohno, A. (1996) *Tetrahedron Letters*, **37**, 1629. (b) Nakamura, K. and Matsuda, T. (1998) *Journal of Organic Chemistry*, **63**, 8957. (c) Matsuda, T., Harada, T., Nakajima, N. and Nakamura, K. (2000) *Tetrahedron Letters*, **41**, 4135. (d) Hamada, H., Miura, T., Kumobayashi, H., Matsuda, T., Harada, T. and Nakamura, K. (2001) *Biotechnology Letters*, **23**, 1603. (e) Matsuda, T., Nakajima, Y., Harada, T. and Nakamura, K. (2002) *Tetrahedron: Asymmetry*, **13**, 971. (f) Nakamura, K., Matsuda, T., Itoh, T. and Ohno, A. (1996) *Tetrahedron Letters*, **37**, 5727. (g) Nakamura, K., Matsuda, T., Shimizu, M. and Fujisawa, T. (1998) *Tetrahedron*, **54**, 8393. (h) Matsuda, T., Harada, T., Nakajima, N., Itoh, T. and Nakamura, K. (2000) *Journal of Organic Chemistry*, **65**, 157. (i) Nakamura, K., Matsuda, T. and Harada, T. (2002) *Chirality*, **14**, 703.

20 (a) Nakamura, K., Kawai, Y. and Ohno, A. (1990) *Tetrahedron Letters*, **31**, 267. (b) Nakamura, K., Inoue, K., Ushio, K. and Oka, S. and Ohno, A. (1987) *Chemistry Letters*, **17**, 679. (c) Nakamura, K., Kawai, Y., Oka, S. and Ohno, A. (1989) *Tetrahedron Letters*, **30**, 2245. (d) Dahl, A.C., Fjeldberg, M. and Madsen, J. (1999) *Tetrahedron: Asymmetry*, **10**, 551. (e) Nakamura, K., Kawai, Y., Oka, S. and Ohno, A. (1989) *Bulletin of the Chemical Society of Japan*, **62**, 875. (f) Forni, A., Moretti, I., Prati, F. and Torre, G. (1994) *Tetrahedron*, **50**, 11995. (g) Kim, J.-H. and Oh, W.-T. (1992) *Bulletin of the Korean Chemical Society*, **13**, 2. (h) Hayakawa, R., Nozawa, K., Shimizu, M. and Fujisawa, T. (1998) *Tetrahedron Letters*, **39**, 67. (i) Ushio, K., Hada, J., Tanaka, Y. and Ebara, K. (1993) *Enzyme and Microbial Technology*, **15**, 222. (j) Hayakawa, R., Shimizu, M. and Fujisawa, T. (1997) *Tetrahedron: Asymmetry*, **8**, 3201. (k) Nakamura, K., Kawai, Y., Nakajima, N. and Ohno, A. (1991) *Journal of Organic Chemistry*, **56**, 4778.

21 Li, F., Cui, J., Qian, X., Ren, W. and Wang, X. (2006) *Chemical Communications*, 865.
22 Nakamura, K., Inoue, Y., Matsuda, T. and Misawa, I. (1999) *Journal of the Chemical Society, Perkin Transactions I*, 2397.
23 (a) Nakamura, K., Fujii, M. and Ida, Y. (2000) *Journal of the Chemical Society, Perkin Transactions I*, 3205. (b) Kometani, T., Toide, H., Daikaiji, Y. and Goto, M. (2001) *Journal of Bioscience and Bioengineering*, **91**, 525. (c) Andrade Conceição, G.J., Moran, P.J.S. and Rodrigues, J.A.R. (2003) *Tetrahedron: Asymmetry*, **14**, 43.
24 (a) Matsuda, T., Harada, T. and Nakamura, K. (2000) *Chemical Communications*, 1367. (b) Matsuda, T., Watanabe, K., Kamitanaka, T., Harada, T. and Nakamura, K. (2003) *Chemical Communications*, 1198.
25 (a) Howarth, J., James, P. and Dai, J.F. (2001) *Tetrahedron Letters*, **42**, 7517. (b) Eckstein, M., Villela, M., Liese, A. and Kragl, U. (2004) *Chemical Communications*, 1084.
26 (a) Itoh, N., Morihama, R., Wang, J., Okada, K. and Mizuguchi, N. (1997) *Applied and Environmental Microbiology*, **63**, 3783. (b) Kita, K., Fukura, T., Nakase, K., Okamoto, K., Yanase, H., Kataoka, M. and Shimizu, S. (1999) *Applied and Environmental Microbiology*, **65**, 5207. (c) Howell, J.A.S., Palin, M.G., Jaouen, G., Top, S., Hafa, H.E. and Cense, J.M. (1993) *Tetrahedron: Asymmetry*, **4**, 1241. (d) Howell, J.A.S., Palin, M.G., Hafa, H.E., Top, S. and Jaouen, G. (1992) *Tetrahedron: Asymmetry*, **3**, 1355. (e) Baldoli, C., Buttero, P.D., Maiorana, S., Ottolina, G. and Riva, S. (1998) *Tetrahedron: Asymmetry*, **9**, 1497.

27 Nakamura, K., Takenaka, K., Fuji, M. and Ida, Y. (2002) *Tetrahedron Letters*, **43**, 3629.
28 (a) Ema, T., Moriya, H., Kofukuda, T., Ishida, T., Maehara, K., Utaka, M. and Sakai, T. (2001) *Journal of Organic Chemistry*, **66**, 8682. (b) Paizsa, C., Tosa, M., Majdik, C., Moldovan, P., Novák, L., Kolonits, P., Marcovici, A., Irimie, F.-D. and Poppe, L. (2003) *Tetrahedron: Asymmetry*, **14**, 1495.
29 (a) Itoh, N., Matsuda, M., Mabuchi, M., Dairi, T. and Wang, J. (2002) *European Journal of Biochemistry/FEBS*, **269**, 2394. (b) Stampfer, W., Kosjek, B., Faber, K. and Kroutil, W. (2003) *Journal of Organic Chemistry*, **68**, 402.
30 (a) Nakamura, K., Miyai, T., Ushio, K., Oka, S. and Ohno, A. (1988) *Bulletin of the Chemical Society of Japan*, **61**, 2089. (b) Züger, M.F., Giovannini, F. and Seebach, D. (1983) *Angewandte Chemie (International Ed. in English)*, **22**, 1012. (c) Seebach, D., Züger, M.F., Giovannini, F., Sonnleitner, B. and Fiechter, A. (1984) *Angewandte Chemie (International Ed. in English)*, **23**, 151. (d) Matzinger, P.K. and Leuenberger, H.G.W. (1985) *Applied Microbiology and Biotechnology*, **22**, 208.
31 (a) Hasegawa, J., Ogura, M., Tsuda, S., Maemoto, S., Kutsuki, H. and Ohashi, T. (1990) *Agricultural and Biological Chemistry*, **54**, 1819. (b) Nakamura, K., Inoue, Y., Matsuda, T. and Ohno, A. (1995) *Tetrahedron Letters*, **36**, 6263. (c) Nakamura, K., Fujii, M. and Ida, Y. (2001) *Tetrahedron: Asymmetry*, **12**, 3147. (d) Allan, G.R. and Carnell, A.J. (2001) *Journal of Organic Chemistry*, **66**, 6495. (e) Takemoto, M. and Achiwa, K. (1998) *Phytochemistry*, **49**, 1627. (f) Nie, Y., Xu, Y., Mu, X.Q., Tang, Y., Jiang, J. and Sun, Z.H. (2005) *Biotechnology Letters*, **27**, 23.

Part Three
Imino Reductions

10
Imine Hydrogenation
Carmen Claver and Elena Fernández

10.1
Recent Advances in the Asymmetric Hydrogenation of Imines

Chiral aromatic amines are extremely interesting products because of their applications in the pharmaceutical, agrochemical and fine chemical industries. The enantioselective hydrogenation of C=N double bonds using chiral transition metal complexes as catalysts is one of the most useful methods for preparing optically active amines [1]. This reaction has drawbacks, however: coordination can take place through both the nitrogen donor atom and the double bond, and the substrate and catalyst intermediates are unstable under catalytic conditions. The C=N substrates, for instance, are often sensitive to hydrolysis. Homogeneous catalysts can complex with both the imine and the amine product or can be deactivated by the formation of trimers or oligomers. In consequence, catalytic activity is often low. Furthermore, in the case of the acyclic imines one of the major problems for achieving high enantiomeric excess is the equilibrium between the *E* and *Z* isomer of the imine, which makes it difficult for the catalyst to convert all stereoisomers in selective manner. Many ruthenium-, rhodium- and iridium-based catalytic systems are excellent in hydrogenating functionalized olefins and ketones but are much less efficient with imine substrates.

In spite of these problems, an Ir-catalyzed imine hydrogenation is the key step in the largest scale enantioselective catalytic process used in industry, which is also one of the fastest homogeneous catalytic systems: the industrial production of the chiral herbicide (*S*)-metolachlor [*N*-(1′-methyl-2′-methoxyethyl)-*N*-chloroacetyl-2-ethyl-6-methylaniline] in amounts greater than 10 000 tonnes per annum at about 80% ee [2]. The catalyst is a combination of [Ir(COD)Cl]$_2$ (COD = 1,5-cyclooctadiene), the chiral ferrocenyl diphosphine Xyliphos, tetrabutylammonium iodide (TBAI) and sulfuric acid. 2-Methyl-6-ethylphenyl-1′-methyl-2′-methoxyethylimine (MEA(monoethanolamine)-imine) is hydrogenated under 80 bar (8 MPa) hydrogen pressure at 323 K and at a substrate/catalyst ratio exceeding 10^6 to yield MEA-amine in 79% ee and with an initial turnover frequency that is said to exceed 1.8 × 10^6 h^{-1} (Scheme 10.1). This process has been extensively studied and

Modern Reduction Methods. Edited by Pher G. Andersson and Ian J. Munslow.
Copyright © 2008 WILEY-VCH Verlag GmbH & Co. KGaA, Weinheim
ISBN: 978-3-527-31862-9

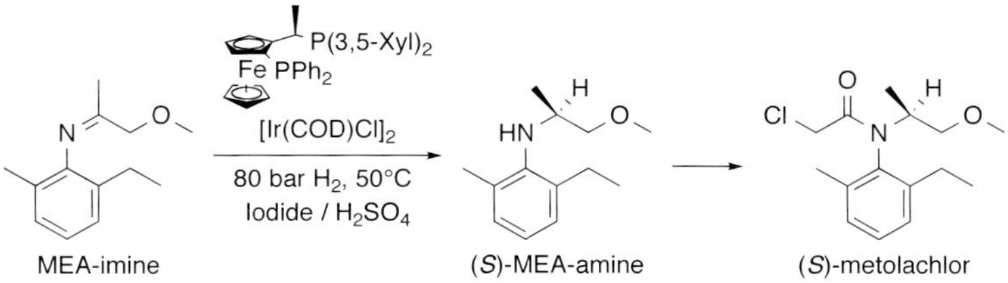

Scheme 10.1

documented [1–4], which has increased the interest of both academic and industrial research groups in asymmetric imine hydrogenation.

The state of the art and the most important results in asymmetric hydrogenation of imines have been described in several reviews, in particular by H.-U. Blaser et al. in 1999 [1a], and more recently by X. Zhang et al. in 2003 [1d]. Both reviews focus on the most efficient catalytic systems, and point out the various problems for enantioselective hydrogenation. Despite the catalytic efficiency of Ti and Rh catalysts, only limited success has been reported for acyclic N-alkylimines. However, good to excellent enantioselectivity has been obtained for acyclic N-arylimines, including MEA-imine for the production of (S)-metolachlor and cyclic imines. The scope of the catalysts is in general very limited because, as the reviews reflected, hydrogen pressures are high and the catalysts were deactivated.

In recent years, however, it has been found that several new systems based on metals such as Ir, with phosphine-based chiral auxiliaries, can provide high to excellent enantioselectivity in the hydrogenation of imines, even at low hydrogen pressures. This chapter describes several new strategies. It deals first with the iridium catalytic systems, classified according to the nature of the ligand, and then considers other metals (Rh, Ru, Ti, Zr and Au).

10.1.1
Iridium Catalysts

In general, two types of precursors are used in the Ir-catalyzed hydrogenation of imines: (a) cationic derivatives bearing a chiral chelating ligand and a diolefin coligand such as COD. The counter anion A^- can play an important role in these systems; (b) neutral precursors which provide catalytic systems prepared *in situ* from [Ir(COD)Cl]$_2$ and a stoichiometric amount of chiral ligand.

10.1.1.1 Iridium / P-P Ligands
Most of the early effective catalysts in the asymmetric hydrogenation of imines contain chiral chelating bisphosphine ligands. In general, these systems are limited in substrate scope, since they require high pressures, long reaction times and additives. The importance of additive effects on Ir-catalyzed imine hydrogenation has been known since the first reports on the subject [1]. In the case of the asymmetric hydrogenation of 2,3,3-trimethylindolenine catalyzed by an iridium

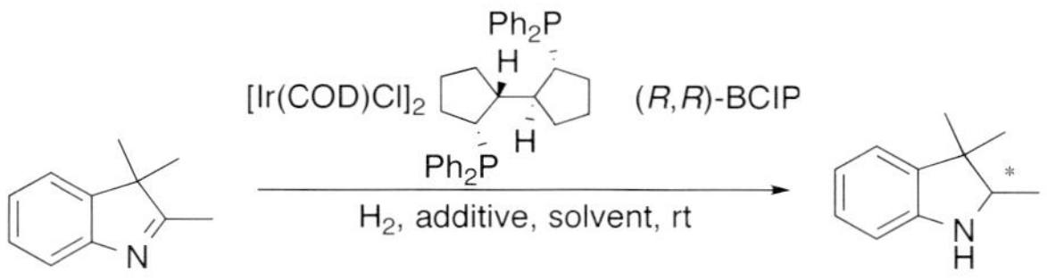

10.1 Recent Advances in the Asymmetric Hydrogenation of Imines

Scheme 10.2

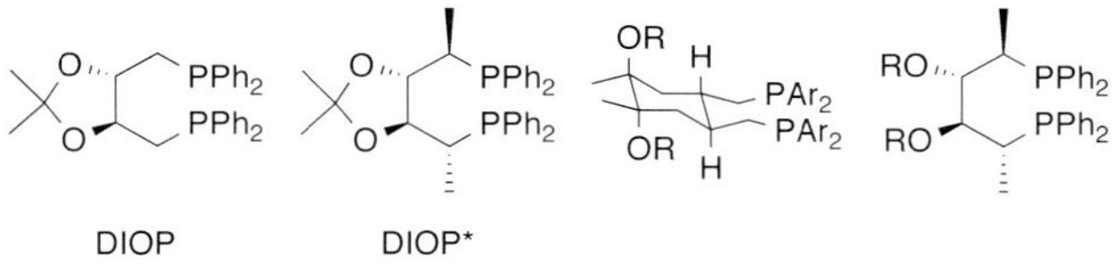

Figure 10.1 1,4-Bisphosphine ligands based on DIOP backbone.

system using a BCIP ligand (Scheme 10.2), the additive effects of imides and 1,2-indandione increase the enantiomeric excess to 95% [5].

DIOP was one of the first ligands to be explored for imine reduction, although ee values were only moderate [6]. Several modified DIOP derivatives have been prepared, which have been applied in Ir-catalyzed asymmetric hydrogenations of acyclic imine (Figure 10.1) [7]. In the hydrogenation of cyclic imines, with $[Ir(COD)Cl]_2$ / DIOP complex generated *in situ*, and 10 mol% of I_2 as an additive, 85% ee was obtained at 0 °C in CH_2Cl_2 which is better than previously reported with DIOP [6]. The conformational mobility of the backbone of this type of ligand plays an important role in both the reactivity and the selectivity. Isolating a pure precatalyst is another key factor in obtaining high catalytic activity. In general, 1,4-bisphosphine ligands have been found to be effective for the Ir-catalyzed hydrogenation of imines [7].

Chiral (2S, 4S)-2,4-bis(diphenylphosphino)pentane (BDPP) has also been applied in the Ir-catalyzed imine hydrogenation. The system is active in mild conditions. The $[Ir(COD)(S,S)-(BDPP)]PF_6$ complex hydrogenates N-(α-methyl-p-methoxybenzylidine)benzylamine at 5 bar and 40 °C, although no enantioselectivity was observed. The immobilized catalyst in montmorillonite K-10 (MK-10) achieves 59% ee by using a recycled catalyst, discussed in Section 10.2. [8].

As has often been observed, the chiral phosphane ligand, BINAPHANE, highly effective in the Rh-catalyzed hydrogenation of enamides [9a] is ineffective for the hydrogenation of imines. As an example of the key role of the ligand, however, the related chiral 1,1′-bisphosphanoferrocene, (f-BINAPHANE), is active in the Ir-catalyzed asymmetric hydrogenation of N-(1-phenylethylidene)aniline (84% ee) (Scheme 10.3) [9b]. The enantioselectivity with the neutral precursor $[Ir(COD)Cl]_2$ was higher than with a cationic precursor $[Ir(COD)_2]PF_6$. The weakly coordinating solvent CH_2Cl_2 was more desirable than other solvents such as THF, toluene or methanol. A change in hydrogen pressure has no clear effect on the enantioselectivity, but the conversion was increased under high H_2 pressure. The highest

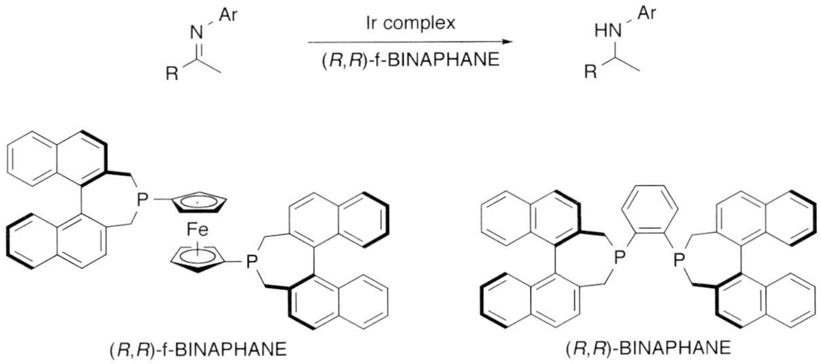

Scheme 10.3

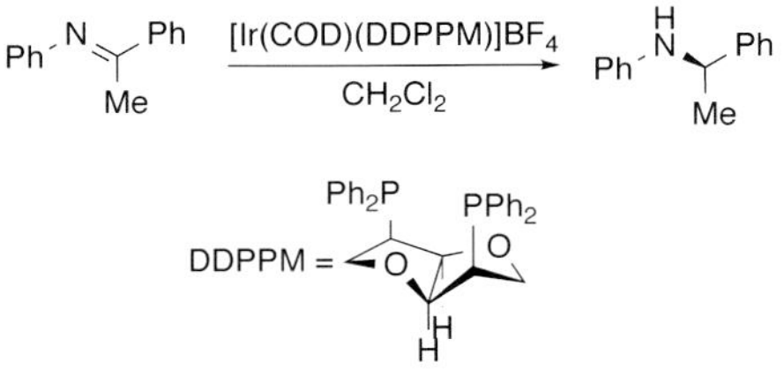

Scheme 10.4

enantioselectivity was ≈99% ee for sterically hindered substrates (Ar′ = 2,6-dimethyl-*N*-phenyl) [9b].

Iridium complexes of the type [Ir(COD)(DDPPM)]X were applied in the asymmetric hydrogenation of imines (Scheme 10.4). Interestingly, the DDPPM complexes performed efficiently even under an atmospheric hydrogen pressure, whereas at higher pressures catalyst activity was drastically reduced. Depending upon the reaction conditions *N*-arylimines were hydrogenated to the corresponding secondary amines in high yields and enantioselectivities (80–94% ee). In contrast to noncoordinating anions such as $[BF_4]^-$ and $[PF_6]^-$, coordinating anions like chlorides did not form active catalysts. The cationic Ir-DDPPM hydrogenation system performed well in chlorinated solvents, whereas coordinating solvents deactivated the system. Dimeric and trimeric Ir(III) polyhydride complexes were formed from the reaction of [Ir(COD)(DDPPM)]PF_6 with molecular hydrogen at atmospheric pressure and were found to inhibit catalytic activity [10].

Recently, attempts to selectively reduce acyclic aromatic *N*-aryl imines under 1 bar of H_2 with Ir(I)-(*S*)-BINAP systems failed for a neutral iridium complex and for cationic complexes. However, the complex containing tetrakis(3,5-bis(trifluoromethyl)phenyl)borate (BARF$^-$), promoted the reduction considerably even under 1 bar of hydrogen pressure to give the product in 93% yield after 1.5 hours, although with low ee value(16%) [11]. The pronounced rate acceleration effect of BARF$^-$ has also been observed in Ir-catalyzed reduction of olefins. In the same report [11], the ligand (*S*,*S*)-1,2-bis(*t*-butylmethylphosphino)ethane was used to

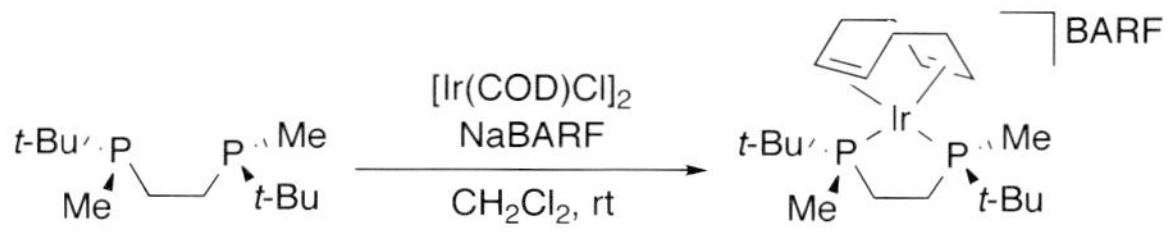

Scheme 10.5

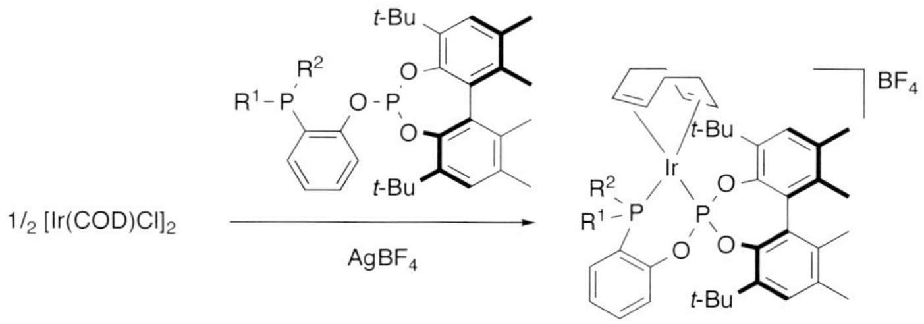

Scheme 10.6

form a cationic complex, containing BARF⁻ (Scheme 10.5). This catalytic system catalyzes the hydrogenation of acyclic aromatic *N*-aryl imines under 1 bar of H_2 at room temperature to give the corresponding optically active secondary amines at up to 99% ee.

10.1.1.2 Iridium / Phosphine–Phosphite Ligands

Phosphine–phosphite ligands, P-OP (Scheme 10.6), are an interesting family of ligands because of their particular electronic properties, due to the presence of two strongly coordinating phosphorus functionalities, one of them a good π-acceptor group. A family of modular chiral P-OP ligands containing differently modified biaryl fragments has been prepared and applied in the Ir-catalyzed hydrogenation of *N*-aryl imines. Ligand screening clearly identified the nature of the backbone as a critical variable in this reaction. Ligand optimization has pointed to 2-[(S)-(*o*-anisil)phenylphosphino]ethyl-(S)-3-3′-di-*t*-butyl-5,5′,6,6′-tetramethylbiphenyl-2,2′-diylphosphite as the best ligand in the series, 84% ee in the hydrogenation of imine PhN=CMePh. Similar enantioselectivities of 72–85% ee have been obtained with this ligand in the reduction of several *N*-aryl imines. As usual in the iridium catalytic systems, both the cationic and neutral catalytic systems have been studied. Neutral precatalysts seem to be better in this case than cationic ones [12].

10.1.1.3 Iridium / Diphoshite, Diphosphinite and Phosphinite–Phosphite Ligands

Phosphorous ligands derived from carbohydrates have been used in the Ir-catalyzed hydrogenation of imines. The advantage of these catalytic systems is that the enantiomerically pure ligands are easy to prepare. Iridium complexes incorporating xylose diphosphinite and diphosphite ligands (Figure 10.2) are active catalysts for the hydrogenation of imines, although providing only moderate ee

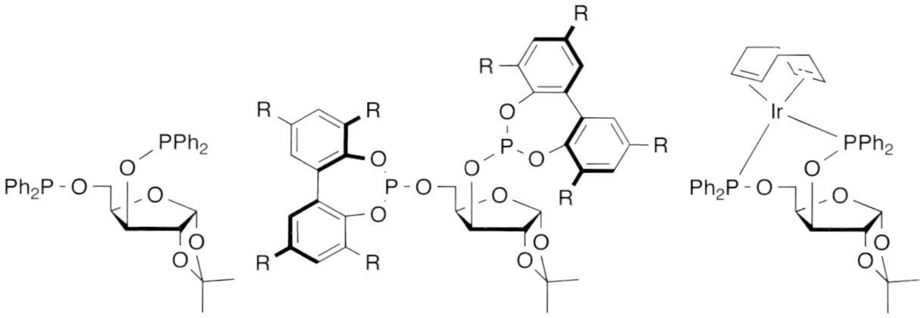

Figure 10.2 Xylose diphosphinite and diphosphite ligands.

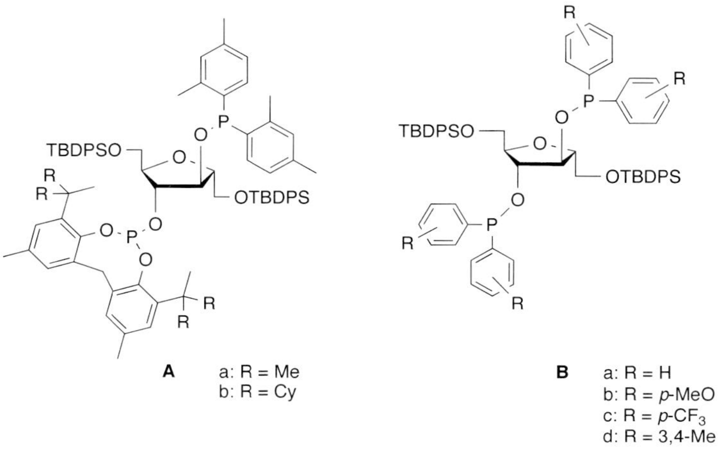

Figure 10.3 Phosphinite–phosphite ligands.

A a: R = Me
       b: R = Cy

B a: R = H
       b: R = p-MeO
       c: R = p-CF$_3$
       d: R = 3,4-Me

values. The important fact is that the enantioselectivity depends on the fine-tuning of the structural parameters of the ligand. These catalytic systems were active at 50 bar of H$_2$ and 25 °C. In the asymmetric hydrogenation of N-aryl imines, results were poorly reproducible. However, in the hydrogenation of N-(phenylethylidene)aniline using the complex [Ir(COD)(xylose diphosphinite)]BF$_4$ as catalyst precursor, ee values up to 57% were achieved, optimum at 10 bar H$_2$ [13].

The diphosphinite ligands having different electron-donating or electron-withdrawing groups in the aryl group, and the phosphinite–phosphite ligands, (Figure 10.3), which are directly prepared from glucosamine, have been used in combination with [Ir(COD)Cl]$_2$. The cationic complexes formed from [Ir(COD)$_2$]BF$_4$ and the diphosphinites or phosphinite–phosphite ligands provided hydrogenation of N-(phenylethylidene)benzylamine with conversions between 70% and 100%. The use of additives was, in general, detrimental to both the conversion and the enantioselectivity. In the hydrogenation of N-(phenylethylidene)aniline the results were best with ligand **(A)** (R = Me) (76% ee), but in the hydrogenation of N-(phenylethylidene)benzylamine results were best with ligand **(B)** (R = p-MeO), (70% ee) [14].

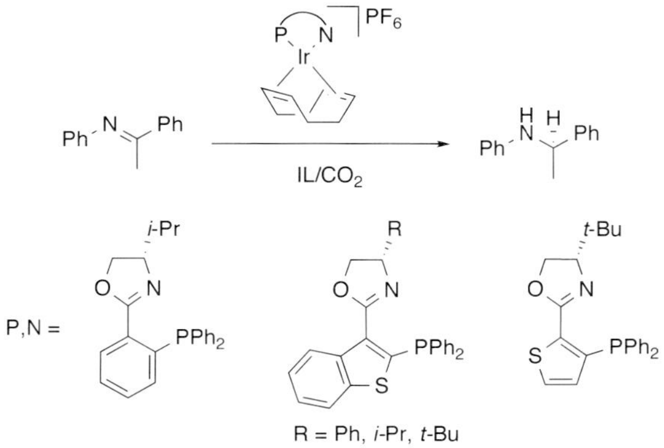

Scheme 10.7

10.1.1.4 Iridium / P,N-Ligands

P,N ligands, and in particular phosphine-oxazolines [15] and related ligands, emerged at the end of the 1990s as a new alternative class for the Ir-catalyzed enantioselective reduction of imines [16] and olefins [17]. As far as the hydrogenation of asymmetric imines is concerned, Pfaltz et al. reported that iridium complexes containing chiral phosphinooxazolines (PHOX ligands) (Scheme 10.7), provide high enantioselectivity in the hydrogenation of N-(phenylethylidene)aniline (89% ee) and related imines [16]. This is one of the most successful examples of cationic precursors in the asymmetric hydrogenation of imines.

Cationic Ir(I) complexes with chiral phosphino dihydrooxazoles, modified with perfluoroalkyl groups in the ligand, have shown to be efficient catalysts for the hydrogenation of N-(1-phenylethylidene)aniline in supercritical carbon dioxide ($scCO_2$). Both the side-chains and the lipophilic anions increased the solubility, but the anion also had a dramatic effect on the enantioselectivity, with BARF$^-$ leading to the highest enantiomeric excess obtained (up to 81%), [16b].

Although the reported P,N-oxazolines include structures where the P-atom is attached to the ring through C-2 or C-4, the corresponding five- or six-membered rings can be formed on coordination to a metal. Therefore, (4S,5S)-2-R-4-diphenylphosphino(methyl)-5-phenyl-1,3-oxazolines (R = Me, Et, Ph), where the P-atom is tethered to the oxazoline ring at C-4 can be prepared through a simple synthetic route (Figure 10.4). These phosphine-oxazoline ligands have been applied in the Ir-catalyzed asymmetric hydrogenation of PhCH$_2$N7=C(Me)Ph [18]. The [Ir(COD)(P,N-oxazoline)]PF$_6$ complex catalyzes the hydrogenation of N-(1-phenylethylidene)benzylamine, in CH$_2$Cl$_2$, to the corresponding amine with up to 63% ee when R = Et in the ligand.

Other chiral P,N ligands such as chiral aminophosphine-oxazoline [19], phosphine-imidazoline [20] and oxazoline-thioether [21] ligands have been used in relation to phosphine-oxazolines. The chiral aminophosphine-oxazoline auxiliaries (Figure 10.5) of general formula [Ir(COD)L]X [X = PF$_6$ and BARF$^-$] have been

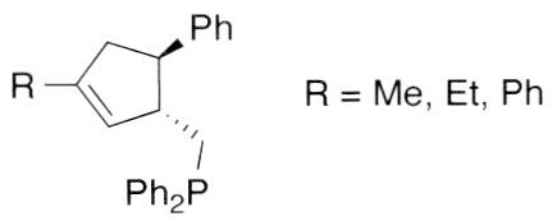

Figure 10.4 Phosphine-oxazoline ligands.

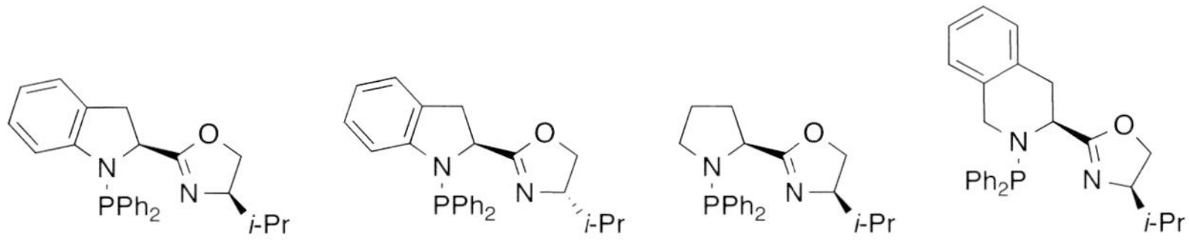

Figure 10.5 Chiral aminophosphine-oxazoline auxiliaries.

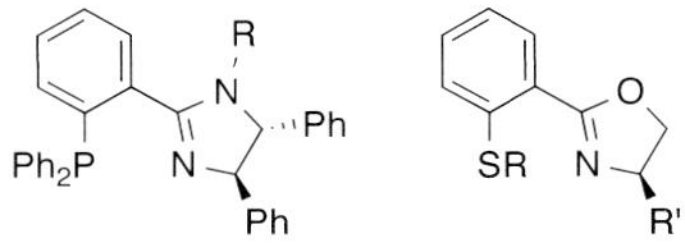

Figure 10.6 Phosphine-imidazoline and oxazoline-thioether ligands.

applied to the asymmetric hydrogenation of two imines: N-(phenylethylidene)aniline, and N-(phenylethylidene)benzylamine, providing the corresponding chiral amines in 90% and 82% ee respectively [20]. The Ir-catalyst containing a ligand with (R) configuration on the oxazoline ring and (S) on the aminophosphine residue, induced a higher selectivity (80% ee) than its diastereomeric complex containing a ligand with (S) configuration on the oxazoline unit (14% ee). In addition, the amines with opposite configuration are obtained with the two diastereomeric chiral auxiliaries, clearly showing that the stereogenic center of the oxazoline unit has a considerable impact on the selectivity of the reaction.

Generally, high conversions and good-to-high enantioselectivities (over 68% ee) are obtained under 50 bar of H_2 at room temperature.

Phosphine-imidazoline [20] and oxazoline-thioether [21] have also been used with iridium precursors, although the enantioselectivities obtained are generally moderate (Figure 10.6). They were able to reduce acyclic and cyclic imines providing moderate enantioselectivity (up to 50%). In general, additives had a negative effect on the selectivity. For the reduction of acyclic imines the catalytic systems prepared *in situ* and based on $[Ir(COD)_2]BF_4/L$ provided better activity than the precursors based on the neutral dinuclear $[Ir(COD)Cl]_2/L$.

One of the most successful P,N systems in the Ir-catalyzed hydrogenation of imines consists of ligands derived from 2-azanorbornan-3-ylmethanol, as recently reported by Andersson *et al.*. Functionalization of 2-azanorbornane-oxazoline with phosphine leads to a novel class of phosphine-oxazoline ligands (Scheme 10.8), which have been applied in the Ir-catalyzed hydrogenation of acyclic N-arylimines [22]. Hydrogenation of N-(1-phenylethylidene)aniline revealed optimal results

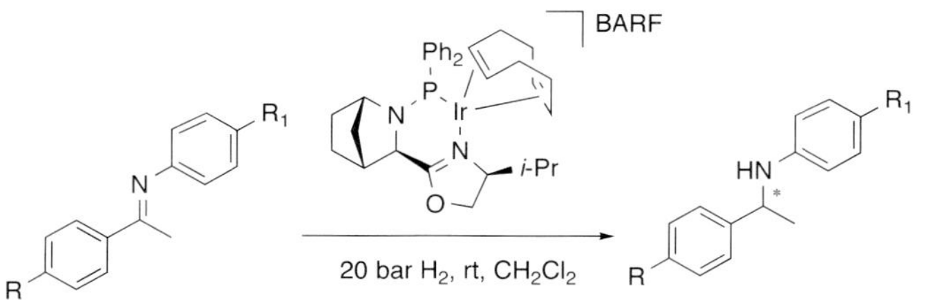

Scheme 10.8

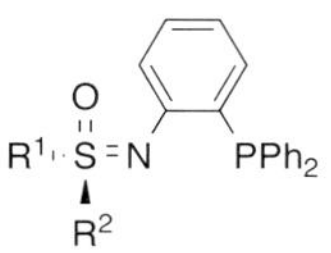

Figure 10.7 Sulfoximine derivative ligands.

when the reaction was performed in CH_2Cl_2 at 20 bar H_2 with a catalyst loading of 0.5 mol%. The corresponding (R)-N-phenyl-N-(1-phenylethyl) amine was obtained in 90% ee and 98% conversion in 2 hours.

Sulfoximine derivatives have also been successfully used recently as effective catalysts for the asymmetric hydrogenation of imines (Figure 10.7). In the hydrogenation of the acetophenone-derived imine (p-MeO)PhN=CMe(Ph), the catalyst generated by mixing [Ir(COD)Cl]$_2$ with the sulfoximine (where R^1 = Ph, R^2 = Me) and then adding iodine, provides 79% ee. It has been observed that the ligand structure clearly affects catalyst performance. Increasing the steric bulk of the alkyl substituent of the sulfoximine, α to the sulfur atom, drastically lowered both the activity and the enantioselectivity of the catalysts. However, with the isobutyl-substituted sulfoximine (where R^1 = Ph, R^2 = i-Bu), conversion of arylimines was excellent after only 4 hours, and the enantioselectivity of the resulting amine reached an unprecedented 96% ee. Hydrogenations of related substrates showed that the aryl group on the imine nitrogen atom had a strong effect on the reactivity and selectivity of the catalyst [23].

10.1.1.5 Iridium / N-Ligands

Other Ir-catalysts containing a chiral monodentate phosphoramidite and various N-donor ligands have been used for asymmetric imine hydrogenation. The commercially available chiral monodentate phosphoramidite (S)-Monophos was used in combination with N-donor ligands to prepare the complexes [Ir(S)-Monophos(COD)(L)]BARF, which have proved to be efficient catalysts for the asymmetric hydrogenation of 2,3,3-trimethylindolenine (Scheme 10.9) [24]. The hydrogenation of this imine using [Ir(S)-Monophos(COD)(L)]X (X = SbF$_6$, BARF; L = 3-methylisoquinoline) showed that when methanol solvent is used the BARF salts of these catalysts were superior to the corresponding SbF$_6$ salts as previously described, [11]. The complex [Ir(S)-Monophos(COD)(imine)]BARF catalyzed the

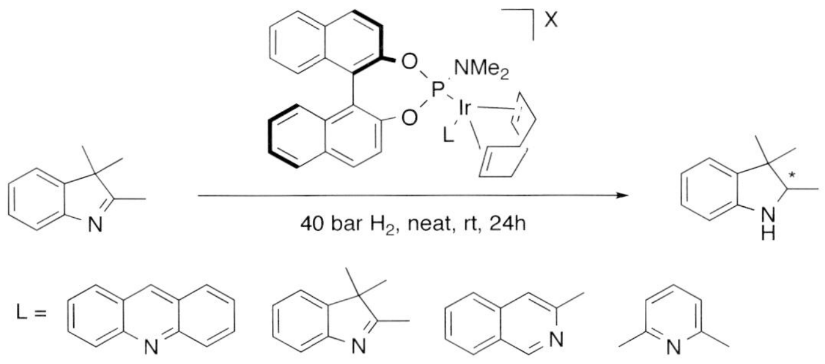

Scheme 10.9

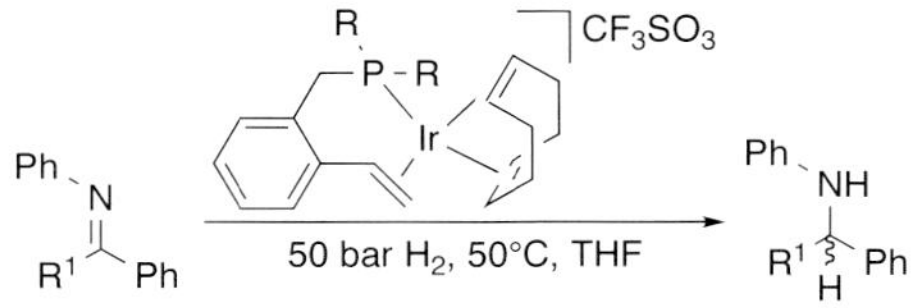

Scheme 10.10

reduction of 2,3,3-trimethylindolenine with similar enantioselectivity (44% ee) to complexes with a pyridine-type ligand but at a greater rate (75% conversion in 24 hours). Only when L = 2,6-lutidine did the iridium complex show a similar rate, with 80% conversion in 24 hours. This may be because the pyridine ligands compete for binding with the substrate so the rate decreases, while 2,6-lutidine is more a sterically hindered ligand and so competes poorly for binding.

10.1.1.6 Other Iridium / Phosphorous Systems

Ir(I) complexes containing P-olefin ligands, such as (5H-dibenzo[a,d]cyclohepten-5-yl)-phosphane (TROPP^R; R = phosphorus-bound substituent = Ph, Cyc) as a rigid, concave-shaped, mixed phosphane olefin ligand (Scheme 10.10) have been used as catalyst precursors in the hydrogenation of imines. With the complex [Ir(COD) (TROPP^{Cyc})]OTf, values of turnover frequency (TOF) of >6000 h^{-1} were reached in the hydrogenation of N-phenylbenzylidenamine. Lower activities (TOF > 80 h^{-1}) are observed with N-phenyl-(1-phenylethylidene)amine. The best ee value (86%) was obtained with PhN=CMePh as substrate and the R,R form of the (10-menthyloxy-5H-dibenzo[a,d]cyclohepten-5-yl)diphenylphosphane ligand [25].

Secondary phosphine oxides easily prepared in a two-step one-pot procedure from readily available starting materials have been used as monodentate enantio-pure ligands in the Ir-catalyzed asymmetric hydrogenation of acetophenone-based imines, [26]. The results obtained with [Ir(COD)Cl]$_2$ and [Ir(COD)$_2$]BF$_4$, precursors showed that neutral chloride-containing Ir-precursors gave both the highest rates and the highest enantioselectivity. Enantioselectivities were as high as 76% ee where L/Ir = 2. Addition of pyridine (Pyr/Ir = 1:2) raised the value to 83% ee.

As well as the iridium approaches described above, a number of transition metal-based catalysts, such as those containing Rh, Ru, Pd, Ti, Zr and Au have been applied to the asymmetric hydrogenation of imines.

10.1.2
Rhodium and Palladium Catalysts

Rh(I)-complexes bearing ancillary P-ligands can also be useful for the asymmetric hydrogenation of imines. N-(1-Phenylethylidene)benzylamine was hydrogenated under H_2 pressure of 50 bar. Chelating diphosphines forming seven-membered rings had higher activities than catalysts based on smaller chelate rings. Dialkylphosphines as ligands are less active than diarylphosphines. The reactivity of diphosphine catalysts was increased by the addition of p-toluenesulfonic acid [27]. Rh-complexes based on chiral diphosphinites and a diphosphite also rapidly converted the substrate to the desired amine. One of the most interesting systems is the Rh(I) complex bearing an electron-deficient ligand, the diphosphinite [1,2-O-dihydroxyethane-bis(diphenylphosphinite)] (DPOE), which hydrogenates this imine rather effectively with complete reduction with no additives within 5 hours and 71% ee [27].

The Duphos derivatives, 1,2-bis(phospho-1-ano)benzenes (Me-, Et-, and i-Pr-Duphos), are very efficient as ligands in the Rh-catalyzed hydrogenation of olefins and have also been used in the Rh-catalyzed asymmetric hydrogenation of N-acylhydrazones. Hydrogenation of the N-benzoylhydrazone of acetophenone proceeded readily under mild conditions using [Rh(COD)(Duphos)]CF_3SO_3. Of the three Duphos ligands, Et-Duphos proved to be superior in terms of enantioselectivity, providing the product N-benzoylhydrazine in 88% ee [28]. N-tosylimines are interesting substrates for these reactions since they are relatively stable and can be easily obtained from the corresponding ketones exclusively as the (E)-isomer [29]. In addition, the strongly electron-withdrawing character of the tosyl group reduces the inhibitory effect of the reduction product on the catalysts, which can lead to higher reactivity. Rhodium catalysts based on the [Rh(COD)$_2$]BF_4 complex with a chiral diphosphane such as (S,S,R,R)-Tangphos, (R_P,S_C)-Duanphos, S_P-BINAPINE, (S)-C_3-Tunephos, (Scheme 10.11) provides ee values of 80–94% for several substrates. The Rh-catalysts with the ligands (R,R)-Et-Duphos or (R,R)-Me-Duphos show high conversions; however, the ee values of the products were relatively low (63% and 68%, respectively).

Rhodium systems have recently been applied in the asymmetric hydrogenation of α-acyl imino esters. The [Rh(S,S,R,R)-TangPhos(COD)]BF_4 provided ee values between 90% and 99% in the hydrogenation of several α-imino esters with different substituents on the aromatic ring of the ester group [30].

Interestingly, palladium systems, Pd(OCOCF$_3$)$_2$/TangPhos hydrogenate N-tosylimines with enantioselectivities of up to 99% ee and conversions of more than 99%. However, relatively high hydrogen pressure and catalytic loading are important limitations.

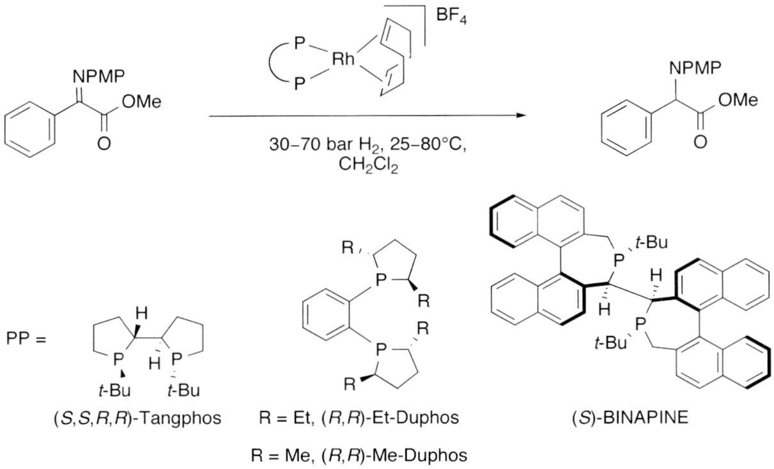

Scheme 10.11

10.1.3
Ruthenium Catalysts

Noyori's Ru(II) dichloride(diphosphine)(diamine) complexes have proved to be very efficient at reducing ketones with very high enantioselectivities even at extremely low catalyst loadings. These have also been applied in the asymmetric hydrogenation of imines [31a]. When the same precatalysts are used, imine reduction requires higher pressure and temperature conditions than ketone reduction, and the enantioselectivities are in general lower [31b]. Large libraries of structurally diverse diphosphine and diamine ligands have been used to rapidly generate an array of catalysts with very different stereoelectronic properties. For a given imine the most appropriate diphosphine/diamine combination is best identified by extensive screening and it seems difficult to predict.

The (R,R)-Et-Duphos/(R,R)-DACH (1,2-diaminocyclohexane) combination provides the best enantioselectivity (ee up to 92%) [31c]. Ru-BINAP catalysts hydrogenate N-tosylimines, and the best result reported so far is the 84% ee reported by Charette [32].

10.1.4
Titanium and Zirconium Catalysts

Chiral titanocenes have shown excellent levels of enantioselectivity in the hydrogenation of several types of imines, particularly cyclic ones [33, 34].

Two different patterns have been proposed for cyclic and acyclic imines. The enantiomeric excesses for the hydrogenation of cyclic imines (up to 98%) were essentially insensitive to changes in reaction conditions. However, for acyclic imines the ee values (up to 76%) were dependent on several variables, most significantly hydrogen pressure. This phenomenon has been explained on the basis of the interconversion of the *syn* and *anti* isomers of acyclic imines during the hydrogenation.

A mechanistic study of the asymmetric titanocene-catalyzed imine hydrogenation has been discussed elsewhere in this chapter [35].

Enantiopure biphenyl-bridged titanocene and zirconocene complexes, obtained by an asymmetric thermal transformation of the binaphthol complexes formed from the metallocene racemates and subsequent transformation to the corresponding dichlorides, were applied as catalysts in the asymmetric hydrogenation of cyclic and acyclic imines [36]. The results were similar to those obtained for the titanocene catalysts [33].

10.1.5
Gold Catalysts

Gold complexes as catalysts have been the object of considerable interest in recent years and they have been applied to several catalytic process. The complex $\{(AuCl)_2[(R,R)\text{-Me-Duphos}]\}$ can be easily prepared by reacting the diphosphine with two equivalents of [AuCl(tht)] and provides high activity and 75% ee at 4 bar and 20 °C, in the hydrogenation of *anti*-*N*-benzyl(1-phenylethylidene)imine [37].

10.2
Green Approaches

The increasing demand for environmentally friendly methods from academia and industry, has recently encouraged the scientific community to develop benign and economically viable homogeneously catalyzed processes [38]. Considerable effort has been made to conduct the asymmetric hydrogenation of imines under safe conditions, emphasizing simple product separation and the possibility of recycling the catalytic system. In fact, because of the expense of the chiral ligand in asymmetric versions, only reactions with extremely high turnover numbers have been commercialized. To this end, some of the green approaches developed so far have been collected under the following conceptual methodologies: (i) aqueous–organic two-phase solvent systems, (ii) catalyst immobilization on insoluble materials and (iii) carbon dioxide/ionic liquid media.

10.2.1
Aqueous–Organic Two-Phase Solvent Systems

This method enables the catalyst to be easily recycled by reusing the aqueous phase after separating it from the product, which remains in the organic phase. One of the main strategies used to prepare water-soluble catalysts is functionalizing the chiral ligand with sulfonated groups.

Interestingly, the degree of sulfonation on a ligand influences not only catalyst recovery but also enantioselectivity. Sinou [39], Bakos [40] and de Vries [41] studied the influence of total or partial sulfonation on BDPP in the hydrogenation of the N-benzylacetophenone imine, under 70 bar H_2 at room temperature. The *in situ* catalyst formed from [Rh(COD)Cl]$_2$ and sulfonated BDPP proved to be most effective for hydrogenating the imine when the degree of sulfonation of the phosphine was below 2 and the reaction was performed in a H_2O-AcOEt two-phase solvent system, (Scheme 10.12). Despite the high ee values observed (94–96%), no recycling of the catalyst is mentioned. However, the lower the degree of sulfonation, the lower, or null, was the solubility of the catalytic system in the aqueous media.

An extractable Ir-Xyliphos (Xyliphos = chiral ferrocenyldiphosphine) catalytic system has recently been applied in the enantioselective hydrogenation of MEA-imine to produce the (S)-MEA-amine as an intermediate in the synthesis of (S)-metolachlor [42]. In the presence of acetic acid and tetrabutylammonium iodide (TBAI) under 80 bar H_2 at 25–30 °C, the extractable catalytic system provided total conversion with ratio of moles of substrate/moles of Ir = 120 000 and TOF = 36 000 h^{-1}, (Scheme 10.13). The enantioselectivity achieved (79%) was

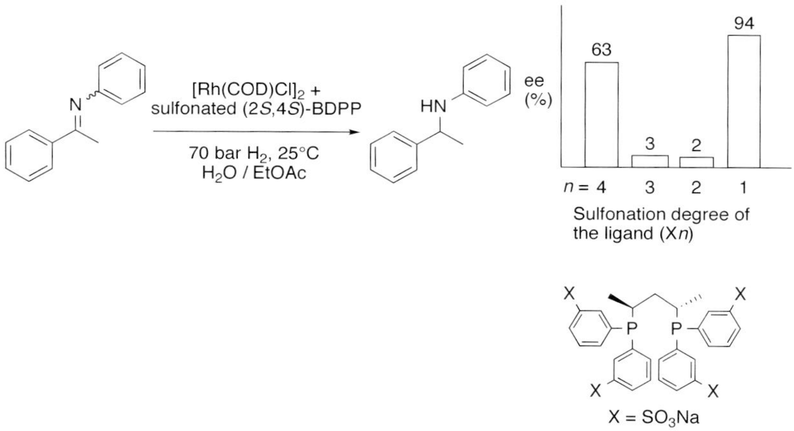

Scheme 10.12

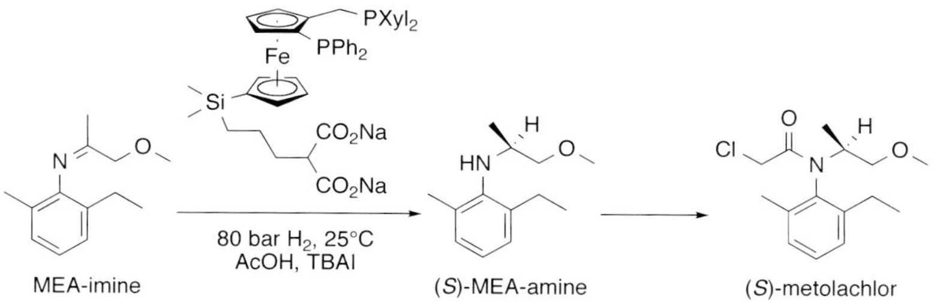

Scheme 10.13

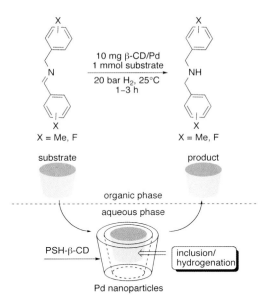

Scheme 10.14

close to that of the unfunctionalized Ir-Xyliphos. Although the catalytic system was separated with >90% efficiency by extraction into NaOH, no mention was made of reuse.

In order to overcome some of the problems associated with the aqueous–organic biphasic catalytic hydrogenation of imines, such as the lack of phase-transfer efficiency of hydrophobic substrates to the catalytic species in the aqueous media, and the limitation to water-soluble organometallic catalytic species, Török and Liu have developed an alternative aqueous biphasic reaction catalyzed by metal nanoparticles. They have introduced the novel concept of using an artificial host immobilized on metal nanoparticles as an efficient phase-transfer catalyst for aqueous biphasic media (Scheme 10.14) [43]. A water soluble β-cyclodextrin-modified Pd-nanoparticle was used as a multifunctional nanocatalyst for the chemoselective hydrogenation of a series of arylimines under 20 bar of H_2 at 25 °C. The presence of electron-withdrawing substituents was shown to lower the yields slightly. No asymmetric induction was explored at this stage even though a precedent on a supported Pd catalyst had been found to be an effective and selective catalyst for preparing chiral 2,2,2-trifluoro-1-phenylethylamines through the diastereoselective hydrogenation of the corresponding N-phenylethyl amines and the subsequent hydrogenolysis of the methylbenzyl group [44]. The recovered water-soluble nanoparticle catalyst was not contaminated by the organic phase (actually composed of the substrate) at the end of the reaction and was recycled.

10.2.2
Catalyst Immobilization on Insoluble Materials

Immobilizing complex catalysts on nonsoluble structures makes it possible to conduct solution-like reactions in the solid state and minimize many of the barriers associated with the unremovable homogeneous soluble catalyst. An early approach involved immobilization on such smectite clay structures as montmorillonite, which combine properties such as cation exchange, intercalation and swelling properties [45]. Cationic Ir- and Rh(I)-complexes were immobilized by absorption onto the external surface of montmorillonite K-10 and lithium hectorite, but by intercalation by ion-exchange into bentonite (Scheme 10.15) [46]. Their application in the hydrogenation of the N-benzylidene aniline showed that the activities in the first run of the heterogenized systems were comparable to those of their homogenous counterparts. The recovered catalyst immobilized onto montmorillonite K-10 was reused in consecutive hydrogenations with high activity (>98% conversion) for at least 13 runs. Interestingly the immobilization process prevented the formation of catalytically inactive species produced by oligomerization reactions [47].

A potential industrial application of these immobilized systems involved transforming primary amine 4-aminediphenylamine to N-phenyl-N-isopropyl-p-phenylenediamine (IPPD) or p-phenyldiamine to N,N'-di(1,4-dimethylpentyl)-p-phenylenediamine (DMPPD) through the catalytic hydrogenation of the imine intermediate [48], in a single step under milder and solvent-free conditions (Scheme 10.16). The catalytic system [Rh(COD)(PPh$_3$)$_2$]BF$_4$ immobilized onto montmorillonite K-10 provided activity that was similar to that of the homogeneous counterpart (>96% conversion), but the selectivity on the dialkyl amine diminished slightly on reuse after five consecutive runs.

Further studies have been made of the separation of the immobilized catalytic systems onto insoluble supports to perform asymmetric hydrogenation of imines. The most representative approaches in the 1990s were the Ir-diphosphine complex immobilized on silicates, metal oxides or polymers [49], and the Rh-phosphine complexes immobilized on silica gel for the stereoselective hydrogenation of folic

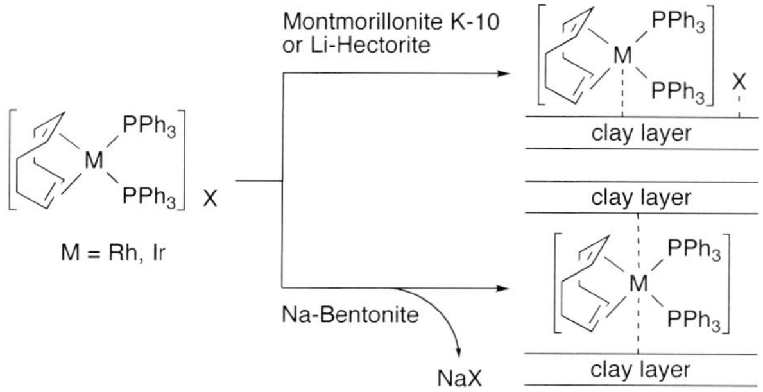

Scheme 10.15

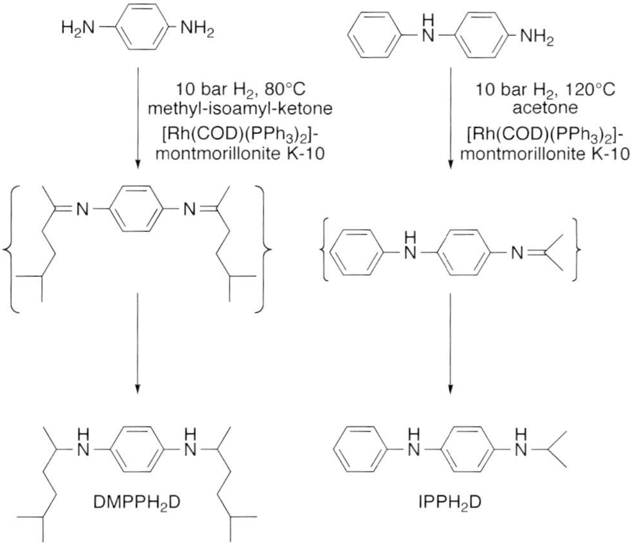

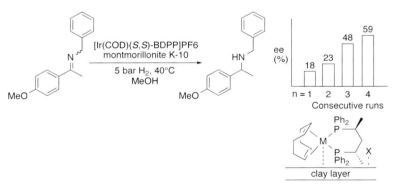

Scheme 10.16

Scheme 10.17

acid [50]. Alternatively, the immobilization of [Ir(S,S)-BDPP(COD)]PF$_6$ on montmorillonite K-10 allowed the hydrogenation of N-(α-methyl-p-methoxybenzylidene)benzylamine at 5 bar H$_2$ and 40 °C [8]. While conversion was comparable to that of the homogeneous catalytic system, the heterogenized system induced enantiomerically enriched mixtures of amine product. Remarkably, when the recovered supported catalyst was used in further consecutive hydrogenation reactions, the catalyst became less active but more enantioselective (up to 59% ee in the fourth run) (Scheme 10.17). The immobilization also prevented the catalytic system from decomposing or deactivating on exposure to air.

A modular concept for preparing immobilized enantioselective catalysts has recently been developed. It consists of a functionalized Xyliphos ligand [3b],

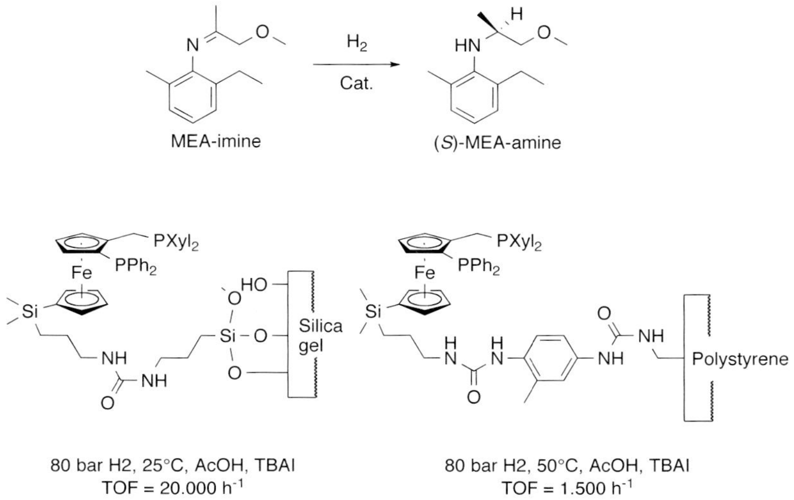

Scheme 10.18

covalently attached to an insoluble support via a linker [51]. Its application in the hydrogenation of the hindered N-arylimine MEA in the presence of acetic acid and iodide, has provided the best heterogenized catalyst, with turnover number (TON) > 95 000 and TOF up to $20 000 h^{-1}$, when the catalyst was supported on silica (Scheme 10.18). While these catalysts were easily and efficiently separated by filtration and the enantioselectivity remained high, the activities diminished consecutively on reuse as a result of catalyst deactivation by irreversible dimer formation. The Ir-BPPM and Ir-DIOP catalysts immobilized on silica gel had a similar effect [2c, 52]. The polystyrene-bound catalytic system turned out to be much less active. This was attributed to slower mass transport in the polymer matrix.

To avoid mass transport through the insoluble structures, an alternative approach was to immobilize the less expensive palladium and nickel metallic species modified with chiral salen ligands, on mesoporous silica structures such as MCM-41, ITQ-2, ITQ-6 delaminated zeolites and amorphous silica (Scheme 10.19). These heterogenized catalysts have been tested in the hydrogenation of anti-N-benzyl-(1-phenylethylidene)imine and 5-phenyl-3,4-dihydro-2H-pyrrole, 4–5 bar H_2 at 40 °C, with a substrate/catalyst ratio = 100 000 [52]. The immobilized catalyst on the mesoporous molecular sieve MCM-41 and the delaminated zeolites provided higher activities (TOF up to $150 000 h^{-1}$) than the homogeneous counterparts and the accessibility and adsorption properties of the solids. This activity was even increased by introducing acidity on the support (TOF up to $200 000 h^{-1}$). The stability of the catalyst toward recycling was good, and no decrease in activity was observed after it had been reused four times. However, the observed enantioselectivity was low (10–15%), which was attributed to the planarity of the metal complex.

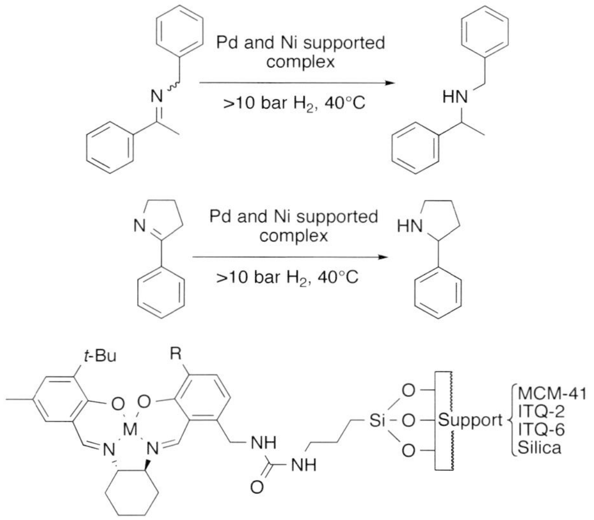

Scheme 10.19

10.2.3
Carbon Dioxide / Ionic Liquid Media

The use of scCO$_2$ as a solvent is particularly good in enhancing reaction rates [53], and improving regio- [54] and enantioselectivity [55], in reactions where hydrogen is one of the reactants. Recent advances, inspired by the advantageous miscibility of scCO$_2$ with many gases to avoid potential mass-transfer limitations, have led to efficient enantioselective hydrogenations of prochiral imines under this toxicologically and environmentally benign medium [16b]. Enantiopure phosphinodihydrooxazoles were successfully used as ligands [16a] in the Ir-catalyzed enantioselective hydrogenation of the imines N-(1-phenylethylidene)aniline and N-(1-phenylethylidene)benzylamine. To increase the solubility of transition metal catalysts in scCO$_2$, it was suggested that perfluorinated side-chains could be incorporated either directly in the ligands or in the anions when cationic complexes were used (Scheme 10.20).

It has been demonstrated that the anion is of key importance for solubilizing the catalytically active species, and the levels of asymmetric induction (about 80% ee) were high only with catalysts containing the BARF anion. This catalytic system was much more efficient in the scCO$_2$ medium than conventional organic solvents because of a different rate profile, not just a simple increase in the overall rate of the process. scCO$_2$ was also used in an integrated process as the medium for reaction and separation: the pure product was easily isolated and the catalyst efficient recovered. However, while the levels of enantioselectivity remained constant in subsequent runs (>70%), longer reaction times were required for quantitative conversions. This catalyst deactivation was attributed to the sensitivity

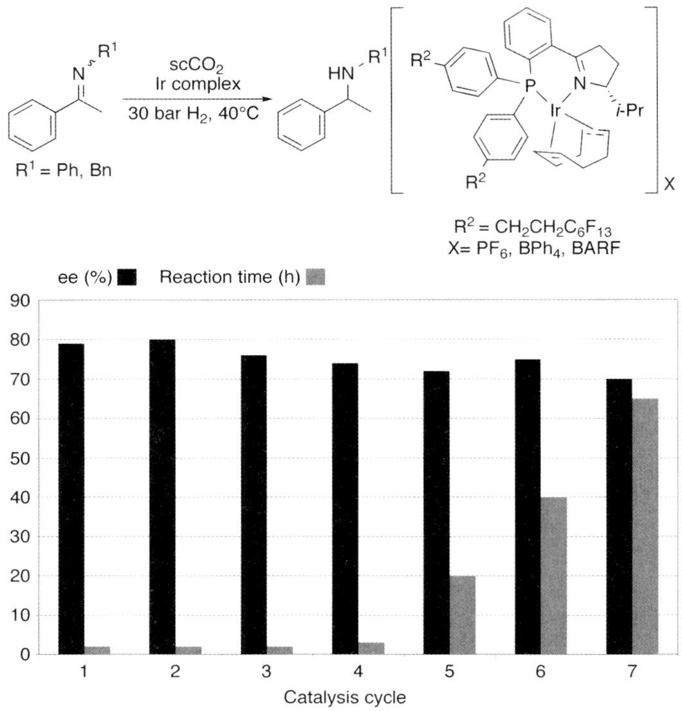

Scheme 10.20

of the active metal intermediates to adventitious oxygen during the batch-wise recycling.

In the attempt to prevent deactivation of the catalyst during the recycling process, it was subsequently demonstrated [56] that the combination of ionic liquids (ILs) and scCO$_2$ not only overcomes this problem but also fully exploits the benefits of both systems through the combination of the molecular interaction of the stationary IL phase with the catalyst and the mass transfer properties of the mobile CO$_2$ phase (Scheme 10.21) [57]. It was found that the cationic Ir-complex modified with the enantiomerically pure phosphinodihydrooxazole ligand leads to activation, tuning and immobilization of the catalytic system in the hydrogenation of N-(1-p henylethylidene)aniline. In particular it was observed first that the presence of CO$_2$ was beneficial for efficient hydrogenation in the IL media, and second that dissolving the catalyst in IL led to activation by anion exchange, which allowed the *in situ* use of catalytic systems. The choice of the anion in the IL greatly influenced the selectivity of the catalyst. In addition, the IL led to greatly enhanced long-term stability of the iridium catalyst, as it was found that IL solutions of the catalyst were significantly less air-sensitive than analogue solutions in conventional organic solvents. However, catalyst solutions of [EMIM][BTA]-Ir complex were found to be

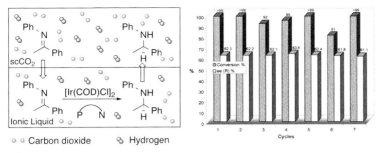

Scheme 10.21

remarkably stable even when exposed to air (75% conversion and 56% ee after 20 hours in air). Finally, the products were readily removed and isolated from the catalyst solution by CO_2 extraction without cross contamination of the IL or catalyst. Although the relative amount of CO_2 used for the isolation of a certain amount of product was quite high in the bath procedure, the authors suggested that this could be greatly reduced under continuous-flow operation.

To this end, the first application of a continuous-flow strategy has been described for the catalytic hydrogenation of imines with high chemoselectivity. This application uses a continuous flow-reactor with a mixed hydrogen–liquid flow stream, [58].

10.3
Mechanistic Insights

Despite impressive advances in the catalytic asymmetric hydrogenation of imines, some aspects are still controversial, particularly in comparison with the state of the art in alkene hydrogenation chemistry. A well-understood mechanistic cycle to explain the sequential steps for converting imines into amines is still a challenge, particularly when describing the enantiofacial differentiation of the substrate. In addition, slight modifications in the substrate, additives, solvent and acidic medium also have a substantial effect on the catalytic activity. Thus, the mechanism and the role of each reactant need to be deeply and widely understood if imine hydrogenation is to be applied technologically. The literature contains numerous mechanistic proposals for C=N hydrogenation, although the acceptance of one mechanism or another is fairly arbitrary. The criterion used in this chapter for selecting the most representative mechanistic approaches is based on the following conceptual features: (i) homolytic and heterolytic H_2-metal activation; (ii) H^-/H^+ transfer to the imine in the inner or outer coordination sphere; (iii) neutral or ionic mechanism; (iv) ligand-assisted mechanism; and (v) enantio-differentiation steps.

10.3.1
Homolytic and Heterolytic H$_2$-Metal Activation

It has been assumed that the H$_2$ molecule coordinates to a metal complex at a vacant site as a η^2-dihydrogen ligand [59–62]. Then, the first key step in the catalytic hydrogenation of the C=N polar bond is the activation of H$_2$ by transition metal species via homolytic or heterolytic cleavage. However, electron-rich metal complexes seem to favor the homolytic cleavage of H$_2$ by changing the oxidative state of the metal [63]. In contrast, electron-poor metal centers should preferentially lead to heterolytic cleavage of the H–H bond, with the dihydrogen ligand acting as a Brønsted acid (Scheme 10.22) [64, 65].

It has generally been thought that catalytic systems based on iridium, rhodium and ruthenium complexes, M(I), undergo homolytic addition of H$_2$ to give M(III) metal species [66]. A general schematic mechanism for the [Ir(COD)(PR$_3$)$_2$]PF$_6$-catalyzed hydrogenation of N-(β-naphthyl methylene)aniline has been proposed in accordance with the experimentally determined rate law (Scheme 10.23) [67]. Therefore, the precursor of catalyst (**A**) can be rapidly hydrogenated to form octahedral dihydro-bisimine intermediate (**B**), which reversibly undergoes intramolecular hydride transfer to give a coordinated amine species (**C**). This intermediate then oxidatively adds dihydrogen as the rate-determining step to yield a dihydrido species containing both coordinated imine and amine compounds [68]. Displace-

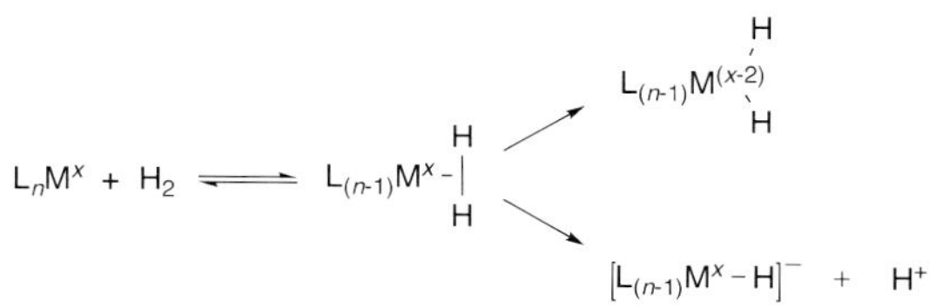

Scheme 10.22

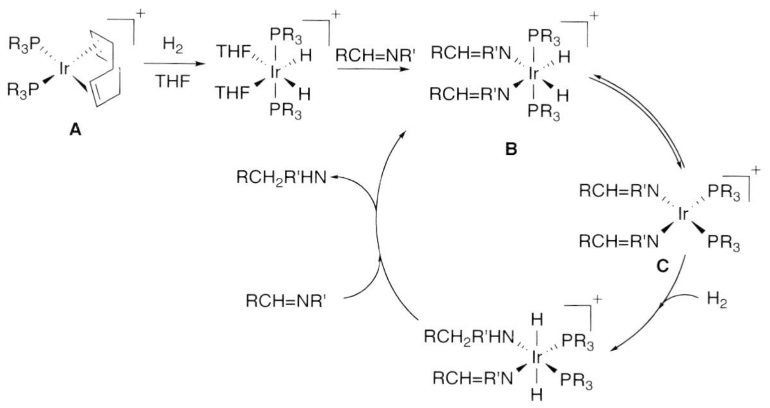

Scheme 10.23

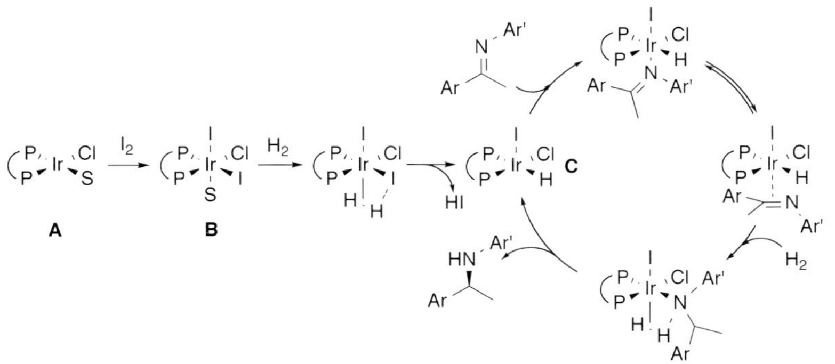

Scheme 10.24

ment of the product by a new incoming imine molecule regenerates the active species and completes the cycle.

Alternative iridium complexes, such as $[Ir(P-P)HI_2]_2$, seem to facilitate the hydrogenation of *N*-arylimines and the oxidative state of Ir(III) remains unchanged throughout the catalytic cycle [69]. It was proposed that the heterolytic cleavage of H_2 by the Ir-N species could generate an amine and an Ir(III)-H species. Further studies have demonstrated that the *addition of* I_2 leads to an Ir(III)-catalyzed pathway, and provide evidence for its beneficial role [70]. A similar hypothesis involves the oxidative addition of I_2 to Ir(I) precursor **(A)** [Ir(F-BINAPHANE)Cl(S)] (S = solvent or substrate), generating the Ir(III) complex **(B)** (Scheme 10.24) [9b, 71]. The heterolytic cleavage of H_2 in the presence of amine can form the Ir(III)-H species ©. An imine substrate is coordinated in the inner sphere, followed by migratory insertion into the Ir(III)-H bond to form an Ir(III)-amide complex. Heterolytic cleavage of H_2 provide the amine product and regenerates the Ir(III)-H active species.

10.3.2
H⁻/H⁺ Transfer to the Imine in the Inner or Outer Coordination Sphere

Another relevant key step in the catalytic process is to assume that the hydride attacks the imine substrate in the inner or outer coordination sphere [72]. The catalytic behavior in these cases is very different, because in the inner coordination sphere the coordination of the C=N substrate allows the electrophilic activation of the carbon on the imine group by the metal ion, so that a *cis* hydride ligand can migrate to this β-carbon. In the outer coordination sphere the imine is not coordinated to the metal and usually has a lower hydride affinity, so that electrophilic activation might be required by an external electrophile, intermolecularly (Scheme 10.25).

The great majority of imine hydrogenation mechanistic studies have used late transition metal catalysts. However, some early transition metal complexes can effectively hydrogenate imines, even asymmetrically, and mechanistic approaches

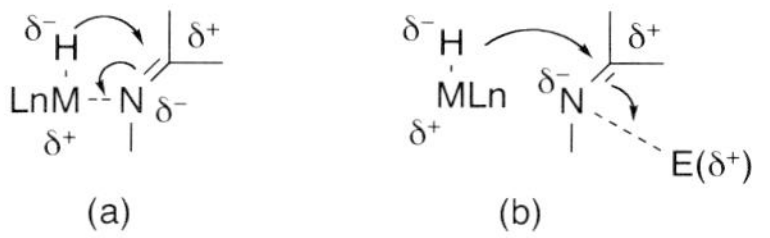

Scheme 10.25

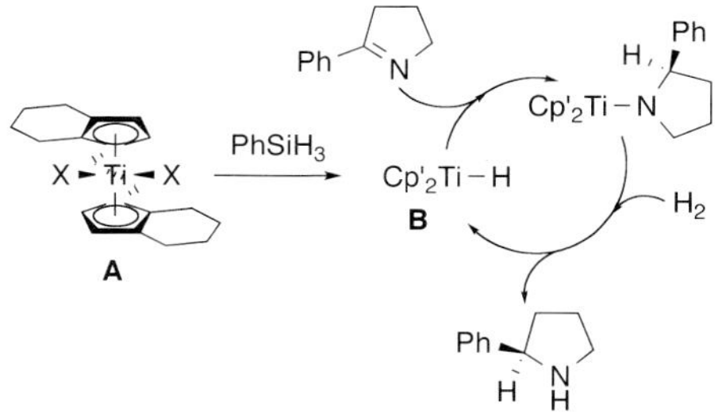

(a) LnM–N with δ−H···δ+ interactions

(b) MLn–N with δ−H and E(δ+)

Scheme 10.26

Catalytic cycle showing conversion of A (X–Ti–X with indenyl ligands) via PhSiH₃ to Cp′₂Ti–H (B), reaction with Ph–N imine, formation of Cp′₂Ti–N intermediate, release of H₂, and product Ph–CH(H)–N(H) pyrrolidine.

have been developed. Thus, the titanocene complex [ethylene-1,2-bis(η^5-4,5,6,7,-tetrahydro-1-indenyl)]titanium 1,1′-binaphthyl 2,2′-diolate (**A**), can be reduced under reaction conditions to give the Ti(III)-H species (**B**) (Scheme 10.26) [33a, 34, 35]. Fast insertion of the imine into a titanium hydride in the inner sphere, slow β-H elimination of the resulting titanium-amide intermediate and slow hydrogenolysis of the titanium amide to form the amine and regenerate the Ti(III)-H species are assumed to be the most plausible steps of the catalytic cycle. The same authors have demonstrated that the *syn* isomer of the imines reacts faster than the *anti* isomer, and that the *syn* and *anti* imines interconvert under the reaction conditions. Similar mechanistic pathways, where the H⁻/H⁺ is transferred to the imine in the inner sphere, have been used for organolanthanide-catalyzed imine hydrogenations, where the insertion of the imine into the Ln-H intermediates is a rapid and irreversible step, while hydrogenolysis is turnover-limiting. Small amounts of PhSiH₃ accelerate the imine hydrogenation process, possibly because of the deamidation of the lanthanide center (Scheme 10.27) [73].

Alternatively, it has been suggested that a rapid and reversible protonation of the substrate in the outer sphere of the catalytic metal center is the key to catalytic performance for the hydrogenation of imines. In these cases, hydrogenation rates increase with the acidity and are first-order in substrate. In particular, the Ru-catalyzed hydrogenation of iminium cations has been shown to proceed by coordinating H₂ to the Ru(P-P)CpX precursor of the catalyst, and then transferring the H⁺ to the amine product [74]. Both processes have been shown to be fast and their rate has little impact on the turnover frequency. However, transfer of hydride to the iminium cation has been rationalized as the turnover-limiting step (Scheme 10.28a), although it can be faster when the chelate ring in the metal species is smaller or when more electron-donating substituents are introduced into the phosphine. Similarly, it has been suggested that the selective hydrogenation of

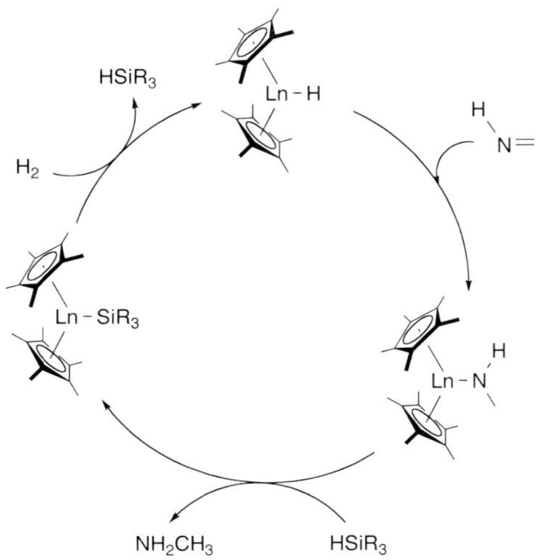

Scheme 10.27

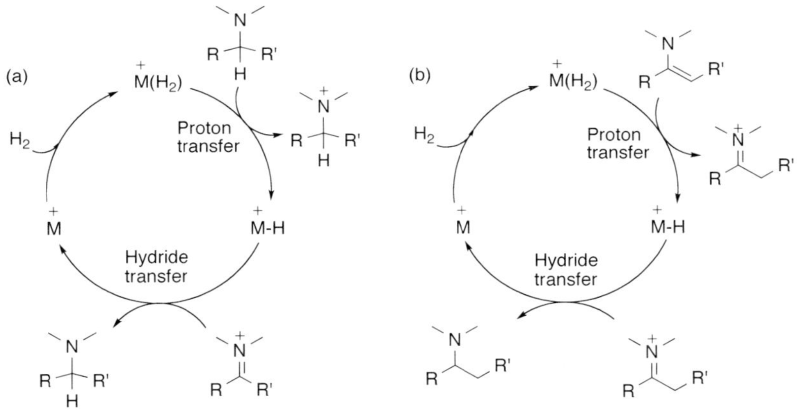

Scheme 10.28

enamines to amines proceeds by intermolecular H⁺ transfer to the substrate and consecutive hydride transfer to the iminium intermediate (Scheme 10.28b).

10.3.3
Neutral or Ionic Mechanisms

Hydrogenation by an ionic mechanism [75] involves heterolytic cleavage of H_2 and separate transfer of H⁺ and H⁻ to the substrate, unlike hydrogenation by a neutral concerted pathway in which homolytic cleavage of H_2 on the metal (normally by oxidative addition), is followed by imine insertion in M-H and reductive elimination sequences [8–11, 15]. However, conceptually the hydrogenation of imines by an ionic mechanism has such advantages as compatibility with ionizing solvents (water and alcohols) and selectivity for polar C=N bonds over nonpolar C=C bonds.

Calculations made with model complexes and isolated intermediates indicate that the effective hydrogenation of N-benzylideneaniline under mild conditions with [IrH$_2$(η^6-C$_6$H$_6$)(P^i-Pr$_3$)]BF$_4$ may follow an ionic mechanism [76]. A distinctive feature of this catalytic system is that H$^+$ and H$^-$ can be transferred simultaneously to the imine, and also that the H$^+$ is transferred intermolecularly from the NH moiety of the amine hydrogenated product. Therefore, the hydrogenation of the C=N has been postulated to proceed in the outer metal sphere where the effectiveness of the hydrogen bonding plays an important role, and solvent properties are crucial. Donor substituents can compete with imine and amine for the reactive sites, but they can also strongly stabilize catalytic intermediates. The mechanism has been formulated in terms of the transformation of precursor **(A)** into the resting state **(B)** by amine interaction, followed by formation of catalytic intermediate **(C)** by another amine interaction (Scheme 10.29). Homolytic cleavage of H$_2$ through oxidative addition generates intermediate **(D)**, and then the hydrogen-

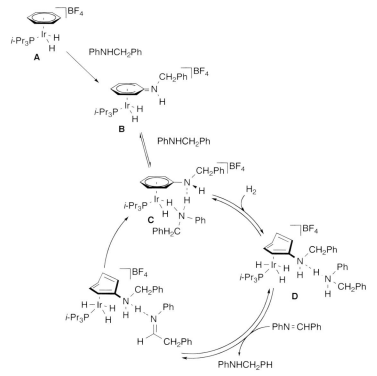

Scheme 10.29

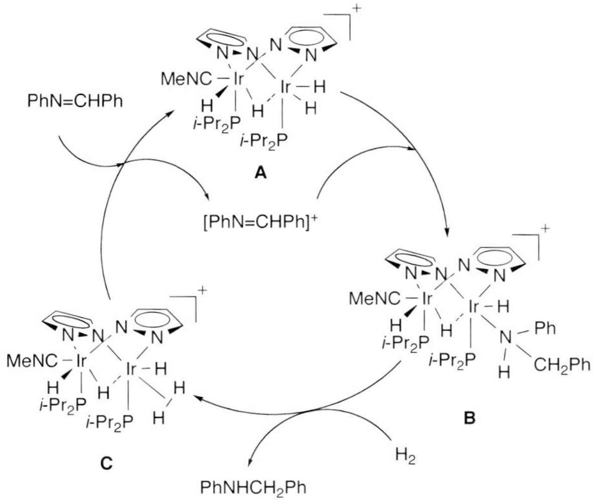

Scheme 10.30

bonded amine is replaced by imine. Finally, the hydrogenated product is produced by simultaneous H^+/H^- transfer to the imine from the NH moiety of the arene-coordinated amine ligand and the metal-H, respectively. A close ionic outer-sphere mechanism has also been postulated by the same authors for the dinuclear [Ir$_2$(μ-H)(μ-Pz)$_2$H$_3$(NCMe)(P-i-Pr$_3$)$_2$] in an attempt to explain the efficiency of the hydrogenation of imines in a versatile dinuclear environment [77]. Despite the apparent coincidence of mechanistic features between the mononuclear and the dinuclear catalytic precursor, the postulated catalytic cycles have no elementary step in common, which illustrates the variety of mechanistic resources available for the ionic hydrogenation of imines. In this case, complex **(A)** becomes an active catalyst in the hydrogenation of *N*-benzylideneaniline when catalytic amounts of protic acid are added to provide intermediate **(B)**. The rate determining substitution of the amine ligand in **(B)** by H$_2$ to give **(C)**, followed by fast elementary steps of non-simultaneous H^+/H^- transfer to the imine, permits the selective hydrogenation of C=N, even in the presence of readily hydrogenated alkenes (Scheme 10.30). Although this mechanism involves the participation of a single metal center from the dinuclear complex, ligand modification at the spectator metal atom causes significant changes in the hydrogenation rate.

Testing the catalytic hydrogenation of several imines with the series of complexes [Re(CO)$_n$(PMe$_3$)$_{5-n}$(solvent)]$^+$ has demonstrated for the first time that the solvent-stabilization effect plays an important role in an ionic mechanistic pathway with heterolytic cleavage of H$_2$ [78]. For optimum catalysis, there must be a subtle balance between the counteracting effects of the strength of back-bonding of the H$_2$ ligand and its acidity.

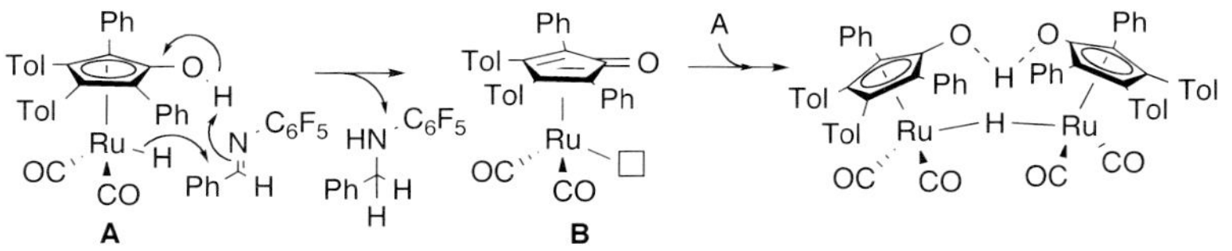

Scheme 10.31

10.3.4
Ligand-assisted Mechanisms

The catalytic system can have an ancillary ligand *cis* to the hydride that assists in the H⁻ transfer to the imine step. In certain cases H_2 can be heterolytically split to give a M-hydride and a protonated ancillary ligand (containing an NH or OH functional group). These catalytic systems have been referred to as "metal–ligand bifunctional catalysts" [79], and are characterized by an ancillary ligand that provides a H⁺ that can be transferred when the hydride is also transferred to the substrate both in the inner or in the outer metal sphere.

One significant example of the ligand-assisted hydrogenation of imines is the tolyl derivative of Shro's hydroxycyclopentadienyl ruthenium hydride system [80], which can transfer a H⁺ from the ligand to the imine simultaneously with the H⁻ transfer from the metal center in the outer coordination sphere (Scheme 10.31). The reaction begins with net *trans* addition of a proton and hydride from intermediate (A) to the imine and formation of coordinatively unsaturated intermediate (B). This study has led to a greater understanding of the reduction mechanism and has demonstrated that the rate-limiting step changes as a function of imine basicity. For electron-deficient substituted imines, the simultaneous H⁺/H⁻ transfer to the imine is the rate-limiting step. However, for electron-rich imines the rate-limiting step is the coordination of nitrogen to ruthenium and reversible hydrogen transfer leads to imine isomerization.

The hydride complexes *trans*-RuHCl(diamine)(diphosphine) [81] and $RuCl_2$(diamine)(diphosphine) [31c], have been shown to be active precatalysts in the asymmetric hydrogenation of imines in the presence of a base, even in the asymmetric version. Presumably catalytic cycles involve H⁺/H⁻ transfer to the imine in the outer coordination sphere of the metal center, with the amine ligand-assisted protocol (Scheme 10.32).

10.3.5
Enantiodifferentiation Steps

Proposed models based on the crystal structures of catalysts and the isolation of catalyst-substrate adducts have been the most important source of data about the stereochemical trends in the asymmetric hydrogenation of imines.

The mode of coordination of the imine in the metal inner sphere may explain the highly efficient catalytic performance of Ir-Siphox for the hydrogenation of

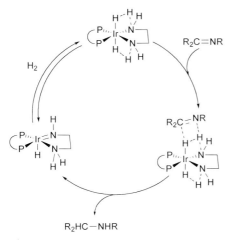

Scheme 10.32

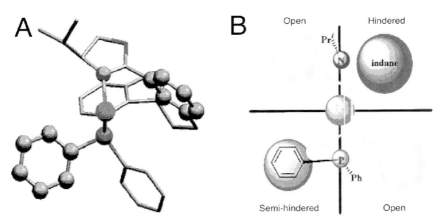

Figure 10.8 Proposed models based on the crystal structure of Ir-Siphox.

acyclic *N*-aryl ketimines with ee values up to 97% [82]. The ketimine substrate must coordinate to iridium with the *Si* face because the two bulky aryl groups can be located in the two open quadrants (Figure 10.8). To account for the observed absolute configuration of the products in the asymmetric titanocene-catalyzed hydrogenation of imines [33a, 34, 35], stereochemical models that take into account the electronic structure of the complex and the *anti*- and *syn*-imine interactions with the titanium-hydride species have demonstrated that the substituents on the N-atom of the imine have the greatest influence on the stereochemical outcome of the reaction and the *syn* and *anti* isomers of the imine should react to give opposite enantiomers of the product.

On the other hand, other mechanistic studies have assumed that the migratory insertion of the hydride into the imine is the enantioselective step. The isolation of catalyst–substrate adducts for chiral Xyliphos catalytic systems has demonstrated that while coordination of the imine is not diastereoselective, migratory insertion leads to the postulated and kinetic intermediates [3b].

References

1 (a) Blaser, H.-U. and Spindler, F. (1999) *Hydrogenation of Imino Groups*, in *Comprehensive Asymmetric Catalysis*, Vol. 1 (eds E.N. Jacobsen, A. Pfaltz and H. Yamamoto), Springer, Berlin, pp. 247–65. (b) Kobayashi, S. and Ishitani, H. (1999) *Chemical Reviews*, **99(5)**, 1069. (c) Okuma, T. and Noyoriin, R. (2004) *Hydrogenation of Imino Groups*, in *Comprehensive Asymmetric Catalysis*, Suppl. 1, Vol. 1 (eds E.N. Jacobsen, A. Pfaltz and H. Yamamoto), Springer, Berlin, pp. 43–53. (d) Tang, W. and Zhang, X. (2003) *Chemical Reviews*, **103**, 3029.

2 (a) Blaser, H.-U. and Spindler, F. (1997) *Chimia*, **51**, 297. (b) Blaser, H.-U., Buser, H.-P., Coers, K., Hanreich, R., Jalett, H.-P., Jelsch, E., Pugin, B., Schneider, H.-D., Spindler, F. and Wegmann, A. (1999) *Chimia*, **53**, 275. (c) Blaser, H.-U. (2002) *Advanced Synthesis and Catalysis*, **344**, 17.

3 (a) Spindler, F., Pugin, B. and Blaser, H.-U. (1990) *Angewandte Chemie (International Ed. in English)*, **29**, 558. (b) Dorta, R., Broggini, D., Stoop, R., Rhegger, H., Spindler, F. and Togni, A. (2004) *Chemistry – A European Journal*, **10**, 267. (c) Blaser, H.-U., Hanreich, R., Schneider, H.-D., Spindler, F. and Steinacher, B. (2004) *The quiral switch of Metolachlor: The development of a large-scale enantioselective catalytic process*, in *Asymmetric Catalysis on Industrial Scale* (eds H.-U. Blaser and E. Schmidt), Wiley-VCH Verlag GmbH, Weinheim.

4 (a) Dorta, R., Broggini, D., Kissner, R. and Togni, A. (2004) *Chemistry – A European Journal*, **10**, 4546. (b) Dorta, R., Broggini, D., Stoop, R., Rüegger, H., Spindler, F. and Togni, A. (2004) *Chemistry – A European Journal*, **10**, 1312.

5 Zhu, G. and Zhang, X. (1998) *Tetrahedron: Asymmetry*, **9**, 2415.

6 (a) Morimoto, T., Nakajima, N. and Achiwa, K. (1995) *Synlett*, **7**, 748, (b) Sablong, R. and Osborn, J.A. (1996) *Tetrahedron: Asymmetry*, **7**, 3059.

7 Liu, D., Li, W. and Zhang, X. (2004) *Tetrahedron: Asymmetry*, **15**, 2181.

8 Margalef-Catalá, R., Claver, C., Salagre, P. and Fernandez, E. (2000) *Tetrahedron: Asymmetry*, **11**, 1469.

9 (a) Xiao, D., Zhang, Z. and Zhang, X. (1999) *Organic Letters*, **10**, 1679. (b) Xiao, D. and Zhang, X. (2001) *Angewandte Chemie (International Ed. in English)*, **40**, 3425.

10 Dervisi, A., Carcedo, C. and Ooi, L. (2006) *Advanced Synthesis and Catalysis*, **348**, 175.

11 Imamoto, T., Iwadate, N. and Yoshida, K. (2006) *Organic Letters*, **8**, 2289.

12 (a) Vargas, S., Rubio, M., Suárez, A. and Pizzano, A. (2005) *Tetrahedron Letters*, **46**, 2049. (b) Vargas, S., Rubio, M., Suárez, A., del Rio, D., Alvarez, E. and Pizzano, A. (2006) *Organometallics*, **25**, 961.

13 Guiu, E., Muñoz, B., Castillón, S. and Claver, C. (2003) *Advanced Synthesis and Catalysis*, **345**, 169.

14 Guiu, E., Aghmiz, M., Díaz, Y., Claver, C., Meseguer, B., Militzer, C. and Castillón, S. (2006) *European Journal of Organic Chemistry*, **3**, 627.

15 Helmchen, G. and Pfaltz, A. (2000) *Accounts of Chemical Research*, **33**, 336.

16 (a) Schnider, P., Koch, G., Prétòt, R., Wang, G., Bohnen, F.M., Krüger, C. and Pfaltz, A. (1997) *Chemistry – A European Journal*, **3**, 887. (b) Kainz, S., Brinkmann, A., Leitner, W. and Pfaltz, A. (1999) *Journal of the American Chemical Society*, **121**, 6421.

17 (a) Lightfoot, A., Schnider, P. and Pfaltz, A. (1998) *Angewandte Chemie*, **110**, 3047. (b) Lightfoot, A., Schnider, P. and Pfaltz, A. (1998) *Angewandte Chemie (International Ed. in English)*, **37**, 2897. (c) Blackmond, D.G., Lightfoot, A., Pfaltz, A., Rosner, T.,

Schnider, P. and Zimmermann, N. (2000) *Chirality*, **12**, 442.
18 Ezhovaa, M.B., Patricka, B.O., James, B.R., Wallerb, F.J. and Ford, M.E. (2004) *Journal of Molecular Catalysis A: Chemical*, **224**, 71.
19 Blanc, C., Agbossou-Niedercorna, F. and Nowogrocki, G. (2004) *Tetrahedron: Asymmetry*, **15**, 2159.
20 Guiu, E., Claver, C., Benet-Buchholz, J. and Castillón, S. (2004) *Tetrahedron: Asymmetry*, **15**, 3365.
21 Guiu, E., Claver, C. and Castillón, S. (2004) *Journal of Organometallic Chemistry*, **689**, 1911.
22 Trifonova, A., Diesen, J.S., Chapman, C.J. and Andersson, P.G. (2004) *Organic Letters*, **6**, 3825.
23 Moessner, C. and Bolm, C. (2005) *Angewandte Chemie (International Ed. in English)*, **44**, 7564.
24 Faller, J.W., Milheiro, S.C. and Parr, J. (2006) *Journal of Organometallic Chemistry*, **691**, 4945.
25 Maire, P., Deblon, S., Breher, F., Geier, J., Bohler, C., Ruegger, H., Schonberg, H. and Grutzmacher, H. (2004) *Chemistry–A European Journal*, **10**, 4198.
26 Jiang, X., Minnaad, A.J., Hessen, B., Feringa, B.L., Duchateau, A.L.L., Andrien, J.G.O., Boogers, J.A.F. and de Vries, J.G. (2003) *Organic Letters*, **5**, 1503.
27 (a) Tararov, V.I., Kadyrov, R., Riermeier, T.H., Holz, J. and Börner, A. (1999) *Tetrahedron: Asymmetry*, **10**, 4009. (b) Tararov, V.I. and Börner, A. (2005) *Synlett*, **2**, 203.
28 Burk, M.J. and Feaster, J.E. (1992) *Journal of the American Chemical Society*, **114**, 6266.
29 Yang, Q., Shang, G., Gao, W., Deng, J. and Zhang, X. (2006) *Angewandte Chemie (International Ed. in English)*, **45**, 3832.
30 Shang, G., Yang, Q. and Zhang, X. (2006) *Angewandte Chemie (International Ed. in English)*, **45**, 6360.
31 (a) Ohkuma, T., Ooka, H., Hashiguchi, S., Ikariya, T. and Noyori, R. (1995) *Journal of the American Chemical Society*, **117**, 2675. (b) Uematsu, N., Fujii, A., Hashiguchi, S., Ikariya, T. and Noyori, R. (1996) *Journal of the American Chemical Society*, **118**, 4916. (c) Cobley, Ch.J. and Henschke, J.P. (2003) *Advanced Synthesis and Catalysis*, **345**, 195.
32 Charette, A.B. and Giroux, A. (1996) *Tetrahedron Letters*, **37**, 6669.
33 (a) Willoughby, C.A. and Buchwald, S.L. (1992) *Journal of the American Chemical Society*, **114**, 7562. (b) Willoughby, C.A. and Buchwald, S.L. (1993) *Journal of Organic Chemistry*, **58**, 7627.
34 Willoughby, C.A. and Buchwald, S.L. (1994) *Journal of the American Chemical Society*, **116**, 8957.
35 Willoughby, C.A. and Buchwald, S.L. (1994) *Journal of the American Chemical Society*, **116**, 11703.
36 (a) Ringwald, Z.M., StLrmer, R. and Brintzinger, H.H. (1999) *Journal of the American Chemical Society*, **121**, 1524. (b) Hansen, M.C. and Buchwald, S.L. (2000) *Organic Letters*, **2**, 713.
37 González-Arellano, C., Corma, A., Iglesias, M. and Sánchez, F. (2005) *Chemical Communications*, 3451.
38 DeVos, D.E., Vankelecom, I.F.J. and Jacobs, P.A. (2000) *Chiral Catalyst Immobilization and Recycling*, Wiley-VCH Verlag GmbH, Weinheim.
39 Amrani, Y., Lecomte, L., Sinou, D., Bakos, J., Coth, I. and Heil, B. (1989) *Organometallics*, **8**, 542.
40 (a) Bakos, J., Orosz, A., Heil, B., Laghmari, M., Lhoste, P. and Sinou, D. (1991) *Chemical Communications*, 1684. (b) Bakos, J. (1994) Aqueous Organometallic Chemistry and Catalysis, Vol. 3, High Technology NATO ASi Series, 15, p. 231.
41 (a) Lensink, C. and de Vries, J.G. (1992) *Tetrahedron: Asymmetry*, **3**, 235. (b) Lensink, C., Rijnberg, E. and de Vries, J.G. (1997) *Journal of Molecular Catalysis A: Chemical*, **116**, 199.
42 Pugin, B., Landert, H., Spindler, F. and Blaser, H.-U. (2002) *Advanced Synthesis and Catalysis*, **344**, 974.
43 Mhadgut, S.C., Palaniappan, K., Thimmaiah, M., Hackney, S.A., Török, B. and Liu, J. (2005) *Chemical Communications*, 3207.
44 Török, B. and Prakash, G.K.S. (2003) *Advanced Synthesis and Catalysis*, **345**, 165.
45 (a) Pinnavaia, T.J. (1983) *Science*, **220**, 365. (b) Choudary, B.M., Vasantha, G., Sharma, M. and Bharathi, P. (1989) *Angewandte Chemie (International Ed. in English)*, **28**, 465. (c) Choudary, B.M., Kumar, K.R. and Kantam, M.L. (1991) *Journal of Catalysis*,

130, 41. (d) Crocker, M., Herold, R.H.M., Beglass, J.G. and Companje, P. (1993) *Journal of Catalysis*, **141**, 70.
46 (a) Margalef-Català, R., Salagre, P., Fernández, E. and Claver, C. (1999) *Catalysis Letters*, **60**, 121. (b) Claver, C., Fernández, E., Margalef-Català, R., Medina, F., Salagre, P. and Sueiras, J.E. (2001) *Journal of Catalysis*, **201**, 70.
47 Crabtree, R.H. (1979) *Accounts of Chemical Research*, **12**, 331.
48 Margalef-Català, R., Claver, C., Salagre, P. and Fernández, E. (2000) *Tetrahedron Letters*, **41**, 6583.
49 (a) Pugin, B. (1997) WO97102232. (b) Pugin, B., Sindler, F. and Müller, M. (1992) USA5,252,751. (c) Jalett, H.-P. and Siebenhaar, B. (1997) WO97105095.
50 Brunner, H., Rublak, P. and Helgel, M. (1997) *Chemische Berichte*, **130**, 55.
51 Blaser, H.-U., Pugin, B., Spindler, F. and Togni, A. (2002) *Comptes Rendus Chimie*, **5**, 379.
52 Ayala, V., Corma, A., Iglesias, M., Rincón, J.A. and Sánchez, F. (2004) *Journal of Catalysis*, **224**, 170.
53 (a) Jessop, P.G., Ikariya, T. and Noyori, R. (1994) *Nature*, **368**, 231. (b) Jessop, P.G., Hsiao, Y., Ikariya, T. and Noyori, R. (1996) *Journal of the American Chemical Society*, **118**, 344. (c) Jessop, P.G., Ikatiya, T. and Noyori, R. (1995) *Chemical Reviews*, **95**, 259. (d) Leitner, W. (1995) *Angewandte Chemie (International Ed. in English)*, **34**, 2207. (e) Koch, D. and Leitner, W. (1998) *Journal of the American Chemical Society*, **120**, 13398.
54 (a) Rathke, J.W., Klingler, R.J. and Krause, T.R. (1991) *Organometallics*, **10**, 1350. (b) Guo, Y. and Akgerman, A. (1997) *Industrial and Engineering Chemistry Research*, **36**, 4581. (c) Bach, I. and Cole-Hamilton, D.J. (1998) *Chemical Communications*, 1463.
55 (a) Kainz, S. and Leintner, W. (1998) *Catalysis Letters*, **55**, 223. (b) Xiao, J., Nefkens, S.C.A., Jessop, P.G., Ikariya, T. and Noyori, R. (1996) *Tetrahedron Letters*, **37**, 2813. (c) Burk, J., Feng, S., Gross, M.F. and Tomas, W. (1995) *Journal of the American Chemical Society*, **117**, 8277.
56 Solinas, M., Pfaltz, A., Lozzi, P.G. and Leitner, W. (2004) *Journal of the American Chemical Society*, **126**, 16142.
57 Blanchard, L.A., Hancu, D., Beckman, E.J. and Brennecke, J.F. (1999) *Nature*, **299**, 28.
58 Saaby, S., Knudsen, K.R., Laslow, M. and Ley, S.V. (2005) *Chemical Communications*, 2909.
59 Kubas, G.J. (2001) *Metal Dihydrogen and Sigma-Bonded Complexes*, Kluwer Academic Publishers, Plenum Press, New York.
60 Esteruelas, M.A. and Oro, L.A. (1998) *Chemical Reviews*, **98**, 577.
61 Jessop, P.G. and Morris, R.H. (1992) *Coordination Chemistry Reviews*, **121**, 155.
62 Heinekey, D.M. and Oldham, W.J.J. (1993) *Chemical Reviews*, **93**, 913.
63 (a) Osborn, J.A., Jardine, F.H., Young, J.F. and Wilkinson, G. (1966) *Journal of the Chemical Society. A*, **4**, 1711. (b) Halpern, J. (1959) *Advances in Catalysis*, **11**, 301.
64 (a) Osakuda, K., Ohshiro, K. and Yamamoto, A. (1991) *Organometallics*, **10**, 404. (b) Grass, V., Lexa, D. and Saveant, J.M. (1997) *Journal of the American Chemical Society*, **119**, 7526.
65 (a) Fong, T.P., Forde, C.E., Lough, A.J., Morris, R.H., Rigo, R., Rocchini, E. and Stephan, T. (1999) *Journal of the Chemical Society, Dalton Transactions*, 4475. (b) Belkova, N.V., Bakhmutova, E.V., Shubina, E.S., Bianchini, C., Peruzzini, M., Bakhmutov, V.I. and Epstein, L.M. (2000) *European Journal of Inorganic Chemistry*, **10**, 2163.
66 James, B.R. (1997) *Catalysis Today*, **37**, 209.
67 (a) Herrera, V., Muñoz, B.K., Landaeta, V.R. and Canudas, N. (2001) *Journal of Molecular Catalysis. A*, **174**, 141. (b) Landaeta, V.R., Muñoz, B.K., Peruzzini, M., Herrera, V., Bianchinni, C. and Sanchez-Delgado, R.A. (2006) *Organometallics*, **25**, 403.
68 (a) Marcazzan, P., Patrick, B.O. and James, B.R. (2004) *Inorganic Chemistry*, **43**, 6838. (b) Marcazzan, P., Abu-Gnim, C., Seneviratne, K.N. and James, B.R. (2004) *Inorganic Chemistry*, **43**, 4820.
69 Ng, Y., Chan, C. and Osborn, J.A. (1990) *Journal of the American Chemical Society*, **112**, 9400.
70 (a) Spindler, F. and Blaser, H.-U. (1999) *Enantiomer*, **4**, 557. (b) Togni, A. (1996) *Angewandte Chemie (International Ed. in English)*, **35**, 1475.
71 Fryzuk, M.D., MacNeil, P.A. and Retting, S.J. (1987) *Journal of the American Chemical Society*, **109**, 2803.

References

72 Clapham, S.E., Hadzovic, A. and Morris, R.H. (2004) *Coordination Chemistry Reviews*, **248**, 2201.

73 Obora, Y., Ohta, T., Stern, Ch.L. and Marks, T.J. (1997) *Journal of the American Chemical Society*, **119**, 3745.

74 (a) Magee, M.P. and Norton, J.R. (2001) *Journal of the American Chemical Society*, **123**, 1778. (b) Guan, H., Iimura, M., Magee, M.P., Norton, J.R. and Zhu, G. (2005) *Journal of the American Chemical Society*, **127**, 7805.

75 (a) Bullock, R.M. (2004) *Chemistry – A European Journal*, **10**, 2366. (b) Song, J.S., Szalda, D.J., Bullock, R.M., Lawrie, C.J.C., Rodkin, M.A. and Norton, J.R. (1992) *Angewandte Chemie (International Ed. in English)*, **31**, 1233.

76 Martín, M., Sola, E., Tejero, S., Andrés, J.L. and Oro, L.O. (2006) *Chemistry – A European Journal*, **12**, 4043.

77 Martín, M., Sola, E., Tejero, S., López, J.A. and Oro, L.O. (2006) *Chemistry – A European Journal*, **12**, 4057.

78 Liu, X.-Y., Venkatesan, K., Schmalle, H.W. and Berke, H. (2004) *Organometallics*, **23**, 3153.

79 Haack, K.J., Hashiguchi, S., Fujii, A., Ikariya, T. and Noyori, R. (1997) *Angewandte Chemie (International Ed. in English)*, **36**, 285.

80 (a) Casey, C.P., Singer, S.W., Powell, D.R., Hayashi, R.K. and Kavana, M. (2001) *Journal of the American Chemical Society*, **123**, 1090. (b) Casey, C.P. and Johnson, J.B. (2005) *Journal of the American Chemical Society*, **127**, 1883.

81 (a) Abdur-Rashid, K., Lough, A.J. and Morris, R.H. (2000) *Organometallics*, **19**, 2655. (b) Abdur-Rashid, K., Lough, A.J. and Morris, R.H. (2001) *Organometallics*, **20**, 1047.

82 Zhu, S.-F., Xie, J.-B., Zhang, Y.-Z., Li, S. and Zhou, Q.-L. (2006) *Journal of the American Chemical Society*, **128**, 12886.

11
Imino Reductions by Transfer Hydrogenation
Martin Wills

11.1
History and Background

In contrast to ketones, and in particular with respect to asymmetric reductions, the reduction of compounds containing C=N bonds has remained relatively underdeveloped. Some significant breakthroughs have, however, been reported in the last ten years. This chapter will review reductions through metal hydride complexes, "Meerwein–Ponndorf–Verley (MPV) type" reductions, and other methods including transfer hydrogenation with organocatalysts and asymmetric Leuckart–Wallach aminations of ketones. Although there will be discussion of mechanistic aspects, the main focus of this review will be on more recent synthetic applications. The reader's attention is also directed to a number of recent useful reviews of this area [1].

11.2
Mechanisms of C=N Bond Reduction by Transfer Hydrogenation

Using either organic or organometallic catalysts, transfer hydrogenation of imines and other C=N bond-containing substrates, is most commonly carried out using isopropanol, formic acid or a formate salt as the hydrogen source. The most commonly, but not exclusively, used metals are ruthenium, rhodium and iridium.

The mechanism of the transfer hydrogenation of imines has been the subject of some extended synthetic, computational and mechanistic studies. The tutorial review by Bäckvall [1f] provides a very good summary of this. An excellent review by Morris [1c] also contains an incisive overview of hydrogenation and transfer hydrogenation and provides a very logical nomenclature for the possible reduction reaction pathway.

The mechanisms that may be considered for imine reduction are the same as for ketone reductions and may be divided into two broad classes (Figure 11.1). In the first of these, the MPV reduction or its reverse (Oppenauer oxidation), a metal

Modern Reduction Methods. Edited by Pher G. Andersson and Ian J. Munslow.
Copyright © 2008 WILEY-VCH Verlag GmbH & Co. KGaA, Weinheim
ISBN: 978-3-527-31862-9

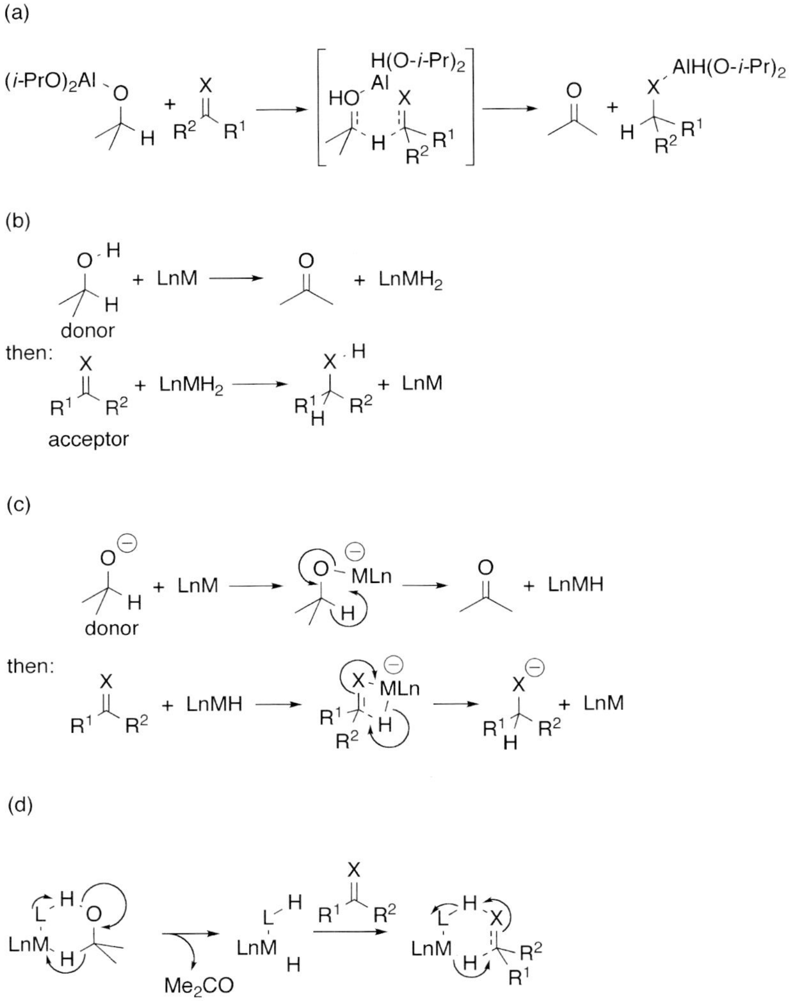

Figure 11.1 Possible mechanisms of hydrogen transfer.
(a) Direct hydrogen transfer (MPV reduction and Oppenauer oxidation). (b) Dihydride route. (c) Monohydride mechanism, inner sphere, no ligand assistance.
(d) Monohydride mechanism, outer sphere, ligand-assisted; the mechanism is as for (c), but with the involvement of a basic group "L" on the ligand.

coordinates to both substrate and hydrogen source (typically isopropanol) and hydrogen transfer takes place via a six-centered transition state (Figure 11.1a). This has been referred to as the "direct hydrogen pathway" [1f] and is typically observed with metal catalysts that cannot readily form hydrides. In the alternative "hydridic route," a metal-hydride complex is involved. Unsurprisingly, metals such as Rh,

Ru and Ir favor this mechanism, that is, metals which can readily form metal hydrides. Again there are several possible routes for hydrogen transfer. In the "dihydride" mechanism, both hydrogen atoms are transferred to the metal to form an intermediate which then transfers the hydrogens to a substrate (Figure 11.1b). In the alternative monohydride mechanism, the substrate interacts directly with the metal center (Figure 11.1c). A hydride is then transferred to the carbon of the C=N bond and the nitrogen atom of the product then engages in a direct interaction with the metal atom. Loss of amine (product) is facilitated by protonation to give the product and regenerate the catalyst which re-enters the catalytic cycle. A variation on this mechanism, involving "metal ligand bifunctional catalysis" [1e] or what Morris has called "ligand assistance" [1c], is the situation where two hydrogen atoms are simultaneously transferred from the catalyst to the substrate in a six-membered transition state (Figure 11.1d). In this mechanism, the substrate does not come into direct contact with the metal, but is reduced through an "outer sphere" mechanism.

Whether a reduction proceeds through the monohydride or dihydride mechanism can be determined by racemizing a sample of chiral alcohol containing a deuterium atom adjacent to the hydroxyl group. If the dihydride mechanism operates then the deuterium level will drop to ~50% upon racemization (the OH gaining a corresponding amount of D label) because the two hydrogen atoms in the metal hydride are equivalent. On the other hand, if the monohydride mechanism is operating then the label will remain on the C atom because the mechanism demands that the donor atom be the same as the recipient atom for the hydride on the metal.

Unlike ketone reductions, which have been the subject of extensive studies and optimizations, the mechanism of imine reduction is rather less fully developed and understood. However, some incisive studies have been completed, and these will be introduced as each class of catalyst is discussed below.

11.3
Asymmetric Reduction of C—N Bonds: Catalysts, Mechanisms and Results

11.3.1
Organometallic Catalysts Based on Ru, Rh, and Ir

The ruthenium complex $Ru_3(CO)_{12}$ was reported in 1987 to be capable of catalyzing the transfer of hydrogen from isopropanol to benzylideneanilines [2]. The active catalyst was isolated and shown by X-ray crystallography to be a complex of the imine substrate and a trimetallic hydride.

Bäckvall reported, in 1992, a racemic imine reduction using the Ru complex $[RuCl_2(PPh_3)]$ as catalyst and isopropanol as hydrogen source (Scheme 11.1) [3]. Base was required for the reaction to be successful. It was speculated that the ruthenium hydride was first formed from the chloride with the aid of the base. In later studies, this was confirmed and the hydride $[RuH_2(PPh_3)]$ was isolated and

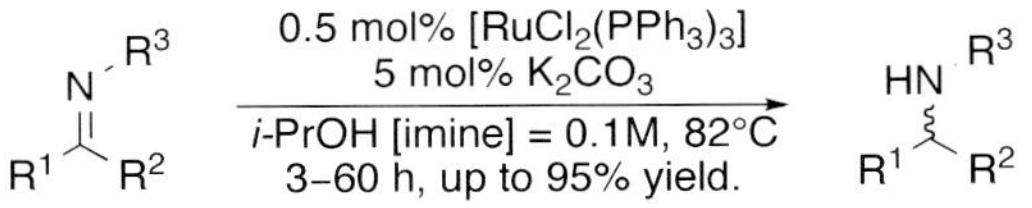

Scheme 11.1 Transfer hydrogenation of imines using a Ru(II) catalyst.

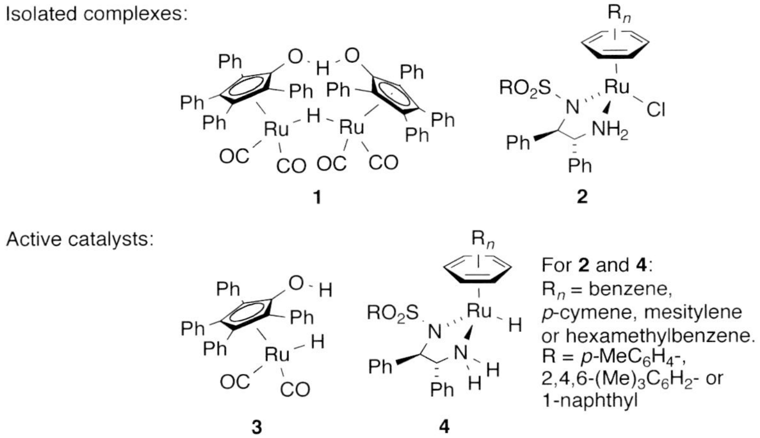

Isolated complexes:

1

2

Active catalysts:

3

4

For **2** and **4**:
R_n = benzene,
p-cymene, mesitylene
or hexamethylbenzene.
R = p-MeC$_6$H$_4$-,
2,4,6-(Me)$_3$C$_6$H$_2$- or
1-naphthyl

Figure 11.2 Complexes used for imine transfer hydrogenation.

was demonstrated to be an effective catalyst for C=N bond reduction without a requirement to add base [4]. Mechanistic studies have indicated that this reaction proceeds through the hydride mechanism without ligand assistance, as illustrated in Figure 11.1b. Ruthenium-catalyzed imine reduction by isopropanol can be accelerated by microwave irradiation [5]. Iridium/diphosphine catalysts are effective in the reduction of imines formed through aza-Wittig reactions [6], although elevated temperatures (110 °C) are required.

Formic acid (FA), together with triethylamine (TEA), in a 5:2 (molar) azeotropic mixture (FA/TEA), has frequently been used in asymmetric transfer hydrogenation (ATH) of imines. The reductive amination of ketones, in the presence of formic acid, is known as the Leuckart–Wallach reaction. An asymmetric variant of this reaction has been developed, and will be described later in this review. K. Wagner provides a very useful summary of early uses of FA/TEA in reductive amination [7].

A number of other catalysts have been used for imine transfer hydrogenation, most notably the Shvo diruthenium catalyst **(1)** [8–11, 12] and the Noyori ruthenium (II)-based catalyst **(2)** [13, 14] (Figure 11.2). Although the Shvo catalyst **(1)** is highly effective and utilizes isopropanol as the hydrogen source, the Noyori catalyst **(2)** provides a means for highly enantioselective imine reduction and requires formic acid/triethylamine for success. Diruthenium complex **(1)** is itself a precatalyst which is converted into two molecules of the active catalyst **(2)**, at the outset of the reaction. Likewise, complex **(3)** is converted into the active hydride

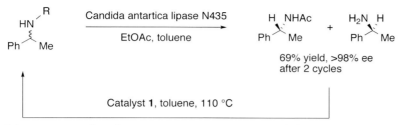

Scheme 11.2 Coupling of amine racemization to enzymic amide formation.

complex (**4**) upon exposure to a hydrogen source such as isopropanol or formic acid in the presence of a base. Bis-bidentate Schiff base complexes of Ru(III) have also been reported to reduce C=N bonds in racemic form using isopropanol as the source of hydrogen [15].

Bäckvall [8–10, 12] and Casey [11] have reported extensively on the use of dimer (**1**) in both ketone and imine reduction. The catalyst is highly active, giving high values of turnover frequency (TOF) even at very low loadings. A cosolvent is generally beneficial to the reaction. The mechanism of reduction has also been studied in detail.

Dimeric catalyst (**1**) is competent in the racemization of amines, which is difficult due to hydrolysis of the intermediate imines [16]. This is a useful process because it can be combined with an enzyme-catalyzed acylation to furnish enantiomerically enriched amides (Figure 11.3). Although the process requires the separation of amide from amine at each cycle, amides of high enantioselectivity (>98% ee) can be generated within a few cycles.

The first report of the use of Ru(II) complex (**2**) was published in 1996 by Noyori, using the TsDPEN system (Figure 11.3) [13]. This provided a method for the highly enantioselective reduction of imines under very mild conditions; typically at room temperature and without the need for the use of high pressures of hydrogen gas. It should be noted that reduction cannot be carried out in isopropanol, which is so commonly used for ketone reduction, using the TsDPEN catalysts. Also unlike the analogous ketone reductions, the addition of polar co-solvents such as acetonitrile, DMF, DMSO and DCM improves the rate of the reactions. In neat FA/TEA the reactions are slower.

The structure of the arene ring on the complex has an important effect on the enantioselectivity, as has the exact structure of the "ArSO$_2$" component on the asymmetric ligand in the Ru(II) complex. The examples in Figure 11.3 underline the catalyst requirements for successful reduction of each substrate.

Cyclic imines bearing alkyl and benzyl groups are generally reduced in high enantioselectivity, providing a very effective and general route to the asymmetric synthesis of isoquinoline alkaloids in particular; several of the reductions provided direct access to natural products. It is noteworthy, however, that similar substrates with aryl groups adjacent to the C=N bond are reduced with lower selectivity. As will be discussed later, a detailed study of aryl-substituted cyclic imines has revealed

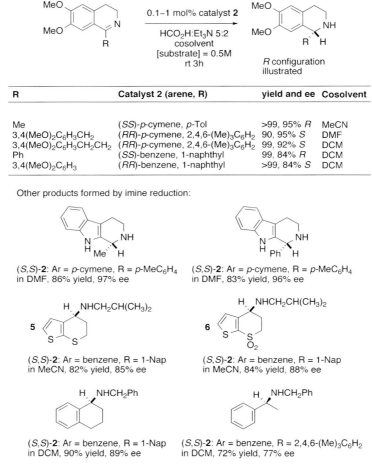

Figure 11.3 Asymmetric imine reduction with Ru(II) catalysts.

that these reactions can be substantially improved through the optimization of reaction conditions. In addition to the isoquinolines, cyclic imines fused to indoles also give products in high enantiomeric excess.

Acyclic imines are generally reduced in a lower selectivity, possibly due to the involvement of a mixture of imine configurational isomers. The observation of a higher ee value for the reduction of the imine of benzylamine with tetralone (89% ee) over the analogous acetophenone-derived imine (only 77% ee) suggests that fixation of this configurational mixture may be advantageous. This observation proved valuable in the synthesis of MK-0417 precursors **(5)** and **(6)** by reduction of the precursor acyclic imines.

An interesting feature of the reaction, as noted by Noyori, is the high speed of the imine reduction relative to the ketone; in the region of >1000-fold, and higher than that observed with NaBH$_3$CN. This may reflect protonation of the imine under the reaction conditions.

Rhodium (III) / pentamethylcyclopentadienyl (Cp') complexes of TsDPEN and other bidentate ligands are isoelectronic with and analogous in structure to the Ru(II)/arene complexes. This similarity has been exploited extensively by Avecia in its development of the Rh(III)-based "CATHy" reduction system. However the first published report on the application of Rh(III) complexes to imine reduction was by Baker in 1999 (Scheme 11.3) [17]. To a large extent the results mirror those obtained with the Ru(II) system although the ee values are generally slightly lower, and are particularly low for substrates containing aromatic substituents adjacent to the C=N group. This trend suggests that an unfavorable secondary interaction may be interfering with the control of the substrate approach to the catalyst. The reaction times, however, appear generally shorter for the rhodium(III) catalysts than for the ruthenium(II) analogues.

Avecia have developed Rh(III)-based asymmetric transfer hydrogenation reductions of ketones and of imines, using both amino alcohols and monotosylated diamines as ligands. They have assigned the trade marked name "CATHy" (catalytic asymmetric transfer hydrogenation) to the Rh catalysts they have developed. Although the majority of this work is featured in patents [18], Blacker *et al.* have described the application of Rh(III) reduction systems to *N*-phosphinoyl

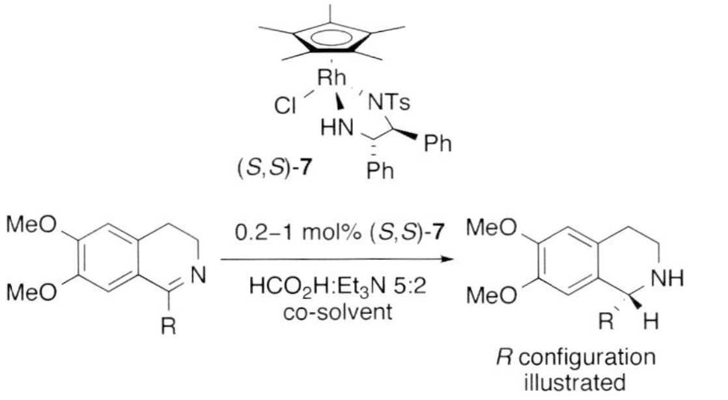

R	time/min	mol% catalyst	yield and ee	Cosolvent
Me	10	0.5	96, 89% ee R	MeCN
Et	10	0.5	93, 83% ee R	DMF
i-Pr	10	0.5	96, 99% ee R	DCM
Cyclohexyl	10	0.5	94, 97% ee R	DCM
Ph	180	1	90, 4.4% ee R	DCM
3,4-(MeO)$_2$C$_6$H$_3$	180	1	89, 3.2% ee R	DCM

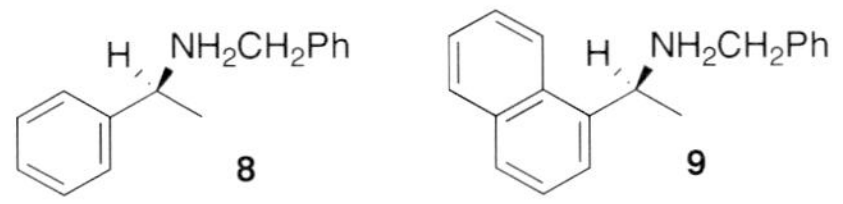

8
(*S*,*S*)-**7**: in DCM, 10 min, 88% yield, 8.4% ee (*S*)

9
(*S*,*S*)-**7**: in DCM, 10 min 85% yield, 8.4% ee (*S*)

Scheme 11.3 Asymmetric imine reduction with Rh(III) catalysts.

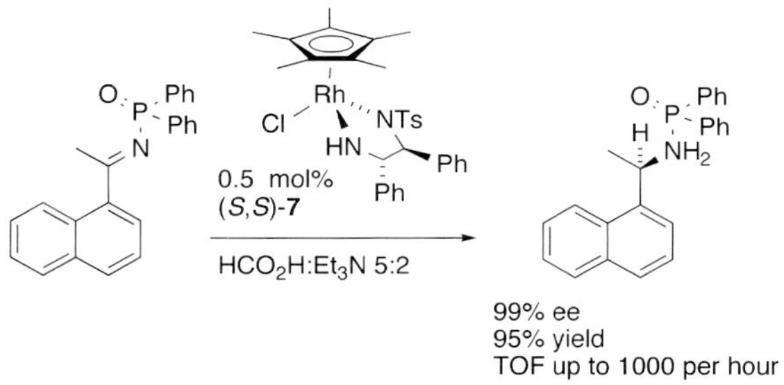

Scheme 11.4 Asymmetric N-phosphoroyl imine reduction.

imine reduction (Scheme 11.4) [19]. The result shown in Scheme 11.4 conceals an extensive series of detailed studies which have led to the optimization of the process. During the course of the work, Avecia have prepared a large number of catalysts and have investigated the reaction conditions in detail, including the effects of co-solvents. Some interesting observations were made, for example that the turnover frequency (TOF) can be increased (up to 1000 per hour) by bubbling nitrogen through the solution and agitating vigorously. Through the use of such modifications, the reduction can be carried out on a very large scale.

The asymmetric reduction of N-phosphinoyl imines has also been attempted using a multiple dendrimer supported Ru(II)/TsDPEN catalyst, but was reported to be inactive and decomposed under the reduction conditions [20].

11.3.1.1 Mechanistic Discussion

Bäckvall and Casey have focused many research investigations on the Shvo catalyst **(1)** [8–11, 12, 16]. This complex catalyzes both the racemization of enantiomerically pure amines [16] and the reduction of imines using isopropanol as the hydrogen source [8]. There is now convincing evidence that the mechanism of the reaction involves coordination of the imine to the metal followed by hydride transfer, that is, the monohydride mechanism of Figure 11.1. Although an intramolecular hydrogen bond between the —OH of the Cp ring may assist the reaction, the mechanism appears to be a clear example of an "inner" rather than an "outer" sphere" process [9–11, 12]. This is somewhat in contrast to the "outer sphere" mechanism now firmly established for ketone reductions by the same complexes. It is noteworthy that analogous ionic mechanisms of imine reduction by *hydrogenation* have been reported [21]. An excellent discussion of the very complex mechanistic picture is presented in a recent tutorial review by Bäckvall [1f].

Although very little mechanistic information is available on the Noyori Ru(II) system **(2)** in imine reductions, a 2006 paper [14] presented evidence for the operation of an ionic pathway in analogy to catalyst **(1)**. In particular, it was demonstrated that the addition of acid to the reaction was essential for a rapid reaction to be observed, suggesting that protonation of the imine is required. Although the full details remain to be determined, this observation suggests that the imine reduction mechanism is not analogous to the ligand-assisted mechanism of ketone reduction by the same catalysts (Figure 11.1d). The means by which the enantioselectivity of imine reduction is controlled also remains a matter of speculation at the present time. Although it may be controlled by the same factors that control the ketone reduction, there is no direct evidence to support this.

Recent kinetic studies have been carried out on the Rh(III) catalyst system by Blackmond [22, 23] which have served to shed further light on the mechanism of the reduction of imines. The study revealed that the resting state of the catalyst was as the Rh-hydride species, that is, where the chloride of the "precatalyst" is replaced by a hydride. The successful operation of the catalyst is shown to depend upon the acid–base equilibria, and this observation has led to a highly valuable, improved experimental protocol in which formic acid is added slowly to the solution rather than as a single dose at the start [22]. The observed importance of the solution pH mirrors Xiao's observations in the aqueous ketone reduction process [24].

11.3.2
Asymmetric Reductive Aminations Using Ammonium Formate to Give the Primary Amine Directly (the Leuchart–Wallach Reaction)

Although highly desirable, this is a very difficult and challenging process. Nevertheless, some useful recent breakthroughs have been achieved [25, 26]. A group at Nagoya reported on the use of several metal complexes for the reaction, of which the best proved to be the commercially available [Cp*Rh(III)Cl$_2$]$_2$. Although this gave only racemic products, the conversions were good. A remarkably effective asymmetric version was reported by Kadyrov and Riermeier, who found that several ruthenium and rhodium catalysts worked well in the conversion of acetophenone derivatives to enantiomerically enriched amines (Scheme 11.5). Of the catalysts studied, the best was [Ru-(R)-(TolBINAP)Cl$_2$], which was formed *in situ*. Amine products with ee values of up to 95% were formed, although a hydrolysis step was required in the work-up to convert the formate coproduct (anywhere up to 94% of the total product) into the amine. The table in Scheme 11.5 gives the

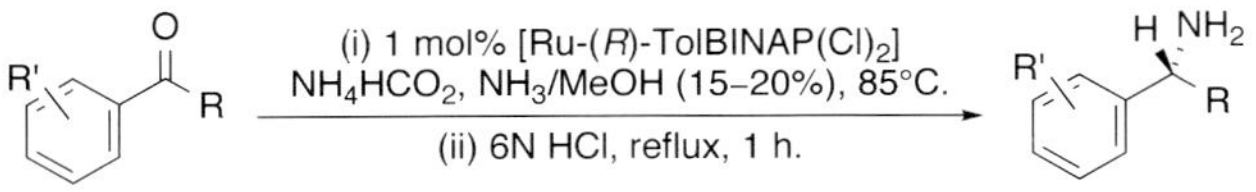

Aryl group	R	Yield and ee
Ph	Me	92, 95% ee R
Ph	Et	89, 95% ee R
3-MeC$_6$H$_4$	Me	74, 89% ee R
4-MeC$_6$H$_4$	Me	93, 93% ee R
4-ClC$_6$H$_4$	Me	93, 92% ee R
4-(NO$_2$)C$_6$H$_4$	Me	92, 95% ee R

Scheme 11.5 Asymmetric Leuckart–Wallach reactions with a Ru/TolBINAP catalyst.

total yield of amine after hydrolysis, but the paper gives full details of the relative amounts of formamide and amine that were formed with each substrate.

11.3.3
MPV Type Reductions

The MPV reduction reaction is a very venerable method for the reduction of ketones through the use of aluminum (III) salts. The accepted mechanism for MPV reduction involves a six-membered transition state with hydride transfer from the alcohol and concomitant formation of a ketone (that is acetone, if isopropanol is used). Unfortunately the development of the reaction in an asymmetric sense has been hindered by the requirement for the use of stoichiometric quantities of metal and, therefore, of catalyst. Solutions have been found to this problem, however, and several truly catalytic methods for ketone reductions have been reported [27]. The literature on MPV-type reductions of imines is quite limited, but a very effective system based on Al(III) BINOL complexes has been reported (Scheme 11.6) [28].

Although the BINOL/Al(III) process requires stoichiometric quantities of ligands and metal, both are reusable after the reaction and are inexpensive, making this a viable and apparently very versatile process. The authors suggest a transition state model in which the BINOL chelates the Al(III) and the phosphinoyl group

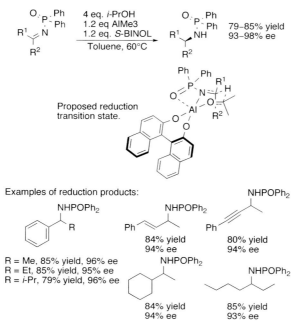

Scheme 11.6 MPV reduction of N-phosphinoyl imines using BINOL/Al complexes.

11.3 Asymmetric Reduction of C—N Bonds: Catalysts, Mechanisms and Results

also interacts in order to provide further stabilization. The use of BINOL is essential for catalysis; however, the use of too much (above 1.8 equivalents) results in dramatic reduction in the rate of reduction.

11.3.4
Carbene-based Catalysts

Another interesting development in recent years has been the development of *N*-heterocyclic carbene (NHC) complexes for the reduction of imines (Scheme 11.7) [29–32]. Although the methods are generally not asymmetric, the potential for future applications is clear from the TONs achieved. Most of the popular metals are represented, and the favored ligands tend to be bidentate, with large groups on the peripheral nitrogen atoms. These carbenes have also been successfully employed for ketone reductions (TON up to 19 000 in the case of **(11)** [31]. The combination of nickel (0) with NHC complexes also works effectively [29] although rather higher catalyst loadings are required compared to the bidentate versions. In the case of iridium complex **(12)** [32a], the neopentyl group is favored at the "wingtip" position as this improves the stability of the catalysts by preventing the possibility of a Hoffmann elimination.

A ruthenium complex of an *N*-heterocyclic carbene has recently been demonstrated to be capable of both the hydrogenation (i.e. use of hydrogen gas) and transfer hydrogenation of imines and ketones [33]. Isopropanol is used as the hydrogen source, with a catalyst loading of 2 mol%.

11.3.5
Organocatalytic Methods

Asymmetric reduction of imines can be achieved using a Hantzsch ester as the hydrogen source. When coupled with a suitable homochiral hydrogen-bonding

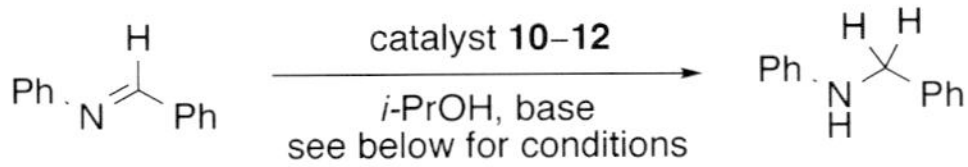

Using catalyst **10**: 0.015 mol% , 20 h, 55°C, TON: 4200 (63% conversion)
Using catalyst **11**: 0.1 mol%, 10 h, reflux, >98% yield, TON ca. 1000
Using catalyst **12**: 0.1 mol%, 10 min, 100% yield, TON 6000
Using Ni(0)/IMes: 5 mol%, 1.5 h,100°C, 89% yield

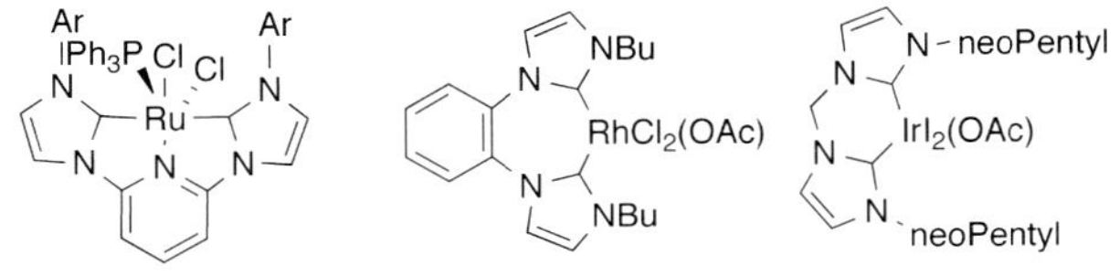

10 Ar = 2,6-(*i*-Pr)$_2$C$_6$H$_3$ [30]   **11** [31]   **12** [32]

Scheme 11.7 Carbene complexes used in transfer hydrogenation of imines.

catalyst, such as a BINOL-derived phosphate, this can be a powerful organocatalytic method. Some of the most recent examples are featured in Schemes 11.8 and 11.9 [34–39]. Both Rueping [35] and List [36] published the use of very similar catalysts for the reduction of isolated imines, to good effect. Both groups published the reductions of a wide range of representative substrates, with some of the highest enantiomeric excesses being obtained for imines with *ortho*-substituents. Both groups used a range of catalysts and in each case the best BINOL derivative proved to be an example with very bulky group adjacent to the oxygen atoms. The List catalyst has some advantages in terms of lower catalyst loadings and higher enantioselectivity, and was also reported to work on an aliphatic example; the imine of 3-methylbutan-2-one was reduced in 90% ee [36].

In later work, Rueping disclosed the highly enantioselective reductions of a series of cyclic imines (Scheme 11.9a) [37]. Careful optimization of the catalyst and conditions led to a process for benzoxazine reduction in >99% ee in one case and of at least 98% ee in the optimized examples. Remarkably, the catalyst loading could be dropped to as low as 0.01 mol%, at a cost of only 3% ee (decreased from 96% to 93% ee). Benzoxazinone reduction was also successful (Scheme 11.9a) and products of high enatioselectivity were again obtained, providing a very effective route for the synthesis of aryl-substituted amino acids following hydrolysis.

A reductive amination of ketones by aromatic amines was recently reported by Macmillan, who employed related BINOL-derived phosphate catalysts, thus avoiding the need to purify the imine substrates. Several ketones were used and the ee values were in the range of 83–97% (Scheme 11.9b). Reductive amination of butanone gave a product of 83% ee in 71% yield, which, alongside several other nonaromatic examples, demonstrated the applicability of the method beyond acetophenone derivatives [38]. Several heteroaromatic amines were also successfully employed in this asymmetric reductive amination.

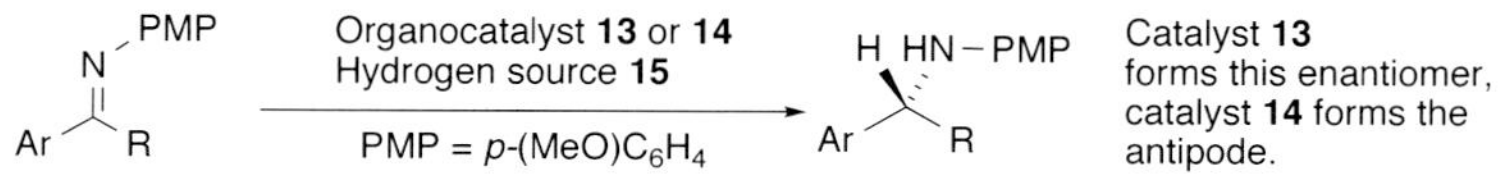

Comparative example:
Ar = 2-Napth, R = Me; 59% yield, 70% ee (20 mol% catalyst **13**), benzene, 60°C
Ar = 2-Napth, R = Me; 85% yield, 84% ee (1 mol% catalyst **14**), toluene, 35°C
Highest examples:
Ar = 2-FC$_6$H$_4$, R = Me, 82% yield, 84% ee (20 mol% catalyst **13**) benzene, 60°C
Ar = 2-MeC$_6$H$_4$, R = Me, 91% yield, 93% ee (1 mol% catalyst **14**) toluene, 35°C

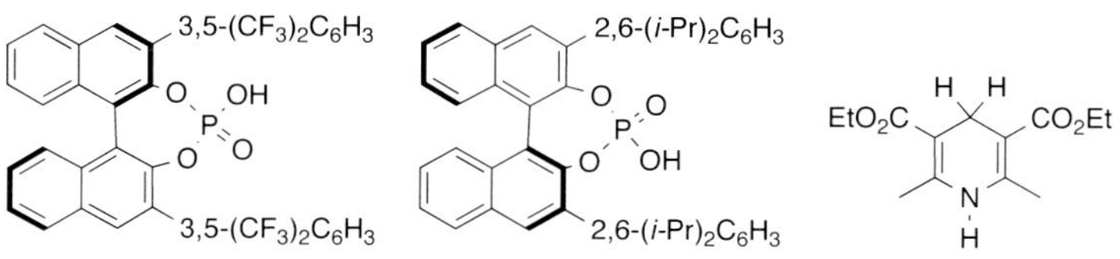

Scheme 11.8 Organocatalysis of imine reduction.

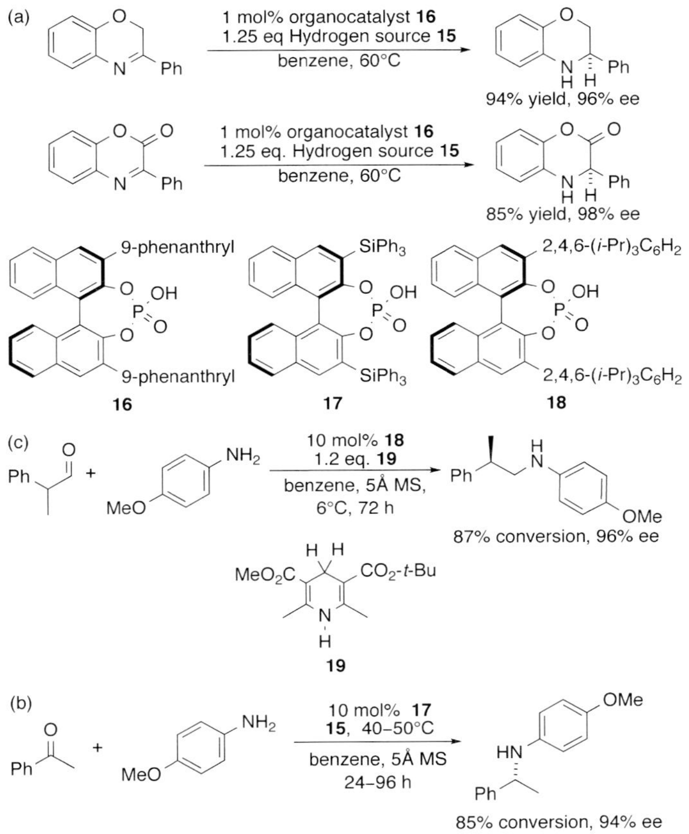

Scheme 11.9 (a) Organocatalysis of cyclic imine reduction. (b) Organocatalysis of reductive amination. (c) Reductive amination coupled to dynamic kinetic resolution.

The combination of asymmetric reductive amination with dynamic kinetic resolution (DKR) was also successfully investigated [39]. List demonstrated that racemic aldehydes could be successfully converted to branched-chain secondary amines in excellent enantiomeric excess Again the use of a highly hindered phosphate was essential, and intriguingly the best results required the very specific use of a particular Hantzsch ester, in this case **(19)** (Scheme 11.9c).

11.4
Specific Synthetic Applications

11.4.1
Sultams

Enantioselective imine reduction as a method for forming sultams represented one of the very first applications of the Noyori Ru(II)/TsDPEN methodology

(Scheme 11.10). Ahn published a paper in 1997 describing the specific application of the method to two substrates (Scheme 11.10) [40]. This was followed in 1999 by the Rh(III) paper of Mao and Baker in which the same TsDPEN ligand was used, but on an extended range of sultam precursors [17]. The most significant observation was the much higher activity of the Rh(III) range compared to the earlier Ru(II) catalysts. In the case of the *meta*-chlorophenyl example, the use of the equivalent Ru(II)/TsDPEN catalyst gave a product of just 69% ee after 180 minutes, compared to 81% ee within 30 minutes for the Rh(III)/TsDPEN. At the time of publication, the absolute configurations were assumed rather than proven in two cases.

The enantioselective asymmetric synthesis of sultams by C=N bond reduction has also been achieved using dendrimer supported catalysts (that is one catalyst per dendrimer) [41] and multiple dendritic complexes (that is several catalysts on one dendrimer) [20] (Scheme 11.10). In almost all cases DCM was used as the co-solvent of choice. In the case of the multiple dendrimer work, the use of unsupported Ru(II)/TsDPEN gave improved results (up to 96% ee) over the earliest reports for this reaction, presumably as a result of optimization of the conditions.

In a very recent paper, a sulfonated, and therefore water-soluble, TsDPEN derivative was used as the basis of a very effective Ru(II) catalyst for reduction of iminium salts and imines under aqueous conditions, representing the first report of this process [42]. In this case, sodium formate was used as the hydrogen source and cetyltrimethylammonium bromide was required as an additive (Scheme 11.10).

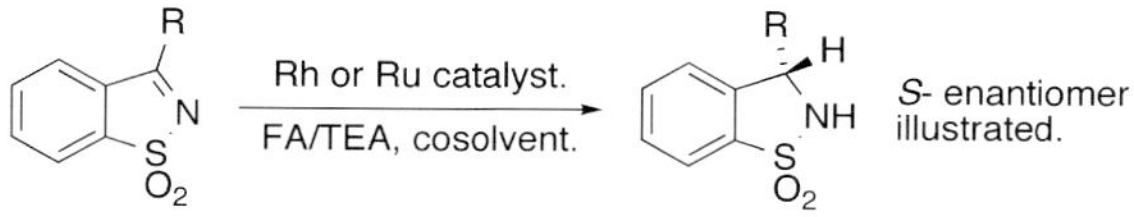

Using Ru(II)/(S,S)-TsDPEN catalyst 2: [40]
R = Bn, up to 91% ee (S) at S/C = 200, DCM 25°C, 17 h
R = t-Bu, up to 91% ee (S) at S/C = 200, DCM 25°C, 17 h
(one recrystallization gave optically pure product in 75% yield)

Using Rh(III)/(R,R)-TsDPEN catalyst 7: [17]
R = Me, 68% ee (R) at S/C = 200, DCM 20°C, 20 min
R = Bu, 67% ee (S presumed) at S/C = 200, DCM 20°C, 30 min
R = Bn, 68% ee (S) at S/C = 200, DCM 20°C, 40 min [(S,S) catalyst used]
R = m-ClC$_6$H$_4$, 81% ee (S presumed) at S/C = 200, DCM 20°C, 30 min

Using a dendrimer-supported Ru(II)/(R,R)-TsDPEN catalyst [41]
R = t-Bu, 95.5% ee (R) at S/C = 100, 28°C, DCM, 10 h

Using a multiple dendritic Ru(II)/(R,R)-TsDPEN catalyst: [20]
R = Bn, up to 96% ee (R) at S/C = 100, 28°C, DCM, 2 h
R = t-Bu, up to 98.4% ee (R) at S/C = 100, 28°C, DCM, 6 h

Using a water soluble Ru(II)/sulfonated (R,R)-TsDPEN catalyst): [42]*
R = Me, up to 65% ee (R) at S/C = 100, 28°C, 6 h
R = t-Bu, up to 94% ee (R) at S/C = 100, 28°C, 10 h
*with added cetyltrimethylammonium bromide

Scheme 11.10 Asymmetric synthesis of sultams.

Scheme 11.11 Asymmetric reduction of acyclic N-sulfonyl imines.

Using a multiple dendrimer Ru(II)/(R,R)-TsDPEN catalyst: [20]
51.8% ee (R), at S/C = 100, 28°C, DCM, 25 h
Using [Ru(p-cymene)Cl-(R,R)-TsDPEN] catalyst 2: [20]
45.0% ee (R), at S/C = 100, 28°C, DCM, 25 h

Using a water soluble Ru(II)/sulfonated TsDPEN catalyst: [42]*
No reduction observed!
*with added cetyltrimethylammonium bromide.

The reduction of acyclic N-sulfonylated imines is far less developed, but in one report on the use of multiple dendrimer catalysts, some useful, but not outstanding, enantioselectivities were achieved. However, the performance of the dendrimer was similar to that of the unsupported catalyst (Scheme 11.11) [20]. The attempted use of a sulfonated TsDPEN to induce an aqueous-phase transformation of the same substrate resulted in no reduction being observed [42].

11.4.2
Tetrahydroisoquinolines and Tetrahydro-β-carbolines

Probably the largest group of compounds which have been prepared by transfer hydrogenation of imines are those based on the tetrahydroisoquinoline structure, since the precursor is an ideal structure for a highly enantioselective reaction, following the precedent set by Noyori (Figure 11.3).

Tetrahydroisoquinoline reduction where there is an aryl group adjacent to the nitrogen carbon is difficult. In some early work, Vedejs was able to obtain some excellent results through reaction optimization but also found that the reaction was highly dependent on the structure of the substrate (Scheme 11.12) [43]. The desired product of Vedejs' study was the *ortho*-amino-substituted (20), for use as a chiral hydrogen donor. The use of the Ru(II)/TsDPEN catalyst (2) (arene = benzene, R = *p*-Tol) on a number of potential substrates was tested. Imine (21a), bearing a nitro group, could not be reduced efficiently, and (21b), containing a NHTs group, was not reduced at all, presumably due to product inhibition. Both (21c) and (21d) were reduced efficiently, but associated problems with protecting group removal made the synthesis undesirable and eroded both ee value and yield. It was possible to reduce the *ortho*-NH_2 analogue (not shown), but this material needed to be very pure for the reduction to be successful.

The best solution was found through the use of *ortho*-bromo compound (22) as the substrate. Following optimization, and again the use of DCM co-solvent, reduction was achieved with an ee of 98.3% (rising to 98.7% if the 1-napthylsulfonyl ligand was used instead) and >99% ee upon recrystallization. Although yields were not quantitative, this method had the best blend of selectivity and practicality. Displacement of the bromide with an amine furnished the desired product (20). A number of useful observations were made during the course of the work, in

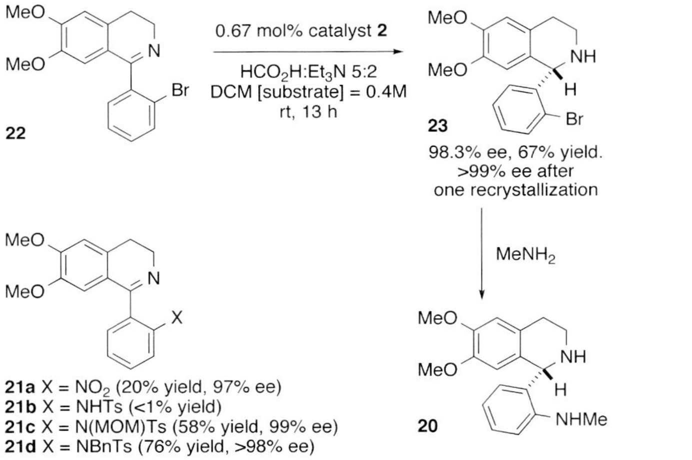

Scheme 11.12 Asymmetric reduction of aryl-substituted dihydroisoquinolines.

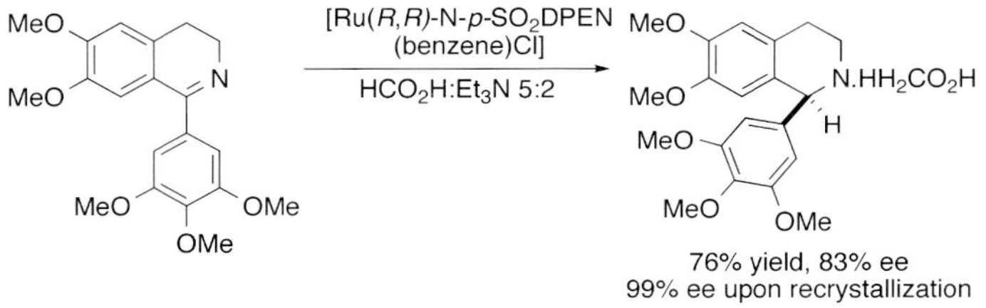

Scheme 11.13 Asymmetric tetrahydroisoquinoline synthesis.

particular it was noted that regular venting of the gases from the reaction was highly beneficial to the yield.

There are very few other reports of the reduction of dihydroisoquinolines containing aryl substituents at C1; however, an approach to S-(+)-cryptosyline using this method has been disclosed. Although the crude product was of 83% ee, it could be raised to 99% ee by recrystallization (Scheme 11.13) [44].

Probably the largest class of tetrahydroquinolines which have been prepared by ATH of imines are those containing a benzylic group on the imine, presumably because these provide a route to morphine-type alkaloids. Sheldon et al. published two reductions (Scheme 11.14) which were directed towards this end. In one case the ee value was 86% but for the second substrates (following the "Rice" route) the reduction proceeded in a remarkable 99% ee [45]. A similar reduction was also published by Boros and colleagues at GSK (Scheme 11.15) [46] for the synthesis of cis-dibenzoquinone chlorofumarate derivatives, which contain two tetrahydroisoquinoline rings.

The asymmetric synthesis of a series of emetines, which also contain two tetrahydroisoquinoline rings, was reported by Tietze et al. Asymmetric transfer hydro-

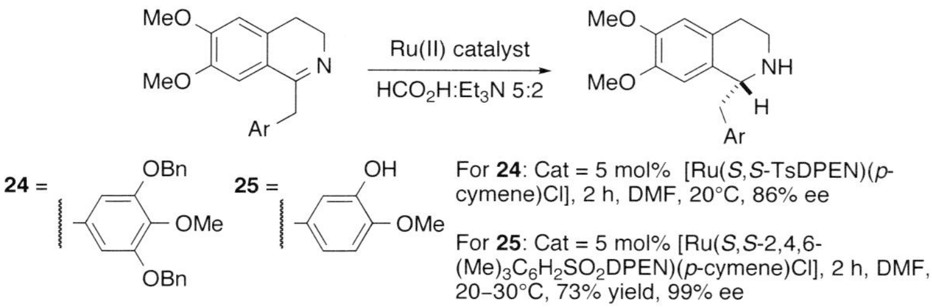

Scheme 11.14 Synthetic approach to a morphine precursor.

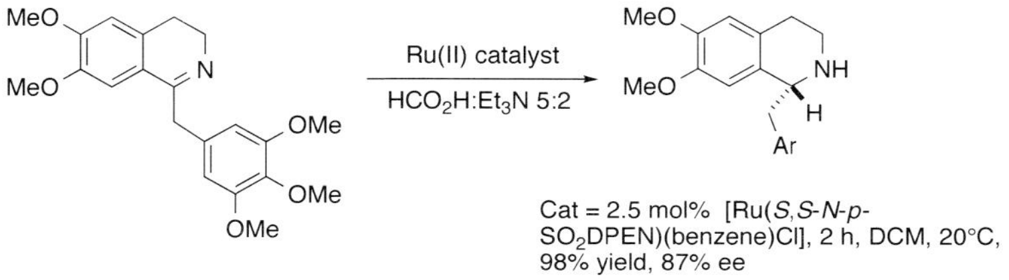

Scheme 11.15 cis-Dibenzoquinone chlorofumarate synthesis.

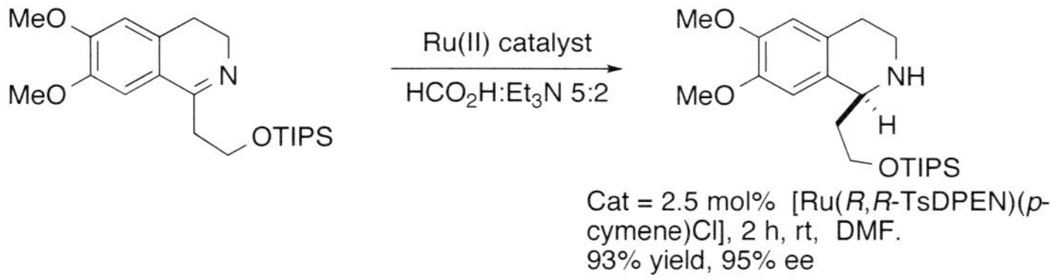

Scheme 11.16 Emetine synthesis ring 1.

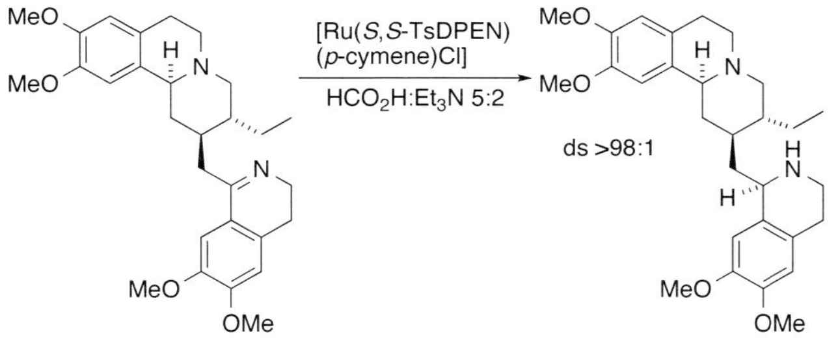

Scheme 11.17 Emetine synthesis ring 2.

genation was used to prepare each ring; Schemes 11.16 and 11.17 illustrate the key reduction steps [47]. The second reduction involves the reduction of an advanced intermediate and proceeds well, despite the complicating functional groups. This last step was also employed in a synthesis of the same target by Itoh et al. [48]. Other studies have revealed that esters (Scheme 11.18) [49] and

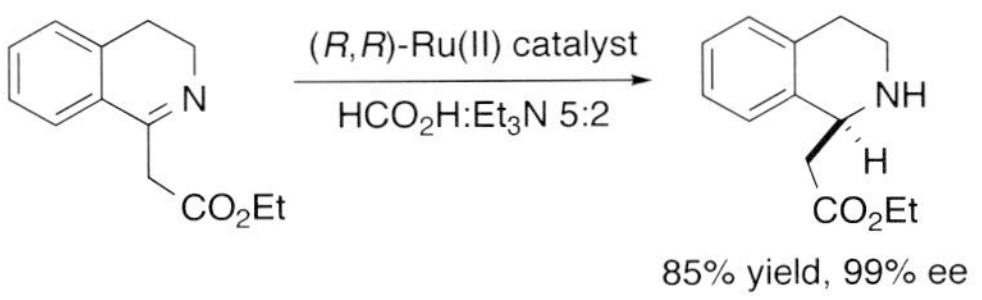

Scheme 11.18 Ester-substituted tetrahydroisoquinolines.

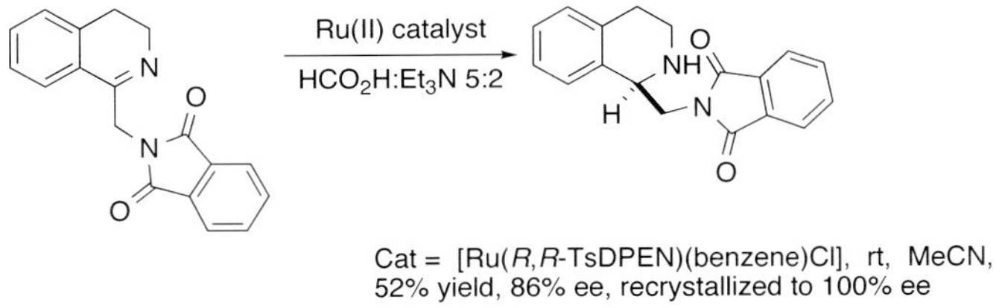

Scheme 11.19 Amino-substituted tetrahydroisoquinolines.

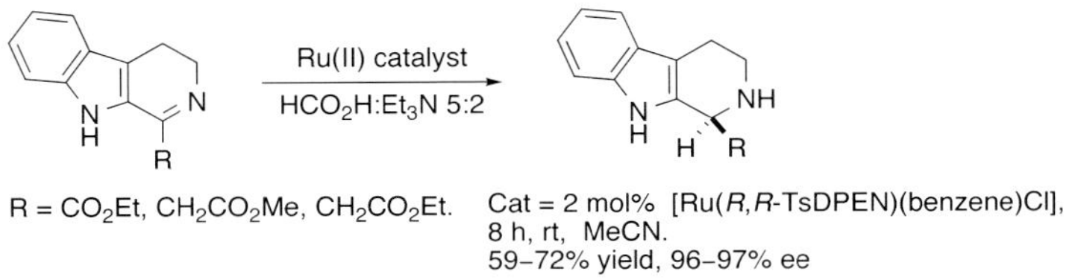

Scheme 11.20 Tetrahydro-β-carbolines 1.

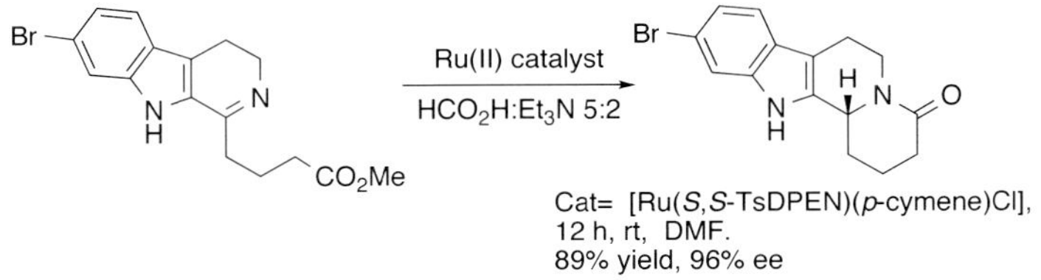

Scheme 11.21 Tetrahydro-β-carbolines 2.

phthalimides (Scheme 11.19) [50] could also be tolerated in the asymmetric reduction. The latter reduction represents a key step in the synthesis of (R)-(−)-praziquantel.

The imine precursors of tetrahydro-β-carbolines, by virtue of their cyclic structure, also represent a highly "matched" substrate for imine ATH. Several examples of reductions have been reported (Schemes 11.20–11.22) [49, 51]. Tietze *et al.* investigated the reductions of substrates containing ester substituents, to good effect [49]. An ideal application of this chemistry proved to be to the asymmetric synthesis of arborescidines A–C, for which the transformations in Schemes 11.21

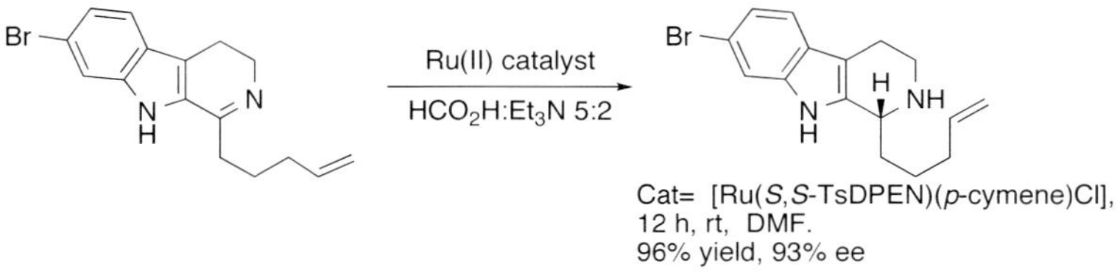

Scheme 11.22 Tetrahydro-β-carbolines 3.

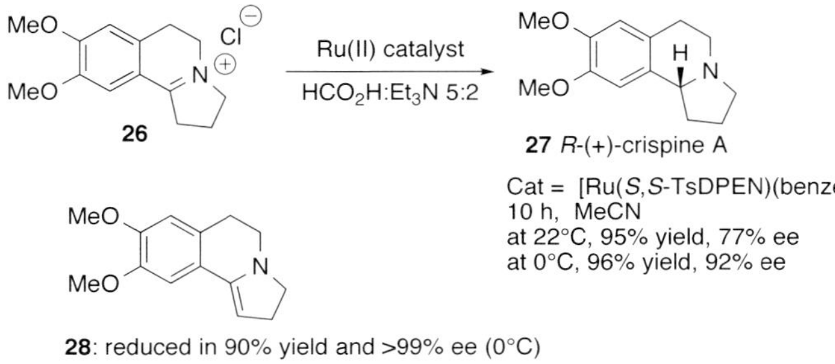

Scheme 11.23 Iminium salt reduction.

and 11.22 [51] are pivotal. This work served to establish the absolute configuration of these natural products. Of the cosolvents which have been used in imine reduction, DMF, MeCN and DCM are the most popular, and can have dramatic effects. For the reaction shown in Scheme 11.22, for example, the use of MeCN in place of DMF resulted in no reduction being observed. In other applications (for example Scheme 11.19), MeCN is a suitable solvent.

The direct reduction of iminium salts is also a productive process, as demonstrated by Czarnocki *et al.*, who have employed this reaction in a concise synthesis of (*R*)-(+)-crispine A (**27**) (Scheme 11.23) from iminium (**26**) [52]. Although the iminiums were reduced in excellent enantiomeric excess, a remarkably high ee value was obtained for the reduction of the related enamine substrate (**28**) (>99% ee). The reasons for this are not yet clear, but may prove to be of significant interest to workers in the area. This group also attempted the reaction with amino alcohol ligands, which are popular in ketone ATH. However the yields were very low (3–5%), reflecting the general observation that monotosylated diamines are virtually always used in ATH of imines with Ru(II) catalysts. It is likely that the corresponding amino alcohol/Ru(II) complexes are decomposed by the presence of formic acid. However an exception is detailed below.

The reduction of cyclic imines by ATH with Ru/TsDPEN complexes has also been reported under aqueous conditions [42]. Scheme 11.24 illustrates some of the products and the conditions under which they have been prepared, generally in excellent enantiomeric excess.

Although most work has focused on C=N reductions in five-membered or larger rings, the methodology may also be applied to the synthesis of aziridines by

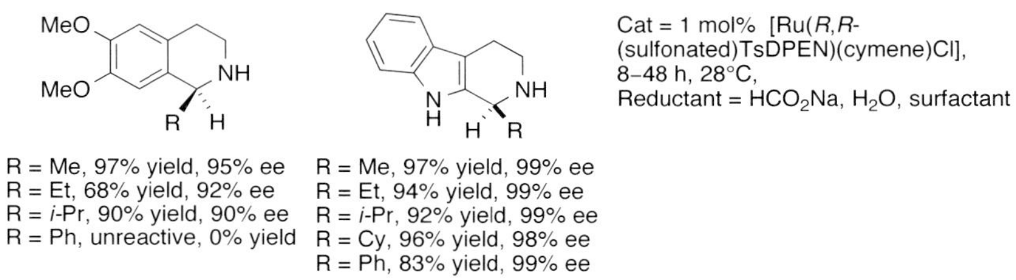

Scheme 11.24 Asymmetric C=N reduction products under aqueous conditions.

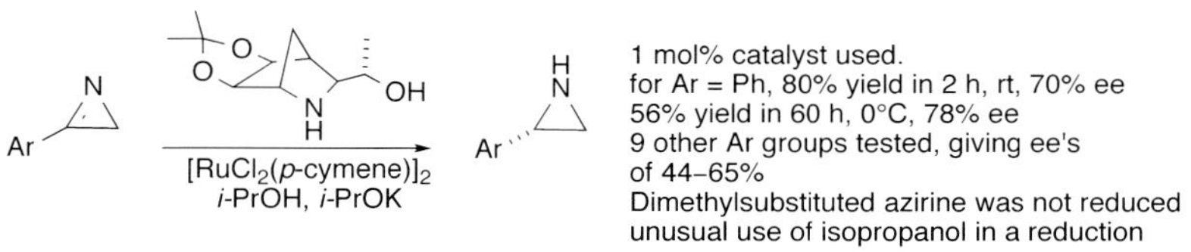

Scheme 11.25 Asymmetric C=N bond reduction in an azirine.

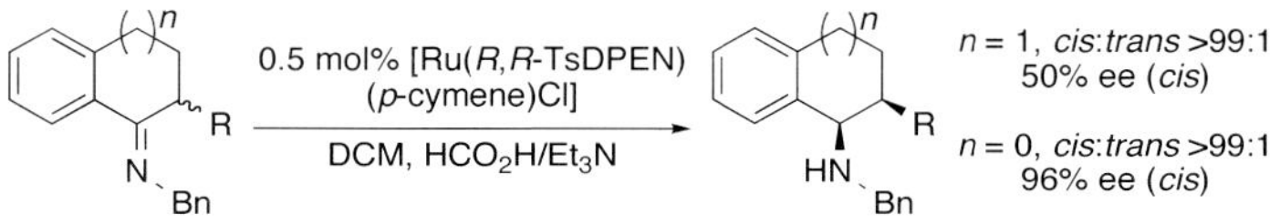

Scheme 11.26 DKR/asymmetric transfer hydrogenation of imines.

reduction of starting materials containing strained ring C=N bonds (Scheme 11.25) [53]. This is probably the only known example of ATH of an imine using an amino alcohol/Ru(II) complex in isopropanol. Generally, formic acid/triethylamine is required, in which amino alcohol/Ru(II) complexes are unstable. The dimethyl substituted azirine, that is with two methyl groups on the β-carbon, was not reduced under the conditions used in this investigation.

In a very detailed report, Lassaletta *et al.* described the clever combination of imine ATH with racemization to create a dynamic kinetic resolution reaction (Scheme 11.26) [54]. In certain cases, products were generated with high de and ee values in a single step, and the use of dichloromethane as a co-solvent was favored. This asymmetric reduction of an acyclic imine is one of a relatively small group, as most studies have been on cyclic substrates. Although the Ru(II) catalyst gave the best result for the example shown, Rh(III) and Ir(III) catalysts were also tested, and the Rh(III)/TsDPEN catalysts worked well with cyclohexyl-derived substrates (lacking the fused aromatic ring).

The formation of C—N bonds can be achieved in a one-pot process by combining reductive intramolecular amination with reduction (Schemes 11.27 and 11.28) [55, 56]. This involves the combination of a *t*-Boc-containing substrate with formic acid alone for a period, then the addition of catalyst. The tetrahdyroisoquinoline synthesis illustrated in Scheme 11.27 gives similar results to those generated by

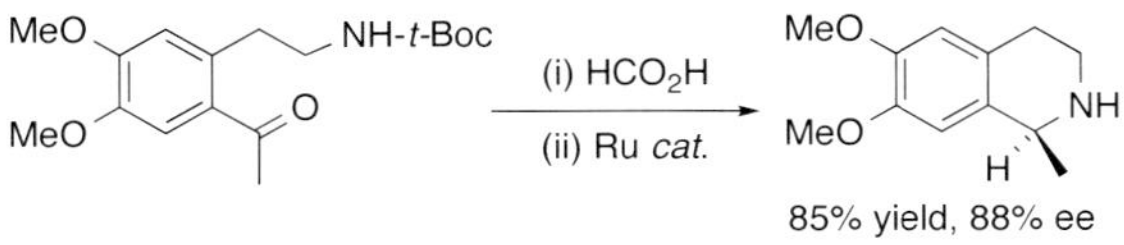

Scheme 11.27 One-pot reductive amination.

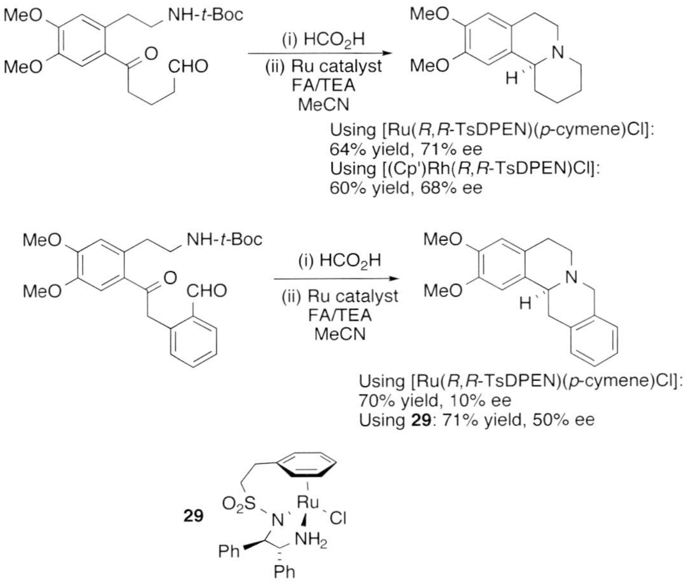

Scheme 11.28 Formation of several rings in a one-pot reduction.

the reduction of the isolated imine [55]. The method can be used to generate products containing multiple rings through a sequence of condensations and reductions (Scheme 11.28). In the most complex case, the maximum ee value was only 50% and the reaction required a tethered version of the Ru(II)/TsDPEN catalyst (**29**) for success.

Although dihydroisoquinolines are excellent substrates, quinoline itself and isoquinolium salts can be reduced by ruthenium and rhodium catalysts to the tetrahydroquinoline and tetrahydroisoquinolines respectively [57] using formic acid as a hydrogen source. These transformations may become the subject of future asymmetric research efforts.

11.5
Conclusion

In conclusion, transfer hydrogenation reactions of C=N bonds are valuable synthetic transformations that can be catalyzed by a multitude of organometallic catalysts and organocatalysts. In each case the mechanisms depend on the nature of the catalyst, with transition-metal catalysts favoring a metal-hydride intermediate and Al(III) catalysts favoring a "direct hydrogen transfer." The asymmetric reductions are not as well studied or developed as the corresponding ketone reductions, although a number of valuable applications have already been identified, particularly to the synthesis of alkaloid natural products. Asymmetric transfer hydrogenation of acyclic imines represents an ongoing challenge for researchers in this area. The most popular catalysts for ATH of imines are the Ru(II)/TsDPEN catalysts invented by Noyori, which are generally used in formic acid/triethylamine in the presence of a co-solvent, although Rh(III) and Ir(III) catalysts have also been investigated. This promises to be an important future research area.

References

1 (a) Palmer, M.J. and Wills, M. (1999) Asymmetric transfer hydrogenation of C=O and C=N bonds. *Tetrahedron: Asymmetry*, **10**, 2045. (b) Noyori, R. and Hashiguchi, S. (1997) Asymmetric Transfer Hydrogenation Catalyzed by Chiral Ruthenium Complexes. *Accounts of Chemical Research*, **30**, 97. (c) Clapham, S.E., Hadzovic, A. and Morris, R.H. (2004) Mechanisms of the H_2-hydrogenation and transfer hydrogenation of polar bonds catalyzed by ruthenium hydride complexes. *Coordination Chemistry Reviews*, **248**, 2201. (d) Gladiali, S. and Alberico, E. (2006) Asymmetric transfer hydrogenation: chiral ligands and applications. *Chemical Society Reviews*, **35**, 226. (e) Ikariya, T., Murata, K. and Noyori, R. (2006) Bifunctional transition metal-based molecular catalysts for asymmetric syntheses. *Organic and Biomolecular Chemistry*, **4**, 393. (f) Samec, J.S.M., Bäckvall, J.-E., Andersson, P.G. and Brandt, P. (2006) Mechanistic aspects of transition metal-catalyzed hydrogen transfer reactions. *Chemical Society Reviews*, **35**, 237. (g) Gladiali, S. and Alberico, E. (2004) Transfer hydrogenations. *Transition Metals for Organic Synthesis*, 2nd edn, Vol. **2**, Wiley-VCH Verlag GmbH, Weinheim, pp. 145–66. (h) Gladiali, S. and Mestroni, G. (1998) Transfer hydrogenations. *Transition Metals for Organic Synthesis*, **2**, 97–119. (i) Jacobsen, E.N., Pfaltz, A. and Yamamoto, H. (eds) (1999) *Comprehensive Asymmetric Catalysis*, Springer, Berlin. (j) Kitamura, M. and Noyori, R. (2004) Hydrogenation and transfer hydrogenation, in *Ruthenium in Organic Synthesis* (ed. Shun-Ichi Murahashi), Wiley-VCH Verlag GmbH, Weinheim, pp. 3–52. (k) Bullock, R.M. (2004) Catalytic Ionic Hydrogenations. *Chemistry – A European Journal*, **10**, 2366.

2 Basu, A., Bhaduri, S., Sharma, K. and Jones, P.G. (1987) Isolation and X-ray structure of [(μ-H)Ru$_3$(CO)$_9$(μ$_3$-PhNCPh)], the catalytically active cluster on Ru$_3$(CO)$_{12}$-catalysed transfer hydrogenation of benzylideneaniline. *Chemical Communications*, 1126.

3 Wang, G.-Z. and Bäckvall, J.-E. (1992) Ruthenium-catalysed transfer hydrogenation of imines by propan-2-ol. *Chemical Communications*, 980.

4 (a) Mizushima, E., Yamaguchi, M. and Yamagishi, T. (1999) Effective transfer hydrogenation of unsaturated compounds by ruthenium dihydride complex in propan-2-ol. *Journal of Molecular Catalysis A: Chemical*, **148**, 69. (b) Aranyos, A., Csjernyik, G., Kalman, K.J. and Bäckvall, J.-E. (1999) Evidence for a ruthenium dihydride species as the active catalyst in the RuCl$_2$(PPh$_3$)-catalyzed hydrogen transfer reaction in the presence of base. *Chemical Communications*, 351.

5 Samec, J.S.M., Mony, L. and Bäckvall, J.-E. (2005) Efficient ruthenium catalyzed

transfer hydrogenation of functionalized imines by isopropanol under controlled microwave heating. *Canadian Journal of Chemistry*, **83**, 909.

6 Cami-Kobeci, G. and Williams, J.M.J. (2004) Conversion of alcohols into N-alkyl anilines via an indirect aza-Wittig reaction. *Chemical Communications*, 1072.

7 (a) Wagner, V.K. (1970) Reactions with addition compounds containing activated formic acid. *Angewandte Chemie (International Ed in English)*, **9**, 50. (b) Narita, K. and Sekiya, M. (1977) Vapor-liquid equilibrium for the formic acid-triethylamine system examined by the use of a modified still. Formic acid-trialkylamine azeotropes. *Chemical and Pharmaceutical Bulletin*, **25**, 135.

8 Samec, J.S.M. and Bäckvall, J.-E. (2002) Ruthenium-catalysed transfer hydrogenation of imines by propan-2-ol in benzene. *Chemistry – A European Journal*, **8**, 2955.

9 Ell, A.H., Johnson, J.B. and Bäckvall, J.-E. (2003) Mechanism of ruthenium-catalyzed hydrogen transfer reactions. Evidence for a stepwise transfer of CH and NH hydrogens from an amine to a (cyclopentadienone)ruthenium complex. *Chemical Communications*, 1652.

10 Samec, J.S.M., Ell, A.H. and Bäckvall, J.-E. (2004) Mechanism of hydrogen transfer to imines from a hydroxycyclopentadienyl ruthenium hydride. Support for a stepwise mechanism. *Chemical Communications*, 2748.

11 Casey, C.P. and Johnson, J.B. (2005) Isomerisation and deuterium scrambling evidence for a change in the rate-determining step during imine hydrogenation by Shvo's hydroxycyclopentadienyl ruthenium hydride. *Journal of the American Chemical Society*, **127**, 1883.

12 Samec, J.S.M., Ell, A.H., Aberg, J.B., Privalov, T., Eriksson, L. and Bäckvall, J.-E. (2006) Mechanistic study of hydrogen transfer to imines from a hydroxycyclopentadienyl ruthenium hydride. Experimental support for a mechanism involving coordiantion of imine to ruthenium prior to hydrogen transfer. *Journal of the American Chemical Society*, **128**, 14293.

13 Uematsu, N., Fujii, A., Hashiguchi, S., Ikariya, T. and Noyori, R. (1996) Asymmetric Transfer Hydrogenation of Imines. *Journal of the American Chemical Society*, **118**, 4916.

14 Aberg, J.B., Samec, J.S.M. and Bäckvall, J.-E. (2006) Mechanistic investigation into the hydrogenation of imines by [p-(Me$_2$CH)C$_6$H$_4$Me]RuH(NH$_2$CHPhCHPhNSO$_2$C$_6$H$_4$-p-CH$_3$). Experimental support for an ionic pathway. *Chemical Communications*, 2771.

15 Venkatachalam, G. and Ramesh, R. (2006) Ruthenium(III) bis-bidentate Schiff base complexes mediated transfer hydrogenation of imines. *Inorganic Chemistry Communications*, **9**, 703.

16 (a) Pamies, O., Ell, A.H., Samec, J.S.M., Hermanns, N. and Bäckvall, J.E. (2002) An efficient and mild ruthenium-catalyzed racemization of amines: application to the synthesis of enantiomerically pure amines. *Tetrahedron Letters*, **43**, 4699.
(b) Paetzold, J. and Bäckvall, J.-E. (2005) Chemoenzymatic Dynamic Kinetic Resolution of Primary Amines. *Journal of the American Chemical Society*, **127**, 17620.

17 Mao, J. and Baker, D.C. (1999) A chiral rhodium complex for rapid asymmetric transfer hydrogenation of imines with high enantioselectivity. *Organic Letters*, **1**, 841.

18 (a) Blacker, A.J. and Mellor, B.J. (1997) Preparation of chiral arylalkanols by transfer hydrogenation using chiral metal cyclopentadiene complex catalysts. WO9842643B1, Avecia Ltd, 26 March 1997. (b) Blacker, A.J. and Campbell, L.A. (2000) Transfer hydrogenation process. EP1117627, Avecia Ltd. (c) Campbell, L.A. and Martin, J. (2001) Transfer hydrogenation process. EP1210305, Avecia Ltd.

19 Blacker, A.J. and Martin, J. (2004) Scale up studies in asymmetric transfer hydrogenation, in *Asymmetric Catalysis on an Industrial Scale: Challenges, Approaches and Solutions* (eds H.U. Blaser and E. Schmidt), John Wiley & Sons, Inc., New York, pp. 201–16.

20 Chen, Y.-C., Fei, T.-W., Deng, J.-G., Liu, H., Cui, X., Zhu, J., Jiang, Y.-Z., Choi, M.C.K. and Chan, A.S.C. (2002) Multiple

dendritic catalysts for asymmetric transfer hydrogenation. *Journal of Organic Chemistry*, **67**, 5301.

21 (a) Guan, H., Iimura, M., Magee, M.P., Norton, J.R. and Zhu, G. (2005) Ruthenium-Catalyzed Ionic Hydrogenation of Iminium Cations. Scope and Mechanism. *Journal of the American Chemical Society*, **127**, 7805. (b) Magee, M.P. and Norton, J.R. (2001) Stoichiometric, Catalytic, and Enantioface-Selective Hydrogenation of C:N Bonds by an Ionic Mechanism. *Journal of the American Chemical Society*, **123**, 1778. (c) Schlaf, M., Ghosh, P., Fagan, P.J., Hauptman, E. and Bullock, R.M. (2001) Metal-catalyzed selective deoxygenation of diols to alcohols. *Angewandte Chemie (International Ed in English)*, **49**, 3887.

22 Blackmond, D., Ropic, M. and Stefinovic, M. (2006) Kinetic Studies of the Asymmetric Transfer Hydrogenation of Imines with Formic Acid Catalyzed by Rh-Diamine Catalysts. *Organic Process Research and Development*, **10**, 457.

23 Richards, S., Ropic, M., Blackmond, D. and Walmsley, A. (2004) Quantitative determination of the catalysed asymmetric transfer hydrogenation of 1-methyl-6,7-dimethyxy-3,4-dihydroisoquinole using in situ FTIR and multivariate curve resolution. *Analytica Chimica Acta*, **519**, 1.

24 Wu, X., Li, X., King, F. and Xiao, J. (2005) Insight into and practical application of pH-controlled asymmetric transfer hydrogenation of aromatic ketones in water. *Angewandte Chemie (International Ed in English)*, **44**, 3407.

25 Kadyrov, R. and Riermeier, T.H. (2003) Highly enantioselective hydrogen-transfer reductive amination: catalytic asymmetric synthesis of primary amines. *Angewandte Chemie (International Ed in English)*, **42**, 5472.

26 Kitamura, M., Lee, D., Hayashi, S., Tanaka, S. and Yoshimura, M. (2002) Catalytic Leuckart-Wallach-Type reductive amination of ketones. *Journal of Organic Chemistry*, **7**, 8685.

27 Nishide, K. and Node, M. (2002) Recent development of asymmetric syntheses based on the Meerwein-Ponndorf-Verley reduction. *Chirality*, **14**, 759.

28 Graves, C.R., Scheidt, K.A. and Nguyen, S.T. (2006) Enantioselective MSPV reduction of ketimines using 2-propanol and (BINOL)Al(III). *Organic Letters*, **8**, 1229.

29 Kuhl, S., Schneider, R. and Fort, Y. (2003) Transfer Hydrogenation of Imines Catalyzed by a Nickel(0)/NHC Complex. *Organometallics*, **22**, 4184.

30 Danopoulos, A.A., Winston, S. and Motherwell, W.B. (2002) Stable N-functionalized "pincer" bis carbene ligands and their ruthenium complexes; synthesis and catalytic studies. *Chemical Communications*, 1376.

31 Albrecht, M., Crabtree, R.H., Mata, J. and Peris, E. (2002) Helating bis-carbene rhodium(III) complexes in transfer hydrogenation of ketones and imines. *Chemical Communications*, 32.

32 (a) Miecznikowski, J.R. and Crabtree, R.H. (2004) Transfer hydrogenation reduction of ketones, aldehydes and imines using chelated iridium(III) N-heterocyclic bis-carbene complexes. *Polyhedron*, **23**, 2857. (b) Gnanamgari, D., Moores, A., Rajaseelan, E. and Crabtree, R.H. (2007) Transfer Hydrogenation of Imines and Alkenes and Direct Reductive Amination of Aldehydes Catalyzed by Triazole-Derived Iridium(I) Carbene Complexes. *Organometallics*, **26**, 1226.

33 Burling, S., Whittlesey, M.K. and Williams, J.M.J. (2005) Direct and transfer hydrogenation of ketones and imines with a ruthenium N-heterocyclic carbene complex. *Advanced Synthesis and Catalysis*, **347**, 591.

34 Singh, S. and Batra, U.K. (1989) Asymmetric reductions of imines using NADH models. *Indian Journal of Chemistry, Section B: Organic Chemistry Including Medicinal Chemistry*, **28**, 1.

35 Reuping, M., Sugiono, E., Azap, C., Theissmann, T. and Bolte, M. (2005) Enantioselective Bronsted acid catalysed transfer hydrogenation: Organocatalytic reduction of imines. *Organic Letters*, **7**, 3781.

36 Hoffmann, S., Seayad, A.M. and List, B. (2005) A powerful Bronsted acid catalyst

for the organocatalytic asymmetric transfer hydrogenation of imines. *Angewandte Chemie (International Ed in English)*, **44**, 7424.
37. Reuping, M., Antonchick, A.P., Theissmann, T. (2006) Remarkably low catalyst loading in Bronsted acid catalysed transfer hydrogenations: Enantioselective reduction of benzooxazines, benothiazines and benzooxazinones. *Angewandte Chemie (International Ed in English)*, **45**, 6751.
38. Storer, R.I., Carrera, D.E., Ni, Y. and MacMillan, D.C. (2006) Enantioselective organocatalytic reductive amination. *Journal of the American Chemical Society*, **128**, 84.
39. Hoffmann, S., Nicoletto, M. and List, B. (2006) Catalytic asymmetric reductive amination of aldehydes via dynamic kinetic resolution. *Journal of the American Chemical Society*, **128**, 13074.
40. Ahn, K.H., Ham, C., Kim, S.-K. and Cho, C.-W. (1997) Practical synthesis of chiral sultam auxiliaries: 3-substituted-1,2-benzisothiazoline 1,1-dioxides. *Journal of Organic Chemistry*, **62**, 7047.
41. Chen, Y.-C., Wu, T.-W., Jiang, L., Deng, J.-G., Liu, H., Zhu, J. and Jiang, Y.-Z. (2005) Synthesis of Dendritic Catalysts and Application in Asymmetric Transfer Hydrogenation. *Journal of Organic Chemistry*, **70**, 1006.
42. Wu, J., Wang, F., Ma, Y., Cui, X., Cun, L., Deng, J. and Yu, B. (2006) Asymmetric trasnfer hydrogenation of imines and iminiums catalysed by a water-soluble catalyst in water. *Chemical Communications*, 1766.
43. Vedejs, E., Trapencieris, P. and Suna, E. (1999) Substituted isoquinolines by Noyori transfer hydrogenation: Enantioselective synthesis of chiral diamines containing an aniline subunit. *Journal of Organic Chemistry*, **64**, 6724.
44. Samano, V., Ray, A.J., Thomson, J.B., Mook, R.A., Jr, Jung, D.K., Koble, C.S., Martin, M.T., Bigham, E.C., Regitz, C.S., Feldman, P.L. and Boros, E.E. (1999) Synthesis of untra-short-acting neuromuscular blocker GW 0430: A remarkably stereo- and regioselective synthesis of mixed tetrahydroisoquinolinium chlorofumarates. *Organic Letters*, **1**, 1993.
45. Meuzelaar, G.J., van Vliet, M.C.A., Maat, L. and Sheldon, R.A. (1999) Chemistry of Opium Alkaloids, 45: Improvements in the total synthesis of morphine. *European Journal of Organic Chemistry*, 2315.
46. Kaldor, I., Feldman, P.L., Mook, R.A., Jr, Ray, J.A., Samano, V., Sefler, A.M., Thompson, J.B., Travis, B.R. and Boros, E.R. (2001) Stereocontrolled synthesis of cis-dibenzoquinolizine chlorofumarates: Curare-like agents of ultrashort duration. *Journal of Organic Chemistry*, **66**, 3495.
47. Tietze, L.F., Rackelmann, N. and Muller, I. (2004) Enantioselective total syntheses of the ipecacuanha alkaloid emetine, the Alangium alkaloid tubulosine and a novel benzoquinolizidine alkaloid by using a domino process. *Chemistry – A European Journal*, **10**, 2722.
48. Itoh, T., Miyazaki, M., Fukuoka, H., Nagata, K. and Ohsawa, A. (2006) Formal total synthesis of (-)-emetine using catalytic asymmetric allylation of cyclic imines as a key step. *Organic Letters*, **8**, 1295.
49. Tietze, L.F., Zhou, Y. and Topken, E. (2000) Synthesis of simple enantiopure tetrahydro-β-carbolines and tetrahydroisoquinolines. *European Journal of Organic Chemistry*, 2247.
50. Roszkowski, P., Maurin, J.K. and Czarnocki, Z. (2006) Enantioselective synthesis of (R)-(−)-praziquantel (PZQ). *Tetrahedron: Asymmetry*, **17**, 1415.
51. Santos, L.S., Pilli, R.A. and Rawal, V.H. (2004) Enantioselective toral synthesis of (+)-arborescidine A, (−)-arborescidine B and (−)-arborescidine C. *Journal of Organic Chemistry*, **69**, 1283.
52. (a) Szawkalo, J., Zawadzka, A., Wojtasiewicz, K., Leniewski, A., Drabowicz, J. and Czarnocki, Z. (2005) First enantioselective synthesis of the antitumor alkaloid (+)-crispine A and determination of its enantiomeric purity by ^{1}H NMR. *Tetrahedron: Asymmetry*, **16**, 3619. (b) Szawkalo, J. and Czarnocki, Z. (2005) Enantioselective synthesis of some tetracyclic isoquinolone alkaloids by asymmetric transfer hydrogenation catalysted by a chiral ruthenium complex. *Monatshefte fuer Chemie*, **1136**, 1619.

53 Roth, P., Andersson, P.G. and Somfai, P. (2002) Asymmetric reduction of azirines; a new route to chiral aziridines. *Chemical Communications*, 1752.

54 Ros, A., Magriz, A., Dietrich, H., Ford, M., Fernández, R. and Lassaletta, J.M. (2005) Transfer hydrogenation of α-branched ketimines: enantioseelcitve synthesis of cycloalkylamines via dynamic kinetic resolution. *Advanced Synthesis and Catalysis*, **347**, 1917.

55 Williams, G.D., Pike, R.A., Wade, C.E. and Wills, M. (2003) A One-Pot Process for the Enantioselective Synthesis of Amines via Reductive Amination under Transfer Hydrogenation Conditions. *Organic Letters*, **5**, 4227.

56 Williams, G.D., Wade, C.E. and Wills, M. (2005) One-pot formation of nitrogen-containing heterocyclic ring systems using a deprotection-cyclization-asymmetric reduction sequence. *Chemical Communications*, 4735.

57 (a) Wu, J., Liao, J., Zhu, J. and Deng, J. (2006) Transfer hydrogenation of isoquinolinium salts catalyzed by a rhodium complex. *Synlett*, 2059. (b) Watanabe, Y., Ohta, T., Tsuji, Y., Hiyoshi, T. and Tsuji, Y. (1984) Ruthenium catalyzed reduction of nitroarenes and azaaromatic compounds using formic acid. *Bulletin of the Chemical Society of Japan*, **57**, 2440.

12
Hydroboration and Diboration of Imines
Stephen A. Westcott and R. Thomas Baker

12.1
Introduction

Amines are a remarkably important class of compounds that have widespread applications as solvents, textile additives, rubber chemicals, raw material for resins, agricultural chemicals, water treatment, disinfectants and corrosion inhibitors and in the manufacture of detergents. Amines are also used as key intermediates in organic synthesis and as bases in synthetic transformations and are important building blocks in many common polymers such as Nylon. Owing to their vast synthetic importance, and the observation that over 75% of all drugs and drug candidates include nitrogenous functionalities [1], numerous methods have been developed for their production. Two of the most common methods of generating amines are via the reductive amination of carbonyl compounds and the hydrogenation or reduction of nitriles [2–4].

The production of amines from nitriles is mechanistically similar to reductive amination as both have an intermediate imine functionality. The reductive amination of aldehydes or ketones generally proceeds via initial condensation of the carbonyl compound and the amine to give a carbinolamine intermediate (Scheme 12.1), which subsequently dehydrates to generate the imine or iminium ion. Reduction of the imine group affords the corresponding amine, where the choice of the reducing agent is critical to the success of the reaction. Although the catalytic hydrogenation reaction is an effective way of reducing imines, reactions can give mixtures of products and low yields depending upon the nature of the imine. Another limitation to this method involves reducing substrates containing divalent sulfur, which may inhibit and eventually deactivate the catalyst [5]. An alternative method for the reduction of imines involves the use of hydride reducing agents, especially boron hydrides such as $NaBH_4$. While the latter known for some time [6–9], this chapter highlights some of the latest developments in the use of boron hydrides and related diboron compounds for the reduction of imines.

Modern Reduction Methods. Edited by Pher G. Andersson and Ian J. Munslow.
Copyright © 2008 WILEY-VCH Verlag GmbH & Co. KGaA, Weinheim
ISBN: 978-3-527-31862-9

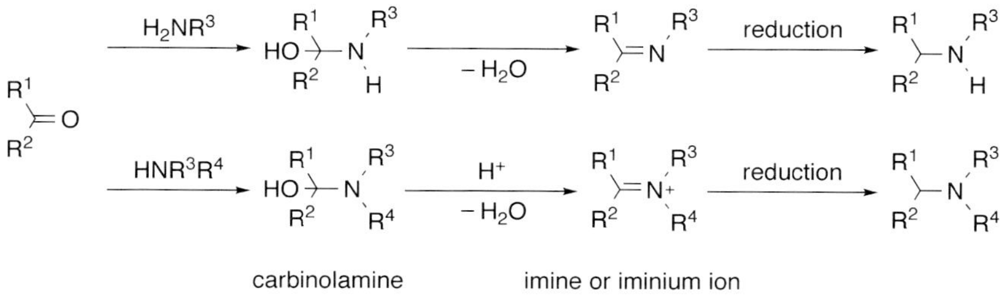

Scheme 12.1 General reductive amination pathway.

12.2
Uncatalyzed Reactions

In 1963 Schellenberg [10] reported that sodium borohydride (NaBH$_4$) could be used effectively in the reductive amination of a wide range of substrates. However, these reactions suffered from poor selectivities and low yields in the case of sterically hindered imines. As a result, a considerable amount of research has subsequently concentrated on the development of new borohydride reducing agents for these important reactions. Borch and coworkers found that sodium cyanoborohydride (NaBH$_3$CN) [11] was a much more selective reducing agent than NaBH$_4$ for the reductive amination of ketones and aldehydes. Unfortunately, the major drawbacks of using this expensive reagent include its high toxicity, the observation that reactions are carried out using a fivefold excess of amine at pH 6–8, and the fact that these reactions generate dangerously toxic by-products such as HCN and NaCN. Since then, sodium triacetoxyborohydride [NaBH(OAc)$_3$] (STAB-H) has been identified as a safer alternative than NaBH$_3$CN [12, 13]. However, solvents such as methanol or water are not suitable for reactions with STAB-H due to decomposition of the reducing agent. A copper(I) tetrahydroborate complex in the presence of sulfamic acids has also been used recently for the direct reductive elimination of carbonyl compounds [14]. As several excellent reviews and papers on borohydride reducing agents are currently available [15–21], readers are encouraged to consult these publications for a more complete survey of the development of this area. The rest of this chapter will therefore focus on the advantages of using neutral boranes (including adducts) and diboranes for the reduction of imines.

12.2.1
Imines

Although borohydride reagents are still commonly employed for the reduction of imines [22–36], studies have shown that polyhedral boranes can also be used as reducing agents [37–39]. For instance, early work in this area by Gaines and Coons [37] found that reactions of the iminium salt [Me$_2$NCH$_2$]I with salts of B$_5$H$_8^-$,

1-Et-B$_5$H$_7^-$, and 1-Br-B$_5$H$_7^-$ gave the μ-[(dimethylamino)methyl]pentaborane derivatives μ-(Me$_2$NCH$_2$)B$_5$H$_8$, 1-Et-μ-(Me$_2$NCH$_2$)B$_5$H$_7$, and 1-Br-μ-(Me$_2$NCH$_2$)B$_5$H$_7$, respectively, in high yields. A few years later, Sneddon and coworkers reported that reaction of *nido*-6-SB$_9$H$_{11}$ with *N-t*-butylformaldimine (H$_2$C=N-*t*-Bu) afforded the novel zwitterionic compound 9-[(*t*-Bu)NH$_2^+$CH$_2$]-*nido*-6-SB$_9$H$_9^-$. This unique product was purported to arise from an initial hydroboration of the imine at the B9 cage position followed by proton transfer from a cage-bridging site to the nitrogen atom [38].

More relevant to organic synthesis, however, are the recent advances made using simple boranes such as BH$_3$·THF, BH$_3$·SMe$_2$, BH$_3$·pyridine, and derivatives [40–45]. While great strides have been made using these important reagents, these adducts are not without certain disadvantages. Specifically, the low concentration of reducing agent in commercially available BH$_3$·THF limits its applications to only one solvent and the complex itself is not stable over long periods [46]. As borane-dimethylsulfide (BMS) does not suffer from these drawbacks, its use in organic synthesis has been relatively common. For instance, *N*-alkylimines derived from cyclobutanones have been stereoselectively reduced to the corresponding *cis*-3-arylcyclobutylamines without formation of the *trans*-isomers using BH$_3$·SMe$_2$ [47]. Remarkably, attempts to reduce these *N*-(cyclobutylidene)amines with NaBH$_4$ did not afford any of the desired product but led to mixtures containing the dehydrochlorinated *N*-(cyclobutenylidene)amine (Scheme 12.2). The less alkaline properties of the borane were reported to be responsible for the unique selectively in these reactions, whereas the ionic nature of the borohydride resulted in products arising from dehydrochlorination.

Interestingly, addition of borane to a 2,2,4,4-unsubstituted *N*-(cyclobutylidene) amine, did not give the expected reduction product, but rather a stable complex between the imine and the borane, which could even be purified by flash chromatography on silica gel (Scheme 12.3). While heating of the imine-borane adduct gave the secondary amine, treatment with aqueous HCl or aqueous NaOH gave unwanted 3-phenylcyclobutanone and *cis*-3-phenylcyclobutanol, respectively. Although BH$_3$·SMe$_2$ has also been used in the hydroboration of poly(ethylene imine)s [48], and is often considered superior to its THF counterpart in terms of

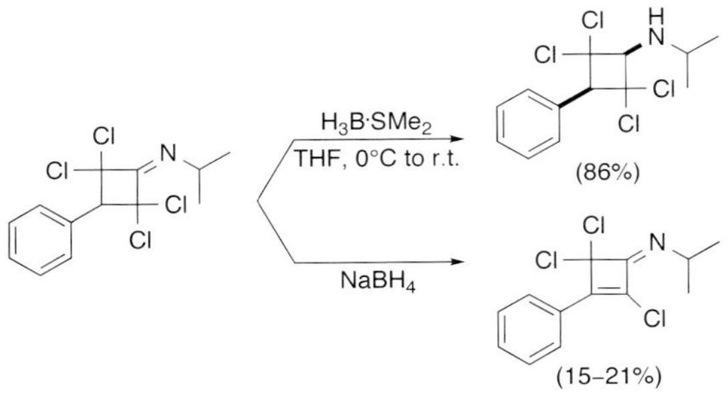

Scheme 12.2 Addition of different reducing agents to an *N*-(cyclobutylidene)amine.

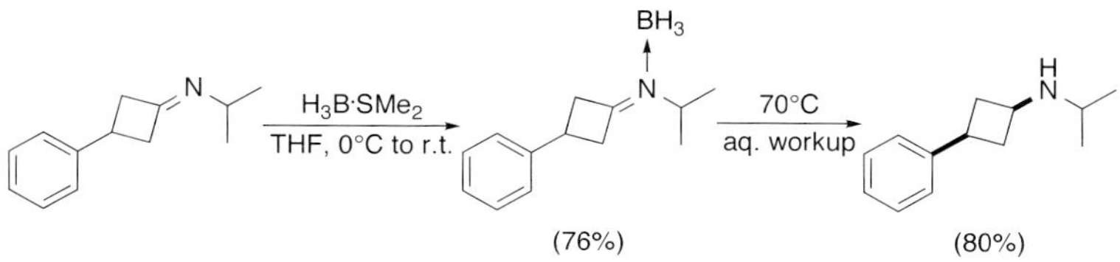

Scheme 12.3 Formation of a stable imine-borane adduct.

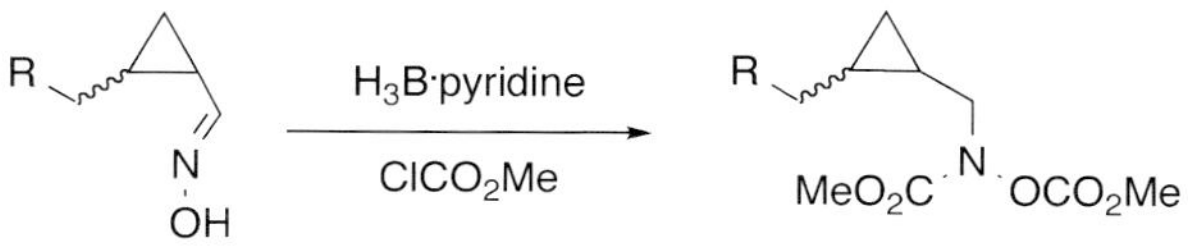

Scheme 12.4 Reduction and acylation of cyclopropyloximes.

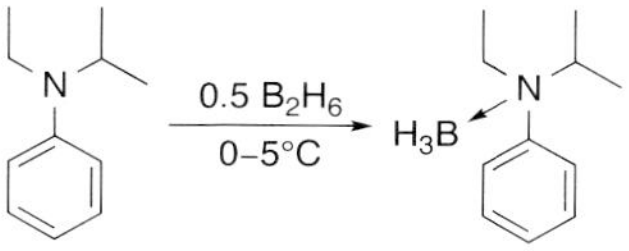

Scheme 12.5 Synthesis of N-ethyl-N-isopropylaniline-borane.

stability [49], use of this reagent has been limited as the odorous DMS (dimethyl sulfide) group is generated as a toxic and volatile side-product.

A considerable effort has thus focused on alternatives to BMS. Indeed, N-alkylamino phthalimides have been synthesized via a reductive amination reaction of N-amino phthalamides and aldehydes using pyridine-borane in acetic acid [42]. The presence of acid is required to activate the relatively stable pyridine-borane adduct. The same reagent has also been used effectively to reduce cyclopropyloximes [50], which, after acylation, yielded diacylated hydroxyamines (Scheme 12.4). Beside the unwanted health risks associated with pyridine and the deleterious environmental effects, this reagent is relatively unstable at elevated temperatures and distillations at reduced pressures can lead to violent decompositions [51]. This leads to problems concerning large-scale use in industrial applications as well as difficulties associated with storage over extended periods.

Borane-amine reagents have been found to be particularly attractive for large-scale industrial reactions, as many are air-stable liquids or solids. Brown and coworkers have examined several such reagents and found N-ethyl-N-isopropylaniline to be a highly reactive and environmentally benign borane carrier for a variety of substrates, including imines and nitriles (see below) [52]. N-Ethyl-N-isopropylaniline-borane is readily prepared in high yield by passing a slight excess of diborane gas into the neat amine at low temperature (Scheme 12.5) and is stable at room temperature indefinitely under nitrogen. While reactions of N-ethyl-N-isopropylaniline-borane with N-benzylidene-butylamine occur rapidly

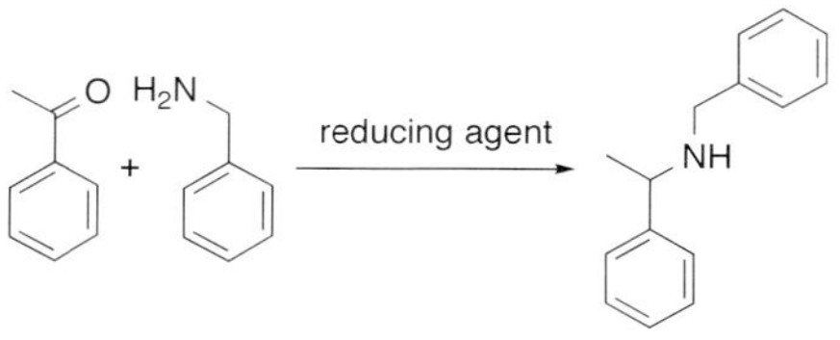

Reducing agent	Yield (%)
BH$_3$·pyridine (AcOH)	0
BH$_3$·pyridine (molecular sieves)	10
NaBH(OAc)$_3$	55
BH$_3$·picoline	87

Scheme 12.6 The use of α-picoline borane in the reductive amination of acetophenone and benzylamine.

to give the corresponding amine upon work-up, reductions of N-benzylidene-aniline take considerably longer and are complete after 4–8 hours.

More recently, commercially available α-picoline-borane has been used expeditiously in one-pot reductive aminations of a wide range of aldehydes and ketones with amines [51]. Reactions with this crystalline solid are carried out in MeOH, H$_2$O, or neat with a small amount of AcOH with yields superior to those from other common reducing agents (Scheme 12.6).

In another elegant study, Connolly and coworkers have illustrated the advantages of using borane-*t*-butylamine in the synthesis of methyl(3-nitrobenzyl)amine [53]. They found that this borane reagent could be activated with methanesulfonic acid and used for the selective reductive amination of 3-nitrobenzaldehyde with methylamine to give the corresponding amine in high yield and purity, where no partial reduction of the nitro group had occurred. This reduction has been demonstrated on scales up to 5 kg. Conversely, the sodium borohydride-mediated reductive amination proceeded with partial reduction of the nitro group to generate a complex mixture of potentially dangerous azo- and azoxy-containing products.

Isolation of the product amine using these borane-amine adducts is complicated by the presence of the original carrier amine. If the carrier amine has a significantly different boiling point or polarity from the product amine, the reduction product can be isolated by distillation or column chromatography. If the product amine is strongly complexed with any remaining borane, which occurs frequently for less-hindered amines, the stable product is treated with dilute HCl. The more hindered carrier amine can be extracted into the aqueous phase, separated, and recovered by alkalization. Addition of boron trifluoride-etherate to the borane adduct generates the desired product amine.

To avoid complicated work-ups associated with using sodium salts or adducts, catecholborane (HBcat, cat = 1,2-O$_2$C$_6$H$_4$) has been used as a gentle and effective reagent to reduce preformed boronated benzaldimines to the corresponding amines. Reactions occurred at room temperature where addition of the B—H bond of HBcat to the imine gave a putative borylamine intermediate (Scheme 12.7a). Although NaBH$_4$ could also be used in these reactions, yields of isolated products were higher in reactions employing HBcat [54]. This methodology has recently been extended to reduce phosphinoimines, where HBcat selectively added to the C=N double bond without interference from the basic phosphine moiety [55]. In an earlier study, addition of HBcat to *o*-(HO)-C$_6$H$_4$CH=NPh rapidly gave a novel heterocyclic compound which was characterized by multinuclear NMR spectros-

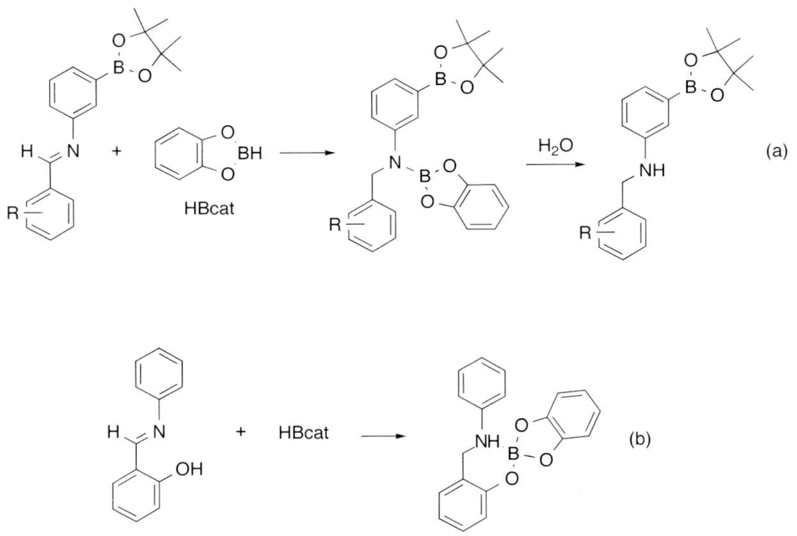

Scheme 12.7 Reduction of benzaldimines using catecholborane (HBcat).

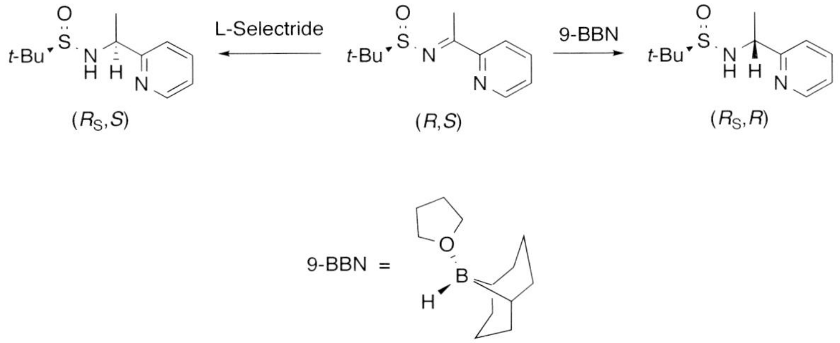

Scheme 12.8 The reduction of enantiopure N-t-butanesulfinyl ketimines derived from pyridyl ketones.

copy as well as a single-crystal X-ray diffraction study (Scheme 12.7b). The formation of this heterocycle was rationalized via an initial hydroboration of the C=N functionality followed by B—O bond formation and hydrogen transfer [56].

N-t-Butanesulfinyl amines have recently been prepared in high yields and diastereoselectivities from the reduction of enantiopure N-t-butanesulfinyl ketimines derived from pyridyl ketones (Scheme 12.8) using a number of different reducing agents [57]. While reactions with NaBH$_4$ gave high yields of the desired amines, low diastereoselectivities limited its use in this study. Interestingly, reactions with borabicyclo[3.3.1]nonane (9-BBN) reduced the sulfinyl ketimines in high yields and with diastereoselectivities up to 98% de. The opposite diastereoisomer was obtained in reactions using L-Selectride (tri-sec-butylborohydride). An excess of the

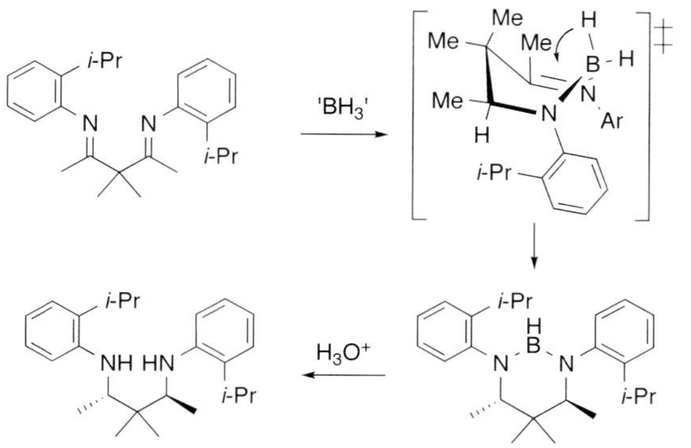

Scheme 12.9 Synthesis of a C_2-symmetric diazaborinane from the dihydroboration of a bulky 1,3-diiimine.

reducing agent (up to four equivalents) was required for all these reactions, presumably due to interaction of the electron-poor borane with the nucleophilic nitrogen of the pyridine group.

12.2.2
Diimines

In an interesting study, Mair and coworkers found that either diborane, generated *in situ* in dichloromethane from a tetraalkylammonium borohydride [58], or $BH_3 \cdot THF$ can be added to sterically hindered 1,3-diimines to give stable diazaborinanes [59]. Interestingly, similar chloro-1,3,2-diazaborolidine products are obtained from the addition of BCl_3 to 1,2-diimines [60]. These reactions occur stereoselectively to give the *l*-diastereomers after one crystallization with de values in excess of 95%. This high level of diastereocontrol has been rationalized by an intramolecular reduction in which the chirality of the first-reduced imine group is transferred efficiently to the second by chelation-control. A boat-like transition state has been proposed where the first asymmetric center directs a methyl group in an equatorial orientation to avoid bulky isopropyl groups on the aryl ring. Although reactions with 2,6-diisopropylaniline derivatives afforded a remarkably stable diazaborinane, addition of acid to the 2-isopropylaniline analogue afforded the corresponding diamine in good yield (Scheme 12.9). Conversely, reductions using $AlCl_3/LiAlH_4$ mixtures gave lower (de 75%) and opposite selectivities in favor of the *u*-isomers.

In a related study, reactions of 2,2′-methylenebis(oxazolines) (BOXs) with catecholborane gave the corresponding C_2-symmetric boron complexes, which were used as active catalysts in the enantioselective reduction of prochiral ketones (Scheme 12.10). The boron-bis(oxazolinate) complexes could be readily prepared in CH_2Cl_2 at room temperature as white solids stable to moisture and oxygen [61].

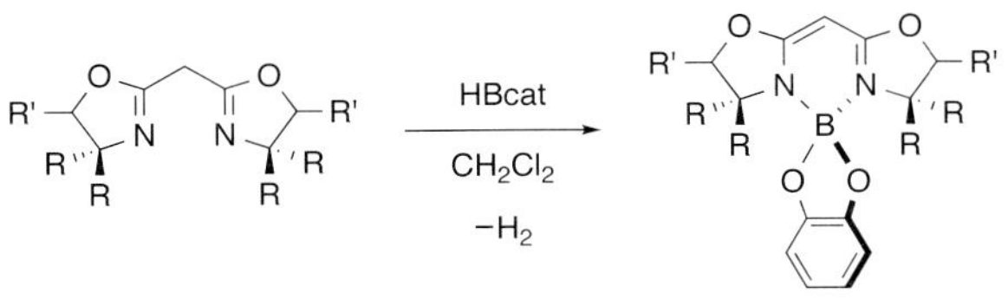

Scheme 12.10 Synthesis of a C_2-symmetrical boron-bis(oxazolinate) complex.

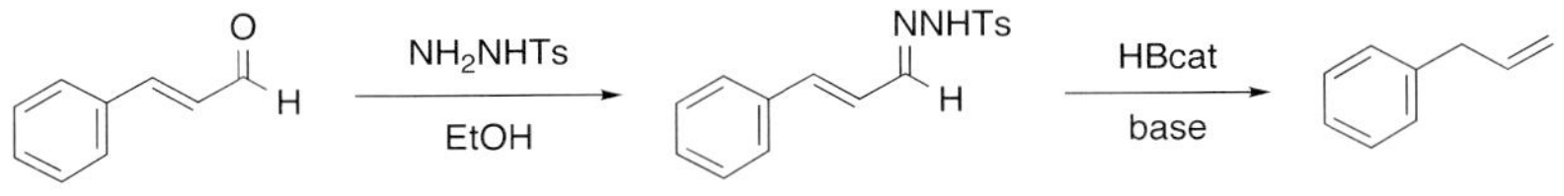

Scheme 12.11 Conversion of *trans*-cinnamaldehyde into allylbenzene using HBcat.

12.2.3
Tosylhydrazones

The ability to convert ketones to the corresponding methylene derivatives is of key importance in organic synthesis. As such, a great deal of effort has been expended on this transformation, with the classical reduction methods using strong acids (Clemmensen) or bases (Wolff–Kishner) to accomplish this feat. Unfortunately, these harsh methods do not allow for the reduction of carbonyl groups in the presence of sensitive functional groups. Recent advances in this area have found that reduction of the corresponding tosylhydrazones with boron hydride reagents offers a mild and effective alternative to the more severe classic routes.

Early work has shown that $NaBH_4$ [62, 63] and $NaBH_3CN$ [64] can be used for these reactions, however, a large excess of borohydride is required and more than one hydrocarbon product is frequently generated. Kabalka and Baker found that HBcat can be used for these reductions under mild reaction conditions to give the desired methylene products in high yields [65]. Reactions can be conducted at room temperature, at neutral pH and using common aprotic solvents. Catecholborane is exceptionally tolerant of other functional groups as only aldehydes are reduced faster than the tosylhydrazones (Scheme 12.11). Kabalka and Summers later found that bis(benzoyloxy)borane, $HB(O_2CPh)_2$, was also effective and selective in these transformations [66].

This methodology has been extended to the synthesis of (*S*)-6-methyldeaminosinefungin, an analogue of the nucleoside *S*-adenosylhomocysteine [67]. One of the key steps in this elaborate synthesis was a deoxygenation of a C-6-ketone which had proved to be sensitive to both acidic and alkaline conditions. Condensation of the ketone with *p*-toluenesulfonohydrazine afforded the corresponding hydrazone in 98% yield and subsequent reduction afforded the methylene analogue in 73% yield (Scheme 12.12). Interestingly, while the catecholborane–tosylhydrazone reaction was also used in the reaction of abietadienal, a diterpene aldehyde, this method was less effective than Zn/NaI reductions of the corresponding protected alcohols [68].

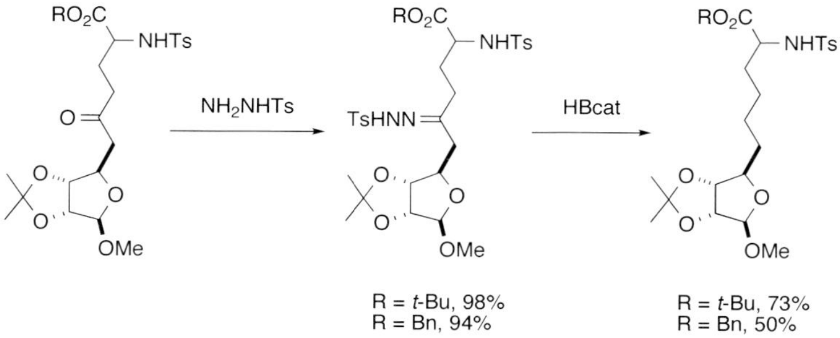

Scheme 12.12 Reduction of a C-6-ketone using the tosylhydrazone-hydroboration method.

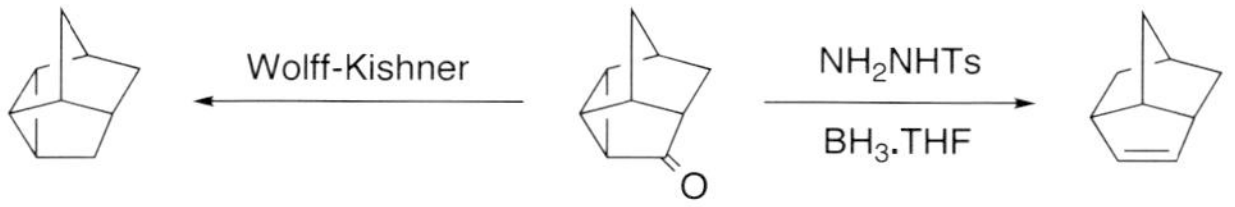

Scheme 12.13 Reduction of 5-oxo-2,4-didehydrobrendane.

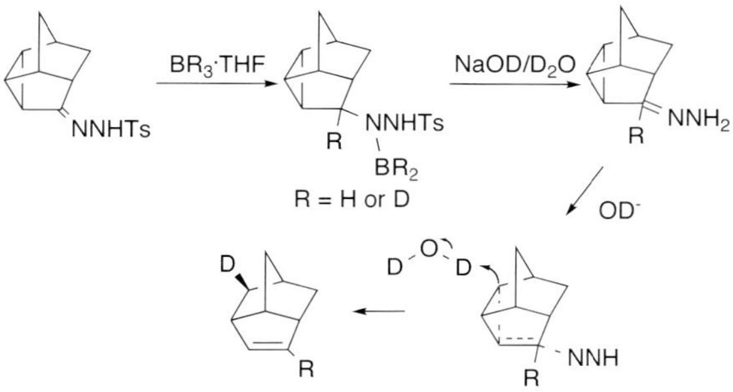

Scheme 12.14 Proposed mechanism for the hydroboration of the tosylhydrazone derivative of 5-oxo-2,4-didehydrobrendane.

Different products have been obtained in the Wolff–Kishner reduction of cyclopropyl ketone 5-oxo-2,4-didehydrobrendane compared to the tosylhydrazone–hydroboration method [69]. The classical Wolff–Kishner reduction proceeded without rearrangement, while the borane-mediated reduction afforded the rearranged 4-brendene (Scheme 12.13). Results obtained from deuterium labeling experiments suggest that the borane-mediated reduction proceeded via a concerted mechanism of diazene decomposition and cyclopropyl–homoallyl rearrangement (Scheme 12.14). Similar chemistry was reported for analogous reactions with 9-oxo-2,8-didehydronoradamantane.

12.2.4
Nitriles

A considerable amount of effort has focused on the hydroboration of nitriles in both organic [66] and inorganic [70, 71] synthesis. It has long been known that addition of borane to nitriles affords the corresponding substituted borazines, where the electron-deficient boron group has added to the more electronegative nitrogen of the nitrile group. Indeed, early work in this area has shown that addition of diborane to acetonitrile gave ethylborazine, (EtNBH)$_3$, in 35–40% yield (Scheme 12.15) [72, 73] along with a number of carboraza bicyclic systems [74]. More recently, computational [75] and ^{11}B NMR spectroscopy [76] studies have investigated the mechanism of the reduction of nitriles using boranes, indicating an overall associative pathway for the case of acrylonitrile.

Conversion of the borazines to the corresponding amines occurs upon hydrolysis [77–79]. An early study used this methodology to prepare ^{11}C-labeled norepinephrine hydrochloride for its use as a potential radiopharmaceutical [80]. Since then a considerable body of work in this area has appeared in the literature [52, 81–84]. Noteworthy is the development of the odorless borane carrier 2-(perfluorooctyl)ethyl methyl sulfide–borane adduct, which permits hydroborations of nitriles to occur in a dichloromethane/perfluorinated hydrocarbon mixture with subsequent recycling of the sulfide by fluorous extraction [84].

It is also well known that addition of a monosubstituted (H$_2$BR) or disubstituted (HBR$_2$) borane to a nitrile (R'CN) affords an aldiminoborane (R'CH=NBHR or R'CH=NBR$_2$, respectively) [85], which dimerizes subsequently to give four-membered heterocycles. For instance, Nicoud and coworkers have recently found that hydroboration of 4-iodobenzonitrile with mesitylborane gave the corresponding aldiminoborane, which rapidly dimerized to give a bis(4-iodophenyl)cyclodiborazane core [86]. The resulting cyclodiborazane, later transformed to a two-photon-absorbing chromophore, can exist as four potential stereoisomers, considering the different substituents in the core (Scheme 12.16). The aryl groups can be in *syn* or *anti* configurations and the mesityl groups can be *cis* or *trans* to one another. However, only two steroisomers were observed in this study, the *trans-syn* and *trans-anti* isomers, as computational results suggested that a *cis* arrangement of

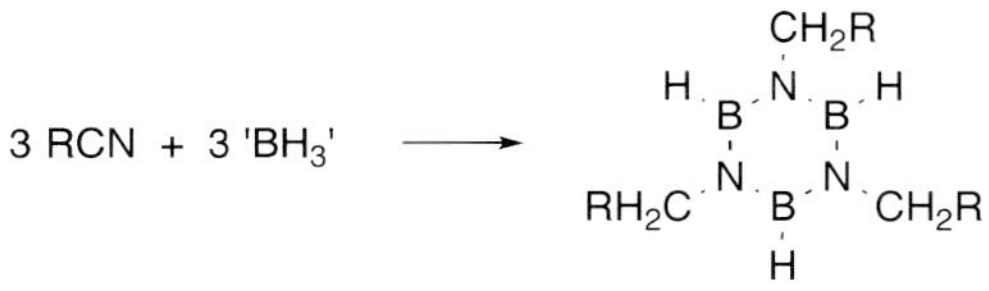

Scheme 12.15 Synthesis of borazines from the hydroboration of nitriles.

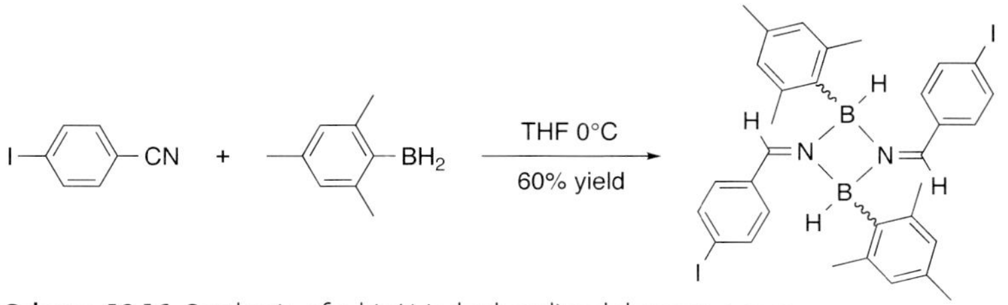

Scheme 12.16 Synthesis of a bis(4-iodophenyl)cycloborazane core.

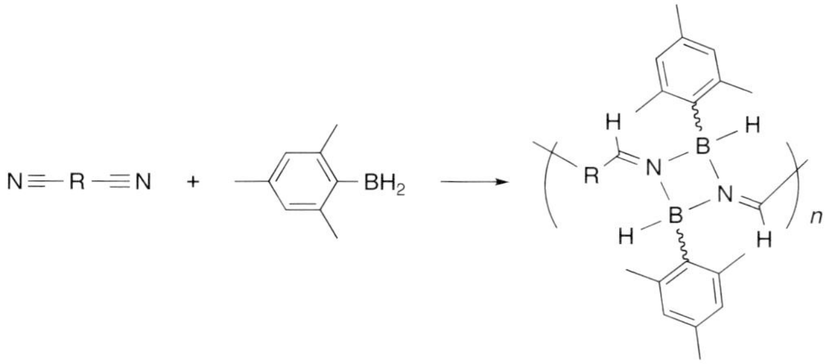

Scheme 12.17 Synthesis of poly(cyclodiborazane)s by the hydroboration–dimerization–polymerization of dicyano compounds.

bulky mesityl groups was disfavored. While the molecular structure of the *trans-anti* isomer was provided in that work, in a related study Lesley and coworkers isolated, and characterized structurally, the *trans-syn* isomer from the hydroboration of 4-methoxybenzonitrile with thexylborane [87]. This hydroboration-dimerization methodology has been extended to dicyano compounds in an extensive effort [88–93] to generate poly(cyclodiborazanes) (Scheme 12.17).

Also of interest is the report by Sneddon and coworkers that addition of nitriles to *nido*-5,6-$C_2B_8H_{11}^-$ or isoelectronic *nido*-$B_{10}H_{13}^-$ proceeded via initial nucleophilic attack at the nitrile carbon, followed by hydroboration and then cage insertion to give new azacarborane clusters in good yields [94]. Likewise, a recent report by Kennedy and coworkers details the reaction of *t*-BuNC with *anti*-$B_{18}H_{12}$ to give products derived from a reductive oligomerization of the isocyanide group [95]. Considering the wealth of interest in the hydroboration of nitriles, it is somewhat surprising that the analogous reactivity with isocyanides has received so little attention.

12.3
Catalyzed Reactions

Enantiomerically pure amines and amino acids are of great importance in pharmaceutical, biological and synthetic chemistry. The use of chiral catalysts for the enantioselective reduction of imines using borohydrides is well established. For instance, Yamada and coworkers have published an excellent review on this subject [96], highlighting reactions catalyzed by optically active ketoiminato cobalt(II) complexes (Scheme 12.18).

1,3,2-Oxazaborolidines have been used recently as Lewis acid catalysts for the enantioselective borane reduction of prochiral imines, and two comprehensive reviews have appeared on this subject [97, 98]. Commonly employed B–H oxazaborolidines are prepared by the *in situ* addition of a chiral amino alcohol with an excess of borane, without isolation and characterization of the catalyst precursor.

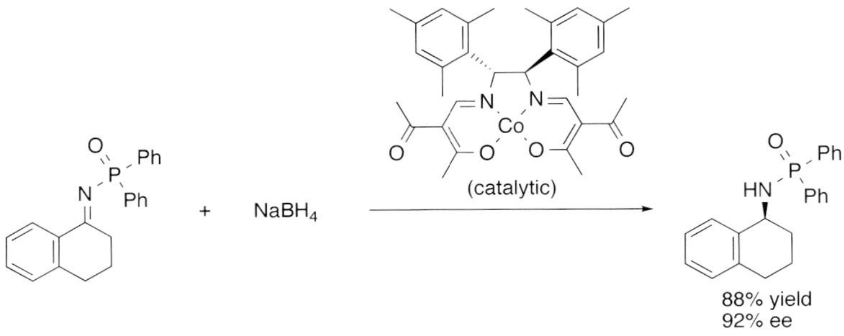

Scheme 12.18 An optically active ketoiminato Co(II) complex for the enantioselective borohydride reduction of imines.

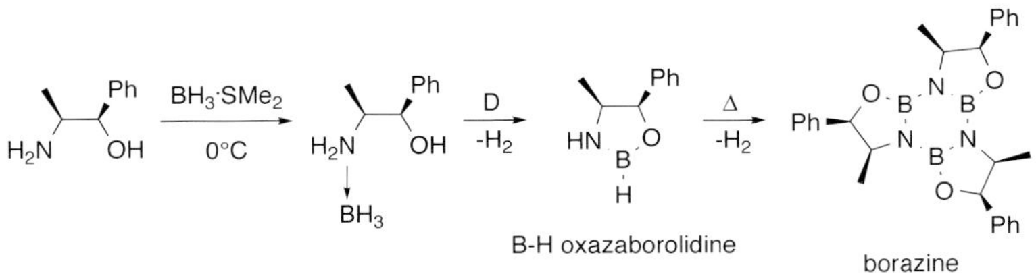

Scheme 12.19 Synthesis of a B-H oxazaborolidine and the corresponding polycyclic borazine.

A recent study [99] has shown that polycyclic borazines are frequently formed via this method and could account for some of the unwanted side-products and low enantioselectivities sometimes observed in the catalyzed reactions. Indeed, borane was found to react initially with norephedrine at 0 °C to give the expected N-borane adduct, as shown by multinuclear NMR spectroscopy. The expected B-H oxazaborolidine was generated upon warming the mixture and the borazine product formed on heating the mixture to 135 °C with loss of dihydrogen (Scheme 12.19).

A report by Field and coworkers examined the effect of the reducing agent on the oxazaborolidine-mediated asymmetric reduction of imines [100]. They found that borane-THF effectively reduced phenyl-(1-phenylethylidene)amine in the presence of Corey's reagent, (S)-methyl-CBS-oxazaborolidine, with high enantio-control (87% ee) in favor of the R-amine. However, when catecholborane was used as a reducing agent, lower enantioselectivites (47% ee) were observed in formation of the S-amine, the opposite outcome to that found in reactions using borane-THF. This change in the sense of oxazaborolidine-mediated asymmetric reduction of imines was rationalized in terms of steric arguments when using the bulkier catecholborane reducing agent. This methodology has been used successfully in the asymmetric synthesis of the herbicide metolachlor [101–103].

Other catalytic systems have also been investigated for the borane reduction of imines. For instance, a family of phenylalanine-substituted fullerenes have been prepared by the diacid-catalyzed borane-THF reduction of the imine precursors

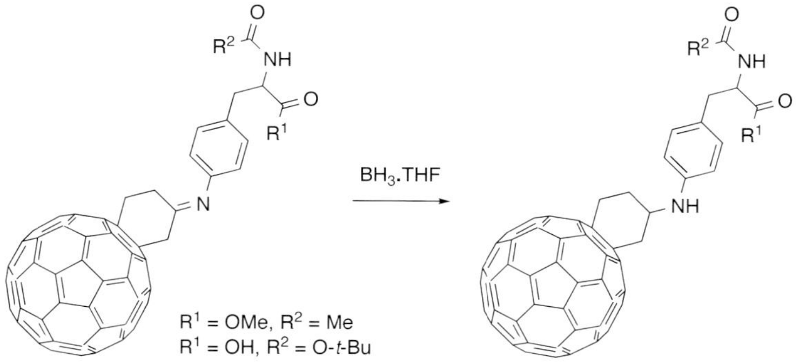

Scheme 12.20 Synthesis of fullerene-based amino acid derivatives.

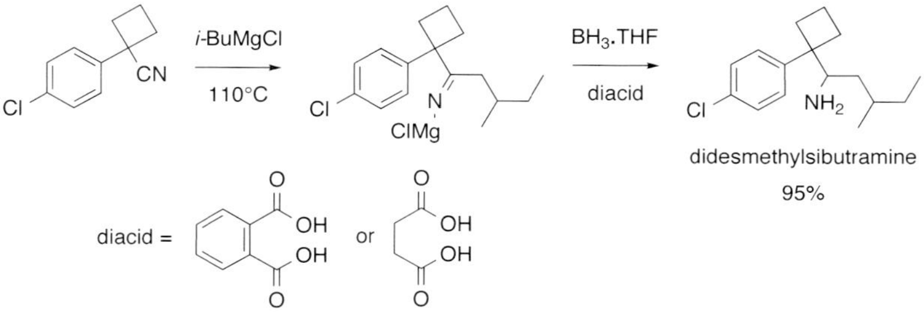

Scheme 12.21 The diacid catalyzed synthesis of didesmethylsibutramine.

(Scheme 12.20). This reaction was completely selective as no competing reduction of the C_{60} group was observed in this study [104]. Diacids were also used [105] in the preparation of didesmethylsibutramine, a major active metabolite of the anti-obesity drug Meridia (sibutramine hydrochloride monohydrate). Interestingly, this study generated an imine intermediate from the addition of a Grignard reagent to a nitrile, whereupon subsequent addition of borane-THF in the presence of one equivalent of a diacid gave the desired amine in 95% overall yield. Without the phthalic acid, however, the reaction was slow, even at ambient temperatures, and many by-products were observed with the amine isolated in only 75% yield. While reactions using $NaBH_4$ also gave the amine in 76% yield, those employing either $NaBH(OAc)_3$ or $NaBH_3CN$ failed to generate any of the desired amine (Scheme 12.21).

A polymer-supported chiral sulfonamide catalyst [106] has been used in the borane reduction of β-keto nitriles to afford optically active 1,3-amino alcohols with moderate to high enantioselectivities (68–96% ee). Reactions were conducted using 30 mol% of the catalyst and $BH_3 \cdot SMe_2$ as the reducing agent of choice, where this methodology was used successfully to generate the antidepressant drugs (R)-fluoxetine and (R)-duloxetine (Scheme 12.22). Likewise, reduction of α-alkyl-β-keto nitriles to *anti*-γ-amino alcohols was also reported using $BH_3 \cdot SMe_2$ and cerium chloride as the Lewis acid catalyst [107].

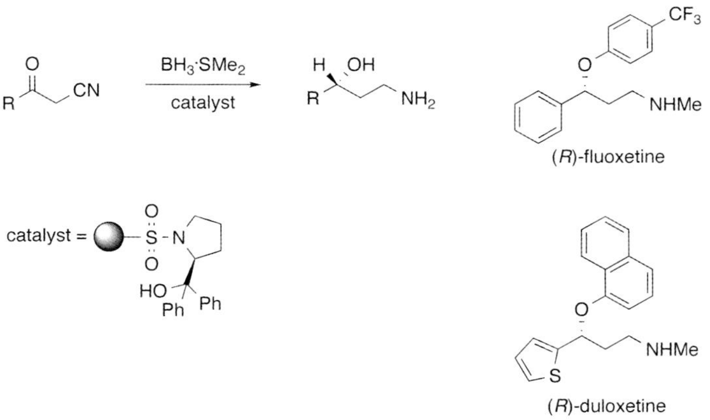

Scheme 12.22 Asymmetric reduction of β-keto nitriles.

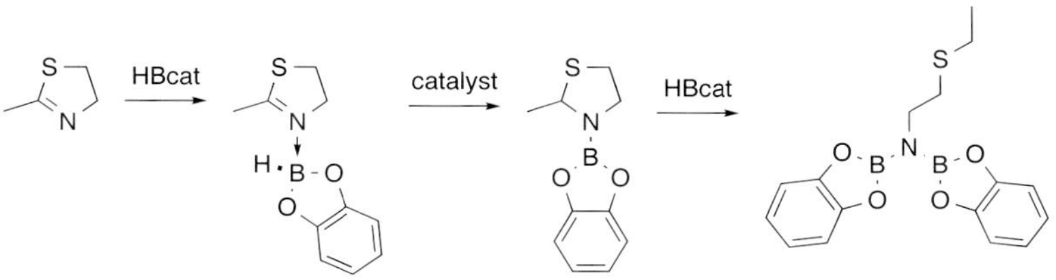

Scheme 12.23 The metal-catalyzed hydroboration of 2-methyl-2-thiazoline.

12.3.1
Transition Metals

That transition metal complexes could be used to catalyze the hydroboration of imines was originally reported in 1995 [56] using catecholborane and a number of coinage metal complexes. In the absence of catalyst, addition of HBcat to benzaldimines initially gave iminoborane adducts, which slowly reduced to the corresponding N-borylamines. No adduct formation or reduction was observed for reactions with the hindered ketimine PhN=CPh$_2$. Interestingly, both copper and gold complexes could be used to facilitate these reactions, showing rate enhancements superior to the more commonly employed Wilkinson's catalyst, RhCl(PPh$_3$)$_3$. The catalyzed hydroboration of several cyclic imines was also reported. For instance, addition of HBcat to 2-methyl-2-thiazoline gave a mixture of Lewis base adduct and the reduction product N-Bcat-2-methylthiazolidine (Scheme 12.23). The imine-borane adduct was reported to be quite stable and did not convert to the amine readily in the absence of catalyst. Two equivalents of HBcat could be used to generate the ring-opened product EtSCH$_2$CH$_2$N(Bcat)$_2$. Two mechanistic pathways for the coinage metal-catalyzed hydroboration of imines have been postulated. One invokes traditional oxidative addition of HBcat to the metal center, as frequently proposed for reactions using rhodium catalysts. The second relies on the Lewis-acidic coinage metals to activate the imine substrate via coordination through the nitrogen atom. Development of a positive charge in the imine carbon via resonance was believed to enhance its reactivity toward the polar B—H bond of HBcat.

A number of different rhodium catalysts have also successfully catalyzed the hydroboration of a series of allyl imines [108] with HBcat, where addition occurred initially at the more reactive imine functionality to give unsaturated borylamines of the type $RCH_2N(Bcat)CH_2CH=CH_2$. Also reported in this study is the first metal-catalyzed hydroboration of imines using pinacolborane, a much less reactive borane than HBcat (Scheme 12.24). The mechanism for these reactions presumably proceeds via initial oxidative addition of the borane to the metal center, followed by coordination of the imine and insertion into either the Rh—H or Rh—B bond. Subsequent reductive elimination would therefore generate the reduced borylamine product. Previous model studies have demonstrated the addition of B—H bonds to an imine at a metal center using pre-formed molybdenum [109] and titanium complexes [110]. An intriguing study by Buranov and Morrill has shown that $RhCl_3$ can be used to catalyze the hydroboration of alkenyl nitriles with borane-THF to give a number of products, including formation of 4-amino-2,6-dimethylpyrimidine [111].

An exciting alternative to the hydroboration reaction, and a practically unexplored area, is the transition metal-catalyzed diboration of imines. We have found [112] that reaction of $PhN=C(Me)Ph$ and $B_2cat'_2$ (cat' = 4-t-Bu-1,2-$O_2C_6H_3$) did not proceed without a catalyst, even at elevated temperatures (90 °C for 1 week). Using 2 mol% $RhCl(PPh_3)_3$ at 25 °C, however, the reaction produced equal amounts of N-borylenamine and N-borylamine products, as ascertained by NMR spectroscopy (Scheme 12.25). The formation of these products was rationalized by a selective β-hydrogen elimination step to give the N-borylenamine along with one equivalent of HBcat', which subsequently added to the unreacted imine to give the reduced amine. A combined ONIOM (Our owN n-layered Integrated molecular Orbital + molecular mechanics Method) and density functional study [113] has confirmed

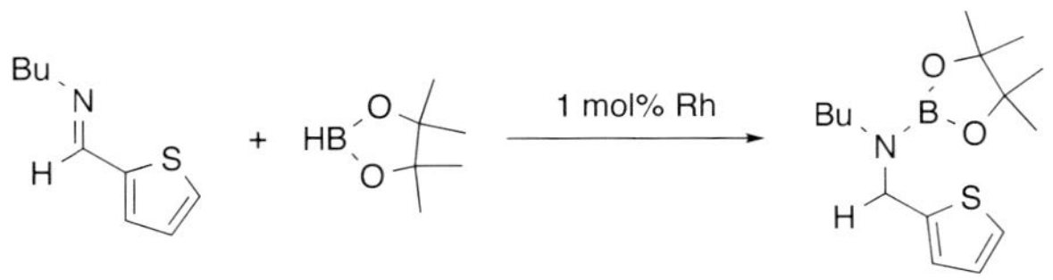

Scheme 12.24 The metal-catalyzed hydroboration of an aldimine using pinacolborane.

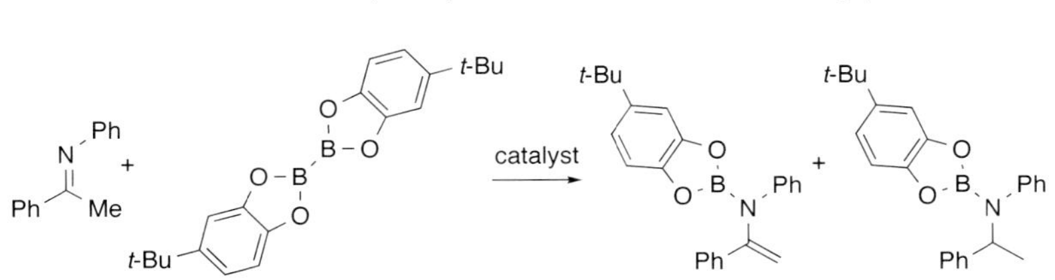

Scheme 12.25 The metal-catalyzed diboration of $PhN=C(Me)Ph$ using $B_2cat'_2$.

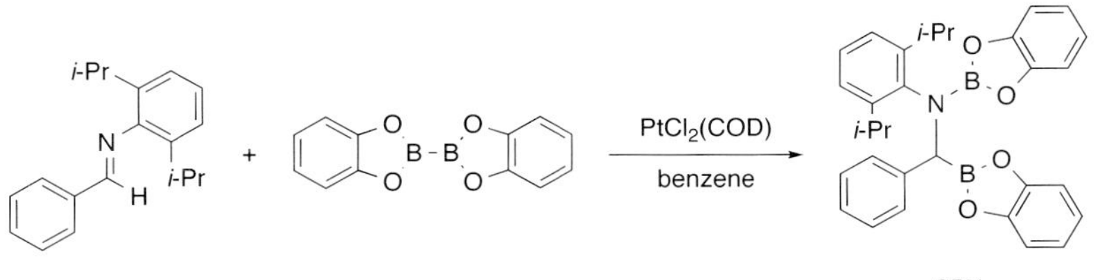

Scheme 12.26 The metal-catalyzed diboration of aldimine using $PtCl_2(COD)$.

that these catalytic imine boration reactions proceed via initial oxidative addition of the B—B bond to the rhodium complex followed by imine coordination and migratory insertion of the imine into the rhodium–boron bond. A β-hydrogen elimination step would give a monoboration product, whereas carbon–boron bond formation would afford the diborated product. Bulky imines appear to facilitate C—H bond activation and retard B—C bond formation, whereas C—B bond formation is preferred in the cases of less bulky substrates.

Interestingly, the diboration of sterically hindered aldimines could be readily achieved (Scheme 12.26) using commercially available $PtCl_2(COD)$ (COD = 1,5-cyclooctadiene) as the catalyst and $B_2cat'_2$ as the diboron source [114]. Reactions were conducted in benzene and mild conditions were required for substrates containing bulky or electron-donating substituents on the nitrogen-bound aryl group with yields up to 95% for the case of *N*-benzylidene-2,6-diisopropylaniline. Lower yields of diborated products were obtained, however, in reactions with *N*-benzylidene-4-anisidine. The effectiveness of this catalytic reaction was dependent upon the platinum-bound halide, the diene and diboron reagent, as lower yields were obtained in reactions using $PtX_2(COD)$ (X = Br, I), $PtCl_2$(dicyclopentadiene) and B_2pin_2. Likewise, donor ligands such as PPh_3, or even THF, seriously inhibited the reaction and drastically reduced the amount of diborated product generated (<5%). Initial attempts to selectively deborate the N—B bond proved unsuccessful.

12.4
Conclusions

Boron hydride reagents are versatile and effective agents for the reductive amination of a wide range of substrates. Although borohydride complexes are frequently used, these reactions can suffer from poor selectivities and problems associated with work-up procedures. Organoborane hydrides provide an interesting alternative to these salts, and reactions are highly dependent upon the nature of the borane employed to facilitate reductions. Finally, metal-catalyzed addition of boranes and diboranes to imines is a relatively unexplored area of research that offers tremendous potential to the field of organic synthesis. The use of chiral metal catalysts will allow for the production of enantiopure amines.

Acknowledgments

First and foremost we would like to thank Professor Pher Andersson and Dr Ian Munslow for the kind invitation to write this review and for their help with the publication. We also thank the Canada Research Chairs Program, Canadian Foundation for Innovation-Atlantic Innovation Fund, Natural Science and Engineering Research Council of Canada, and Mount Allison University and LANL for financial support.

References

1 Gonzalez, A.Z., Canales, E. and Soderquist, J.A. (2006) N-Propargylamides via the asymmetric Michael addition of B-alkynyl-10-TMS-9-borabicyclo[3.3.2]decanes to N-acylimines. *Organic Letters*, **8**, 3331.

2 Hayes, K.S. (2001) Industrial processes for manufacturing amines. *Applied Catalysis A*, **221**, 187.

3 Gomez, S., Peters, J.A. and Maschmeyer, T. (2002) The reductive amination of aldehydes and ketones and the hydrogenation of nitriles: mechanistic aspects and selectivity control. *Advanced Synthesis and Catalysis*, **344**, 1037.

4 Chen, B., Dingerdissen, U., Krauter, J.G.E., Lansink Rotgerink, H.G.J., Möbus, K., Ostgard, D.J., Panster, P., Riermeier, T.H., Seebald, S., Tacke, T. and Trauthwein, H. (2005) New developments in hydrogenation catalysis particularly in synthesis of fine and intermediate chemicals. *Applied Catalysis A*, **280**, 17.

5 Rylander, P.N. (1967) *Catalytic Hydrogenation over Platinum Metals*, Academic Press, New York, p. 21.

6 Schmitt, J., Cornu, P.J., Pluchet, H., Panouse, J.J. and Comoy, P. (1963) Recherches sur les amino-steroides .7. Steroides a restes amines fixes en c 3– preparation par reduction au borohydrure de sodium et au diborane dimines et denamines derivees de ceto-3 steroides. *Bulletin de la Societe Chimique de France*, 816.

7 Dalton, D.R., Miller, S.I., Dalton, C.K. and Crelling, J.K. (1971) The synthesis of 1-(image-nitrobenzyl)-4-hydroxy-6,7-methylenedioxy-1,2,3,4-tetrahydroisoquinoline. A general synthesis of 1-substituted-4-oxygenated-1,2,3,4-tetrahydroisoquinolines. *Tetrahedron Letters*, **12**, 575.

8 Montury, M. and Gore, J. (1975) Hydroboration of imines of 2-cyclohexenones. *Bulletin de la Societe Chimique de France*, 2622.

9 Nose, A. and Kudo, T. (1986) Selective reduction of imines with the diborane-methanol system. *Chemical and Pharmaceutical Bulletin*, **34**, 4817.

10 Schellenberg, K.A. (1963) The synthesis of secondary and tertiary amines by borohydride reduction. *Journal of Organic Chemistry*, **28**, 3259.

11 Borch, R.F., Bernstein, M.D. and Durst, H.D. (1971) Cyanohydridoborate anion as a selective reducing agent. *Journal of the American Chemical Society*, **93**, 2897.

12 Bhanushali, M.J., Nandurkar, N.S., Bhor, M.D. and Bhanage, B.M. (2007) Direct reductive amination of carbonyl compounds using bis(triphenylphosphine) copper(I) tetrahydroborate. *Tetrahedron Letters*, **48**, 1273.

13 Chaux, F., Denat, F., Espinosa, E. and Guilard, R. (2006) An easy route towards regioselectively difunctionalized cyclens and new cryptands. *Chemical Communications*, 5054.

14 Abdel-Magid, A.F. and Mehrman, S.J. (2006) A review on the use of sodium triacetoxyborohydride in the reductive amination of ketones and aldehydes. *Organic Process Research and Development*, **10**, 971.

15 de Souza, M.V.N. and Vasconcelos, T.R.A. (2006) Recent methodologies mediated by sodium borohydride in the reduction of different classes of compounds. *Applied Organometallic Chemistry*, **20**, 798.

16 Kim, J., Suri, J.T., Cordes, D.B. and Singaram, B. (2006) Asymmetric reductions involving borohydrides: A practical asymmetric reduction of ketones mediated by (L)-TarB-NO$_2$: A chiral Lewis acid. *Organic Process Research and Development*, **10**, 949.

17 Pasumansky, L., Goralski, C.T. and Singaram, B. (2006) Lithium aminoborohydrides: powerful, selective, air-stable reducing agents. *Organic Process Research and Development*, **10**, 959.

18 Gribble, G.W. (2006) The synthetic versatility of acyloxyborohydrides. *Organic Process Research and Development*, **10**, 1062.

19 Gribble, G.W. (1998) Sodium borohydride in carboxylic acid media: a phenomenal reduction system. *Chemical Society Reviews*, **27**, 395.

20 Ricci, A. (ed.) (2000) *Modern Amination Methods*, Wiley-VCH Verlag GmbH, Weinheim.

21 Tice, N.C., Parkin, S. and Selegue, J.P. (2007) Synthesis, characterization and crystal structures of boron-containing intermediates in the reductive amination of ferrocenecarboxaldehyde to a bis(ferrocenylmethyl) amine. *Journal of Organometallic Chemistry*, **692**, 791.

22 Razavi, H. and Polt, R. (2000) Asymmetric syntheses of (−)-8-epi-swainsonine and (+)-1,2-Di-epi-swainsonine. Carbonyl addition thwarted by an unprecedented aza-pinacol rearrangement. *Journal of Organic Chemistry*, **65**, 5693.

23 Cha, J.S. and Lee, D.Y. (2002) Reaction of lithium (2,3-dimethyl-2-butyl)-t-butoxyborohydride with selected organic compounds containing representative functional group. *Bulletin of the Korean Chemical Society*, **23**, 856.

24 Periasamy, M. and Thirumalaikumar, M. (2000) Methods of enhancement of reactivity and selectivity of sodium borohydride for applications in organic synthesis. *Journal of Organometallic Chemistry*, **609**, 137.

25 Ramsay, S.L., Freeman, C., Grace, P.B., Redmond, J.W. and MacLeod, J.K. (2001) Mild tagging procedures for the structural analysis of glycans. *Carbohydrate Research*, **333**, 59.

26 Dubber, M. and Lindhorst, T.K. (1998) Synthesis of octopus glycosides: core molecules for the construction of glycoclusters and carbohydrate-centered dendrimers. *Carbohydrate Research*, **310**, 35.

27 Gonschior, M., Kötteritzsch, M., Rost, M., Schönecker, B. and Wyrwa, R. (2000) Synthesis of N,N-bis[2-(2-pyridyl)ethyl]amino steroids and related compounds intended as chiral ligands for copper ions. *Tetrahedron: Asymmetry*, **11**, 2159.

28 Jestin, E., Bultel-Rivière, K., Faivre-Chauvet, A., Barbet, J., Loussouarn, A. and Gestin, J.-F. (2006) A simple and efficient method to label L-fucose. *Tetrahedron Letters*, **47**, 6869.

29 Bhattacharyya, S., Rana, S., Gooding, O.W. and Labadie, J. (2003) Polymer-supported triacetoxyborohydride: A novel reagent of choice for reductive amination. *Tetrahedron Letters*, **44**, 4957.

30 Heinrichs, G., Schellenträger, M. and Kubik, S. (2006) An enantioselective fluorescence sensor for glucose based on a cyclic tetrapeptide containing two boronic acid binding sites. *European Journal of Organic Chemistry*, 4177.

31 Thomas, S., Collins, C.J., Cuzens, J.R., Spiciarich, D., Goralski, C.T. and Singaram, B. (2001) Novel tandem SNAr amination-reduction reactions of 2-halobenzonitriles with lithium N,N-dialkylaminoborohydrides. *Journal of Organic Chemistry*, **66**, 1999.

32 Vogeley, N.J., White, P.S. and Templeton, J.L. (2003) Methyl ethyl differentiation in diastereoselective hydride addition to (Tp'W(CO)(PhCCMe)(NH=CMeEt))(BAr'$_4$). *Journal of the American Chemical Society*, **125**, 12422.

33 Luvino, D., Smietana, M. and Vasseur, J.-J. (2006) Selective fluorescence-based detection of dihydrouridine with boronic acids. *Tetrahedron Letters*, **47**, 9253.

34 Hajipour, A.R., Mohammadpoor-Baltork, I. and Rahu, A. (2000) 1-Benzyl-1-azonia-4-azabicyclo[2.2.2]octane tetrahydroborate (BAAOTB) as a selective reducing agent. *Indian Journal of Chemistry, Section B: Organic Chemistry Including Medicinal Chemistry*, **39**, 239.

35 Hajipour, A.R., Mohammadpoor-Baltork, I. and Noroallhi, M. (2001) Butyltriphenylphosphonium tetrahydroborate (BTPPTB) as a selective reducing agent for the reduction of imines, enamines and oximes and reductive alkylation of aldehydes or ketones with primary amines in methanol or under solid-phase conditions. *Indian Journal of Chemistry, Section B: Organic Chemistry Including Medicinal Chemistry*, **40**, 152.

36 Tokunaga, Y., Ohta, G., Yamauchi, Y., Goda, T., Kawai, N., Sugihara, T. and Shimomura, Y. (2006) Efficient synthesis of rotaxane based on complexation of acetylene and dicobalt hexacarbonyl: introduction of a transformable functional group. *Chemistry Letters*, **35**, 766.

37 Gaines, D.F. and Coons, D.E. (1986) Reactions of boron hydrides with the iminium salt [Me$_2$N:CH$_2$]I. Synthesis and characterization of 1-X-μ-(Me$_2$NCH$_2$)B$_5$H$_7$ (X=H, Et, Br), a new class of bridge-substituted pentaborane derivatives. *Inorganic Chemistry*, **25**, 364.

38 Küpper, S., Carroll, P.J. and Sneddon, L.G. (1992) Reactions of nido-6-SB$_9$H$_{11}$ with imines. Structural characterization of endo-9-((tert-Bu)MeC=NH)-arachno-6-SB$_9$H$_{11}$S. *Inorganic Chemistry*, **31**, 4921.

39 Park, E.-S., Lee, J.-H., Kim, S.-J. and Yoon, C.M. (2003) One-pot reductive amination of acetals with aromatic amines using decaborane (B$_{10}$H$_{14}$) in methanol. *Synthetic Communications*, **33**, 3387.

40 Moormann, A.E. (1993) Reductive amination of piperidines with aldehydes using borane-pyridine. *Synthetic Communications*, **23**, 789.

41 Peterson, M.A., Bowman, A. and Morgan, S. (2002) Efficient preparation of n-benzyl secondary amines via benzylamine-borane mediated reductive amination. *Synthetic Communications*, **32**, 443.

42 Nara, S., Sakamoto, T., Miyazawa, E. and Kikugawa, Y. (2003) A convenient synthesis of 1-alkyl-1-phenylhydrazines from N-aminophthalimide. *Synthetic Communications*, **33**, 87.

43 Johansson, A., Lindstedt, E.-L. and Olsson, T. (1997) A one-pot reductive amination of ketones to primary amines using borane-dimethyl sulfide complex. *Acta Chimica Scandinavica*, **51**, 351.

44 Olsen, C.A., Franzyk, H. and Jaroszewski, J.W. (2005) N-Alkylation and indirect formation of amino functionalities in solid-phase synthesis. *Synthesis*, 2631.

45 Zhou, F., Wang, Z. and Yin, D. (1994) New development in the hydroboration reactions of unsaturated carbon-carbon bonds in organic synthesis. *Hecheng Huaxue*, **2**, 117.

46 Kollonitsch, J. (1961) Reductive ring-cleavage of tetrahydrofurans by diborane. *Journal of the American Chemical Society*, **83**, 1515.

47 Verniest, G., Claessens, S. and De Kimpe, N. (2006) Stereoselective reduction of N-(3-arylcyclobutylidene)-amines. *Tetrahedron Letters*, **47**, 3299.

48 Bergbreiter, D.E. and Xu, G.-F. (1996) Surface selective modification of poly(vinyl chloride) film with lithiated α,ω-diaminopoly(alkene oxide)s. *Polymer*, **37**, 2345.

49 Patra, P.K., Nishide, K., Fuji, K. and Node, M. (2004) Dod-S-Me and methyl 6-morpholinohexyl sulfide (MMS) as new odorless borane carriers. *Synthesis*, 1003.

50 Wu, P.-L., Chen, H.-C. and Line, M.-L. (1997) Homodienyl [1,5]-hydrogen shift of cis- and trans-N-acyl-2-alkylcyclopropylimines. *Journal of Organic Chemistry*, **62**, 1532.

51 Sato, S., Sakamoto, T., Miyazawa, E. and Kikugawa, Y. (2004) One-pot reductive amination of aldehydes and ketones with α-picoline-borane in methanol, in water, and in neat conditions. *Tetrahedron*, **60**, 7899.

52 Brown, H.C., Bhaskar Kanth, J.V. and Zaidlewicz, M. (1998) Molecular addition compounds. 11. N-Ethyl-n-isopropylaniline-borane, a superior

53 Connolly, T.J., Constantinescu, A., Lane, T.S., Matchett, M., McGarry, P. and Paperna, M. (2005) Assessment of a reductive amination route to methyl(3-nitrobenzyl)amine hydrochloride. *Organic Process Research and Development*, **9**, 837.

54 Vogels, C.M., Nikolcheva, L.G., Norman, D.W., Spinney, H.A., Decken, A., Baerlocher, M.O., Baerlocher, F.J. and Westcott, S.A. (2001) Synthesis and antifungal properties of benzylamines containing boronate esters. *Canadian Journal of Chemistry*, **79**, 1115.

55 Halcovitch, N.R., Burford, R.J., Geier, M.J., Geier, S.J., Vogels, C.M., Decken, A. and Westcott, S.A. (2007) 233rd ACS National Meeting, Chicago, IL, INOR-951.

56 Baker, R.T., Calabrese, J.C. and Westcott, S.A. (1995) Coinage metal-catalyzed hydroboration of imines. *Journal of Organometallic Chemistry*, **498**, 109.

57 Chelucci, G., Baldino, S., Chessa, S., Pinna, G.A. and Soccolini, F. (2006) An easy route to optically active 1-substituted-1-pyridyl-methylamines by diastereoselective reduction of enantiopure *N*-tert-butanesulfinyl ketimines. *Tetrahedron: Asymmetry*, **17**, 3163.

58 Brändström, A., Junggren, U. and Lamm, B. (1972) An improved method for the preparation of solutions of diborane. *Tetrahedron Letters*, **13**, 3173.

59 Carey, D.T., Mair, F.S., Pritchard, R.G., Warren, J.E. and Woods, R.J. (2003) Borane and alane reductions of bulky *N,N'*-diaryl-1,3-diimines: structural characterization of products and intermediates in the diastereoselective synthesis of 1,3-diamines. *Dalton Transactions*, 3792.

60 Mair, F.S., Manning, R., Pritchard, R.G. and Warren, J.E. (2001) Imine chloroboration: reaction of boron trichloride with a bulky diazadiene gives not a diazaborolium salt, but a 2,4,5-trichloro-1,3,2-diazaborolidine. *Chemical Communications*, 1136.

61 Bandini, M., Bottoni, A., Cozzi, P.G., Miscione, G.P., Monari, M., Pierciaccante, R. and Umani-Ronchi, A. (2006) Chiral C2-boron-bis(oxazolines) in asymmetric catalysis – a theoretical study of the catalyzed enantioselective reduction of ketones promoted by catecholborane. *European Journal of Organic Chemistry*, 4596.

62 Caglioti, L. (1966) The reduction of tosylhydrazones and of acyl tosylhydrazides. *Tetrahedron*, **22**, 487.

63 Hutchins, R.O. and Natale, N.R. (1978) Sodium borohydride in acetic acid. A convenient system for the reductive deoxygenation of carbonyl tosylhydrazones. *Journal of Organic Chemistry*, **43**, 2299.

64 Hutchins, R.O., Kacher, M. and Rua, L. (1975) Synthetic utility and mechanism of the reductive deoxygenation of α,β-unsaturated *p*-tosylhydrazones with sodium cyanoborohydride. *Journal of Organic Chemistry*, **40**, 923.

65 Kabalka, G.W. and Baker, J.D., Jr (1975) New mild conversion of ketones to the corresponding methylene derivatives. *Journal of Organic Chemistry*, **40**, 1834.

66 Kabalka, G.W. and Summers, S.T. (1981) A mild and convenient conversion of ketones to the corresponding methylene derivatives via reduction of tosylhydrazones by bis(benzoyloxy)borane. *Journal of Organic Chemistry*, **46**, 1217.

67 Peterli-Roth, P., Maguire, M.P., León, E. and Rapoport, H. (1994) Syntheses of 6-deaminosinefungin and (S)-6-methyl-6-deaminosinefungin. *Journal of Organic Chemistry*, **59**, 4186.

68 Lee, H.-J., Ravn, M.M. and Coates, R.M. (2001) Synthesis and characterization of abietadiene, levopimaradiene, palustradiene, and neoabietadiene: Hydrocarbon precursors of the abietane diterpene resin acids. *Tetrahedron*, **57**, 6155.

69 Kragol, G., Benko, I., Muharemsphahić, J. and Mlinarić-Majerski, K. (2003) A cyclopropyl-homoallyl rearrangement accompanying the borane-mediated reduction of tosylhydrazones. *European Journal of Organic Chemistry*, 2622.

70 Hawthorne, M.F. (1962) Amine boranes-X: Alkylideneamino *t*-butylboranes. The

hydroboration of nitriles with trimethylamine *t*-butylborane. *Tetrahedron*, **17**, 117.
71 Lloyd, J.E. and Wade, K. (1964) Reactions between dialkylboranes and methyl cyanide. Ethylideneaminodimethylborane and diethylethylideneaminoborane. *Journal of the Chemical Society*, 1649.
72 Jennings, J.R. and Wade, K. (1968) The diborane–methyl cyanide reaction. Further studies on volatile products. *Journal of the Chemical Society. (A)*, 1946–1950.
73 Stone, F.G.A. and Emeléus, J. (1950) The reaction of diborane with some alkene oxides and vinyl compounds. *Journal of the Chemical Society*, 2755.
74 Coult, R., Fox, M.A., Rand, B., Wade, K. and Westwood, A.V.K. (1997) Convenient direct syntheses of novel fused-ring CB_4N_5 systems by nitrile hydroboration. *Journal of the Chemical Society, Dalton Transactions*, 3411.
75 Cui, R., Sun, Z. and Fan, Z. (1999) Hydroboration reaction of organic nitriles. *Fenzi Kexue Xuebao*, **15**, 86.
76 Jaganyi, D. and Mzinyati, A. (2006) An ^{11}B NMR spectroscopy investigation of the mechanism of the reduction of nitriles by $BH_3 \cdot SMe_2$. *Polyhedron*, **25**, 2730.
77 Brown, H.C. and Subba Rao, B.C. (1957) Selective reductions with diborane, an acidic-type reducing agent. *Journal of Organic Chemistry*, **22**, 1135.
78 Brown, H.C. and Subba Rao, B.C. (1960) Hydroboration. III. The reduction of organic compounds by diborane, an acid-type reducing agent. *Journal of the American Chemical Society*, **82**, 681.
79 Brown, H.C. and Kortnyk, W. (1960) Hydroboration. IV. A study of the relative reactivities of representative functional groups toward diborane. *Journal of the American Chemical Society*, **82**, 3866.
80 Fowler, J.S., MacGregor, R.R., Ansari, A.N., Atkins, H.L. and Wolf, A.P. (1974) Radiopharmaceuticals. 12. A new rapid synthesis of carbon-11 labeled norepinephrine hydrochloride? *Journal of Medicinal Chemistry*, **17**, 246.
81 Beveridge, K.A., McAuley, A. and Xu, C. (1991) Preparation of the macrobicyclic ligand 17-oxa-1,5,8,12-tetraazabicyclo[10.5.2]nonadecane: characterization of copper(II) intermediates in a template synthesis. *Inorganic Chemistry*, **30**, 2074.
82 Brown, H.C., Kanth, J.V.B., Dalvi, P.V. and Zaidlewicz, M. (1999) Molecular addition compounds. 15. Synthesis, hydroboration, and reduction studies of new, highly reactive *tert*-butyldialkylamine-borane adducts. *Journal of Organic Chemistry*, **64**, 6263.
83 Pemberton, N., Åberg, V., Almstedt, H., Westermark, A. and Almqvist, F. (2004) Microwave assisted synthesis of highly substituted aminomethylated 2-pyridones. *Journal of Organic Chemistry*, **69**, 7830.
84 Crich, D. and Neelamkavil, S. (2002) Fluorous dimethyl sulfide: A convenient, odorless, recyclable borane carrier. *Organic Letters*, **4**, 4175.
85 Dorokhov, V.A. and Lappert, M.F. (1969) Cyclic boron compounds. Part IX. Aldiminoboranes and their cyclic dimers. *Journal of the Chemical Society. A*, 433.
86 Hayek, A., Nicoud, J.-F., Bolze, F., Bourgogne, C. and Baldeck, P.L. (2006) Boron containing two-photon absorbing chromophores: electronic interaction through the cyclodiborazane core. *Angewandte Chemie (International Ed. in English)*, **45**, 6466.
87 Lesley, M.J.G., Pineau, M.R., Nelson, D.M. and Crundwell, G. (2007) *syn*-2,4-Bis(4-methoxybenzylidene)-*trans*-1,3-dithexyl-2,4-diaza-1,3-diboracyclobutane. *Acta Crystallographica E*, **63**, o1204.
88 Chujo, Y., Tomita, I. and Saegusa, T. (1993) Hydroboration polymerization of dicyano compounds. *Polymer Bulletin*, **31**, 553.
89 Matsumi, N. and Chujo, Y. (1999) Synthesis of poly(cyclodiborazane)s by hydroboration polymerization of dicyano compounds with tripylborane. *Polymer Bulletin*, **43**, 151.
90 Matsumi, N., Naka, K. and Chujo, Y. (1998) Hydroboration polymerization of dicyanoanthracene using mesitylborane. *Macromolecules*, **31**, 8047.
91 Chujo, Y., Tomita, I. and Saegusa, T. (1994) Hydroboration polymerization of

dicyano compounds. 4. synthesis of stable poly(cyclodiborazane)s from dialkylboranes. *Macromolecules*, **27**, 6714.

92 Chujo, Y., Tomita, I. and Saegusa, T. (1992) Allylboration polymerization. 1. Synthesis of boron-containing polymers by the reaction between triallylborane and dicyano compounds. *Macromolecules*, **25**, 3005.

93 Chujo, Y., Tomita, I., Murata, N., Mauermann, H. and Saegusa, T. (1992) Hydroboration polymerization of dicyano compounds. 1. Synthesis of boron-containing polymers by the reaction between t-BuBH$_2$NMe$_3$ and dicyano compounds. *Macromolecules*, **25**, 27.

94 Wille, A.E., Su, K., Carroll, P.J. and Sneddon, L.G. (1996) New synthetic routes to azacarborane clusters: nitrile insertion reactions of $nido$-5,6-C$_2$B$_8$H$_{11}^-$ and $nido$-B$_{10}$H$_{13}^-$. *Journal of the American Chemical Society*, **118**, 6407.

95 Jelínek, T., Kilner, C.A., Barrett, S.A., Štíbr, B., Thornton-Pett, M. and Kennedy, J.D. (2005) Macropolyhedral borane reaction chemistry: Reductive oligomerisation of ter-BuNC by anti-B$_{18}$H22 to give the boron-coordinated {(ter-BuNHCH){ter-BuNHC(CN)}CH$_2$:} carbene residue. *Inorganic Chemistry Communications*, **8**, 491.

96 Yamada, T., Nagata, T., Sugi, K.D., Yorozu, K., Ohtsuka, Y., Miyazaki, D. and Mukaiyama, T. (2003) Enantioselective borohydride reduction catalyzed by optically active cobalt complexes. *Chemistry – A European Journal*, **9**, 4485.

97 Kobayashi, S. and Ishitani, H. (1999) Catalytic enantioselective addition to imines. *Chemical Reviews*, **99**, 1069.

98 Burkhardt, E.R. and Matos, K. (2006) Boron reagents in process chemistry: Excellent tools for selective reductions. *Chemical Reviews*, **106**, 2617.

99 Stepanenko, V., Ortiz-Marciales, M., Barnes, C.E. and Garcia, C. (2006) Studies on the synthesis of borazines from borane and 1,2-aminoalcohols. *Tetrahedron Letters*, **47**, 7603.

100 Kirton, E.H.M., Tughan, G., Morris, R.E. and Field, R.A. (2004) Rationalising the effect of reducing agent on the oxazaborolidine-mediated asymmetric reduction of N-substituted imines. *Tetrahedron Letters*, **45**, 853.

101 Cho, B.T., Ryu, M.H., Chun, Y.S., Dauelsburg, C., Wallbaum, S. and Martens, J. (1994) A direct comparison study of asymmetric borane reduction of C=N double bond mediated by chiral oxazaborolidines. *Bulletin of the Korean Chemical Society*, **15**, 53.

102 Cho, B.T. and Chun, Y.S. (1990) Asymmetric reduction of N-substituted ketimines with the reagent prepared from borane and (S)-(–)-2-amino-3-methyl-1,1-diphenylbutan-1-ol (itsuno's reagent): enantioselective synthesis of optically active secondary amines. *Journal of the Chemical Society, Perkin Transactions 1*, 3200.

103 Cho, B.T. and Chun, Y.S. (1992) Enantioselective synthesis of optically active metolachlor via asymmetric reduction. *Tetrahedron: Asymmetry*, **3**, 337.

104 Yang, J. and Barron, A.R. (2004) A new route to fullerene substituted phenylalanine derivatives. *Chemical Communications*, 2884.

105 Lu, Z.-H., Bhongle, N., Su, X., Ribe, S. and Senanayake, C.H. (2002) Novel diacid accelerated borane reducing agent for imines. *Tetrahedron Letters*, **43**, 8617.

106 Wang, G., liu, X. and Zhao, G. (2005) Polymer-supported chiral sulfonamide catalyzed one-pot reduction of β-keto nitriles: A practical synthesis of (R)-fluoxetine and (R)-duloxetine. *Tetrahedron: Asymmetry*, **16**, 1873.

107 De Nino, A., Dalpozzo, R., Cupone, G., Maiuolo, L., Procopio, A., Tagarelli, A. and Bartoli, G. (2002) Stereoselective complete reduction of α-alkyl-β-ketonitriles to anti γ-amino alcohols. *European Journal of Organic Chemistry*, 2924.

108 Vogels, C.M., O'Connor, P.E., Phillips, T.E., Watson, K.J., Shaver, M.P., Hayes, P.G. and Westcott, S.A. (2001) Rhodium catalyzed hydroborations of allylamines and allylimines. *Canadian Journal of Chemistry*, **79**, 1898.

109 Pizzano, A., Sánchez, L., Gutiérrez, E., Monge, A. and Carmona, E. (1995) B-H

reactivity of a dihydrobis(pyrazolyl)
borate ligand: products of
intramolecular acyl and iminoacyl
hydroboration. *Organometallics*, **14**, 14.
110 Binger, P., Sandmeyer, F. and Krüger,
C. (1995) Reactivity of bimetallic
compounds containing planar
tetracoordinate carbon toward
unsaturated organic substrates.
Organometallics, **14**, 2969.
111 Buranov, A.U. and Morrill, T.C. (2003)
RhCl$_3$-catalyzed hydroboration of
alkenyl nitriles. *Tetrahedron Letters*, **44**,
6301.
112 Cameron, T.M., Baker, R.T. and
Westcott, S.A. (1998) Metal-catalysed
multiple boration of ketimines. *Chemical
Communications*, 2395.
113 Ananikov, V.P., Szilagyi, R., Morokuma,
K. and Musaev, D.G. (2005) Can steric
effect induce the mechanism switch in
the rhodium catalyzed imines boration
reaction?; A density functional and
ONIOM study. *Organometallics*, **24**,
1938.
114 Mann, G., John, K.D. and Baker, R.T.
(2000) Platinum-catalyzed diboration
using a commercially available catalyst:
Diboration of aldimines to α-
aminoboronate esters. *Organic Letters*, **2**,
2105.

13
Hydrosilylation of Imines
Olivier Riant

13.1
Introduction

The first examples of the production of chiral secondary alcohols and amines by the reduction of a carbonyl or imino group using a silane as a reducing agent and a transition metal catalyst was reported during the mid 1970s by Kagan [1]. Following this first report, the literature showed a growing interest in the discovery of efficient catalytic systems for the hydrosilylation of ketones and aldehydes, while imines only started to show significant advances much later. One of the main reasons is that most of the efficient catalytic systems for the reduction of ketones were rhodium-based catalysts that usually required activated silanes such as bis-aryl silanes. While these systems were active under very mild conditions, they suffered from strong economic drawbacks due to the cost of the catalysts, relatively poor catalytic efficiency compared to hydrogenation and the cost of the silanes. Imines being poorer substrates than aldehydes and ketones suffered even more from these limitations and were neglected as substrates for hydrosilylation reactions until more efficient systems were discovered during the 1990s. Indeed, imines are less electrophilic substrates compared to their carbonyl counterparts and are also susceptible to isomerization to enamines, when they posses one or more α-hydrogen. It is, however, possible to increase the electrophilic character of the imino group by the introduction of an aryl or an electron-withdrawing group on the nitrogen. The most common activating groups are tosyl, Boc or diphenylphosphinoyl. However, it is usually considered that only diphenylphosphinoyl combines ease of synthesis, stability to moisture and simplicity of removal for further transformations in synthesis. When designing a chiral catalyst for enantioselective addition of a nucleophile on an aldimine or a ketimine, another difficulty arises from the possible *cis-trans* isomerization of the C=N double bond.

However, nucleophilic addition to imines remains an attractive method for the synthesis of amines and increasing numbers of examples in the field of asymmetric catalysis have been reported during the since the 1990s [2]. Concerning reduction processes, the costs of the catalytic systems and the silanes have often drawn chemists to develop more economic alternatives, such as hydrogenation.

Modern Reduction Methods. Edited by Pher G. Andersson and Ian J. Munslow.
Copyright © 2008 WILEY-VCH Verlag GmbH & Co. KGaA, Weinheim
ISBN: 978-3-527-31862-9

Indeed, one of the most efficient enantioselective catalytic processes developed in industry uses an iridium-disphosphine-based catalysts for the asymmetric hydrogenation of a prochiral imine, leading to the production of the herbicide metolachlor on a multi-tonne scale [3]. However, the discovery of new catalytic systems which allowed the use of cheaper silanes brought a rebirth to the field of hydrosilylation of C=O and C=N bonds. Therefore, this chapter will focus on reviewing the hydrosilylation processes of imines that have been discovered and developed during the 10–15 years prior to publication [4].

13.2
Rh, Ir, Ru Based Catalysts

As quoted earlier, the first example of the asymmetric hydrosilylation of an imine by a rhodium-DIOP complex was reported in 1973 by the group of Kagan [1]. The use of diphenylsilane in the reduction of N-(α-methylbenzylidene)benzylamine gave the amine in 50% ee. This pioneering work was followed by several other examples, although the enantioselectivities remained moderate at ~60%. The development of new P,N-based chiral ligands by the groups of Uemura and Hidai led to some improvements for the asymmetric hydrosilylation of imines and ketoximes with iridium- and ruthenium-based catalysts, although the reported methods, previously designed for the reduction of ketones to chiral secondary alcohols, proved to be highly substrate dependent in terms of enantioselection.

Cyclic imines such as **(1)** are interesting precursors for the preparation of chiral pyrrolidines by reduction and were investigated by Uemura and Hidai using Ru- and Ir-based chiral catalysts (Scheme 13.1) [5]. The first catalytic system utilized

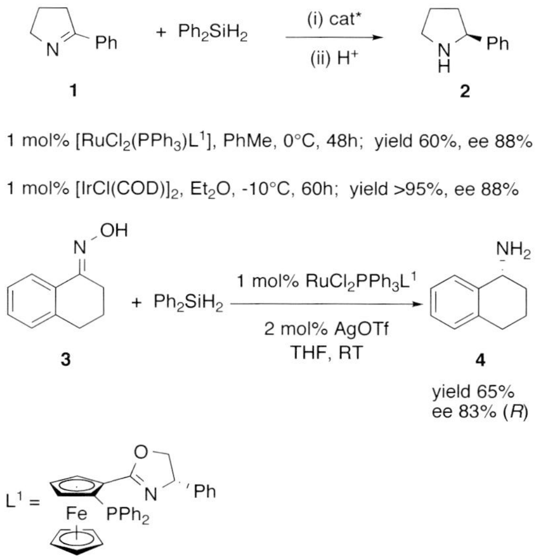

1 mol% [RuCl$_2$(PPh$_3$)L^1], PhMe, 0°C, 48h; yield 60%, ee 88%

1 mol% [IrCl(COD)]$_2$, Et$_2$O, -10°C, 60h; yield >95%, ee 88%

Scheme 13.1 Ru- and Ir-based catalysis utilizing the ferrocenyl phosphinooxazoline ligand L^1.

ruthenium in combination with a chiral ferrocenyl phosphinooxazoline ligand L[1]. The corresponding amine (2) was isolated in 88% ee, albeit in modest yield of 60%. It was later found that an iridium catalyst bearing the same ligand gave a similar enantioselectivity but a quantitative yield of the pyrrolidine. However, when other substrates, such as the corresponding six-membered imine, were investigated, the enantioselectivities dropped to 7%. Another investigation from the same groups focused on the use of ketoximes (such as (3)) as substrates that can lead to the chiral primary amines after reduction [6]. The Ru-based catalyst, when activated by silver triflate, gave a promising 83% ee for amine (4). Unfortunately, the catalysts proved again to be less efficient with other ketoximes, such as dialkyl ketoximes.

While there are still increasing examples for the Rh-, Ir- and Ir-based catalyzed reduction of ketones, the applications of such catalysts for the hydrosilylation of imines remain underdeveloped.

The ruthenium cluster (5) was reported by Kira *et al.* to be reactive toward nitriles, yielding iminosilyl bridged diruthenium complexes (Figure 13.1) [7]. This observation led the authors to examine this cluster as a catalyst for the hydrosilylation of simple ketimines. Full conversion of the imines required the use of bis-aryl silanes and 5 mol% of the catalyst. Recently reported by the group of Messerle is the more efficient catalyst (6) (Figure 13.1) [8]. This cationic iridium complex bearing a bis(pyrazol-1-yl)methano ligand gave high activity for the ketimine derived from acetophenone and aniline when 0.5 mol% of catalyst was used with triethylsilane in methanol. An initial turn over frequency (TOF) of $5000\,h^{-1}$ was measured for this substrate, yielding the corresponding desilylated amine in quantitative yield in less than 10 min. This family of catalysts was later used in an elegant one-pot hydroamination/hydrosilylation process (Scheme 13.2) [9].

^{1}H NMR investigations showed that the cationic complex (6) catalyzes the intramolecular hydroamination of the amino alkyne (7) to give the cyclic imine (8) as an intermediate. The hydrosilylation of the amine occurs in a second step to yield the silylated pyrrolidine (9). Such studies might show the future direction for the use of Rh–Ir-based catalysts for the hydrosilylation of imines and their use in

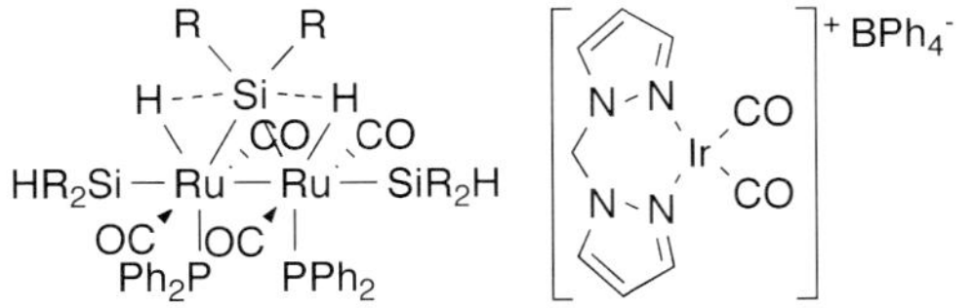

Figure 13.1 Catalysts active toward nitriles.

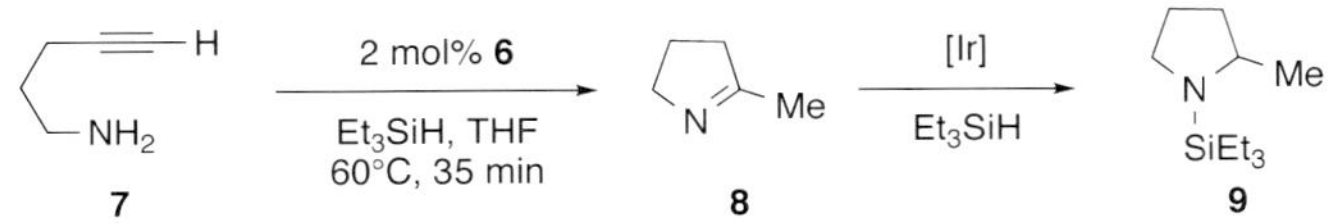

Scheme 13.2 One-pot hydroamination/hydrosilylation.

multistep one-pot processes that may overcome the main disadvantages of such catalytic systems regarding their cost.

Iridium-based catalysts have also been reported by the group of Ishii for the optimization of the one-pot reductive amination of secondary amines with aldehydes, using triethylsilane or poly(methylhydrosiloxane) (PMHS) as reducing agents [10]. The catalytic reaction takes place at 75 °C in dioxane with 2 mol% catalyst, [IrCl(COD)]$_2$, with the corresponding tertiary amines being isolated in good yields (71–90%). Excellent yields were also obtained at 50 °C in THF when the inexpensive PMHS was used as the silylating agent. Mechanistic investigations by deuterium experiments suggested that the main catalytic process goes through the hydrosilylation of an enamine, generated by the condensation of the secondary amine and the aldehyde.

13.3
Titanium-based Catalysts

A major breakthrough occurred in the early 1990s in the field of hydrosilylation of ketones and lactones with the discovery of titanium-based catalysts by the groups of Buchwald, Halterman and Harrod [4]. Buchwald first developed such precatalysts for the enantioselective hydrogenation of imines [11], before the same group optimized new conditions for the asymmetric hydrosilylation of this family of substrates. Titanocene-based complexes were used as precatalysts, which were first activated with organometallic reagents such as butyllithium giving the active Ti(III) hydride catalysts. The same group then developed a more efficient method and applied it to the asymmetric hydrosilylation of prochiral ketones and ketimines using a chiral C_2-symmetric difluorotitanocene precatalyst (Scheme 13.3) [12].

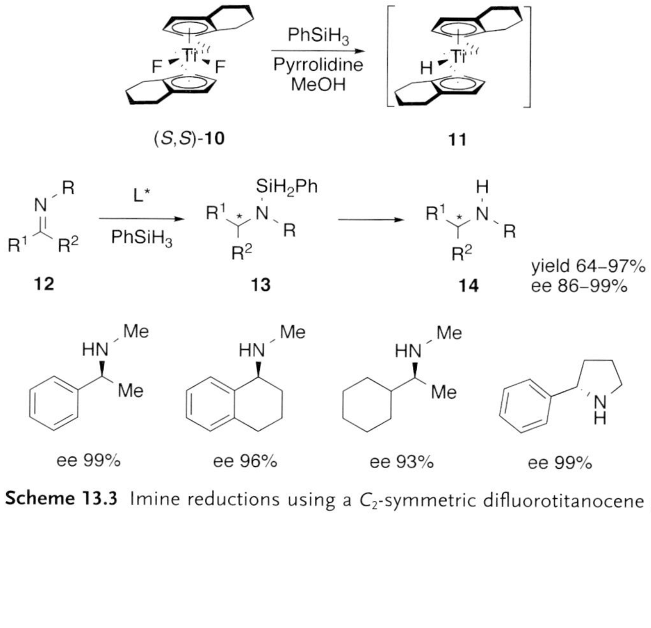

Scheme 13.3 Imine reductions using a C_2-symmetric difluorotitanocene precatalyst.

The difluoro analogue of Brintzinger's catalyst, **(10)**, was activated by a combination of phenylsilane, methanol and pyrrolidine to give the hypothetical Ti(III) hydride catalyst. The role of the methanol and pyrrolidine are postulated to convert phenylsilane into the more reactive Ph(MeO)$_2$SiH reagent, which in turn carries out the reduction of Ti(IV) to Ti(III) and transfers its hydride to titanium by reaction with a fluoride ligand. The first generation of the catalytic system used the highly active, albeit expensive, phenylsilane as the stoichiometric reducing agent. This system gave high selectivities, ee values were over 90% in all cases, and the amount of catalyst could be decreased to 0.02 mol% with no reduction in the enantioselectivity. It is also noteworthy that this catalytic system was used on nonactivated ketimines, bearing simple alkyl substituents on the nitrogen atom. Still, this first-generation catalytic system suffered from two major drawbacks: (i) the system was strongly sensitive to steric bulk on the nitrogen substituent, and was hence limited to methyl groups or cyclic imines; and (ii) phenylsilane has to be used for good reactivity and is expensive for large-scale applications. A re-examination of its proposed mechanistic cycle gave rise to some useful observations that led to optimizations which avoided these drawbacks (Figure 13.2).

The main mechanism (cycle **A**) includes two main elementary processes. The addition of the titanium hydride **(15)** to the imino bond, leading to the titanium amide **(17)**, which is postulated to be the enantiodetermining step, is followed by σ-bond metathesis with phenylsilane (via intermediate **(18)**), leading to the silylated chiral amine **(19)** and regeneration of **(15)**. This last step is assumed to be slow and very sensitive to steric bulk. Therefore, the addition of a less bulky primary

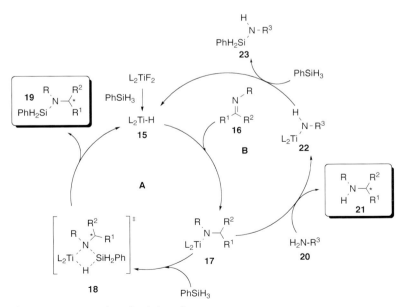

Figure 13.2 Proposed mechanistic cycle.

amine **(20)**, as an additive, could lead to the less-hindered titanium amide **(22)** by transmetallation and release the amine **(21)**. The σ-bond metathesis would then occur more easily on the titanium amide **(22)**, regenerating the active catalyst **(15)** (cycle **B**). This hypothesis was confirmed by experimentation and a survey of various additives. The main advantage of this method is that it also allowed the use of PMHS (a cheap and easy to handle silane) as a replacement for phenylsilane in the procedure (Scheme 13.4) [13a].

This procedure was also applied to the enantioselective hydrosilylation of N-aryl imines (Scheme 13.4), yielding the corresponding chiral anilines in up to 99% ee [13b]. This method proved to be especially efficient with imines derived from dialkyl ketones and unsaturated ketones. Applications in asymmetric synthesis soon followed through the synthesis of calcitemic compound **(22)** [14] and piperidine alkaloids **(23)** and **(24)** (Figure 13.3) [15].

The procedure using phenylsilane was later used for the kinetic resolution of 3-substituted indanones and tetralones with high enantioselectivities [16] and the straightforward synthesis of (1S,4S)-sertraline, marketed as an antidepressant drug, was also described (Scheme 13.5).

The group of Doye has recently reported the use of titanium complexes as catalysts for hydroamination/hydrosilylation sequences (Scheme 13.6) [17].

Addition of a primary alkyl or aryl amine occurs on the least-hindered carbon of the alkyne **(28)** in the presence of dimethyl titanocene as a catalyst to give an intermediate imine. The subsequent addition of a stoichiometric amount of phe-

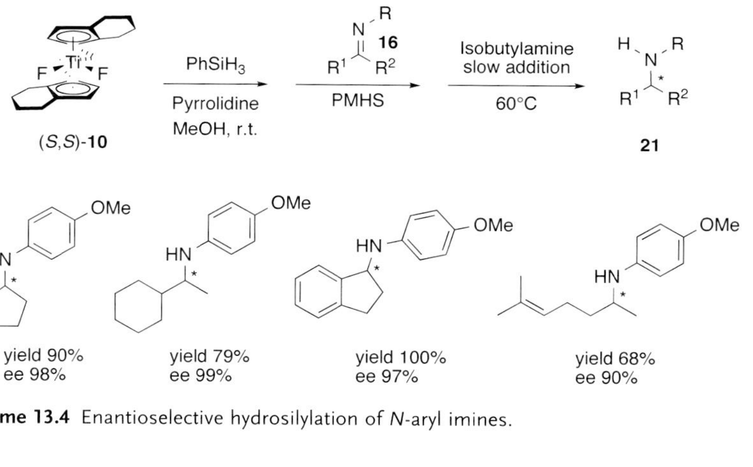

Scheme 13.4 Enantioselective hydrosilylation of N-aryl imines.

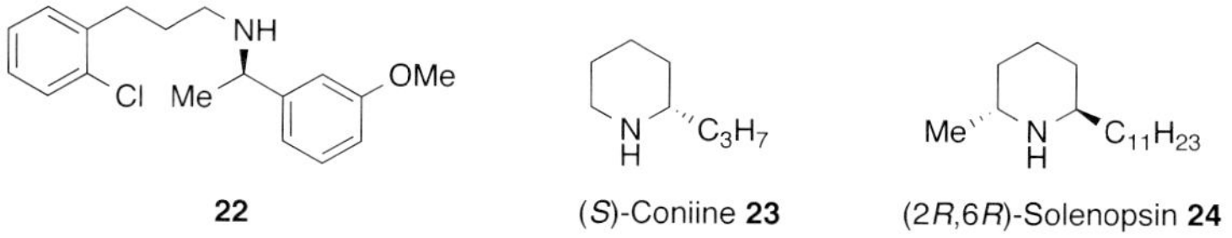

Figure 13.3 Applications of imine hydrosilylation.

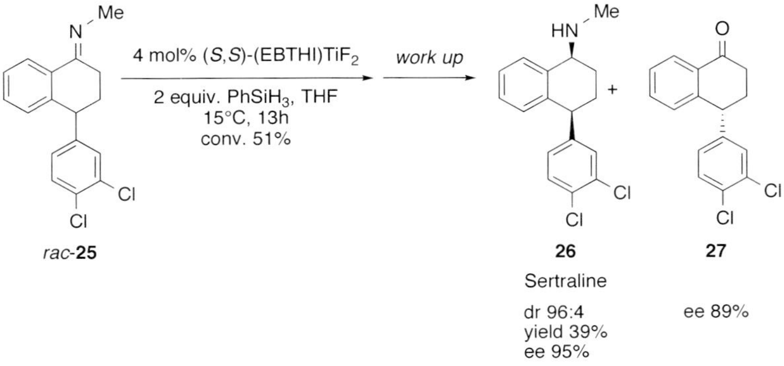

Scheme 13.5 Synthesis of (1S,4S)-sertraline via a kinetic resolution of rac-(25).

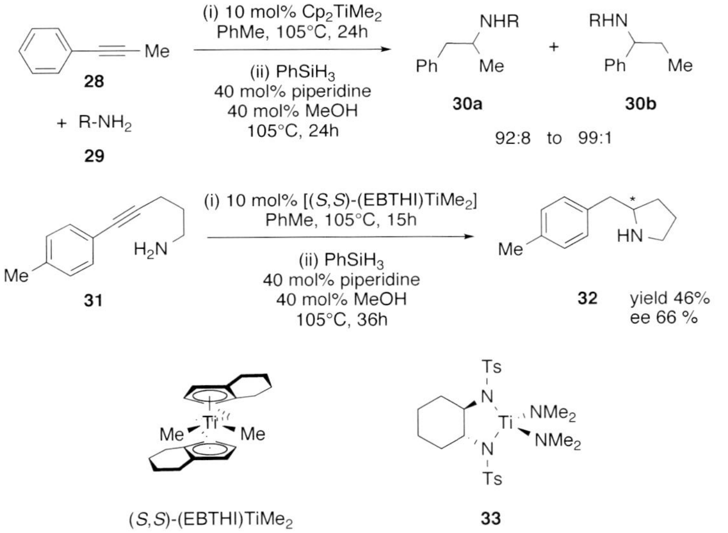

Scheme 13.6 Ti-complexes as catalysts for hydroamination/hydrosilylation.

nylsilane, piperidine and methanol as activating agents leads to the *in situ* hydrosilylation of the imine giving secondary amines **(30a,b)**. Using a chiral catalyst in a tandem process allowed for the intramolecular formation of the chiral pyrrolidine **(32)**. When the dimethyl derivative of Brintzinger's catalyst was used, amine **(32)** was recovered in 66% ee. The analogous piperidine ring was also tested with the easily available chiral complex **(33)**, which gave a promising 55% ee for the piperidine heterocycle. Tillack *et al.* reported the survey of various titanocene catalysts for the hydrosilylation of various aldimines with diphenylsilane [18]. The highest conversions were obtained with $Cp_2Ti(PhCCTMS)$, with a 30:1 substrate:catalyst ratio.

13.4
Zinc-, Copper-, and Rhenium-based Catalysts

Another major breakthrough in the field of hydrosilylation of carbonyl compounds was reported in 1998 by the group of Mimoun who carried out a survey of various transition metal salts as precursors of metal hydride catalysts for the hydrosilylation of carbonyl compounds [19]. Zinc complexes were found to be efficient precatalysts for the reduction of various chemical functionalities with PMHS, a chiral version was later detailed by the same group [20] and subsequently optimized by the groups of Carpentier, Mortreux and Walsh [21]. One of the catalytic systems for the reduction of ketones reported by Carpentier was also applied to the hydrosilylation of imines (Scheme 13.7) [22].

The combination of diethylzinc with a simple diamine ligand proved efficient to carry out the reduction of imines (34), when the reaction was performed in a protic solvent such as methanol. However, an excess (two to four equivalents) was required for full conversion as competitive dehydrogenative silylation of the protic solvent occurred. A mechanistic investigation was later described by the same group for the hydrosilylation of ketones [23] that indicated the role of the alcohol cosolvent in the production of the active catalyst. These observations and the work carried out by the groups of Mimoun, Carpentier and Walsh on the enantioselective hydrosilylation of ketones allowed Yun et al. to optimize the catalytic protocol for the enantioselective hydrosilylation of imines (Scheme 13.8) [24].

Electrophilic imines (36) bearing an electron-withdrawing diphenylphosphinyl group were used as substrates for the reduction into their corresponding protected amines (37). The authors also used methanol as a cosolvent on the hypothesis that the solvolysis of a strong zinc–nitrogen bond formed by the addition of a zinc hydride species to the imine would accelerate the catalytic process. The use of a chiral ligand, previously described for the asymmetric reduction of ketones, resulted in high enantioselectivities of various aryl (heteroaryl) and alkyl ketimines with excellent yields. In this case, an excess of PMHS (three equivalents) was required for full conversion of the starting ketimines.

While the use of chiral copper hydrides in enantioselective reduction of ketones and electrophilic alkenes has been widely developed, there is only one example of the application of such a catalytic system being employed in enantioselective hydrosilylation of prochiral imines. Lipschutz et al. recently reported a highly

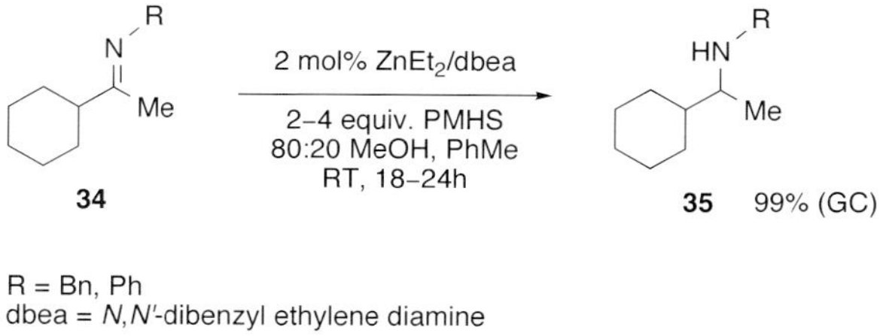

R = Bn, Ph
dbea = N,N'-dibenzyl ethylene diamine

Scheme 13.7 Zinc-catalyzed hydrosilylation of imines.

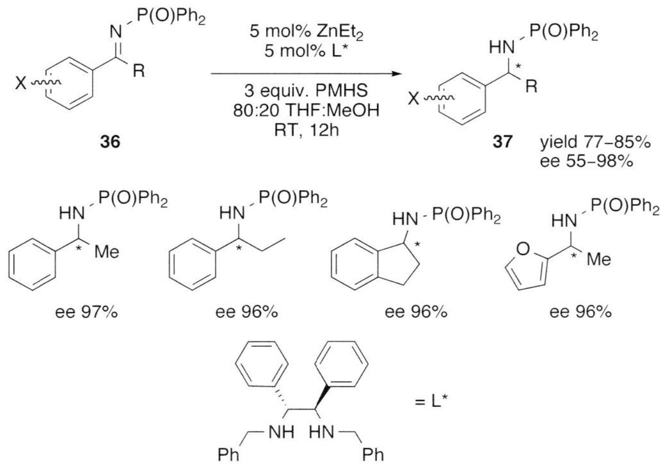

Scheme 13.8 Reduction of imines bearing the electron-withdrawing diphenylphosphinyl group.

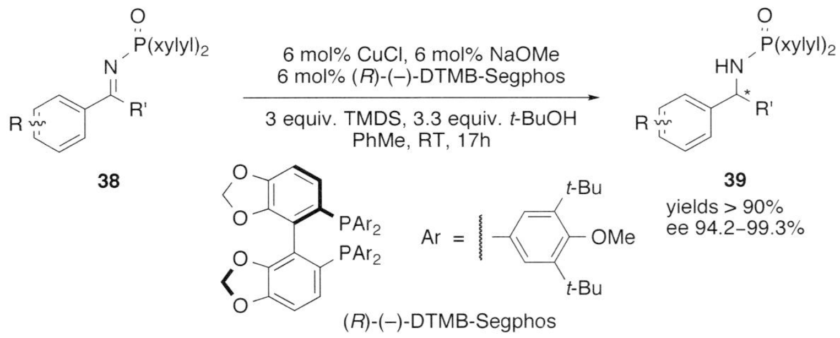

Scheme 13.9 Cu(I)-catalyzed hydrosilylation of imines.

TMDS = tetramethyldisiloxane

enantioselective process for the asymmetric reduction of electrophilic ketimines using a copper (I)/chiral disphosphine catalyst (Scheme 13.9) [25].

This system uses a combination of an *in situ* generated copper alcoxide bearing a hindered chiral diphosphine, from the Segphos family, as a precursor of a chiral copper hydride catalyst. Imines activated by bis-arylphosphinyl groups were also required for good reactivity and the steric hindrance was increased on the electron-withdrawing group by the introduction of xylyl groups on the phosphorus atom. Three equivalents of TMDS (six equivalents of Si—H per imino bond) and *t*-butanol were also required for optimum reactivity. Although this system still requires a fairly high catalyst loading (compared with the catalyst loading usually required for simple ketones and electrophilic olefins), it remains one of the most enantioselective reported to date as all the enantiomeric excesses reported for various aryl alkyl ketimines were in the range of 94.2–99.3%.

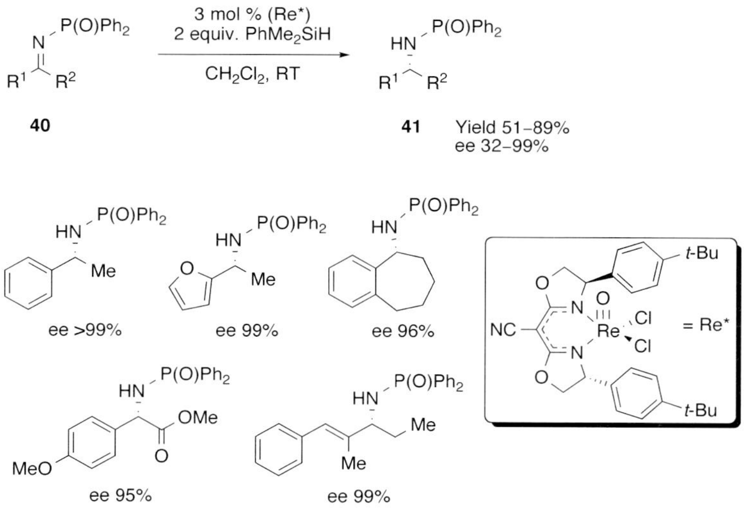

Scheme 13.10 Re(V) catalyst bearing a chiral cyano(bisoxazoline) ligand.

This family of electrophilic imines was also recently used by the group of Toste for asymmetric hydrosilylation using a new Re(V) catalyst bearing a chiral cyano(bisoxazoline) ligand (Scheme 13.10) [26].

Good reactivity was obtained with 3 mol% of chiral catalyst and it is noteworthy that the high oxidation state of the rhenium complex allowed the catalytic reaction to be performed under "open flask" conditions. A wide range of ketimines was successfully tested, including α-imino esters and α,β-unsaturated imines. Only one case of a dialkyl ketimine was reported with a modest enantioselectivity (32% ee).

13.5
Lanthanide-based Catalysts

The group of Takaki has reported the preparation of an ytterbium imine complex (**43**) as a useful catalyst for a variety of catalytic transformations, including imine hydrosilylation [27].

The azaytterbiumcyclopropane complex (**43**) is prepared by reaction of a ketimine (**42**) with an equimolar amount of ytterbium metal in a THF-HMPA solution, and can be isolated as air- and moisture-sensitive red-black crystals. However, it can also be prepared *in situ* for catalytic reactions, such as dehydrogenative silylation of amines, hydrosilylation of alkenes and imines and the hydrophosphination of various unsaturated systems. The proposed mechanism for the hydrosilylation of imines is depicted in Scheme 13.11, and starts with the activation of the imine complex by phenylsilane, which gives the active ytterbium hydride cata-

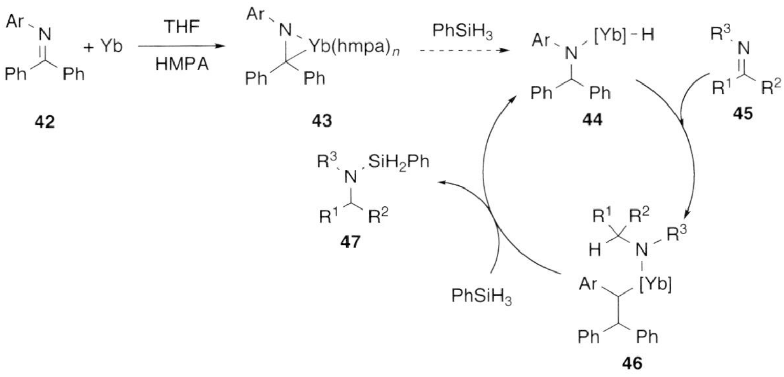

Scheme 13.11 Use of the ytterbium imine complex (43) in imine reductions.

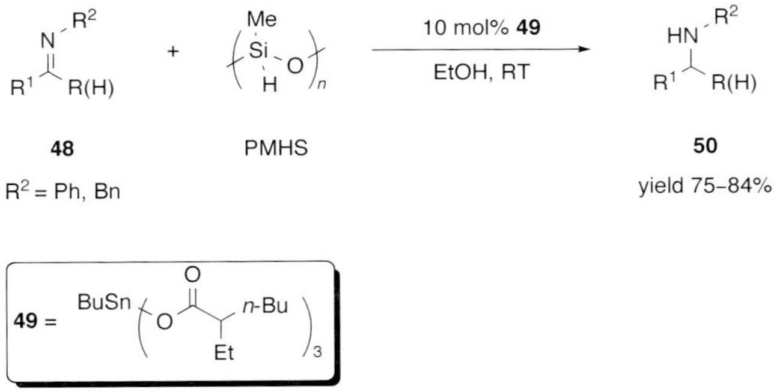

Scheme 13.12 First example of tin-based hydrosilylation of imines.

lyst (44). Hydride transfer to the imine substrate (45) gives the bis-amido intermediate (46), which undergoes σ-bond metathesis with the silane to yield the silylated amine (47) and regenerates the catalyst (44). This catalyst displayed good activity for a range of unactivated aldimines (with 5 mol% catalyst loading and phenylsilane as the reducing agent) but gave disappointing results when ketimines were used as substrates.

13.6
Tin-based Catalysts

In 1997 Lopez and Fu reported the first example of a tin-based catalyst for the direct hydrosilylation of imines (Scheme 13.12) [28].

The tin carboxylate catalyst (49) was successfully used in the hydrosilylation of unactivated aldimines and ketimines. This system also proved particularly interesting for the development of low-cost reduction processes of imines as it combines a cheap catalyst and reducing agent (three equivalents PMHS per imine substrate).

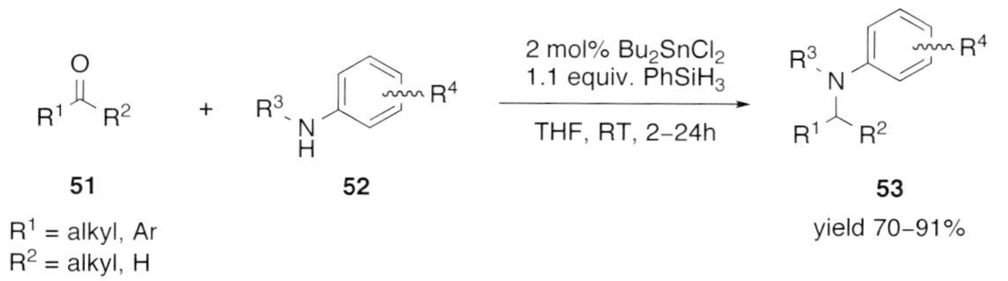

Scheme 13.13 Reductive amination of aromatic amines with aldehydes and ketones.

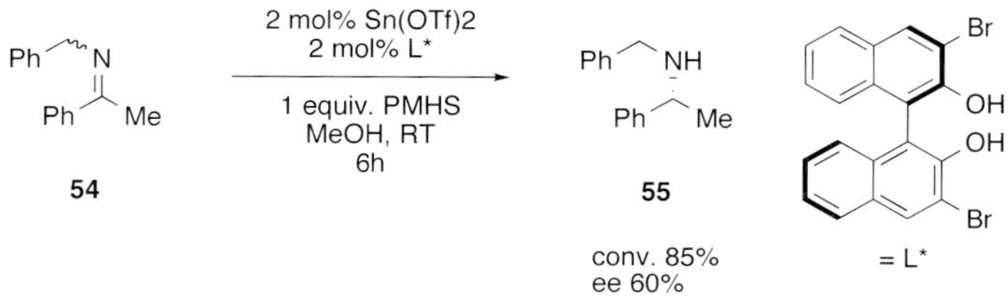

Scheme 13.14 Tin-catalyzed asymmetric hydrosilylation of a prochiral imine.

Further development in this area was later reported by Apodaca and Xiao [29], who showed that such tin catalysts could be used for the direct reducing amination of aromatic amines with aldehydes and ketones (Scheme 13.13).

Optimization of the catalytic system was carried out with 2 mol% of dibutyltin dichloride in THF at room temperature with various silanes. Phenylsilane proved the most active and it was tested with a number of substrates, although it was shown that the use of an excess of PMHS did not improve the yields. The exact role of the tin catalyst is still not clear, although the authors postulated that the *in situ* formation of a tin hydride (Bu$_2$SnClH) could be responsible for the activity. They also proposed that the dibutyltin dichloride was acting as a simple Lewis acid catalyst in the hydrosilylation reaction. As shown in these studies, the use of tin-based catalysts could be particularly interesting for the development of low-cost reductive amination processes with silanes and a promising application lies in the development of chiral tin-based catalysts. There is, however, only one example reported for the asymmetric hydrosilylation of a prochiral imine with a chiral tin catalyst (Scheme 13.14) [30].

High-throughput screening methods were used to find a high-performing catalyst for the hydrosilylation of the imine **(54)**, by mixing metal precursors (such as Sn, Cd, In and Zn salts) with various commercially available chiral ligands. The combination of tin(II) triflate and the modified binol ligand L* gave the best results with 85% conversion of the starting imine and 60% ee for the resulting amine **(55)**. Although no further optimization of the catalytic system or application to other substrates was reported, this example opens a new pathway for the development of low-cost chiral catalysts for the asymmetric hydrosilylation of imines.

13.7
Chiral Lewis Bases as Catalysts

The concept that neutral and anionic Lewis bases can promote the transfer of activated silyl nucleophiles to carbonyl substrates offers a useful alternative to the use of metal-based catalysts and was first used by the group of Kobayashi for the reduction of aldimines with a trichlorosilane/DMF reagent [31]. This concept has also been widely used in the field of asymmetric catalysis for various types of silyl nucleophiles, including allyl-, enol-, cyanide- and hydride-based reagents. The application to asymmetric hydrosilylation of imines came much later, however, and has been developed by the groups of Hosomi, Kočovský, Matsumura and Sun. In a pioneering work, Homosi et al. showed that a binaphthol bis-lithium alcoxide could catalyze the hydride transfer of trimethoxisilane on an arylsulfonyl-activated imine with moderate activity and enantioselectivity (72% ee) [32]. As depicted in the Scheme 13.15 and Table 13.1, the most efficient catalysts for the asymmetric hydrosilylation of imines with trichlorosilanes are chiral bidentate Lewis bases.

All the reported systems used the inexpensive trichlorosilane as the reducing reagent with 10–20 mol% catalyst under various conditions; most reported chiral bases prepared from α-amino-acids. Although the first system reported, by the group of Matsumura, displayed moderate enantioselectivities and substrate scope, this work opened the way to the design of more selective catalysts based on similar structures. The groups of Sun and Kočovský later reported other formamide-derived amino acids with noncyclic and cyclic structures, with good reactivities and improved enantioselectivities. The catalyst (**60**) developed by Sun et al. also gave high enantioselectivities with dialkyl imines and a second-generation catalyst

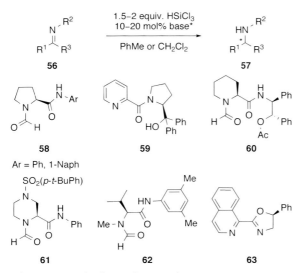

Scheme 13.15 Chiral Lewis base catalysis.

Table 13.1 Chiral Lewis base catalysts.

Group	Year	Catalyst	R^1, R^2	R^3	ee (%)	Reference
Matsumura	2001	(58)	R^1 = Ar R^2 = Ph, Bn	Me	49–66	[33]
Matsumura	2006	(59)	R^1 = Ar R^2 = Ph, Bn	Me, CO$_2$Me	45–80	[34]
Sun	2006	(60)	R^1 = Ar, alkyl, styryl R^2 = Ar	Me	87–96	[35]
Sun	2006	(61)	R^1 = Ar, Cy R^2 = Bn	Me, Alk.	82–97	[36]
Kočovský	2004–2006	(62)	R^1 = Ar, styryl R^2 = Ar	Me, CO$_2$Me	Up to 92	[37]
Kočovský	2006	(63)	R^1 = Ar R^2 = Ar	Me	85–87	[38]

(61) gave improved selectivities and allowed some variations of the alkyl group at the R^3 position. Kočovský *et al.* recently reported an elegant system that showed that pyridine-oxazoline structures such as (63) could also activate trichorosilane and promote the enantioselective reduction of ketones and ketimines with good to high enantioselectivities. In all their reported ligands, the authors underlined the importance of secondary π–π interaction between the substrate and the ligand for optimum enantioselectivities. Although examination of all the reported systems shows that the required loading of the chiral base is still high (10–20 mol%), this methodology uses inexpensive reducing agents and recoverable catalysts.

13.8
Miscellaneous Methods

Fernandes and Romao reported the use of dioxomolybdenum dichloride (MoO$_2$Cl$_2$) for the hydrosilylation of imines with phenylsilane [39]. This system uses 10 mol% catalyst in refluxing THF and was limited to bis-aryl aldimines giving the corresponding amines in moderate to good yields (50–97%).

Tris-pentafluorophenyl borane [B(C$_6$F$_5$)$_3$], a commercially available Lewis acid, has been used as a catalyst in many transformation, including Diels–Alder and aldol reactions, and has been also recently applied to the hydrosilylation of ketones and imines (Scheme 13.16) [40].

This catalysis is attractive as it uses a simple organic Lewis acid as a catalyst and a wide range of aldimines and ketimines can be reduced with excellent yields. The authors postulated a mechanism with the intervention of the active hydride species (66), arising from the activation of the hydride on the silane by the Lewis acid borane reagent. The transfer of the hydride occurs thus on the boron atom while the silane coordinates and activates the imine substrate to facilitate the hydride transfer from the boron atom to the imine.

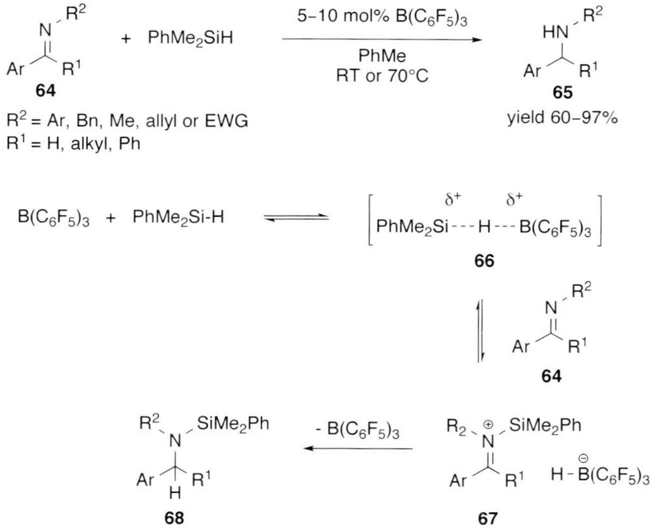

Scheme 13.16 Organic Lewis acid B(C$_6$F$_5$)$_3$ reducing a range of aldimines and ketamines.

Some useful Lewis/Brønsted acid silane combinations have been reported for the reductive amination of amines with carbonyl alkylating agents [41–43]. Dubé and Scholte used trifluoroacetic acid to activate the reductive amination of amides, carbamates and ureas with aliphatic and aromatic aldehydes [41]. This method employs the inexpensive triethylsilane as the reducing agent and was later used by Chen et al. for the N-alkylation of benzodiazepine derivatives [42]. A straightforward "one pot" amination of various aldehydes and ketones with primary and secondary amines was proposed by the group of Chandrasekhar, who used a combination of titanium(IV) isopropoxide and PMHS for the reductive amination [43].

13.9
Conclusion

While there is now a wide range of catalysts available for the hydrosilylation of aldehydes and ketones, this review shows that the number of well performing catalytic systems for the more challenging substrates that are ketimines is still limited. However, some recent reports in this field, with the discovery of zinc, copper and rhenium catalysts as well as chiral organocatalysts, indicate rapid development in the area of the production of chiral amines by asymmetric hydrosilylation of imines. Preliminary results shows that future advances will probably be focused on the development of reductive amination processes using organosilanes as reducing agents, for which chiral catalysts remain to be found.

References

1. (a) Langlois, N., Dang, T.P. and Kagan, H.B. (1973) *Tetrahedron Letters*, **49**, 4865. (b) Dumont, W., Poulain, J.-C., Dang, T.P. and Kagan, H.B. (1973) *Journal of the American Chemical Society*, **95**, 8295.
2. (a) Vilaivan, T., Bhanthumnavin, W. and Sritana-Anant, Y. (2005) *Current Organic Chemistry*, **9**, 1315. (b) Kobayashi, S. and Ishitani, H. (1999) *Chemical Reviews*, **99**, 1069.
3. (a) Blaser, H.U., Spindler, F. and Studer, M. (2001) *Applied Catalysis A*, **221**, 119. (b) Blaser, H.U. (2002) *Advanced Synthesis and Catalysis*, **344**, 17.
4. (a) Nishiyama, H. (1999) Chapter 6.3, in *Comprehensive Asymmetric Catalysis* (eds E.N. Jacobsen, A. Pfaltz and H. Yamamoto), Springer, Berlin. (b) Carpentier, J.F. and Bette, V. (2002) *Current Organic Chemistry*, **6**, 913. (c) Brunel, J.M. (2003) *Recent Research Developments in Synthetic Organic Chemistry*, **7**, 155. (d) Riant, O., Mostefaï, N. and Courmarcel, J. (2004) *Synthesis*, 2943.
5. (a) Takei, I., Nishibayashi, Y., Arikawa, Y., Uemura, S. and Hidai, M. (1999) *Organometallics*, **18**, 2271. (b) Nishibayashi, Y., Takei, I., Uemura, S. and Hidai, M. (1998) *Organometallics*, **17**, 3420.
6. Takei, I., Nishibayashi, Y., Ishii, Y., Misobe, Y., Uemura, S. and Hidai, M. (2001) *Chemical Communications*, 2360.
7. Hashimoto, H., Aratani, I., Kabuto, C. and Kira, M. (2003) *Organometallics*, **22**, 2199.
8. Field, L.D., Messerle, B.A. and Rumble, S.L. (2005) *European Journal of Organic Chemistry*, 2881.
9. Field, L.D., Messerle, B.A. and Wren, S.L. (2003) *Organometallics*, **22**, 4393.
10. Mizuta, T., Sakagushi, S. and Ishii, Y. (2005) *Journal of Organic Chemistry*, **70**, 2195–9.
11. (a) Willoughby, C.A. and Buchwald, S.L. (1992) *Journal of the American Chemical Society*, **114**, 7562. (b) Willoughby, C.A. and Buchwald, S.L. (1994) *Journal of the American Chemical Society*, **116**, 8952.
12. Verdaguer, X., Lange, U.E.W., Reding, M.T. and Buchwald, S.L. (1996) *Journal of the American Chemical Society*, **118**, 6784.
13. (a) Verdaguer, X., Lange, U.E.W. and Buchwald, S.L. (1998) *Angewandte Chemie (International Ed. in English)*, **37**, 1103. (b) Hansen, M.C. and Buchwald, S.L. (2000) *Organic Letters*, **2**, 713.
14. Hansen, M.C. and Buchwald, S.L. (1999) *Tetrahedron Letters*, **40**, 2033.
15. Reding, M.T. and Buchwald, S.L. (1998) *Journal of Organic Chemistry*, **63**, 6344.
16. Yun, J. and Buchwald, S.L. (2000) *Journal of Organic Chemistry*, **65**, 767.
17. Heutling, A., Pohlki, F., Bytschkov, I. and Doye, S. (2005) *Angewandte Chemie (International Ed. in English)*, **44**, 2951.
18. Tillack, A., Lefeber, C., Peulecke, N., Thomas, D. and Rosenthal, U. (1997) *Tetrahedron Letters*, **38**, 1533.
19. (a) Mimoun, H. (1995) WO 96/12694. (b) Mimoun, H. (1998) WO 99/12877.
20. (a) Mimoun, H. (1999) *Journal of Organic Chemistry*, **64**, 2582. (b) Mimoin, H., de Saint Laumer, J.Y., Giannini, L., Scopelliti, R. and Floriani, C. (1999) *Journal of the American Chemical Society*, **121**, 6158.
21. (a) Mastranzo, V.M., Quintero, L., de Parodi, C.A., Juastiri, E. and Walsh, P.J. (2004) *Tetrahedron*, **60**, 1781. (b) Bette, V., Mortreux, A., Savoia, D. and Carpentier, J.-F. (2004) *Tetrahedron*, **60**, 2837. (c) Bette, V., Mortreux, A., Ferioli, F., Martelli, G., Savoia, D. and Carpentier, J.-F. (2004) *European Journal of Organic Chemistry*, 3040.
22. Bette, V., Mortreux, A., Lehmann, C.W. and Carpentier, J.-F. (2003) *Chemical Communications*, 332.
23. Bette, V., Mortreux, A., Savoia, D. and Carpentier, J.-F. (2005) *Advanced Synthesis and Catalysis*, **347**, 289.
24. Park, B.-M., Mun, S. and Yun, J. (2006) *Advanced Synthesis and Catalysis*, **348**, 1029.
25. Lipshutz, B.H. and Shimizu, H. (2004) *Angewandte Chemie (International Ed. in English)*, **43**, 2227.
26. Nolin, K.A., Ahn, R.W. and Toste, F.D. (2005) *Journal of the American Chemical Society*, **127**, 12462.

27 (a) Takaki, K., Kamata, T., Miura, Y., Shishido, T. and Takehira, K. (1999) *Journal of Organic Chemistry*, **64**, 3891. (b) Takaki, K., Komeyama, K. and Takehira, K. (2003) *Tetrahedron*, **59**, 10381.
28 Lopez, R.M. and Fu, G. (1997) *Tetrahedron*, **48**, 16349.
29 Apodaca, R. and Xiao, W. (2001) *Organic Letters*, **3**, 1745.
30 Ireland, T., Fontanet, F. and Tchao, G.-G. (2004) *Tetrahedron Letters*, **45**, 4383.
31 Kobayashi, S., Yasuda, M. and Hashiya, I. (1996) *Chemistry Letters*, 407.
32 Nishikori, H., Yoshihara, R. and Hosomi, A. (2003) *Synlett*, 561.
33 Iwasaki, F., Onomura, O., Mishima, K., Kanematsu, T., Maki, T. and Matsumura, Y. (2001) *Tetrahedron Letters*, **42**, 2525.
34 Onomura, O., Kouchi, Y., Iwasaki, F. and Matsumura, Y. (2006) *Tetrahedron Letters*, **47**, 3751.
35 Wang, Z., Ye, X., Wie, S., Wu, P., Zhang, A. and Sun, J. (2006) *Organic Letters*, **8**, 999.
36 Wang, Z., Cheng, M., Wu, P., Wie, S. and Sun, J. (2006) *Organic Letters*, **8**, 3045.
37 (a) Malkov, A.V., Mariani, A., MacDougall, K.N. and Kocovsky, P. (2004) *Organic Letters*, **6**, 2253. (b) Malkov, A.V., Stoncius, S., MacDougall, K.N., Mariani, A., McGeoch, G.D. and Koovský, P. (2006) *Tetrahedron*, **62**, 264.
38 Malkov, A.V., Stewart Liddon, A.J.P., Ramirez-López, P., Bendová, L., Haigh, D. and Kočovský, P. (2006) *Angewandte Chemie (International Ed. in English)*, **45**, 1432.
39 Fernandes, A.C. and Romào, C.C. (2005) *Tetrahedron Letters*, **46**, 8881.
40 Blackwell, J.M., Sonmor, E.R., Scoccitti, T. and Piers, W.E. (2000) *Organic Letters*, **2**, 3921.
41 Dubé, D. and Scholte, A.A. (1999) *Tetrahedron Letters*, **40**, 2295.
42 Chen, B.-C., Sundeen, J.E., Guo, P., Bednarz, M.S. and Zhao, R. (2001) *Tetrahedron Letters*, **42**, 1245.
43 Chandrasekhar, S., Raji Reddy, Ch. and Moinuddin, A. (2000) *Synlett*, 1655.

Part Four
Miscellaneous Reductions

14
Alkene and Imino Reductions by Organocatalysis

Hans Adolfsson

14.1
Introduction

The reduction of organic compounds is a well-studied transformation and currently a multitude of methods exists for the direct reduction of functional groups such as alkenes, carbonyls and imines. The majority of today's protocols rely on the use of either stoichiometric hydride reagents or molecular hydrogen combined with a suitable transition metal for a successful outcome of the transformation. In particular, in the field of asymmetric reductions, hydrogenation dominates as the standard procedure from small-scale reductions performed in academic laboratories to multi-tonne industrial applications [1, 2]. The increasing interest in compounds that demonstrate biological activity and contain a hydrogen substituent at the stereogenic center has stimulated the development of efficient enantioselective reductions and this particular transformation has dominated the area of asymmetric catalysis. The pioneering discoveries in the 1960s by Knowles and coworkers, who revealed that chiral rhodium complexes based on the Wilkinson catalyst were able to catalyze the hydrogenation of unsaturated carboxylic acids with a slight enantiomeric excess under homogeneous conditions, opened the door for an intensive search for more efficient and selective catalysts [3]. A number of milestones have been reached since the initial experiments were performed, and today we are equipped with a multitude of different methods for the enantioselective reduction of unsaturated organic compounds. Until now, the methods developed for the reduction of organic compounds have been dominated by the use of metal catalysts surrounded by proper stereo-discriminating chiral ligands. Highly efficient catalysts based on rhodium(I) or ruthenium(II) complexes containing chiral diphosphines were introduced for the enantioselective olefin reduction of enamines and unsaturated carboxylic acids with molecular hydrogen [1, 4] More recently, iridium complexes possessing chiral P,N ligands have been used for the enantioselective reduction of nonfunctionalized olefins [5–8]. The asymmetric reductions of ketones and imines are commonly performed using molecular hydrogen and chiral Ru(II) catalysts [4]. A mild alternative for the latter

Modern Reduction Methods. Edited by Pher G. Andersson and Ian J. Munslow.
Copyright © 2008 WILEY-VCH Verlag GmbH & Co. KGaA, Weinheim
ISBN: 978-3-527-31862-9

reductions is to perform the reactions under hydrogen transfer conditions [9, 10]. Metal-catalyzed hydrogen transfer reactions using isopropanol or formic acid as hydrogen sources have been extensively studied during the last decade [11]. Highly enantioselective processes, in particular for the reduction of ketones, have been established using catalysts based on vicinal amino alcohols, diamines or pseudo-dipeptides in combination with Ru(II)- and Rh(III)-arene precursors [11–13]. The common denominator in the above systems is that the principal center of reactivity is positioned on a (transition) metal hydride or dihydride.

Recently, alternative methods have appeared for the selective reduction of unsaturated organic compounds, and these organocatalytic methods are presented and discussed in this chapter. Ever since the re-discovery by List and coworkers that cyclic amino acids like proline efficiently and selectively catalyze the aldol reaction, the use of small organic compounds as catalysts has attracted enormous attention [14, 15]. Currently organocatalytic protocols have been developed for a multitude of different organic transformations [16–18].

14.2
Reducing Agents

The stoichiometric reducing agents in traditional reduction protocols are main-group metal hydrides, molecular hydrogen or various hydride sources such as 2-propanol or formic acid. However, for organocatalyzed reductions different hydrogen sources are needed, and there are currently only two classes of reducing agents that are compatible with the reaction conditions. These two classes of hydrogen donors are represented by either different N-heterocycles such as dihydropyridines (Hantzsch esters) or silanes, in particular trichlorosilane.

14.2.1
N-Heterocyclic Hydrogen Donors

Biological oxidations and reductions are made in cascade reactions involving metalloenzymes and organic electron- or hydride-transferring cofactors such as flavin–adenine dinucleotide ($FADH_2$) or nicotinamide–adenine dinucleotide (NADH) [19]. Model studies of such enzymatic reduction systems have revealed that structurally simpler esters or amides of 1,4-dihydropyridines act as efficient hydride donors (Figure 14.1) [20]. The first example of using such hydride donors in organocatalyzed reductions was made by List and coworkers, who demonstrated that the readily prepared Hantzsch diester **(1a)** (R = Et) in combination with catalytic amounts of ammonium salts of secondary amines effectively accomplished the chemoselective saturation of the C—C π-bond in α,β-unsaturated aldehydes (see below) [21]. Different Hantzsch esters have thereafter been widely employed within the field of organocatalytic reductions.

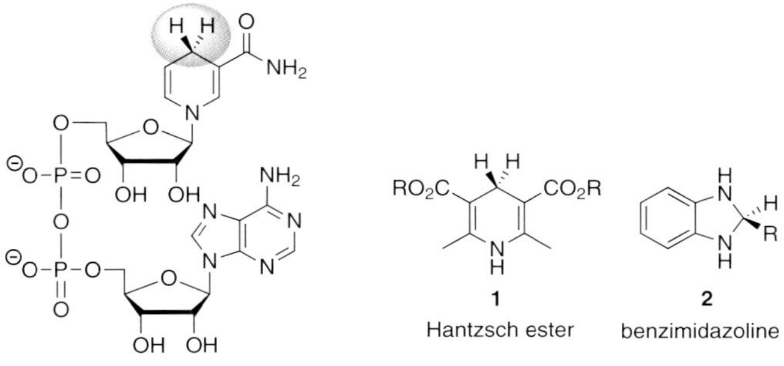

Figure 14.1 Natural and synthetic hydride donors.

A powerful alternative to the Hantzsch esters was recently introduced by Ramachary and Reddy, who discovered that *in situ*-generated benzimidazolines (**2**) can act as efficient hydride donors in transfer hydrogenation reactions of activated alkenes [22].

14.2.2
Silanes

Various silane derivatives are often used as hydride donors in combination with transition metal catalysts for the hydrosilylation of unsaturated organic compounds such as alkenes, imines and ketones. Silanes are normally too unreactive for a direct hydride transfer to an unsaturated compound and the addition of a suitable mediator is necessary in order increase the reactivity. This is in particular true for the *trichlorosilane* (**3**) (see Figure 14.3) which requires activation by a Lewis base to release its hydride [23, 24]. Matsumura and coworkers found that activating trichlorosilane with catalytic amounts of *N*-formylpyrrolidine derivatives resulted in the chemoselective reduction of imines [25, 26]. Additionally, the use of L-proline-derived formamides as catalysts allowed for the first stereoselective organocatalyzed reduction of imines. Several developments within this field have thereafter been accomplished by the use of other organocatalysts (see below).

14.3
Alkene Reduction

The reduction of alkenes using organocatalysis is currently restricted to activated compounds, where α,β-unsaturated carbonyl compounds are the dominant substrate class. Nevertheless, the use of organocatalysis often allows for highly chemoselective olefin reduction in such compounds.

14.3.1
Alkene Reduction by Transfer Hydrogenation of α,β-Unsaturated Aldehydes and Ketones

Inspired by how natural systems perform reduction reactions, List and coworkers recently reported that Hantzsch ester (1a) worked as a good NADH mimic in the hydride transfer reaction to the iminium ion formed from 2-nitrocinnamyl aldehyde and a catalytic amount of dibenzylammonium trifluoroacetate (Scheme 14.1) [21]. The reaction proved to be highly chemoselective, yielding the saturated aldehyde (4) in high chemical yield (94%).

In a series of experiments they showed that different ammonium salts of secondary amines, including pyrrolidine and piperidine, efficiently catalyzed the transformation. Screening different enal substrates using a catalytic amount of dibenzylammonium trifluoroacetate and the NADH-mimic (1a) they demonstrated that a wide range of β-mono- or β,β-disubstituents (aryl or alkyl) were tolerated and gave the corresponding saturated aldehydes in good yields. However, enals containing an additional substituent in the α-position were not reduced under these conditions. The successful results using ammonium salts of cyclic secondary amines as catalysts indicated that chiral cyclic amines like the commonly used organocatalyst (5) (Scheme 14.2) [27] would catalyze the reaction and perhaps be able to induce asymmetry in the hydride transfer step. List and coworkers were indeed rewarded since, in the reduction of enal (6) using a substoichiometric amount of the HCl salt of (5) as catalyst, product (7) was obtained in 81% yield with 81% ee. In further studies on iminium-catalyzed asymmetric hydride transfers to olefins, the groups of MacMillan and List almost simultaneously reported that β,β-disubstituted aldehydes could be reduced in good yields and excellent

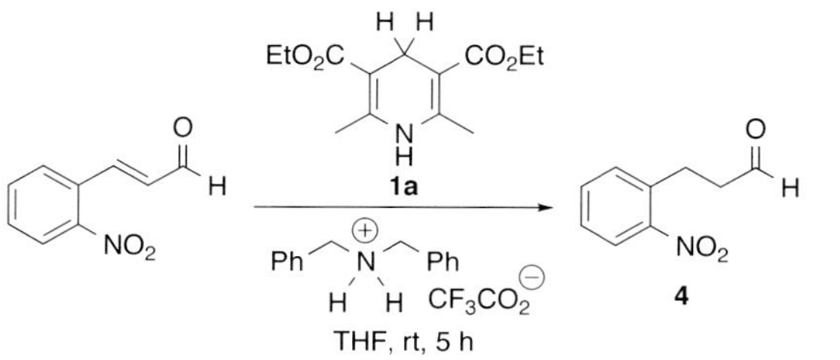

Scheme 14.1 Organocatalytic reduction of α,β-unsaturated aldehydes.

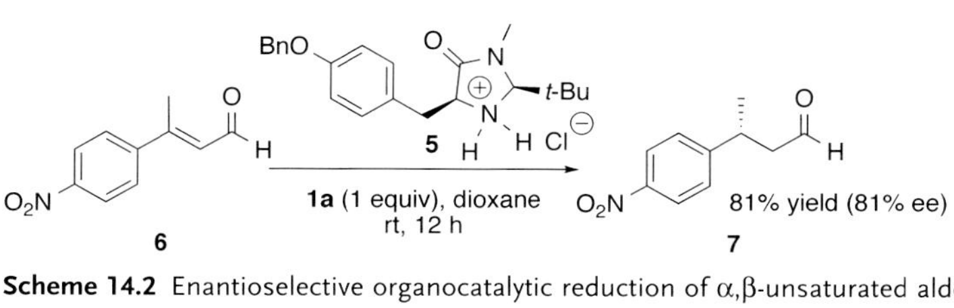

Scheme 14.2 Enantioselective organocatalytic reduction of α,β-unsaturated aldehydes.

enantioselectivity (Table 14.1 and Table 14.2) [28, 29]. The catalysts used by both groups were based on the same imidazolidinone skeleton differing only in the ring substituents.

The group of List reported that compound (**8**) was the optimum catalyst for the enantioselective reduction of β-aryl-substituted α,β-unsaturated aldehydes using the Hantzsch ester (**1b**) as the hydride donor (Table 14.1). Performing the reaction in dioxane at 13 °C using a slight excess of (**1b**) (1.02 equivalents) gave the saturated aldehydes in high yields (77–90%) with up to 96% ee. MacMillan, on the other hand, reported that superior levels of enantiomeric excess were obtained employing the amine salt (**9**)-TFA in chloroform at −30 °C [28]. Under these conditions a number of different trisubstituted α,β-unsaturated aldehydes were reduced in high yields (74–95%) with enantioselectivity up to 97% ee (Table 14.2).

Most interestingly, regardless of whether an (*E*)- or a (*Z*)-olefin was used as substrate, they both converged into the same (*S*)-enantiomer of the product. This result is in direct contrast to many metal-catalyzed hydrogenations where the olefin geometry in general dictates the outcome of the reaction. The reason for the observed stereoconvergence is believed to origin from a fast *E*–*Z* isomerization reaction mediated by the catalyst prior to the reduction reaction.

In accordance with the results reported by MacMillan and coworkers, the List group observed similar enantioconvergence when performing the reduction on *E*/*Z*-substrate mixtures. From a practical point of view, *E*/*Z*-isomeric mixtures obtained from either Wittig-type reactions or olefin metathesis reactions can be used directly in the reductions without initial separations.

The reaction is believed to proceed via the mechanism depicted in Scheme 14.3. The iminium ion is initially formed in a reaction between the imidazolidinone

Table 14.1 Organocatalytic reduction of β-aryl-substituted α,β-unsaturated aldehydes.

Entry	Ar	Yield (%)	ee (%)
1	Phenyl	77[a]	90
2	4-Cyanophenyl	89	96
3	4-Nitrophenyl	83	94
4	4-Bromophenyl	90	94
5	4-(Trifluoromethyl)phenyl	85	94
6	2-Naphthyl	86	92

a Isolated as the 2,4-dinitrophenylhydrazone derivative.

14 Alkene and Imino Reductions by Organocatalysis

Table 14.2 Organocatalytic reduction of β,β-disubstituted α,β-unsaturated aldehydes.

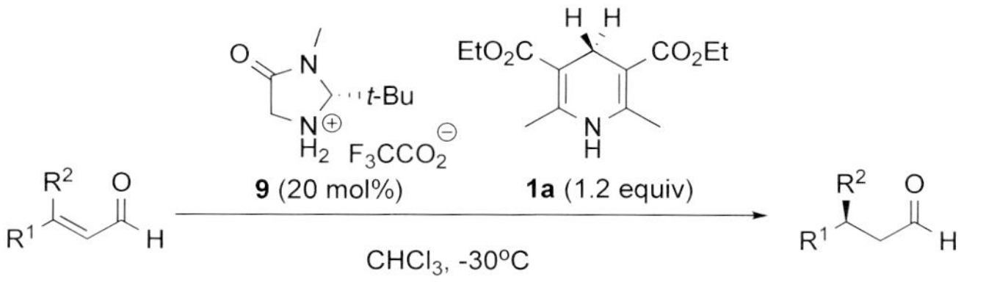

Entry	R^1	R^2	Time (h)	Yield (%)	ee (%)
1[a]	Phenyl	Methyl	23	91	93
2	Phenyl	Ethyl	16	74	94
3	3,4-Dichlorophenyl	Methyl	16	92	97
4	Cyclohexyl	Methyl	10	91	96
5[b]	Cyclohexyl	Ethyl	23	95	91
6[c]	MeO_2C-	Methyl	26	83	91
7	$TIPSOCH_2$-	Methyl	72	74	90
8[c,d]	tert-Butyl	Methyl	0.5	95	97

a Reaction performed at –45 °C.
b Reaction performed using 10 mol% catalyst.
c Reaction performed using 5 mol%.
d Reaction performed at –50 °C.

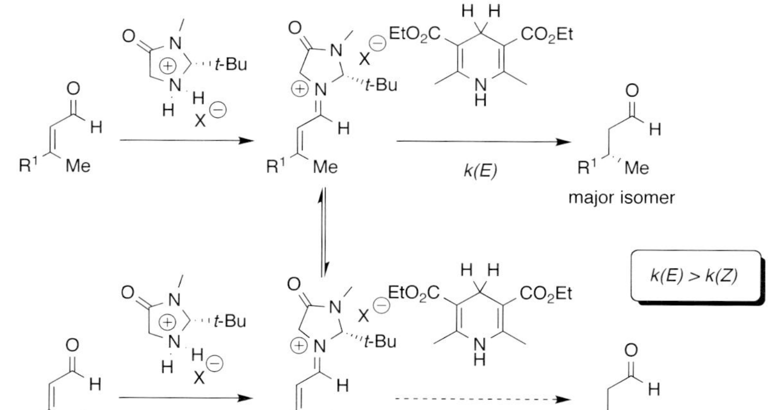

Scheme 14.3 Enantioconvergence in the reduction of E/Z-substrates.

catalyst and the enal. Depending on the geometry of the olefin starting material, E and Z-configured ions are formed. The reason for the excellent stereoconvergence observed in the reaction is explained by a rapid interconversion of the two iminium ions prior to the rate-determining hydride attack from the dihydropyri-

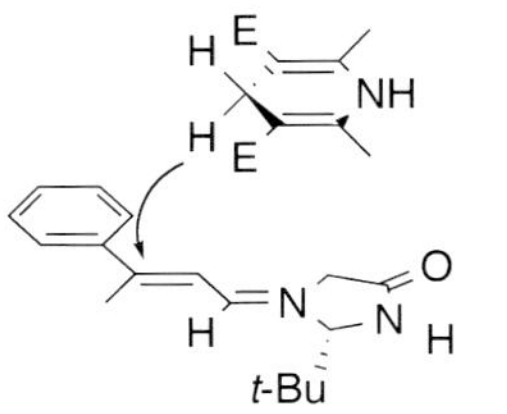

Figure 14.2 Model for the origin of enantiocontrol in the transfer-hydrogenation of enals.

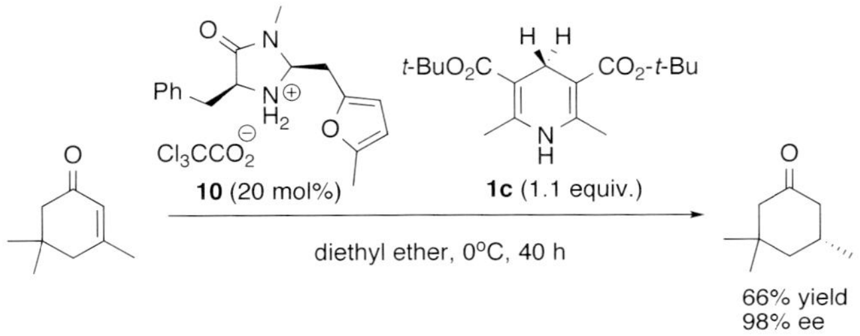

Scheme 14.4 Enantioselective reduction of 3,5,5-trimethylcyclohex-2-enone.

dine. The hydride is then selectively transferred to the *E*-configured olefin from the least sterically hindered face, producing the (*S*)-isomer of the product (Figure 14.2).

In a recent extension of the above-described protocol, MacMillan and coworkers have developed a highly enantioselective method for the reduction of cyclic enones [30]. They investigated a series of previously employed imidazolidinone salts as catalysts for the conjugate reduction of 3-phenyl-2-cyclopentenone with Hantzsch ester **(1a)** as hydride donor and obtained only poor or no conversion to the corresponding cyclopentanone product. However, replacing the organocatalyst with the furyl imidazolidinone **(10)**, a compound previously successfully employed in the organocatalyzed enantioselective Diels–Alder reaction with cyclic enones [31], 3-phenyl-2-cyclopentenone was reduced in high yield and with moderate enantioselectivity. Optimizing the reaction conditions and replacing the Hantzsch ester **(1a)** with the corresponding *t*-butyl ester **(1c)** resulted in a general protocol for the enantioselective reduction of cyclic enones containing various substituents in the ring-systems (Scheme 14.4 and Table 14.3).

The typical organocatalysts used in the chemo- and enantioselective reduction of enals are different imidazolidinone derivatives. However, Zhao and Córdova recently reported that a combination of prolinol derivative **(11)** and benzoic acid (10 mol% of each) works as an efficient catalyst cocktail for the enantioselective reduction of β-aryl-substituted α,β-unsaturated aldehydes [32]. Using the Hantzsch ester **(1a)** as hydride source they obtained the saturated aldehyde products in good yields (58–81%) and in enantioselectivities ranging up to 97% ee (Scheme 14.5). This protocol was later coupled to the Mannich reaction, and enantioselective

Table 14.3 Organocatalytic reduction of β-substituted cyclic enones.

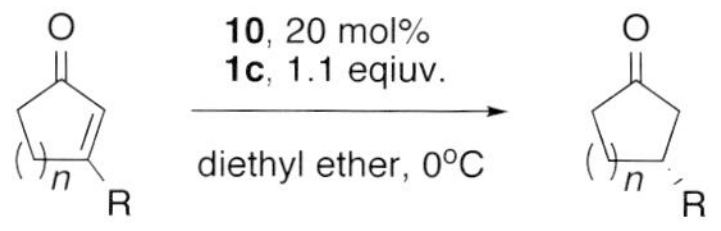

Entry	R	n	Time (h)	Yield (%)	ee (%)
1	Methyl	1	9	72	95
2	tert-Butyl	1	6	81	96
3	Benzyl	1	11	78	90
4[a]	Benzyloxy	1	13	89	91
5	Phenyl	1	9	73	91
6[b]	Cyclohexyl	1	8	85	96
7	COMe	1	1	78	91
8	CO$_2$Me	1	1	83	90
9	Butyl	2	25	82	90
10	Cyclohexyl	2	24	71	88
11	Butyl	3	9	70	92

a Reaction performed with 1.3 equivalents of (**1c**).
b Reaction performed with 1.1 equivalents of the ethyl Hantzsch ester (**1a**).

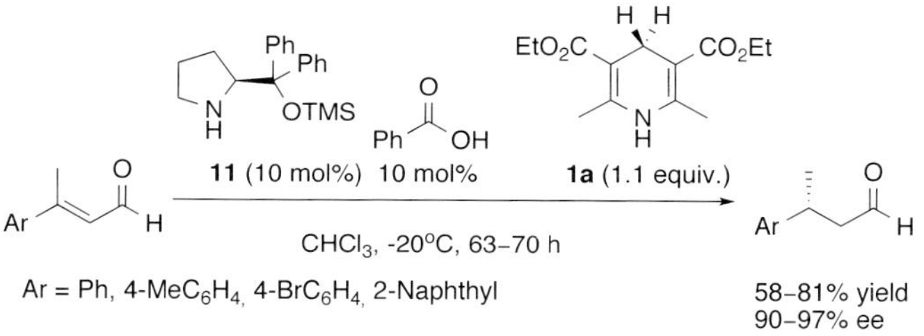

Scheme 14.5 Enantioselective reduction of enals using a prolinol-derived catalyst.

reductive Mannich-type reactions were studied in a one-pot tandem process (see below).

14.3.2
Alkene Reduction in Organocatalytic Tandem Processes

Using a combination of the Knoevenagel condensation and organocatalytic reduction, Ramachary and coworkers have developed an efficient method for the direct formation of the corresponding saturated condensation products [33]. Proline was employed as catalyst for the one-pot three-component reaction between aldehydes, different active methylenes like Meldrum's acid and Hantzsch ester (**1a**), and the products were in general obtained in high yields (Scheme 14.6). The reaction is

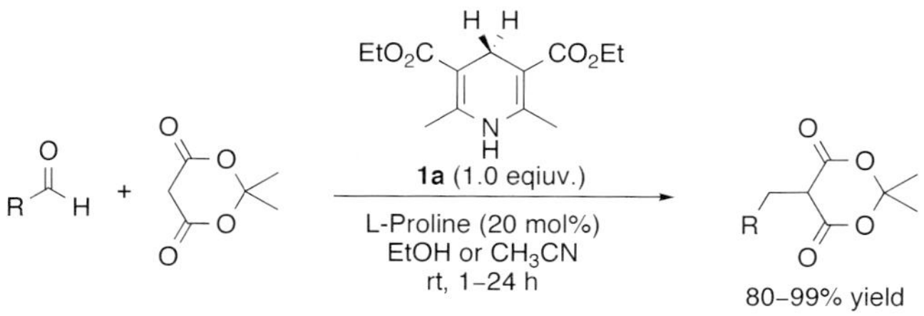

Scheme 14.6 One-pot three-component reaction.

believed to proceed via an initial condensation reaction between the aldehyde and the β-dicarbonyl derivative, which after dehydration is reduced by the Hantzsch ester. Furthermore, the reaction was demonstrated to be mediated by either the Hantzsch ester or proline, although the addition of catalytic amounts of the amino acid resulted in higher reaction rates.

In addition to Meldrum's acid, a number of different active methylenes were employed in the reaction to generate a library of reduced condensation products. A subsequent alkylation process was realized by addition of base (K_2CO_3) and alkyl or allyl halides to the reaction mixture after the initial formation of the reduced condensation products. The Knoevenagel condensation/reduction protocol was later extended to encompass cyclic and noncyclic ketones, which were efficiently reacted with active methylenes, in particular ethyl cyanoacetate, and Hantzsch ester (**1a**) to form the corresponding reduced condensation products [34]. Similar to the aldehyde protocol, *in situ* alkylations of the products formed in the ketone reactions generated the corresponding quaternary carbon compounds. It should be pointed out that despite the use of L-proline as catalyst in the above reactions, no enantiomeric enrichment of the products was reported.

Along the same lines, Ramachary and Reddy reported an interesting one-pot condensation/reduction procedure where an alternative *in situ*-formed hydride donor was introduced. The combination of 1,2-diaminobenzene, ethyl cyanoacetate and two equivalents of an aldehyde resulted in the formation of benzimidazoline (**2**) along with the above-described condensation product [22]. In a subsequent reaction, benzimidazoline (**2**) transfers a hydride to the condensation product to generate the reduced condensation product and a 2-substituted benzimidazole (Scheme 14.7). This development indicates that other hydride donors than the popular Hantzsch esters might dominate the area of future organocatalytic reductions.

Another example of an efficient tandem process involving organocatalytic reduction was introduced by Zhao and Córdova, who combined the prolinol-catalyzed reduction of β-methylcinnamic aldehydes described above with a subsequent Mannich-type reaction [32]. In this protocol, the initial enal-reduction using Hantzsch ester (**1a**) and catalytic amounts of the prolinol derivative (**11**) and benzoic acid is directly followed by the addition of an *N-p*-methoxyphenyl (PMP)-protected ethyl α-iminoglyoxylate which facilitates the Mannich-type coupling (Scheme 14.8). Notably, the opposite diastereomer of the condensation product

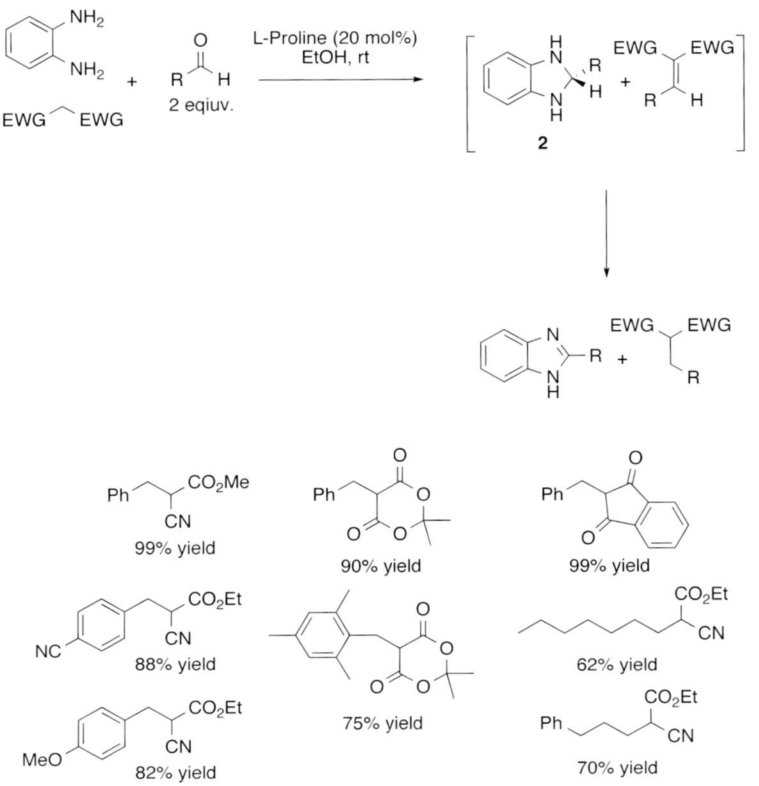

Scheme 14.7 One-pot condensation/reduction synthesis of 2-substituted benzimidazoles. EWG = electron-withdrawing group.

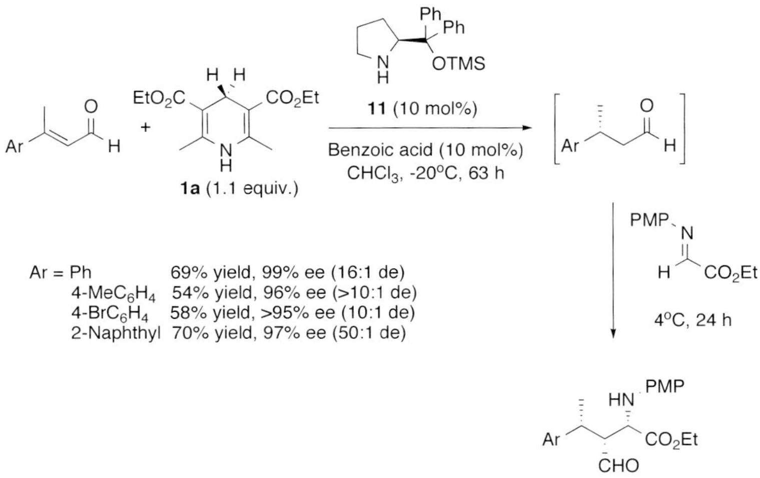

Ar = Ph   69% yield, 99% ee (16:1 de)
4-MeC$_6$H$_4$ 54% yield, 96% ee (>10:1 de)
4-BrC$_6$H$_4$ 58% yield, >95% ee (10:1 de)
2-Naphthyl   70% yield, 97% ee (50:1 de)

Scheme 14.8 Tandem process involving organocatalytic reduction.

was formed in 80% yield and in 96% ee (5:1 dr) when (R)-proline (35 mol%) was added along with the imine in the second reaction step.

14.4
Imine Reductions

14.4.1
Enantioselective Reductions of Ketimines Using Trichlorosilane as Reducing Agent

The reduction of imines represents a most useful route for the preparation of functionalized amino compounds. In comparison to their corresponding carbonyl derivatives, imines usually require more acidic reducing agents since imine activation often is necessary prior to the attack of the hydride. In the area of organocatalyzed reductions, Matsumura and coworkers reported a highly chemoselective imine-reduction employing trichlorosilane as hydride donor and N-formyl pyrrolidines as catalysts [25]. A number of ketimines derived from substituted acetophenones were reduced in good yields in the presence of catalytic amounts of N-formylpyrrolidine (12), although the enantioselectivity in all cases was modest (Scheme 14.9). A slight change in the catalyst structure, replacing the phenyl ring with the 1-naphthyl counterpart in the amide functionality, resulted in a moderate increase of the enantioselectivity (66% ee) in the reduction of N-phenyl acetophenone imine.

Inspired by the above-described findings of Matsumura, the group of Malkov and Kočovský investigated the use of N-formyl L-valine amides as organocatalysts for the reduction of ketimines with trichlorosilane [35]. In a screening process using a number of different valine based catalysts they found that electron-releasing substituents on the aryl ring in the catalyst had a slight positive influence on the enantioselectivity of the reduction reaction (Table 14.4).

When the reaction was performed at −20 °C slightly higher enantioselectivity was obtained. A number of different ketimines derived from acetophenones were reduced using the above protocol and in general enantioselectivities around 90% ee were observed. The reduction of ketimines derived from alkyl alkyl ketones resulted in significantly lower enantioselectivity. The important interactions

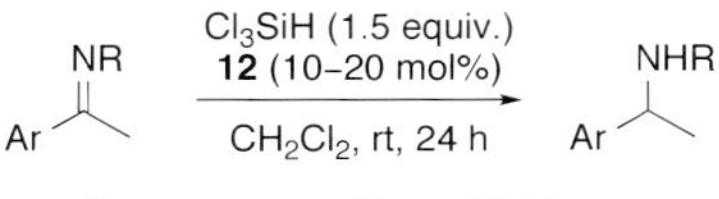

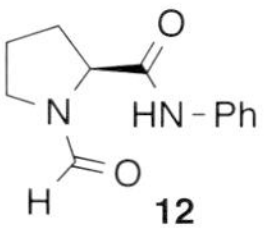

Ar	R	Yield	ee
phenyl	phenyl	91%	55%
phenyl	benzyl	97%	55%
4-nitrophenyl	phenyl	>99%	49%
4-chlorophenyl	phenyl	95%	54%
2-naphthyl	phenyl	56%	49%

Scheme 14.9 Imine reductions.

14 Alkene and Imino Reductions by Organocatalysis

Table 14.4 Enantioselective reduction of imines catalyzed by N-formyl-protected amino acid amides.

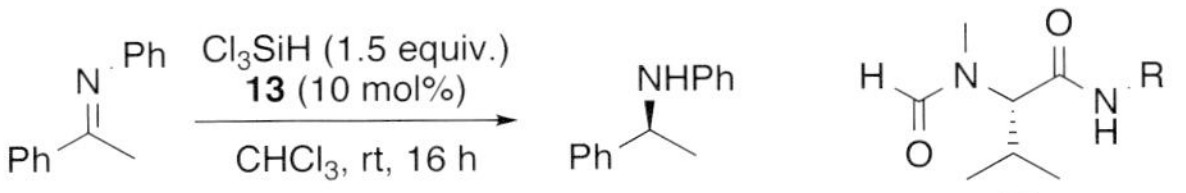

Entry	Catalyst (13), R =	Yield (%)	ee (%)
1	Phenyl	79	86
2	4-Methoxyphenyl	62	85
3	3,5-Dimethoxyphenyl	81	82
4	3,5-Dimethylphenyl	70	89
5	3,5-Di(trifluoromethyl)phenyl	88	53
6	3,5-Dichlorophenyl	35	56

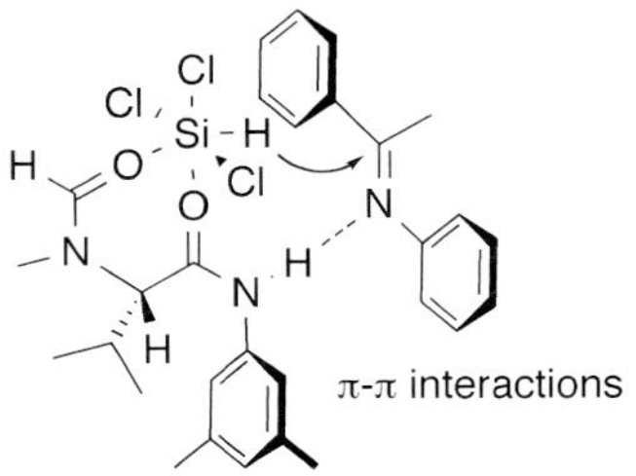

Figure 14.3 Model for the enantioselective reduction of ketones catalyzed by (13).

between the catalyst (3) and the substrate were explained as depicted in Figure 14.3. The amino acid-derived catalyst coordinates to the reagent in a bidentate fashion through the two amide carbonyls. This interaction activates the silane for a subsequent hydride transfer. The enantiocontrol is believed to arise from a combination of hydrogen-bonding interaction between the catalyst and the ketimine nitrogen, and through π–π interactions between the aryl-groups of the catalyst and the substrate. It should be pointed out that amino acid amides lacking an aryl substituent were unable to catalyze the reaction.

The group of Malkov and Kočovský extended this study to encompass other amino acids than valine; however, from a catalytic point of view the original catalysts with the general structure (13) appear to be among the best [36]. A general trend regarding the key features of the catalyst and the substrate for optimum selectivity and activity was found and is depicted in Figure 14.4.

In a recent extension to the above concept, Kočovský and coworkers have studied the effect of introducing fluorous tags on the valine-based catalysts [37]. They found that asymmetric reduction of ketimines with trichlorosilane was readily performed using these catalysts and the products were formed with high enanti-

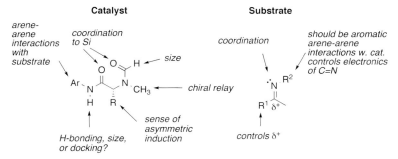

Figure 14.4 Key prerequisites of the catalyst and the substrate in the reduction of ketimines using amino acid amides and trichlorosilane **(3)**.

oselectivity (up to 92% ee). The fluorous tag was attached to the *N*-phenyl substituent on the valine amide and this catalyst modification efficiently simplified product isolation and catalyst recovery.

On the basis of the catalyst concept introduced by the Matsumura and Kočovský groups, Sun and coworkers have examined the catalytic activity of *N*-formyl derivatives of pipecolinic acid amides [38]. They found that catalysts based on pipecolinic acid exhibited significantly better reactivity and selectivity than the corresponding proline derivatives. Screening a set of different *N*-formyl amides revealed that catalyst **(14)** displayed high catalytic activity and selectivity in the trichlorosilane reduction of *N*-phenylacetophenone ketimine. The scope of the reaction was examined and they found that catalyst **(14)** showed high tolerance towards different functionalized ketimines (Table 14.5). As can be seen in Table 14.5, in comparison to the organocatalysts developed by Matsumura and Kočovský, the Sun pipecholinic acid derivatives display better activity and selectivity for the ketimine reduction.

In a subsequent study, the Sun group discovered that *S*-chiral sulfinamides act as efficient and enantioselective organocatalysts for the trichlorosilane reduction of ketimines [39]. Screening a number of different sulfinamides, based on the commercially available Ellman's chiral auxiliary [40], and with the general structure of **(15)**, they found that catalyst **(15a)** showed highest efficacy in the reduction of *N*-phenylacetophenone ketimine (Equation 14.1).

$$\underset{\textbf{15}}{t\text{-Bu}\overset{\overset{\text{O}}{\|}}{\underset{}{\text{S}}}\underset{\text{H}}{\text{N}}{}^{R}} \qquad \underset{\textbf{15a}}{t\text{-Bu}\overset{\overset{\text{O}}{\|}}{\underset{}{\text{S}}}\underset{\text{H}}{\text{N}}\diagdown\underset{\text{F}}{\bigcirc}\text{-OH}} \qquad (14.1)$$

Sulfinamide **(15a)** was thereafter employed as organocatalyst in the reduction of a whole range of differently substituted aryl alkyl ketimines and the corresponding amine products were obtained in yields ranging from 78% to 98% and with enantioselectivities from 74% to 93% ee.

Table 14.5 Asymmetric reduction of ketimines catalyzed by (14).

Entry	R^1	R^2	Yield (%)	ee (%)
1	Phenyl	Phenyl	97	95
2	4-Methoxyphenyl	Phenyl	95	93
3	4-Nitrophenyl	Phenyl	96	95
4	4-Trifluoromethylphenyl	Phenyl	85	96
5	2-Naphthyl	Phenyl	92	93
6	Phenyl	4-Methoxyphenyl	98	92
7	Phenyl	4-Methylphenyl	90	95
8	Phenyl	4-Chlorophenyl	98	93
9[a]	Cyclohexyl	Phenyl	81	95
10[b]	Isopropyl	Phenyl	86	91

a The ketimine was added as a 10/1 E/Z-mixture.
b The ketimine was added as a 6/1 E/Z-mixture.

14.4.2
Enantioselective Reductions of Ketimines Using Hantzsch Esters as Reducing Agents

An alternative approach for the organocatalyzed reduction of ketimines uses Hantzsch esters as hydride donors. An early example of this type of reaction was described in 1989, when Singh and Batra reported that chiral organic Brønsted acids mediated the reduction process [41, 42]. They found that α-amino acid hydrochlorides, for example L-cysteine-HCl and L-serine-HCl, mediated the reduction of N-phenylacetophenone ketimine in moderate to good yields and with selectivity up to 62% ee. In addition to amino acid salts, they investigated other chiral organic acids such as camphorsulfonic acid and tartaric acid as catalysts for the reaction, but obtained lower enantioselectivity of the formed N-phenyl-α-phenethylamine.

More recently Rueping and coworkers rediscovered this process and reported that Brønsted acids (for example diphenyl phosphate) were able to catalyze the reduction of ketimines with a Hantzsch dihydropyridine as hydrogen source [43]. When diphenyl phosphate was replaced by the chiral phosphoric acid (16), a chiral Brønsted acid developed by Akiyama and coworkers [44], the reduction of ketimines proceeded to yield the chiral amine in modest yields and enantioselectivity (up to 62%). After optimization of the reaction conditions, the group of Rueping managed to increase the selectivity of the reaction, but in most cases examined, the enantiomeric excess never reached higher than 80% (Scheme 14.10). This

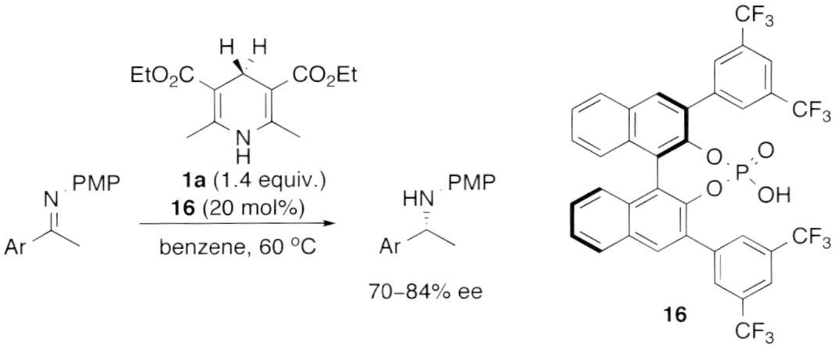

14.4 Imine Reductions

Scheme 14.10 Chiral Brønsted (16) reduction of ketimines.

Table 14.6 Asymmetric ketimine reduction catalyzed by phosphoric acid (17).

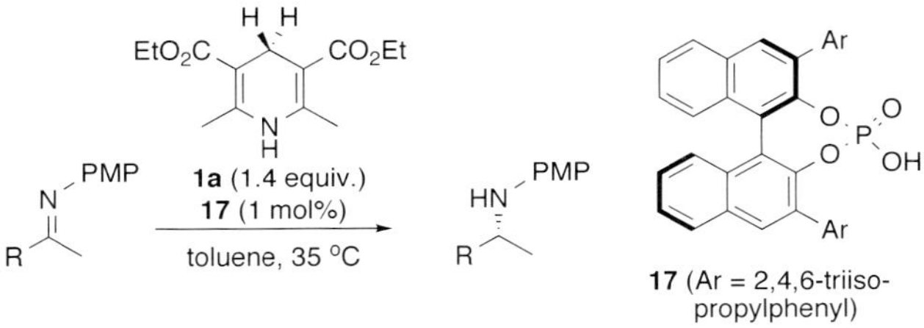

Entry	R	Time (h)	Yield (%)	ee (%)
1	Phenyl	45	96	88
2	4-Cyanophenyl	42	87	80
3	4-Nitrophenyl	42	96	80
4	4-Methylphenyl	42	98	88
5	2-Methylphenyl	71	91	93
6	2,4-Dimethylphenyl	71	88	92
7	2-Naphthyl	42	85	84
8	Isopropyl	60	80	90

Brønsted acid-catalyzed reduction process is believed to be initiated by protonation of the ketimine by the chiral acid, whereafter a hydride transfer occurs from the Hantzsch ester to the formed iminium ion. The chiral anion formed after imine protonation is believed to provide the asymmetric environment necessary for chiral induction in the addition step.

Almost simultaneously with the work of Rueping, List and coworkers reported an imine reduction system employing the same Akiyama-type of phosphoric acids [45]. The protocol developed by List differed only slightly from the system presented by Rueping, but List's group screened a larger number of acids and found that compound (17) was a superior catalyst for the ketimine reduction. Using only 1 mol% of this acid catalyst, a series of aryl alkyl ketimines were reduced in good yields and in enantioselectivities between 80 and 93% ee (Table 14.6).

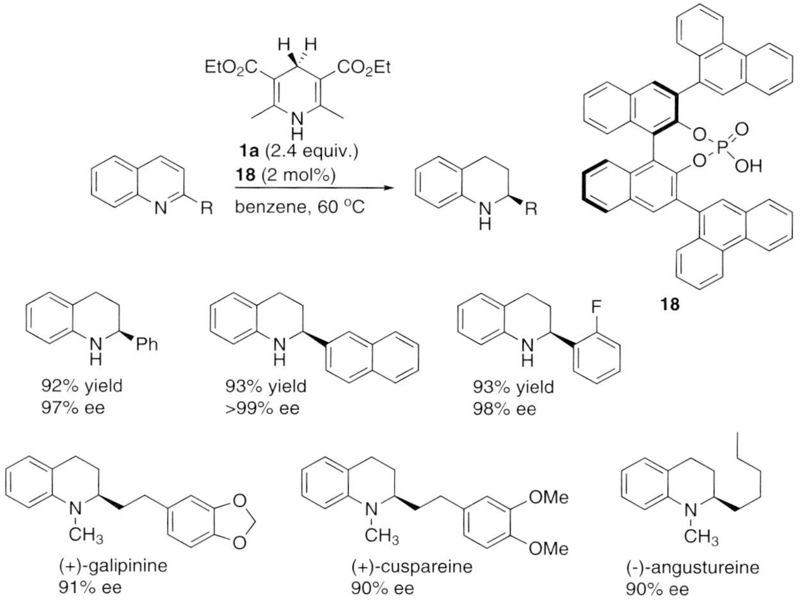

Scheme 14.11 Partial reduction of quinolines.

Another highly interesting organocatalyzed imine reduction was recently presented by the group of Rueping. Based on the same concept as presented above, they demonstrated that Brønsted acids could be employed as catalysts for the partial reduction of quinolines under transfer hydrogenation conditions (Scheme 14.11) [46]. A wide range of substituted quinolines were reduced under mild conditions and in excellent yields to the corresponding tetrahydroquinolines using 1 mol% diphenyl phosphate and the standard Hantzsch diethyl ester. When the reaction was performed using the Akiyama-type phosphoric acid (**18**) as catalyst, excellent enantioselectivity (90% to >99% ee) was obtained for a great number of quinoline derivatives (Scheme 14.11) [47]. In addition, the protocol was employed for the preparation of the biologically active tetrahydroquinoline alkaloids galipinine, cusparein and angustureine. In a two-step procedure (asymmetric reduction followed by reductive methylation) the natural products were formed and isolated in good yields (79–89%) with high enantiomeric excess (90–91% ee).

Mechanistically the reduction of quinolines is proposed to be initiated by protonation of the substrate, followed by a hydride transfer from the Hantzsch ester to the 4-position of the heterocycle, leading to an intermediate enamine. Further protonation of the enamine and a subsequent transfer of the second hydride will facilitate the formation of the tetrahydroquinoline skeleton.

14.4.3
Organocatalytic Reductive Amination of Aldehydes and Ketones

The general theme in the above-described ketimine reductions is that only preformed imine substrates were employed as starting material. However, the reductive amination protocol, wherein primary amines and aldehydes or ketones are joined under reductive conditions, is a widely used and most powerful method for the direct formation of substituted amines [48]. This process takes advantage of the often swift *in situ* imine formation which is favored under slightly acidic conditions if the formed water is removed efficiently. Using the concept that an acidic catalyst with the possibility to take part in H-bonding can facilitate the *in situ* imine formation and subsequently direct an incoming nucleophile, the MacMillan group presented an excellent study in which they demonstrated that highly enantioselective reductive aminations can be performed efficiently [49]. In a screening process using various known hydrogen-bonding catalysts, MacMillan and coworkers examined the reaction between acetophenone, *p*-anisidine and the ethyl Hantzsch ester, and found that the binol-derived phosphoric acids introduced by Akiyama were superior catalysts. Further elaboration of the binaphthyl structure led to compound **(19)**, which showed high catalytic activity and excellent enantiocontrol in the reductive coupling reaction. The removal of water liberated in the imine formation step proved to be most important for achieving high catalytic activity and selectivity, and the solution was to add 5 Å molecular sieves to the reaction mixture. With this protocol in hand, the scope of the reaction was investigated using catalyst **(19)** and they found that a huge number of different ketones and aromatic amines could be merged together efficiently to form secondary amines with high enantiomeric excess (Table 14.7 and Figure 14.5).

The enantioselective reductive amination of ketones provides a direct route into chiral α-branched amines. The corresponding reaction using aldehydes as substrates simply led to achiral compounds; however, if enolizable α-branched aldehydes are used in the process, kinetic resolution can be achieved and provides a direct route to enantiomerically enriched β-branched amines. Such a protocol was developed by List and coworkers, who demonstrated that α-branched aldehydes combined with *p*-anisidine underwent reductive amination in the presence of Hantzsch esters and phosphoric acid catalysts [50]. Screening a number of different Akiyama-type phosphoric acids, including the triphenylsilyl derivative introduced by MacMillan, the List group found that the 2,4,6-triisopropylphenyl derivative was a superior catalyst for the enantioselective reductive amination of aldehydes. A highly beneficial feature of this system is that the acid catalyst facilitated rapid racemization of the starting aldehyde, which allowed a *dynamic* kinetic resolution process to occur. The reaction turned out to be rather general and a series of α-branched aldehydes were subjected to the reduction protocol with good to excellent results (Scheme 14.12).

In a similar albeit achiral protocol, Menche and Arikan have demonstrated that thiourea acts as an efficient organocatalyst for the reductive amination of

Table 14.7 Enantioselective reductive amination of ketones.

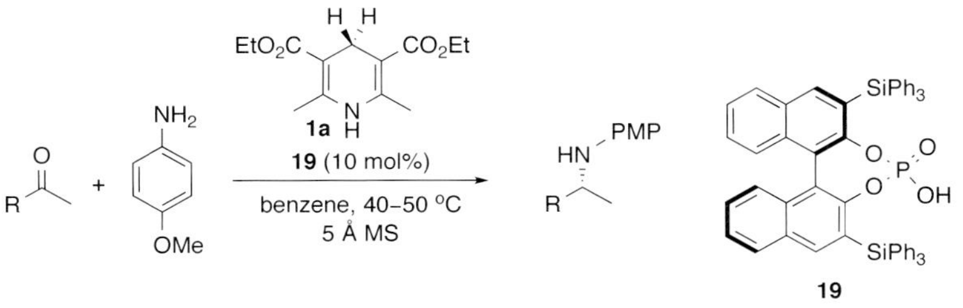

Entry	R⁽¹⁾	Time (h)	Yield (%)	ee (%)
1	Phenyl	24	87	94
2	4-Chlorophenyl	72	75	95
3	4-Nitrophenyl	72	71	95
4	3-Fluorophenyl	72	81	95
5	2-Naphthyl	72	73	96
6	Hexyl	96	72	91
7	2-Phenylethyl	72	75	94
8	Cyclohexyl	96	49	86

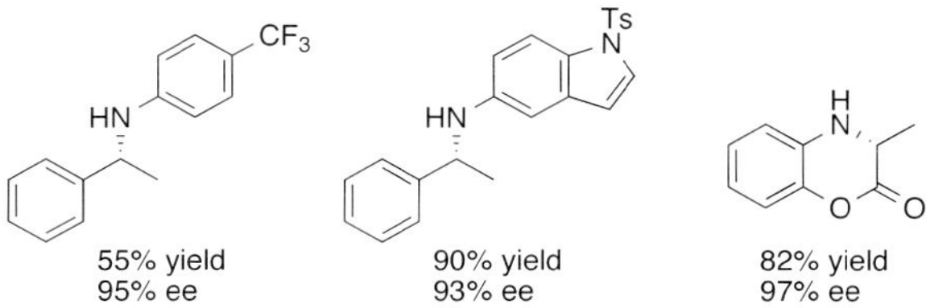

Figure 14.5 Other secondary imines.

aldehydes using aromatic amines and the ethyl Hantzsch ester [51]. Using the conditions depicted in Scheme 14.13, benzaldehyde was reacted with different anilines to yield the corresponding N-benzylanilines in yields ranging from 72% to 93%. Substituted benzaldehydes as well as two aliphatic aldehydes were reacted with p-anisidine and the secondary amines formed were isolated in good yields.

14.5
Concluding Remarks

Whereas transition metal-catalyzed reduction of organic compounds is an area that has been examined for several decades and thus has had a chance to mature, the corresponding organocatalytic reductions are still in the early years. We have

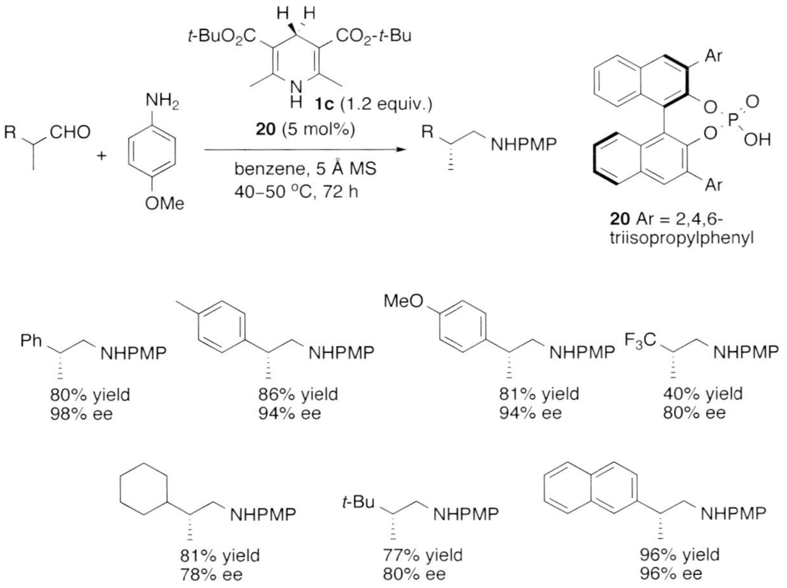

Scheme 14.12 Dynamic kinetic resolution.

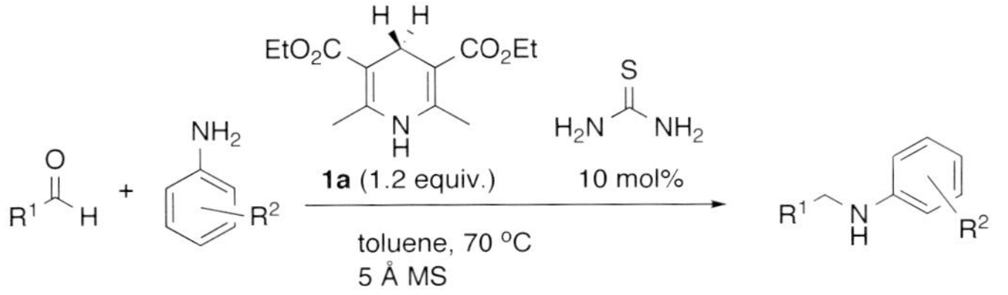

Scheme 14.13 Thiourea-catalyzed reductive amination of aldehydes.

witnessed enormous progress in the field of organocatalysis in recent years and the contributions in the area of reductions presented in this chapter are probably only the beginning. Regarding the reduction of alkenes, so far only substrates activated by conjugation to aldehydes or ketones are able to undergo organocatalytic reductions. These substrates react readily under iminium activation conditions, but other α,β-unsaturated compounds such as esters or amides cannot be activated in this fashion. The introduction of catalysts which do not require covalent interaction with the substrate, for example chiral acids, phase-transfer agents or H-bonding catalysts, has broadened the scope of reactions mediated by organocatalysts. It can therefore be foreseen that such compounds will be employed as catalysts in future alkene reduction using organocatalysis. The choice of reducing agents is currently limited to either Hantzsch dihydropyridines or to trichlorosilane. We can expect that new reagents with suitable properties for organocatalytic reductions will be developed and introduced in the future.

References

1 Blaser, H.-U., Malan, C., Pugin, B., Spindler, F., Steiner, H. and Studer, M. (2003) *Advanced Synthesis and Catalysis*, **345**, 103.
2 Noyori, R., Kitamura, M. and Ohkuma, T. (2004) *Proceedings of the National Academy of Sciences of the United States of America*, **101**, 5356.
3 Knowles, W.S. (2002) *Angewandte Chemie*, **114**, 2096; (2002) *Angewandte Chemie (International Ed. in English)*, **41**, 1998.
4 Noyori, R. (2002) *Angewandte Chemie*, **114**, 2108; (2002) *Angewandte Chemie (International Ed. in English)*, **41**, 2008.
5 Lightfoot, A., Schnider, P. and Pfaltz, A. (1998) *Angewandte Chemie*, **110**, 3047; (1998) *Angewandte Chemie (International Ed. in English)*, **37**, 2897.
6 Bell, S., Wuestenberg, B., Kaiser, S., Menges, F., Netscher, T. and Pfaltz, A. (2006) *Science*, **311**, 642.
7 Källström, K., Hedberg, C., Brandt, P., Bayer, A. and Andersson, P.G. (2004) *Journal of the American Chemical Society*, **126**, 14308.
8 Hedberg, C., Källström, K., Brandt, P., Hansen, L.K. and Andersson, P.G. (2006) *Journal of the American Chemical Society*, **128**, 2995.
9 Zassinovich, G., Mestroni, G. and Gladiali, S. (1992) *Chemical Reviews*, **92**, 1051.
10 Gladiali, S. and Alberico, E. (2006) *Chemical Society Reviews*, **35**, 226.
11 Palmer, M.J. and Wills, M. (1999) *Tetrahedron: Asymmetry*, **10**, 2045.
12 Noyori, R. and Hashiguchi, S. (1997) *Accounts of Chemical Research*, **30**, 97.
13 Västilä, P., Zaitsev, A.B., Wettergren, J., Privalov, T. and Adolfsson, H. (2006) *Chemistry – A European Journal*, **12**, 3218.
14 List, B., Lerner, R.A. and Barbas, C.F., III (2000) *Journal of the American Chemical Society*, **122**, 2395.
15 Hajos, Z.G. and Parrish, D.R. (1974) *Journal of Organic Chemistry*, **39**, 1615.
16 Dalko, P. and Moisan, L. (2001) *Angewandte Chemie (International Ed. in English)*, **40**, 3726.
17 Dalko, P. and Moisan, L. (2004) *Angewandte Chemie (International Ed. in English)*, **43**, 5138.
18 Berkessel, A. and Gröger, H. (2005) *Asymmetric Organocatalysis. From Biomimetic Concepts to Applications in Asymmetric Synthesis*, Wiley-VCH Verlag GmbH, Weinheim.
19 Westheimer, F.H. (1987) *Mechanism of action of the pyridine nucleotides*, in *Pyridine Nucleotide Coenzyme, Part A* (eds D. Dolphin, R. Poulson and O. Avramovic), Wiley-Interscience, New York, p. 253.
20 Murakami, Y., Kikuchi, J.-L., Hisaeda, Y. and Hayashida, O. (1996) *Chemical Reviews*, **96**, 721.
21 Yang, J.W., Hechavarria Fonseca, M.T. and List, B. (2004) *Angewandte Chemie (International Ed. in English)*, **43**, 6660.
22 Ramachary, D.B. and Reddy, G.B. (2006) *Organic and Biomolecular Chemistry*, **4**, 4463.
23 Benkeser, R.A. and Snyder, D.C. (1982) *Journal of Organometallic Chemistry*, **225**, 107.
24 Kobayashi, S., Yasuda, M. and Hachiya, I. (1996) *Chemistry Letters*, 407.
25 Iwasaki, F., Onomura, O., Mishima, K., Kanematsu, T., Maki, T. and Matsumura, Y. (2001) *Tetrahedron Letters*, **42**, 2525.
26 For ketone reduction, see: Iwasaki, F., Onomura, O., Mishima, K., Maki, T. and Matsumura, Y. (1999) *Tetrahedron Letters*, **40**, 7507.
27 Ahrendt, K.A., Borths, C.J. and MacMillan, D.W.C. (2000) *Journal of the American Chemical Society*, **122**, 4243.
28 Ouellet, S.G., Tuttle, J.B. and MacMillan, D.W.C. (2005) *Journal of the American Chemical Society*, **127**, 32.
29 Yang, J.W., Hechavarria Fonseca, M.T. and List, B. (2005) *Angewandte Chemie (International Ed. in English)*, **44**, 108.
30 Tuttle, J.B., Oullet, S.G. and MacMillan, D.W.C. (2006) *Journal of the American Chemical Society*, **128**, 12662.
31 Northrup, A.B. and MacMillan, D.W.C. (2002) *Journal of the American Chemical Society*, **124**, 7894.
32 Zhao, G.-L. and Córdova, A. (2006) *Tetrahedron Letters*, **47**, 7417.

33 Ramachary, D.B., Kishor, M. and Ramakumar, K. (2006) *Tetrahedron Letters*, **47**, 651.
34 Ramachary, D.B., Kishor, M. and Reddy, G.B. (2006) *Organic and Biomolecular Chemistry*, **4**, 1641.
35 Malkov, A.V., Mariani, A., MacDougall, K.N. and Kočovský, P. (2004) *Organic Letters*, **6**, 2253.
36 Malkov, A.V., Stončius, S., MacDougall, K.N., Mariani, A., McGeoch, G.D. and Kočovský, P. (2006) *Tetrahedron*, **62**, 264.
37 Malkov, A.V., Figlus, M., Stončius, S. and Kočovský, P. (2007) *Journal of Organic Chemistry*, **72**, 1315.
38 Wang, Z., Ye, X., Wei, S., Wu, P., Zhang, A. and Sun, J. (2006) *Organic Letters*, **8**, 999.
39 Pei, D., Wang, Z., Wei, S., Zhang, Y. and Sun, J. (2006) *Organic Letters*, **8**, 5913.
40 Ellman, J.A., Owens, T.D. and Tang, T.P. (2002) *Accounts of Chemical Research*, **35**, 984.
41 For a recent review on chiral Brønsted acid catalysis, see: Bolm, C., Rantanen, T., Schiffers, I. and Zani, L. (2005) *Angewandte Chemie (International Ed. in English)*, **44**, 1758.
42 Singh, S. and Batra, U.K. (1989) *Indian Journal of Chemistry, Section B: Organic Chemistry Including Medicinal Chemistry*, **28**, 1.
43 Rueping, M., Sugiono, E., Azap, C., Theissmann, T. and Bolte, M. (2005) *Organic Letters*, **7**, 3781.
44 Akiyama, T., Itoh, J., Yokota, K. and Fuchibe, K. (2004) *Angewandte Chemie (International Ed. in English)*, **43**, 1566.
45 Hoffmann, S., Seayad, A.M. and List, B. (2005) *Angewandte Chemie (International Ed. in English)*, **44**, 7424.
46 Rueping, M., Theissmann, T. and Antonchick, A.P. (2006) *Synlett*, 1071.
47 Rueping, M., Antonchick, A.P. and Theissmann, T. (2006) *Angewandte Chemie (International Ed. in English)*, **45**, 3683.
48 Tararov, V.I. and Börner, A. (2005) *Synlett*, 203.
49 Storer, R.I., Carrera, D.E., Ni, Y. and MacMillan, D.W.C. (2006) *Journal of the American Chemical Society*, **128**, 84.
50 Hoffmann, S., Nicoletti, M. and List, B. (2006) *Journal of the American Chemical Society*, **128**, 13074.
51 Menche, D. and Arikan, F. (2006) *Synlett*, 841.

15
Alkyne Reductions
Ian J. Munslow

15.1
Introduction

Alkynes are highly versatile building blocks for organic chemists due to the ease with which they can be modified. The selective semi-hydrogenation of alkynes is an important process for the fine chemical industry, manufacturing of insect sex pheromones and vitamins [1], and for industrial polymerization processes, where removal of alkynes from alkene feedstocks is crucial [2]. The hydrometallation of alkynes is perhaps the most straightforward method for the production of vinyl-metal intermediates. Vinylmetal reagents are important organic intermediates that can function in oxidation and reduction reactions as nucleophiles or in Pd-catalyzed cross-coupling reactions.

This chapter covers three methods for alkyne reductions: hydrogenation [3, 4], hydroboration [5] and hydrosilylation [6]. For the hydroalumination and hydromagnesiation of alkynes, the reader is directed to the relevant sections in Chapter 18. Their reduction with dissolved metals is discussed in Chapter 17. Also of interest, but not discussed here, are the hydrometallation reactions including hydrostannation [7], hydrogermylation [8] and hydroselenation [9].

15.2
Hydrogenation

The addition of molecular hydrogen (H_2) to carbon–carbon multiple bonds has been studied extensively and is a technique frequently used by chemists [10]. The partial reduction of alkynes gives alkenes, whereas total reduction gives alkanes. Total reduction can be simply achieved by catalytic hydrogenation. Alkanes are easily obtained with palladium [11], Raney nickel [12] or Adams' platinum oxide [13] under low pressures of hydrogen. A number of alkynes have been fully reduced with the complexes [RhCl(PPh$_3$)$_3$] and [Ru(OAc)(PPh$_3$)$_3$] [14]. The electron-rich titanium complexes [Cp$_2$Ti(PhC≡CPh)(PMe$_3$)], [(MeCp)$_2$Ti(PhC≡CPh)(PMe$_3$)]

Modern Reduction Methods. Edited by Pher G. Andersson and Ian J. Munslow
Copyright © 2008 WILEY-VCH Verlag GmbH & Co. KGaA, Weinheim
ISBN: 978-3-527-31862-9

and [Cp(C$_5$Me$_5$)Ti(PhC≡CPh)], have also been reported to be catalyst precursors for the hydrogenation of alkynes to alkanes under an atmospheric pressure of hydrogen [15], though in the presence of excess PMe$_3$ only semi-hydrogenation occurs and *cis*-olefins are selectively obtained.

15.2.1
Semi-hydrogenation

The semi-hydrogenation of alkynes is an important process in organic chemistry and can be carried out using a number of catalysts both heterogeneously [3] and homogeneously [4], as well as by noncatalytic chemical methods [16] (for example with diimide [17], dissolved metals [18], or hydroalumination [19]).

Conversion to (Z)-alkenes has traditionally been achieved via heterogeneous catalyst systems. Of these, the Lindlar catalyst [20], nickel boride [21], dispersed nickel on graphite [22] and palladium immobilized on clay or borohydride exchange resin [23] are perhaps the most efficient and well known. Partial isomerization of the (Z)-alkene to the (E)-alkene, double-bond shift and over-reduction to the alkane are notable issues, especially with the Lindlar catalyst [3]. However, a general and selective method of semi-hydrogenation to (E)-alkenes remains a challenging problem for chemists.

Although Lindlar's catalyst has successfully been used in the semi-hydrogenations of alkynes for a wide range of functionalities [24], substrates containing amines are generally protected beforehand [25]. Campos *et al.* reported that when using Lindlar's catalyst in the semi-hydrogenation of alkynes possessing a primary amine, over-reduction becomes a problem (Table 15.1, entry 1) [26]. However, they reported that the addition of one equivalent of ethylenediamine

Table 15.1 Amino alkyne reductions with Lindlar's catalyst.

Entry	Additive	Product distribution (%)		
		(Z)-alkene	(E)-alkene	Alkane
1	None	0	0	100
2	Pyridine	70	20	10
3	Quinoline	80	13	7
4	Ethylenediamine	97.6	2.0	0.4
5	Propylenediamine	96.7	2.5	0.7

15.2 Hydrogenation

efficiently affords the semi-hydrogenation of amino alkynes with minimal E/Z isomerization and over-reduction (entry 4).

The recent use of *in situ*-generated Ni(0) nanoparticles and molecular hydrogen has been reported by Yus and coworkers as a highly efficient method for the formation of *cis*-alkenes (Table 15.2) [27].

The Ni(0) nanoparticles were generated from the reducing system $NiCl_2 \cdot Li \cdot DTBB(cat.) \cdot ROH$ (DTBB = 4,4'-di-*t*-butylbiphenyl), where $NiCl_2$ acts as the source of the Ni(0) nanoparticles, lithium reduces the Ni(II) to Ni(0) and generates H_2 *in situ* via reaction with the alcohol (ROH). A catalytic amount of DTBB acts as an electron carrier from the Li to $NiCl_2$. The system was found to reduce both internal and terminal alkynes in high yields (62–99%) and selectivities, while tolerating a variety of functional groups. Some over-reduction (10–15%) was observed with terminal alkynes (entries 6 and 7).

Only a few homogeneous catalyst systems are known that show good selectivity for a wide range of substrates, namely palladium [28], rhodium [29] and $Cr(CO)_3$ catalysts [30]. Recently, the homogeneous Pd(0) complex **(1)** (Figure 15.1), reported by van Laren and Elsevier, has shown good selectivity in the formation of (*Z*)-alkenes as well as tolerating a variety of functional groups (esters, carboxylic and nitro groups) [31]. Only small amounts of (*E*)-alkenes (<1–6%) and over-reduction to alkanes (<1–13%) were observed in a number of cases.

However, the hydrogenation of alkynes to (*E*)-alkenes has been reported for only a small number of complexes. The reduction of acetylenedicarboxylic acid dimethyl ester and diphenylacetylene to their respective *trans*-alkenes has been reported with $[RhH_2(O_2COH)(P\text{-}i\text{-}Pr_3)_2]$ [32]. This complex was also shown to facilitate the isomerization of *cis*-stilbene to *trans*-stilbene, though the process was around eight

Table 15.2 Alkyne semi-hydrogenation with Ni(0) nanoparticles

Entry	Alkyne	Product	Hydrogen source	Yield (%)
1	*n*-Pr—≡—*n*-Pr	*n*-Pr / *n*-Pr	EtOH	99
2	Et—≡—OH	Et / OH	EtOH	83
3	*n*-Hex—≡—OMe	*n*-Hex / OMe	EtOH	85
4	*n*-Hex—≡—NEt₂	*n*-Hex / NEt₂	EtOH	90
5	*n*-Hex—≡	*n*-Hex /	*i*-PrOH	85
6	BnO—≡	BnO /	*i*-PrOH	74
7	Ph-NH-≡	Ph-NH- /	*i*-PrOH	79

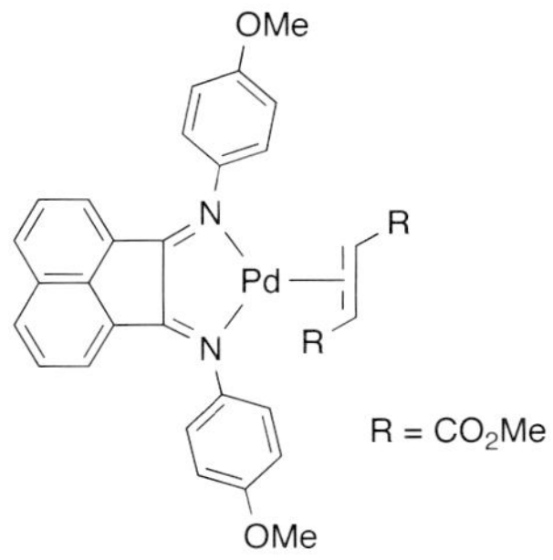

Figure 15.1 Homogeneous Pd(0) complex (**1**).

times slower than for the hydrogenation of alkynes. *trans*-Stilbene was also the initial major product, leading the authors to postulate that the *E/Z* isomerization was achieved mainly through a dipolar intermediate (M$^+$=CR–$^-$CHR).

Dialkyl and diaryl alkynes have been hydrogenated to (*E*)-alkanes with the dinuclear catalyst precursor [(μ-H)Rh{P(O-*i*-Pr$_3$)$_2$}$_2$]$_2$ [33], but the system is limited by a slow reaction rate (approximately 1 turnover min^{-1} at 20 °C) and very short catalyst lifetime (approximately 5 minutes under catalytic conditions). Bargon and coworkers have reported that [Cp*Ru(η^4-CH$_3$CH=CHCH=CHCO$_2$H)][CF$_3$SO$_3^-$] also reduces internal alkynes to (*E*)-alkenes [34]. However, no activity was observed with terminal alkynes, possibly due to the formation of stable vinylidene complexes. A mechanism proceeding via a binuclear complex was postulated.

15.3
Hydroboration

The hydroboration of carbon–carbon multiple bonds allows for the practical synthesis of organoboranes, which are highly versatile organometallic reagents for organic synthesis [5]. For this reason, hydroboration has been the focus of much attention and a large range of hydroborating agents has been developed (Figure 15.2) and has been reviewed at length [5a, 35]. Although the reaction usually requires no catalyst, when the hydroborating agent is bound to heteroatoms the uncatalyzed reaction requires elevated temperatures. In the mid-1980s, the discovery that some transition metal complexes facilitated the reaction at room temperature (RT) and affected chemo-, regio- and stereoselectivity opened up a new field of chemistry [35].

Reactions between borane·THF (BH$_3$·THF) and terminal alkynes generally result in complicated mixtures of products [5a, 36]. The equivalent reaction with internal alkynes is more controlled, and reasonable yields of the corresponding *cis*-trialkenylboranes can be obtained (Scheme 15.1) [37]. However, unsymmetrical internal alkynes react with only modest regioselectivity [38].

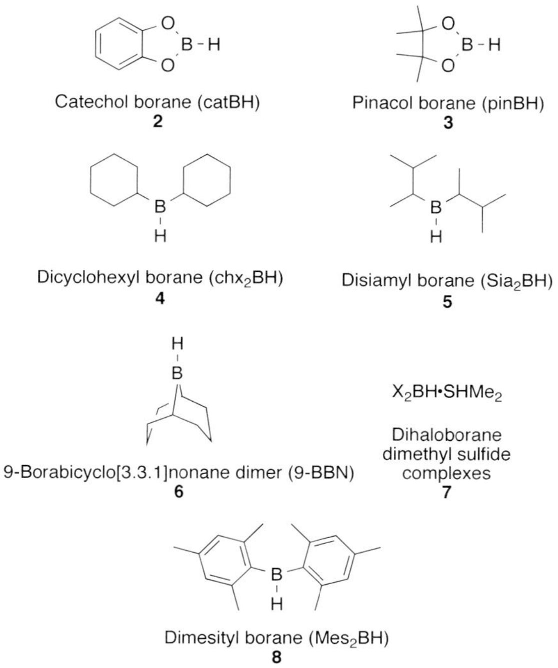

15.3 Hydroboration

Catechol borane (catBH)
2

Pinacol borane (pinBH)
3

Dicyclohexyl borane (chx$_2$BH)
4

Disiamyl borane (Sia$_2$BH)
5

9-Borabicyclo[3.3.1]nonane dimer (9-BBN)
6

X$_2$BH·SHMe$_2$

Dihaloborane
dimethyl sulfide
complexes
7

Dimesityl borane (Mes$_2$BH)
8

Figure 15.2 Borane reagents.

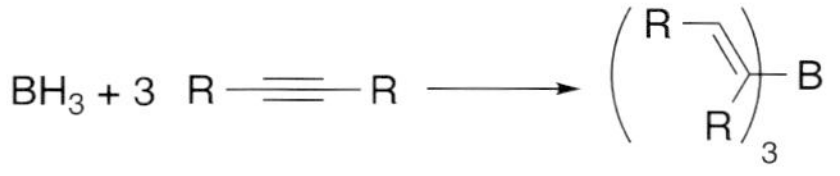

$$BH_3 + 3\ R\text{—}\!\!\equiv\!\!\text{—}R \longrightarrow \left(\begin{array}{c} R \\ \\ R \end{array}\!\!=\!\!\begin{array}{c} \\ B \\ \end{array}\right)_3$$

9

Scheme 15.1 Borane reaction with internal alkynes.

Both steric and electronic factors affect the regioselectivity of monohydroboration (Figure 15.3). Whereas hydroboration with 9-BBN is more susceptible to electronic factors and less to steric ones than Sia$_2$BH (disiamyl borane) [39], steric effects dominate the reactivity of highly hindered disubstituted boranes such as Mes$_2$BH. Indeed, the regioselectivity with Mes$_2$BH in the hydroboration of unsymmetrical alkynes is greater than that of any other known borane reagent (Figure 15.3) [40]. The stoichiometric reaction of boranes with alkynes has been well investigated [5b], and reagents that give vinyl boranes or diboryl alkanes and which are also sensitive to the steric environment of the carbon–carbon triple bond have been developed.

Arase and coworkers have reported the synthesis of some isomerically pure α,β-unsaturated nitriles via hydroboration of alkynes (Scheme 15.2) [41]. For example, the hydroboration of hex-1-yne with Sia$_2$BH and subsequent treatment of (E)-hex-1-enylbis(1,2-dimethylpropyl)borane with excess CuCN and Cu(OAc)$_2$

	Ph≡Me		Et≡Me		n-Pr≡Me		Ph≡H	
BH₃·THF	74	26	40	60	40	60		
9-BBN	65	35	22	78	22	78		
pinBH	15	85			7	93	0	100
catBH	27	73			40	60	0	100
Sia₂BH	19	81	39	61	39	61		
Br₂BH·SMe₂	64	36			25	75		
Mes₂BH	2	98			10	90		

Figure 15.3 Regioselectivity (%) of borane additions.

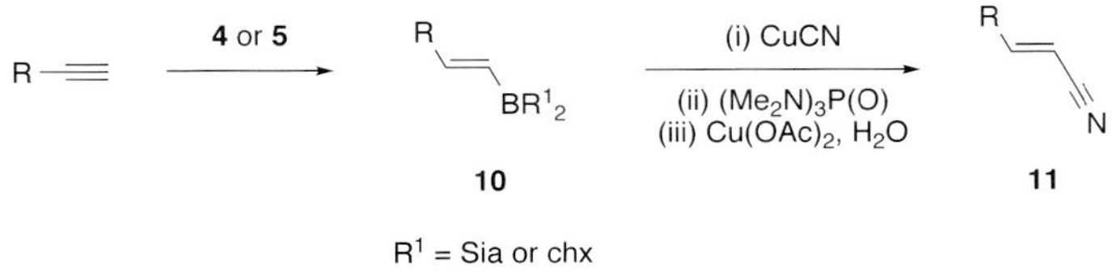

Scheme 15.2 α,β-Unsaturated nitriles via hydroboration of alkynes.

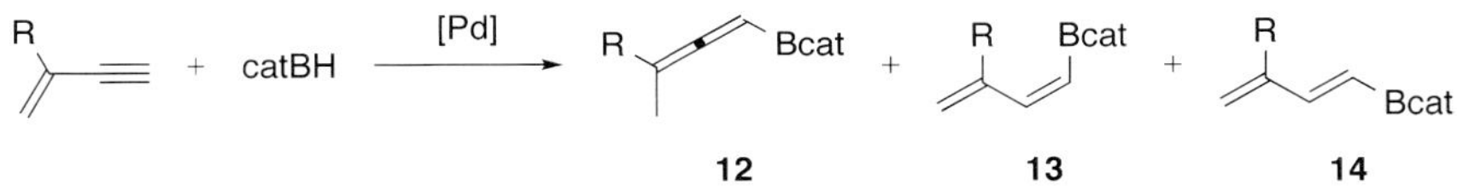

Scheme 15.3 Palladium-catalyzed hydroboration of 1,3-enynes.

in the presence of $(Me_2N)_3P(O)$ and H_2O gave the desired product, (E)-1-cyanohex-1-ene, in 88% yield.

15.3.1
Catalysis

15.3.1.1 Palladium
Suzuki and coworkers briefly reported that the hydroboration of 1,3-enynes with Pd(PPh₃)₄ gives predominately the allenyl borane (12) (R = Me, Scheme 15.3) [42]. Hayashi and coworkers further investigated this reaction with a number of Pd-phosphine complexes [43]. They found that the bidentate phosphine ligands 1,2-bis(diphenylphosphino)ethane (dppe), 1,4-bis(diphenylphosphino)butane (dppb) and 1,1′-bis(diphenylphosphino)ferrocene (dppf) catalyzed the 1,2-addition to the 1,3-enyne giving (E)-dienyl boranes (14). Monodentate phosphines gave predominately the allenyl borane (12).

15.3.1.2 Rhodium
Burgess and coworkers have studied the Rh-catalyzed (Wilkinson's catalyst) hydroboration of phenylethyne with two equivalents of catecholborane (catBH) [44].

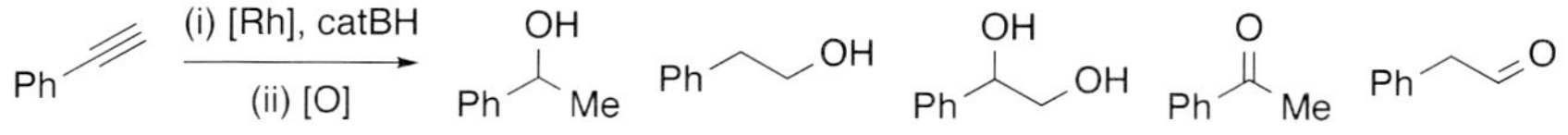

15.3 Hydroboration

Scheme 15.4 Rhodium-catalyzed hydroboration of phenylethyne.

Table 15.3 Comparison of Wilkinson's catalyst and RhCl(CO)(PPh$_3$)$_2$ in the catalytic hydroboration of terminal alkynes.

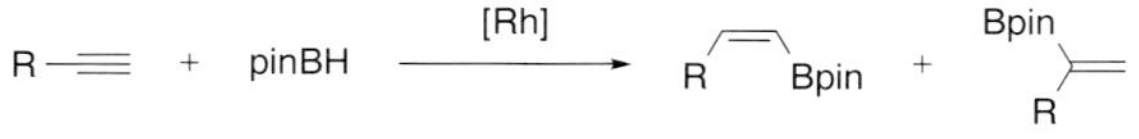

Entry	[Rh]	R	Product distribution (%)		Yield (%)
			(20)	(21)	
1	RhCl(PPh$_3$)$_3$	n-hex	71	29	99
2	RhCl(CO)(PPh$_3$)$_2$	n-hex	99	1	99
3	RhCl(PPh$_3$)$_3$	i-Pr	79	21	99
4	RhCl(CO)(PPh$_3$)$_2$	i-Pr	99	1	99
5	RhCl(PPh$_3$)$_3$	t-Bu	99	1	99
6	RhCl(PPh$_3$)$_3$	Ph	48	52	99
7	RhCl(CO)(PPh$_3$)$_2$	Ph	98	2	99
8	RhCl(PPh$_3$)$_3$	Cl(CH$_2$)$_3$	40	60	99
9	RhCl(CO)(PPh$_3$)$_2$	Cl(CH$_2$)$_3$	99	1	98
10	RhCl(PPh$_3$)$_3$	MeOCH$_2$	30	70	99
11	RhCl(CO)(PPh$_3$)$_2$	MeOCH$_2$	99	1	99

After oxidative work-up, carbonyl compounds, alcohols and/or diols were obtained, with the product distribution depending on reaction conditions. Short reaction times favored the production of isomeric carbonyl compounds ((**18**) and (**19**), Scheme 15.4), whereas extended reaction times favored the formation of alcohols (**15**), (**16**) and (**17**). Diol (**17**) was postulated to be formed from the catalytic hydroboration of the vinylboronate ester intermediate.

Pereira and Srebnik have also investigated the hydroboration of a number of alkynes with pinacolborane (pinBH) catalyzed by Wilkinson's catalyst [45]. Their results, like those of Burgess [44], showed that RhCl(PPh$_3$)$_3$ was unselective, except where the R group was a bulky substituent such as t-Bu (Table 15.3, entry 5). However, changing one phosphine ligand on Rh to a CO ligand, yielding RhCl(CO)(PPh$_3$)$_2$, greatly affected the regioselectivity, and the reaction gave essentially pure *anti*-Markovnikov product (**20**).

Miyaura and coworkers have recently reported the use of [Rh(COD)Cl]$_2$ in the hydroboration of terminal alkynes [46]. They found that adding one equivalent of

Table 15.4 [Rh(COD)Cl]$_2$ in the hydroboration of terminal alkynes.

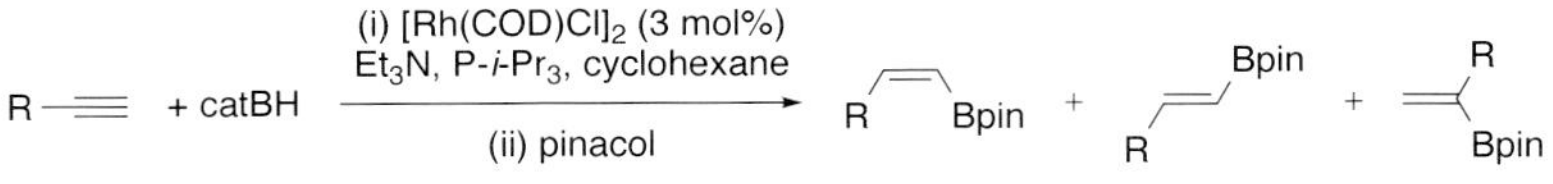

$$R-\!\!\!\equiv\;+\;catBH\;\xrightarrow[\text{(ii) pinacol}]{\substack{\text{(i) [Rh(COD)Cl]}_2\text{ (3 mol\%)}\\ \text{Et}_3\text{N, P-}i\text{-Pr}_3\text{, cyclohexane}}}\;R\!\!\diagup\!\!\diagdown\text{Bpin}\;+\;\text{Bpin}\!\!\diagup\!\!\diagdown R\;+\;\underset{\text{Bpin}}{\overset{R}{=\!\!=}}$$

$$\qquad\qquad\qquad\qquad\qquad\qquad\qquad 22 \qquad\qquad 23 \qquad\qquad 24$$

Entry	Alkyne	Product distribution (%)			Yield (%)
		(22)	(23)	(24)	
1	Me(CH$_2$)$_7$C≡CH	99	1	0	79
2	TBDMSO(CH$_2$)$_3$C≡CH	98	2	0	72
3	TBDMSOCH$_2$C≡CH	>99	<1	0	71
4	Me(TBDMSO)CHC≡CH	98	2	0	70
5	t-BuC≡CH	89	11	0	62
6	TMSC≡CH	98	2	0	70
7	PhC≡CH	99	1	0	60
8	Me(CH$_2$)$_5$C≡CH	99	1	0	74
9[a]	Me(CH$_2$)$_5$C≡CH	18	65	17	60

a   No NEt$_3$ was present.

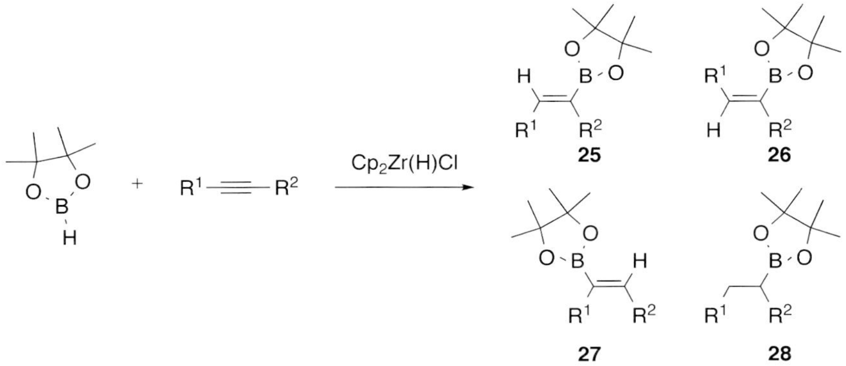

Scheme 15.5 Zirconium-catalyzed hydroboration of internal alkynes with pinBH.

Et$_3$N was essential for obtaining *cis*-alkenylboronates (Table 15.4). For easier analysis, the catecholborate esters were converted into their corresponding pinacolborate esters. Omitting Et$_3$N resulted in a mixture of all isomers (entry 9). The use of P-*i*-Pr$_3$ was also found to be crucial to the reaction. While the corresponding PCy$_3$ complex also resulted in comparable selectivities, complexes of PPh$_3$, PMePh$_2$, PMe$_3$, P-*n*-Pr$_3$ and P-*t*-Bu$_3$ yielded mixtures of all the possible isomers.

15.3.1.3 Zirconium

The hydroboration of alkynes with pinBH has been catalyzed by the zirconium complex Cp$_2$Zr(H)Cl (Scheme 15.5) [47, 48]. The reaction with terminal alkynes

Table 15.5 Zirconium-catalyzed hydroboration of internal alkynes with pinBH.

Entry	R¹	R²	Product distribution (%)				Yield (%)
			(25)	(26)	(27)	(28)	
1	n-Hex	H	98	2	0	0	93
2	Cl(CH$_2$)$_3$	H	96.8	2.4	0	0.8	94
3	TMS	H	90.3	2.2	3.2	4.3	88
4	Cyclopentyl	H	97.5	0.7	0	1.8	93.7
5	t-Bu	H	100	0	0	0	95.2
6	Ph	H	97.2	0.8	0.7	0.9	74.7
7	Ph(CH$_2$)$_3$	H	98.3	1.7	0	0	87.2
8	i-Pr	Me	96.9	2.2	0	0.9	93.8
9	t-Bu	Me	100	0	0	0	91.5
10	Et	Et	100	0	0	0	92.6
11	(EtO)$_2$CH	H	81.9	10.8	7.3	0	82.2
12	MeOCH$_2$	H	95	2.5	2.5	0	86.8
13	TMSC≡CCH$_2$	H	96.3	2.5	0.8	0.4	81.5

gave excellent yields and regioselectivities; the *anti* addition product **(26)** was obtained only in small quantities. Hydrogenation product **(28)**, presumably due to hydrolysis of the *gem*-borazirconocene alkanes [49], was detected only occasionally and in small amounts (Table 15.5).

The lowest regioselectivity was observed when the alkyne bore an acetal substituent (Table 15.5, entry 11); only 82% of the desired *cis* product was obtained, along with 11% *trans* and 7% of the other the regioisomer **(27)**. This discrepancy was postulated to be due to coordination of the hydroborating reagent with an oxygen atom of the acetal [47].

15.3.1.4 Titanium

Although dimethyltitanocene (Cp$_2$TiMe$_2$) is a highly efficient and selective catalyst for the hydroboration of alkenes [50], it is inferior to Cp$_2$Ti(CO)$_2$, **(29)**, as it gives rise to double additions of catBH to alkynes and exhibits lower regioselectivity. The hydroboration of alkynes with **(29)** proceeds in an *anti*-Markovnikov fashion. Despite giving excellent yields (Table 15.6), the reaction produced only low selectivities with unsymmetrical internal alkynes (entry 4) [51].

15.3.2 Mechanism

Miyaura and coworkers proposed a mechanism for the Ir- and Rh-catalyzed hydroboration of terminal alkynes with catBH and pinBH (Scheme 15.6) [46]. Deuterium experiments using 1-deutero-1-octyne suggested that the mechanism did not fit those previously proposed for the catalyzed *trans*-hydrometallation of terminal alkynes [52]. It was found that the β-hydrogen in the *cis*-product origi-

Table 15.6 Titanium-catalyzed hydroboration of alkynes.

Entry	Alkyne	Product(s)	Ratio	Yield (%)
1	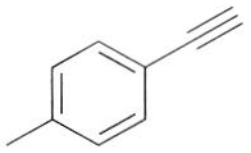 4-methylphenyl-C≡CH	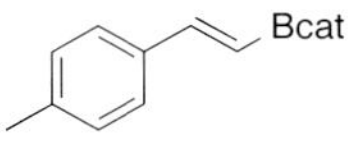 (E)-4-MeC$_6$H$_4$-CH=CH-Bcat	–	96
2	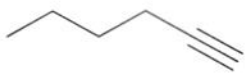 n-BuC≡CH	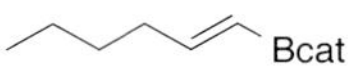 n-Bu-CH=CH-Bcat	–	97
3	Ph—≡—Ph	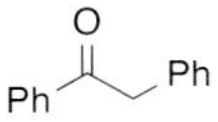 PhC(O)CH$_2$Ph	–	9[a]
4	Ph—≡—	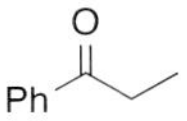 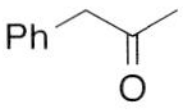 PhC(O)CH$_2$CH$_3$ / PhCH$_2$C(O)CH$_3$	67:33	89[a,b]

a Isolated yield after oxidation to ketone.
b Yield of both regioisomers.

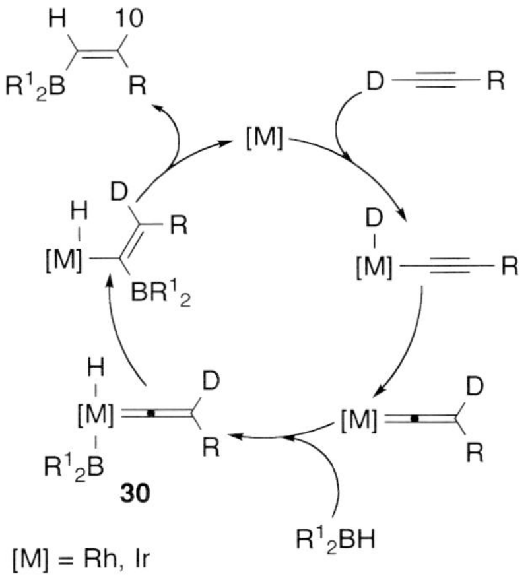

[M] = Rh, Ir

Scheme 15.6 Ir- and Rh-catalyzed hydroboration of terminal alkynes.

nated not from the borane reagent but selectively from the deuterium label on the terminal carbon. A mechanism involving isomerization of the *trans*-product to the *cis*-isomer was therefore discounted. To account for the deuterium migration and the *anti*-addition of the B–H bond, the mechanism was instead postulated to proceed via vinylidene complex **(30)**.

The Zr-mediated hydroboration of alkynes is believed to proceed in a manner similar to that proposed by Marks for lanthanide-catalyzed hydroboration of alkenes with catBH [53]. The proposed pathway is believed to involve alkyne insertion, σ-bond metathesis, and finally vinyl-halide exchange. Wang *et al*. speculated that the intramolecular interaction between the zirconium center and oxygen atom

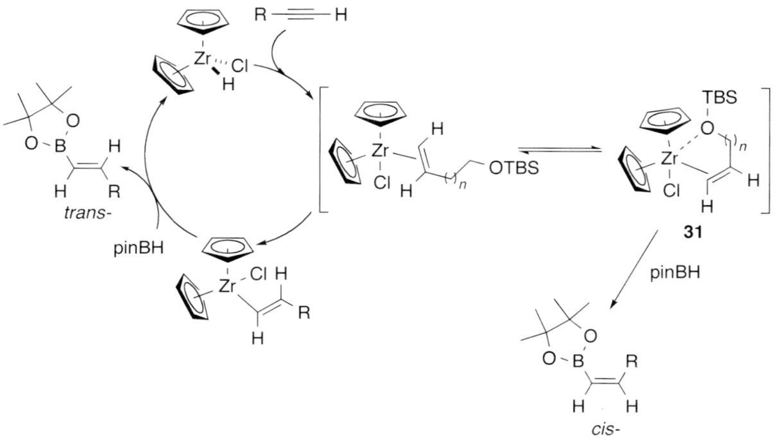

Scheme 15.7 Mechanism of zirconium-catalyzed hydroboration of terminal alkynes.

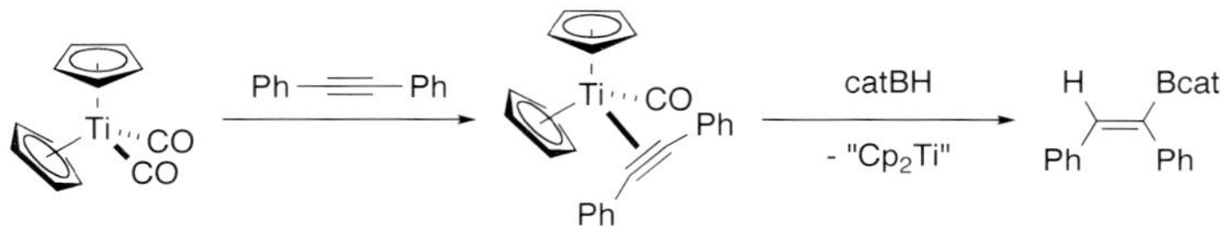

Scheme 15.8 Titanium-catalyzed hydroboration of 1,2-diphenylacetylene.

of the substrate favors the formation of a pseudo-Z intermediate (**31**) and subsequent *cis*-product (Scheme 15.7) [47].

The $Cp_2Ti(CO)_2$ system reported by Hartwig and Muhoro is highly efficient in the hydroboration of alkynes [50]. Although the mechanism of alkyne reductions with $Cp_2Ti(CO)_2$ is not known, the electronics of group IV d^0 complexes suggest that it likely involves σ-bond metathesis between a Ti-alkyne complex and the borane. Hartwig and Muhoro showed that alkynes react with $Cp_2Ti(CO)_2$, displacing one carbonyl group. They also showed that $Cp_2Ti(CO)(PhC≡CPh)$ reacts with catBH, giving quantitative production of (Z)-PhCH=CPh(catB) (Scheme 15.8) [50, 51].

15.4 Hydrosilylation

Vinyl silanes are useful building blocks that can be used in a number of synthetic processes including Pd-catalyzed cross-couplings [54], electrophilic substitutions [55] and Tamao–Fleming oxidations [56]. Vinyl silanes, unlike vinyl boranes, can be easily carried through other synthetic operations [57]. The catalytic hydrosilylation of alkynes is perhaps the most straightforward way to access vinyl silanes.

15.4.1
Terminal Alkynes

The hydrosilylation of terminal alkynes gives three isomeric products: *syn*- and *anti*-β-addition products and the α-isomer (Scheme 15.9) [6b]. The ratio of products varies dramatically depending on substrate, choice of catalyst (metal and ligands) and to a lesser extent reaction conditions.

The reagent tris(trimethylsilyl)silane (HSi(TMS)$_3$) simply adds to alkynes in a *trans*-manner forming (Z)-β-vinyl silanes, in typically >95:5 (Z/E) selectivity [58]. Hydrosilylation with HSi(TMS)$_3$ to give *anti*-addition is well accepted to be a radical-initiated process [59]. When activated by Et$_3$B/O$_2$, reactions can be conducted at temperature as low as 0 °C [58]. The hydrosilylation of terminal alkynes with platinum catalysts gives *syn*-addition products, leading to (E)-β-vinyl silanes and the α-regioisomer (Table 15.7) [60].

Oro and coworkers showed that the known Ru complex (**35**) [61] reacts with a stoichiometric amount of HSiEt$_3$ to give (E)-β-PhCH=CH(SiEt$_3$) and (**34**) (Scheme 15.10) [62]. Complex (**34**) also efficiently catalyzes the hydrosilylation of phenylacetylene with HSiEt$_3$ [62]. However, in contrast to the stoichiometric reaction the formation of (Z)-β-vinyl silanes is observed. Although, under the reaction conditions, complex (**35**) is the main species, catalysis proceeds via initial interaction of the catalyst with the silane.

Ozawa and coworkers have shown that the hydrosilylation of terminal alkynes (RC≡CH; R = Ph, *p*-tolyl, Cy and *n*-hexyl) with a number of hydrosilanes can be affected in a stereodivergent manner by small changes in the Ru-catalyst structure (Scheme 15.11) [63]. Catalytic amounts of RuHCl(CO)(PPh$_3$)$_3$ gave (E)-β-vinyl silanes (>99% selectivity) whereas Ru(SiMe$_2$Ph)Cl(CO)(P-*i*-Pr$_3$)$_2$ gave (Z)-β-vinyl silanes (99% selectivity).

In 2000, Na and Chang reported that [Ru(*p*-cymene)Cl$_2$]$_2$ was a highly regio- and stereoselective catalyst for the hydrosilylation of terminal alkynes (Z/E>95/5) [64]. Although the selectivities are some of the best reported, even for α-branched substrates (Scheme 15.12), it is limited to trialkyl and triphenyl silanes.

Interestingly, Na and Chang reported a dramatic directing effect for substrates with appropriately positioned hydroxyl groups when using [RuCl$_2$(*p*-cymene)]$_2$ as

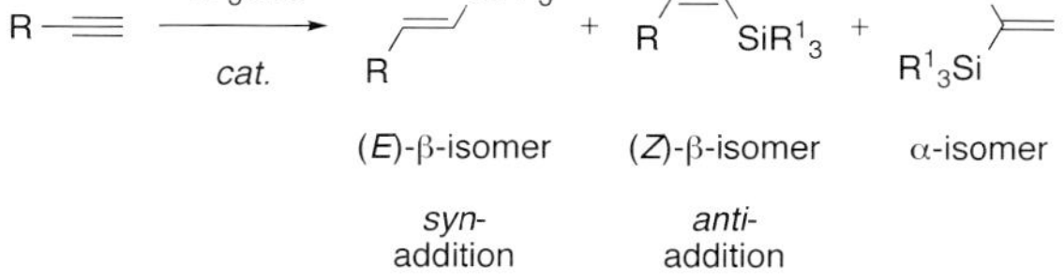

$$R-\!\!\equiv \quad \xrightarrow[\text{cat.}]{R^1_3SiH} \quad \underset{R}{\diagup\!\!\!\diagdown}\!SiR^1_3 \;+\; \underset{R}{\diagup\!\!\!\diagdown}\!SiR^1_3 \;+\; \underset{R^1_3Si}{\diagup\!\!\!\diagdown}\!R$$

(E)-β-isomer   (Z)-β-isomer   α-isomer

syn-addition   *anti*-addition

Scheme 15.9 Hydrosilylation of terminal alkynes.

15.4 Hydrosilylation

Table 15.7 Platinum-catalyzed hydrosilylation of terminal alkynes.

$$R\!\!-\!\!\equiv \xrightarrow[\text{HSiR}^1_3]{[\text{Pt}]\ 0.01\ \text{mol\%}} \underset{\alpha\text{-isomer}}{R\!\!\diagup\!\!\overset{\text{SiR}^1_3}{\diagdown}} + \underset{(E)\text{-}\beta\text{-isomer}}{R\!\!\diagdown\!\!\diagup\!\!\text{SiR}^1_3}$$

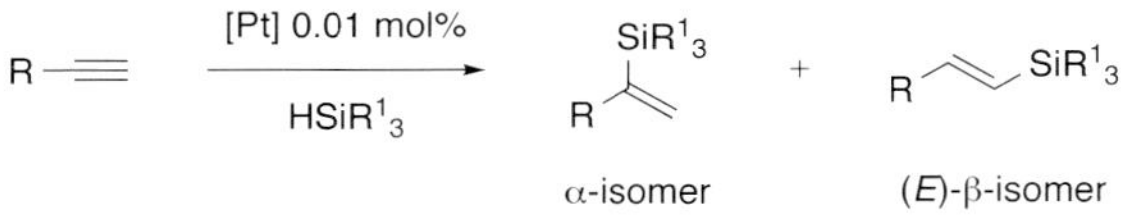

Entry	R	HSiR$^1_3$	[Pt]^a	Ratio ($\alpha:\beta$)	Yield (%)
1	Et	HSiEt$_3$	A	8:92	88
2	Et	HSiCl$_2$Me	A	14:86	80
3	Et	HSiCl$_3$	B	10:90	88
4	n-Pr	HSiEt$_3$	A	4:96	92
5	n-Pr	HSiCl$_2$Me	A	5:95	86
6	n-Pr	HSiCl$_3$	A	4:96	88
7	n-Bu	HSiEt$_3$	A	4:96	93
8	n-Bu	HSiCl$_2$Me	A	6:94	97
9	n-Bu	HSiCl$_3$	A	5:95	94
10	n-Pent	HSiEt$_3$	A	5:95	97
11	n-pent	HSiCl$_2$Me	A	4:96	95
12	n-Pent	HSiCl$_3$	A	4:96	93
13	cyclo-C$_5$H$_9$	HSiEt$_3$	A	0:100	97
14	cyclo-C$_5$H$_9$	HSiCl$_2$Me	A	0:100	92
15	cyclo-C$_5$H$_9$	HSiCl$_3$	A	0:100	90
16	Ph	HSiEt$_3$	C	11:89	90
17	Ph	HSiCl$_2$Me	A	0:100	81
18	Ph	HSiCl$_3$	B	0:100	80

a [Pt] catalysts: A = {Pt(μ-H)[Si(CH$_2$Ph)Me](PCy$_3$)}$_2$, B = [Pt(μ-H)(SiCl$_3$)(PCy$_3$)]$_2$, C = [Pt(μ-H)(SiEtMe$_2$)(PCy$_3$)]$_2$.

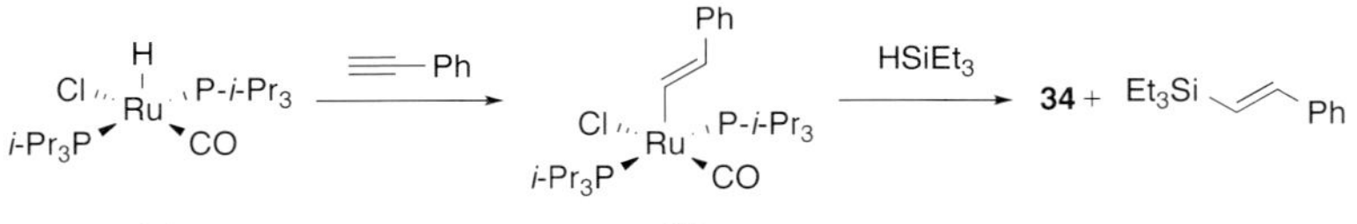

Scheme 15.10 Synthesis of (**35**) and its stoichiometric reaction with HSiEt$_3$.

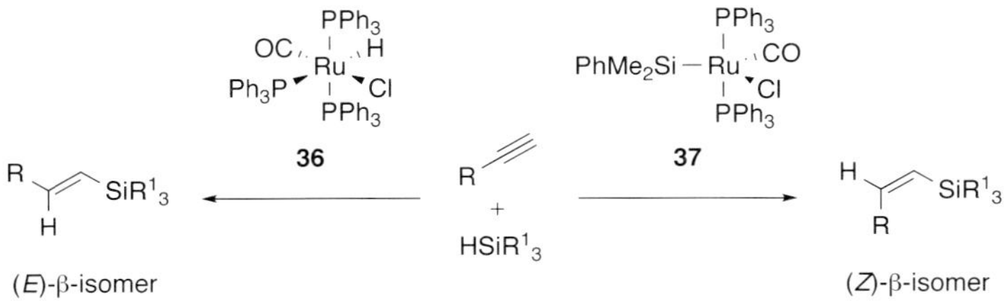

Scheme 15.11 Hydrosilylation of terminal alkynes catalyzed by (**36**) or (**37**).

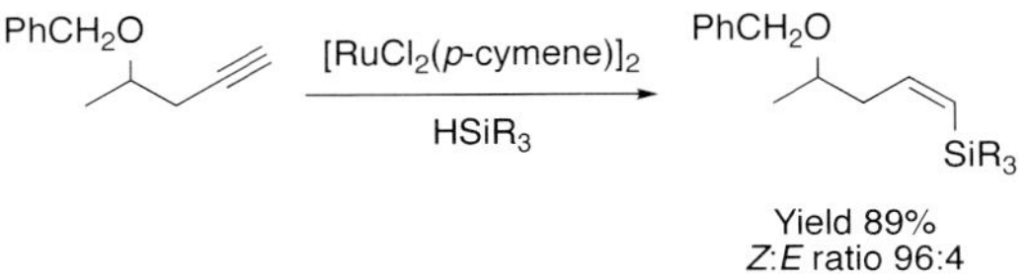

Scheme 15.12 Ruthenium-catalyzed hydrosilylation of terminal alkynes.

Table 15.8 Hydroxyl-group-directing effects on the ruthenium-catalyzed hydrosilylation of terminal alkynes.

Entry	Alkyne	Product (major)	β-(Z):α	Yield (%)
1	 HO-alkyne	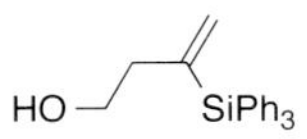 HO-CH₂-C(=CH₂)SiPh₃	2:98	47
2	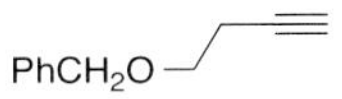 PhCH₂O-alkyne	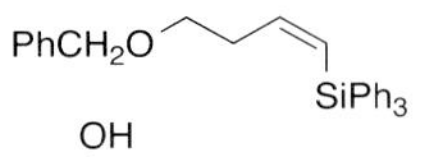 PhCH₂O-CH₂CH₂-CH=CH-SiPh₃ (Z)	>99:1	89
3	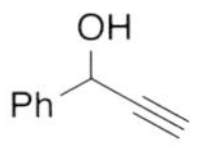 Ph-CH(OH)-alkyne	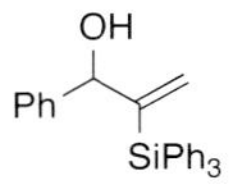 Ph-CH(OH)-C(=CH₂)SiPh₃	13:87	60
4	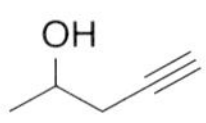 CH₃-CH(OH)-CH₂-alkyne	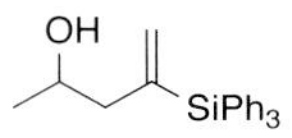 CH₃-CH(OH)-CH₂-C(=CH₂)SiPh₃	2:98	59
5	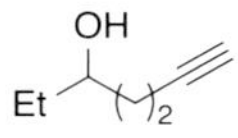 Et-CH(OH)-(CH₂)₂-alkyne	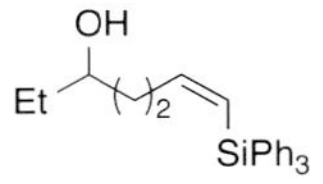 Et-CH(OH)-(CH₂)₂-CH=CH-SiPh₃	92:8	53
6	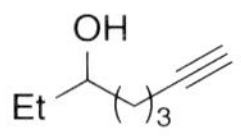 Et-CH(OH)-(CH₂)₃-alkyne	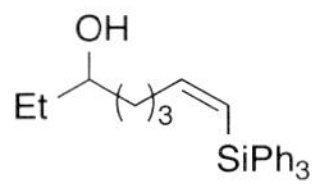 Et-CH(OH)-(CH₂)₃-CH=CH-SiPh₃	96:4	61
7	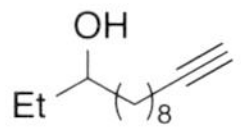 Et-CH(OH)-(CH₂)₈-alkyne	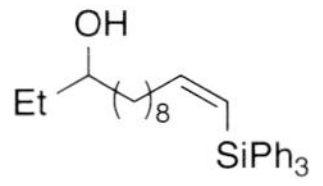 Et-CH(OH)-(CH₂)₈-CH=CH-SiPh₃	98:2	86

the hydrosilylation catalyst (Table 15.8). For example, the hydrosilylation of 3-butyn-1-ol (Table 15.8, entry 1) gave the α-regioisomer in excellent selectivity, 98:2, whereas the benzyl-protected alkyne provided the β-(Z)-vinyl silane as the major regioisomer (entry 2). Increasing the chain length between the hydroxyl and alkyne groups significantly reduced this affect (entries 5–7). These results imply that the mechanism proceeds through coordination of the carbinol oxygen to a Ru intermediate.

The cyclopentadienyl-ruthenium complexes **(38)–(40)** have also been used in an alternative strategy for α-vinyl silane formation (Figure 15.4).

Yamamoto and coworkers showed that the trihydride-Ru complex **(38)** gave preferentially α-vinyl silanes with a number of terminal alkynes (Table 15.9) [65].

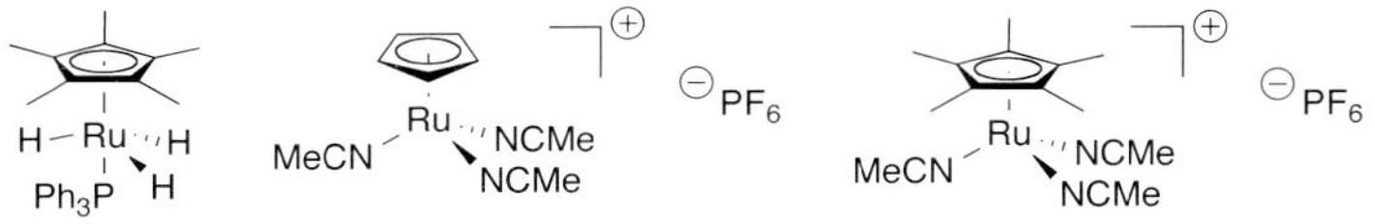

Figure 15.4 Cyclopentadienyl-ruthenium complexes that have been used in the formation of α-vinyl silanes.

Table 15.9 Hydrosilylation of terminal alkynes with complex (38).

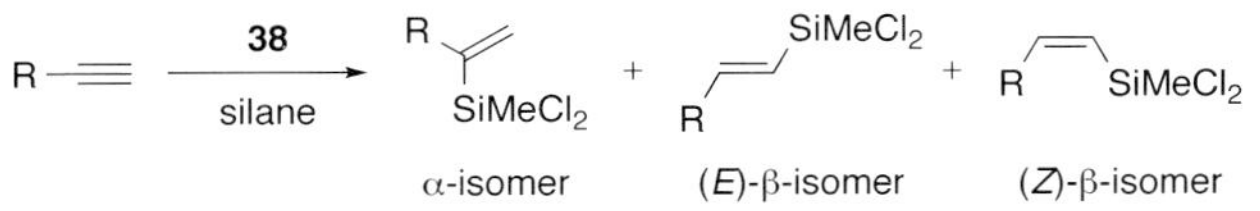

Entry	Alkyne	Silane	α:β (E:Z)	Yield (%) (α+β)
1	$n\text{-}C_5H_{11}$	$HSiMeCl_2$	85:15 (1:9)	65
2	$n\text{-}C_6H_{13}$	$HSiMeCl_2$	92:8 (0:1)	89
3	$AcO(CH_2)_3$	$HSiMeCl_2$	93:7 (0:1)	68
4	$AcO(CH_2)_2$	$HSiMeCl_2$	97:3 (0:1)	83
5	$PhOCH_2$	$HSiMeCl_2$	>99:1	78
6	Ph	$HSiMeCl_2$	69:31 (1:1)	81
7	t-Bu	$HSiMeCl_2$	17:83 (1:0)	82
8	TMS	$HSiMeCl_2$	71:29 (1:0)	18

They also reported that the pentamethylcyclopentadienyl (Cp*) was essential to obtain the α-regioisomer in high selectivity. Trost and Ball have reported the use of the cationic cyclopentadienylruthenium complexes (39) and (40) in hydrosilylation reactions to afford predominately α-vinyl silanes (Figure 15.4) [66, 67]. These catalysts are compatible with a number of functional groups, such as alcohols, acids, protected amines and internal alkynes, as well as functioning with alkyl-, aryl-, halo- and alkoxy-silanes [67].

The iridium complex [IrH(H$_2$O)(bq)(PPh$_3$)$_2$] (bq = 7,8-benzoquinolinato) has been demonstrated by Jun and Crabtree to give good selectivities for (Z)-β-vinylsilanes with HSiMePh$_2$ [68]. However, significant deterioration in selectivities were observed with bulky alkynes or silanes with electron-withdrawing groups. Iyoda and coworkers reported that the Ir-complex [Ir(η^5-C$_9$H$_7$)(COD)] (η^5-C$_9$H$_7$ = η^5-indenyl), in the presence of 4,4′,5,5′-tetramethylbiphosphinine (tmbp) efficiently gives (E)-β-vinylsilanes (Table 15.10) [69]. In the absence of tmbp, (Z)-β-vinylsilanes were preferentially formed: the hydrosilylation of 1-octyne gave 67% of the (Z)-β-isomer, 21% (E)-β-isomer and 12% of the α-isomer (cf. entry 5).

Table 15.10 Hydrosilylation with the Ir complex [Ir(η⁵-C₉H₇)(COD)] in the presence of tmbp.

$R\!\equiv\!\equiv$ + HSiMe₂Ph $\xrightarrow{\text{[Ir] / tmbp}}$ (E)-β-isomer + (Z)-β-isomer + α-isomer

tmbp = (structure shown)

Entry	R	Product ratio			Yield (%)
		(E)-β	(Z)-β	α	
1	Ph	84	16	0	88
2	p-Cl-C₆H₄	85	13	2	84
3	p-MeO-C₆H₄	91	7	2	86
4	t-Bu	100	0	0	59
5	n-C₆H₁₃	94	4	2	96
6	TMS	100	0	0	59
7	HOCH₂	93	1	6	44
8	HO(CH₂)₂	97	0	3	91
9	HO(CH₂)₃	95	1	4	92

15.4.2
Internal Alkynes

The hydrosilylations of internal alkynes are considerably less developed than those of terminal alkynes. Symmetrical internal alkynes can be hydrosilylated to yield a single product, but unsymmetrical internal alkynes rarely react with high levels of regioselectivity [70].

Platinum-based catalysts exhibit similar selectivities in the hydrosilylation of internal and terminal alkynes, giving *syn*-addition, but show reduced reaction rates [71]. In the case of unsymmetric internal alkynes, the observed trend of regioselectivity has been attributed to the silyl group preferring the alkyne carbon that has the least bulky substituent (Table 15.11) [60].

The hydrosilylation of internal alkynes has also been efficiently catalyzed by a titanium species generated *in situ* from titanocene dichloride (Cp₂TiCl₂) and two equivalents of *n*-BuLi [72]. High *syn*-addition was observed; for example, hex-3-yne was hydrosilylated to (*E*)-3-(methylphenylsilyl)hex-3-ene in 96% yield. The catalyst was also highly selective for the one unsymmetric alkyne substrate that was reported, namely 1-(trimethylsilyl)propyne. Again, only the (*E*)-regioisomer was observed, with the silyl group adding to the least-hindered alkyne carbon atom. A limitation of the procedure is that high catalyst loading (20 mol%) was required.

Recently ruthenium carbene complexes have been demonstrated to act as alkyne hydrosilylation catalysts, affording mainly (*Z*)-alkenes via *trans* addition

Table 15.11 Platinum-catalyzed hydrosilylation of internal alkynes.

$$R\!\!\equiv\!\!R^1 \xrightarrow[\text{HSiX}_2]{[\text{Pt}]} \underset{R}{\overset{SiX_3}{\diagdown}}\!\!=\!\!R^1 \;+\; \underset{R}{\overset{X_3Si}{\diagdown}}\!\!=\!\!R^1$$

41   42

Entry	R	R′	HSiX$_3$	[Pt]^a	Product ratio (41)	Product ratio (42)	Yield (%)
1	Me	Me	HSiEt$_3$	(43)	–	–	92
2	Me	Me	HSiCl$_2$Me	(44)	–	–	87
3	Me	Me	HSiCl$_3$	(43)	–	–	87
4	Ph	Ph	HSiEt$_3$	(43)	–	–	95
5	Ph	Ph	HSiCl$_2$Me	(43)	–	–	82
6	Ph	Ph	HSiCl$_3$	(43)	–	–	76
7	Me	Et	HSiEt$_3$	(43)	35	65	97
8	Me	Et	HSiCl$_2$Me	(43)	35	65	96
9	Me	Et	HSiCl$_3$	(43)	30	70	94
10	Me	n-Pr	HSiEt$_3$	(43)	40	60	96
11	Me	n-Pr	HSiCl$_2$Me	(43)	28	72	95
12	Me	n-Pr	HSiCl$_3$	(43)	32	68	92

a [Pt] catalysts: **(43)** = {Pt(μ-H)[Si(CH$_2$Ph)Me](PCy$_3$)}$_2$, **(44)** = [Pt(μ-H)(SiCl$_3$)(PCy$_3$)]$_2$.

[73]. Although only one internal alkyne was presented, *trans* addition was observed in that case [73a].

Complex **(45)** has recently been reported to catalyze the formation vinyl silanes via *syn* or *anti* hydrosilylation [74]. The reaction of diphenylethyne with HSiEt$_3$ afforded the (*E*)-isomer exclusively in 60% yield (Table 15.12, entry 1). Interestingly, using triethoxysilane resulted in reversed regioselectivity, and predominantly the (*Z*)-isomer was observed (entries 4–6). However, for unsymmetrical internal alkynes, exclusive *syn* hydrosilylation was observed for all cases, though only moderate yields and regioselectivity were obtained.

Perhaps the most successful catalyst for the hydrosilylation of unsymmetrical internal alkynes is Cp*$_2$YCH$_3$·THF [75, 76]. Single stereo- and regioisomers were obtained in high yields for the hydrosilylation of unsymmetrical alkynes (Table 15.13, entries 2–11) [75]. Although the reaction of non-2-yne afforded two regioisomers (4.2:1, *E/Z*) under the reaction conditions, this was improved to 7.2:1 through the use of the catalyst Cp*$_2$Y[CH(TMS)$_2$]. Entries 6–11 demonstrate that the catalyst is tolerant of a wide range of functional groups. However, highly hindered alkynes (entry 5) require longer reaction times and suffer from reduced yields, presumably due to dehydrogenative polymerization. More recently the hydrosilylation of alkynyl silanes has been reported with Cp*$_2$YCH$_3$·THF [76].

Table 15.12 Cobalt-catalyzed hydrosilylation of internal alkynes.

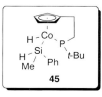

Entry	R	R^1_3SiH	Product ratio		Yield (%)
			(E)-	(Z)-	
1	Ph	$HSiEt_3$	100	0	60
2	Et	$HSiEt_3$	100	0	71
3	CO_2-t-Bu	$HSiEt_3$	100	0	66
4	Ph	$HSi(OEt)_3$	9	91	69
5	Et	$HSi(OEt)_3$	8	92	77
6	CO_2-t-Bu	$HSi(OEt)_3$	5	95	55

Table 15.13 $Cp^*_2YCH_3·THF$-catalyzed hydrosilylation of internal alkynes.[a]

Entry	Alkyne		Product (major)	Yield (%)
1	R = n-Butyl	R^1 = n-Butyl		83
2	R = Cyclohexyl	R^1 = Me		74
3	R = s-Butyl	R^1 = Me		74
4	R = s-Butyl	R^1 = n-Pentyl		93
5	R = t-Butyl	R^1 = n-Decyl		28[b]
6	R = Me	R^1 = Me		82
7	R = Me	R^1 = n-Decyl		89
8	R = H	R^1 = Me		23
9	X = Cl			79
10	X = OTHHP			84
11	X = NMe_2			73

a Reaction time 24 h.
b Reaction time 48 h.

15.4.3
Mechanism

The Chalk–Harrod mechanism for catalytic hydrosilylation was first proposed in 1965 [77]. Insertion of a coordinated alkyne (or alkene) into a metal–hydrogen bond (hydrometallation) is followed by a reductive elimination of alkenyl (or alkyl) and silyl ligands giving *syn* addition products (Scheme 15.13, pathway **A**). This mechanism, originally derived from studies on cobalt-catalyzed hydrosilylation of olefins, has been successfully applied to platinum-catalyzed hydrosilylation of alkenes and alkynes. However, the mechanism has been modified to address the observations of (Z)-β-vinylsilanes from *trans* addition, α-vinylsilanes and dehydrogenative silylation [78]. In this modified Chalk–Harrod mechanism, the coordinated alkyne (or alkene) inserts into the metal–silicon (silylmetallation) bond. This is followed by reductive elimination of the β-silylalkenyl (or β-silylalkyl) and hydride ligands, giving *syn* addition products (Scheme 15.13, pathway **B**) [79].

To account for *anti* addition, both Crabtree [52c] and Ojima [52a] have proposed that following insertion of the alkyne into the M–Si bond, E/Z isomerization of the intermediate vinyl complex takes place (Scheme 15.14), either through a

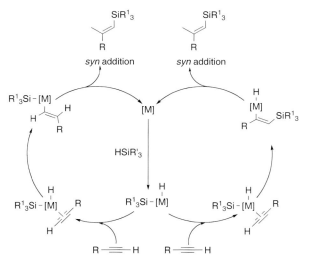

A: Chalk-Harrod   **B**: modified Chalk-Harrod

Scheme 15.13 The Chalk–Harrod and modified Chalk–Harrod mechanisms.

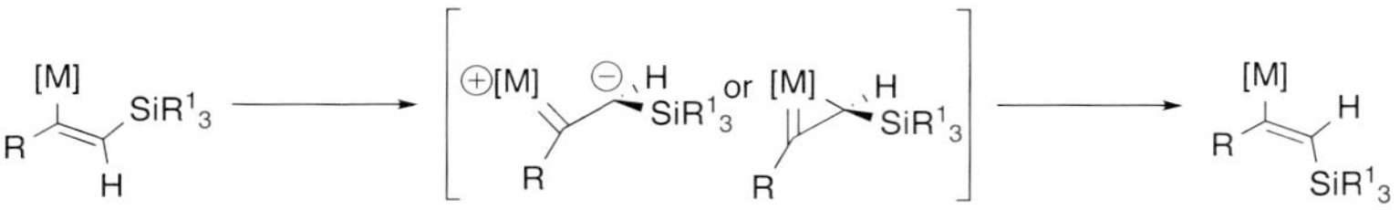

Scheme 15.14 E/Z isomerization of a vinylmetal intermediate.

zwitterionic carbene **(46)** or metallocyclopropane intermediate **(47)**. Subsequent reductive elimination of the β-silylalkenyl gives the *anti* addition product.

15.5
Conclusions

Alkyne reductions have a long history, and chemists have developed many novel techniques for controlling their regio-, chemo- and stereoselectivity. However, there remain many problems. The hydrogenation of alkynes to (*E*)-alkenes has been reported for only a small number of catalysts with limited substrate scope. The uncatalyzed hydroboration of alkynes can be carried out with excellent yields and regioselectivity, but the catalytic hydroboration of internal alkynes has no current solutions.

Further understanding of the associated mechanisms will hopefully reduce current selectivity limitations.

References

1 (a) Moil, K. (1981) The synthesis of insect pheromones, in *The Total Synthesis of Natural Products* (ed. J. Apsimon), Wiley Interscience, New York. (b) Bartlett, P.A. (1980) *Tetrahedron*, **36**, 1. (c) Ackroyd, J. and Scheinmann, F. (1982) *Chemical Society Reviews*, **11**, 321.

2 Molnár, Á., Sárkány, A. and Varga, M. (2001) *Journal of Molecular Catalysis A: Chemical*, **173**, 185.

3 For heterogeneous see: Siegel, S. (1991) *Heterogeneous catalytic hydrogenation of C=C and C≡C*, in *Comprehensive Organic Synthesis*, Vol. 8 (eds B.M. Trost and I. Fleming), Pergamon, Oxford, pp. 417–42.

4 For homogeneous see: Takaya, H. and Noyori, R. (1991) *Homogeneous catalytic hydrogenation of C=C and C≡C*, in *Comprehensive Organic Synthesis*, Vol. 8 (eds B.M. Trost and I. Fleming), Pergamon, Oxford, pp. 443–69.

5 (a) Pelter, A., Smith, K. and Brown, H. (1988) *Borane Reagents*, Academic Press Inc., San Diego, CA. (b) Smith, K. and Pelter, A. (1991) *Hydroboration of C=C and C≡C*, in *Comprehensive Organic Synthesis*, Vol. 8 (eds B.M. Trost and I. Fleming), Pergamon, Oxford, pp. 703–31.

6 (a) Ojima, I., Li, Z. and Zhu, J. (1998) *Hydrosilylation of C=C and C≡C*, in *The Chemistry of Organosilicon Compounds*, Vol. 2 (eds Z. Rappoport and Y. Apeloig), John Wiley & Sons, Ltd, Chichester, pp. 1687–792. (b) Hiyama, T. and Kusumoto, T. (1991) Xxxx, in *Comprehensive Organic Synthesis*, Vol. 8 (eds B.M. Trost and I. Fleming), Pergamon, Oxford, pp. 763–92.

7 Trost, B.M. and Ball, Z.T. (2005) *Synthesis*, **6**, 853.

8 (a) Corriu, R.J.P. and Moreau, J.J.E. (1972) *Journal of Organometallic Chemistry*, **40**, 97. (b) Nozaki, K., Ichinose, Y., Wakamatsu, K., Oshima, K. and Utimoto, K. (1990) *Bulletin of the Chemical Society of Japan*, **63**, 2268. (c) Bernardoni, S., Lucarini, M., Pedulli, G.F., Valgimigli, L. Gevorgyan, V. and Chatgilialoglu, C. (1997) *Journal of Organic Chemistry*, **62**, 8009. (d) Schwier, T. and Gevorgyan, V. (2005) *Organic Letters*, **7**, 5191.

9 (a) Kuniyasu, H., Ogawa, A., Sato, K., Ryu, I. and Sonoda, N. (1992) *Tetrahedron Letters*, **33**, 5525. (b) Kamiya, I., Nishinaka, E. and Ogawa, A. (2005) *Journal of Organic Chemistry*, **70**, 696.

10 (a) Rylander, P.N. (1985) *Hydrogenation Methods*, Academic Press, London. (b) Marvell, E.N. and Li, T. (1973) *Synthesis*,

11, 457. (c) Henrick, C.A. (1977) *Tetrahedron*, **33**, 1845.

11 Hershberg, E.B., Oliveto, E.P., Gerold, C. and Johnson, L. (1951) *Journal of the American Chemical Society*, **73**, 5073.

12 Campbell, K.N. and O'Connor, M.J. (1939) *Journal of the American Chemical Society*, **61**, 2897.

13 Rylander, P.N. (1979) *Cataytic Hydrogenation in Organic Synthesis*, Academic Press, New York.

14 (a) Osborn, J.A., Jardine, F.H., Young, J.F. and Wilkinson, G. (1966) *Journal of the Chemical Society. A*, **12**, 1711. (b) Mitchell, R.W., Ruddick, J.D. and Wilkinson, G. (1971) *Journal of the Chemical Society. A*, **20**, 3224.

15 Demerseman, B., Le Coupanec, P. and Dixneuf, P.H. (1985) *Journal of Organometallic Chemistry*, **287**, C35.

16 Pasto, D.J. (1991) *Reduction of C=C and C≡C by noncatalytic chemical methods*, in *Comprehensive Organic Synthesis*, Vol. 8 (eds B.M. Trost and I. Fleming), Pergamon, Oxford, pp. 471–88.

17 Corey, E.J., Mock, W.L. and Pasto, D.J. (1961) *Tetrahedron Letters*, **2**, 347.

18 (a) Brandsma, L., Nieuwenhuizen, W.F., Zwikker, J.W. and Mäeorg, U. (1999) *European Journal of Organic Chemistry*, **4**, 775. (b) Kaufman, D., Johnson, E. and Mosher, M.D. (2005) *Tetrahedron Letters*, **46**, 5613.

19 Ohmori, K., Hachisu, Y., Suzuki, T. and Suzuki, K. (2002) *Tetrahedron Letters*, **43**, 1031.

20 (a) Lindlar, H. (1952) *Helvetica Chimica Acta*, **35**, 446. (b) Lindlar, H. and Dubuis, R. (1973) *Organic Synthesis Collection* **5**, 880.

21 Choi, J. and Yoon, N.M. (1996) *Tetrahedron Letters*, **35**, 1057.

22 Savoia, D., Tagliavini, E., Trombini, C. and Umano-Ronchi, A. (1981) *Journal of Organic Chemistry*, **46**, 5340.

23 Choudary, B.M., Vasantha, G., Sharma, M. and Bharathi, P. (1989) *Angewandte Chemie (International Ed. in English)*, **28**, 465.

24 (a) Noyori, R. and Suzuki, M. (1984) *Angewandte Chemie (International Ed. in English)*, **23**, 847. (b) Zhou, W.-S. and Shen, Z.-W. (1991) *Journal of the Chemical Society, Perkin Transactions I*, **11**, 2827. (c) Yanagisawa, A., Habaue, S. and Yamamoto, H. (1992) *Tetrahedron*, **48**, 1969. (d) Nicolaou, K.C., Xu, J.-Y., Kim, S., Ohshima, T., Hosokawa, S. and Pfefferkorn, J. (1997) *Journal of the American Chemical Society*, **119**, 11353. (e) Rzasa, R.M., Shea, H.A. and Romo, D. (1998) *Journal of the American Chemical Society*, **120**, 591. (f) Evans, D.A. and Fitch, D.M. (2000) *Angewandte Chemie (International Ed. in English)*, **39**, 2536.

25 (a) Altenbach, H.-J. and Himmeldirk, K. (1995) *Tetrahedron: Asymmetry*, **6**, 1077. (b) Hamprecht, D., Josten, J. and Steglich, W. (1996) *Tetrahedron*, **52**, 10883. (c) Fujita, M., Chiba, K., Nakano, J., Tominaga, Y. and Matsumoto, J. (1998) *Chemical and Pharmaceutical Bulletin*, **46**, 631. (d) Knight, D.W. and Little, P.B. (1998) *Tetrahedron Letters*, **39**, 5105.

26 Campos, K.R., Cai, D., Journet, M., Kowal, J.J., Larsen, R.D. and Reider, P.J. (2001) *Journal of Organic Chemistry*, **66**, 3634.

27 Alonso, F., Osante, I. and Yus, M. (2006) *Advanced Synthesis and Catalysis*, **348**, 305.

28 (a) Pelagatti, P., Venturini, A., Leporati, A., Carcelli, M., Costa, M., Bacchi, A., Pelizzi, G. and Pelizzi, C. (1998) *Journal of the Chemical Society, Dalton Transactions*, **16**, 2715. (b) Bacchi, A., Carcelli, M., Costa, M., Pelagatti, P., Pelizzi, C. and Pelizzi, G. (1994) *Gazzetta Chimica Italiana*, **124**, 429.

29 Schrock, R.R. and Osborn, J.A. (1976) *Journal of the American Chemical Society*, **98**, 2143.

30 Sodeoka, M. and Shibasaki, M. (1985) *Journal of Organic Chemistry*, **50**, 1147.

31 van Laren, M.W. and Elsevier, C.J. (1999) *Angewandte Chemie (International Ed. in English)*, **38**, 3715.

32 Yoshida, T., Youngs, W.J., Sakaeda, T., Ueda, T., Otsuka, S. and Ibers, J.A. (1983) *Journal of the American Chemical Society*, **105**, 6273.

33 (a) Burch, R.R., Shusterman, A.J., Muetterties, E.L., Teller, R.G. and Williams, J.M. (1983) *Journal of the American Chemical Society*, **105**, 3546. (b) Burch, R.R., Muetterties, E.L., Teller, R.G. and Williams, J.M. (1982) *Journal of the American Chemical Society*, **104**, 4257.

34 Schleyer, D., Niessen, H.G. and Bargon, J. (2001) *New Journal of Chemistry*, **25**, 423.
35 Beletskaya, I. and Pelter, A. (1997) *Tetrahedron*, **53**, 4957.
36 Zweifel, G. and Arzoumanian, H. (1967) *Journal of the American Chemical Society*, **89**, 291.
37 Brown, H.C. and Zweifel, G. (1961) *Journal of the American Chemical Society*, **83**, 3834.
38 Zweifel, G., Clark, G.M. and Polston, N.L. (1971) *Journal of the American Chemical Society*, **93**, 3395.
39 Brown, H.C., Scouten, C.G. and Liotta, R. (1979) *Journal of the American Chemical Society*, **101**, 96.
40 Pelter, A., Singaram, B. and Brown, H.C. (1983) *Tetrahedron Letters*, **24**, 1433.
41 Masuda, Y., Hoshi, M. and Arase, A. (1991) *Journal of the Chemical Society, Chemical Communications*, 748.
42 Satoh, M., Nomoto, Y., Miyaura, N. and Suzuki, A. (1989) *Tetrahedron Letters*, **30**, 3789.
43 Matsumoto, Y., Naito, M. and Hayashi, T. (1992) *Organometallics*, **11**, 2732.
44 Burgess, K., van der Donk, W.A., Westcott, S.A., Marder, T.B., Baker, R.T. and Calabrese, J.C. (1992) *Journal of the American Chemical Society*, **114**, 9350.
45 Pereira, S. and Srebnik, M. (1996) *Tetrahedron Letters*, **37**, 3283.
46 Ohmura, T., Yamamoto, Y. and Miyaura, N. (2000) *Journal of the American Chemical Society*, **122**, 4990.
47 Wang, Y.D., Kimball, G., Prashad, A.S. and Wang, Y. (2005) *Tetrahedron Letters*, **46**, 8777.
48 Pereira, S. and Srebnik, M. (1995) *Organometallics*, **14**, 3127.
49 (a) Zheng, B. and Srebnik, M. (1994) *Journal of Organometallic Chemistry*, **474**, 49. (b) Deloux, L., Skrzypczak-Jankun, E., Cheesman, B.V., Srebnik, M. and Sabat, M. (1994) *Journal of the American Chemical Society*, **116**, 10302.
50 Hartwig, J.F. and Muhoro, C.N. (2000) *Organometallics*, **19**, 30.
51 He, X. and Hartwig, J.F. (1996) *Journal of the American Chemical Society*, **118**, 1696.
52 (a) Ojima, I., Clos, N., Donovan, R.J. and Ingallina, P. (1990) *Organometallics*, **9**, 3127. (b) Tanke, R.S. and Crabtree, R.H. (1990) *Journal of the American Chemical Society*, **112**, 7984. (c) Jun, C.-H. and Crabtree, R.H. (1993) *Journal of Organometallic Chemistry*, **447**, 177. (d) Maruyama, Y., Yamamura, K., Nakayama, I., Yoshiuchi, K. and Ozawa, F. (1998) *Journal of the American Chemical Society*, **120**, 1421.
53 Harrison, K.N. and Marks, T.J. (1992) *Journal of the American Chemical Society*, **114**, 9220.
54 (a) Hatanaka, Y. and Hiyama, T. (1991) *Synlett*, **12**, 845. (b) Denmark S.E. and Sweis, R.F. (2002) *Accounts of Chemical Research*, **35**, 835. (c) Hiyama, T. (1998) *Organosilicon Compounds in Cross–Coupling Reactions*, in *Metal-Catalyzed Cross-Coupling Reactions* (eds F. Diederich and P.J. Stang), Wiley-VCH Verlag GmbH, Weinheim.
55 (a) Fleming, I., Dunogues, J. and Smithers, R. (1989) *Organic Reactions*, **37**, 57. (b) Blumenkopf, T.A. and Overman, L.E. (1986) *Chemical Reviews*, **86**, 857.
56 (a) Fleming, I., Henning, R. and Plaut, H. (1984) *Journal of the Chemical Society, Chemical Communications*, 29. (b) Tamao, K., Ishida, N., Tanaka, T. and Kumada, M. (1983) *Organometallics*, **2**, 1694.
57 Trost, B.M. and Ball, Z.T. (2005) *Journal of the American Chemical Society*, **127**, 17644.
58 Kopping, B., Chatgilialoglu, C., Zehnder, M. and Giese, B. (1992) *Journal of Organic Chemistry*, **57**, 3994.
59 Chatgilialoglu, C. (1992) *Accounts of Chemical Research*, **25**, 188.
60 (a) Tsipis, C.A. (1980) *Journal of Organometallic Chemistry*, **187**, 427. (b) Green, M., Spencer, J.L., Stone, F.G.A. and Tsipis, C.A. (1977) *Journal of the Chemical Society, Dalton Transactions*, **61**, 1525.
61 Werner, H., Esteruelas, M.A. and Otto, H. (1986) *Organometallics*, **5**, 2295.
62 Esteruelas, M.A., Herrero, J. and Oro, L.A. (1993) *Organometallics*, **12**, 2377.
63 Katayama, H., Taniguchi, K., Kobayashi, M., Sagawa, T., Minami, T. and Ozawa, F. (2002) *Journal of Organometallic Chemistry*, **645**, 192.
64 Na, Y. and Chang, S. (2000) *Organic Letters*, **2**, 1887.

65 Kawanami, Y., Sonoda, Y., Mori, T. and Yamamoto, K. (2002) *Organic Letters*, **4**, 2825.
66 Trost, B.M. and Ball, Z.T. (2001) *Journal of the American Chemical Society*, **123**, 12726.
67 Trost, B.M. and Ball, Z.T. (2005) *Journal of the American Chemical Society*, **127**, 17644.
68 Jun, C.-H. and Crabtree, R.H. (1993) *Journal of Organometallic Chemistry*, **447**, 177.
69 Miyake, Y., Isomura, E. and Iyoda, M. (2006) *Chemistry Letters*, **35**, 836.
70 Stork, G., Jung, M.E., Colvin, E. and Noel, Y. (1974) *Journal of the American Chemical Society*, **96**, 3684.
71 Tsipis, C.A. (1980) *Journal of Organometallic Chemistry*, **188**, 53.
72 Takahashi, T., Bao, F., Gao, G. and Ogasawara, M. (2003) *Organic Letters*, **5**, 3479.
73 (a) Maifeld, S.V., Tran, M.N. and Lee, D. (2005) *Tetrahedron Letters*, **46**, 105. (b) Aricó, C.S. and Cox, L.R. (2004) *Organic and Biomolecular Chemistry*, **2**, 2558.
74 Yong, L., Kirleis, K. and Butenschön, H. (2006) *Advanced Synthesis and Catalysis*, **348**, 833.
75 Molander, G.A. and Retsch, W.H. (1995) *Organommetallics*, **14**, 4570.
76 Molander, G.A., Romero, J.A.C. and Corrette, C.P. (2002) *Journal of Organometallic Chemistry*, **647**, 225.
77 (a) Harrod, J.F., Chalk, A.J. and Davies, N.R. (1965) *Nature*, **205**, 280. (b) Chalk, A.J. and Harrod, J.F. (1965) *Journal of the American Chemical, Society*, **87**, 16. (c) Harrod, J.F. and Chalk, A.J. (1965) *Journal of the American Chemical, Society*, **87**, 1133. (d) Chalk, A.J. and Harrod, J.F. (1967) *Journal of the American Chemical Society*, **89**, 1640.
78 Schroeder, M.A. and Wrighton, M.S. (1977) *Journal of Organometallic Chemistry*, **128**, 345.
79 Vicent, C., Viciano, M., Mas-Marzá, E., Sanaú, M. and Peris, E. (2006) *Organometallics*, **25**, 3713.

16
Metal-Catalyzed Reductive Aldol Coupling

Susan A. Garner, Soo Bong Han and Michael J. Krische

16.1
Introduction – Reductive Generation of Enolates from Enones

Enolate-mediated C—C bond formations rank among the most fundamental and broadly utilized processes in synthetic organic chemistry. Indeed, the earliest examples of enolate coupling vis-à-vis aldol addition extend back nearly 150 years and were instrumental in defining organic chemistry as a distinct field of research [1]. From their inception, enolate chemistry and organic chemistry were closely linked, and they continue to shape one another to this very day.

A significant development in the field of enolate chemistry is represented by the advent of methods for the preformation of structurally defined enolates, which conferred heightened levels of control in additions to diverse C-electrophiles. Enolate preformation was first accomplished by Hauser in 1951, who employed lithium amide in ammonia as a base for the stoichiometric deprotonation of esters [2]. It was not until 1963 that Wittig, in a seminal paper, described the use of lithium diisopropylamide (LDA) for the stoichiometric deprotonation of aldimines, in a reaction now known as the "Wittig directed aldol condensation." [3]. Later in 1970, Rathke utilized lithium hexamethyldisilazane for the preformation of ester enolates [4], which led to the use of lithium *N*-isopropylcyclohexylamide (LICA) for the generation of ester enolates in 1971 [5] and, ultimately, the use of LDA [6] The ability to direct stereoselective enolization in kinetically controlled deprotonations mediated by lithium amides, as described by Ireland's model (1976) [6b] or under thermodynamic conditions, combined with the observation that *Z*- and *E*-enolates react stereospecifically with aldehydes through closed "Zimmerman–Traxler" transition structures [7] to provide *syn*- and *anti*-addition products [8] ultimately led to a renaissance in the area of aldol chemistry in the 1980s.

With these advances in enolate generation, certain limitations were brought to light, including the challenge of regioselective enolate formation. For nonsymmetric ketones that possess different degrees of substitution at the α-positions, as in the case methylcyclohexanone, regioselective enolate formation may be achieved by conducting the deprotonation under kinetic or thermodynamic conditions [9].

Modern Reduction Methods. Edited by Pher G. Andersson and Ian J. Munslow.
Copyright © 2008 WILEY-VCH Verlag GmbH & Co. KGaA, Weinheim
ISBN: 978-3-527-31862-9

However, for these substrates high levels of regioselection are exceedingly difficult to attain. A classical example involves the deprotonation of cholesterane-3-one, which exhibits an overwhelming thermodynamic preference for formation of the Δ_3-enolate. The Δ_2-enolate cannot be formed cleanly via deprotonation under kinetic or thermodynamic conditions. Conversely, introduction of 7,8-unsaturation results in an overwhelming thermodynamic preference for formation of the Δ_2-enolate, and the Δ_3-enolate cannot be formed selectivity under kinetic or thermodynamic conditions for deprotonation (Figure 16.1) [10, 11]

Seminal studies performed by Stork in the 1960s address the challenge of regiocontrolled enolate formation by establishing the use of enones as enolate precursors [12]. Specifically, dissolving-metal reduction (Li/NH$_3$) of conjugated enones was found to promote regiospecific enolate generation, enabling access to enolate isomers, and adducts thereof, that cannot be formed by way of base-mediated deprotonation (Scheme 16.1) [11c].

Subsequent to Stork's findings, a variety of metal catalysts enabling conjugate reduction of α,β-unsaturated carbonyl compounds were reported, and enantiose-

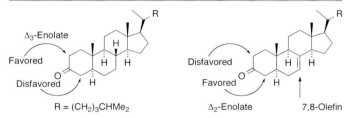

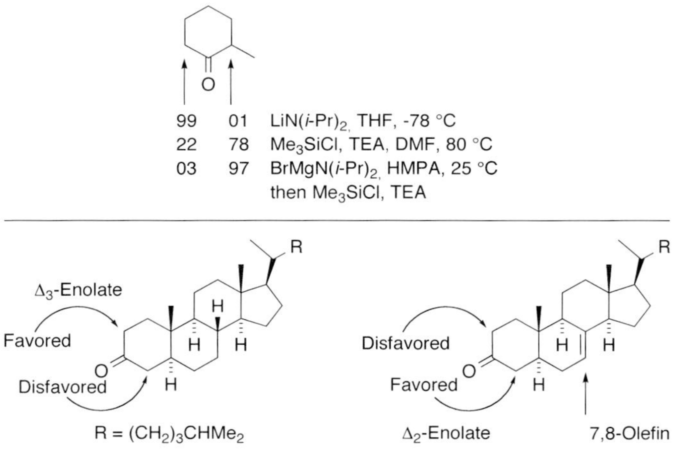

Figure 16.1 Regioselective enolization of nonsymmetric ketones via base-mediated deprotonation.

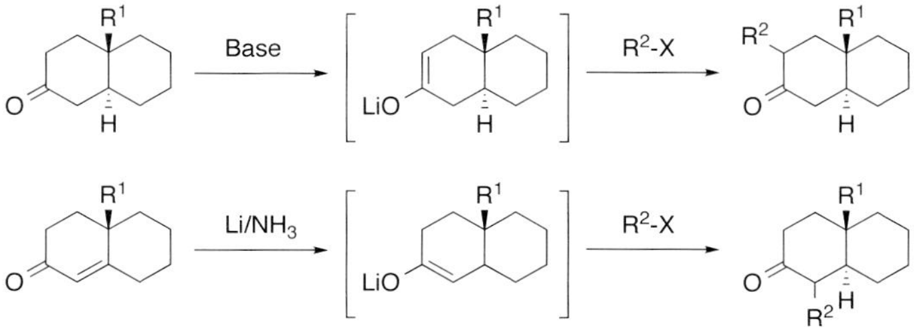

Scheme 16.1 Regiospecific enolization via dissolving-metal reduction—enones as latent enolates.

lective variants of these transformations soon followed [13, 14]. Beyond asymmetric hydrogenation [14], highly enantioselective hydrosilane-mediated (Rh [15] and Cu [13c, 16]), borohydride-mediated (Co [17]) and, most recently, ethanol-mediated (Pd [18]) conjugate reductions have been developed. The silane-mediated conjugate reduction methodologies enable stoichiometric preformation of enol derivatives, which may be exploited in tandem 1,4-reduction–electrophilic trapping sequences [19]. More recently, related processes have emerged, wherein solutions of α,β-unsaturated carbonyl compound are reduced directly in the presence of electrophiles to furnish products of C—C coupling. A challenge inherent to the development of such transformations relates to the design of catalytic systems that exhibit high levels of chemoselectivity for conjugate reduction, thereby mitigating competitive reduction of the electrophile. By meeting this challenge, one dispenses altogether with the requirement of stoichiometric enolate preformation.

In this account, methods for direct metal-catalyzed reductive C—C coupling of α,β-unsaturated carbonyl compounds to aldehydes to provide aldol products are catalogued. Focus is maintained on couplings that proceed through the intermediacy of enolates. Related enone–aldehyde couplings that involve homo-enolates or oxy-π-allyls as reactive intermediates, and therefore result in functionalization of the enone β-position, are not covered [20]. Content is organized on the basis of reaction type, and is categorized further on the basis of metal catalyst. This review covers literature reported up to the end of January 2007.

16.2
The Reductive Aldol Reaction

The issues of selectivity posed by the aldol reaction continue to inspire development of increasingly effective protocols for stereocontrolled aldol addition [21, 22]. The advent of metallic [23] and organic [24] catalysts for direct enantioselective aldol addition of unmodified carbonyl compounds represents an especially significant milestone, as these processes herald a departure from the use of chiral auxiliaries and preformed enol(ate) derivatives. Regioselective enolization in direct enantioselective aldol additions of nonsymmetric ketones generally favors activation of the less substituted enolizable position. For example, direct aldol couplings of 2-butanone catalyzed by L-proline [25, 26] or the heterobimetallic catalyst LaLi$_3$-tris(binaphthoxide) (LLB) furnish linear aldol adducts [27]. Complementary regiochemistry may be achieved through the catalytic reductive coupling of α,β-unsaturated carbonyl compounds and aldehydes to form aldol products. By exploiting enones as regiochemical control elements, one gains access to the branched aldol adducts relevant to polypropionate synthesis (Scheme 16.2).

Following seminal studies by Revis (1987) [28a], the reductive aldol reaction has become the topic of intensive investigation [29]. To date, catalysts for reductive aldol coupling based on rhodium [28, 30], cobalt [31], iridium [32], ruthenium [33], palladium [34], copper [35, 36], nickel [37] and indium [38, 39] have been devised. A survey of these catalytic systems is given below.

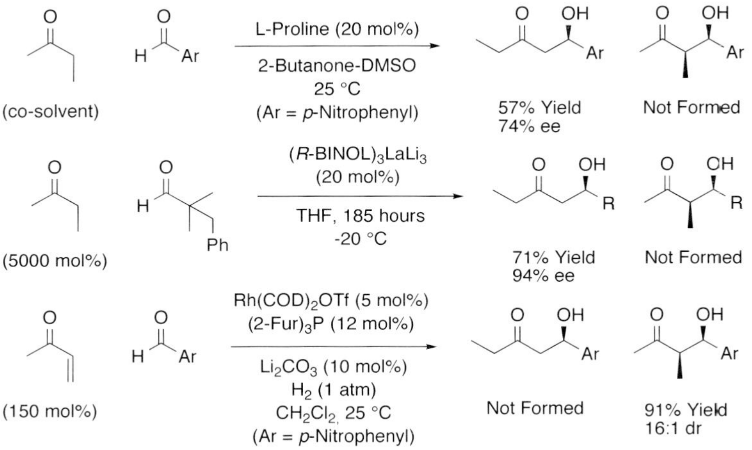

Scheme 16.2 Complementary regiocontrol in direct versus reductive catalytic aldol couplings.

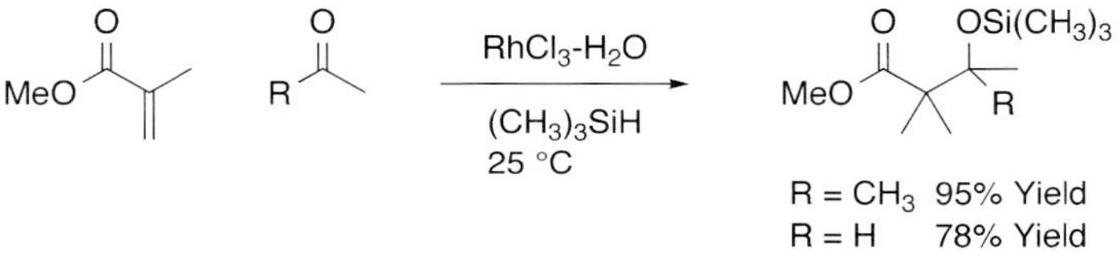

Scheme 16.3 First reported reductive aldol reaction described by Revis in 1987.

16.2.1
Rhodium

In 1987, Revis reported the first metal-catalyzed reductive coupling of α,β-unsaturated esters with carbonyls to form aldol products, termed the "reductive aldol reaction." [28a]. In this transformation, $RhCl_3 \cdot 3H_2O$ serves as precatalyst and trimethylsilane serves as terminal reductant. At ambient temperature, α,β-unsaturated esters and lactones were found to couple to both aldehydes and ketones. Notably, as demonstrated by the coupling of methyl methacrylate to acetone, two contiguous tetrasubstituted carbon centers may be created in this transformation. The silyl ketene acetal derived from methyl methacrylate does not couple to acetone upon exposure to the reaction conditions, suggesting intervention of a rhodium enolate in the reductive catalytic process (Scheme 16.3).

Attempted coupling of methyl vinyl ketone to acetone under the conditions reported by Revis does not provide the aldol adduct, but delivers the corresponding enol silane. It is likely that rhodium enolates derived from vinyl ketones are not sufficiently nucleophilic to engage ketone electrophiles in intermolecular aldol coupling. Consistent with this hypothesis, the reductive aldol coupling of vinyl ketones to aldehydes reported by Matsuda in 1990 proceeds readily using $Rh_4(CO)_{12}$

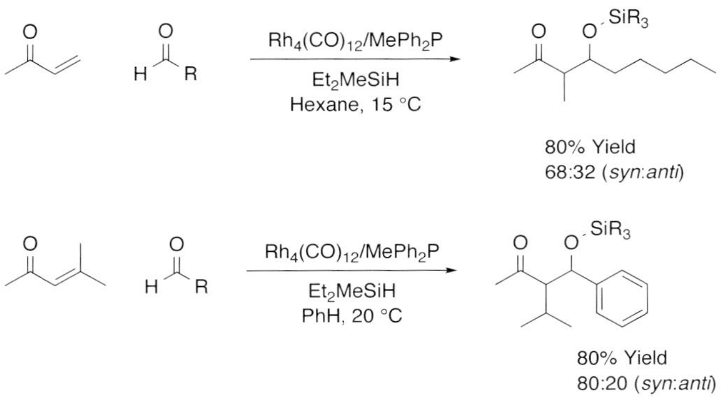

Scheme 16.4 Rhodium-catalyzed reductive aldol coupling employing $Rh_4(CO)_{12}$ as catalyst precursor.

as catalyst precursor and Et_2MeSiH as terminal reductant [28b]. As demonstrated by the coupling of methyl vinyl ketone and mesityl oxide, variable degrees of β-substitution are tolerated in the enone partner. Again, control experiments involving introduction of preformed enol silane to the reaction conditions does not provide an aldol product, suggesting intermediacy of rhodium enolates in the reductive catalytic process. This interpretation is supported by the fact that structurally related O-bound rhodium enolates have been isolated and are found to react with benzaldehyde to furnish corresponding rhodium aldolates (Scheme 16.4) [40].

In 1999, the first diastereoselective reductive aldol reaction was discovered by Morken using an arrayed catalyst evaluation protocol [28c]. Evaluation of 192 independent catalytic systems revealed a strong interdependence of reaction variables. Optimal conditions identified for the reductive aldol coupling of methyl acrylate and benzaldehyde, which involve use of $[Rh(cod)Cl]_2$ dimer as the precatalyst, Me-Duphos as ligand and Cl_2MeSiH as terminal reductant, provide the aldol coupling product in good yield with exceptional *syn*-diastereoselectivity. When these conditions are applied to other substrate combinations, good levels of *syn*-diastereoselectivity persist although isolated yields are diminished in the case of enolizable aldehyde partners. Despite the use of a chiral ligand, asymmetric induction is not observed. It was later found that reactions employing Cl_2MeSiH as reductant give rise to silyl ketene acetals that spontaneously engage aldehydes in carbonyl addition, thus precluding involvement of the chirally modified catalyst in the enantiodetermining event [28d]. In contrast, using Et_2MeSiH as the terminal reductant, high levels of enantioselection may be achieved [28e]. However, diminished yields again were observed in connection with the use of enolizable aldehyde partners (Scheme 16.5).

Two plausible mechanistic pathways for the rhodium-catalyzed reductive aldol coupling of acrylates and aldehydes have been proposed (Scheme 16.6). Both mechanism involve initial oxidative addition of hydrosilane to $L_nRh(I)$ to form $L_nRh(H)(SiR_3)$ **(1)**. In one possible scenario, catalytic cycle **(A)**, the aldehyde inserts

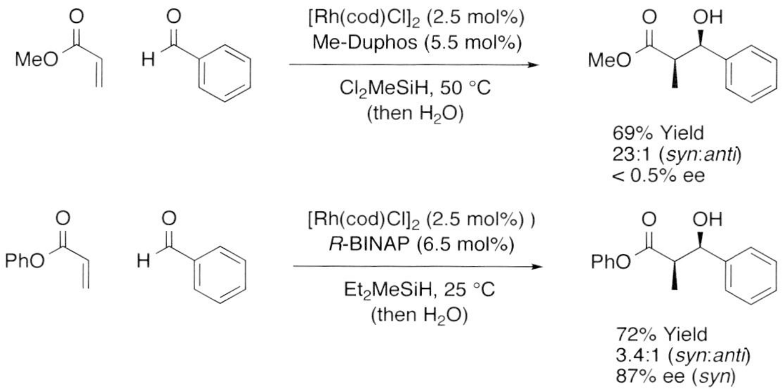

16 Metal-Catalyzed Reductive Aldol Coupling

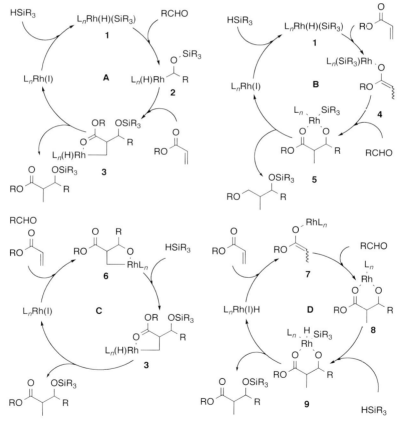

Scheme 16.5 First highly diastereo- and enantioselective reductive aldol couplings.

Scheme 16.6 Plausible mechanistic pathways for reductive aldol coupling as corroborated by deuterium labeling.

into the Rh—Si bond to form complex **(2)**, followed by acrylate insertion into the newly formed Rh—C bond to form intermediate **(3)**. Finally, C—H reductive elimination delivers the aldol as the silyl ether along with $L_nRh(I)$. A related mechanism has been proposed for the rhodium-catalyzed silylformylation of the aldehydes [41]. Alternatively, as depicted in catalytic mechanism **(B)**, $L_nRh(H)(SiR_3)$ **(1)** may insert the acrylate to furnish rhodium enolate **(4)**. Carbonyl addition provides the rhodium aldolate **(5)**, which upon O—Si reductive elimination delivers the silyl-protected aldol and $L_nRh(I)$. This mechanism finds support in the aldol addition of stoichiometrically preformed rhodium enolates [40]. For the enantioselective reductive aldol coupling depicted in Scheme 16.5, use of phenyldimethylsilyldeuteride as terminal reductant results in partial deuteration at the former acrylate β-position, which is consistent with both catalytic cycles **(A)** and **(B)**. In another experiment, it was found that after 24 hours in the absence of aldehyde less than 5% of the silylketene acetal was detected. Also, if preformed silyl ketene acetal is used instead of the acrylate, less than 5% of the reductive aldol product is formed. These data are consistent with the intermediacy of a rhodium enolate. Intervention of rhodium enolate **(4)** is especially attractive, as partial deuterium incorporation at the former enone β-position would arise as a consequence of reversible acrylate hydrometallation. Two additional mechanistic possibilities that were not enumerated by the authors, but that are consistent with the results of deuterium labeling, and which may be operative in other catalytic systems (see above), are represented by catalytic cycles **(C)** and **(D)**. In catalytic cycle **(C)**, acrylate–aldehyde oxidative coupling mediated by $L_nRh(I)$ provides the oxametallacyclic intermediate **(6)**. σ-Bond metathesis of **(6)** with silane provides intermediate **(3)**, which upon C—H reductive elimination provides the silyl-protected aldol adduct. Finally, as shown in catalytic cycle **(D)**, monohydride-based catalytic cycles involving rhodium(I) enolate **(7)** and rhodium(I) aldolates **(8)** and **(9)** may be envisioned (Scheme 16.6).

Silane-mediated reductive aldol cyclizations employing Wilkinson's complex, $RhCl(PPh_3)_3$, or $RhH(PPh_3)_4$ as precatalyst have been described [28g, h]. Although triphenylphosphine is used as ligand in each case, the stereochemical outcome of the reaction was found to be highly dependent upon the choice of precatalyst. Whereas cyclizations employing $RhCl(PPh_3)_3$ provide the *cis*-isomer with modest levels of stereoselection, cyclizations employing $RhH(PPh_3)_4$ as precatalyst deliver the *anti*-isomer with relatively high levels of stereoselectivity. These data suggest a precatalyst-dependent partitioning of alternate catalytic mechanisms. It is possible that Wilkinson's complex, $RhCl(PPh_3)_3$, promotes catalytic cycle **(B)** involving intermediates **(1)**, **(4)** and **(5)**, while $RhH(PPh_3)_4$ promotes monohydride catalytic cycle **(D)** involving intermediates **(7)**, **(8)** and **(9)** (Scheme 16.6). Notably, attempted cyclization of the corresponding methyl ketone using Wilkinson's complex, $RhCl(PPh_3)_3$, provides an acyclic product that appears to arise by way of olefin isomerization to furnish a conjugated enone that is subject to 1,4-reduction (Scheme 16.7).

Using a Rh(Phebox) catalyst, silane-mediated intermolecular reductive aldol coupling is achieved with excellent levels of *anti*-diastereoselectivity accompanied by exceptional levels of asymmetric induction across a range of substrates [28i].

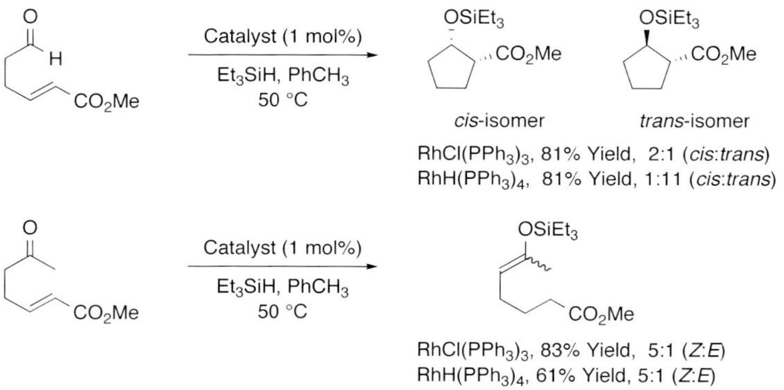

Scheme 16.7 Rhodium-catalyzed reductive aldol cyclization.

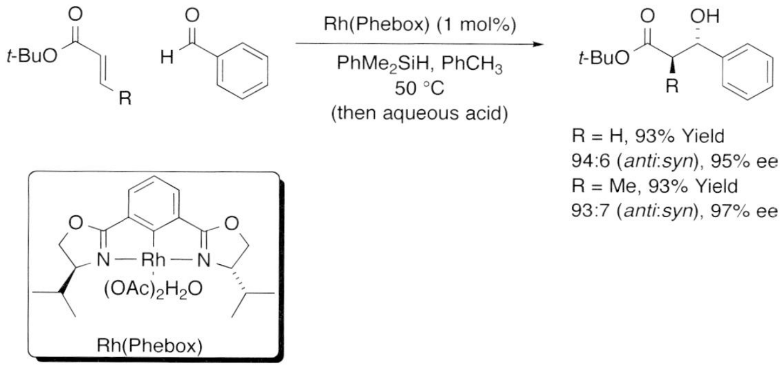

Scheme 16.8 Highly *anti*-diastereoselective and enantioselective reductive aldol coupling catalyzed by Rh(Phebox).

The coupling reactions are conducted at 50 °C with dropwise addition of the hydrosilane to a solution of *t*-butyl acrylate, aldehyde and the rhodium complex. The reaction is believed to involve initial reduction of the Rh^{III}(Phebox) complex to provide Rh^{I}(Phebox), which promotes catalytic cycle **(B)** involving intermediates **(1)**, **(4)** and **(5)** (Scheme 16.6). Intervention of a chair-like transition state and an (*E*)-rhodium enolate are invoked in the stereodetermining step. The reaction is applicable to both aromatic and aliphatic aldehydes, as well as β-substituted pronucleophiles such as *t*-butyl crotonate. Further studies of the Rh(Phebox) catalyst demonstrate that appropriate substitution of the ligand phenyl moiety confers heightened levels of reactivity and selectivity (Scheme 16.8) [28j, k].

An unusual variation of the catalytic reductive aldol reaction involves use of β-sulfido aldehydes as the stoichiometic reductant [28l]. The β-sulfido moiety promotes formation of chelated acyl rhodium hydrides, which engage in enone hydrometallation en route to rhodium enolates. Carbonyl addition to a second molecule of aldehyde followed by C—O reductive elimination from the resulting

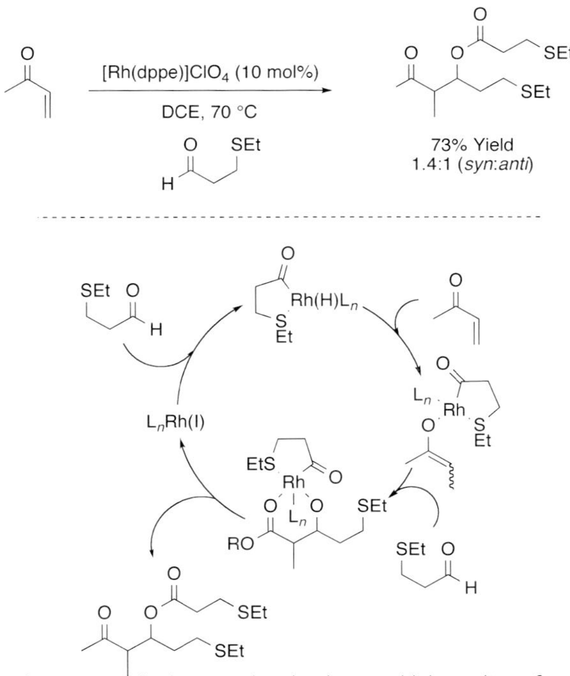

Scheme 16.9 Rhodium-catalyzed reductive aldol coupling of β-sulfido aldehydes and proposed catalytic mechanism.

aldolate delivers the *O*-acylated aldol adduct with concomitant regeneration of the active catalyst. This first generation catalytic system provides aldol adducts in good yields with diastereoselectivities ranging from 1:1 to 3.3:1 in favor of the *syn*-isomer (Scheme 16.9).

In 2002, it was found that reductive aldol coupling may be achieved by hydrogenating enones in the presence of aldehydes [30]. Through the use of tri-2-furylphosphine as ligand [42], the observed levels of *syn*-diastereoselection, which are obtained at ambient temperature, exceed those observed in reactions of lithium enolates conducted at –78 °C [30e]. High *syn*-diastereoselectivity suggests a kinetically controlled process [21], and may be rationalized on the basis of a mechanism involving stereospecific formation of the (Z)-(O)-enolate, as determined by internal hydride delivery to the enone *S-cis* conformer through a six-centered transition structure, and subsequent addition of the (Z)-(O)-enolate to the aldehyde through a Zimmerman–Traxler type transition state [7]. By virtue of the facile kinetic pathway for enone reduction, competitive conventional hydrogenation of "hydrogen-labile" functional groups (alkynes, alkenes, benzylic ethers and nitroarenes) is not observed under the conditions of hydrogen-mediated coupling, thus enabling chemoselective activation of vinyl ketones and even divinyl ketones [30f]. Because hydrogen-mediated aldol coupling occurs under essentially neutral conditions in low dielectric media at ambient temperature, sensitive *N*-Boc-α-aminoaldehydes

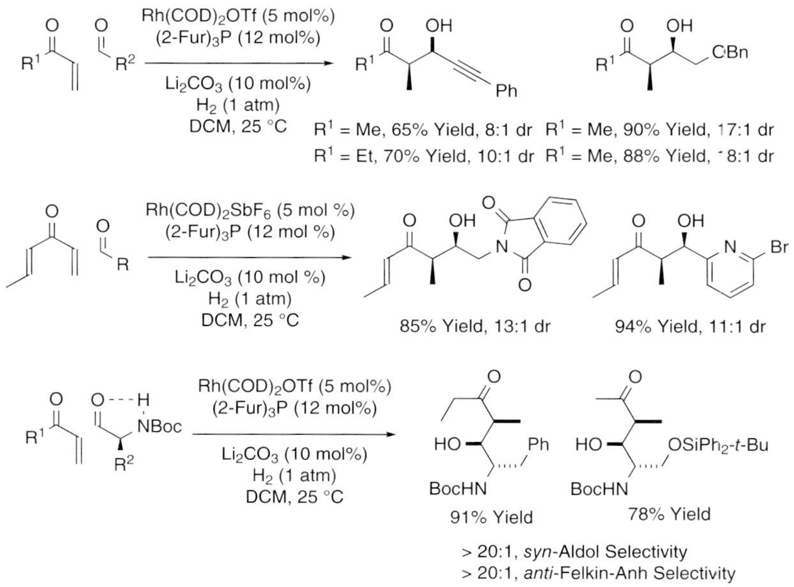

Scheme 16.10 Highly *syn*-diastereoselective rhodium-catalyzed reductive aldol couplings mediated by hydrogen.

may react through hydrogen-bonded chelates without epimerization to furnish aldol adducts that embody exceptional levels of *syn*-aldol diastereoselectivity and *anti*-Felkin–Anh control (Scheme 16.10) [30g].

For the hydrogen-mediated reductive aldol additions, cationic rhodium precatalysts along with substoichiometric quantities of mild basic additives are found to suppress competitive conventional hydrogenation pathways. Whereas neutral rhodium(I) catalysts induce homolytic hydrogen activation [43], corresponding cationic complexes used in the presence of basic additives are known to promote heterolytic hydrogen activation (H_2 + M-X → M-H + HX) [44]. This behavior derives from the enhanced acidity of the cationic dihydrides [45]. Hence, basic additives may disable direct enolate-hydrogen reductive elimination manifolds by deprotonation of the (hydrido)rhodium intermediates $L_nRh^{III}X(H)_2$ or (enolato)$Rh^{III}X(H)L_n$, thereby promoting entry into a monohydride-based catalytic cycle. The results of isotopic labeling studies involving reductive coupling of MVK and *p*-nitrobenzaldehyde performed under an atmosphere of elemental deuterium are consistent with this interpretation. The aldol adduct incorporates a single deuterium atom at the former enone β-position [30e]. This result excludes Morita–Baylis–Hillman pathways *en route* to aldol product, as deuterium incorporation at the α-carbon is not observed. Irreversible enone hydrometallation is suggested by the incorporation of a single deuterium atom. A catalytic cycle involving enone–aldehyde oxidative coupling (mechanism (**C**), Scheme 16.6) cannot be excluded and may account for the fact that easily reduced enones participate in the coupling, whereas α,β-unsaturated esters do not (Scheme 16.11).

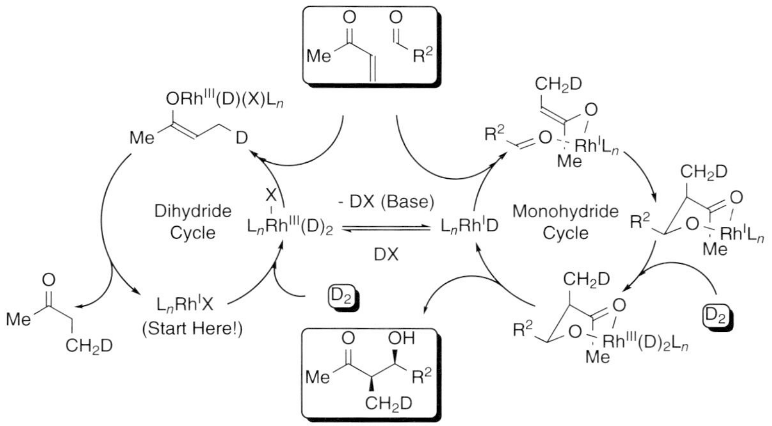

Scheme 16.11 Rhodium-catalyzed hydrogenative aldol coupling under an atmosphere of deuterium: partitioning of carbonyl addition and 1,4-reduction manifolds.

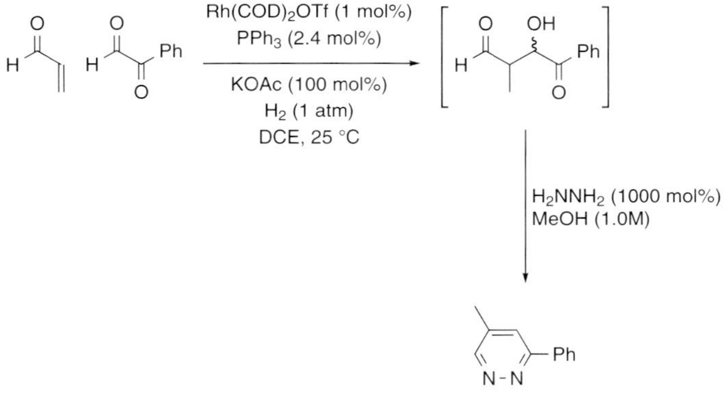

78% Yield, 17:1 dr, R² = p-NO₂Ph
Exclusive monodeuteration
at the former enone β-position

Scheme 16.12 Acrolein as a metallo-aldehyde enolate precursor in hydrogenative aldol additions to α-ketoaldehydes.

The use of metallo-aldehyde enolates in aldol coupling typically suffers from polyaldolization, product dehydration and competitive Tishchenko-type processes [21d, 46]. Indeed, catalytic cross aldolization of aldehyde donors only has been achieved through amine catalysis and the use of aldehyde-derived enol silanes [47]. The first use of metallo-aldehyde enolates in catalytic cross aldolization was achieved by hydrogenating enals in the presence of α-ketoaldehydes [30d]. The resulting β-hydroxy-γ-keto aldehydes are highly unstable, but may be trapped through condensation with hydrazine to afford 3,5-disubsituted pyridazines (Scheme 16.12).

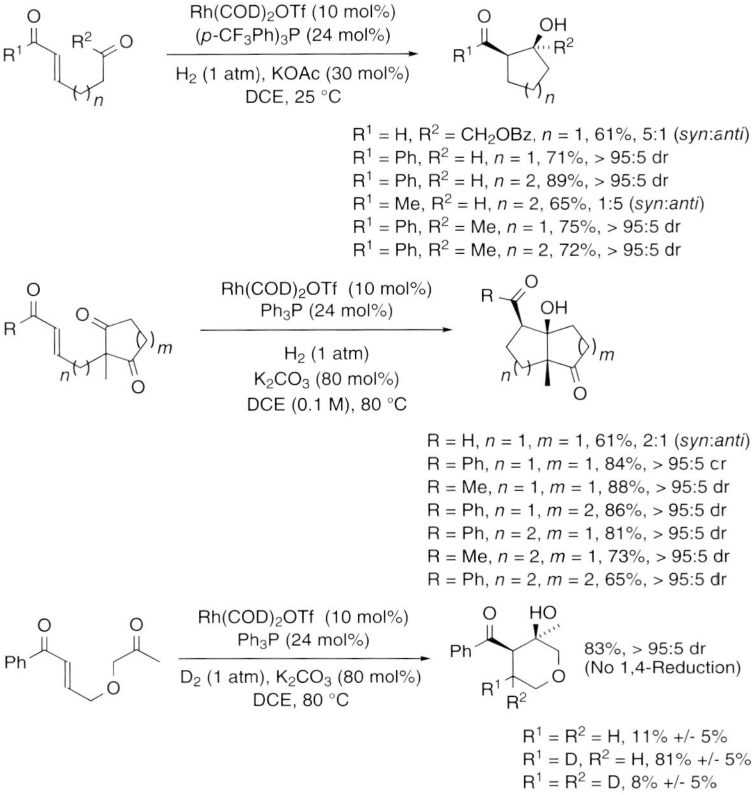

Scheme 16.13 Aldol cycloreduction via rhodium-catalyzed hydrogenation.

Under hydrogenation conditions, the cyclization of enone and enal donors tethered to aldehyde and ketone acceptors is readily achieved with high levels of *syn*-diastereoselectivity [30a, b, c]. The observance of *syn*-aldol adducts is consistent once again with the intermediacy of a *Z*-enolate and a closed Zimmerman–Traxler type transition structure. Reversible aldol addition is suggested by an inversion in diastereoselectivity observed in connection with aliphatic enone pronucleophiles and the results of isotopic labeling. In the latter case, hydrogenative aldol cyclization under an atmosphere of deuterium results in deuterium incorporation exclusively at the former enone β-position, but as a distribution of deuterated products. This result supports reversible enone hydrometallation in the case of ketone acceptors, where reversible aldol addition is anticipated (Scheme 16.13).

16.2.2
Cobalt

In 1989, a cobalt-catalyzed reductive aldol coupling mediated by phenylsilane was reported by Mukaiyama [31a]. The reaction exhibits broad scope and is applicable

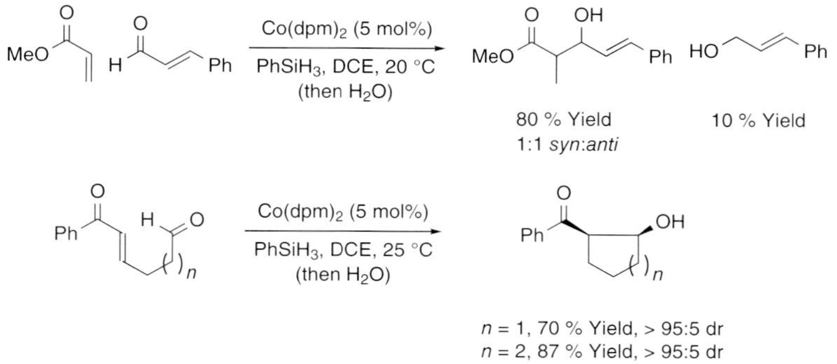

Scheme 16.14 Cobalt-catalyzed intermolecular reductive aldol coupling (dpm = dipivaloylmethanato).

to the coupling of α,β-unsaturated nitriles, amides and esters to a diverse assortment of aldehydes. While good yields of coupling product are obtained, roughly equimolar distributions of diastereomers are observed. Later in 2001, an intramolecular variant of the cobalt-catalyzed reductive aldol reaction was reported [31b, c]. Unlike the intermolecular coupling, the cyclization occurs with high levels of *syn*-diastereoselectivity for aryl- and heteroaryl-substituted enones. α,β-Unsaturated nitriles, amides and esters do not participate in the cyclization, instead providing products of conjugate reduction. This methodology is applicable to the formation of five- and six-membered ring aldol adducts (Scheme 16.14).

The catalytic mechanism proposed for the cobalt-catalyzed reductive aldol coupling takes into account the unique reactivity of Co(dpm)$_2$, a tetrahedral d^7-metal complex with three unpaired electrons. Tetrahedral d^7-metal complexes are known to engage in single-electron oxidative addition through a free radical mechanism [48]. Additionally, Co(II) complexes are known to disproportionate [49]. Hence, two plausible mechanisms may account for generation of the (hydrido)cobalt intermediates required for entry into catalytic cycles involving enone hydrometallation: (i) single-electron oxidative addition of silane to Co(II) or (ii) disproportionation followed by two-electron oxidative addition of silane to Co(I). In either case, a Co(III) hydride is formed, suggesting a Co(I)–Co(III) cycle. Alkene geometry does not influence diastereoselection, which is consistent with a catalytic mechanism involving enone hydrometallation. Additionally, reductive cyclization mediated by PhSiD$_3$ results in the incorporation of a single deuterium atom at the former enone β-position as an equimolar mixture of epimers, as determined by single-crystal neutron diffraction. The incorporation of a single deuterium atom suggests that β-hydride elimination of the metallo-enolate is slow with respect to subsequent steps in the catalytic cycle. The observance of an equimolar mixture of epimers suggests any combination of the following: (i) the enolate resides as the *O*-bound haptomer, (ii) π-facial interconversion of the metallo-enolate is faster than aldehyde addition, or (iii) aldol addition is reversible. The observed *syn*-diastereomers likely arise through addition of the (*Z*)-cobalt enolate to the tethered aldehyde through a chair-like transition structure (Scheme 16.15).

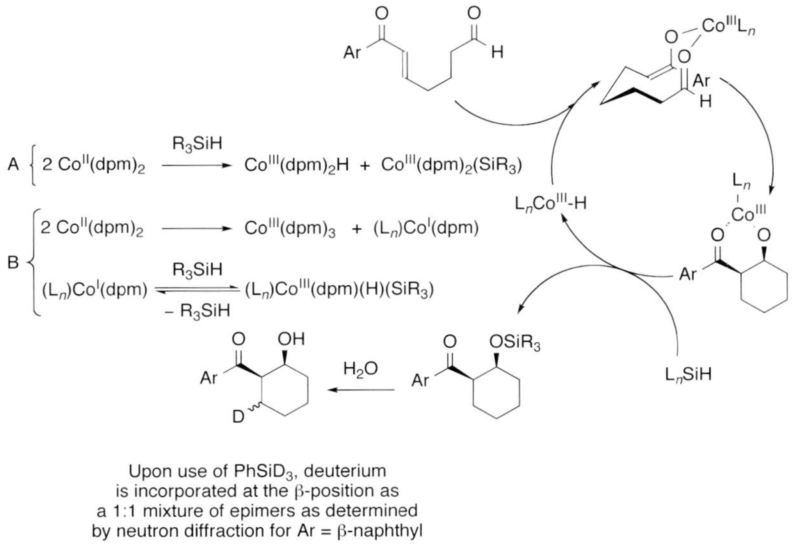

Scheme 16.15 Plausible mechanism for the cobalt-catalyzed reductive aldol cyclization as supported by isotopic labeling studies.

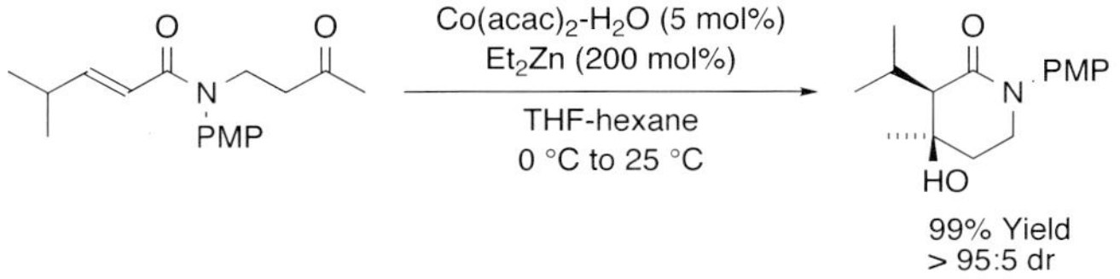

Upon use of PhSiD$_3$, deuterium is incorporated at the β-position as a 1:1 mixture of epimers as determined by neutron diffraction for Ar = β-naphthyl

Scheme 16.16 Cobalt-catalyzed reductive aldol cyclization mediated by diethylzinc.

A cobalt-catalyzed reductive aldol cyclization reaction employing diethylzinc as the stoichiometric reductant was reported recently[31d]. Under optimal conditions, α,β-unsaturated amides are cyclized onto tethered ketones to provide a variety of 4-hydroxypiperdin-2-ones in excellent yield and with exceptional levels of diastereoselectivity (Scheme 16.16). A hydrometallative mechanism involving intermediacy of an amide enolate was proposed. Specifically, it was postulated that the cobalt precatalyst and diethylzinc engage in transmetallation–β-hydride elimination to furnish a cobalt hydride. Hydrometallation of the α,β-unsaturated amide delivers a cobalt enolate that participates in a second transmetallation with diethylzinc to produce a zinc enolate and an ethylcobalt species. Addition of the zinc enolate to the tethered ketone provides the zinc-aldolate of the cyclized product. β-Hydride elimination of the ethylcobalt species regenerates the cobalt hydride along with ethylene (Scheme 16.17).

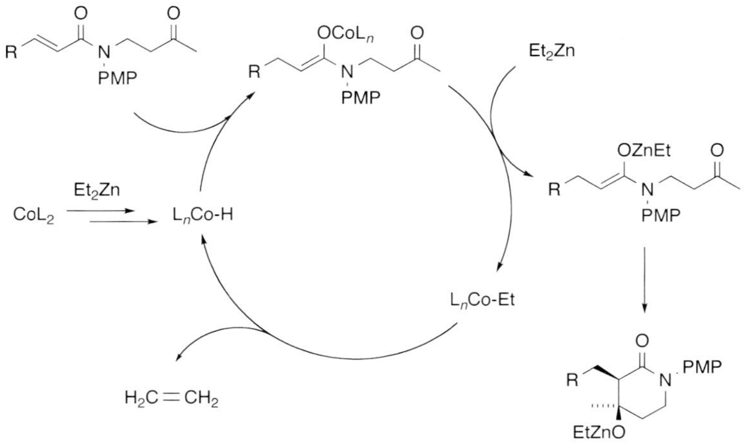

Scheme 16.17 Mechanism proposed for the cobalt-catalyzed reductive aldol cyclization mediated by diethylzinc.

16.2.3
Iridium

In 2001, an Ir(pybox) complex was reported to catalyze highly diastereo- and enantioselective reductive aldol couplings mediated by silane [32]. Simple aliphatic aldehydes are not sufficiently reactive to participate in the coupling. However, aldehydes bearing heteroatoms at the α- and β-positions, which benefit from σ-inductive effects, provide moderate yields of coupling product. For the indicated α-chiral aldehyde, highly stereoselective reductive aldol addition is observed in the matched case. In the mismatched case, aldehyde reduction is observed. For the indicated β-chiral aldehyde, erosion in the level of *syn*-diastereoselectivity is observed. For both enantiomers of the β-chiral aldehyde, the ratio of the major and minor *syn*-aldol isomers is nearly identical, indicating good levels of catalyst-directed asymmetric induction (Scheme 16.18).

16.2.4
Ruthenium

Recently, a RuHCl(CO)(PPh$_3$)$_3$-catalyzed reductive aldol dimerization of α,β-unsaturated aldehydes to provide α-hydroxymethyl ketones was reported that employs 2-propanol as the terminal reductant [33]. The first step in the proposed catalytic mechanism involves enal hydrometallation to provide a ruthenium enolate. Addition of the enolate to a second molecule of aldehyde provides a ruthenium aldolate, which upon β-hydride elimination delivers an intermediate 1,3-dicarbonyl compound with regeneration of ruthenium hydride. Sequential hydrogen transfer from 2-propanol reduces the formyl moiety and gives an unsaturated dimer, which upon further reduction provides the saturated dimer (Scheme 16.19).

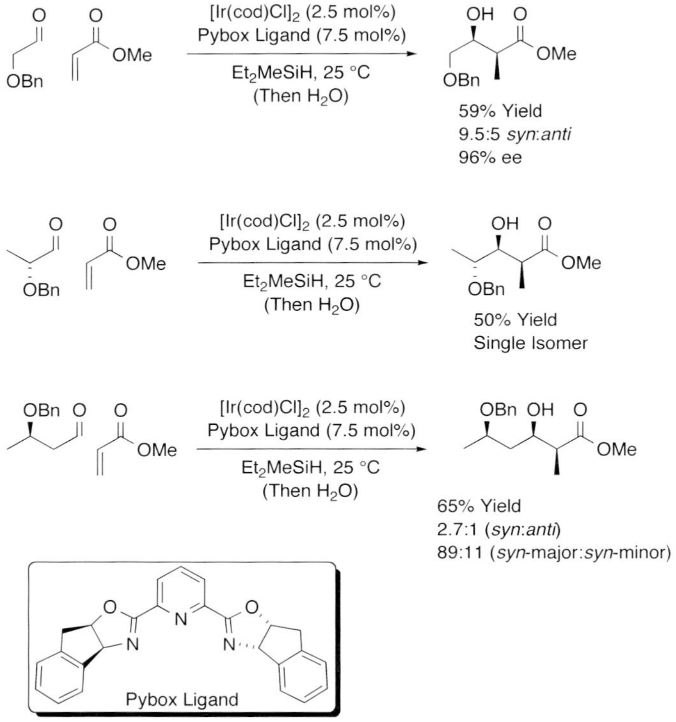

Scheme 16.18 Diastereo- and enantioselective reductive aldol coupling catalyzed by Ir(pybox).

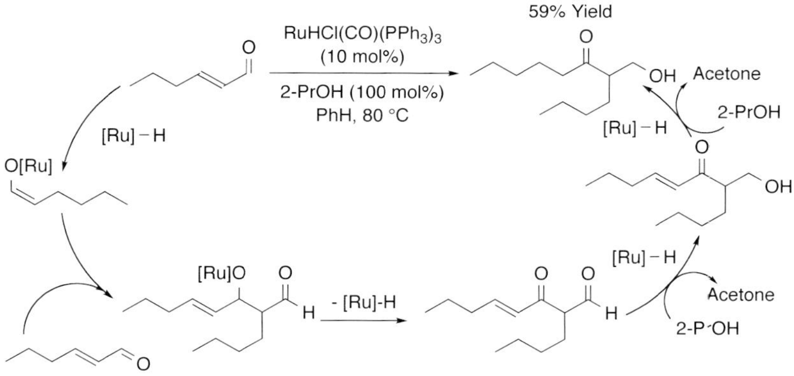

Scheme 16.19 Ruthenium-catalyzed reductive dimerization of enals mediated by 2-propanol.

16.2.5
Palladium

A singular example of palladium-catalyzed reductive aldol coupling was disclosed in 1998 [34]. This trichlorosilane-mediated coupling employs tetrakis(triphenylphosphine)palladium as the catalyst precursor. Under the reported reaction conditions, the aldol product is obtained in good yield with modest levels of *anti*-diastereoselectivity. The first steps in the catalytic mechanism are postulated to involve oxidative addition of the trichlorosilane to palladium(0), followed by hydrometallation of the α,β-unsaturated compound. The remainder of the mechanism is unclear. It is not known whether the aldehyde addition occurs through the intermediacy of the palladium enolate or whether O—Si reductive elimination occurs to provide an enol silane that reacts with the aldehyde spontaneously (Scheme 16.20).

16.2.6
Copper

Stryker's reagent, $[PPh_3CuH]_6$, has been used stoichiometrically in reductive aldol cyclization [35]. Soon thereafter, it was found that substoichiometric quantities of Stryker's reagent catalyze the conjugate reduction of α,β-unsaturated ketones to furnish enol silanes, which participate in Lewis acid-catalyzed Mukaiyama aldol additions [19b]. To date, true reductive aldol processes that circumvent stoichiometric enol(ate) preformation based on Stryker's reagent have not been developed. However, related reductive cyclizations of α,β-acetylenic ketones onto tethered ketones catalyzed by Stryker's reagent and mediated by silane have been described (Scheme 16.21) [35d].

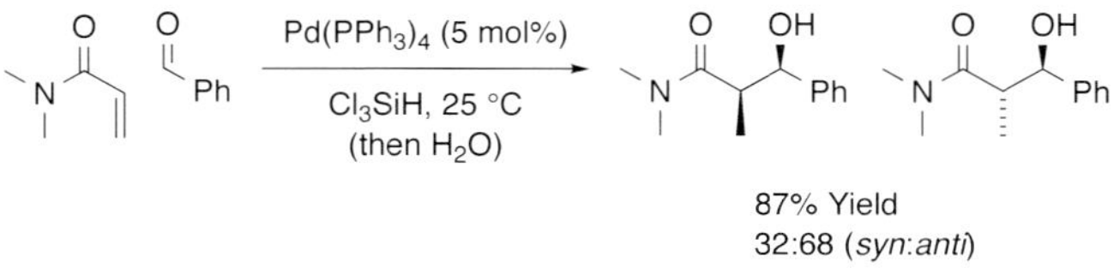

Scheme 16.20 Palladium-catalyzed reductive aldol coupling mediated by silane.

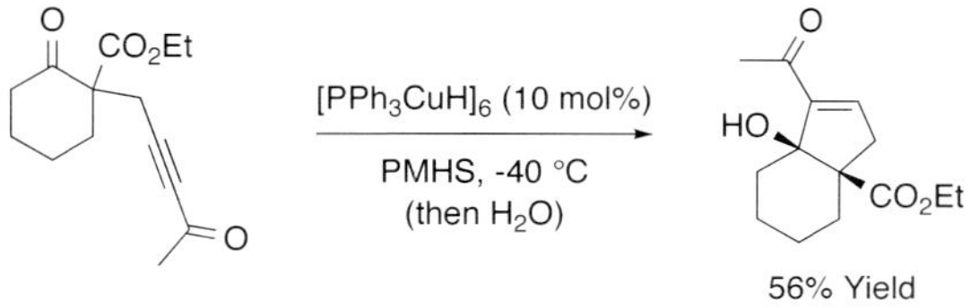

Scheme 16.21 Reductive cyclization of zyxoneoneonexyz,β-acetylenic ketones onto tethered ketones catalyzed by Stryker's reagent.

The first catalytic reductive aldol coupling to employ a copper catalyst was reported in 1999[36a]. Cuprous chloride is postulated to initiate the addition of Bu₃SnH to the enone substrate through a radical mechanism, thereby generating a tin enolate. The intervention of radical intermediates finds indirect support by the fact that related tributylstannane-mediated halide reductions are arrested upon introduction of the radical inhibitor galvinoxyl. The authors propose that the copper catalyst not only promotes formation of the tin enolate but also catalyzes subsequent aldehyde addition by acting as a Lewis acid. This method is applicable to the coupling of vinyl ketones to aliphatic, aromatic and α,β-unsaturated aldehydes. The aldol products are obtained in good isolated yield with modest levels of *syn*-diastereoselection (Scheme 16.22).

In 2005, a diastereo- and enantioselective copper(II)-catalyzed reductive aldol cyclization mediated by 1,1,3,3-tetramethylhydroxysiloxane (TMDS) was developed by Lam and applied to the synthesis of β-hydroxy lactones [36b]. The substrates for this transformation are α,β-unsaturated esters of hydroxy ketones. The products form as single diastereomers, as determined by ¹H NMR analysis of the unpurified reaction mixtures. In an effort to develop an asymmetric variant of the cyclization, several chiral ligands were screened, revealing moderate levels of enantioselection. This methodology was extended to the synthesis of 4-hydroxypiperdin-2-ones, which are formed through the cycloreduction of corresponding α,β-unsaturated keto-amides. Again, exceptional levels of diastereoselection are observed (Scheme 16.23) [36c].

The authors propose a catalytic mechanism wherein silane-mediated reduction of Cu(OAc)₂·H₂O provides a copper (I)-bisphosphine hydride complex, which initiates the catalytic cycle through hydrometallation of the α,β-unsaturated ester. The resulting copper enolate undergoes addition to the tethered ketone to give a copper aldolate, which upon σ-bond metathesis with siloxane delivers the product with concomitant regeneration of the copper (I)-bisphosphine hydride complex to close the cycle. The observed stereoselectivity is postulated to arise through aldehyde addition of the (Z)-copper enolate in a closed transition structure (Scheme 16.24).

The first highly diastereo- and enantioselective intermolecular reductive aldol coupling to simple ketones was reported by Riant using [CuF(PPh₃)₃]·2MeOH as precatalyst and phenylsilane as the stoichiometric reductant [36d]. Entry into the catalytic cycle is achieved through the reaction of hydrosilane and

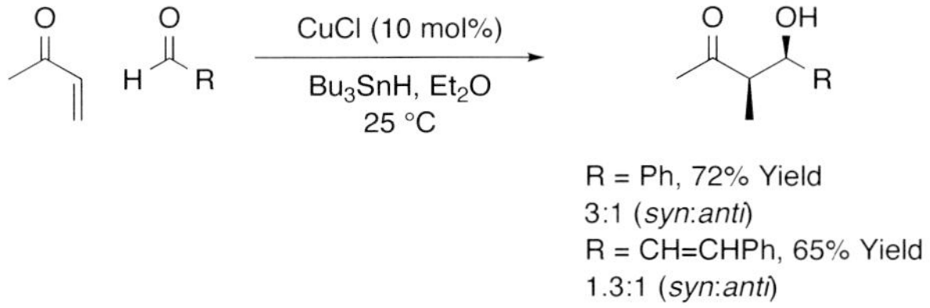

Scheme 16.22 Copper-initiated radical addition of Bu₃SnH to enones resulting in reductive aldol coupling.

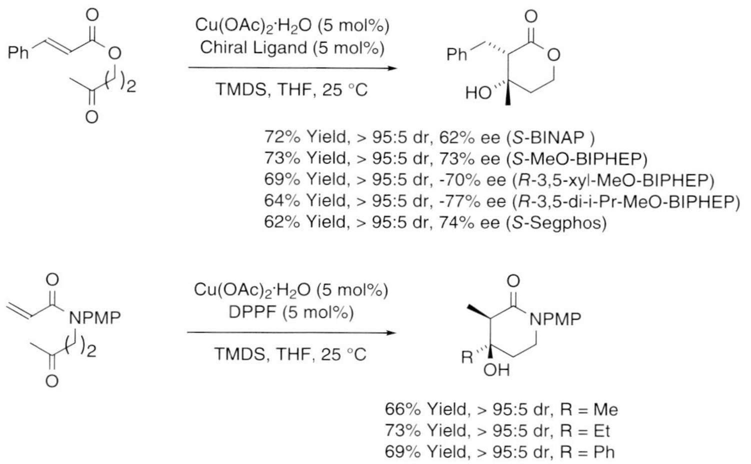
Cu(OAc)2·H2O (5 mol%)
Chiral Ligand (5 mol%)
TMDS, THF, 25 °C
72% Yield, > 95:5 dr, 62% ee (S-BINAP)
73% Yield, > 95:5 dr, 73% ee (S-MeO-BIPHEP)
69% Yield, > 95:5 dr, -70% ee (R-3,5-xyl-MeO-BIPHEP)
64% Yield, > 95:5 dr, -77% ee (R-3,5-di-i-Pr-MeO-BIPHEP)
62% Yield, > 95:5 dr, 74% ee (S-Segphos)
Cu(OAc)2·H2O (5 mol%)
DPPF (5 mol%)
TMDS, THF, 25 °C
66% Yield, > 95:5 dr, R = Me
73% Yield, > 95:5 dr, R = Et
69% Yield, > 95:5 dr, R = Ph

Scheme 16.23 Diastereo- and enantioselective copper(II)-catalyzed reductive aldol cyclizations mediated by silane.

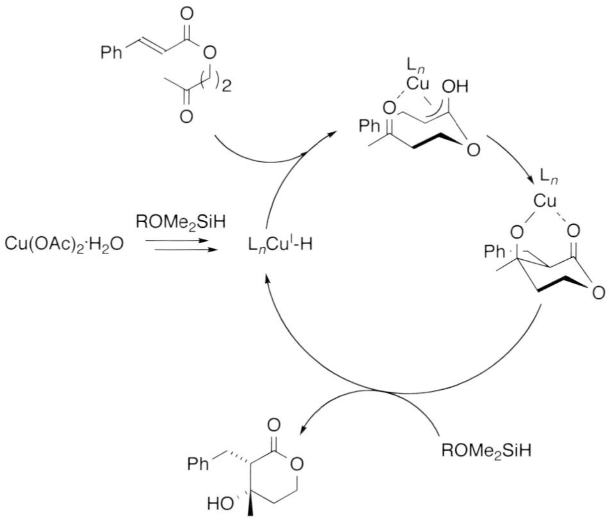
Cu(OAc)2·H2O
ROMe2SiH
LnCuI-H
ROMe2SiH

Scheme 16.24 Proposed catalytic cycle for the copper-catalyzed reductive aldol cyclization mediated by silane.

$[CuF(PPh_3)_3]\cdot 2MeOH$ to generate a copper(I) hydride, which initiates a hydrometallative mechanism akin to that previously described (Scheme 16.24). Under optimum conditions employing a "Taniaphos" ligand, methyl acrylate couples to methyl ketones such as acetophenone with remarkable levels of *anti*-diastereoselectivity and absolute stereocontrol. Aldehyde couplings performed under these conditions exhibit modest levels of diastereoselection and highly variable levels of enantioselection [36e]. Very similar conditions for intermolecular reductive aldol

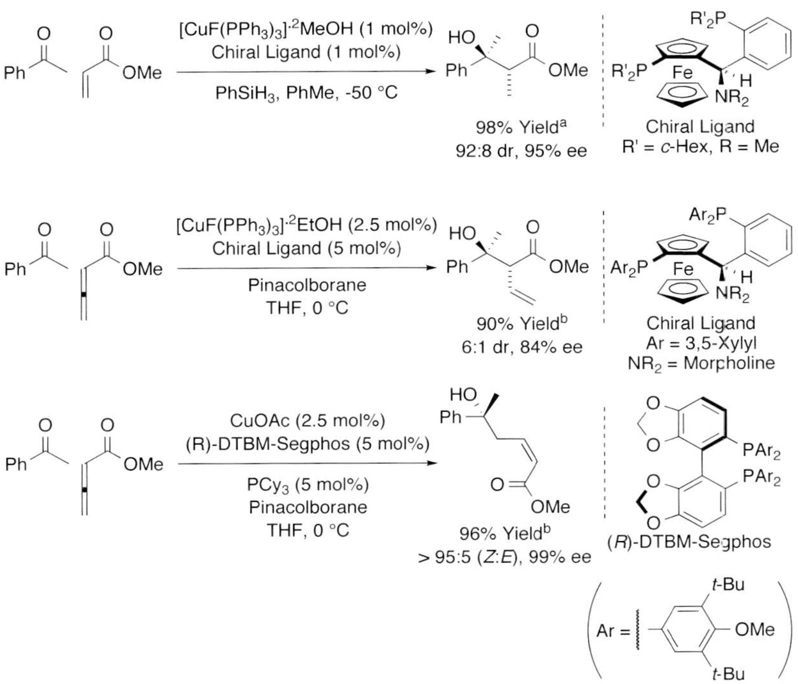

Scheme 16.25 Diastereo- and enantioselective reductive aldol additions to ketones catalyzed by copper.[a] Yield of isolated product. [b] Yield was determined by ^{1}H NMR of the crude reaction mixture using an internal standard.

coupling of acrylates, β-substituted acrylates and allenic esters to simple ketones using [CuF(PPh$_3$)$_3$]·2EtOH as precatalyst and triethoxysilane as the stoichiometric reductant were concurrently reported by Shibasaki and Kanai [36f]. In these reactions, which employ (R)-tol-BINAP as ligand, modest levels of relative and absolute stereocontrol were observed. Later it was found that using a Taniaphos ligand with pinacolborane as reductant, selectivities in reductive aldol couplings of allenic esters to acetophenone were improved [36g]. Furthermore, through the use of P(c-Hex)$_3$ as an additive, highly enantioselective alkylation is directed to the γ-position of the purported dienolate intermediate (Scheme 16.25).

16.2.7
Nickel

A nickel-catalyzed reductive aldol addition of *t*-butyl acrylate to aromatic and aliphatic aldehydes mediated by trialkylborane has been reported [37]. Interestingly, an aryl iodide is required to initiate the catalytic process, though it is not incorporated into the reaction product. In the absence of phenyl iodide no reaction occurs and starting materials are recovered. It was found that the reaction requires

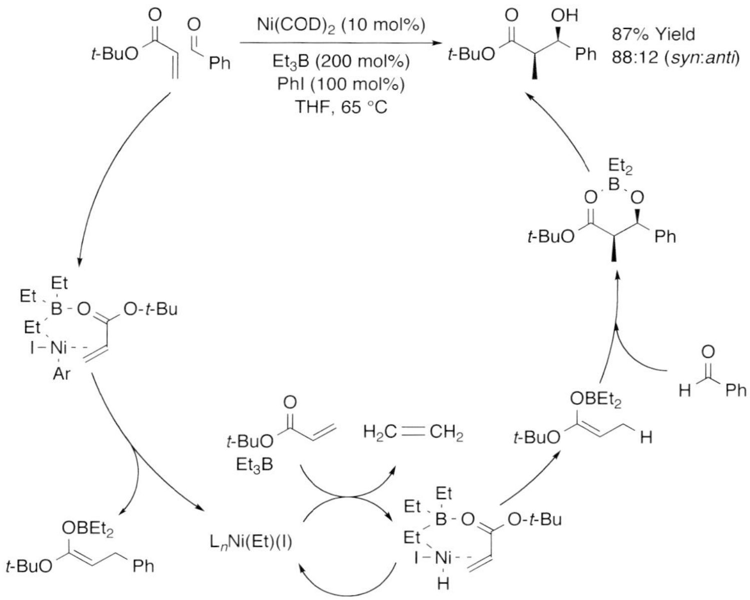

Scheme 16.26 Nickel-catalyzed reductive aldol coupling initiated by aryl iodides and proposed catalytic mechanism.

sterically unencumbered aryl iodides, which led the authors to propose that phenyl iodide undergoes oxidative addition to the nickel(0) precatalyst. In agreement with this hypothesis, direct use of nickel(II) precatalysts results in an alternate reaction pathway involving ethyl transfer to the acrylate in advance of aldol addition. Based on the collective data, the following catalytic mechanism was proposed. Oxidative addition of phenyl iodide to the nickel(0) precatalyst is followed by coordination of acrylate. The resulting complex is transformed to an ethyl(iodo)nickel species, along with an arylated boron enolate that spontaneously reacts with aldehyde to furnish trace quantities of tandem conjugate addition-aldol coupling product. The ethyl(iodo)nickel species, which represents the first point of entry into the actual catalytic cycle, eliminates ethylene and then coordinates another molecule of acrylate and triethylborane to generate the indicated nickel hydride complex. Acrylate hydrometallation provides a boron enolate and regenerates the ethyl(iodo)nickel species. The boron enolate is presumed to react spontaneously with the aldehyde to deliver the aldol product (Scheme 16.26).

16.2.8
Indium

In 2002, a reductive aldol coupling employing stoichiometric quantities of indium(III) bromide and tributylstannane was reported [38]. In 2004, it was found that use of triethylsilane as terminal reductant enables corresponding catalytic

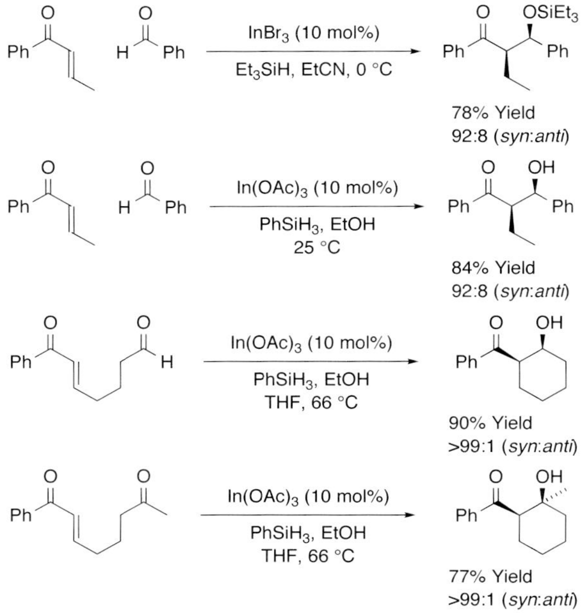

Scheme 16.27 Indium-catalyzed reductive aldol coupling.

processes [39a]. This protocol is applicable to the highly *syn*-diastereoselective coupling of enones to aldehydes. Entry into the catalytic mechanism is postulated to occur through transmetallation between $InBr_3$ and Et_3SiH to generate $HInBr_2$ and Et_3SiBr. The indium hydride is proposed to initiate a hydrometallative mechanism similar to that previously described (mechanism **(D)**, Scheme 16.6), but involving σ-bond metathesis between the intermediate indium aldolate and silane. It has been suggested that the hydrometallation event occurs through a six-centered transition structure involving the *S-cis* conformer of the enone, stereospecifically delivering the (*Z*)-indium enolate [50]. Aldehyde addition of the (*Z*)-indium enolate through a Zimmerman–Traxler type transition structure accounts for the observed *syn*-diastereoselectivity [7]. Concurrently, an indium-catalyzed reductive aldol coupling employing indium(III) acetate as the catalyst and $PhSiH_3$ as reductant was reported [39b]. This catalyst system enables *syn*-diastereoselective intermolecular and intramolecular reductive coupling (Scheme 16.27).

16.3
Conclusion

Following to the pioneering work of Revis in 1987, research in the area of catalytic reductive aldol coupling has flourished. These methods herald a departure from the use of stoichiometrically preformed enol(ate) derivatives in aldol coupling and support the feasibility of the developing related C–C couplings that circumvent stoichiometric usage of metallated reagents. Although remarkable progress has been made, there remain many unresolved challenges in this burgeoning area of research. For example, catalytic systems that employ cost-effective reductants that minimize or entirely avoid generation of stoichiometric by-products should be identified. In this regard, a significant advance resides in the development of reductive aldol couplings mediated by hydrogen gas. The development of related couplings mediated by feedstock reductants such as formate, carbon monoxide and 2-propanol would represent valuable contributions to the ever-growing arsenal of methods for catalytic reductive aldol coupling.

References

1 Though largely attributed to Würtz, the aldol reaction was reported first by Borodin: (a) von Richter, V. (1869) V. von Richter, aus St. Petersburg am 17. October 1869. *Berichte der Deutschen Chemischen Gesellschaft*, **2**, 552 (Borodin's earliest results are cited in this article). (b) Würtz, A. (1872) Sur un Aldéhyde-Alcool. *Bulletin de la Societe Chimique de Paris*, **17**, 436. (c) Borodin, A. (1873) Ueber einen neuen Abkömmling des Valerals. *Berichte der Deutschen Chemischen Gesellschaft*, **6**, 982. See also: Kane, R. (1838) *Annual Review of Physical Chemistry, Series 2*, **44**, 475.

2 (a) Hauser, C.R. and Puterbaugh, W.H. (1951) β-Hydroxy Esters From Ketones and Esters. *Journal of the American Chemical Society*, **73**, 2972. (b) Hauser, C. R. and Puterbaugh, W.H. (1953) Aldol Condensation of Esters with Ketones or Aldehydes to Form β-Hydroxy Esters by Lithium Amide. Comparison with the Reformatsky Reaction. *Journal of the American Chemical Society*, **75**, 1068. (c) Hauser, C.R. and Lindsay, J.K. (1955) Aldol Condensation of Ethyl Acetate with Ketones to Form β-Hydroxy Esters by Lithium Amide. *Journal of the American Chemical Society*, **77**, 1050. (d) Dunnavant, W.R. and Hauser, C.R. (1960) Synthesis of β-Hydroxy Esters from Ethyl Acetate and Ketones or Aldehydes by Means of Lithium Amide. Some Results with Other Esters. *Journal of Organic Chemistry*, **25**, 503.

3 (a) Wittig, G., Frommeld, H.D. and Suchanek, P. (1963) Über gezielte Aldolkondensation. *Angewandte Chemie*, **75**, 978. (b) Wittig, G. and Reiff, H. (1968) Directed Aldol Condensations. *Angewandte Chemie (International Ed. in English)*, **7**, 7. (c) Wittig, G. (1967) Directed Aldol Aondensations. *Record of Chemical Progress*, **28**, 45.

4 Rathke, M.W. (1970) Preparation of Lithio Ethyl Acetate. Procedure for the Conversion of Aldehydes and Ketones to .beta.-Hydroxy Esters. *Journal of the American Chemical Society*, **92**, 3222.

5 Rathke, M.W. and Lindert, A. (1971) Reaction of Lithium N-Isopropylcyclohexylamide with Esters. Method for the Formation and Alkylation of Ester Enolates. *Journal of the American Chemical Society*, **93**, 2318.

6 (a) Posner, G.H. and Loomis, G.L. (1972) A Useful Method for α-Methylation of γ-Butyrolactones. *Journal of the Chemical Society Chemical Communications*, 892. (b) Ireland, R.E., Meuller, R.H. and Willard, A.K. (1976) The Ester Enolate Claisen Rearrangement. Stereochemical Control through Stereoselective Enolate Formation. *Journal of the American Chemical Society*, **98**, 2868.

7 Zimmerman, H.E. and Traxler, M.D. (1957) The Stereochemistry of the Ivanov and Reformatsky Reactions. I. *Journal of the American Chemical Society*, **79**, 1920.

8 (a) Dubois, J.-E. and Dubois, M. (1967) Role des Interactions Dans les etats de Transition de L'aldolisation. Controle

Cinetique et Thermodynamique de la Cetolisation Mixte. *Tetrahedron Letters*, **8**, 4215. (b) Dubois, J.-E. and Fort, J.-F. (1972) Dynamic Stereochemistry of Aldolization-XX: Study of the Stereochemical Composition Versus Time; Quantitative Determination of Kinetic and Thermodynamic Stereoselectivities. *Tetrahedron*, **28**, 1653. (c) Dubois, J.-E. and Fort, J.-F. (1972) Dynamic Stereochemistry of Aldolization-XXI: Definition of the "Restoring Energy" of a System of Reversible Competitive Reactions. *Tetrahedron*, **28**, 1665. (d) Heathcock, C. A., Buse, C.T., Kleschick, W.A., Pirrung, M.C., Sohn, J.E. and Lampe, J. (1980) Acyclic Stereoselection. 7. Stereoselective Synthesis of 2-Alkyl-3-hydroxy Carbonyl Compounds by Aldol Condensation. *Journal of Organic Chemistry*, **45**, 1066.

9 (a) House, H.O., Czuba, L.J., Gall, M. and Olmstead, H.D. (1969) Chemistry of Carbanions. XVIII. Preparation of Trimethylsilyl Enol Ethers. *Journal of Organic Chemistry*, **34**, 2324. (b) Kraft, M. E. and Holton, R.A. (1983) Regiospecific Preparation of Thermodynamic Silyl Enol Ethers Using Bromomagnesium Dialkylamides. *Tetrahedron Letters*, **24**, 1345.

10 For a review, see: Velluz, L., Valls, J. and Nomine, G. (1965) Recent Advances in the Total Synthesis of Steroids. *Angewandte Chemie (International Ed. in English)*, **4**, 181.

11 (a) Corey, E.J. and Sneen, R.A. (1955) Calculation of Molecular Geometry by Vector Analysis. Application to Six-membered Alicyclic Rings. *Journal of the American Chemical Society*, **77**, 2505. (b) Berkoz, B., Chavez, E.P. and Djerassi, C. (1962) Steroids. Part CLXXII. Factors Controlling the Direction of Enol Acetylation of 3-Oxo-steroids. *Journal of the Chemical Society*, 1323. (c) Mazur, Y. and Sondheimer, F. (1958) Synthesis and Reactions of Ring A Methylated Saturated Steroids. *Journal of the American Chemical Society*, **80**, 5220. (d) Mazur, Y. and Sondheimer, F. (1958) Synthesis of 4α-Methyl-Δ⁷-steroiäs. The Interrelationship of Cholesterol, Citrostadienol and Lophenol. *Journal of the American Chemical Society*, **80**, 6296.

12 (a) Stork, G., Rosen, P. and Goldman, N.L. (1961) The α-Alkylation of Enolaters from the Lithium-Ammonia Reduction of α, β-Unsaturated Ketones. *Journal of the American Chemical Society*, **83**, 2965. (b) Stork, G., Rosen, P., Goldman, N., Coombs, R.V. and Tsuji, J. (1965) Alkylation and Carbonation of Ketones by Trapping the Enolates from the Reduction of α, β-Unsaturated Ketones. *Journal of the American Chemical Society*, **87**, 275.

13 For selected reviews encompassing metal-catalyzed conjugate reduction mediated by silanes and borohydrides, see: (a) Keinan, E. and Greenspoon, N. (1991) Partial Reduction of Enones, Styrenes and Related Systems, in *Comprehensive Organic Synthesis*, Vol. 8 (eds B.M. Trost and I. Fleming), Pergamon Press, Oxford, p. 523. (b) Sibi, M.P. and Manyem, S. (2000) Enantioselective Conjugate Additions. *Tetrahedron*, **56**, 8033. (c) Shibata, I. (2004) Recent Advances in Conjugated Reduction of Unsaturated Carbonyl Compounds. *Organometallic News* **2**, 53. (d) Rendler, S. and Oestreich, M. (2007) Polishing a Diamond in the Rough: "Cu-H" Catalysis with Silanes. *Angewandte Chemie (International Ed. in English)*, **46**, 498.

14 For selected reviews encompassing metal-catalyzed conjugate reduction mediated by hydrogen, see: (a) Haskel, A. and Keinan, E. (2002) Palladium-Catalyzed 1, 4-Reduction (Conjugate Reduction), in *Handbook of Organopalladium Chemistry for Organic Synthesis*, Vol. 2 (ed. E.-I. Negishi), John Wiley & Sons, Inc., New York, p. 2767. (b) Farina, V., Reeves, J.T., Senanayake, C.H. and Song, J.J. (2006) Asymmetric Synthesis of Active Pharmaceutical Ingredients. *Chemical Reviews*, **106**, 2734. (c) Jäkel, C. and Paciello, R. (2006) High-Throughput and Parallel Screening Methods in Asymmetric Hydrogenation. *Chemical Reviews*, **106**, 2912.

15 For enantioselective conjugate reductions of unsaturated carbonyl compounds catalyzed by rhodium, see: (a) Ojima, I. and Kogure, T. (1974) Asymmetric

Reduction of α-Keto Esters via Hydrosilylation Catalyzed by a Rhodium(I) Complex with Chiral Phosphin Ligands. *Tetrahedron Letters*, **15**, 1889. (b) Ojima, I. and Kogure, T. (1975) Selective Asymmetric Reduction of α, β-Unsaturated Ketones via Hydrosilylation Catalyzed by Rhodium(I) Complexes with Chiral Phosphine Ligands. *Chemistry Letters*, 985. (c) Hayashi, T., Yamamoto, K. and Kumada, M. (1975) Asymmetric Hydrosilylation of α, β-Unsaturated Carbonyl Compounds. *Tetrahedron Letters*, **16**, 3. (d) Tsuchiya, Y., Kanazawa, Y., Shiomi, T., Kobayashi, K. and Nishiyama, H. (2004) Asymmetric Conjugate Reduction of α, β-Unsaturated Esters with Chiral Rhodium(bisoxazoliny lphenyl) Catalysts. *Synlett*, **14**, 2493. (e) Kanazawa, Y. and Nishiyama, H. (2006) Conjugate Reduction of α, β-Unsaturated Aldehydes with Rhodium (bisoxazolinylphenyl) Catalysts. *Synlett*, **19**, 3343. (f) Ito, J.-I., Shiomi, T. and Nishiyama, H. (2006) Efficient Preparation of New Rhodium- and Iridium-[Bis(oxazolinyl)-3,5-dimethylphenyl] Complexes by C-H Bond Activation: Applications in Asymmetric Synthesis. *Advanced Synthesis and Catalysis*, **348**, 1235. (g) Kanazawa, Y., Tsuchiya, Y., Kobayashi, K., Shiomi, T., Itoh, J., Kikuchi, M., Yamamoto, Y. and Nishiyama, H. (2006) Asymmetric Conjugate Reduction of, α, β-Unsaturated Ketones and Esters with Chiral Rhodium(2,6-bisoxazolinylphenyl) Catalysts. *Chemistry – A European Journal*, **12**, 63.

16 For enantioselective conjugate reductions of unsaturated carbonyl compounds catalyzed by copper, see: (a) Appella, D. H., Moritani, Y., Shintani, R., Ferreira, E.M. and Buchwald, S.L. (1999) Asymmetric Conjugate Reduction of α, β-Unsaturated Esters Using a Chiral Phosphine-Copper Catalyst. *Journal of the American Chemical Society*, **121**, 9473. (b) Moritani, Y., Appella, D.H., Jurkauskas, V. and Buchwald, S.L. (2000) Synthesis of β-Alkyl Cyclopentanones in High Enantiomeric Excess via Copper-Catalyzed Asymmetric Conjugate Reduction. *Journal of the American Chemical Society*, **122**, 6797. (c) Jurkauskas, V. and Buchwald, S.L. (2002) Dynamic Kinetic Resolution via Asymmetric Conjugate Reduction: Enantio- and Diastereoselective Synthesis of 2,4-Dialkyl Cyclopentanones. *Journal of the American Chemical Society*, **124**, 2892. (d) Hughes, G., Kimura, M. and Buchwald, S.L. (2003) Catalytic Enantioselective Conjugate Reduction of Lactones and Lactams. *Journal of the American Chemical Society*, **125**, 11253. (e) Lipshutz, B.H. and Servesdo, J.M. (2003) CuH-Catalyzed Asymmetric Conjugate Reductions of Acyclic Enones. *Angewandte Chemie (International Ed. in English)*, **42**, 4789. (f) Rainka, M.P., Aye, Y. and Buchwald, S.L. (2004) Copper-Catalyzed Asymmetric Conjugate Reduction as a Route to Novel β-Azaheterocyclic Acid Derivitives. *Proceedings of the National Academy of Sciences of the United States of America*, **101**, 5821. (g) Lipshutz, B.H., Servesdo, J.M., Petersen, T.B., Papa, P.P. and Lover, A.A. (2004) Asymmetric 1, 4-Reductions of Hindered β-Substituted Cycloalkenones Using Catalytic SEGPHOS-Ligated CuH. *Organic Letters*, **6**, 1273. (h) Lipshutz, B.H., Servesdo, J.M. and Taft, B.R. (2004) Asymmetric 1,4-Hydrosilylations of α, β-Unsaturated Esters. *Journal of the American Chemical Society*, **126**, 8352. (i) Rainka, M.P., Milne, J.E. and Buchwald, S.L. (2005) Dynamic Kinetic Resolution of α, β-Unsaturated Lactones through Asymmetric Copper-Catalyzed Conjugate Reduction: Application to the Total Synthesis of Eupomatilone-3. *Angewandte Chemie (International Ed. in English)*, **44**, 6177. (j) Lipshutz, B.H. and Frieman, B.A. (2005) CuH in a Bottle: A Convenient Reagent for Asymmetric Hydrosilylations. *Angewandte Chemie (International Ed. in English)*, **44**, 6345. (k) Lipshutz, B.H., Tanaka, N., Taft, B.R. and Lee, C.-T. (2006) Chiral Silanes via Asymmetric Hydrosilylation with Catalytic CuH. *Organic Letters*, **8**, 1963. (l) Lipshutz, B.H., Frieman, B.A. and Tomaso, A.E. (2006) Copper-in-Charcoal (Cu/C): Heterogeneous, Copper-Catalyzed Asymmetric Hydrosilylations. *Angewandte Chemie (International Ed. in English)*, **45**,

1259. (m) Lee, D., Kim, D. and Yun, J. (2006) Highly Enantioselective Conjugate Reduction of β, β-Disubstituted α, β-Unsaturated Nitriles. *Angewandte Chemie (International Ed. in English)*, **45**, 2785.

17 For enantioselective conjugate reductions of unsaturated carbonyl compounds catalyzed by cobalt, see: (a) Leutenegger, U., Madin, A. and Pfaltz, A. (1989) Enantioselective Reduction of α, β-Unsaturated Carboxylates with $NaBH_4$ and Catalytic Amounts of Chiral Cobalt Semicorrin Complexes. *Angewandte Chemie (International Ed. in English)*, **28**, 60. (b) Misun, M. and Pfaltz, A. (1996) Enantioselective Reduction of Electrophilic C=C Bonds with Sodium Tetrahydroborate and 'Semicorrin' Cobalt Catalysts. *Helvetica Chimica Acta*, **79**, 961. (c) von Matt, P. and Pfaltz, A. (1991) Enantioselective Conjugate Reduction of α, β-Unsaturated Carboxamides with Semicorrin Cobalt Catalysts. *Tetrahedron: Asymmetry*, **2**, 691. (d) Yamada, T., Ohtsuka, Y. and Ikeno, T. (1998) Enantioselective Borohydride 1, 4-Reduction of α, β-Unsaturated Carboxamides Using Optically Active Cobalt(II) Complex Catalysts. *Chemistry Letters*, 1129. (e) Ohtsuka, Y., Ikeno, T. and Yamada, T. (2003) Catalytic Enantioselective Protonation of Cobalt–Enolate Equivalents Generated by 1, 4-Reduction with Borohydride. *Tetrahedron: Asymmetry*, **14**, 967. (f) Geiger, C., Kreitmeier, P. and Reiser, O. (2005) Cobalt(II)-Azabis(oxazoline)-Catalyzed Conjugate Reduction of α, β-Unsaturated Carbonyl Compounds. *Advanced Synthesis and Catalysis*, **347**, 249.

18 For enantioselective conjugate reductions of unsaturated carbonyl compounds catalyzed by palladium, see: Tsuchiya, Y., Hamashima, Y. and Sodeoka, M. (2006) A New Entry to Pd-H Chemistry: Catalytic Asymmetric Conjugate Reduction of Enones with EtOH and a Highly Enantioselective Synthesis of Warfarin. *Organic Letters*, **8**, 4851.

19 For selected examples of the stoichiometric preformation of enol derivatives via metal-catalyzed conjugate reduction with subsequent by capture of enol derivative through its reaction with electrophilic reagents, see: (a) Evans, D.A. and Fu, G.C. (1990) Conjugate reduction of α,β-unsaturated carbonyl compounds by catecholborane. *Journal of Organic Chemistry*, **55**, 5678. (b) Chrisman, W., Nosson, K., Papa, P., Sclafani, J.A., Vivian, R.W., Keith, J.M. and Lipshutz, B.H. (2000) Copper Hydride-Catalyzed Tandem 1,4-Reduction/Alkylation Reactions. *Tetrahedron*, **56**, 2779. (c) Yun, J. and Buchwald, S.L. (2001) One-Pot Synthesis of Enantiomerically Enriched 2,3-Disubstituted Cyclopentanones via Copper-Catalyzed 1,4-Reduction and Alkylation. *Organic Letters*, **3**, 1129. (d) Lipshutz, B.H. and Papa, P. (2002) Copper-Catalyzed Reductive Alkylations of Enones: A Novel Transmetalation Protocol. *Angewandte Chemie (International Ed. in English)*, **41**, 4580. (e) Chae, J., Yun, J. and Buchwald, S.L. (2004) One-Pot Sequential Cu-Catalyzed Reduction and Pd-Catalyzed Arylation of Silyl Enol Ethers. *Organic Letters*, **6**, 4809.

20 For reviews that encompass examples of catalytic reductive enone couplings that result in functionalization of the β-position, see: (a) Ikeda, S.-I. (2000) Nickel-Catalyzed Intermolecular Domino Reactions. *Accounts of Chemical Research*, **33**, 511. (b) Montgomery, J. (2004) Nickel-Catalyzed Reductive Cyclizations and Couplings. *Angewandte Chemie (International Ed. in English)*, **43**, 3890.

21 For selected reviews encompassing core physical organic and stereochemical principles associated with the aldol reaction, see: (a) Heathcock, C.H. (1981) Acyclic Stereocontrol Through the Aldol Condensation. *Science*, **214**, 395. (b) Heathcock, C.H. (1982) Acyclic Stereoselection via the Aldol Condensation. *ACS Symposium Series*, **185**, 55. (c) Evans, D.A., Nelson, J V. and Taber, T.R. (1982) Stereoselective Aldol Condensations. *Topics in Stereochemistry*, **13**, 1. (d) Heathcock, C.H. (1984) The Aldol Addition Reaction. *Asymmetric Synthesis*, **3**, 111. (e) Heathcock, C.H. (1991) The Aldol Reaction: Group I and Group II Enolates, in *Comprehensive Organic Synthesis*, Vol. 2 (eds B.M. Trost, I.

Fleming and C.H. Heathcock), Pergamon Press, New York, p. 181.

22 For selected reviews encompassing recent developments in catalytic aldol coupling, see: (a) Machajewski, T.D. and Wong, C.-H. (2000) The Catalytic Asymmetric Aldol Reaction. *Angewandte Chemie (International Ed. in English)*, **39**, 1352. (b) Palomo, C., Oiarbide, M. and García, J.M. (2002) The Aldol Addition Reaction: An Old Transformation at Constant Rebirth. *Chemistry – A European Journal*, **8**, 36. (c) Palomo, C., Oiarbide, M. and García, J.M. (2004) Current Progress in the Asymmetric Aldol Addition Reaction. *Chemical Society Reviews*, **33**, 65. (d) Schetter, B. and Mahrwald, R. (2006) Modern Aldol Methods for the Total Synthesis of Polyketides. *Angewandte Chemie (International Ed. in English)*, **45**, 7506.

23 For a recent review on the use of metallic catalysts for direct enantioselective aldol additions, see: Shibasaki, M., Matsunaga, S. and Kumagai, N. (2004) Direct Catalytic Asymmetric Aldol Reaction Using Chiral Metal Complexes, in *Modern Aldol Reactions*, Vol. 2 (ed. R. Mahrwald), Wiley-VCH Verlag GmbH, Weinheim.

24 For a recent review on the use of organic catalysts for direct enantioselective aldol addition, see: List, B. (2004) Amine-Catalyzed Aldol Reactions, in *Modern Aldol Reactions*, Vol. 1 (ed. R. Mahrwald), Wiley-VCH Verlag GmbH, Weinheim.

25 Sakthivel, K., Notz, W., Bui, T. and Barbas, C.F., III (2001) Amino Acid Catalyzed Direct Asymmetric Aldol Reactions: A Bioorganic Approach to Catalytic Asymmetric Carbon-Carbon Bond-Forming Reactions. *Journal of the American Chemical Society*, **123**, 5260.

26 Using amides of L-proline, direct catalyzed aldol coupling of 2-butanone to *p*-nitrobenzaldehyde affords mixtures of regioisomeric products: Tang, Z., Yang, Z.-H., Chen, X.-H., Cun, L.-F., Mi, A.-Q., Jiang, Y.-Z. and Gong, L.-Z. (2005) A Highly Efficient Organocatalyst for Direct Aldol Reactions of Ketones with Aledydes. *Journal of the American Chemical Society*, **127**, 9285.

27 Yoshikawa, N., Yamada, Y.M.A., Das, J., Sasai, H. and Shibasaki, M. (1999) Direct Catalytic Asymmetric Aldol Reaction. *Journal of the American Chemical Society*, **121**, 4168.

28 For rhodium-catalyzed reductive aldol reaction mediated by silane or other reductants see: (a) Revis, A. and Hilty, T.K. (1987) Novel Synthesis of β-Siloxy Esters by Condensation of Carbonyls and Trimethylsilane with α,β-Unsaturated Esters Catalyzed by $RhCl_3$. *Tetrahedron Letters*, **28**, 4809. (b) Matsuda, I., Takahashi, K. and Sato, S. (1990) Rhodium Catalyzed Direct Coupling of α, β-Unsaturated Ketone, Aldehyde, and Trialkylsilane: An Easy Access to Regio-Defined Aldol Derivatives. *Tetrahedron Letters*, **31**, 5331. (c) Taylor, S.J. and Morken, J.P. (1999) Catalytic Diastereoselective Reductive Aldol Reaction: Optimization of Interdependent Reaction Variables by Arrayed Catalyst Evaluation. *Journal of the American Chemical Society*, **121**, 12202. (d) Zhao, C.-X., Bass, J. and Morken, J.P. (2001) Generation of (*E*)-Silylketene Acetals in a Rhodium-DuPhos Catalyzed Two-Step Reductive Aldol Reaction. *Organic Letters*, **3**, 2839. (e) Taylor, S.J., Duffey, M.O. and Morken, J.P. (2000) Rhodium-Catalyzed Enantioselective Reductive Aldol Reaction. *Journal of the American Chemical Society*, **122**, 4528. (f) Fuller, N.O. and Morken, J.P. (2005) Direct Formation of Synthetically Useful Silyl-Protected Aldol Adducts via the Asymmetric Reductive Aldol Reaction. *Synlett*, 1459. (g) Emiabata-Smith, D., McKillop, A., Mills, C., Motherwell, W.B. and Whitehead, A.J. (2001) Some Preliminary Studies on a Novel Rhodium(I)-Catalysed Tandem Hydrosilylation-Intramolecular Aldol Reaction. *Synlett*, **1302**, (h) Freiría, M., Whitehead, A.J., Tocher, D.A. and Motherwell, W.B. (2004) Further Observations on the Rhodium (I)-Catalyzed Tandem Hydrosilylation-Intramolecular Aldol Reaction. *Tetrahedron*, **60**, 2673. (i) Nishiyama, H., Shiomi, T., Tsuchiya, Y. and Matsuda, I. (2005) High Performance of Rh(Phebox) Catalysts in Asymmetric Reductive Aldol

Reaction: High Anti-Selectivity. *Journal of the American Chemical Society*, **127**, 6972. (j) Ito, J.-I., Shiomi, T. and Nishiyama, H. (2006) Efficient Preparation of New Rhodium- and Iridium-[Bis(oxazolinyl)-3,5-dimethylphenyl] Complexes by C-H Bond Activation: Applications in Asymmetric Synthesis. *Advanced Synthesis and Catalysis*, **348**, 1235. (k) Shiomi, T., Ito, J.-I., Yamamoto, Y. and Nishiyama, H. (2006) 4-Substituted-Phenyl(bisoxazoline)-Rhodium Complexes: Efficiency in the Catalytic Asymmetric Reductive Aldol Reaction. *European Journal of Organic Chemistry*, 5594. (l) Willis, M.C. and Woodward, R.L. (2005) Rhodium-Catalyzed Reductive Aldol Reactions Using Aldehydes as the Stoichiometric Reductants. *Journal of the American Chemical Society*, **127**, 18012.

29 For reviews encompassing the topic of reductive aldol coupling, see: (a) Motherwell, W.B. (2002) Curiosity and Simplicity in the Invention and Discovery of New Metal-Mediated Reactions for Organic Synthesis. *Pure and Applied Chemistry Chimie Pure et Appliquee*, **74**, 135. (b) Huddleston, R.R. and Krische, M.J. (2003) Enones as Latent Enolates in Catalytic Processes: Catalytic Cycloreduction, Cycloaddition and Cycloisomerization. *Synlett*, 12. (c) Jang, H.-Y., Huddleston, R.R. and Krische, M.J. (2003) Nucleophilic Activation of Enones via Homogeneous Catalytic Hydrogenation: Catalytic Reductive C-C Bond Formation under Hydrogenation Condition. *Chemtracts*, **16**, 554. (d) Jang, H.-Y. and Krische, M.J. (2004) Catalytic Hydrogen-Mediated Cross-Coupling of Enones and Carbonyl Compounds: Aldol Condensation by Hydrogenation. *European Journal of Organic Chemistry*, **19**, 3953. (e) Jang, H.-Y. and Krische, M.J. (2004) Catalytic C-C Bond Formation via Capture of Hydrogenation Intermediates. *Accounts of Chemical Research*, **37**, 653. (f) Chiu, P. (2004) Organosilanes in Copper-Mediated Conjugate Reductions and Reductive Aldol Reactions. *Synthesis*, 2210. (g) Nishiyama, H. and Shiomi, T. (2007) Reductive Aldol, Michael and Mannich Reactions. *Topics in Current Chemistry*, **279**, 105.

30 For rhodium-catalyzed reductive aldol reaction mediated by hydrogen, see: (a) Jang, H.Y., Huddleston, R.R. and Krische, M.J. (2002) Reductive Generation of Enolates from Enones Using Elemental Hydrogen: Catalytic C-C Bond Formation under Hydrogenative Conditions. *Journal of the American Chemical Society*, **124**, 15156. (b) Huddleston, R.R. and Krische, M.J. (2003) Enolate Generation under Hydrogenation Conditions: Catalytic Aldol Cycloreduction of Keto-Enones. *Organic Letters*, **5**, 1143. (c) Koech, P.K. and Krische, M.J. (2004) Catalytic Addition of Metallo-Aldehyde Enolates to Ketones: A New C-C Bond-Forming Hydrogenation. *Organic Letters*, **6**, 691. (d) Marriner, G.A., Garner, S.A., Jang, H.Y. and Krische, M.J. (2004) Metallo-Aldehyde Enolates via Enal Hydrogenation: Catalytic Cross Aldolization with Glyoxal Partners As Applied to the Synthesis of 3, 5-Disubstituted Pyridazines. *Journal of Organic Chemistry*, **69**, 1380. (e) Jung, C.K., Garner, S.A. and Krische, M.J. (2006) Hydrogen-Mediated Aldol Reductive Coupling of Vinyl Ketones Catalyzed by Rhodium: High *Syn*-Selectivity through the Effect of Tri-2-Furylphosphine. *Organic Letters*, **8**, 519. (f) Han, S.B. and Krische, M.J. (2006) Reductive Aldol Coupling of Divinyl Ketones via Rhodium-Catalyzed Hydrogenation: syn-Diastereoselective Construction of β-Hydroxyenones. *Organic Letters*, **8**, 5657. (g) Jung, C.K. and Krische, M.J. (2006) Asymmetric Induction in Hydrogen-Mediated Reductive Aldol Additions to α-Amino Aldehydes Catalyzed by Rhodium: Selective Formation of *syn*-Stereotriads Directed by Intramolecular Hydrogen-Bonding. *Journal of the American Chemical Society*, **128**, 17051.

31 For cobalt-catalyzed reductive aldol reaction, see: (a) Isayama, S. and Mukaiyama, T. (1989) Cobalt(II) Catalyzed Coupling Reaction of α, β-Unsaturated Compounds with Aldehydes by the Use of Phenylsilane. New Method for Preparation of β-Hydroxy Nitriles, Amides, and Esters. *Chemistry Letters*, 2005. (b) Baik, T.G., Luis, A.L., Wang, L.C. and Krische, M.J. (2001)

Diastereoselective Cobalt-Catalyzed Aldol and Michael Cycloreductions. *Journal of the American Chemical Society*, **123**, 5112. (c) Wang, L.C., Jang, H.-Y., Roh, Y., Lynch, V., Schultz, A.J., Wang, X. and Krische, M.J. (2002) Diastereoselective Cycloreductions and Cycloadditions Catalyzed by Co(dpm)$_2$-Silane (dpm = 2,2,6,6-tetramethylheptane-3,5-dionate): Mechanism and Partitioning of Hydrometallative versus Anion Radical Pathways. *Journal of the American Chemical Society*, **124**, 9448. (d) Lam, H.W., Joensuu, P.M., Murray, G.J., Fordyce, E.A.F., Prieto, O. and Luebbers, T. (2006) Diastereoselective Cobalt-Catalyzed Reductive Aldol Cyclizations Using Diethylzinc as the Stoichiometric Reductant. *Organic Letters*, **8**, 3729.

32 For iridium-catalyzed reductive aldol reaction, see: Zhao, C.X., Duffey, M.O., Taylor, S.J. and Morken, J.P. (2001) Enantio- and Diastereoselective Reductive Aldol Reactions with Iridium-Pybox Catalysts. *Organic Letters*, **3**, 1829.

33 For ruthenium catalyzed reductive aldol reaction, see: Doi, T., Fukuyama, T., Minamino, S. and Ryu, I. (2006) RuHCl(CO)(PPh$_3$)$_3$-Catalyzed Reductive Dimerization of α, β-Unsaturated Aldehydes Leading to α-Hydroxymethyl Ketones. *Synlett*, **18**, 3013.

34 For palladium catalyzed reductive aldol reaction, see: Kiyooka, S.I., Shimizu, A. and Torii, S. (1998) A Mild Aldol Reaction of Aryl Aldehydes through Palladium-Catalyzed Hydrosilation of α,β-Unsaturated Carbonyl Compounds with Trichlorosilane. *Tetrahedron Letters*, **39**, 5237.

35 For copper-promoted reductive aldol reaction, see: (a) Chiu, P., Chen, B. and Cheng, K.F. (1998) A Conjugate Reduction-Intramolecular Aldol Strategy toward the Synthesis of Pseudolaric Acid A. *Tetrahedron Letters*, **39**, 9229. (b) Chiu, P. (2004) Organosilanes in Copper-Mediated Conjugate Reductions and Reductive Aldol Reactions. *Synthesis*, **13**, 2210. For copper-promoted reductive intramolecular Henry reaction, see: (c) Chung, W.K. and Chiu, P. (2005) Reductive Intramolecular Henry Reactions Induced by Stryker's Reagent. *Synlett*, **1**, 55. For copper-promoted and catalyzed reductive cyclizations of α,β-acetylenic ketones tethered to ketones, see: (d) Chiu, P. and Leung, S.K. (2004) Stoichiometric and Catalytic Reductive Aldol Cyclizations of Alkynediones Induced by Stryker's Reagent. *Chemical Communications*, 2308.

36 For copper-catalyzed reductive aldol reaction, see: (a) Ooi, T., Doda, K., Sakai, D. and Maruoka, K. (1999) Unique Property of Copper(I) Chloride as a Radical Initiator as well as a Lewis Acid: Application to CuCl-Catalyzed Aldol Reaction of α, β-Unsaturated Ketones with Bu$_3$SnH. *Tetrahedron Letters*, **40**, 2133. (b) Lam, H.W. and Joensuu, P.M.A. (2005) Cu(I)-Catalyzed Reductive Aldol Cyclizations: Diastereo- and Enantioselective Synthesis of β-Hydroxylactones. *Organic Letters*, **7**, 4225. (c) Lam, H.W., Murray, G.J. and Firth, J.D. (2005) Diastereoselective Synthesis of 4-Hydroxypiperidin-2-ones via Cu(I)-Catalyzed Reductive Aldol Cyclization. *Organic Letters*, **7**, 5743. (d) Deschamp, J., Chuzel, O., Hannedouche, J. and Riant, O. (2006) Highly Diastereo- and Enantioselective Copper-Catalyzed Domino Reduction/Aldol Reaction of Ketones with Methyl Acrylate. *Angewandte Chemie (International Ed. in English)*, **45**, 1292. (e) Chuzel, O., Deschamp, J., Chauster, C. and Riant, O. (2006) Copper(I)-Catalyzed Enantio- and Diastereoselective Tandem Reductive Aldol Reaction. *Organic Letters*, **8**, 5943. (f) Zhao, D., Oisaki, K., Kanai, M. and Shibasaki, M. (2006) Catalytic Enantioselective Intermolecular Reductive Aldol Reaction to Ketones. *Tetrahedron Letters*, **47**, 1403. (g) Zhao, D., Oisaki, K., Kanai, M. and Shibasaki, M. (2006) Dramatic Ligand Effect in Catalytic Asymmetric Reductive Aldol Reaction of Allenic Esters to Ketones. *Journal of the American Chemical Society*, **128**, 14440. (h) Welle, A., Diez-Gonzalez, S., Tinant, B., Nolan, S.P. and Riant, O. (2006) A Three-Component Tandem Reductive Aldol Reaction Catalyzed by N-Heterocyclic Carbene-Copper Complexes. *Organic Letters*, **8**, 6059.

37 For nickel-catalyzed reductive aldol reaction, see: Chrovian, C.C. and Montgomery, J. (2007) Surprising Role of Aryl Halides in Nickel-Catalyzed Reductive Aldol Reactions. *Organic Letters*, **9**, 537.

38 For a reductive aldol coupling employing stoichiometric quantities of indium reagent, see: Inoue, K., Ishida, T., Shibata, I. and Baba, A. (2002) Remarkable Dependence of Diastereoselectivity on Anhydrous or Aqueous Solvent in the Indium Hydride Promoted Reductive Aldol Reaction of α, β-Unsaturated Ketones. *Advanced Synthesis and Catalysis*, **344**, 283.

39 For indium-catalyzed reductive aldol reaction, see: (a) Shibata, I., Kato, H., Ishida, T., Yasuda, M. and Baba, A. (2004) Catalytic Generation of Indium Hydride in a Highly Diastereoselective Reductive Aldol Reaction. *Angewandte Chemie (International Ed. in English)*, **43**, 711. (b) Miura, K., Yamada, Y., Tomita, M. and Hosomi, A. (2004) Indium(III) Acetate-Catalyzed 1,4-Reduction and Reductive Aldol Reactions of α-Enones with Phenylsilane. *Synlett*, 1985.

40 Slough, G.A., Bergman, R.C. and Heathcock, C.H. (1989) Synthesis of .eta.1 Oxygen-Bound Rhodium Enolates. Applications to Catalytic Aldol Chemistry. *Journal of the American Chemical Society*, **111**, 938.

41 Wright, M.E. and Cochran, B.B. (1993) Rhodium-Catalyzed Silylformylation of Aldehydes: a Mild and Efficient Catalytic Route to .alpha.-Silyloxyaldehydes. *Journal of the American Chemical Society*, **115**, 2059.

42 For tri-2-furylphosphine and triphenylarsine effects in metal-catalyzed reactions, see: (a) Farina, V. and Krishnan, B. (1991) Large Rate Accelerations in the Stille Reaction with Tri-2-Furylphosphine and Triphenylarsine as Palladium Ligands: Mechanistic and Synthetic Implications. *Journal of the American Chemical Society*, **113**, 9585. (b) Farina, V. (1996) New Perspectives in the Cross-Coupling Reactions of Organostannanes. *Pure and Applied Chemistry Chimie Pure et Appliquée*, **68**, 73. (c) Anderson, N.G. and Keay, B.A. (2001) 2-Furyl Phosphines as Ligands for Transition-Metal-Mediated Organic Synthesis. *Chemical Reviews*, **101**, 997.

43 For mechanistic studies on alkene hydrogenation catalyzed by neutral rhodium complexes, see: (a) Tolman, C.A., Meakin, P.Z., Lindner, D.L. and Jesson, J.P. (1974) Triarylphosphine, Hydride, and Ethylene Complexes or Rhodium(I) Chloride. *Journal of the American Chemical Society*, **96**, 2762. (b) Halpern, J. and Wong, C.S. (1973) Hydrogenation of Tris(triphenylphosphine)chlororhodium(I). *Chemical Communications*, 629. (c) Halpern, J., Okamoto, T. and Zakhariev, A. (1976) Mechanism of the Chlorotris (triphenylphosphine) Rhodium(I)-Catalyzed Hydrogenation of Alkenes. The Reaction of Chlorodihydridotris(triphenyl-phosphine)rhodium(III) with Cyclohexene. *Journal of Molecular Catalysis*, **2**, 65.

44 Monohydride catalytic cycles initiated via deprotonation of cationic rhodium dihydrides have been postulated and their veracity further established through the stoichiometric conversion of cationic rhodium(I) complexes to the corresponding rhodium(I) monohydrides via expsoure to hydrogen in the presence of a tertiary amine base: (a) Schrock, R.R. and Osborn, J.A. (1976) Catalytic Hydrogenation Using Cationic Rhodium Complexes. I. Evolution of the Catalytic System and the Hydrogenation of Olefins. *Journal of the American Chemical Society*, **98**, 2134. (b) Schrock, R.R. and Osborn, J.A. (1976) Catalytic Hydrogenation Using Cationic Rhodium Complexes. II. The Selective Hydrogenation of Alkynes to Cis Olefins. *Journal of the American Chemical Society*, **98**, 2143. (c) Schrock, R.R. and Osborn, J.A. (1976) Catalytic Hydrogenation Using Cationic Rhodium Complexes. 3. The Selective Hydrogenation of Dienes to Monoenes. *Journal of the American Chemical Society*, **98**, 4450.

45 For a review of the acidity of metal hydrides, see: Norton, J.R. (1992) Acidity of hydrido transition metal complexes in solution, in *Transition Metal Hydrides* (ed. A. Dedieu), John Wiley & Sons, Inc., New York.

46 Alcaide, B. and Almendros, P. (2003) The Direct Catalytic Asymmetric Cross-Aldol Reaction of Aldehydes. *Angewandte Chemie (International Ed. in English)*, **42**, 858.

47 (a) Denmark, S. and Ghosh, S.K. (2001) The First Catalytic, Diastereoselective, and Enantioselective Crossed-Aldol Reactions of Aldehydes. *Angewandte Chemie (International Ed. in English)*, **40**, 4759. (b) Northrup, A.B. and MacMillan, D.W.C. (2002) The First Direct and Enantioselective Cross-Aldol Reaction of Aldehydes. *Journal of the American Chemical Society*, **124**, 6798. (c) Pidathala, C., Hoang, L., Vignola, N. and List, B. (2003) Direct Catalytic Asymmetric *Enolexo* Aldolizations. *Angewandte Chemie (International Ed. in English)*, **42**, 2785.

48 For a review encompassing single electron oxidative addition, see: Halpern, J. (1970) Oxidative-Addition Reactions of Transition Metal Complexes. *Accounts of Chemical Research*, **3**, 386.

49 Socol, S.M. and Verkade, J.G. (1986) Steric and Electronic Effects in Cobalt(II) Disproportionation with Phosphorus Ligands. *Inorganic Chemistry*, **25**, 2658 and references therein.

50 (a) Boldrini, G.P., Mancini, F., Tagliavini, E., Trombini, C. and Umani-Ronchi, A. (1990) A New Approach to (*Z*)-Vinyloxyboranes *via* 1, 4-Hydroboration of (*E*)-α, β-Unsaturated Ketones. Synthesis of *syn* Aldols. *Chemical Communications*, 1680. (b) Boldrini, G.P., Bortolotti, M., Mancini, F., Tagliavini, E., Trombini, C. and Umani-Ronchi, A. (1991) A New Protocol for Regio- and Stereocontrolled Aldol Reactions Through the Conjugate Addition of Dialkylboranes to .alpha.,.beta.-Unsaturated Ketones. *Journal of Organic Chemistry*, **56**, 5820.

17
Dissolving Metals
Miguel Yus and Francisco Foubelo

17.1
Introduction

Dissolving metals have been used extensively as reducing agents for more than a century but today they have been partially displaced from the central field of organic synthesis by the use of other more selective methodologies, such as metal hydrides [1] and catalytic hydrogenations [2]. However, dissolving metals are still of interest for the selective reduction of specific polar functional groups (such as hindered cyclic ketones) and the reductive cleavage of some activated bonds [3].

Regarding the mode of action of electropositive metals in reduction processes, it is generally assumed that an initial single-electron transfer (SET) from the metal to the organic substrate takes place to give a radical anion [4]. Depending on the organic substrate and on the reaction conditions (mostly on the solvent), the highly reactive resulting radical anion can then decompose via a number of different routes. For instance, for compounds **(I)** with multiple C=X bonds (X = CR_2, O, NR, S), the radical anion **(II)** could be protonated in the presence of a proton source (the solvent) to give a new radical **(III)**, which after a further SET from the metal, and further protonation of the resulting anion **(IV)**, would lead to the reduced product **(V)**. In the absence of a proton source, coupling of two radical anions **(II)** can occur to give first the dianionic species **(VI)** and, after hydrolysis, compounds of type **(VII)**. Organic substrates **(VIII)** with activated bonds X—Y (X = CR_3, OR, SR, NR_2; Y = CR_3, OR, SR, NR_2, Hal (F, Cl, Br, I)) also undergo reductive cleavage by means of electropositive metals through a SET mechanism. In this case, the initially formed radical anion **(IX)** suffers the activated X—Y bond scission to generate a radical **(X)** and an anion **(XI)**. Radical **(X)** would be converted to an anion after a new electron transfer from the metal, and final protonation would yield the corresponding reduced products **(XII)** and **(XIII)**, respectively (Scheme 17.1).

Most of the reductions by means of dissolving metals have been performed using highly electropositive metals, such as Li, Na and K, in liquid NH_3 as solvent. Less electropositive metals (Mg, Ca, Sr, Ba, Ni, Cu, Zn, In, Sn, Pb) have also been used, so by choosing the appropriate metal, solvent and cosolvent it is possible to

Modern Reduction Methods. Edited by Pher G. Andersson and Ian J. Munslow
Copyright © 2008 WILEY-VCH Verlag GmbH & Co. KGaA, Weinheim
ISBN: 978-3-527-31862-9

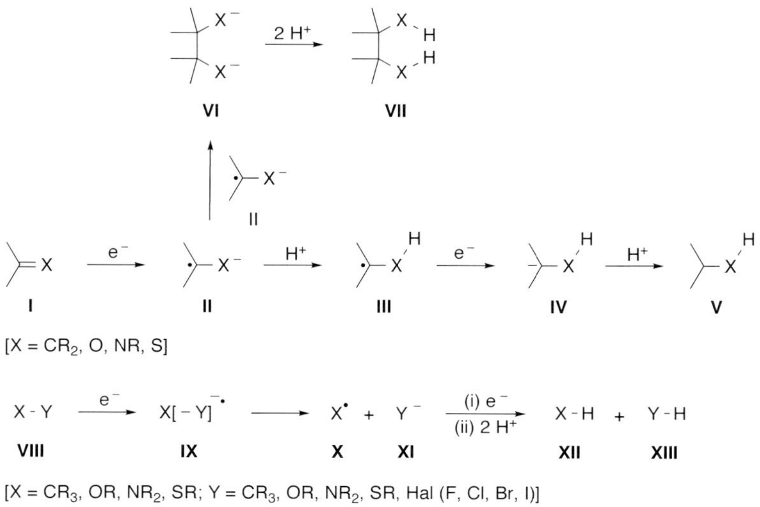

Scheme 17.1 Assumed mode of action of electropositive metals in reduction processes.

modulate the reductive power of the reducing agent. Regarding the chemoselectivity of these reduction processes, several low-valent ions, which have also been used recently as a source of electrons operating in SET mechanisms [5], will not be considered here. A study of the reduction of different functional groups using dissolving metals follows.

17.2
Reduction of Compounds with C=X Bonds

17.2.1
Reduction of Carbonyl Compounds

Metals in alcoholic and aqueous media reduce carbonyl compounds effectively, and this procedure has been known for many years. The most common reductive mixture consists of Na in EtOH, but Na amalgam has also been employed, as well as Zn or Fe in acetic acid. Reductions under strongly basic conditions lead to epimerization, due to enolate formation, for carbonyl compounds bearing a stereogenic center at the α-position [6].

The reduction of carbonyl compounds in liquid NH_3 in the absence of another proton source usually yields equimolecular amounts of the corresponding alkoxide and enolate, so after hydrolysis the starting ketone and the corresponding alcohol are obtained. Better yields of the expected alcohols are obtained by using an excess of the metal, particularly for cyclic ketones [7]. Other solvents than alcohols and liquid NH_3 have been employed in the reduction of carbonyl compounds with dissolving metals, for instance diethyl ether, tetrahydrofuran (THF) (yields are increased under sonication in ethereal solvents), hexamethylphosphoramide (HMPA), ethylenediamine and *N*-ethylpyrrolidone, usually in the presence of a proton source [8]. In these cases, the results are similar to those obtained using liquid NH_3 as solvent. A great number of electropositive metals have been used: Na and K in alcohols [9] or Li, Na, K, Rb, Cs, Ca, Sr, Ba [4b, 7a] and Yb in NH_3 [10]. Amalgams of Li [7a] and Al [11] have also been found to be effective for the reduction of carbonyl compounds. Concerning the chemoselectivity of these processes, the main reaction products are the corresponding alcohols ((**V**), X = O, in Scheme 17.1); however, and depending on the reaction conditions, alcohols, pinacols ((**VII**), X = O, in Scheme 17.1) and conversion of the carbonyl group to a methylene group can occur with aromatic ketones, such as in the case of ketone (**1**) to give compound (**2**) (Scheme 17.2) [12]. The Clemmensen reduction of carbonyl compounds also yields methylene groups [13] with the use of zinc powder in acetic anhydride or ether saturated with hydrogen chloride. Under these acidic reaction conditions, several functional groups are tolerated, such as cyano, acetoxy, phenol ether and alcoxycarbonyl. The mechanism of the Clemmensen reduction is not well understood but it is assumed that the corresponding alcohols are not reaction intermediates.

The stereochemical course of reduction of substituted cyclic ketones has been widely studied. A mixture of epimeric alcohols is obtained in these processes and there is controversy concerning the thermodynamic or kinetic control of these reactions. From experimental results, it has been found that cyclohexanones with one substituent or without substituents at the α-position lead to equatorial alcohols [6, 14]. For instance, the reaction of (−)-menthone (**3**) with Li in liquid NH_3 at −78 °C gave menthol (**4**) as the main reaction product together with a small amount of isomenthol (**5**) (Scheme 17.3) [6]. Reductions of these substrates with metals in alcoholic solvents also gave equatorial alcohols as the major products. This method has been used in the synthesis of enantiomerically pure alcohols starting from epimeric ketones [15] as well as in the synthesis of cholanic acid derivatives starting from the corresponding ketone [9]. It is noteworthy that the same reduction with metal hydrides leads to mixtures of equatorial/axial alcohols, giving predominantly the axial epimers depending upon the position of the substituents.

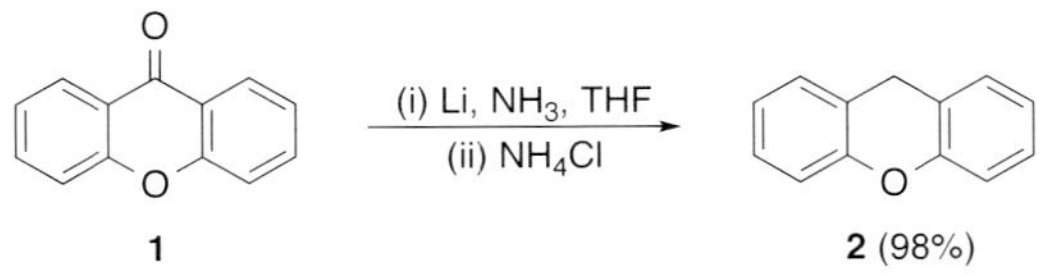

Scheme 17.2 Carbonyl conversion to a methylene group with an aromatic ketone.

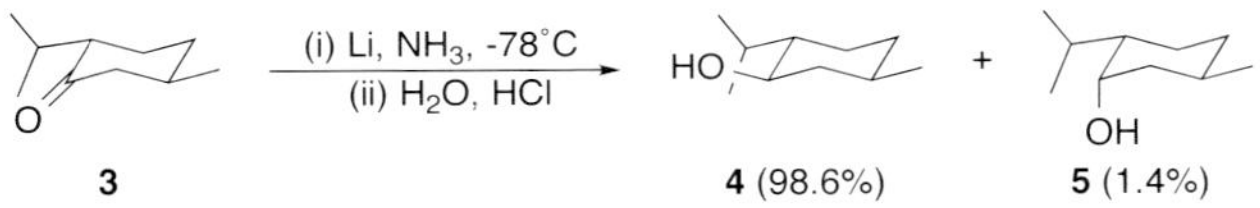

Scheme 17.3 Reduction of (−)-menthone (**3**) with Li in liquid NH_3 at −78 °C.

Sterically hindered cyclohexanones are also reduced with dissolving metals to give, the thermodynamically most stable, equatorial alcohols, such as in the case of 11-keto steroids using Na or Li in liquid NH_3 or alcoholic solvents [16]. These sterically hindered ketones are not reduced effectively by means of metal hydrides, meaning that dissolving metals are the method of choice. However, 12-keto steroids [7b] as well as 1-decalones [7c], which are less sterically hindered than 11-keto steroids, always gives mixtures of epimeric alcohols upon reduction with Li in liquid NH_3 or Na in alcoholic solvents. High stereoselectivity has also been found in the dissolving metal reduction of bicyclo[2.2.1]heptan-2-ones with alkali metals in liquid NH_3 in the presence of a proton source, the *endo*-alcohol being in all cases the major reaction product, regardless of the higher stability of the *exo*-epimer [7d, 14, 17]. Although the reduction of cyclopentanones has been less studied, due to the biological activity of steroid derivatives, 16- and 17-keto steroids have been reduced with dissolving metals. In general, 16-keto steroids give 16-α-hydroxy derivatives [18], meanwhile the 17-keto steroids provide the 17-β-ols with considerably higher stereoselectivities [19].

17.2.2
Reduction of Imines

Imines are reduced by means of dissolving metals to give amines, the mechanism being thought to be similar to that for the reduction of ketones shown in Scheme 17.1 [20]. The most common combinations are Na, Al and Mg in alcoholic solvents and Zn–NaOH [21]. More recently, the treatment of *N*-benzylideneaniline with In powder in ethanolic aqueous ammonium chloride gave a mixture of all the diamine isomers in good yields [22], resulting from the homocoupling of two radical anion intermediates of type **(II)**. Attempted extension of the reaction to the imine **(6)** resulted in the simple reduction of the C=N bond to give the benzylic amine **(7)** (Scheme 17.4). However, the imine coupling could be carried out intramolecularly. Thus, treatment of the bis-imine **(8)** with In in THF in the presence of acetic acid gave decahydroquinoxaline **(9)** in good yield as a single diastereomer (all equatorial substituents) (Scheme 17.4) [22]. Indium metal is an attractive reducing agent not only because it is stable toward air and water but also because its first ionization potential (5.8 eV) is close to that of Na (5.1 eV). However, its relatively high price

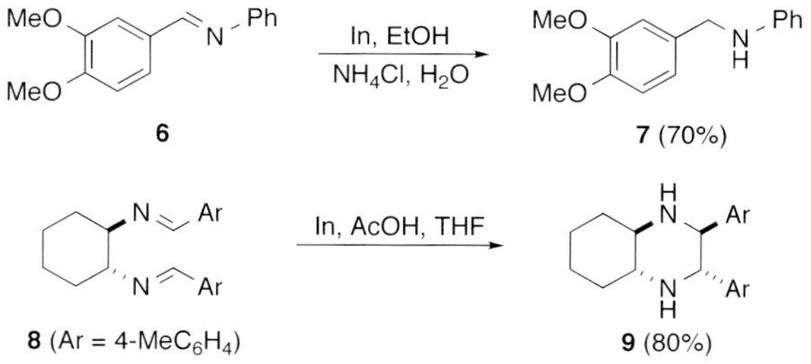

Scheme 17.4 Indium reductions of imines.

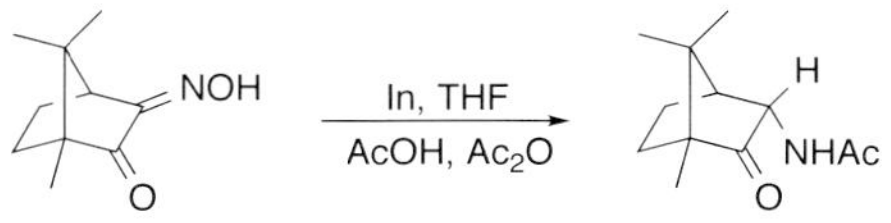

Scheme 17.5 The chemoselective reduction of α-oximino carbonyl compound (**10**).

and low availability are limitations to more extensive use of this metal as a reducing agent in organic synthesis [23].

As with imines, dissolving metals reduce oximes to primary amines, the combination of Na–alcohol being the most commonly used [21]. The same process can be carried out using lithium aluminum hydride (LAH) in THF giving, in general, better yields. The chemoselective reduction of α-oximino carbonyl compounds has been carried out in THF with In metal in the presence of 4 equivalents of AcOH and 2.5 equivalents of Ac_2O to give *N*-acetyl amines in good yields. Regarding the stereochemistry of the process, the single *endo*-diasteromer acetamide (**11**) was obtained from the camphor derived oxime (**10**); notably, the carbonyl group was unaffected in all cases (Scheme 17.5) [22].

17.3
Reduction of Carboxylic Acids and Their Derivatives

Primary alcohols cannot be prepared effectively by reduction of carboxylic acids with dissolving metals. Instead, aliphatic carboxylic acids are transformed into aldehydes using an excess of Li in $MeNH_2$ or $EtNH_2$ at high temperatures [24]. Similarly, acyl halides and acyclic anhydrides are not reduced to alcohols under these reaction conditions. The reduction of carboxylic esters using an alkali metal in an alcoholic solvent yields two alcohols as reaction products (Bouveault–Blanc reduction). This procedure was the standard method for the reduction of aliphatic esters to alcohols prior to the development of metal hydrides as reducing agents. The typical reaction conditions are Na in EtOH although other alkali metals and Na amalgam have also been used. As solvents, other alcohols can replace EtOH while the mixture of $EtOH/NH_3$ is generally more efficient in theses processes than EtOH alone. The acyloin condensation is an alternative dissolving-metal reduction reaction of carboxylic esters [25]. The reduction takes place via two sequential one-electron transfer steps, an aldehyde being involved as a reaction intermediate. For instance, the treatment of compound (**12**) with Na in NH_3 gives the lactone (**13**) after acidic hydrolysis, a selective reduction of the ester group occurring without altering the carboxylic acid functionality (Scheme 17.6) [26].

Although lactones can be reduced under Bouveault–Blanc reaction conditions to their corresponding diols, depending on their ring size they can be transformed into the corresponding lactols. For instance, arabinose (**15**) was prepared from the lactone (**14**) upon treatment with Na amalgam in aqueous H_2SO_4 at pH 3–3.5, yields being considerably lower at higher pH values (Scheme 17.7) [27].

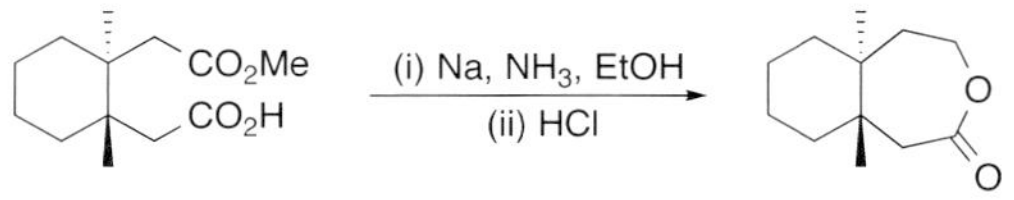

Scheme 17.6 Lactone formation via selective ester reduction.

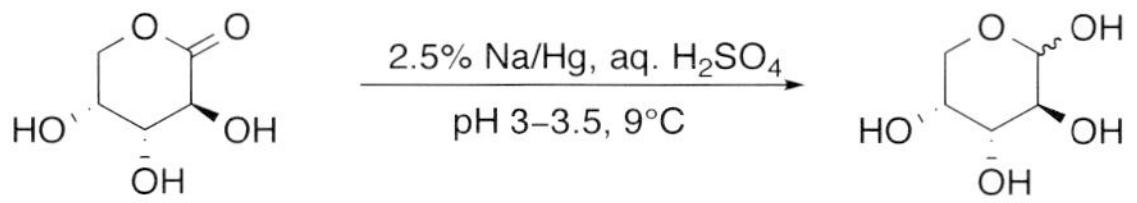

Scheme 17.7 Synthesis of arabinose **(15)**.

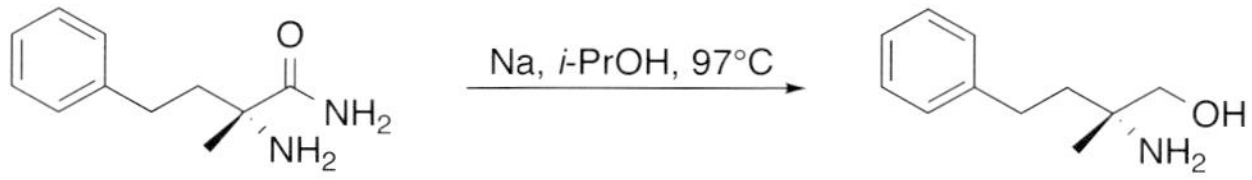

Scheme 17.8 Reduction of the α,α-disubstituted amino amide **(16)**.

The reduction of N,N-disubstituted amides with Li or Na in NH_3 or $MeNH_2$, in the presence of EtOH, AcOH or H_2O, yields aldehydes. Primary alcohols are obtained from amides upon reduction with an excess of alkali in NH_3 or HMPA and in the presence of a proton source. Finally, amines can be prepared by the reduction of amides with dissolving metals in alcoholic solutions [28]. As an example, the reduction of the α,α-disubstituted amino amide **(16)** with Na in i-PrOH gave the amino alcohol **(17)** (Scheme 17.8) [29].

The treatment of nitriles with alkali metals in alcoholic solvents or HMPA, as well as in NH_3 and t-BuOH, yields mainly the decyanation product [30]. However, amines can be prepared in high yields using Ca in NH_3 [31].

17.4
Reduction of functional groups bearing N, O and S

17.4.1
Reduction of Sulfoxides

The reduction of sulfoxides has been reviewed extensively [32–34]. Many low-valent transition metals (Mo(II), Mo(III), Ti(II), V(II), W(II)) have been found to be effective in the deoxygenation of sulfoxides. Deoxygenation of sulfoxides **(18)** was also achieved using a combination of Zn and Me_2SiCl_2 in acetone as the reducing combination to give sulfides **(19)**. In this case, the basic thermodynamic

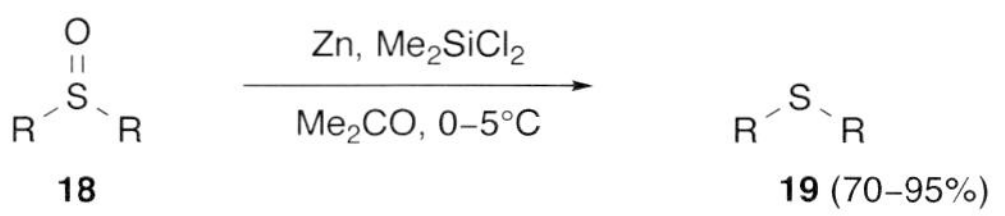

Scheme 17.9 Deoxygenation of sulfoxides.

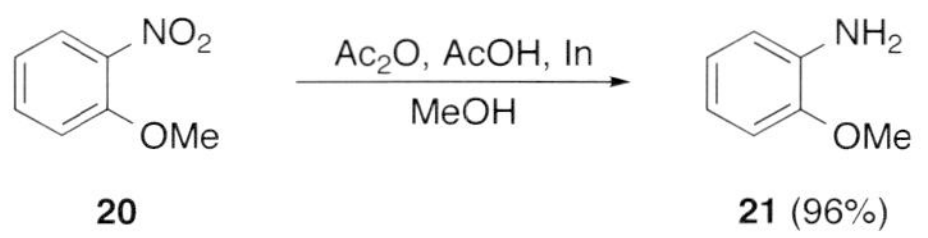

Scheme 17.10 Reduction of nitro arenes.

driving force is the formation of a strong Si—O bond by breaking a weaker Si—Cl bond (Scheme 17.9) [35].

17.4.2
Reduction of Nitro Compounds

The reduction of aromatic nitro compounds usually gives mixtures of nitroso, hydroxylamine and sometimes azo compounds, the combination of Zn and NaOH being the most frequently used for the conversion of nitro compounds into the corresponding symmetric azo derivatives [36, 37]. A selective reduction of nitro arenes to aromatic amines was achieved by using In metal as the reducing reagent. Thus, treatment of a range of aromatic nitro compounds with In powder in aqueous ethanolic ammonium chloride resulted in selective reduction of the nitro group, other substituents such as ester, nitrile, amide, carboxylic acid and halide groups remaining unaffected [38]. More recently, *N*-aryl acetamides were prepared in excellent yields from nitroarenes in the presence of acetic anhydride, acetic acid and In in a one-pot procedure. As an example of the use this method, the nitro arene (**20**) was transformed into the amine (**21**) in high yield (Scheme 17.10) [39].

Reductive heterocyclization of 2-nitroacylbenzenes or 2-nitroiminobenzenes with In metal in the presence of 2-bromo-2-nitropropane (BNP) was also performed in a mixture of H_2O and MeOH (2:1) to give benzisoxazoles in high yields [40].

17.4.3
Reduction of Compounds with N—X Bonds (X = N, O, S)

The reductive cleavage of N—X bonds (X = N, O, S) is of interest in synthetic organic chemistry because, among others, amines, diamines and amino alcohols can be obtained as reaction products. In this way, the amino group is introduced in an organic substrate by the use of the azide ion, hydroxylamines or hydrazides as nucleophiles. Six-membered heterocycles with the N—N unit are accessible

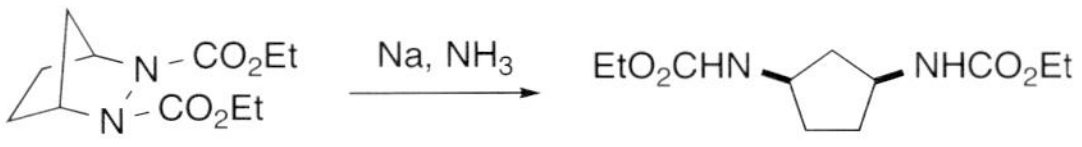

Scheme 17.11 Reductive cleavage of N—N bonds.

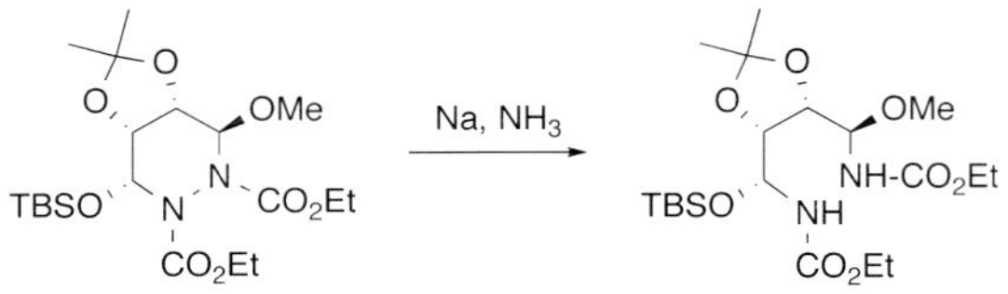

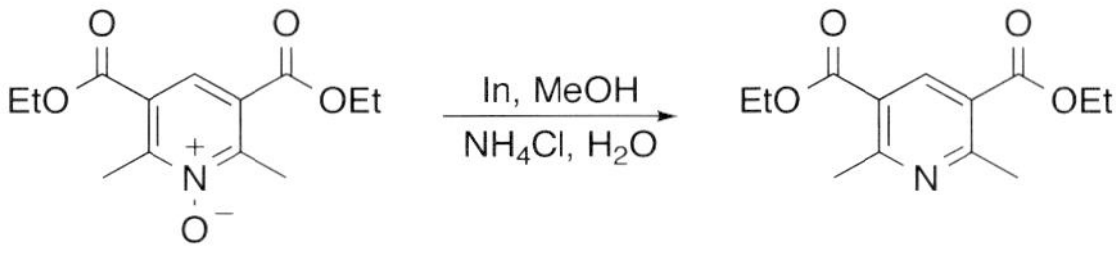

Scheme 17.12 Indium metal reduction of pyridine oxide (**26**).

through a hetero Diels–Alder reaction of an azo compound and a dienophile. The N—N bond of *N,N*-diacyl-2,3-diazanorbornane (**22**) is cleaved with Na in liquid NH_3 to give the 1,3-diaminocyclopentane derivative (**24**) (Scheme 17.11) [41]. Under the same reaction conditions, dicarbamate (**23**) yields the lyxose derivative (**25**) (Scheme 17.11) [42].

The mixture of Na and liquid NH_3 was also used for the reductive cleavage of the N—S and N—O bonds in β-lactam sugar derivatives [43] and β-thiolactams [44] and to transform isoxazolidines into 1,3-hydroxyamines through a reductive ring opening [45]. The deoxygenation of amine *N*-oxides is an important transformation in the synthesis of nitrogen-containing heterocycles [46], and several aromatic and aliphatic amine *N*-oxides were deoxygenated to their corresponding amines in good to quantitative yields using In metal in neutral aqueous media. As an example, the pyridine oxide (**26**) was converted into the pyridine (**27**) (Scheme 17.12) [47]. Other functional groups such as alkenes, halides, esters, ethers, nitriles, amides and sulfones are unaffected under these reaction conditions.

17.5
Reduction of C=C and C≡C Bonds

17.5.1
Reduction of C=C Bonds

The reduction of nonactivated C=C bonds to the corresponding alkanes is not an easy and routine task. Whitesides and Ehmann have described the reduction of alkenes to alkanes using Na in HMPA in the presence of *t*-BuOH. These reductions seem to be quite stereoselective, and the thermodynamically most stable products **(30)** and **(31)** are obtained starting from the alkanes **(28)** and **(29)**, respectively (Scheme 17.13) [48].

Sakai and coworkers reported the selective reduction of the *exo* methylene moiety in the compound **(32)** with Li-EtNH$_2$ to give the spiroalkene **(33)**. Subsequent oxidation of this compound with SeO$_2$ afforded (±)-acorenone B **(34)**, a sesquiterpene isolated from *Bothriochloa intermedia* (Scheme 17.14) [49].

Activated C=C bonds are easily reduced with dissolving metals. Stork proposed that the reduction of α,β-unsaturated ketones **(XIV)** with Na in liquid NH$_3$ proceed by formation of a metallated oxyallyl anion **(XV)** (Scheme 17.15), which undergoes protonation at the β-carbon atom [50]. On the other hand, the reduction of α,β-

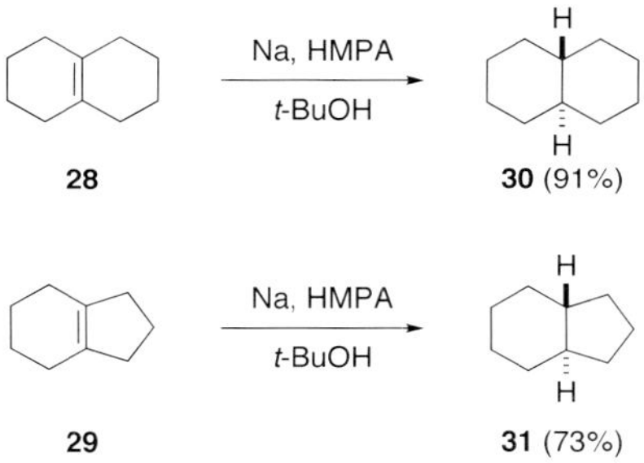

Scheme 17.13 Reduction of nonactivated C=C bonds.

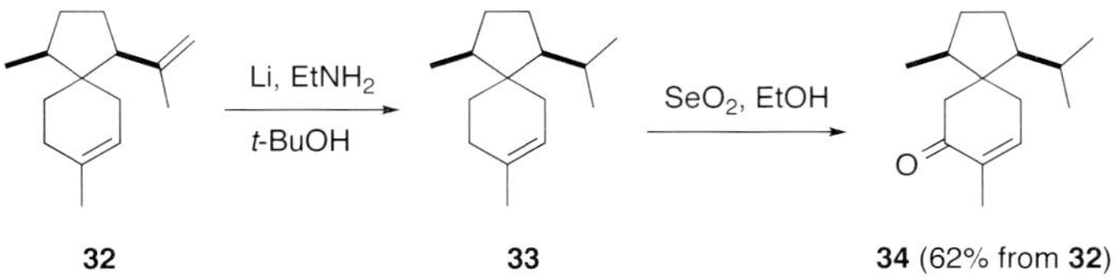

Scheme 17.14 Synthesis of (±)-acorenone B **(34)**.

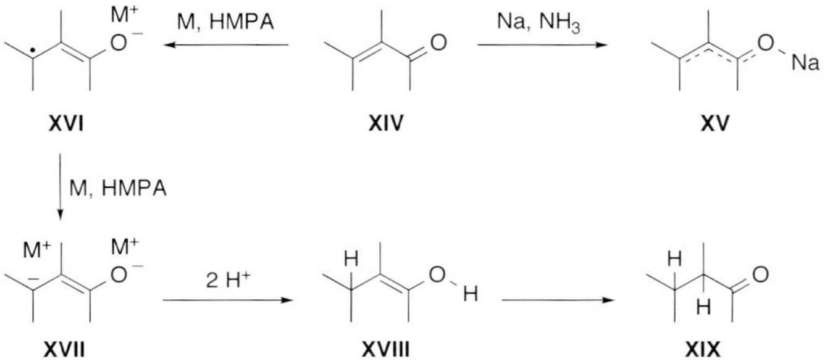

Scheme 17.15 Proposed reduction of α,β-unsaturated ketones.

unsaturated ketones with alkali metals in HMPA in the absence of a proton source has been proposed to proceed via the formation of the dianion (**XVII**) through the corresponding radical anion (**XVI**). Protonation during work-up results in the formation of the enol (**XVIII**) and the final reduced product (**XIX**) (Scheme 17.15) [51].

The reduction of α,β-unsaturated ketones with dissolving metals to give saturated ketones has been widely used in the synthesis of natural products; these processes proceed with high stereoselectivity. Thus, reduction of the enone (**35**) with Li in liquid NH_3 affords predominantly the ketone (**36**) with C_2-symmetry, instead of the *meso* isomer, which is the main reaction product through catalytic hydrogenation (Scheme 17.16) [52]. As another example of these kinds of reductions, enone (**37**) can be converted to the saturated ketone (**38**) in high yield when treated with 3 equivalents of Li in NH_3-THF (Scheme 17.16) [53]. Paquette and coworkers reported the dissolving-metal reduction of unsaturated ketone (**39**) to give the carbotricyclic decahydro-*as*-indacene portion (**40**) of the macrocyclic antibiotic (+)-ikarugamycin in high yield and with *cis* stereochemistry for the B/C rings (Scheme 17.16) [54]. The reduction of the enone (**41**) with Li in liquid NH_3 gave, after treatment of the resulting enolate with $PhNTf_2$, the vinyl triflate (**42**), a precursor of 11-*nor*-Δ^9-tetrahydrocannabinol-9-carboxylic acid, a principal human metabolite of Δ^9-tetrahydrocannabinol (Scheme 17.16) [55].

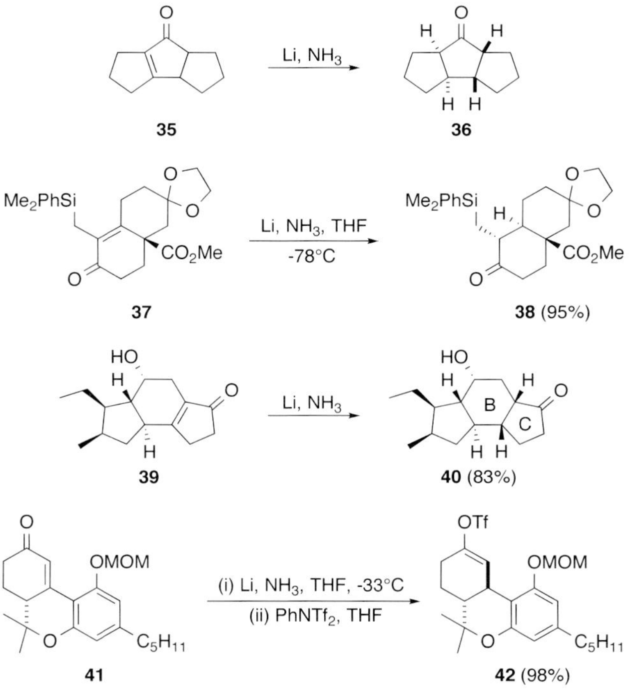

Scheme 17.16 Reduction of α,β-unsaturated ketones with dissolving metals to give saturated ketones.

In one of the steps of the synthesis of (+)-perrottetianal A, Hagiwara and Uda reported the reductive allylation of the optically pure monoacetal (+)-(43) to give the allylated *trans*-decalone (44) in quantitative yield (Scheme 17.17) [56]. When the enone (45) was subjected to reduction with Li in liquid NH_3, the corresponding *cis*-isomer (46) was produced (Scheme 17.17) [57]. The cinnamic ester derivative (47) was reduced with Mg turnings dissolving in methanol without affecting the cyano or ethoxy carbonyl group; subsequently both groups were hydrolyzed by adding aqueous KOH to the reaction mixture and heating to give the diacid (48) (Scheme 17.17) [58]. The ketal (49) was reductively silylated to give the *trans*-silyl enol ether (50) (Scheme 17.17) [59].

Finally, alkenes conjugated to aromatic rings are also easily reduced by means of dissolving metals to their corresponding alkyl arenes [60], similar to conjugated dienes. Thus, the reduction of 1,3-butadiene (51) with Na in liquid NH_3 leads to a mixture of *cis*-(52) and *trans*-(53), the ratio depending on the temperature of the reaction: whereas a 1:1 ratio was found at −78 °C, working at −33 °C the ratio was 7:1 in favor of the *trans* isomer (Scheme 17.18) [61].

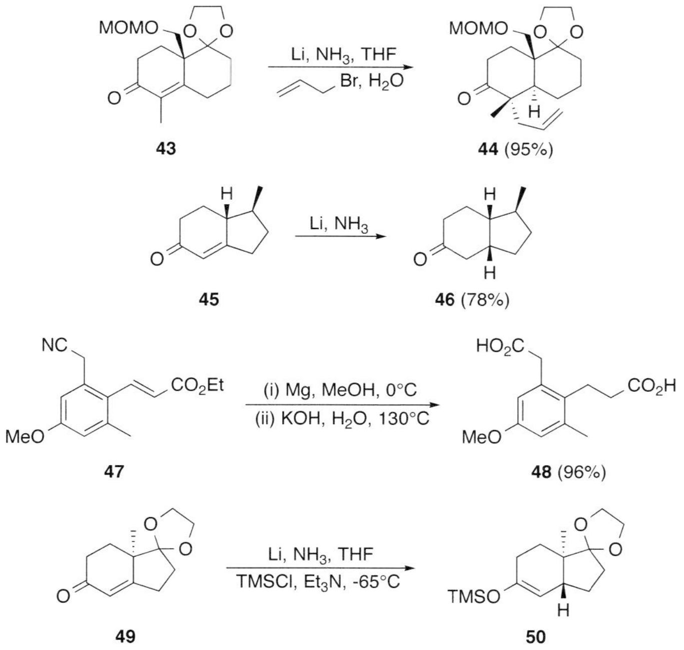

Scheme 17.17 Reduction of α,β-unsaturated ketones.

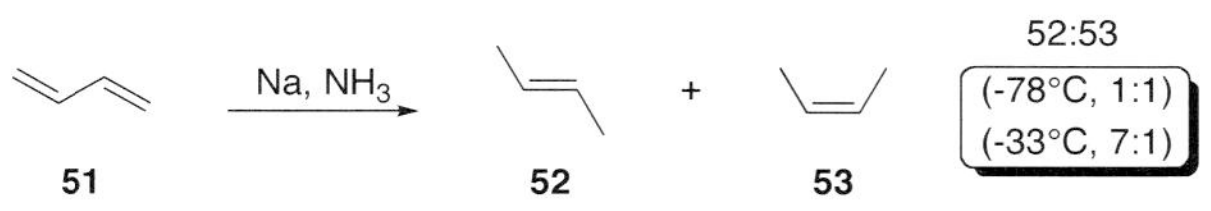

Scheme 17.18 Reduction of conjugated dienes.

17.5.2
Reduction of C≡C Bonds

The mechanism of the alkali metal reduction of alkynes **(XX)** in HMPA in the presence of a proton source is illustrated in Scheme 17.19. The addition of an electron to the triple bond gives the organometallic vinyl radical **(XXI)** which, after protonation (to give the intermediate **(XXII)**) followed by a second reduction gives the *trans*-vinyl organometallic intermediate **(XXIII)**, which after final protonation affords the *trans*-alkene **(XXIV)** (Scheme 17.19) [62]. The same reductions in the absence of a proton source result in a mixture of *trans*- and *cis*-isomers after hydrolysis. However, the reduction with the Zn/Cu couple in MeOH leads predominantly to *cis*-alkenes through a different mechanism, which has not yet been determined [63].

The reduction of tetrahydropyranyl (THP) or *tert*-butyldimethylsilyl (TBS) ethers derived from 3-decyl-1-ol **(54)** with Na in liquid NH_3/THF results in extensive hydrogenolysis of the C—O bond and concomitant bond migration with a low yield of the desired (*E*)-homoallyl ether **(55)**. However, the reduction in the presence of *t*-BuOH led to excellent yields of the desired (*E*)-dec-3-en-1-ol ether **(55)** (Scheme 17.20) [64].

The Zn dissolving-metal reduction of ethyl phenylpropiolate **(56)** to its corresponding cinnamate can be stereochemically controlled by changing the proton source in the reaction. The results of this study, while not fully understood, may imply that surface phenomena, together with a relatively slow-forming equilibrium between the intermediate *cis* and *trans* anion or radical anion, competes with the rate-determining protonation step in the mechanism. Thus, whereas in the presence of HCl as the acid, the *trans* isomer **(57)** is the major reaction product, the corresponding *cis* isomer **(58)** is obtained when AcOH is used as the proton source (Scheme 17.21) [65].

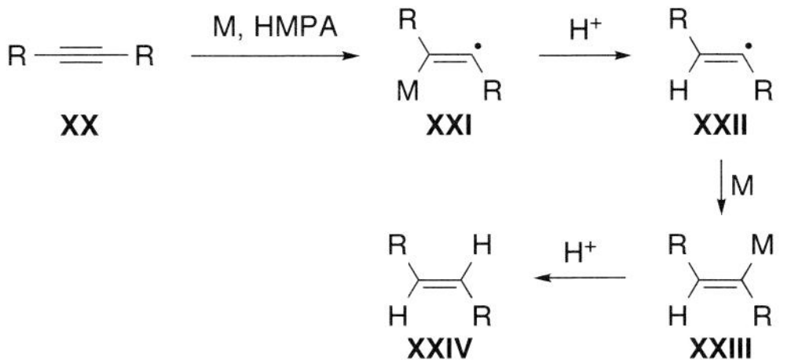

Scheme 17.19 Mechanism of alkali metal reduction of alkynes in HMPA.

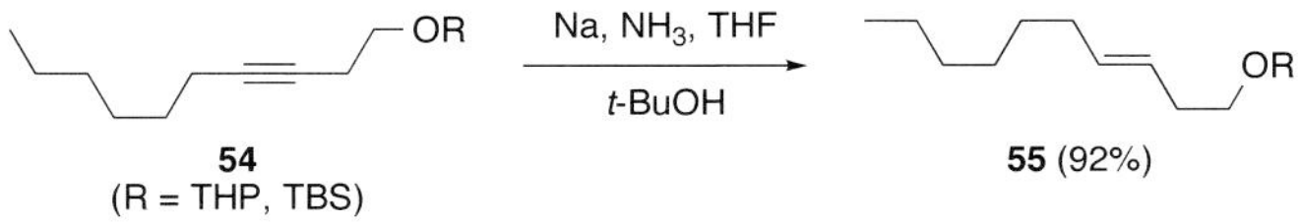

Scheme 17.20 Synthesis of (*E*)-dec-3-en-1-ol ether **(55)**.

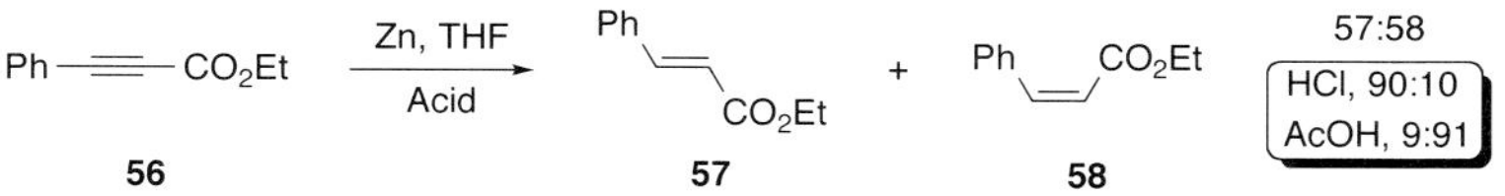

Scheme 17.21 Zn metal reduction of ethyl phenylpropiolate (**56**).

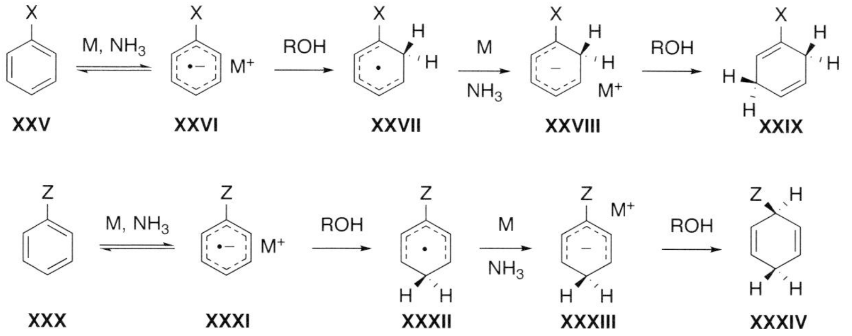

Scheme 17.22 The Birch reduction of aromatic compounds.

Highly selective reduction of terminal alkynes to alkenes in aryl propargyl ethers, amines and esters has been achieved using In metal in aqueous ethanol without formation of undesired side-reactions such as over-reduction of the double bond formed [66].

17.6
Partial Reduction of Aromatic and Heteroaromatic Rings

17.6.1
The Birch Reduction of Aromatic Compounds

The conversion of aromatic compounds to their dihydroderivatives, using a mixture of liquid NH_3, an alcohol (EtOH, i-PrOH or t-BuOH) and a co-solvent (Et_2O and THF) with an alkali metal, was first developed by Wooster [67] and, later and more extensively, by Birch [68]. Instead of NH_3, low-molecular-weight amines can be used in these reductive processes. The mechanism of the Birch reduction for substrates bearing electron-donating (**XXV**) or electron-withdrawing groups (**XXX**) is illustrated in Scheme 17.22. In both cases, the 1,4-dienes (**XXIX**) and (**XXXIV**) are obtained as the main reaction products. The nature of the substituent plays an important role in the structure of the resulting 1,4-dienes. Thus, electron-donating groups (X) direct reduction to unsubstituted 2,5-positions, while electron-withdrawing groups (Z) direct reduction to 1,4-positions. Pentadienyl radicals (**XXVII**) and (**XXXII**) are formed after addition of an electron (giving species (**XXVI**) and (**XXXI**), respectively) followed by protonation. Further addition of a second electron yields the pentadienyl anions (**XXVIII**) and (**XXXIII**), whose protonation occurs predominantly at the central carbon atom [69] to form the nonconjugate dienes **(XXIX)** and **(XXXIV)**, which do not undergo further reduction. If protonation takes place at a terminal atom of the pentadienyl anions, overreduction of the resulting conjugate 1,3-diene may occur, giving cyclohexenes. The Birch reduction of alkylbenzenes, aromatic ethers, acids and amides and bicyclic and polycyclic aromatic compounds has found a wide application in organic synthesis. Some significant examples of these reductions follow.

Corey and Lee reported the Birch reduction of the tetralin **(59)** to give the 1,4-diene **(60)**, an intermediate in the enantioselective total synthesis of pentacyclic triterpenes such as oleanolic acid, erythrodiol and β-amyrin (Scheme 17.23) [70]. In the case of 4-(trimethylsilylmethyl)anisole **(61)**, after Birch reduction and subsequent acidic work-up of **(62)**, 4-methylenecyclohexanone **(63)** was obtained in good yield (Scheme 17.23) [71].

Aromatic carboxylic acids are reduced by alkali metals in a NH_3 solution more readily than are aromatic ethers and alkyl-substituted aromatic compounds. In most cases the 1,4-dihydroisomers are obtained. In the case of 4-(4-chlorobutyl)benzoic acid **(64)**, intramolecular reductive alkylation leads to the spiro compound **(65)** (Scheme 17.24) [72]. It is also possible to perform an asymmetric Birch alkylation of appropriate substrates [73]. Treatment of chiral benzamides **(66)** with K in a mixture of liquid NH_3 and THF in the presence of one equivalent of t-BuOH gives the dienolates **(67)**, which were alkylated with MeI to yield 1,4-cyclohexadiene derivatives **(68)** in high *de* (Scheme 17.24) [74].

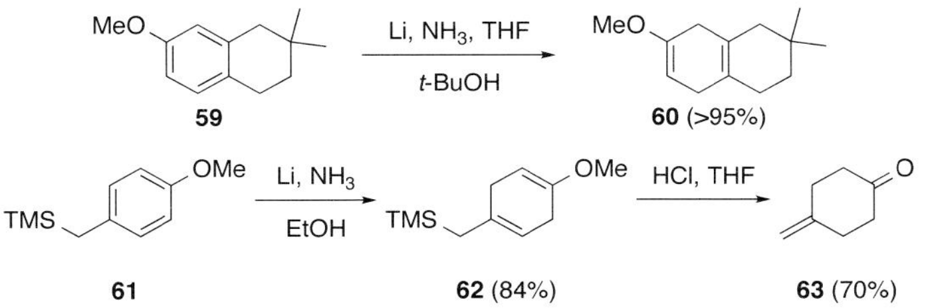

Scheme 17.23 Birch reduction of tetralin **(59)** and 4-(trimethylsilylmethyl)anisole **(61)**.

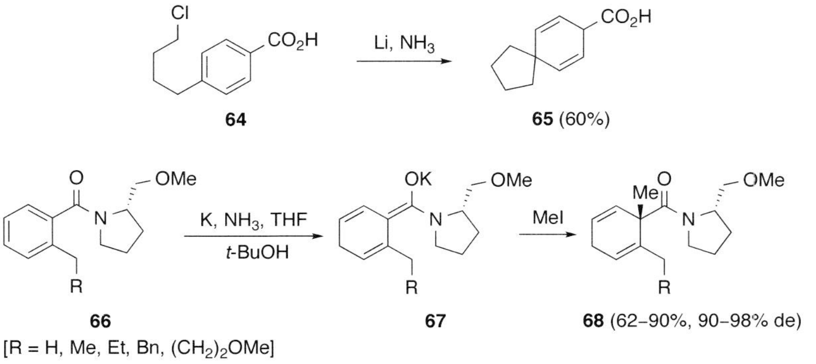

Scheme 17.24 Birch reduction of aromatic carboxylic acids and asymmetric Birch reductions.

Naphthalene reacts with Na in liquid NH_3 to give, after quenching with MeOH, 1,4-dihydronaphthalene [75]. In the case of biphenyl, 1,4-dihydrobiphenyl is also initially obtained [76]. More recently it has been found that the treatment of fullerenes with reducing metals and a proton source leads to the formation of hydrogenated fullerenes. For example, only two isomers of $C_{60}H_6$ (6:1 ratio) are obtained in the Zn(Cu) reduction of C_{60}. Three isomers of $C_{70}H_{10}$ are produced, in ratios that are dependent on the reaction time [77].

17.6.2
Partial Reduction of Heteroaromatic Rings

Five-membered heterocycles, the so called π-exceeding heterocycles, are inherently more difficult to reduce than the π-deficient heterocycles. Thus, the Birch reduction (alkali metals in liquid NH_3) of pyrrole is unknown, but pyrrole is reduced to 2,5-dihydropyrrole as the major reaction product using Zn/HCl [78]. In the case of furan, it is reduced by $Li/MeNH_2$ under forcing conditions to give a mixture of reaction products, but furoic acids under Birch reduction conditions are transformed into 2,5-dihydrofurans [79]. Although thiophenes undergo reductive ring cleavage with alkali metals in liquid NH_3, thiophene-2-carboxylic acid could be reduced to give the 2,5-dihydrothiophene derivative. For instance, the acylthiophene (69) is reduced to give compound (70) in high yield under controlled reaction conditions (Scheme 17.25) [80].

Indoles, benzofurans and benzothiophenes cannot be reduced by means of dissolving metals to their corresponding 2,3-dihydro derivatives without generally disturbing the aromaticity of the benzene ring [81]. In some cases high chemoselectivity can be achieved depending on the reaction conditions. Thus, whereas indoles are reduced by Mg/MeOH to indolines [82], the Birch reduction of 1,3-diphenylbenzo[c]furan leads to a mixture of dihydrobenzofurans [83]. On the other hand, benzothiophenes undergo reductive ring cleavage of the five-membered ring under Birch reaction conditions. However, compound (71) gives the 2,3-dihydro derivative (72) in high yield by reduction with Mg in MeOH (Scheme 17.26) [84]. Other examples of π-exceeding heterocycle reductions are the transformations of 1,2-benzisoxazoles into 2-hydroxyaryl alkyl amines with Na in alcoholic solvents or the conversion of 4-benzyl-3-methyl-4H- [1,2,4]-oxadiazol-5-one (73) into acetamidine (74) with Zn in acetic acid (Scheme 17.26) [85]. In both cases, reductive ring cleavage of the single N—O bonds takes place.

For reduction by dissolving metals of π-deficient heterocycles, Na in liquid NH_3 reduces quinoline to 1,2-dihydroquinoline [86] and isoquinoline to tetrahydroisoquinoline [87]. More recently, Donohoe and coworkers reported the partial

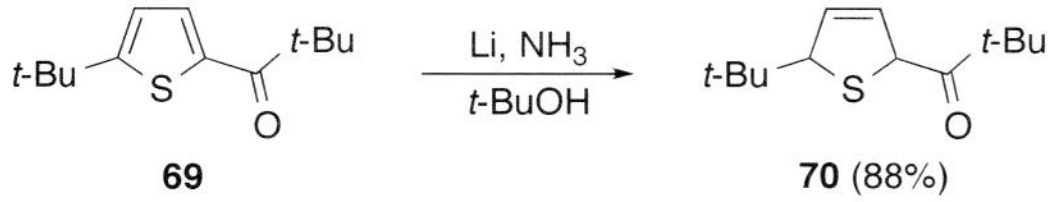

Scheme 17.25 Partial reduction of thiophenes.

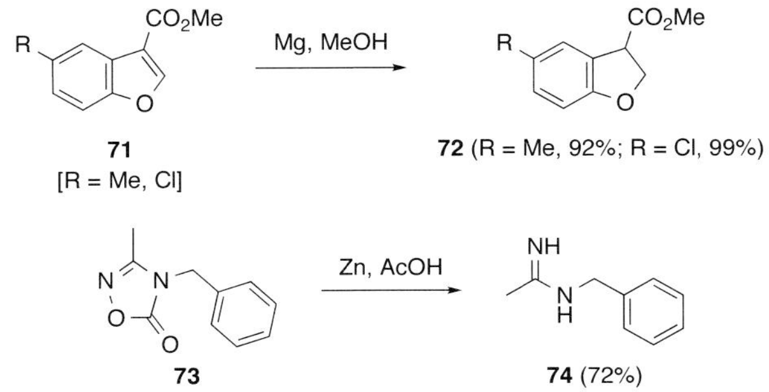

Scheme 17.26 Reduction of five-membered heteroaromatic rings.

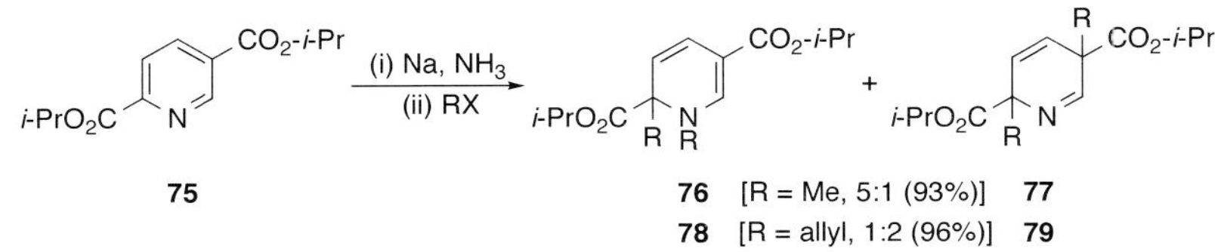

Scheme 17.27 Reductive methylation of the pyridine diester (**75**).

reduction of a series of electron-deficient pyridines to give both 1,2- and 2,5-dihydropyridines. For instance, reductive methylation of the pyridine diester (**75**) gave a mixture of compounds (**76**) and (**77**) in a 5:1 ratio. However, the reaction of compound (**75**) with Na in NH_3 followed by quenching with allyl bromide gave a mixture of the allylated compounds (**78**) and (**79**) in a 1:2 ratio (Scheme 17.27) [88].

Indium metal in aqueous EtOH and in the presence of NH_4Cl selectively reduces the heterocyclic ring of quinolines, isoquinolines and quinoxalines to their corresponding tetrahydro derivatives [89].

17.7
Reduction of Compounds with C—X Bonds

17.7.1
Reduction of α-Functionalized Carbonyl Compounds

Reduction of α-halogenated ketones is effected readily by means of Zn in acetic acid, a zinc enolate probably being the initial reaction intermediate. The reduction of the same substrates in the presence of deuterated solvents yields α-deuterated ketones. Stereoelectronic factors control the approach of the proton donor to the enolate so, for instance, the reduction of the ketone (**80**) with Zn in DOAc gave the deuterated product (**81**) with high stereoselectivity (Scheme 17.28) [90].

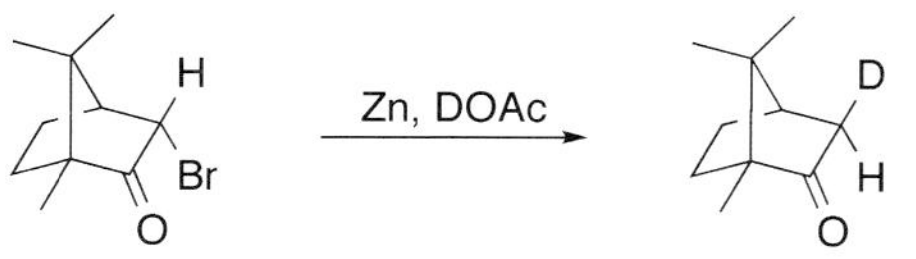

80 **81**

Scheme 17.28 Reduction of α-halogenated ketones.

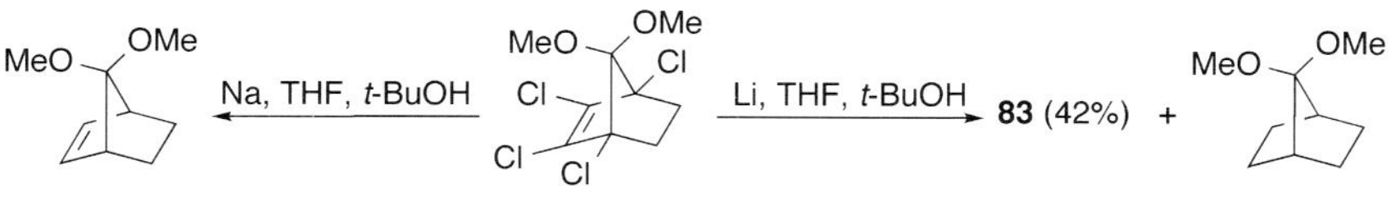

83 (63%) **82** **84** (22%)

Scheme 17.29 Reduction of C—Hal to C—H bonds.

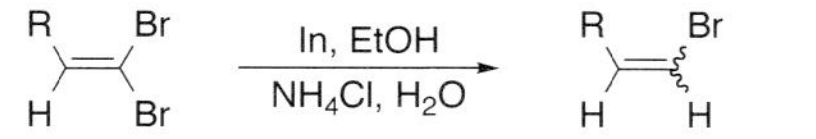

85 **86** (70–95%)

[R = alkyl, aryl, heteroaryl] (*E*/*Z* : 50:50–95:5)

Scheme 17.30 Vinyl bromides via reduction of aryl-substituted *gem*-dibromides.

Different metals, such as Li, Ca and Ba in NH_3 or Zn in acetic acid, effect the reductive cleavage of α-ketols and α-ketol acetates to the corresponding ketones [91]. It is also possible to perform the reduction of C—S bonds of α-alkylsulfanyl ketones under similar reaction conditions [92, 93]. Sulfinyl and sulfonyl groups attached at the α-position of a carbonyl group are also easily reduced with Al amalgam [94].

17.7.2 Reduction of C—Hal to C—H Bonds

Organic chlorides, bromides and iodides are easily reduced with Li, Na, K, Mg, Ca and Ba in the presence of a proton source. The reduction of alkyl fluorides is much more difficult and needs to be performed under stronger reaction conditions, for instance, using K in the presence of a crown ether [95]. Sometimes, these reductions are often accompanied by side-reactions, such as reduction of double bonds to yield the corresponding saturated compound. In the case of the tetrachloronorbornene derivative **(82)**, both allylic and vinylic chlorine atoms are replaced by hydrogen using Na or Li in *t*-BuOH, but Na does not reduce the double bond, giving compound **(83)** whereas Li affords a mixture of bicycles **(83)** and **(84)** (Scheme 17.29) [96].

A stereoselective reduction of aryl-substituted *gem*-dibromides **(85)** to their corresponding vinyl bromides **(86)** has been performed by means of In metal in a mixture of methanol and saturated aqueous ammonium chloride solution (Scheme 17.30) [97].

17.7.3
Reduction of C—O to C—H Bonds

Dissolving metals have also been found to be effective in the deoxygenation of different organic substrates. For instance, a toluene radical anion generated from K metal and toluene with the assistance of a crown ether proved effective in the reductive demethylation of methoxybenzene derivatives [98]. On the other hand, Li and Na in liquid NH_3 reduce allylic and benzylic alcohols, ethers or acetals [99], due to the highly reactive allylic or benzylic C—O bond. Thus, the benzylic acetal **(87)** is reduced to the compound **(88)** without over-reduction using Na in NH_3 and methanol [100], but under the same reaction conditions the allylic alcohol **(89)** leads to bicycle **(90)**, after rearrangement (Scheme 17.31) [101].

The benzyl group is often employed for the protection of alcohols in carbohydrate, peptide and nucleotide chemistry. This protecting group can easily be removed with Na in a mixture of liquid NH_3 and THF [102]. The trityl (triphenylmethyl) group has also been extensively used as a protecting group of alcohols and it can easily be removed by acidic hydrolysis. However, it has been reported that the reaction of alkyl (primary or secondary) allylic and benzylic trityl ethers **(91)** with Li powder and a catalytic amount of naphthalene produces the reductive cleavage of the trityl–oxygen bond, affording the corresponding alcohols **(92)** in good to excellent yields under mild reaction conditions (Scheme 17.32) [103].

Allylic ethers and acetals also undergo reductive cleavage of allylic C—O bonds by means of dissolving metals. Kraft and coworkers reported that in one of the steps toward the synthesis of compounds containing the patchoulol skeleton, the acetate **(93)** was transformed into the deoxygenated methyl enol ether **(94)** by

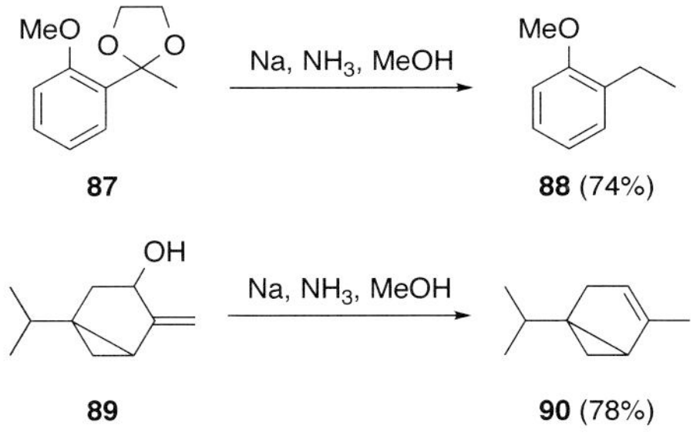

Scheme 17.31 Reduction of C—O to C—H bonds.

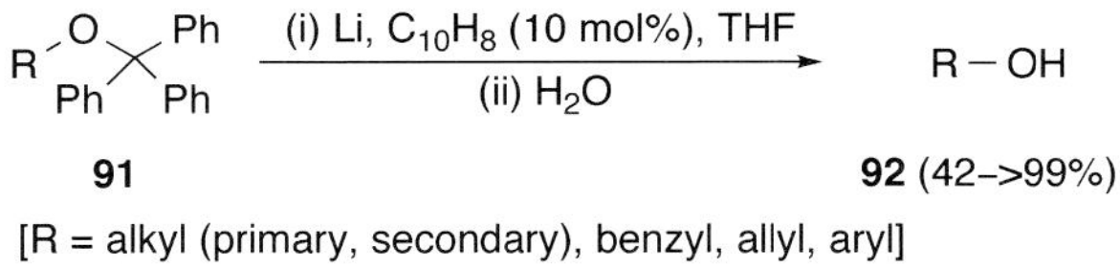

Scheme 17.32 Reductive cleavage of the trityl–oxygen bond.

means of Li in liquid NH_3 in the presence of *t*-BuOH (Scheme 17.33) [104]. Takano and coworkers have developed a facile route to the aromatic bisabolane sesquiterpenes by employing a hydrogenolytic cleavage of the ether bond at the homobenzylic-homoallylic position of the 1,3-dioxepine derivative **(95)**, under dissolving-metal reduction conditions, to give a mixture of the aldehyde **(96)** and the primary alcohol **(97)** (Scheme 17.33) [105].

Deoxygenation of simple alkyl alcohols is not an easy task. Several methodologies have been developed for the reductive cleavage of the C—O bond in different carboxylic, phosphoric or sulfonic esters derived from aliphatic alcohols. Ireland and coworkers reported that *N,N,N′,N′*-tetramethylphosphorodiamidates were cleanly deoxygenated with alkali metals in liquid NH_3 or volatile amines [106]. Mori and Tamura reported the conversion of phosphorodiamidate **(98)** into the compound **(99)** with Li in ethylamine in one of the steps toward the synthesis of ambrein and ambrox (Scheme 17.34) [107].

Similarly to phosphorodiamidates, diethyl vinyl phosphates can also be reduced to their corresponding alkenes by a hydrogenolytic cleavage of the vinylic C—O bond. Thus, treatment of compound **(100)** with an excess of Li in ethylamine and 5 equivalents of *t*-BuOH gave compound **(101)** after reduction of both the enol phosphate and the isopropenyl group (Scheme 17.35) [108].

Epoxides undergo reductive cleavage with Li and Na in amine solvents to give alcohols. Regarding the regiochemistry of the process, the ring opening of epoxides is governed by the stability of the initially formed organometallic intermediate, prior to protonation. Thus, alkyl-substituted epoxides lead to more substituted alcohols, as opposed to for aryl-substituted epoxides [109]. More recently, the reductive opening of epoxides with Li powder in the presence of a stoichiometric

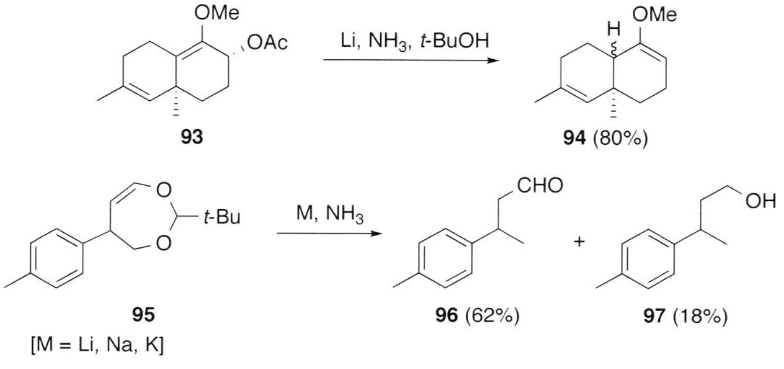

Scheme 17.33 Reductive cleavage of allylic C—O bonds in allylic ethers and acetals.

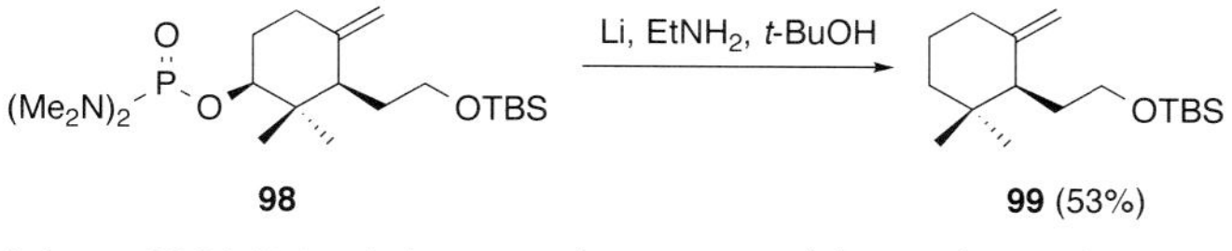

Scheme 17.34 Li in ethylamine reduction toward the synthesis of ambrein and ambrox.

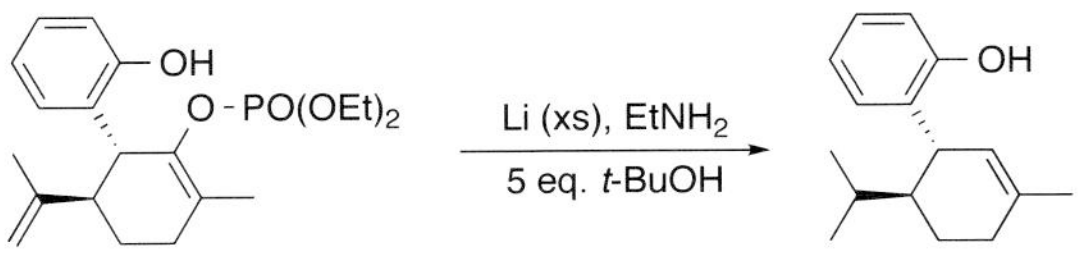

100 **101** (72%)

Scheme 17.35 Hydrogenolytic cleavage of vinylic C—O bonds.

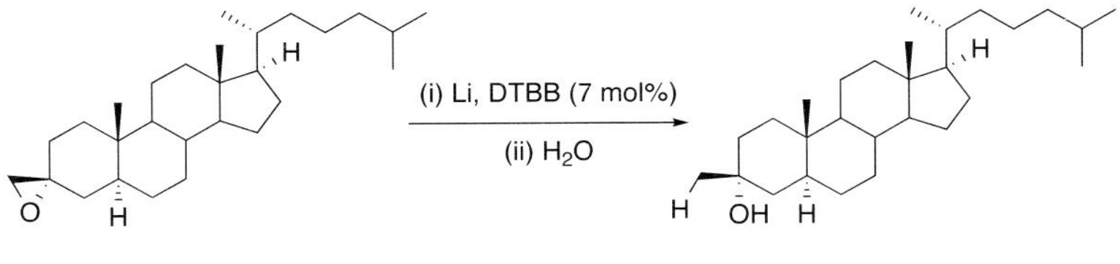

102 **103** (99%)

Scheme 17.36 Reductive cleavage of epoxides.

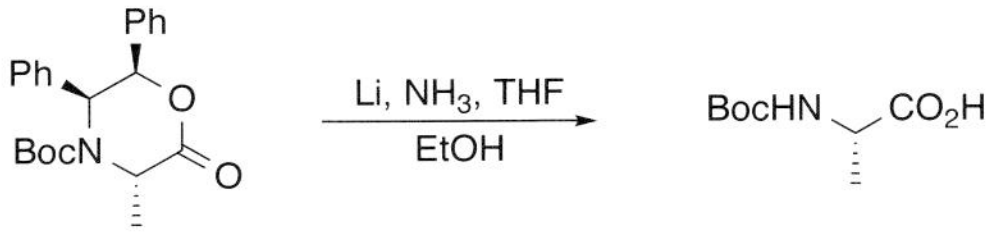

104 **105** (54%, >97% ee)

Scheme 17.37 *N*-Boc-protected α-amino acid via reduction of oxazinone **(104)**.

[110] or catalytic [111] amount of an arene and the reaction of the resulting organolithium compounds with electrophiles has been reported. Thus, the treatment of the cholestanone derivative **(102)** with and excess of Li in the presence of catalytic amount of 4,4′-di-*tert*-butylbiphenyl (DTBB) gave, after hydrolysis with water, the corresponding alcohol **(103)** (Scheme 17.36) [112].

17.7.4 Reduction of C—N to C—H Bonds

As for the C—O bond, dissolving metals are effective in the reductive cleavage of highly reactive benzylic C—N bonds. The reaction of aliphatic and aromatic secondary and tertiary *N*-tritylamines with Li powder and a catalytic amount of naphthalene leads to reductive detritylation, affording the corresponding amines in good to excellent yields [113]. Williams and coworkers developed different methodologies for the asymmetric synthesis of α-amino acids using 3-bromo-5,6-diphenyl-2,3,5,6-tetrahydro-1,4-oxazin-2-ones, either as electrophilic glycine templates [114] or as glycine enolate equivalent [115]. Thus, the dissolving-metal reduction of oxazinone **(104)** allows the direct preparation of *N*-Boc-protected α-amino acid **(105)** (Scheme 17.37) [115].

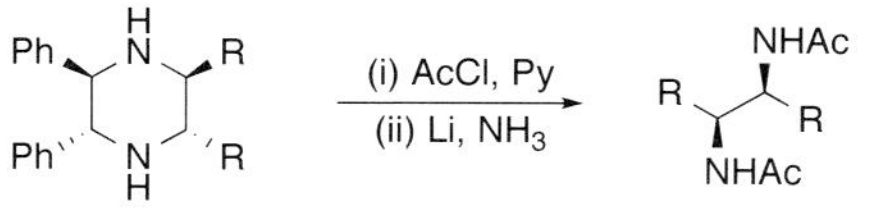

106 107 (80–85%)

[R = Me, *i*-Pr, *i*-Bu]

Scheme 17.38 Enantioselective synthesis of vicinal diamines.

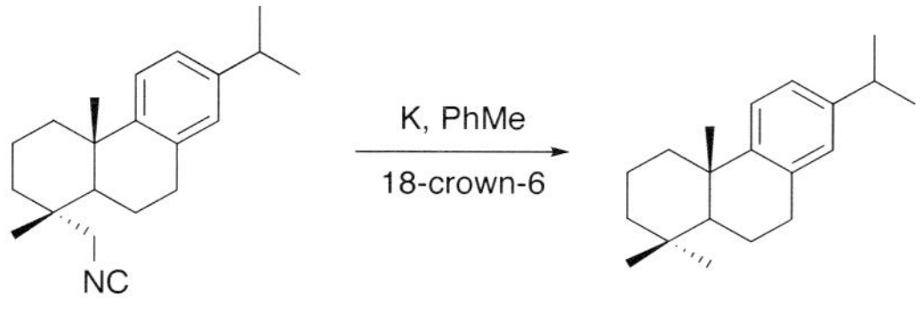

108 109 (97%)

Scheme 17.39 Reductive deamination of aliphatic amines.

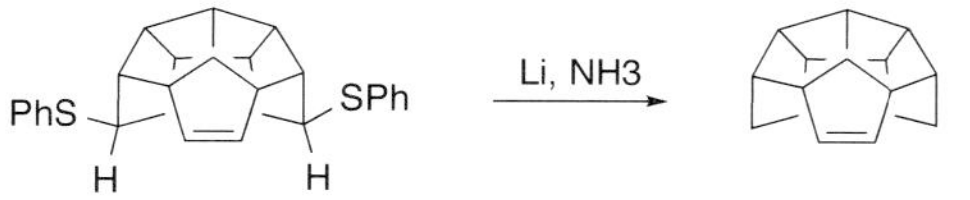

110 111 (88%)

Scheme 17.40 Reductive desulfurization of diphenyl disulfide **(110)**.

Nantz and coworkers have reported an enantioselective synthesis of vicinal diamines. Thus, the treatment of the bisacetamide piperazine derivatives **(106)** (easily accessible from the corresponding 1,2-dicarbonyl compound and 1,2-diphenyl-1,2-diaminoethane) with Li in liquid NH_3 liberated the diamine as its bisacetamide **(107)** (Scheme 17.38) [116].

The treatment of isocyanides (synthesized in high yields from the corresponding amines) with the radical anion generated from K and toluene and the assistance of a crown ether at room temperature for several hours yields the corresponding compounds resulting from the removal of the isocyano group, the whole process representing a reductive deamination of aliphatic amines. As a significant example, compound **(108)** was converted into the tricycle **(109)** in high yield (Scheme 17.39) [117].

17.7.5
Reduction of C—S to C—H Bonds

Aryl alkyl sulfides are easily desulfurized with Li, Na and K in either ethereal solvents or hydrocarbons [118]. The reductive ability of alkali metals may be improved by the presence of aromatic compounds such as naphthalene and 1-(dimethylami no)naphthalene [119]. For instance, the synthesis of the tetrahydro-C_{16}-hexaquinacene **(111)** was successfully achieved by reductive desulfurization of diphenyl disulfide **(110)** with Li in liquid NH_3 (Scheme 17.40) [120].

17.7.6
Reduction of C—C to C—H Bonds

The reductive cleavage of C—C bonds takes place only in highly strained compounds, for instance cyclopropanes, and under special reaction conditions. Thus, stereoelectronic factors present in the compound **(112)** allowed the fully regiocontrolled cleavage of a single cyclopropane bond under dissolving-metal reaction conditions to give a mixture of compounds **(113)** and **(114)** in a 1 : 2 ratio, respectively (Scheme 17.41) [121]. Mehta and Subrahmanyam also reported the synthesis of a [3.3.3]propellanone **(116)** by initial reduction of the ketone **(115)** with Li in liquid NH_3 and further oxidation with pyridinium chlorochromate (PCC) of the resulting epimeric mixture of alcohols (Scheme 17.41) [122].

Maercker and coworkers have also studied the reaction of strained carbocycles bearing phenyl substituents, such as **(117)** and **(120)**, with Li metal. Cleavage of a single C—C bond belonging to the strained carbocycle next to the activating phenyl group was always observed. The reaction of the resulting dilithium compounds **(118)** and **(121)** with MeOD led to the regioselective deuterated products **(119)** and **(122)**, respectively (Scheme 17.42) [123].

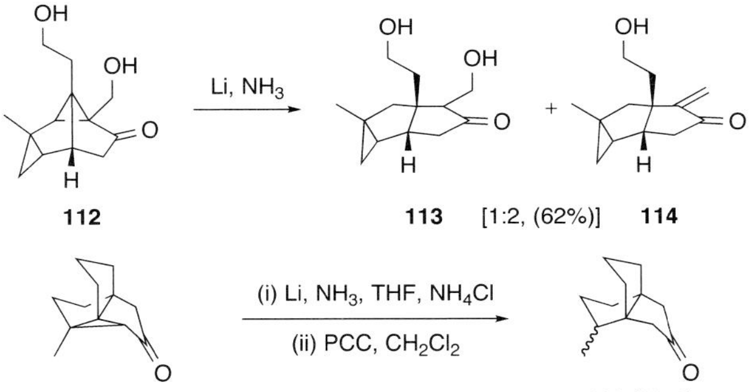

112 **113** [1:2, (62%)] **114**

115 **116** (60%)

Scheme 17.41 Reductive cleavage of cyclopropanes.

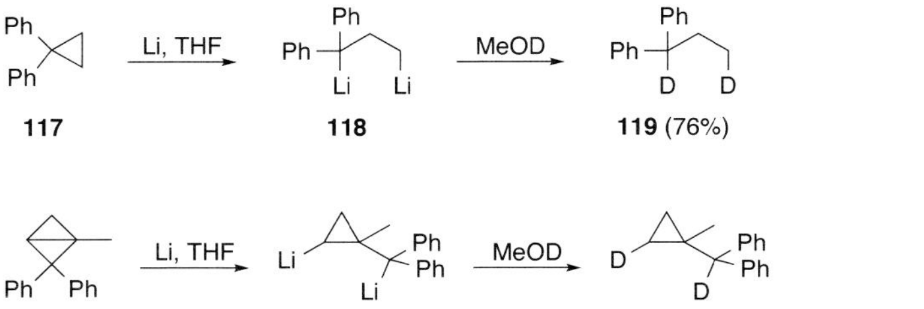

117 **118** **119** (76%)

120 **121** **122** (33%)

Scheme 17.42 Reduction of strained carbocycles bearing phenyl substituents.

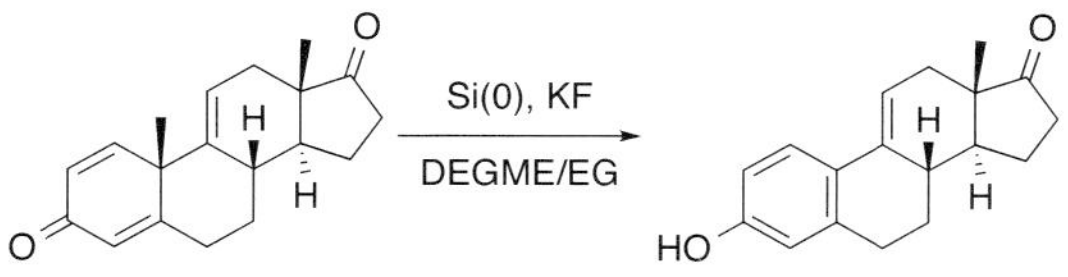

Scheme 17.43 Si(0)/KF in a mixture of DEGME; for example, as a strong, inexpensive and environmentally friendly and safe reagent.

Pearlman and coworkers recently reported the use of the combination of Si(0)/KF in a mixture of diethyleneglycol monoethyl ether (DEGME) and ethylene glycol (EG) as a strong, inexpensive and environmentally friendly and safe reagent for dissolving-metal-type demethylation of androsta-1,4,9-triene-3,17-dione **(123)** to give Δ^9-estrone **(124)** in high yield (Scheme 17.43) [124]. The same transformation has previously been performed with Zn in water and pyridine [125].

References

1. For a review, see: Senda, Y. (2002) *Chirality*, **14**, 110.
2. For recent reviews, see: (a) Glorius, F. (2005) *Organic and Biomolecular Chemistry*, **3**, 4171. (b) Roessler, F. (2003) *Chimia*, **57**, 791. (c) Genet, J.-P. (2003) *Accounts of Chemical Research*, **36**, 908.
3. (a) Mander, L.N. (1991) Partial reduction of aromatic rings by dissolving metals and by other methods, in *Comprehensive Organic Synthesis*, Vol. 8 (eds B.M. Trost and I. Fleming), Pergamon, Oxford. (b) Rabideau, P.W. (1989) *Tetrahedron*, **45**, 1579. (c) Rassat, A. (1977) *Pure and Applied Chemistry Chimie Pure et Appliquée*, **49**, 1049.
4. For reviews, see: (a) Huffman, J.W. (1983) *Accounts of Chemical Research*, **16**, 399. (b) Pradhan, S.K. (1986) *Tetrahedron*, **42**, 6351.
5. Kagan, H.B., Namy, J.L. and Girard, P. (1981) *Tetrahedron*, **9**, 175.
6. Solodar, J. (1976) *Journal of Organic Chemistry*, **41**, 3461.
7. (a) Murphy, W.S. and Sullivan, D.F. (1972) *Journal of the Chemical Society, Perkin Transactions I*, 999. (b) Huffman, J.W. and Copley, D.J. (1977) *Journal of Organic Chemistry*, **42**, 3811. (c) Grieco, P.A., Burke, S., Metz, W. and Nishizawa, M. (1979) *Journal of Organic Chemistry*, **44**, 152. (d) Rautenstrauch, V., Willhalm, B., Thommen, W. and Burger, U. (1981) *Helvetica Chimica Acta*, **64**, 2109.
8. (a) Larcheveque, M. and Cuvigny, T. (1973) *Bulletin de la Société Chimique de France*, 1445. (b) Sowinski, A.F. and Whitesides, G.M. (1979) *Journal of Organic Chemistry*, **44**, 2369.
9. Giordano, C., Perdoncin, G. and Castaldi, G. (1985) *Angewandte Chemie (International Ed. in English)*, **24**, 499.
10. (a) White, J.D. and Larson, G.L. (1978) *Journal of Organic Chemistry*, **43**, 4555. (b) Hou, Z., Takamine, K., Fujiwara, Y. and Taniguchi, H. (1987) *Chemistry Letters*, 2061.
11. Hulce, M. and LaVaute, T. (1988) *Tetrahedron Letters*, **29**, 525.
12. Hall, S.S., Lipsky, S.D., McEnroe, F.J. and Bartels, A.P. (1971) *Journal of Organic Chemistry*, **36**, 2588.
13. For reviews, see: (a) Buchanan, J.G.S.C. and Woodgate, P.D. (1969) *Quarterly Reviews of the Chemical Society*, **23**, 522. (b) Vedejs, E. (1975) *Organic Reactions*, **22**, 401.
14. Huffman, J.W. and Charles, J.T. (1968) *Journal of the American Chemical Society*, **90**, 6486.

15 Ensley, H.E., Parnell, C.A. and Corey, E.J. (1978) *Journal of Organic Chemistry*, **43**, 1610.

16 (a) Heusser, H., Anliker, R. and Jeger, O. (1952) *Helvetica Chimica Acta*, **35**, 1537. (b) Sondheimer, F., Yashin, R., Rosenkranz, G. and Djerassi, C. (1952) *Journal of the American Chemical Society*, **74**, 2696.

17 Rautenstrauch, V. (1986) *Journal of the Chemical Society, Chemical Communications*, 1558.

18 Pradhan, S.K. and Sohani, S.V. (1981) *Tetrahedron Letters*, **22**, 4133.

19 Pradhan, B.P., Hassan, A. and Shoolery, J.N. (1984) *Tetrahedron Letters*, **25**, 865.

20 Smith, J.G. and Ho, I. (1972) *Journal of Organic Chemistry*, **37**, 653.

21 (a) Harada, K. (1970) *Additions to the azomethine group*, in The Chemistry of the Carbon Nitrogen Double Bond (ed. S. Patai), Wiley-Interscience, London, p. 279. (b) Tramontini, M. (1982) *Synthesis*, 605.

22 Pitts, M.R., Harrison, J.R. and Moody, C.J. (2001) *Journal of the Chemical Society, Perkin Transactions I*, 955.

23 For reviews, see: (a) Cintas, P. (1995) *Synlett*, 1087. (b) Ranu, B.C. (2000) *European Journal of Organic Chemistry*, 2347. (c) Chauhan, K.K. and Frost, C.G. (2000) *Journal of the Chemical Society, Perkin Transactions I*, 3015.

24 Bedenbaugh, A.O., Bedenbaugh, J.H., Bergin, W.A. and Adkins, J.D. (1970) *Journal of the American Chemical Society*, **92**, 5774.

25 Bloomfield, J.J., Owsley, D.C. and Nelke, J.M. (1976) *Organic Reactions*, **23**, 259.

26 Huang, F.-C., Lee, L.F.H., Mittal, R.S.D., Ravikumar, P.R., Chan, J.A., Sih, C.J., Caspi, E. and Eck, C.R. (1975) *Journal of the American Chemical Society*, **97**, 4144.

27 Sperber, N., Zaugg, H.E. and Sandstrom, W.M. (1947) *Journal of the American Chemical Society*, **69**, 915.

28 (a) Birch, A.J., Cymerman-Craig, J. and Slaytor, M. (1955) *Australian Journal of Chemistry*, **8**, 512. (b) Benkeser, R.A., Watanabe, H., Mels, S.J. and Sabol, M.A. (1970) *Journal of Organic Chemistry*, **35**, 1210. (c) Bedenbaugh, A.O., Payton, A.L. and Bedenbaugh, J.H. (1979) *Journal of Organic Chemistry*, **44**, 4703.

29 Moody, H.M., Kaptein, B., Broxterman, Q.B., Boesten, W.H.J. and Kamphuis, J. (1994) *Tetrahedron Letters*, **35**, 1777.

30 (a) Van Tamelen, E.E., Rudler, H. and Bjorklund, C. (1971) *Journal of the American Chemical Society*, **93**, 7113. (b) Marshall, J.A. and Bierenbaum, R. (1977) *Journal of Organic Chemistry*, **42**, 3309. (c) Savoia, D., Tagliavini, E., Trombini, C. and Umani-Ronchi, A. (1980) *Journal of Organic Chemistry*, **45**, 3227. (d) Ohsawa, T., Kobayashi, Y., Mizuguchi, Y., Saitoh, T. and Oishi, T. (1985) *Tetrahedron Letters*, **26**, 6103.

31 Doumaux, A.R., Jr (1972) *Journal of Organic Chemistry*, **38**, 508.

32 (a) Dabrowicz, J., Numata, T. and Oae, S. (1977) *Organic Preparations and Procedures International*, **9**, 63. (b) Dabrowicz, J., Togo, H., Milokajczyk, M. and Oae, S. (1984) *Organic Preparations and Procedures International*, **16**, 171.

33 Grossert, J.S. (1988) *Reduction of Sulfoxides. A review*, in The Chemistry of Sulphones and Sulphoxides (eds S. Patai, Z. Rappoport and C.J.M. Stirling), John Wiley& Sons, Ltd, Chichester, p. 925.

34 Madesclair, M. (1988) *Tetrahedron*, **44**, 6537.

35 (a) Schmidt, A.H. and Russ, M. (1981) *Chemische Berichte*, **114**, 1099. (b) Nagasawa, K., Yoneta, A., Umezawa, T. and Ito, K. (1987) *Heterocycles*, **26**, 2607.

36 Schmitt, J., Langlois, M., Perrin, C. and Callet, G. (2004) *Bulletin de la Société Chimique de France*, 1969.

37 (a) Corey, E.J., Nicolaou, K.C., Balanson, R.D. and Machida, Y. (1975) *Synthesis*, 590. (b) Kunesch, G. (1983) *Tetrahedron Letters*, **24**, 5211.

38 Moody, C.J. and Pitts, M.R. (1998) *Synlett*, 1028.

39 Kim, B.H., Han, R., Piao, F., Jun, Y.M., Baik, W. and Lee, B.M. (2003) *Tetrahedron Letters*, **44**, 77.

40 Kim, B.H., Jin, Y., Jun, Y.M., Han, R., Baik, W. and Lee, B.M. (2000) *Tetrahedron Letters*, **41**, 2137.

41 Mellor, J.M. and Smith, N.M. (1984) *Journal of the Chemical Society, Perkin Transactions I*, 2927.

42 Schmidt, R.R., III (1986) *Accounts of Chemical Research*, **19**, 250.
43 Kaluza, Z., Fudong, W., Belzecki, C. and Chmielewski, M. (1989) *Tetrahedron Letters*, **30**, 5171.
44 Nieschalk, J. and Schaumann, E. (1996) *Justus Liebigs Annalen der Chemie*, 141.
45 Wityak, J., Gould, S.J., Hein, S.J. and Keszler, D.A. (1987) *Journal of Organic Chemistry*, **52**, 2179.
46 Ochiai, E.J. (1953) *Journal of Organic Chemistry*, **18**, 534.
47 Yadav, J.S., Reddy, B.V.S. and Reddy, M.M. (2000) *Tetrahedron Letters*, **41**, 2663.
48 Whitesides, G.M. and Ehmann, W.J. (1970) *Journal of Organic Chemistry*, **35**, 3565.
49 Nagumo, S., Suemune, H. and Sakai, K. (1990) *Journal of the Chemical Society, Chemical Communications*, 1778.
50 (a) Stork, G. and Tsuji, J. (1961) *Journal of the American Chemical Society*, **83**, 2783. (b) Stork, G. and Darling, S.D. (1960) *Journal of the American Chemical Society*, **82**, 1512.
51 Angibeaud, P., Larcheveque, M., Normant, H. and Tchoubar, B. (1968) *Bulletin de la Société Chimique de France*, 595.
52 McIntosh, J.M. and Cassidy, K.C. (1991) *Tetrahedron: Asymmetry*, **2**, 1053.
53 Kim, S., Emeric, G. and Fuchs, P.L. (1992) *Journal of Organic Chemistry*, **57**, 7362.
54 Paquette, L.A., Romine, J.L., Lin, H.-S. and Wright, J. (1990) *Journal of the American Chemical Society*, **112**, 9284.
55 Huffman, J.W., Zhang, X., Wu, M.-J. and Joyner, H.H. (1989) *Journal of Organic Chemistry*, **54**, 4741.
56 Hagiwara, H. and Uda, H. (1990) *Journal of the Chemical Society, Perkin Transactions I*, 1901.
57 Tori, M., Sono, M., Nishigaki, Y., Nakashima, K. and Asakawa, Y. (1991) *Journal of the Chemical Society, Perkin Transactions I*, 435.
58 Rucker, M. and Brückner, R. (1997) *Synlett*, 1187.
59 Lajunen, M. and Koskinen, A. (2000) *Journal of the Chemical Society, Perkin Transactions I*, 1439.
60 Johnson, W.S., Vredenburgh, W.A. and Pike, J.E. (1960) *Journal of the American Chemical Society*, **82**, 3409.
61 Bauld, N.L. (1962) *Journal of the American Chemical Society*, **84**, 4345.
62 House, H.O. and Kinloch, E.F. (1974) *Journal of Organic Chemistry*, **39**, 747.
63 Sondengam, B.L., Charles, G. and Akam, T.M. (1980) *Tetrahedron Letters*, **21**, 1069.
64 Doolittle, R.E., Patrick, D.G. and Heath, R.H. (1993) *Journal of Organic Chemistry*, **58**, 5063.
65 Kaufman, D., Johnson, E. and Mosher, M.D. (2005) *Tetrahedron Letters*, **46**, 5613.
66 Ranu, B.C., Dutta, J. and Guchhait, S.K. (2001) *Journal of Organic Chemistry*, **66**, 5624.
67 Wooster, C.B. and Godfrey, K.L. (1937) *Journal of the American Chemical Society*, **59**, 596.
68 (a) Birch, A.J. (1950) *Quarterly Reviews of the Chemical Society*, **4**, 69. (b) Birch, A.J. and Smith, H. (1958) *Quarterly Reviews of the Chemical Society*, **12**, 17.
69 Rabideau, P.W. and Huser, D.L. (1983) *Journal of Organic Chemistry*, **48**, 4266.
70 Corey, E.J. and Lee, J. (1993) *Journal of the American Chemical Society*, **115**, 8873.
71 Coughlin, D.J. and Salomon, R.G. (1979) *Journal of Organic Chemistry*, **44**, 3784.
72 Julia, M. and Malassiné, B. (1972) *Tetrahedron Letters*, 2495.
73 For reviews, see: (a) Schultz, A.G. (1990) *Accounts of Chemical Research*, **23**, 207. (b) Schultz, A.G. (1999) *Chemical Communications*, 1263.
74 Schultz, A.G. and Green, N.J. (1991) *Journal of the American Chemical Society*, **113**, 4931.
75 Hückel, W. and Bretschneider, H. (1939) *Justus Liebigs Annalen der Chemie*, **540**, 157.
76 Grisdale, P.J., Regan, T.H., Doty, J.C., Figueras, J. and Williams, J.L.R. (1968) *Journal of Organic Chemistry*, **33**, 1116.
77 Bergosh, R.G., Meier, M.S., Cooke, J.A.L., Spielmann, H.P. and Weedon, B.R. (1997) *Journal of Organic Chemistry*, **62**, 7667.
78 Hudson, C.B. and Robertson, A.V. (1967) *Tetrahedron Letters*, 4015.
79 (a) Masamune, T., Ono, M. and Matsue, H. (1975) *Bulletin of the Chemical Society*

of Japan, **48**, 491. (b) Kinoshita, T., Miyano, K. and Miwa, T. (1975) *Bulletin of the Chemical Society of Japan*, **48**, 1865. (c) Divanfard, H.R. and Joullié, M.M. (1978) *Organic Preparations and Procedures International*, **10**, 94.

80 Nishino, K., Yano, S., Kohashi, Y., Yamamoto, K. and Murata, I. (1979) *Journal of the American Chemical Society*, **101**, 5059.

81 (a) O'Brien, S. and Smith, D.C.C. (1960) *Journal of the Chemical Society*, 4609. (b) Remers, W.A., Gibs, G.J., Pidacks, C. and Weiss, M.J. (1971) *Journal of Organic Chemistry*, **36**, 279.

82 Fagan, G.P., Chanpleo, C.B., Lane, A.C., Myers, M., Roach, A.G., Smith, C.F.C., Stilling, M.R. and Welbourn, A.P. (1988) *Journal of Medicinal Chemistry*, **31**, 944.

83 Smith, J.G. and McCall, R.B. (1980) *Journal of Organic Chemistry*, **45**, 3982.

84 Boyle, E.A., Mangan, F.R., Markwell, R.E., Smith, S.A., Thomson, M.J., Ward, R.W. and Wyman, P.A. (1986) *Journal of Medicinal Chemistry*, **29**, 894.

85 Moormann, A.E., Wang, J.L., Palmquist, K.E., Promo, M.A., Snyder, J.S., Scholten, J.A., Massa, M.A., Sikorski, J.A. and Webber, R.K. (2004) *Tetrahedron*, **60**, 10907.

86 Watanabe, Y., Ohta, T., Tsuji, Y., Hiyoshi, T. and Tsuji, Y. (1984) *Bulletin of the Chemical Society of Japan*, **57**, 2440.

87 Brown, G.R. and Foubister, A.J. (1982) *Synthesis*, 1036.

88 Donohoe, T.J., McRiner, A.J., Helliwell, M. and Sheldrake, P. (2001) *Journal of the Chemical Society, Perkin Transactions I*, 1435.

89 Moody, C.J. and Pitts, M.R. (1998) *Synlett*, 1029.

90 Sauers, R.R. and Hu, C.K. (1971) *Journal of Organic Chemistry*, **36**, 1153.

91 Rosenfeld, R.S. and Gallagher, T.F. (1955) *Journal of the American Chemical Society*, **77**, 4367.

92 Kurozumi, S., Toru, T., Kobayashi, M. and Ishimoto, S. (1977) *Synthetic Communications*, **7**, 427.

93 Kamata, S., Uyeo, S., Haga, N. and Nagata, W. (1973) *Synthetic Communications*, **3**, 265.

94 Holton, R.A., Crouse, D.J., Williams, A.D. and Kennedy, R.M. (1987) *Journal of Organic Chemistry*, **52**, 2317.

95 Ohsawa, T., Takagaki, T., Haneda, A. and Oishi, T. (1981) *Tetrahedron Letters*, **22**, 2583.

96 Gassman, P.G. and Pape, P.G. (1964) *Journal of Organic Chemistry*, **29**, 160.

97 Ranu, B.C., Samanta, S. and Guchhait, S.K. (2001) *Journal of Organic Chemistry*, **66**, 4102.

98 Ohsawa, T., Hatano, K., Kayoh, K., Kotabe, J. and Oishi, T. (1992) *Tetrahedron Letters*, **33**, 5555.

99 For a review, see: Kaiser, E.M. (1972) *Synthesis*, 391.

100 Pinder, A.R. and Smith, H. (1954) *Journal of the Chemical Society*, 113.

101 Birch, A.J. (1945) *Journal of the Chemical Society*, 809.

102 See, for instance: Wang, Z.-G., Warren, J.D., Dudkin, V.Y., Zhang, X., Iserloh, U., Visser, M., Eckhardt, M., Seeberger, P.H. and Danishefsky, S.J. (2006) *Tetrahedron*, **62**, 4954.

103 Yus, M., Behloul, C. and Guijarro, D. (2003) *Synthesis*, 2179.

104 Kraft, P., Weymuth, C. and Nussbaumer, C. (2006) *European Journal of Organic Chemistry*, 1403.

105 Takano, S., Samizu, K. and Ogasawara, K. (1993) *Synlett*, 393.

106 Ireland, R.E., Muchmore, D.C. and Hengartner, U. (1972) *Journal of the American Chemical Society*, **94**, 5098.

107 Mori, K. and Tamura, H. (1990) *Justus Liebigs Annalen der Chemie*, 361.

108 Agharahimi, M.R. and LeBel, N.A. (1995) *Journal of Organic Chemistry*, **60**, 1856.

109 (a) Brown, H.C., Ikegami, S. and Kawakami, J.H. (1970) *Journal of Organic Chemistry*, **35**, 3243. (b) Kaiser, E.M., Edmonds, C.G., Grubb, S.D., Smith, J.W. and Tramp, D. (1971) *Journal of Organic Chemistry*, **36**, 330. (c) Ayer, W.A., Browne, L.M. and Fung, S. (1976) *Canadian Journal of Chemistry*, **54**, 3276. (d) Benkeser, R.A., Rappa, A. and Wolsieffer, L.A. (1986) *Journal of Organic Chemistry*, **61**, 1.

110 (a) Bartmann, E. (1986) *Angewandte Chemie (International Ed. in English)*, **25**, 653. (b) Dorigo, A.E., Houk, K.N. and

Cohen, T. (1989) *Journal of the American Chemical Society*, **111**, 8976.

111 (a) Bachki, A., Foubelo, F. and Yus, M. (1995) *Tetrahedron: Asymmetry*, **6**, 1907. (b) Bachki, A., Foubelo, F. and Yus, M. (1996) *Tetrahedron: Asymmetry*, **7**, 2997. (c) Soler, T., Bachki, A., Falvello, L.R., Foubelo, F. and Yus, M. (1998) *Tetrahedron: Asymmetry*, **9**, 3939. (d) Soler, T., Bachki, A., Falvello, L.R., Foubelo, F. and Yus, M. (2000) *Tetrahedron: Asymmetry*, **11**, 493.

112 (a) Falvello, L.R., Foubelo, F., Soler, T. and Yus, M. (2000) *Tetrahedron: Asymmetry*, **11**, 2063. (b) Yus, M., Soler, T. and Foubelo, F. (2001) *Tetrahedron: Asymmetry*, **12**, 801.

113 Behloul, C., Guijarro, D. and Yus, M. (2004) *Synthesis*, 1274.

114 (a) Williams, R.M., Sinclair, P.J., Zhai, D. and Chen, D. (1988) *Journal of the American Chemical Society*, **110**, 1547. (b) Williams, R.M. and Fegley, G.J. (1991) *Journal of the American Chemical Society*, **113**, 8796.

115 Williams, R.M. and Im, M.-N. (1991) *Journal of the American Chemical Society*, **113**, 9276.

116 Nantz, M.H., Lee, D.A., Bender, D.M. and Roohi, A.H. (1992) *Journal of Organic Chemistry*, **57**, 6653.

117 Ohsawas, T., Mitsuda, N., Nezu, J.-I. and Oishi, T. (1989) *Tetrahedron Letters*, **30**, 845.

118 (a) Maercker, A. (1987) *Angewandte Chemie (International Ed. in English)*, **26**, 972. (b) Tiecco, M. (1988) *Synthesis*, 749.

119 (a) Screttas, C.G. and Mischa-Screttas, M. (1978) *Journal of Organic Chemistry*, **43**, 1064. (b) Cohen, T., Daniewski, W.M. and Weisenfeld, R.B. (1978) *Tetrahedron Letters*, 4665. (c) Screttas, C.G. and Mischa-Screttas, M. (1979) *Journal of Organic Chemistry*, **44**, 713. (d) Cohen, T., Sherbine, J.P., Matz, J.R., Hutchins, R.R., McHenry, B.M. and Wiley, P.R. (1984) *Journal of the American Chemical Society*, **106**, 3245. (e) Cohen, T., Sherbine, J.P., Mendelson, S.A. and Myers, M. (1985) *Tetrahedron Letters*, **26**, 2965. (f) Ramaiah, M. (1987) *Tetrahedron*, **43**, 3541. (g) Block, E. and Aslam, M. (1988) *Tetrahedron*, **44**, 281.

120 Paquette, L.A. and Galatsis, P. (1989) *Journal of Organic Chemistry*, **54**, 5039.

121 Kang, H.-J., Ra, C.S. and Paquette, L.A. (1991) *Journal of the American Chemical Society*, **113**, 9384.

122 Mehta, G. and Subrahmanyam, D. (1991) *Journal of the Chemical Society, Perkin Transactions I*, 395.

123 Maercker, A., Oeffner, K.S. and Girreser, U. (2004) *Tetrahedron*, **60**, 8245.

124 Lim, C., Evenson, G.N., Perrault, W.R. and Pearlman, B.A. (2006) *Tetrahedron Letters*, **47**, 6417.

125 (a) Tsuda, K., Ohki, E., Nozoe, S. and Ikekawa, N. (1961) *Journal of Organic Chemistry*, **26**, 2614. (b) Tsuda, K., Ohki, E. and Nozoe, S. (1963) *Journal of Organic Chemistry*, **28**, 786.

18
Hydrometallation of Unsaturated Compounds

Usein M. Dzhemilev and Askhat G. Ibragimov

18.1
Introduction

To the number of significant hydrometallating reagents that have been introduced into organic synthesis should be added the stable hydrides of transition metals (Zr) and nontransition metals (B, Al, Mg).

Today, hydrides of the these metals and their derivatives are widely used in the reduction of alkenes, dienes, alkynes, allenes, heteroolefins and carbonyl compounds as well as for the production of the corresponding organometallic compounds and selective alkene functionalization via the synthesis of compounds containing polar metal–carbon (M–C) bonds. These may be involved in reactions with electrophilic or nucleophilic reagents affording functional derivatives.

Generally, the thermal additions of the above hydrides to alkenes proceed via the four-center transition state (**1**), with subsequent cleavage of the M–H bond and metal addition. In this case the corresponding organometallic compounds are formed according to Scheme 18.1.

The use of metal complex catalysts in organic and metalloorganic chemistry allows the chemo-, regio- and stereoselective hydroalumination of unsaturated compounds under mild conditions with high yields.

The role of a transition metal catalyst in these reactions is determined by its ability to activate the starting substrate by coordination to the metal center forming the intermediate hydride complex (**2**). The latter, being a more active hydrometallating reagent, attacks alkenes under mild conditions with subsequent transmetallation of the starting hydride affording the target organic compound (**3**) according to Scheme 18.2.

To date, a large number of reports have appeared in the literature on the application of hydride reagents in both thermal and catalytic hydrometallation of unsaturated compounds. Hydroboration, hydroalumination, hydrozirconation and hydromagnesiation of unsaturated compounds such as alkenes, dienes or alkynes are considered the most effective reactions for producing the related metalloorganic compounds. It is not possible to consider all these reactions within the

Modern Reduction Methods. Edited by Pher G. Andersson and Ian J. Munslow.
Copyright © 2008 WILEY-VCH Verlag GmbH & Co. KGaA, Weinheim
ISBN: 978-3-527-31862-9

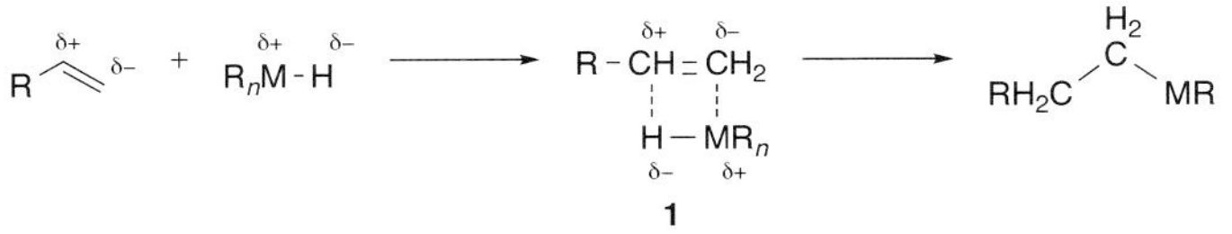

Scheme 18.1 Thermal addition of metal hydrides to alkenes.

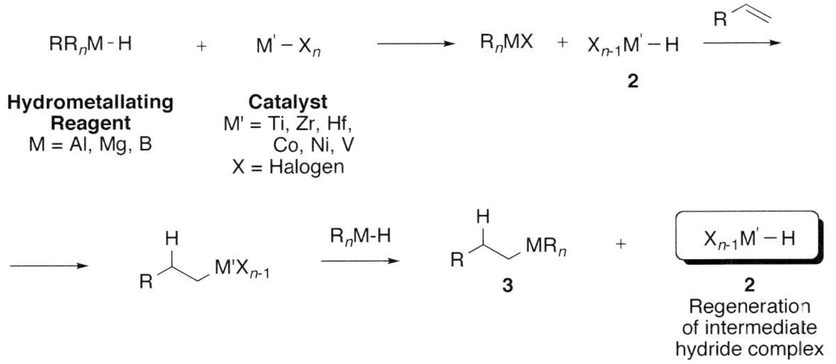

Scheme 18.2 Transition metal-catalyzed hydrometallation of alkenes.

framework of this text. Therefore, our attention will be focused on the hydroalumination and hydromagnesiation of alkenes, dienes and alkynes catalyzed by transition metal complexes, including the most important published data for the 15 years prior to this publication and in some cases the initial reports for comparison of the results and discussion of mechanistic transformations of organometallic reagents containing active metal–carbon bonds.

18.2
Thermal Hydroalumination

18.2.1
Alkenes

Inorganic aluminum hydrides such as AlH_3, $AlHCl_2$, $AlHBr_2$, $LiAlH_4$, $NaAlH_4$ as well as organometallic ones, namely $i\text{-}Bu_2AlH$ (DIBAL), Et_2AlH (DEAH), $i\text{-}Bu_3Al$ (TIBA) and $i\text{-}Bu_2AlCl$ (DIBACh) are widely used in hydroalumination of alkenes. Hydroalumination methods based on DIBAL, TIBA or $LiAlH_4$ are the most commonly used in synthesis. There are reports in the literature on applications of dialkyl amides, cycloalumoxane and diisobutylaluminum phenolate as hydroaluminating reagents (Figure 18.1).

Ziegler and coworkers [1–3] were among the first investigators to report the high reactivity of aluminum hydrides. They showed that the reaction of the simplest α-olefins with AlH_3 in Et_2O produced the corresponding trialkylaluminum derivatives **(4)** in high yields (Scheme 18.3).

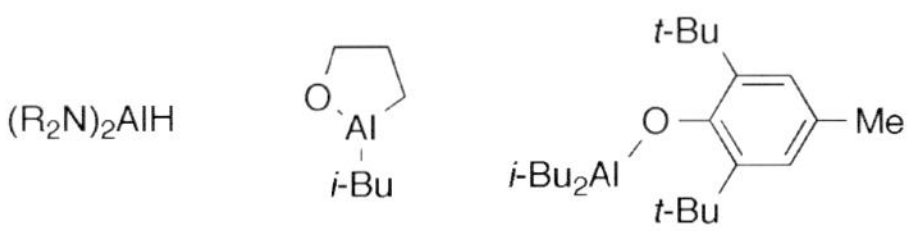

18.2 Thermal Hydroalumination

Figure 18.1 Hydroaluminating reagents.

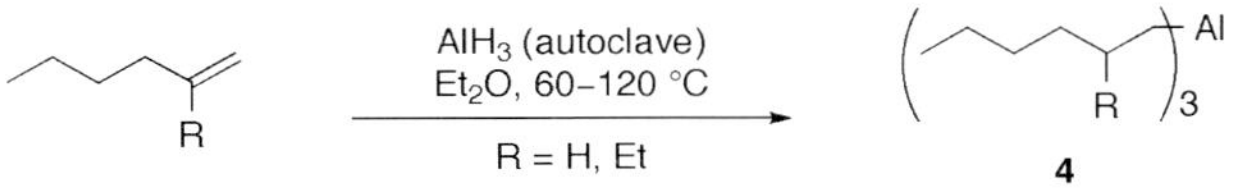

Scheme 18.3 Reactivity of alkylaluminum hydrides with α-olefins.

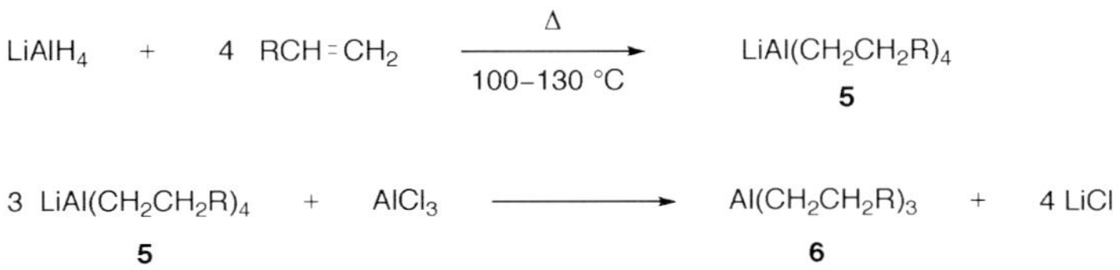

$$LiAlH_4 \;+\; 4\; RCH{=}CH_2 \xrightarrow[100\text{–}130\,°C]{\Delta} \underset{\mathbf{5}}{LiAl(CH_2CH_2R)_4}$$

$$3\; \underset{\mathbf{5}}{LiAl(CH_2CH_2R)_4} \;+\; AlCl_3 \longrightarrow \underset{\mathbf{6}}{Al(CH_2CH_2R)_3} \;+\; 4\; LiCl$$

Scheme 18.4 Formation of trialkylalanates (**6**).

Ziegler further studied, in detail, the thermal reactions involving AlH_3 and α-olefins [4, 5]. Since 1946 when Schlesinger [6] discovered and synthesized for the first time the complex lithium hydride ($LiAlH_4$), it has been used in synthetic practice as a commercial reagent for alkene hydroalumination [7]. One should note that $LiAlH_4$ was the first commercially available complex hydride.

In accordance with the reported data [1, 8] and the investigations of Zakharkin [9], $LiAlH_4$ hydroaluminates α-olefins at higher temperatures (100–130 °C), compared with AlH_3, producing the corresponding lithium tetraalkylalanates (**5**), which further react with $AlCl_3$ yielding trialkylalanates (**6**) (Scheme 18.4).

Detailed research has shown that the use of Zn, Al or Fe salts may increase the rate of hydroalumination of alkenes with $LiAlH_4$ [10–14]. Unfortunately, this reagent has a number of disadvantages that limit its application, namely the hydroalumination of alkenes needs to be carried out in ethereal solvents and at elevated temperatures. Moreover, this reaction is not suitable for all types of alkenes, as the presence of functional groups complicates the reaction.

In spite of the great number of complex aluminum hydrides that have been described in the literature [15], DEAH and DIBAL are the most attractive for alkene hydroalumination due to their commercial availability and low price. In addition, they allow for hydroalumination to be conducted in hydrocarbon solvents.

Wilke has investigated the stereochemistry of thermal hydroalumination, using as an example the reduction of β-pinene with DIBAL under high- and low-temperature conditions (130 °C and 40–60 °C), in which diisobutyl-*trans*-myrtanylaluminum (**7a**) and diisobutyl-*cis*-myrtanylaluminum (**7b**), respectively, were obtained (Scheme 18.5) [16].

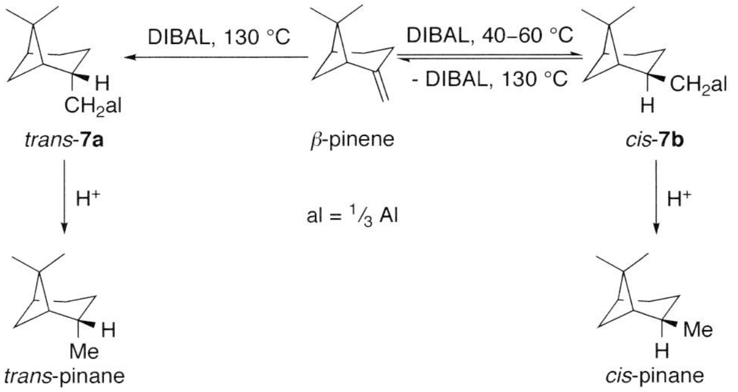

Scheme 18.5 Stereochemistry of the thermal hydroalumination of β-pinene.

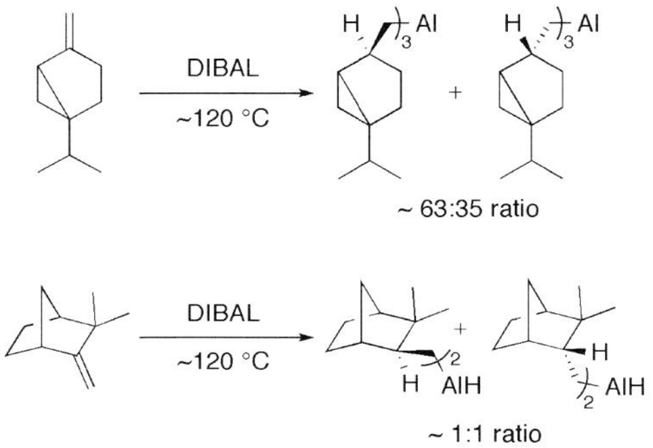

Scheme 18.6 Thermal hydroalumination of sabinene and camphene.

It can be seen clearly that at low temperature hydroalumination occurs stereoselectively via the addition of DIBAL to the carbon–carbon double bond in β-pinene from the sterically less hindered side, giving the thermodynamically less stable *cis*-product **(7b)**. The latter at 130 °C easily loses DIBAL with subsequent formation of the *trans*-isomer **(7a)**.

Thermal hydroalumination of sabinene and camphene with DIBAL or TIBA at 120 °C in hydrocarbon solvents led to isomeric organoaluminum compounds (OACs) (Scheme 18.6) [17]. Examples of hydroalumination of monoterpenes with DIBAL show that the reaction (thermal) is characterized, in most cases, by low stereoselectivity.

In general, the corresponding higher trialkyl-substituted OACs are obtained after treatment of α-olefins or 1,1-disubstituted olefins with DIBAL or TIBA, containing alkyl or *iso*-alkyl substituents related to the starting unsaturated hydrocarbons. One is able to prepare the higher trialkyalanes **(6)** and **(8)** from DIBAL or TIBA and an excess of the appropriate alkene (Scheme 18.7) [16, 18–32].

Thermal hydroalumination of α-olefins, methylene alkanes as well as methylene cycloalkanes with DIBAL or TIBA is characterized by high regioselectivity. The addition of alane hydrides has been found to occur predominately in a

Scheme 18.7 Preparation of the higher alkenes **(6)** and **(8)**.

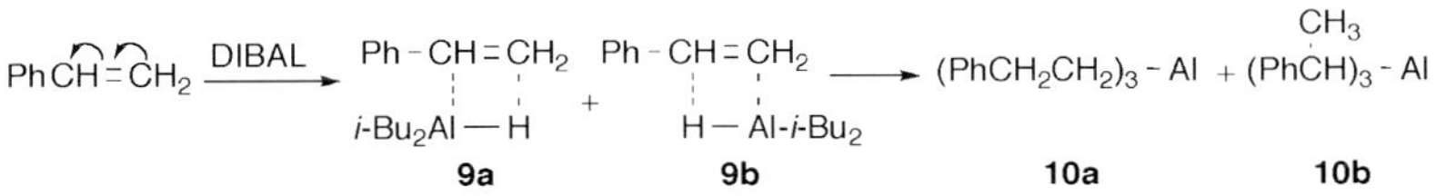

Scheme 18.8 Hydroalumination of styrene.

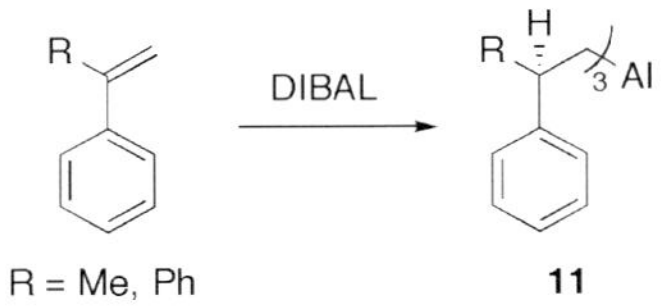

Scheme 18.9 Reaction of α-methylstyrene and α-phenylstyrene with DIBAL.

Markovnikov fashion. However, styrene is one of the few olefins with a terminal double bond for which this reaction occurs with altered regioselectivity. This may be explained by delocalization of the double bond π-electrons over the aromatic ring. As a result of styrene hydroalumination, generation of isomeric complexes responsible for the formation of corresponding OACs **(10a)** and **(10b)** occurs. Structures **(9a)** and **(9b)** indicate these two regioisomeric product intermediates (Scheme 18.8) [33, 34]. In contrast to styrene, α-methylstyrene and α-phenylstyrene react with DIBAL regioselectively giving OAC **(11)** (Scheme 18.9) [35].

Hydroalumination of methylenecyclopentane and its derivatives was found to proceed without skeletal isomerization, giving the corresponding dialkylcyclopentylmethylaluminum in yields of 55–90% (Scheme 18.10) [37–39].

Of interest is methylene cyclobutane, which upon reaction with DEAH subsequently undergoes regioselective hydroalumination and then *in situ* skeletal isomerization accompanied by cleavage of the cyclobutane ring to give diethyl-4-pentenylaluminum **(12)** (Scheme 18.11) [36].

Together with dialkylalanes, LiAlH$_4$ has been used in hydroalumination reactions for cyclic olefins. Thus, from cyclopentene, cycloheptene and *cis*-cyclooctene the corresponding tetraalkyllithiumcycloalanates **(13)** have been obtained (Scheme 18.12) [1].

Upon heating of cyclohexene with LiAlH$_4$, even at 120 °C for more than 20 hours, one is unable to obtain the corresponding tetracyclohexenyllithium alanate in acceptable yields.

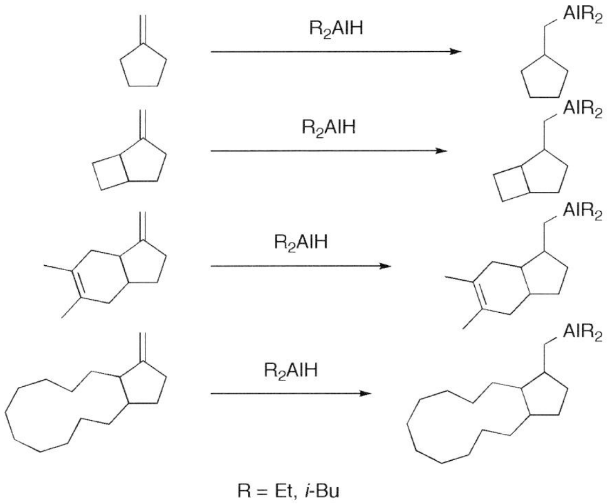

R = Et, *i*-Bu

Scheme 18.10 Hydroalumination of methylenecyclopentane and its derivatives.

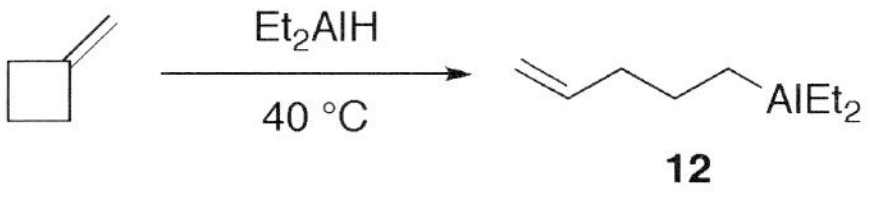

Scheme 18.11 Reaction of methylene cyclobutane with DEAH.

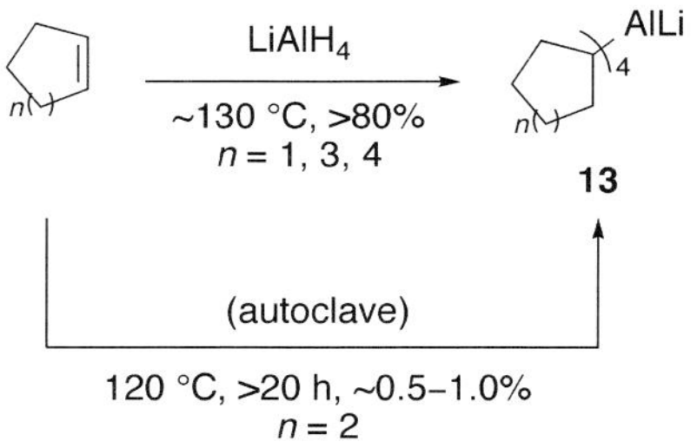

Scheme 18.12 Hydroalumination of cycloalkenes with LiAlH$_4$.

The hydroalumination of C$_5$–C$_8$ cycloolefins with DEAH at ~100 °C for more than 20 hours leads to OAC **(14)** in almost quantitative yields (Scheme 18.13) [1].

18.2.2
Dienes (Unconjugated)

Ziegler and coworkers first reported the hydroalumination of dienes with isolated double bonds [31]. Thus, the hydroalumination of vinylcyclohexene and limonene with DIBAL at 100–120 °C for 3–5 hours leads to the related OACs in quantitative yield (Scheme 18.14).

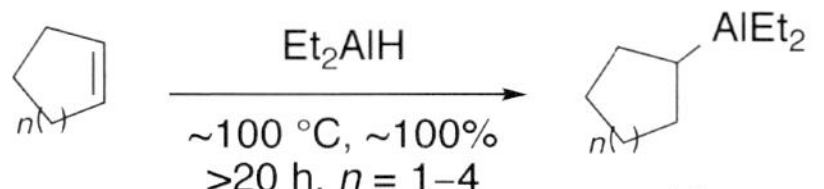

Scheme 18.13 Hydroalumination of cycloolefins with DEAH.

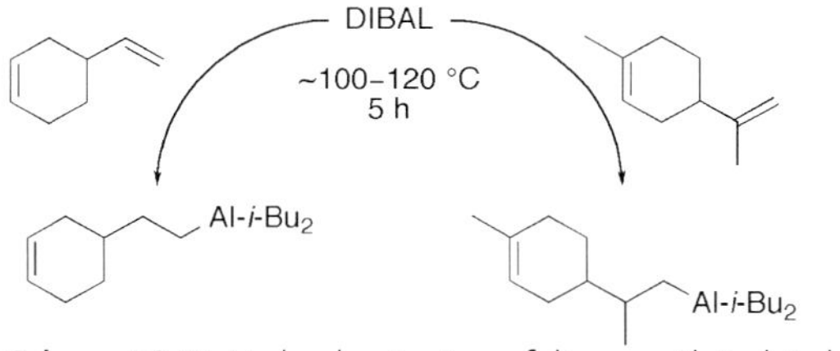

Scheme 18.14 Hydroalumination of dienes with isolated double bonds.

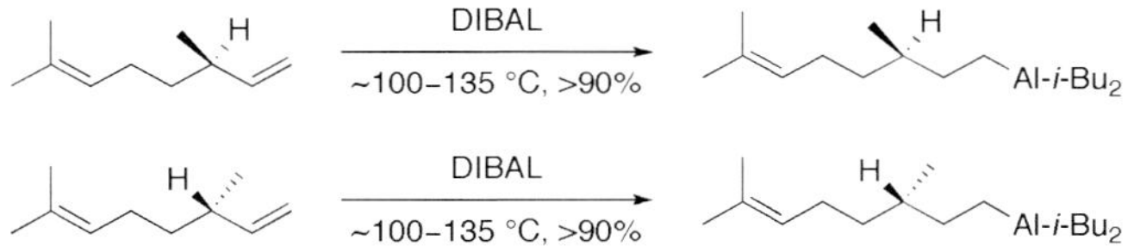

Scheme 18.15 Hydroalumination of acyclic monoterpenes.

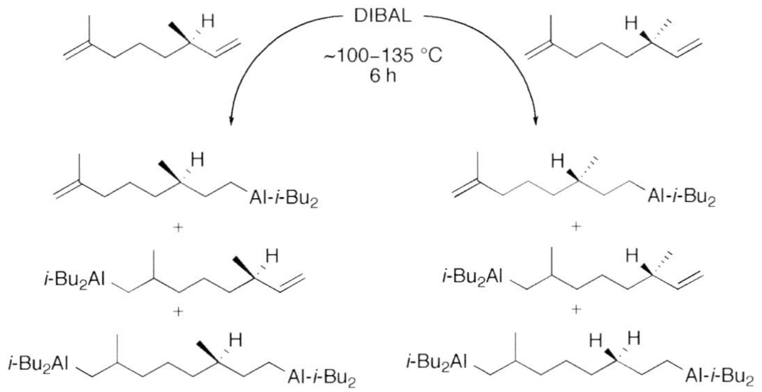

Scheme 18.16 Hydroalumination of (+)- and (−)-2,6-dimethyl-1,7-octadienes.

In an analogous fashion, the hydroalumination of acyclic monoterpenes, for example (+)- and (−)-3,7-dimethyl-1,6-octadiene, has also been performed (Scheme 18.15) [40]. The hydroalumination of (+)- and (−)-2,6-dimethyl-1,7-octadienes, has been found to occur by involving vinyl and methylene groups, and resulted in the formation of mono- and dialuminum compounds (Scheme 18.16) [40]. It should be noted that the thermal hydroalumination of enantiomerically pure acyclic monoterpenes with DIBAL proceeds without racemization.

Aliphatic α,ω-olefins can be selectively converted to their corresponding OACs [37, 41, 42]. The reaction was found to depend upon the concentration of hydroaluminating reagent and the nature of the solvent (Scheme 18.17).

The hydroalumination reaction is rather sensitive to alterations in the stereochemistry of double bonds. Thus, the hydroalumination of (1E,5Z)-cyclodecadiene with DIBAL in refluxing benzene affords the regioisomeric cyclic OAC. Subsequent hydrolysis leads predominantly to (Z)-cyclodecene **(15)**, evidence of higher reactivity the of the *trans* double bond in this reaction (Scheme 18.18) [43].

18.2.3
Dienes (Conjugated)

In contrast to alkenes containing isolated double bonds, the reaction of conjugated dienes with hydridoalanes is more complicated, producing polymeric OAC [44–50]. In most cases, experiments performed with conjugated dienes provided evidence that each of two double bonds in the conjugated system undergoes consecutive hydroalumination reactions giving 1,4-dialuminum compounds of the type **(16)**. Their subsequent transformations led to the polymeric OAC **(17)** (Scheme 18.19).

The hydroalumination of conjugated dienes has been studied in detail for butadiene [45]. As the authors considered, 1,2-hydroalumination first takes place, whereupon the second double bond is involved in the addition reaction forming 1,4-dialuminabutane **(16)**. This then eliminates a molecule of TIBA with simultaneous intramolecular cyclization giving aluminacyclopentane. Further oxidation of **(17)** by molecular oxygen (O_2) produced 1,4-butanediol in more than 60% yield.

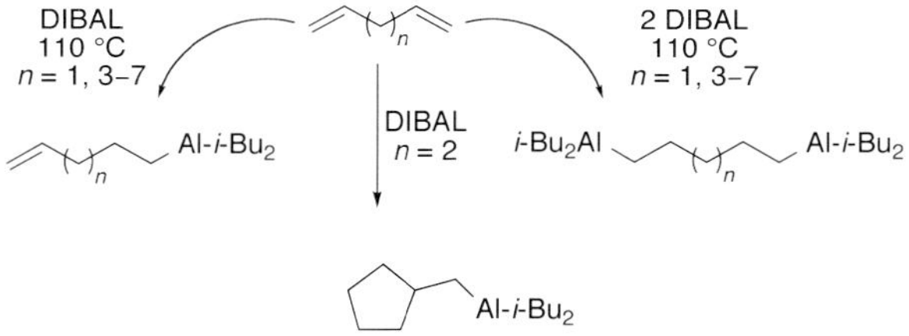

Scheme 18.17 Selective hydroalumination of aliphatic α,ω-olefins.

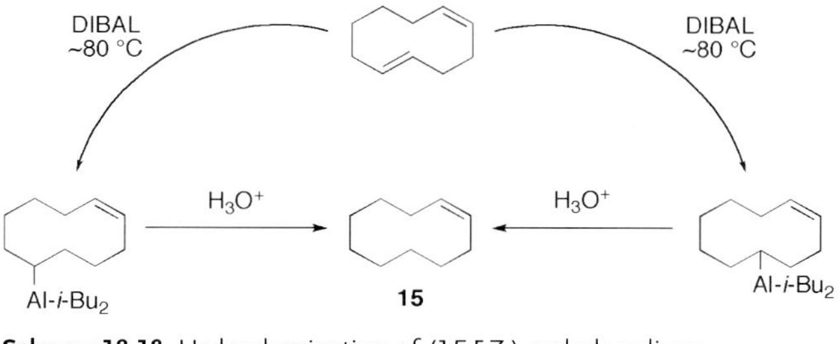

Scheme 18.18 Hydroalumination of (1E,5Z-)-cyclodecadiene.

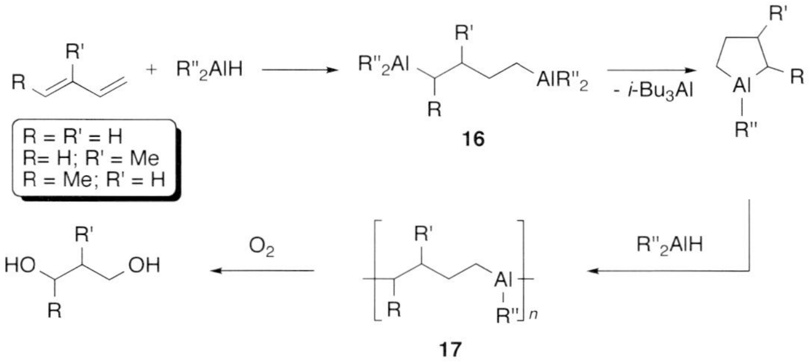

Scheme 18.19 Hydroalumination of conjugated dienes.

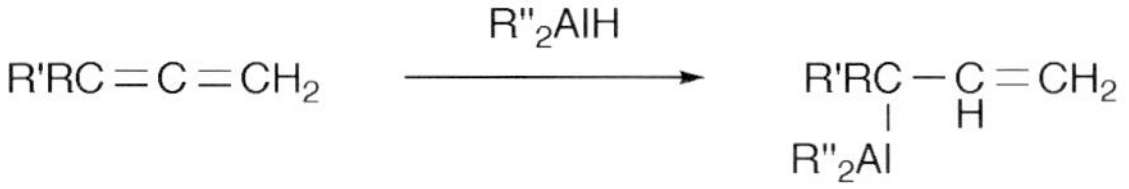

Scheme 18.20 Thermal hydroalumination of 1,2-dienes.

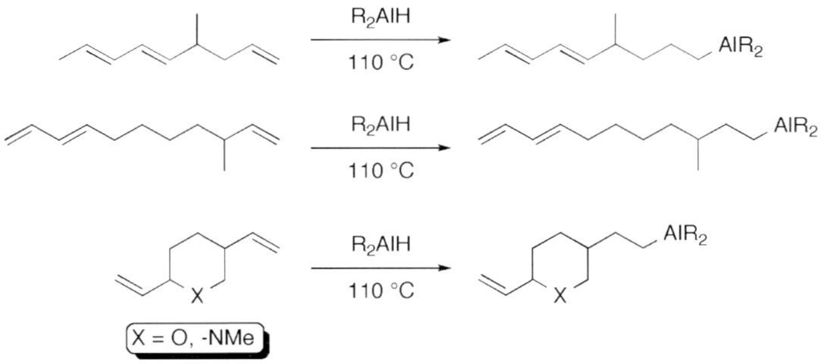

Scheme 18.21 Thermal hydroalumination of acyclic and cyclic polyenes.

Thermal hydroalumination of 1,2-dienes with equimolar amount of hydridoalane was found to proceed via regioselective addition of the R″$_2$Al moiety to the most electronegative carbon, C3, of the diene and the hydrogen atom to carbon C2 forming **(18)** (Scheme 18.20) [49].

There have also been investigations on the thermal hydroalumination of acyclic and cyclic polyenes containing different functional groups such as oxygen, nitrogen or sulfur atoms as well as conjugated double bonds [51–60]. Reactions of polyenes with equimolar amounts of dialkylhydridoalanes indicate that they proceed via initial hydroalumination of the vinyl group, and then the resulting active Al—C bond allows for further functionalization [53] (Scheme 18.21).

18.2.4
Alkynes

There are fewer publications devoted to the hydroalumination of alkynes than to alkenes. Wilke and coworkers were the first to report the thermal hydroalumination of acetylenes [61]. Their investigations revealed that acetylenes react with DIBAL (or TIBA) to produce alkenylalanes **(19)** (Scheme 18.22).

The hydroalumination of monosubstituted alkynes with dialkylhydridoalanes has been studied in detail [62–64]. The authors indicate that under the chosen conditions the hydroalumination of monosubstituted acetylenes affords a mixture of alkyl-, vinyl- and alkynyl-alanes (Scheme 18.23). Disubstituted alkynes rapidly undergo *cis* hydroalumination, depending on the reaction temperature. Upon hydrolysis, they afford the corresponding *cis* and *trans* olefins **(20)**, *trans*-conjugated dienes **(21)** and benzenes **(22)**, see Scheme 18.24 [65–68].

18.3
Catalytic Hydroalumination

18.3.1
Alkenes

The thermal hydroalumination of alkenes with OAC as a synthetic method suffers from a number of disadvantages, considerably restricting its application. Hydroalu-

i-Bu$_2$AlH + HC≡CH ⟶ i-Bu$_2$AlCH=CH$_2$
(i-Bu$_3$Al)

19

Scheme 18.22 Hydroalumination of acetylene.

RC≡CH + R'$_2$AlH ⟶ RCH$_2$CH(AlR'$_2$)$_2$
RCH$_2$=CHAlR'$_2$
RC≡CAlR'$_2$

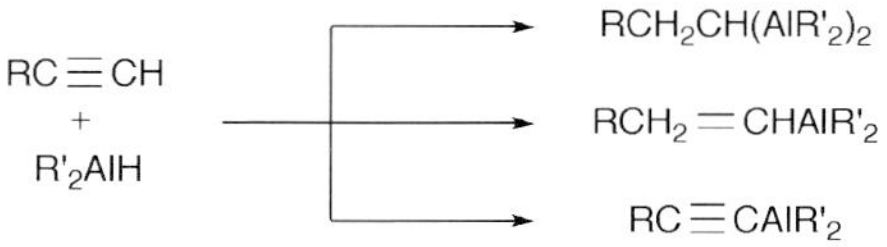

Scheme 18.23 Hydroalumination of terminal alkyne.

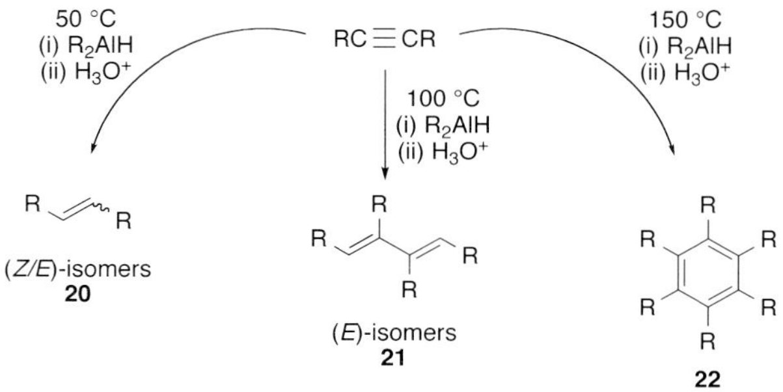

50 °C
(i) R$_2$AlH
(ii) H$_3$O$^+$

RC≡CR

150 °C
(i) R$_2$AlH
(ii) H$_3$O$^+$

100 °C
(i) R$_2$AlH
(ii) H$_3$O$^+$

(Z/E)-isomers
20

(E)-isomers
21

22

Scheme 18.24 Hydroalumination of disubstituted acetylenes.

mination has to be carried out at elevated temperatures (70–150 °C); moreover, this reaction is not suitable for all types of alkenes, due to the presence of functional groups. Usage of a catalyst may reduce these limitations and allow hydroalumination reactions under milder conditions with high conversion rates. In some cases, hydroalumination can be realized only in the presence of a catalyst. In addition, transition metal-based catalysts can govern the regio- and stereochemistry of hydroalumination of unsaturated compounds [69, 70].

The interaction between $LiAlH_4$ and α-alkenes in the presence of Ti- or Zr-complexes is an efficient method of catalytic hydroalumination under mild conditions (0–20 °C) [71]. This procedure allows for the synthesis of lithium tetraalkylalanates in yields of 60–100%. Sato and coworkers [71], who studied catalytic hydroalumination of alkenes with $LiAlH_4$ in the presence of $ZrCl_4$ as a catalyst suggested that this reaction generates the active zirconium hydride complexes **(23)**, which hydrozirconate alkenes to form alkylzirconium intermediates **(24)**. Further transmetallation with $LiAlH_4$ results in regeneration of **(23)** (Scheme 18.25).

Using hydroalumination of hex-1-ene with $LiAlH_4$ as an example, it was shown [72] that $TiCl_4$, VCl_4, $(\eta^5\text{-}Cp)_2TiCl_2$ and $(\eta^5\text{-}Cp)_2ZrCl_2$ result in high catalytic activity compared to $ZrCl_4$. Hydroalumination of terminal alkylalkenes with $LiAlH_4$ in the presence of $[C_5(CH_3)_5]TiCl_2$ [73] or $[C_5(CH_3)_5]ZrCl_2$ [74] as a catalyst was found to proceed at 0 °C in 3–5 hours in yields of 93–99%. In the case of VCl_4 or $(\eta^5\text{-}Cp)_2TiCl_2$, the hydroalumination products were formed together with hex-2-ene resulting from isomerization of the starting material [75]. According to their reactivities in the hydroalumination reaction, aluminum hydrides and initial alkenes can be arranged in the following order: $LiAlH_4 > AlH_3 > AlH_2Cl > AlHCl_2$; $RCH = CH_2 > R_2C = CH_2 > RCH = CHR$ [76].

Co, Ni, Fe and Cr salts [77, 78, 79], as well as transition metal complexes immobilized on inorganic [80] or polymeric supports [80], can be used as catalysts in the hydroalumination reactions. Hydroaluminations of olefins with $LiAlH_4$ in the presence of catalytic amounts of UCl_3 or UCl_4 were performed with quantitative yields [81, 82]. As the authors indicated, the catalytically active species in these reactions was spectroscopically identified as a U(III) compound generated *in situ*, via reduction of UCl_4 by $LiAlH_4$. It is believed that under the reaction conditions a bimetallic complex such as $U(AlH_4)_3$ is the key intermediate, which coordinates the olefin to a uranium center with subsequent addition of the mixed hydrides to a C=C bond.

In contrast with thermal hydroalumination, catalytic hydroalumination affected by Zr and Ti complexes allows the reduction of functionally substituted alkenes such as allylic alcohols and ethers (Scheme 18.26) [83–85].

Obtained *in situ*, lithium tetraalkylalanates are involved in subsequent one-pot conversions. Thus the cross-coupling of lithium tetraalkylalanates with allyl halides

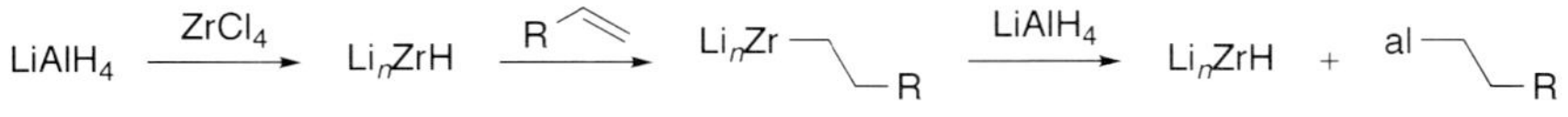

$LiAlH_4 \xrightarrow{ZrCl_4} Li_nZrH \xrightarrow{R \diagdown} Li_nZr\diagdown_R \xrightarrow{LiAlH_4} Li_nZrH + al\diagdown_R$

al = ¼ Al   **23**   **24**   **23**

Scheme 18.25 Catalytic hydroalumination of alkenes.

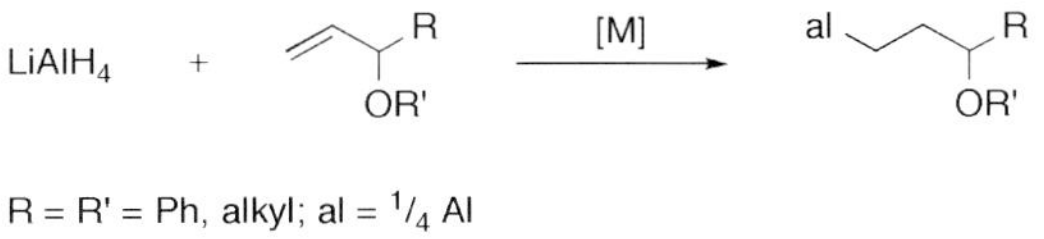

Scheme 18.26 Catalytic hydroalumination of functionally substituted alkenes.

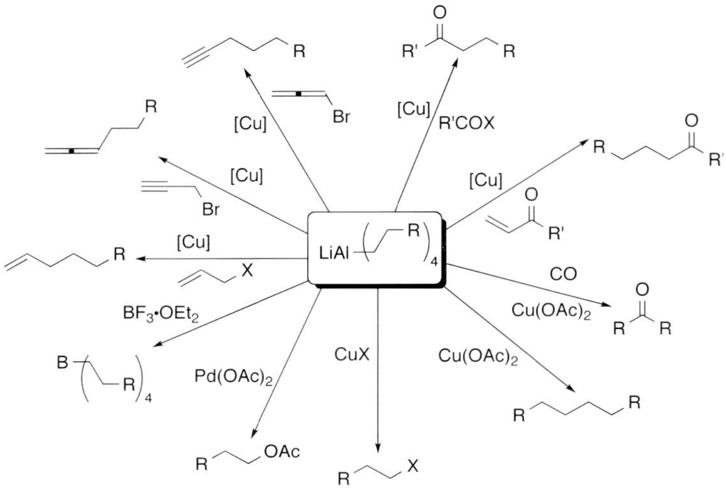

R = R' = Ph, alkyl; al = ¼ Al

Scheme 18.27 Conversion of lithium tetraalkylalanates.

in the presence of copper salts (CuBr, CuI, CuCN, CuCl or Cu(OAc)$_2$) leads to alkenes three carbon atoms longer [86], and the CuCl-catalyzed reaction with propargyl bromide gave terminal allenes in high yields [87]. Later, cross-coupling was successfully extended to allenyl bromide [88], carboxylic acid halides [89], acrolein [90] and methyl vinyl ketone [90, 91]. Catalytic hydroalumination with LiAlH$_4$ and subsequent functionalization of the alkylalanes formed allows conversion of the starting alkenes into their corresponding organic halides [92], acetates [93] or organoboron compounds [94] in one step (Scheme 18.27).

Along with complex aluminum hydrides, readily available alkylaluminum hydrides are also used for catalytic hydroalumination of alkenes. For example, in the course of studies concerning the Ti(OBu)$_4$-catalyzed hydroalumination of a mixture of isomeric alkenes with DIBAL, terminal double bonds were found to react 154 times faster than *cis* disubstituted bonds and 241 times faster than *trans* disubstituted bonds [95].

Hydroalumination of hexa-1,5-dienes with DIBAL has a number of specific features. With Ti(OBu)$_4$ as a catalyst (1.5 mol%) [96], or without it, hydroalumination occurs with subsequent intramolecular cyclization. However, when the reaction was carried out in solvents containing electron-donor atoms (N, O, S, P) no cyclization occurred, and the α,ω-dialuminum compound **(25)** was formed instead (Scheme 18.28) [96].

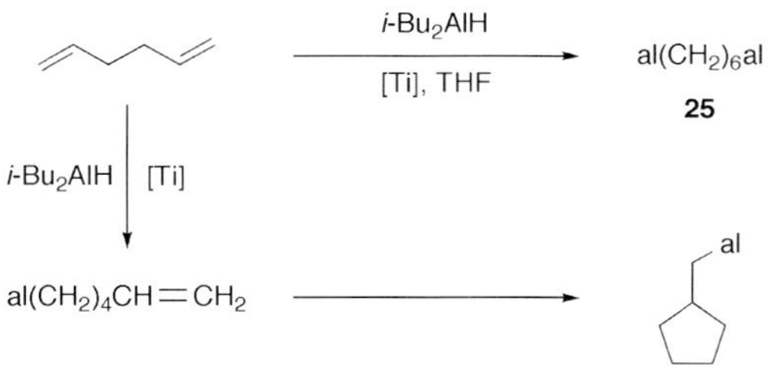

Scheme 18.28 Formation of the α,ω-dialuminum compound (25).

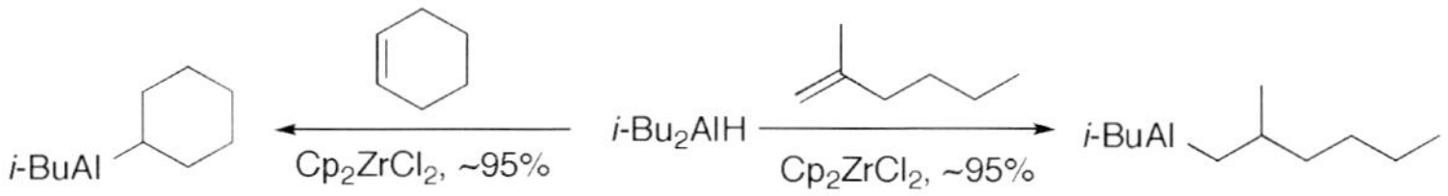

Scheme 18.29 DIBAL hydroalumination of alkyl and cyclic alkenes.

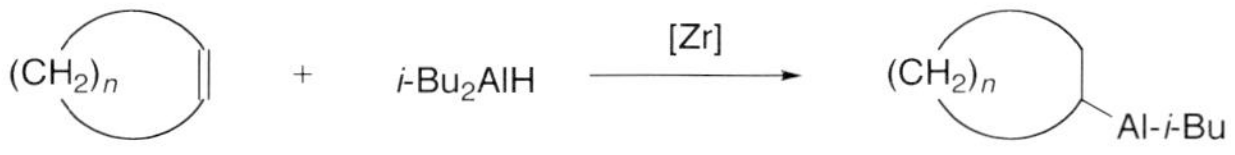

Scheme 18.30 Cycloalkene hydroalumination.

Commercially available DIBAL can be used as an efficient hydroaluminating reagent for linear and cyclic mono-, di- and trienes, and is tolerant to functional groups. Zirconium complexes proved to be the most active catalysts for these reactions [69, 97]. Thus, DIBAL hydroaluminates α-, α,β-disubstituted and cyclic alkenes in high yields in the presence of $ZrCl_4$, even at room temperature (~20 °C) in 3–6 hours [98], while titanium-based catalysts result in the formation of side-products, due to isomerization of the starting alkene [75]. These reactions are not observed in the absence of a catalyst.

An analysis of the literature data on hydrozirconation of α-alkenes [99] prompts the conclusion that hydroalumination yields reactive zirconium hydride intermediates that are able to hydrozirconate alkenes under mild conditions. Further transmetallation of alkyl and cycloalkyl zirconium complexes results in higher OACs (Scheme 18.29) [98]. It should be noticed that the catalytic hydroalumination of cyclohexene, unlike the thermal version, proceeds in practically quantitative yield.

Chlorozirconium alkoxides [$(RO)_nZrCl_{4-n}$] have been shown to be more active than $ZrCl_4$ or $Zr(OBu)_4$ in the hydroalumination reaction [100]. Accordingly, Zr-alkoxides prepared from $ZrCl_4$ and n-BuOH, MeOH, EtOH or n-PrOH have been used to obtain active catalysts. The cycloalkene hydroalumination rate depends upon the ring size and decreases in the order $C_5 > C_6 > C_7 > C_{12} > C_8$ (Scheme 18.30).

Hydroalumination of C_{60} fullerene with i-Bu_2AlH in the presence of $ZrCl_4$ or $(RO)_nZrCl_{4-n}$ was first carried out at ~20 °C in toluene (24 hours, C_{60}^r:[Al] = 1:100–

300) [101, 102]. Data gained from hydrolysis and deuteriolysis of the fullerene-containing OACs suggested that these reaction conditions favor the formation of both hydro-(26) and carboalumination products (27), with a total yield of 90% (Scheme 18.31).

Hydroalumination of oxabicycloalkenes with DIBAL has been achieved in the presence of Ni(COD)$_2$ (COD = cyclooctadiene) or its phosphine derivatives. The regio- and enantioselectivity of the reaction, in this case, are largely dependent on the ligand environment around the catalyst, the amount and ratio of the catalyst components and the DIBAL addition rate [103–106]. Uncertainties still exist concerning the mechanism, such as whether the hydroalumination transformations arise from hydronickelylation or aluminonickelylation of the starting alkenes (Scheme 18.32).

Unlike alkylalanes, bis(dialkylamino)alanes exhibit high reactivity in hydroalumination of alkenes in the presence of Cp$_2$TiCl$_2$ [107, 108]. These compounds hydroaluminate terminal alkenes, methylene alkanes and aliphatic disubstituted alkenes in practically quantitative yields (Scheme 18.33). Cycloalkenes are less reactive in this reaction while tri- and tetrasubstituted alkenes are unreactive.

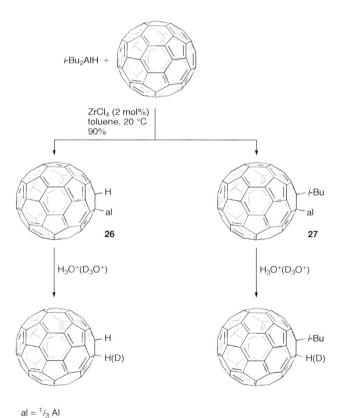

Scheme 18.31 Hydroalumination of C$_{60}$ fullerene with DIBAL.

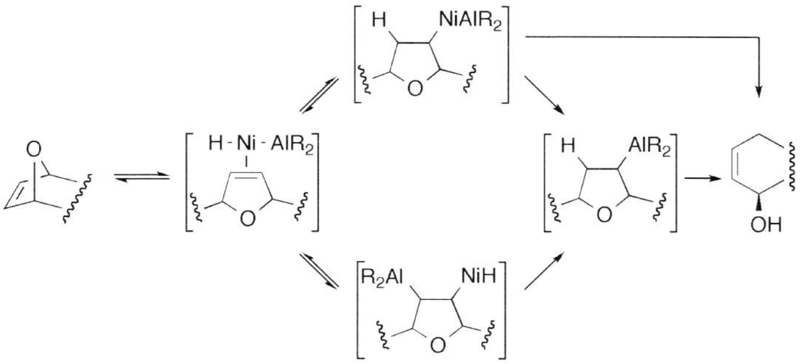

Scheme 18.32 Hydroalumination of oxabicycloalkenes with DIBAL catalyzed by Ni(COD)$_2$.

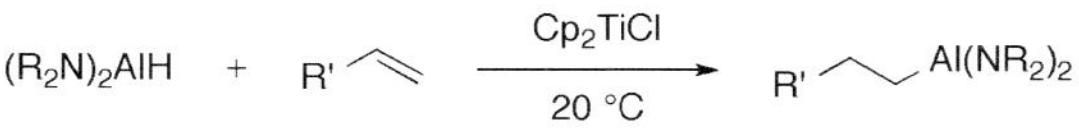

Scheme 18.33 Alkene hydroalumination with bis(dialkylamino)alanes.

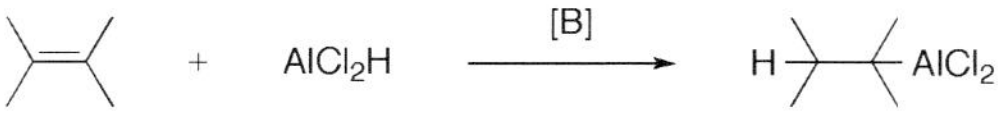

[B] = PhB(OH)$_2$, B(OH)$_3$, BEt$_3$, B(OMe)$_3$, 9-BBN, BF$_3$·OEt$_2$

Scheme 18.34 AlCl$_2$H reacts with α-alkenes in the presence of boron derivatives.

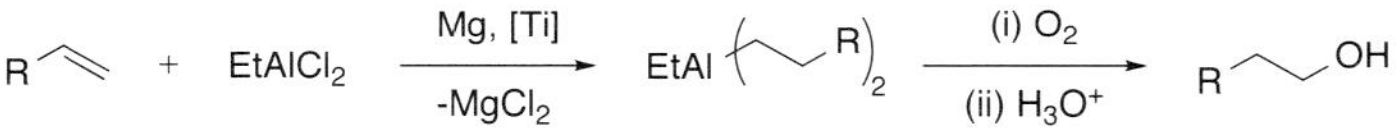

Scheme 18.35 Reaction of terminal olefins with EtAlCl$_2$ in the presence of activated Mg and catalytic [Ti].

As well as aminoalanes, dichloroalane (AlCl$_2$H) reacts with α-alkenes in the presence of boron derivatives giving the corresponding higher alkyl dichloroorganoaluminum compound **(28)**, which is used as a reactive hydroaluminating reagent (Scheme 18.34) [109–112].

Analogous to the procedures mentioned above, the syntheses of dialkyl(ethyl)alanes **(29)** were carried out via the reaction of terminal olefins with EtAlCl$_2$ in the presence of activated Mg and catalytic amounts of Cp$_2$TiCl$_2$ [Ti(O-i-Pr)$_4$, Ti(O-n-Bu)$_4$] in THF (Scheme 18.35). It was suggested that in the presence of low-valent Ti complexes the THF molecule could undergo dehydrogenation to give titanium hydrides as the key intermediates [113].

Hydroalumination of disubstituted alkenes with i-Bu$_3$Al in the presence of Ti^{4+} or Zr^{4+} salts at 110–120 °C gave predominantly OAC **(30)** (Scheme 18.36) [114, 115]. The authors showed that the catalyst induces isomerization of the disubstituted double bond of the starting alkenes to a terminal one, which is then much more readily hydroaluminated with i-Bu$_3$Al.

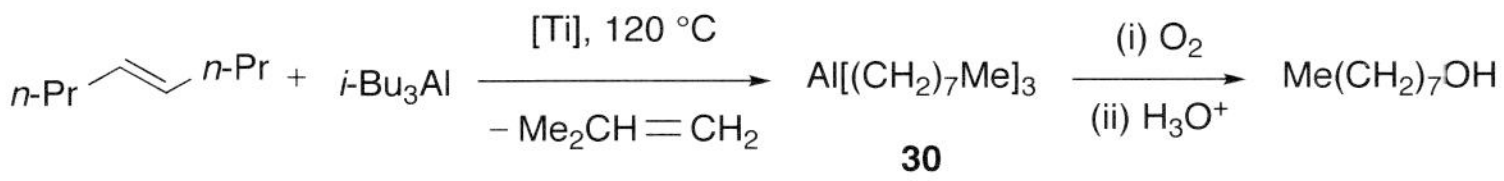

n-Pr$\diagdown$n-Pr + i-Bu$_3$Al $\xrightarrow[-\text{Me}_2\text{CH}=\text{CH}_2]{[\text{Ti}], 120\,°\text{C}}$ Al[(CH$_2$)$_7$Me]$_3$ $\xrightarrow[\text{(ii) H}_3\text{O}^+]{\text{(i) O}_2}$ Me(CH$_2$)$_7$OH
30

Scheme 18.36 Hydroalumination of disubstituted alkenes with i-Bu$_3$Al.

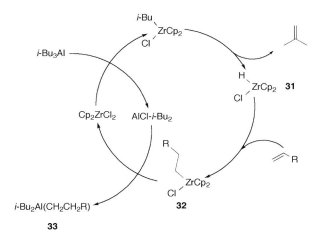

Scheme 18.37 Transmetallation of alkylzirconium complexes resulting in alkylalanes.

In the hydroalumination of alkenes with i-Bu$_3$Al, the yield depends strongly upon the nature of the catalyst [116]. Thus, in the presence of catalytic amounts of Cp$_2$ZrCl$_2$ (0 °C 6 hours), the higher trialkylalanes can be synthesized from terminal alkenes and i-Bu$_3$Al in high yields; no reaction occurs in the absence of catalyst. The high reactivity of i-Bu$_3$Al in the presence of Cp$_2$ZrCl$_2$ is attributed to the formation of zirconium hydrides (**31**) [117, 118]; the latter hydrometallates alkenes under mild conditions giving (**32**). Subsequent transmetallation of alkyl zirconium complexes results in alkylalanes (**33**) (Scheme 18.37). Titanium and zirconium tetrachlorides (TiCl$_4$ and ZrCl$_4$) appear to be less active catalysts. Under these conditions the Cp$_2$TiCl$_2$ complex exhibits no marked activity [119].

Chlorine-containing complexes based on transition metals (Co, Rh, Ni, Pt) exhibit high catalytic activity in the hydroalumination reaction of terminal alkenes with i-Bu$_3$Al. The interaction between dec-1-ene and i-Bu$_3$Al (1.1 molar equivalent) in the presence of a catalyst (2.5–5 mol%) at 25 °C in CH$_2$Cl$_2$ led to decylalane (**34**). Work-up of the latter with molecular iodine (I$_2$) gave 1-iododecane in the following yields: PdCl$_2$(PPh$_3$)$_2$ (90%), Li$_2$PdCl$_4$ (78–86%), K$_2$PtCl$_4$ (86%), Cl$_2$Ni(PPh$_3$)$_2$ (65%), ClCo(PPh$_3$)$_3$ (76%), ClRh(PPh$_3$)$_3$ (79%). However, no reaction occurs in the presence of Pd(OAc)$_2$. These results suggest that the hydroalumination of terminal alkenes under reaction conditions proceeds via formation of intermediate bimetallic complexes of the type Al-Cl-Pd. In this case complexes of Pd(II) catalyze the hydroalumination procedure but are not reduced to Pd(0) [120]. The authors have suggested a sequence of probable transformations (Scheme 18.38) for this reaction.

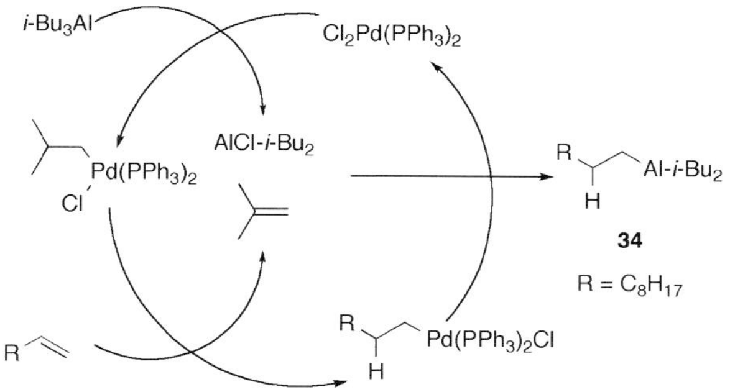

18.3 Catalytic Hydroalumination

Scheme 18.38 Proposed Pd(II)-catalyzed hydroalumination procedure.

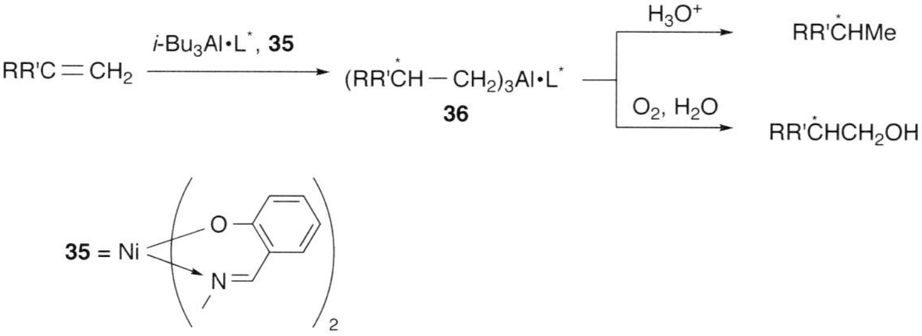

Scheme 18.39 Ni-catalyzed enantioselective hydroalumination.

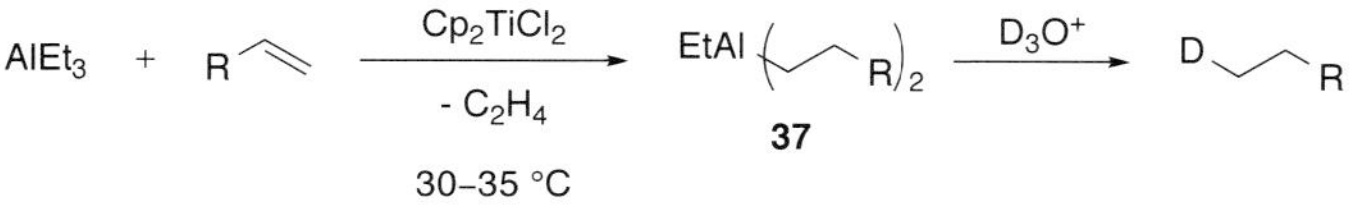

Scheme 18.40 Hydroalumination of terminal alkenes with AlEt$_3$.

Using the molecular complex L*·i-Bu$_3$Al (L* = optically active ligand), enantioselective hydroalumination has been realized in the presence of a nickel(II)-(N-methylsalicylideneamine) complex (35) [121, 122], giving chiral trialkylalanes (36). Subsequent oxidation and hydrolysis of (36) results in optically active alcohols or hydrocarbons. In the reaction (−)-N,N-dimethylmenthylamine, (+)-(S,S)-2,3-dimethoxy-1,4-bis-(N,N-dimethylamino)butane, (+)-N,N-dimethylbornylamine and (+)-(R)-N,N-dimethyl-1-phenylethylamine were used as chiral ligands, which form solvates with i-Bu$_3$Al (Scheme 18.39) [121, 122].

As well as the commonly used synthetic reagents such as DIBAL, DEAH and LiAlH$_4$ for hydroalumination of alkenes, other hydroaluminating reagents such as AlEt$_3$ can react with terminal olefins in the presence of catalytic amounts of Cp$_2$TiCl$_2$ at 30–35 °C giving 1-alkyl-1,1-diethylalanes (37) in 50–90% yields (Scheme 18.40) [123].

The authors of the work [124] have developed an effective preparative method for the synthesis of the higher dialkyl- and cycloalkylaluminum chlorides (38) and

(39). These can be obtained by hydroalumination of acyclic and cyclic alkenes, respectively, with i-Bu$_2$AlCl in the presence of Cp$_2$ZrCl$_2$ under mild conditions (20–40 °C) (Scheme 18.41) [124]; no reaction occurs without catalyst.

This reaction involves nonconjugated di- and trialkenes (including cyclic and functionally substituted ones) [125]. Diisobutylaluminum chloride easily hydroaluminates norbornenes, affording the corresponding *exo*-cycloalkylhalogenoalanes [126]. The reactions involving 4-vinylcyclohexene, norbornene or dicyclopentadiene result in bi- and tricyclic halogenoalanes (Scheme 18.42).

This procedure for Cp$_2$ZrCl$_2$-catalyzed hydroalumination of cyclohexene with i-Bu$_2$AlCl was successfully employed to obtain tricyclohexyltin chloride (40) (Scheme 18.43) [127]. (2,6-Di-t-butyl-4-methylphenoxy)diisobutylaluminum (41) can be used as a regio- and stereoselective reagent for hydroalumination of alkenes and carbonyl compounds in the presence of a two-component catalyst Cp$_2$ZrCl$_2 \cdot i$-Bu$_2$AlH (1 : 1) (Scheme 18.44) [128]. Due to the low reactivity of reagent (41), hydroalumination needs to be performed at elevated temperatures (145 °C). However, the presence of a bulky aryloxy substituent ensures high regioselectivity.

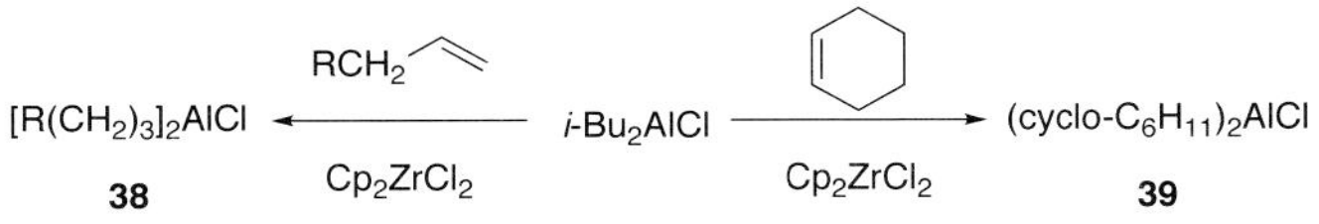

Scheme 18.41 Zr-catalyzed hydroalumination with i-Bu$_2$AlCl.

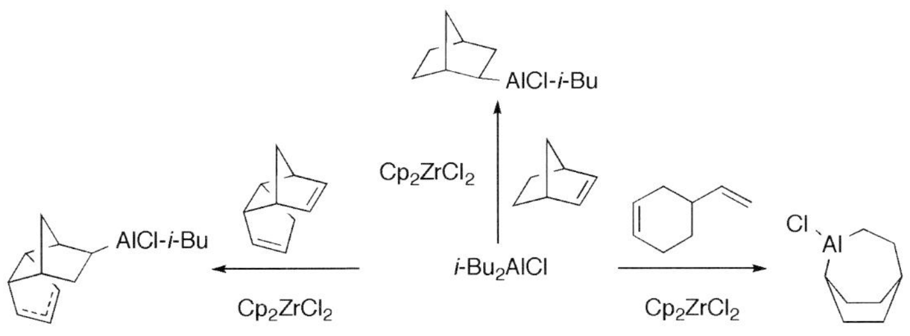

Scheme 18.42 Formation of bi- and tricyclic halogenalanes.

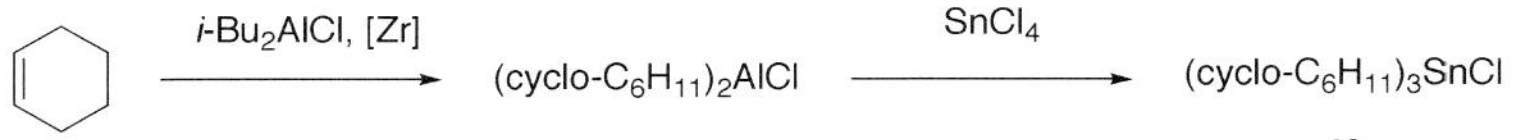

Scheme 18.43 Formation of the tricyclohexyltin chloride (40).

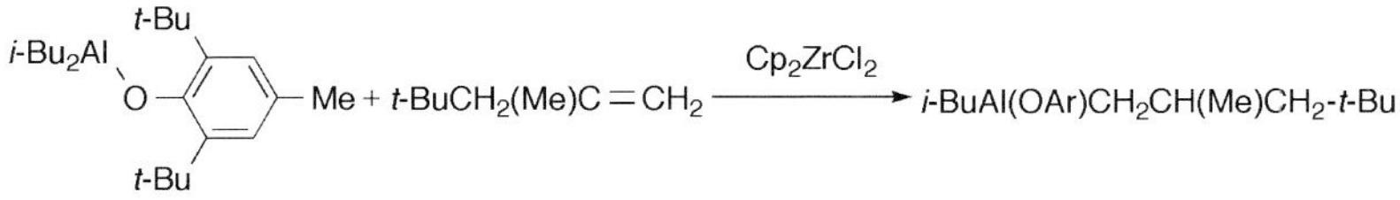

Scheme 18.44 Regio- and stereoselective hydroalumination with (2,6-di-t-butyl-4-methylphenoxy)diisobutylaluminum.

18.3 Catalytic Hydroalumination

The mechanism of catalytic alkene hydroalumination has been a matter of debate for years. Initially, the catalytic activity of Ni-containing catalysts was attributed to colloidal dispersed elemental nickel [129]. Subsequently, hydride Ni complexes such as R-Ni-H or R_2Al-Ni-H have been suggested to play a key role in hydrometallation of alkenes [130]. Studies have led to the conclusion that the active species is a Ni(II) complex, which serves to activate the alkene for the Al—H addition [131]. Further investigations have shown that Ni(II) compounds are rapidly reduced to Ni(0) [132, 133], which react with an olefin and trialkylalane to generate the intermediate complex **(42)**. In a concerted reaction the aluminum and a hydride at the β-carbon of the trialkylalane are transferred to the olefin, giving **(43)** (Scheme 18.45). In these works the authors postulated that intermediate hydride complexes with Al—H or Ni—H bonds are not formed under reaction conditions, and hypothesize that Ni(0) species mostly serve as a template to bring the starting olefin and aluminum compounds into close proximity for their further interaction.

Further work has reported mechanistic studies of the hydroalumination reaction of olefins and acetylenes in which the nickel–aluminum hydride complex **(44)** was generated, under reaction conditions, as the key intermediate (Scheme 18.46). Hydride **(44)** hydrometallates an olefin and the alkyl nickel–aluminum complex **(45)** is obtained, which undergoes reductive elimination releasing alkylalane **(46)** [134].

The direct evidence for an oxidative insertion of Ni(0) into an Al—H bond was provided by the ^{1}H NMR spectrum of the reaction between DIBAL and Ni(PEt$_3$)$_4$, which contained a signal at −12.33 ppm, the region typical for Ni—H signals. In the case of i-Bu$_3$Al, ^{1}H and ^{13}C NMR spectroscopic evidence was found for the oxidative insertion of Ni(0) into the Al—C bond, resulting in only small shifts of the i-Bu groups' resonances [135].

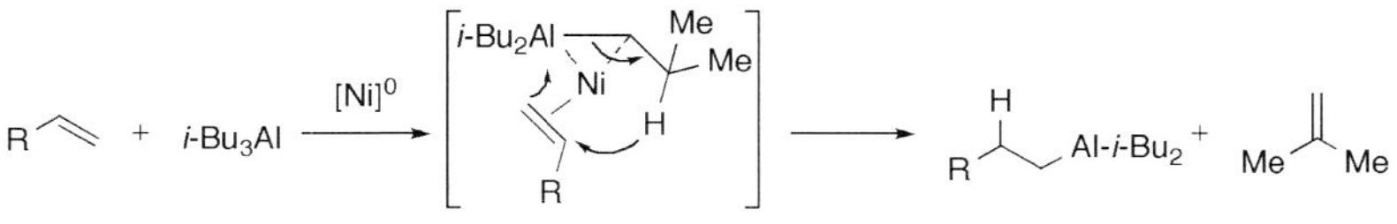

Scheme 18.45 Ni(0) species acting as a template.

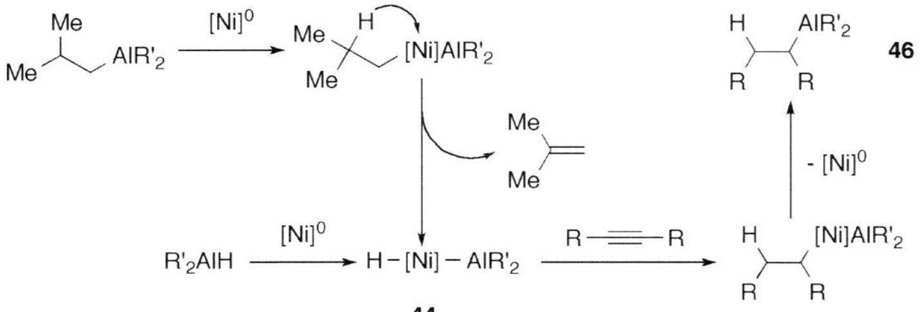

Scheme 18.46 Mechanistic proposals of the Ni-catalyzed hydroalumination.

Dynamic ^{1}H and ^{13}C NMR spectroscopic studies of the hydroalumination reaction of olefins with i-Bu$_2$AlH, i-Bu$_3$Al and i-Bu$_2$AlCl in the presence of Cp$_2$ZrCl$_2$ have identified intermediate complexes responsible for the hydroalumination process [136]. This method uses the characteristic Cp group signal to allow observations of the transformations of in $situ$-generated intermediates.

As a result of these studies, a generalized scheme has been proposed for the mechanism of hydroalumination olefins agents such as i-Bu$_2$AlH, i-Bu$_3$Al and i-Bu$_2$AlCl. Three catalytic systems, Cp$_2$ZrCl$_2$·i-Bu$_2$AlCl, Cp$_2$ZrCl$_2$·i-Bu$_2$AlH and Cp$_2$ZrCl$_2$·i-Bu$_3$Al, were found to initiate the formation of mixed Zr·Al-hydride complexes. But only one of them, that is [Cp$_2$ZrH$_2$·ClAl-i-Bu$_2$]$_2$ complex **(47)** has been identified as an active catalyst able to hydrometallate olefins. The formation of the key intermediate **(47)** was shown to depend on the nature of the starting OAC, which can occur in different ways according to Scheme 18.47.

The kinetics of the elementary reactions of α-olefin hydroalumination with alkylalanes i-Bu$_2$AlX (X = H, Cl, i-Bu) catalyzed by Cp$_2$ZrCl$_2$ have been studied [136]. Each reaction kinetic parameter has been calculated by solving an inverse kinetic problem, and the rate constants as well as activation energies have been determined, allowing the development of a kinetic model for the hydroalumination reaction.

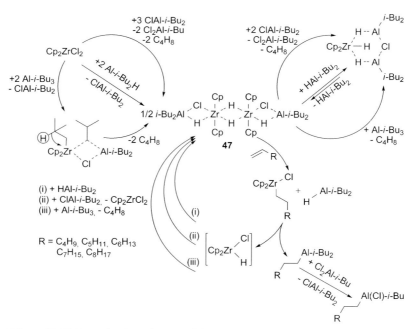

Scheme 18.47 Generalized mechanism of the hydroalumination reaction of olefins with i-Bu$_2$AlH, i-Bu$_3$Al and i-Bu$_2$AlCl in the presence of Cp$_2$ZrCl$_2$.

18.3.2
Dienes

In contrast with terminal alkenes, hydroalumination of 1,3-dienes with LiAlH$_4$ or NaAlH$_4$ in the presence of catalytic amounts of Cp$_2$TiCl$_2$ or TiCl$_4$ is characterized by low selectivity, giving 1,2-, 1,4- or 3,4-adducts (Scheme 18.48) [137, 138]. Catalytic hydroalumination of 1,3-dienes using bis(dialkylamino)alanes also results in low selectivity (Scheme 18.49) [139, 140].

Hydroalumination of 1,2-dienes with LiAlH$_4$ in THF catalyzed by Ti- or Zr-complexes (Cp$_2$TiCl$_2$, TiCl$_4$, TiCl$_3$, ZrCl$_4$, Cp$_2$ZrCl$_2$) are also unselective [141]. In contrast, 1,2-dienes have been reported to be hydroaluminated with high regioselectivity by reaction with dichloroalane (AlHCl$_2$) in the presence of catalytic amounts of organoboranes (Scheme 18.50) [141, 142].

18.3.3
Alkynes

It follows from the previous section that the noncatalyzed hydroalumination of disubstituted acetylenes can be achieved by using DIBAL, DEAH or TIBA under sufficiently mild conditions. However, the reactions are characterized by low selectivity, since the vinylalanes formed in this reaction undergo further transformations [143]. Hydrolysis of the obtained vinylalanes results in 1,2-disubstituted alkenes with Z- or E-configuration depending upon the nature of the starting hydroaluminating reagent [144], the structure of the alkyne [145] and the nature of the solvent (Scheme 18.51) [146].

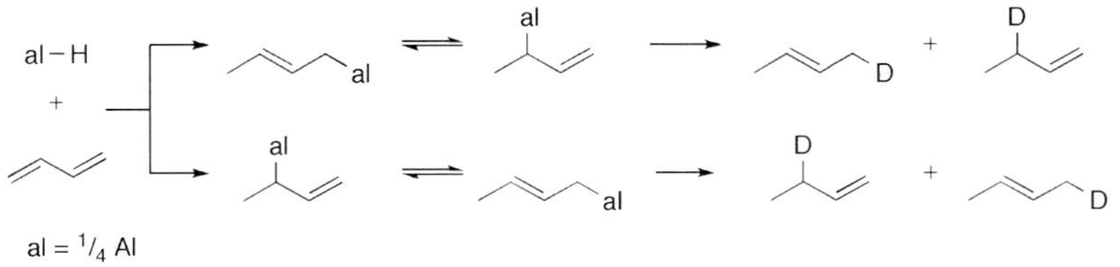

Scheme 18.48 Catalytic hydroalumination of 1,3-dienes with LiAlH$_4$ or NaAlH$_4$.

(i) (i-Pr$_2$N)$_2$AlH, Cp$_2$TiCl$_2$, THF, 12 h, 20 °C
(ii) H$_2$O, 55%

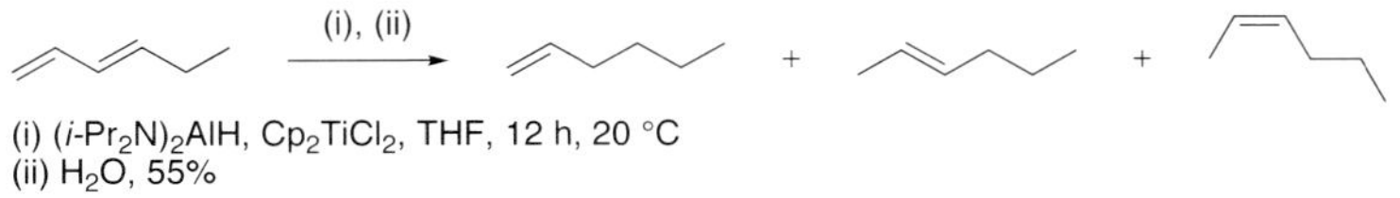

Scheme 18.49 Catalytic hydroalumination of 1,3-dienes with bis(dialkylamino)alanes.

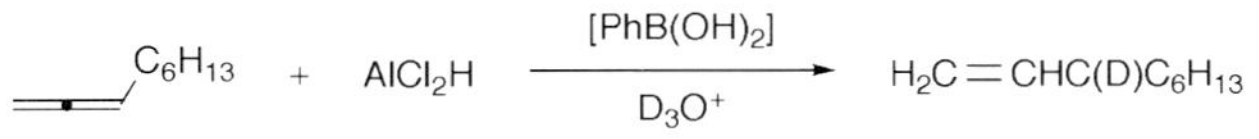

Scheme 18.50 Hydroalumination of 1,2-dienes with catalytic amounts of organoboranes.

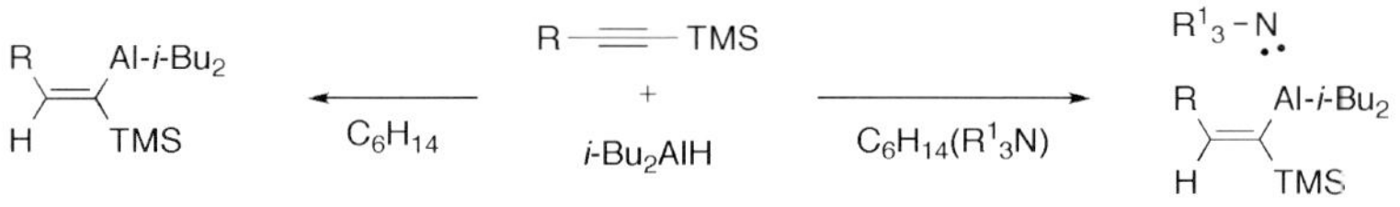

Scheme 18.51 Hydroalumination of alkynes.

Scheme 18.52 Noncatalytic hydroalumination of oct-1-yne with methylcuprate.

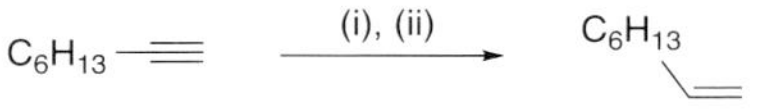

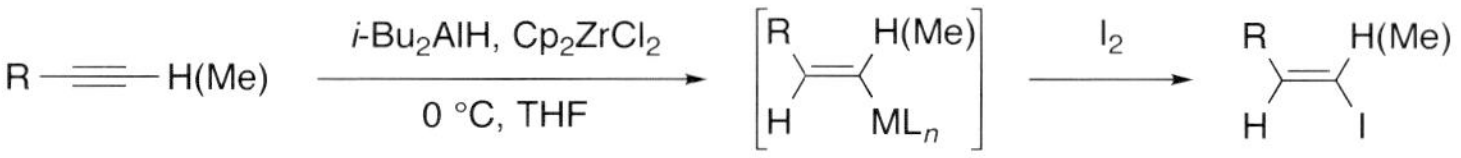

(i) i-Bu$_2$AlH, MeMgBr, CuCl, LiCl, THF
(ii) H$_2$O, 100%

M = Zr, Al

Scheme 18.53 Hydroalumination of terminal and disubstituted alkynes.

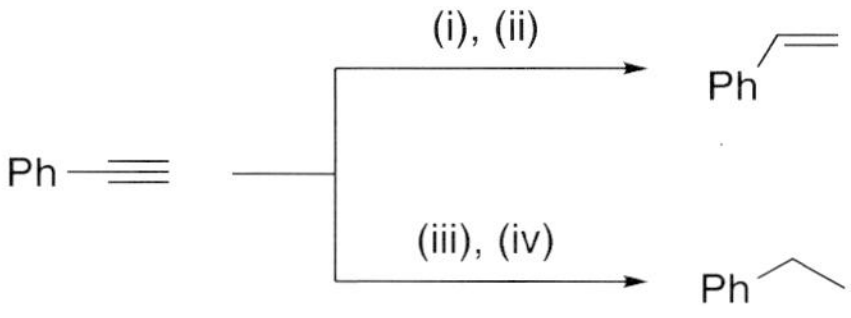

(i) LiAlH$_4$, FeCl$_2$, 20 °C, 24 h; (ii) H$_2$O, 94%

(iii) LiAlH$_4$, NiCl$_2$, 20 °C, 24 h; (iv) H$_2$O, 99%

Scheme 18.54 Unselective hydroalumination of monosubstituted alkynes.

In the presence of transition metal complexes, hydroalumination of alkynes can be carried out with both higher selectivity and higher yields. For example, the reaction of oct-1-yne with DIBAL in the presence of stoichiometric amounts of methyl cuprate (1.25 hours, 20 °C) and subsequent hydrolysis of the reaction mixture gives oct-1-ene in quantitative yield (Scheme 18.52) [147].

Hydroalumination of terminal and disubstituted alkynes with a reagent, obtained *in situ* by the slow dropwise addition of DIBAL to 1 molar equivalent of (η^5-Cp)$_2$ZrCl$_2$, has been carried out with high selectivity (>98%) in THF at 0 °C (Scheme 18.53) [148].

Hydroalumination of monosubstituted alkynes normally proceeds unselectively, yielding a mixture of alkenes and alkanes due to concurrent mono- and dihydroalumination reactions on the triple bond (Scheme 18.54). Selective mono- or dihydroalumination of alkynes can be carried out in the presence of Fe or Ni salts using LiAlH$_4$ [78].

18.3 Catalytic Hydroalumination

Hydroalumination of mono- and disubstituted acetylenes with LiAlH$_4$ in the presence of TiCl$_4$, Cp$_2$TiCl$_2$ [149] or [C$_5$(CH$_3$)$_5$]TiCl$_2$ [73] has been carried out with high selectivity. Thus, the interaction between oct-4-yne or hex-2-yne and aluminum hydrides (LiAlH$_4$, NaAlH$_4$, LiAlMe$_3$H, NaAlMe$_3$H, NaAl(OCH$_2$CH$_2$OCH$_3$)$_2$H$_2$) in the presence of catalytic amounts of Cp$_2$TiCl$_2$ in THF was found to afford alkenylalanes after 2 hours at 20 °C; hydrolysis of these resulted in *cis*-oct-4-ene and *cis*-hex-2-ene, respectively, in approximately 100% yields [149].

Noncatalytic hydroalumination of alkynols with LiAlH$_4$ in diglym at 150 °C was found to yield (*E*)-alkenols [150]. The presence of catalytic amounts of Cp$_2$TiCl$_2$ (10 mol%) provided predominantly (*Z*)-alkenols (*Z/E* ratio ~ 10:1) (Scheme 18.55) [151].

The authors of the work [151] proposed that active complexes [Ti]-H, generated under reaction conditions, probably hydrotitanate the triple bond. Subsequent transmetallation of the resulting alkenyl-Ti complexes **(48)** leads to *E*-alkenylalanes **(49)** and regeneration of the hydride complex [Ti]-H (Scheme 18.56).

(*Z*)-Stilbene has been obtained with high stereoselectivity (more than 90%) in the hydroalumination of tolane with NaAlH$_4$, affected by catalytic amounts of titanium nitride (TiN) (Scheme 18.57) [152]. Without the TiN catalyst (*E*)-stilbene was the major product in the reaction.

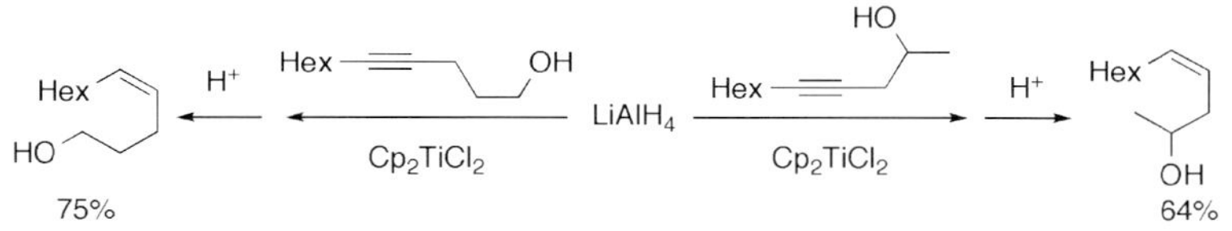

Scheme 18.55 Formation of (*Z*)-alkenols.

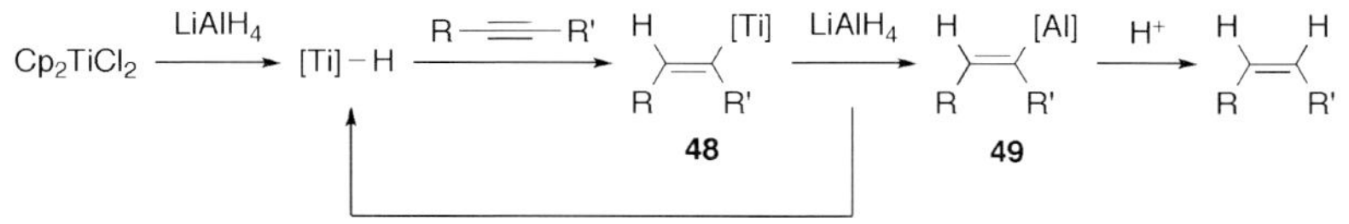

Scheme 18.56 Proposed mechanism for Cp$_2$TiCl$_2$-catalyzed hydroalumination of alkynes.

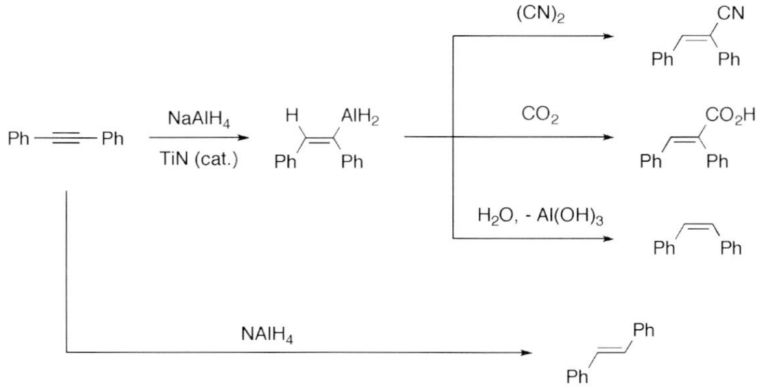

Scheme 18.57 TiN-catalyzed hydroalumination of tolane.

Aminoalanes can also be used in the hydroalumination of alkynes [107]. In this case, hydroalumination of disubstituted alkynes occurs with high selectivity and under milder conditions compared with terminal alkynes. Thus, in the presence of 5 mol% of Cp_2TiCl_2, the reaction of methylphenylacetylene with $(i\text{-}Pr_2N)_2AlH$ gave alkenylalanes in a total yield of >95% (Scheme 18.58) [108]. The reaction of aminoalanes with disubstituted acetylenes occurs predominantly via *cis* addition.

The composition of the reaction mixture depends upon the temperature, the nature of the solvent and the duration of the reaction. Thus, hydroalumination of hex-2-yne with $(i\text{-}Pr_2N)_2AlH$ in benzene at 0°C yields a mixture of alkenylalanes with *trans*- and *cis*-configurations of the substituents. Further hydrolysis of alkenylalanes yields a mixture of isomeric hex-2-enes (Scheme 18.59) [108].

The recently synthesized reagents $i\text{-}Bu_2AlCl\cdot[Cp_2TiCl_2]$ and $Et_3Al\cdot[Cp_2TiCl_2]$, designed for selective hydroalumination of disubstituted acetylenes into their corresponding unsaturated OACs (**50**) and (**51**), deserve special attention (Scheme 18.60) [153, 154]. Hydrolysis of the resulting OAC leads to (Z)-alkenes. Depending upon the structure of the original alkynes and reaction conditions, up to 30% of

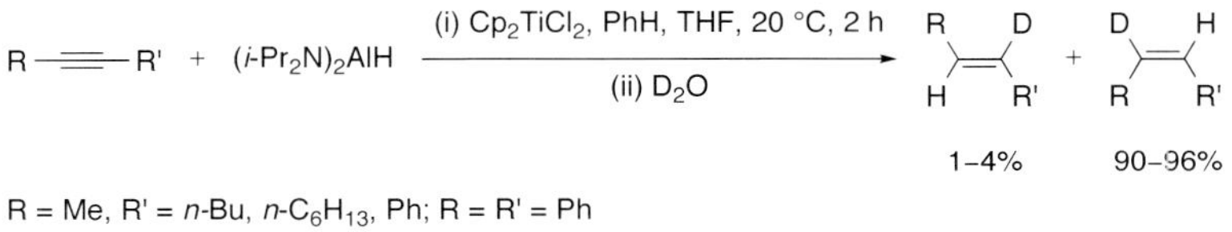

Scheme 18.58 Hydroalumination of alkynes with aminoalanes.

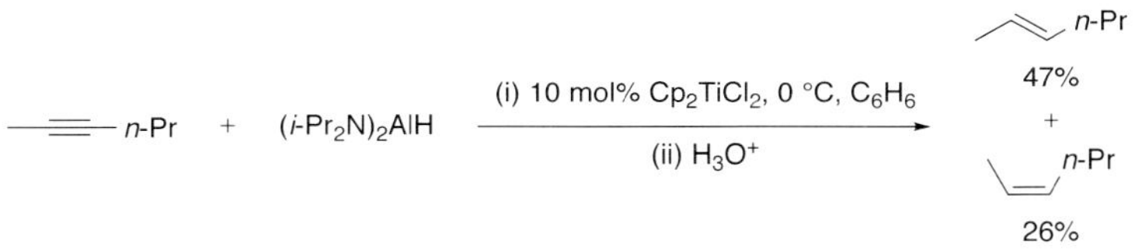

Scheme 18.59 Hydroalumination of hex-2-yne with $(i\text{-}Pr_2N)_2AlH$ at 0°C.

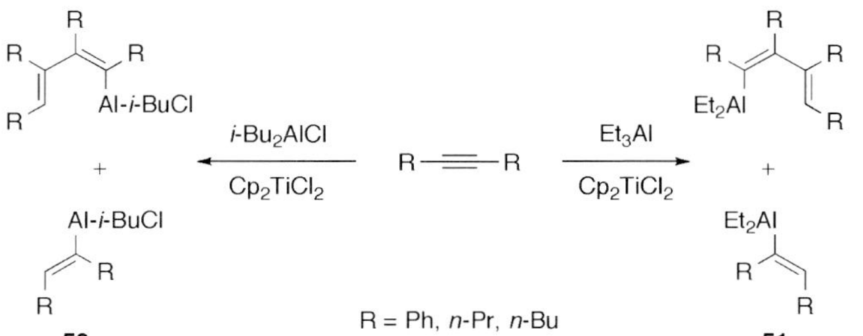

Scheme 18.60 Hydroalumination with $i\text{-}Bu_2AlCl\cdot[Cp_2TiCl_2]$ and $Et_3Al\cdot[Cp_2TiCl_2]$.

alkadienylalanes were also obtained. The structure of the unsaturated OACs **(50)** and **(51)** has been established by spectral methods [155].

Hydroalumination of disubstituted acetylenes with TIBA in the presence of catalytic amounts of Cp_2ZrCl_2 is accompanied by the formation of isomeric alkenylalanes, their deuterolysis affords alkenes (~1:1 ratio) (Scheme 18.61) [156, 157]. Under analogous conditions, terminal alkynes yield a mixture (~1:1 ratio) of mono- and 1,1-di-aluminum compounds in practically quantitative yield (Scheme 18.61) [156].

In contrast to hydroalumination with $Et_3Al \cdot Cp_2TiCl_2$ [154], the reaction of disubstituted acetylenes with $n\text{-}Pr_3Al$ catalyzed by Cp_2ZrCl_2 occurs less selectively and gives a mixture of isomeric alkenylalanes (total yield 63%) and aluminacyclopentadiene (38%) (Scheme 18.62) [156].

It should be noted that Fe or Ni complexes do not selectively hydroaluminate terminal acetylenes with trialkylalanes R_3Al (R = Et, i-Bu, $EtCH(Me)CH_2$). Here the alkenylalanes obtained undergo further transformations resulting, after hydrolytic work-up, in complicated mixtures of olefins, 1,3-dienes and trialkyl-substituted benzenes [158, 159].

Hydroalumination of mono- and disubstituted acetylenes has found wide application in synthetic organic chemistry, particularly in the synthesis of practically useful chemicals such as trisubstituted alkenes [146] and allylic alcohols [160], substituted 1,3-dienes [161] and alk-l-en-3-ynes [162], alkenylsilanes and alkenyl sulfones [163, 164], (*E*)-vinyl phosphonates [165], l-alkyl-2-halogenocyclopropanes [166], (*Z*)-alkenes [167, 168] and others. Thus, in the majority cases, stereoselective hydroalumination of alkynes and allenes occurs under mild conditions. Hydrolysis of the resulting higher alkenylalanes affords (*Z*)-alkenes in high yields.

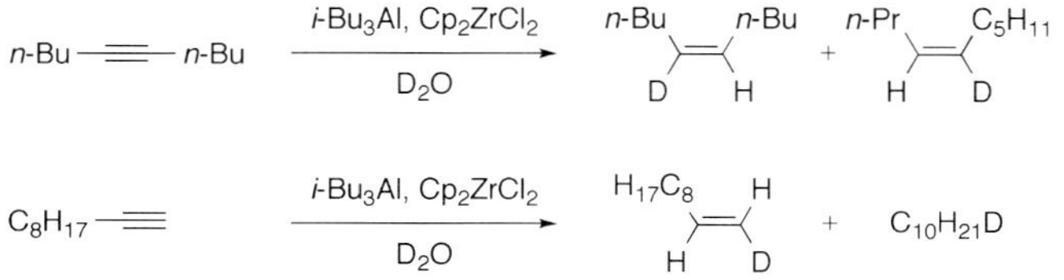

Scheme 18.61 Cp_2ZrCl_2-catalyzed TIBA hydroalumination of di- and mono-substituted alkynes.

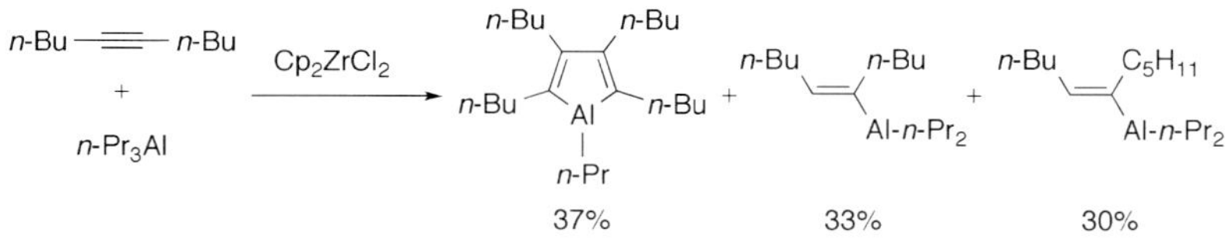

Scheme 18.62 $Et_3Al \cdot Cp_2TiCl_2$-catalyzed hydroalumination of disubstituted alkynes with $n\text{-}Pr_3Al$.

18.4
Catalytic Hydromagnesiation

18.4.1
Alkenes

Thermal hydromagnesiation of unsaturated compounds has not found wide application in the synthesis of organomagnesium compounds (OMCs) due to both low yields and low selectivity [169, 170]. The interaction between magnesium hydrides or alkylmagnesium halogenides and unsaturated compounds (olefins, 1,3-dienes, acetylenes) catalyzed by salts and complexes of titanium and zirconium as well as other transition metals represents a convenient procedure to synthesize OMCs of various structures [171–175]. The diversity of unsaturated compounds as well as the availability of starting reagents enables the synthesis of "nontrivial" OMCs not easily obtained in the conventional way using Grignard reagents.

Finkbeiner and Cooper first reported catalytic hydromagnesiation of α-alkenes with *n*-PrMgBr in the presence of Ti complexes in 1961 [176]. They found that different α-alkenes could be converted into their respective OMCs (**52**) in the presence of TiCl$_4$ (3 mol%) in yields of 20–50% with evolution of an equimolar amount of propylene (Scheme 18.63) [176]. The latter is indirect evidence of the generation, under the reaction conditions, of titanium hydrides, which are responsible for the subsequent hydrometallation of the starting alkenes.

The organomagnesium compounds (**52**) prepared in this way react with aldehydes (R'CHO) or CO$_2$, yielding the respective alcohols (**53**) or carboxylic acids (**54**) (Scheme 18.64).

In addition to TiCl$_4$, compounds such as Ti(O-*i*-Pr)$_4$, Cp$_2$TiCl$_2$, ZrCl$_4$ and VCl$_4$ can also be used as catalysts for this reaction [177, 178]. In contrast to Ti-

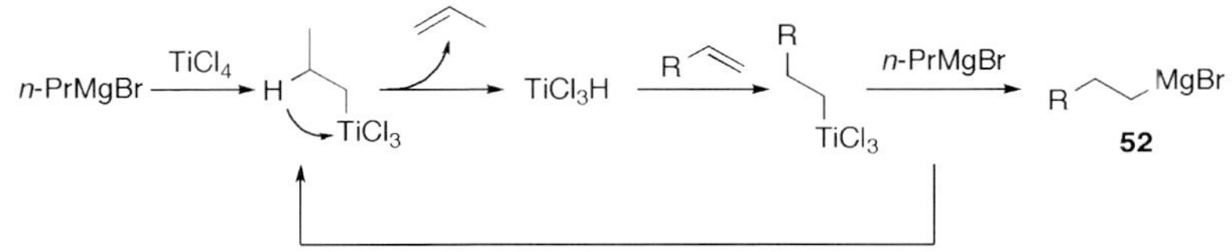

Scheme 18.63 First report of hydromagnesiation of α-alkenes with *n*-PrMgBr.

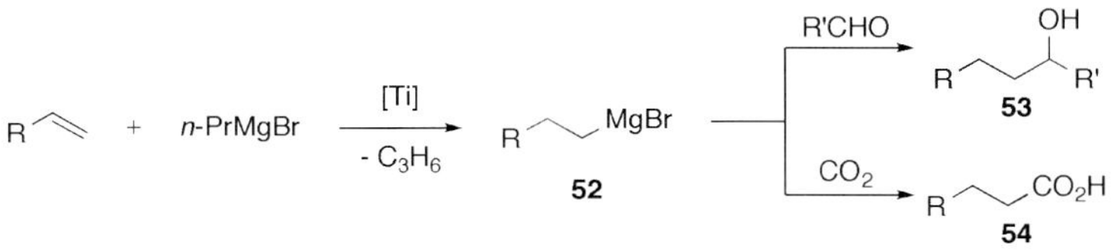

[Ti] = TiCl$_4$; R = alkyl, PhCH$_2$, cyclo-C$_6$H$_{11}$; R' = H, Me

Scheme 18.64 Reactions of (**52**).

containing catalysts, nickel complexes are less selective in the reaction of α-alkenes with RMgX. Thus, for example, both primary (compounds (**52**)) and secondary (compounds (**55**)) OMCs are formed in the presence of catalytic amounts of NiCl$_2$, their ratio being dependent on the structures of the starting alkenes and the nature of the hydrometallation agent (Scheme 18.65) [179, 180].

NiCl$_2$-catalyzed hydromagnesiation of allylic alcohols with *n*-PrMgBr yields the primary OMC (**56**) more selectively [181]. The authors propose that, as in the case of Ti complexes, generation of nickel hydride complexes (**57**), responsible for the hydrometallation of the starting allylic alcohols, is the key step of this reaction (Scheme 18.66).

Cp$_2$TiCl$_2$-catalyzed hydromagnesiation reaction of α-alkenes were studied using a variety of organomagnesiation reagents (EtMgCl, EtMgBr, *n*-PrMgCl, *i*-PrMgBr, *n*-BuMgCl, *sec*-BuMgCl, *i*-BuMgCl, *i*-BuMgBr, *i*-BuMgI, Et$_2$Mg, *n*-Pr$_2$Mg, *i*-Pr$_2$Mg,

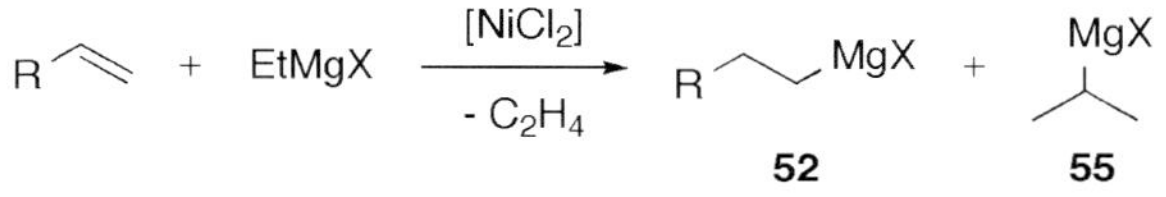

Scheme 18.65 Ni-catalyzed hydromagnesiation of α-alkenes.

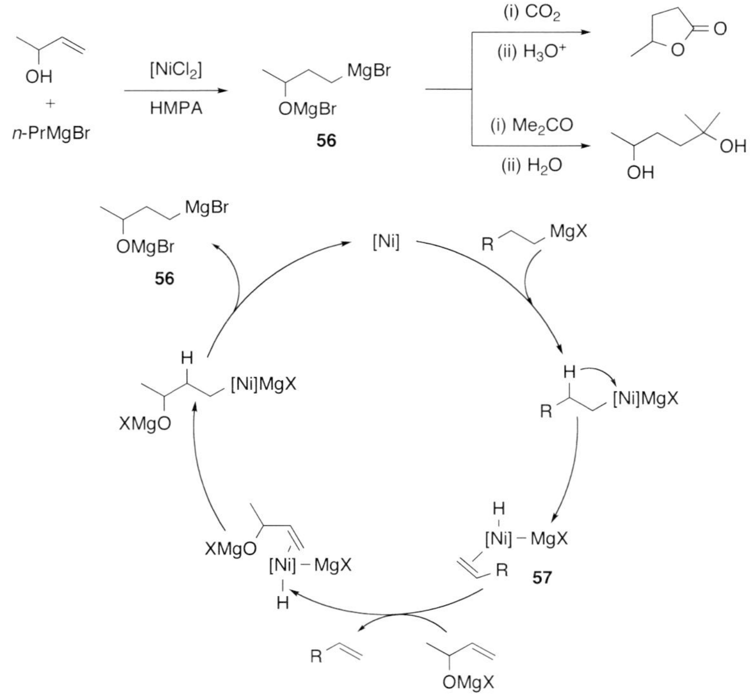

Scheme 18.66 Proposed mechanism of the NiCl$_2$-catalyzed hydromagnesiation of allylic alcohols with *n*-PrMgBr.

n-Bu$_2$Mg, sec-Bu$_2$Mg, t-Bu$_2$Mg) and allow one to obtain higher homologues of the OMCs in satisfactory yields [182].

In addition to alkenes, hydrometallation of imines was accomplished under the action of n-BuMgCl and catalytic amounts of Cp$_2$TiCl$_2$ under mild conditions (Et$_2$O, 25 °C 15 hours) to give magnesium amides. Their subsequent hydrolysis resulted in the corresponding secondary amines (58) (Scheme 18.67) [183]. An efficient method for the hydrogenation of alkenes with hydrogen (1 atm) using a two-component catalyst Cp$_2$TiCl$_2 \cdot$ i-PrMgBr ([Ti]:[Mg]:[alkene] = 1:6:50, 20 °C, 5 hours), in nearly quantitative yield, has also been proposed [184].

Organomagnesium hydrides RMgH (R = H, Hal, Alk) are employed in hydromagnesiation of alkenes. Magnesium hydride, generated *in situ* from metallic magnesium Mg and H$_2$, adds to ethene, isobutene and 1-octene to give the corresponding magnesium dialkyls in yields of 4–13%. Meanwhile, the reaction of α- and 1,1- (or 1,2)-disubstituted alkenes with MgH$_2$ in the presence of catalytic amounts of Cp$_2$TiCl$_2$ in THF, molar ratio alkene:MgH$_2$:[Ti] of 2:3:0.05, under mild conditions (from 20 to 60 °C) was found to afford the magnesium alkyl derivatives in high yields [185]. The system CrCl$_3 \cdot$[Mg-anthracene] (1:1) in THF has been recommended as a promoter for the synthesis of magnesium hydride complexes (HMgCl·MgH$_2$ or 2HMgCl·MgH$_2$) [186, 187]. OMCs (59) were prepared in yields of 93–98% upon hydromagnesiation of α-alkenes with EtMgH in the presence of a Ni- or Zr-containing catalysts in THF (90 °C, 4 hours) (Scheme 18.68) [188].

Bogdanovic and coworkers have employed catalysts based on Zr, Ti, Hf, Cr and Fe halides for the hydromagnesiation of alkenes using magnesium hydride (Scheme 18.69) [189–191].

Hydromagnesiation of ω-N(or O)-containing α-alkenes with MgH$_2$ effected by a ZrCl$_4$ catalyst in THF (70–80 °C) led to OMCs (60) and (61), stabilized by intramolecular coordination of nitrogen or oxygen atoms to the metal center (Scheme 18.70) [192].

Catalytic hydromagnesiation of α-alkenes and subsequent functionalization of the resulting OMC involving the reactive Mg–C bond has been used successfully

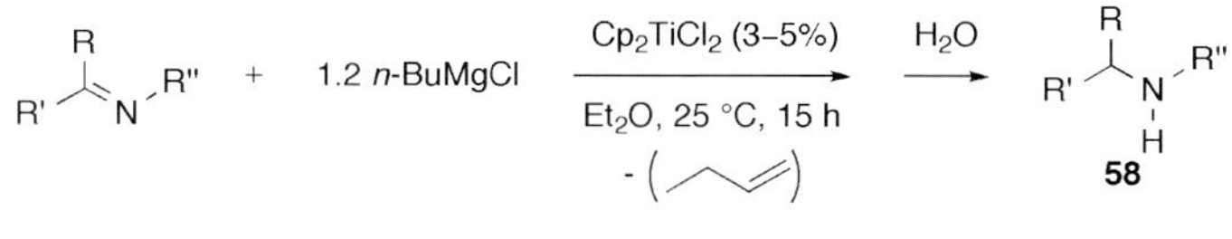

R = H, Me, Et, i-Pr; R' = Et, n-Pr, Ph; R'' = n-Pr, i-Pr, t-Bu, CH$_2$Ph

Scheme 18.67 Hydromagnesiation of imines.

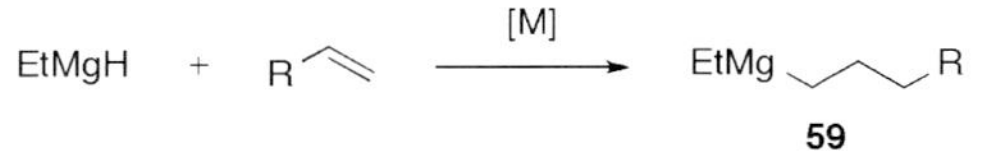

[M] = NiCl$_2$, Cp$_2$ZrCl$_2$; R = alkyl

Scheme 18.68 Formation of OMC (59).

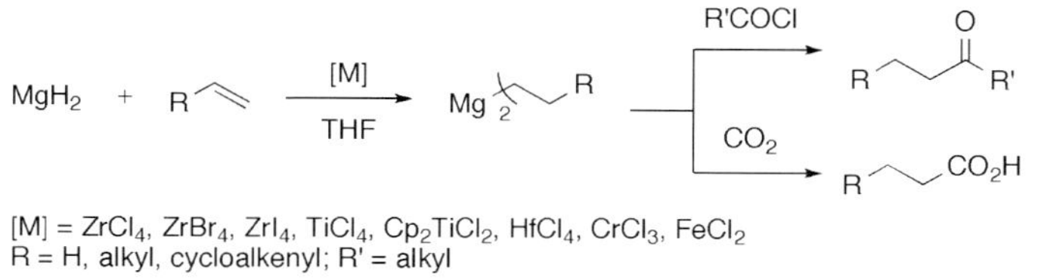
R'COCl
MgH2 + R
[M]
THF
Mg
2
R
CO2
R
R'
O
R
CO2H
[M] = ZrCl4, ZrBr4, ZrI4, TiCl4, Cp2TiCl2, HfCl4, CrCl3, FeCl2
R = H, alkyl, cycloalkenyl; R' = alkyl

Scheme 18.69 Hydromagnesiation of alkenes with MgH_2.

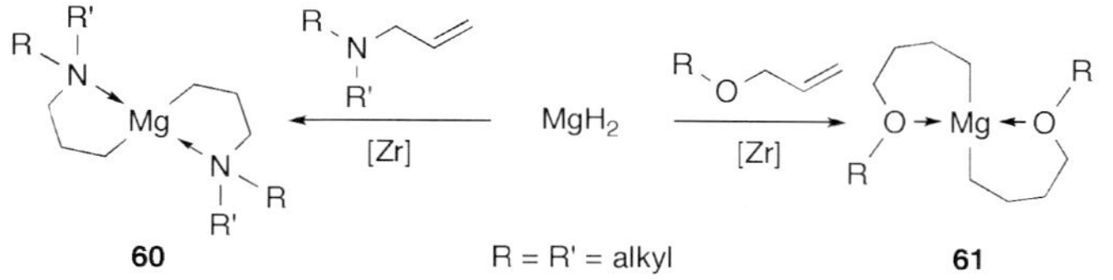
MgH2
[Zr]
[Zr]
Mg
R = R' = alkyl

60 61

Scheme 18.70 Hydromagnesiation of ω-N (or O)-containing α-alkenes with MgH_2.

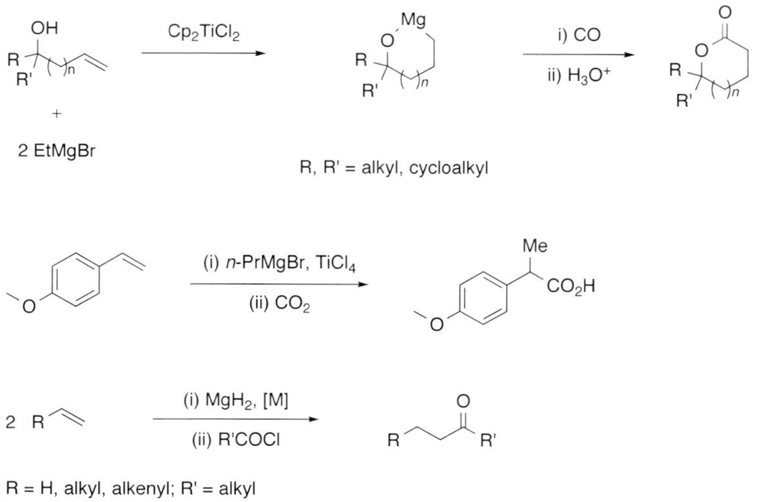
OH
2 EtMgBr
Cp2TiCl2
i) CO
ii) H3O+
R, R' = alkyl, cycloalkyl
(i) n-PrMgBr, TiCl4
(ii) CO2
Me
CO2H
(i) MgH2, [M]
(ii) R'COCl
R = H, alkyl, alkenyl; R' = alkyl

Scheme 18.71 Catalytic hydromagnesiation of α-alkenes and subsequent fuctionalization.

in the synthesis of alcohols [176, 178, 181], steroids [193], γ- and δ-lactones [194], ketones, carboxylic acids [176, 191] and other classes of organic compounds (Scheme 18.71).

18.4.2 Dienes

Catalytic hydromagnesiation of 1,3-dienes with magnesium hydrides is an efficient method for the synthesis of unsaturated OMCs [195, 196]. As in the case of alkenes, Ti-containing catalysts, viz. Cp$_2$TiCl$_2$ and TiCl$_4$, demonstrated the highest catalytic activity in this reaction. With these catalysts, the conversion of the starting diene was approximately 75%. In contrast, the process proceeded nonselectively with Cp$_2$ZrCl$_2$, ZrCl$_4$, HfCl$_4$ or CrCl$_3$, and conversions did not exceed 30% (Scheme 18.72).

Hydromagnesiation of 1,3-dienes (butadiene, isoprene or myrcene) or styrene with n-PrMgBr in the presence of catalytic amounts of Cp$_2$TiCl$_2$ is the simplest procedure for synthesizing unsaturated OMCs [197]. One should note that these studies were preceded by investigations on hydromagnesiation of α-alkenes with propylmagnesium bromide in the presence of TiCl$_4$ [176–178]. However, the attempted use of TiCl$_4$ as a catalyst for hydromagnesiation of 1,3-dienes failed, whereas, in the presence of Cp$_2$TiCl$_2$, regioisomeric allylic OMCs (62a) and (62b) were successfully prepared from 2-substituted 1,3-dienes [197] (Scheme 18.73). It is noteworthy that such reaction is unknown for 1-substituted 1,3-dienes.

The authors [197] proposed the probable mechanism of catalytic hydromagnesiation of 1,3-dienes via titanium organic hydrides (64) generated by reaction of

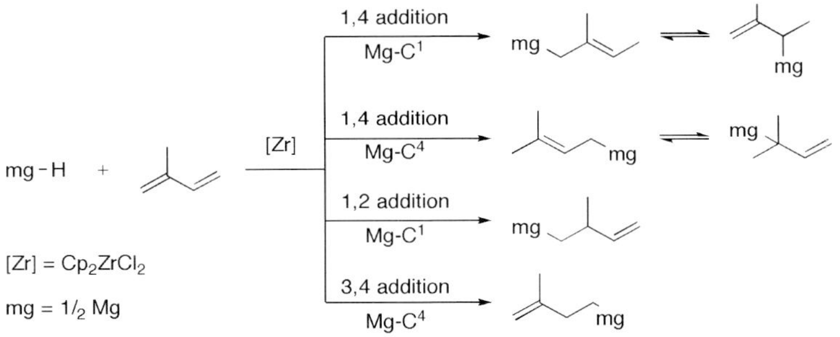

Scheme 18.72 Nonselective hydromagnesiation catalyzed by Cp$_2$ZrCl$_2$.

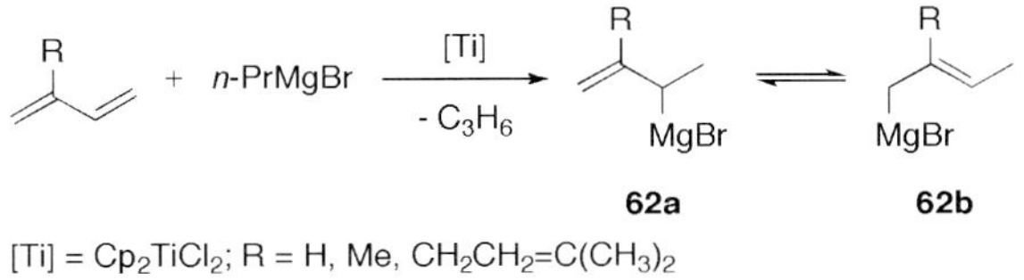

Scheme 18.73 Regioisomeric allylic OMCs (62a, b) prepared from 2-substituted 1,3-dienes.

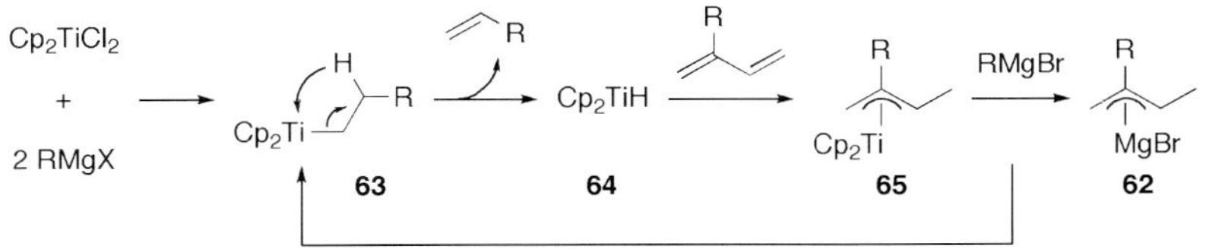

Scheme 18.74 Proposed mechanism of catalytic hydromagnesiation of 1,3-dienes.

Cp$_2$TiCl$_2$ with *n*-PrMgBr, which act as active hydrometallation intermediates in these reactions (Scheme 18.74). Hydrides **(64)** add to 1,3-dienes giving rise to allylic [bis(cyclopentadienyl)]titanium complexes **(65)** [198]. These undergo transmetallation with *n*-PrMgBr (excess) and are converted to regioisomeric allylmagnesium bromide **(62)** with simultaneous regeneration of catalytically active *n*-propylmagnesium complexes **(63)**.

The hydromagnesiation pathway depends on both the nature of the catalyst and the structure of the starting 1,3-diene. Thus, two molecules of butadiene react with *n*-PrMgBr in the presence of catalytic amounts of (PPh$_3$)$_2$NiCl$_2$ to yield 2,7-octadienylmagnesium bromide **(66)**, which undergoes intramolecular cyclization into the OMC **(67)** (Scheme 18.75) [199], while two molecules of isoprene react with *n*-PrMgBr (in ethereal solution) to yield 3,7-dimethyl-2,7-octadienylmagnesium bromide **(68)** (Scheme 18.75) [200].

Hydromagnesiation of 2-alkyl-1,3-dienes (isoprene, myrcene, 2-isopropylbuta-1,3-diene, 2-*t*-butylbuta-1,3-diene) with RMgX (R = Et, Pr, *i*-Pr, *i*-Bu, *sec*-Bu; X = Cl, Br, I) in the presence of nickel catalysts with phosphine oxide or nitrogen-containing ligands was less selective [201, 202]. In this case, allylic-type isomers of monomeric OMCs (**(71)** and **(72)**) were formed in addition to the regioisomeric dimeric hydromagnesiation products **(69)** and **(70)** (Scheme 18.76).

The use of pyridine–nickel complexes as catalysts enabled the hydromagnesiation of 2-alkoxybuta-1,3-dienes [203], 1-alkylbuta-1,3-dienes [204, 205] and di- and polyalkylbuta-1,3-dienes [206] (Scheme 18.77), which could not be effected using Cp$_2$TiCl$_2$ [207].

An efficient method for the synthesis of isoprenoid alcohols or carboxylic acids **(73)** using Grignard reagents has been developed [206–208]. It is based on the reaction of allylic OMCs generated *in situ* by the hydromagnesiation of substituted 1,3-dienes, with CO$_2$ (Scheme 18.78).

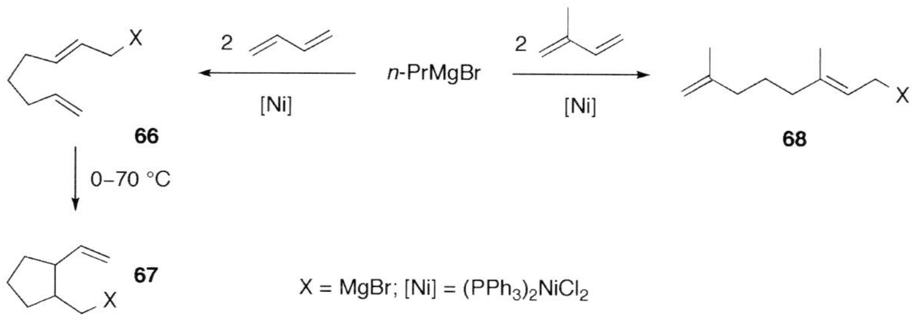

Scheme 18.75 Ni-catalyzed hydromagnesiation of 1,3-dienes.

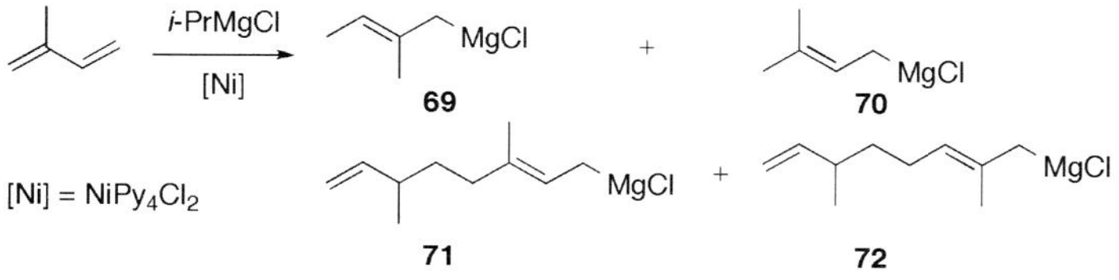

Scheme 18.76 Regioisomeric hydromagnesiation products.

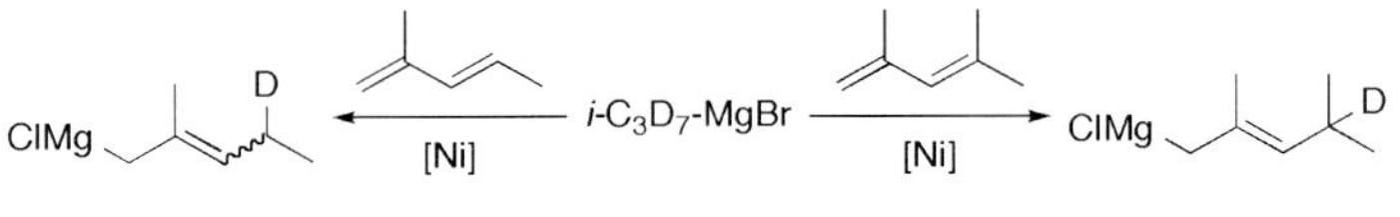

[Ni] = NiPy$_4$Cl$_2$

Scheme 18.77 Ni-catalyzed hydromagnesiation.

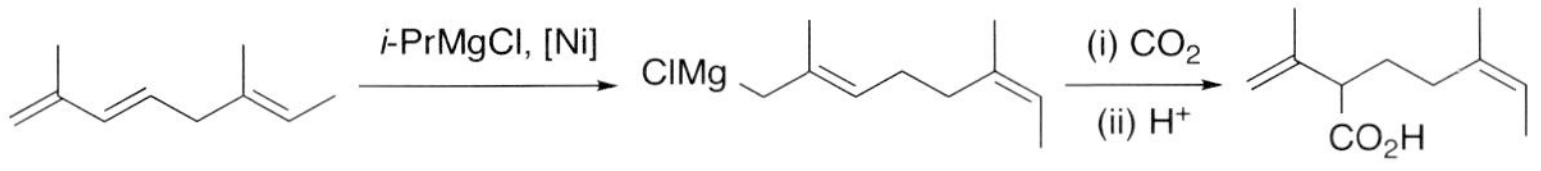

[Ni] = NiPy$_4$Cl$_2$

Scheme 18.78 Synthesis of isoprenoid carboxylic acids.

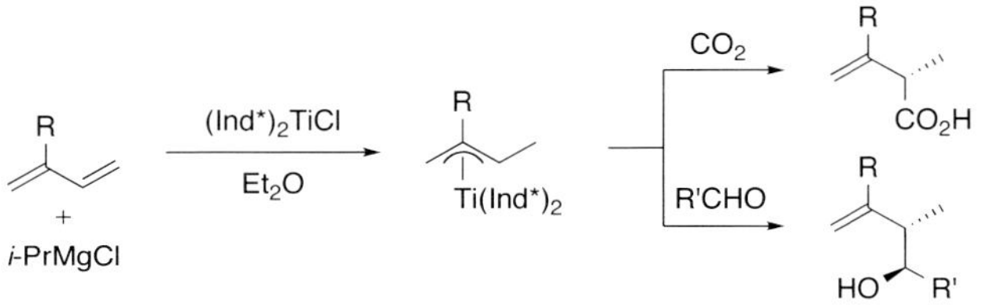

R = C$_9$H$_{19}$, Me$_2$C=CH(CH$_2$)$_3$; R' = Et, i-Pr, t-Bu, Ph

Ind* = η^5-indenyl

Scheme 18.79 Ti-catalyzed optically active carboxylic acids and alcohols.

Hydrometallation of 2-alkylbutadienes with *i*-PrMgCl catalyzed by a chiral indenyl–titanium complex was found to afford allylic titanium complexes; their treatment with CO$_2$ or aldehydes represents a convenient method for the synthesis of optically active β,γ-unsaturated carboxylic acids **(74)** or *threo*-homoallylic alcohols **(75)** with high enantiomeric purity (96%) (Scheme 18.79) [209].

Cross coupling of N-, O- and S-functionalized allylic compounds with dienyl-magnesium reagents catalyzed by transition metal complexes resulted in unsaturated hydrocarbons with an isoprenoid structure in high yields [210]. An analogous scheme was used for the synthesis of polyene sulfides **(76)** and **(77)** as well as thioethers by the reaction of the "isoprenemagnesium" reagent with the allylic electrophiles and accompanied by simultaneous insertion of a sulfur atom (from the S$_8$ molecule) at the Mg–C bond (Scheme 18.80) [211].

Syntheses and transformations of such OMCs of the "non-Grignard" type prepared from 1,3-dienes and highly active metallic Mg are reviewed in more detail elsewhere [173].

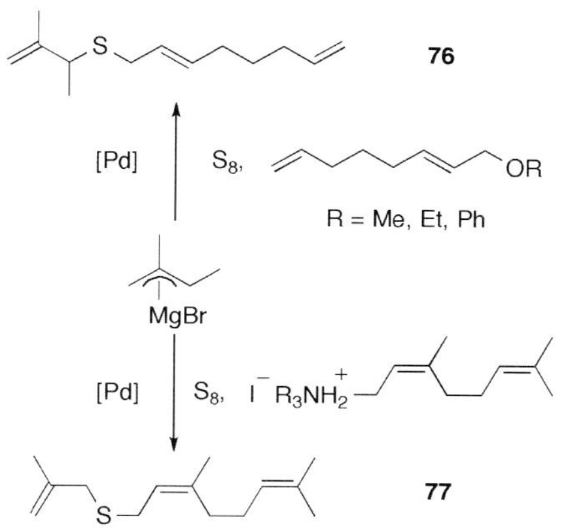

Scheme 18.80 Synthesis of polyene sulfides (**76**) and (**77**).

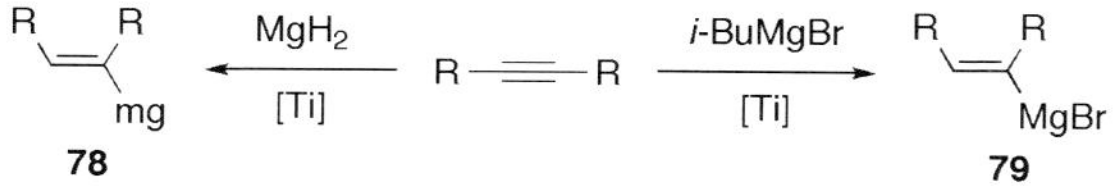

Scheme 18.81 Hydromagnesiation of disubstituted alkynes.

18.4.3
Alkynes

The catalytic hydromagnesiation of alkynes with magnesium hydrides or alkylmagnesium halides in the presence of Ti complexes affords vinylic organomagnesium compounds. The direction and selectivity of this reaction depend on the nature of the catalyst, the structure of the starting alkyne and the hydromagnesiation reagent used. Hydromagnesiation of disubstituted acetylenes with MgH_2 [185], $MgH_2 \cdot CuI$, $MgH_2 \cdot CuO$-t-Bu [212] or i-BuMgHal [171, 213] in the presence of catalytic amounts of Cp_2TiCl_2 was found to yield alkenylmagnesium compounds (**78**) or (**79**) with E-double bonds (Scheme 18.81).

Hydromagnesiation of nonsymmetrical dialkylacetylenes afforded two regioisomeric alkenylmagnesium halides in nearly equal proportions. Hydromagnesiation of alkylarylacetylenes [213] and alkyl(trimethylsilyl)acetylenes [214–219] (20 °C, Et_2O) was more selective. The magnesium atom is preferentially (>95%) attached to the carbon atom bearing the phenyl or silyl substituents [220].

Hydromagnesiation of 3-trimethylsilylprop-2-yn-1-ol with i-BuMgCl in the presence of Cp_2TiCl_2 (ratio [alkyne]:[Mg]:[Ti] = 1:2.4:0.08; 25 °C 6 hours) led, after hydrolysis, to the stereoisomeric alcohols (E)-(**80**) and (Z)-(**80**) in a 95:5 ratio and total yield of 98% (Scheme 18.82) [221].

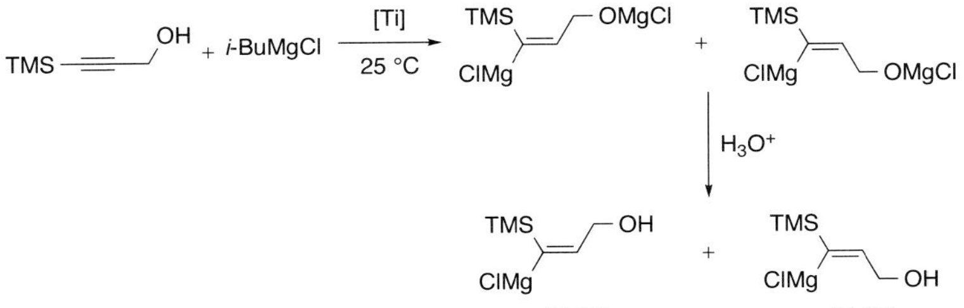

Scheme 18.82 Synthesis of alcohols (E)-(**80**) and (Z)-(**80**).

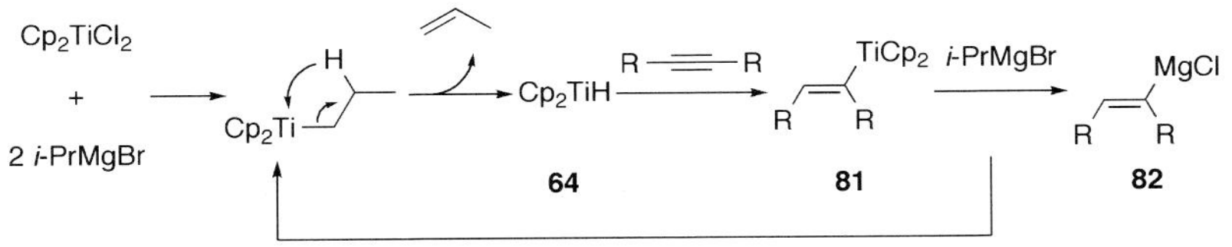

Scheme 18.83 Mechanism of the Ti-catalyzed hydromagnesiation of disubstituted alkynes.

The Cp_2TiCl_2-catalyzed hydromagnesiation of disubstituted acetylenes with organomagnesium compounds of the type $RMgR^1$ has been studied and discussed [222]. It was shown that OMCs containing cyclic and allylic substituents are less reactive than R_2Mg, whereas OMCs with tertiary and vinylic substituents are virtually inert in this reaction.

The hydromagnesiation mechanism of disubstituted acetylenes with Grignard reagents in the presence of Ti complexes implies the generation of reactive titanium hydrides (**64**). These intermediates hydrometallate the starting alkynes, forming alkenyltitanium complexes of the type (**81**). This step determines the regio- and stereoselectivities of the formation of the target products. Transmetallation of complexes (**81**) with RMgX lead to (E)-alkenyl magnesium halides (**82**) (Scheme 18.83) [220].

The regioselectivity of addition of titanium hydride complex (**64**) to nonsymmetrical disubstituted acetylenes is governed by the nature of substituents that determine the electron density on the carbon atoms of the C≡C bond.

Vinylic OMCs synthesized by the hydromagnesiation of disubstituted acetylenes were used in the selective synthesis of (Z)- [185, 222, 223] and (E)-alkenes [224], terpenoids [225], (E)-allylic alcohols [226, 227], (Z)-α,β-unsaturated ketones [228], substituted furans [229, 230], biologically active compounds and fragrances (Scheme 18.84) [215, 231–234].

Cross-coupling of vinylic OMCs with allylic ethers, sulfones, sulfides and quaternary ammonium salts catalyzed by Pd, Ni or Cu complexes led to 1,4-dienes (**83**) with a trisubstituted double bond in E-configuration (Scheme 18.85) [235].

Hydromagnesiation of alkynylsilanes with i-BuMgBr catalyzed by Cp_2TiCl_2 (5 mol%) followed by cross-coupling of in situ-generated vinylic OMCs (**84**) with

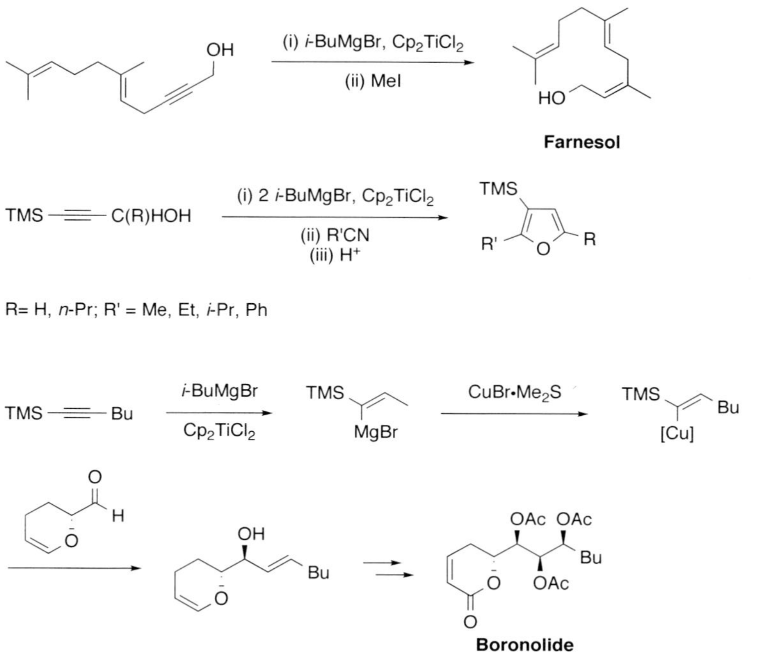

Scheme 18.84 Synthesis of (Z)- and (E)-alkenes, terpenoids, (E)-allylic alcohols, (Z)-α,β-unsaturated ketones, substituted furans, biologically active compounds and fragrances.

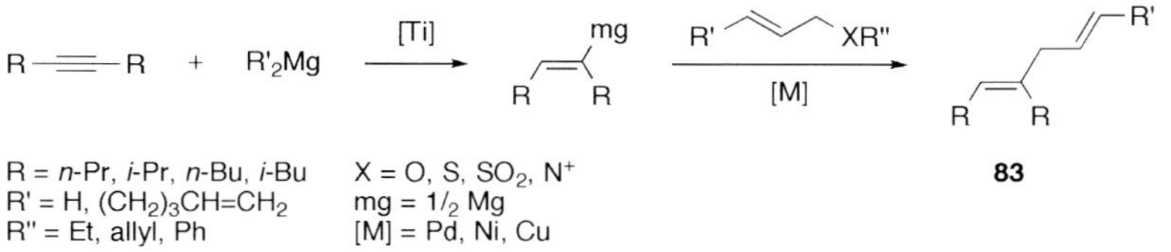

Scheme 18.85 Cross-coupling of vinylic OMCs.

vinyl-, aryltellurenyl- or trialkylstannyl halides represents a real method for the stereoselective synthesis of l,3-dienyl silanes [236], (E)-α-aryltellurenylvinyl silanes [237], (E)-α-silylvinyl stannanes [238], and (E)-1,2-disubstituted vinyl stannanes (Scheme 18.86) [239].

Diynes, in which triple bonds are separated by multiple hydrocarbon bridges, were found to react with i-BuMgBr to yield (Z)-alkenes **(85)** with high regio- and stereoselectivity (Scheme 18.87) [240].

The mixed reagent i-BuMgBr·i-Bu₃Al also provides highly stereoselective hydromagnesiation of alkynes (Scheme 18.88) [240].

Thus, hydromagnesiation of disubstituted alkynes with hydrides or alkylmagnesium halides catalyzed by titanium complexes represents an efficient method for stereoselective conversion of alkynes into (Z)-alkenes.

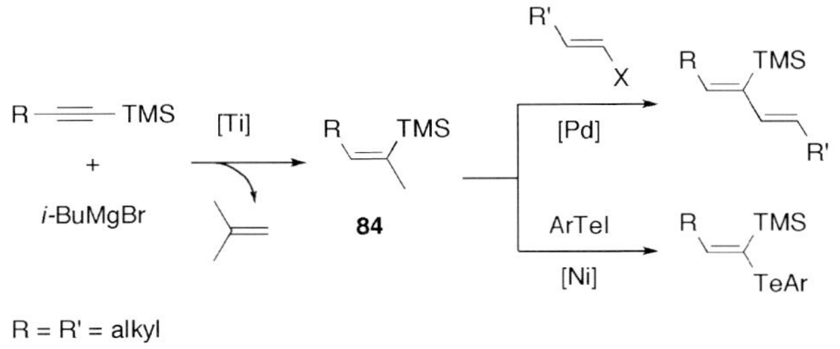

Scheme 18.86 Stereoselective synthesis of vinylsilanes.

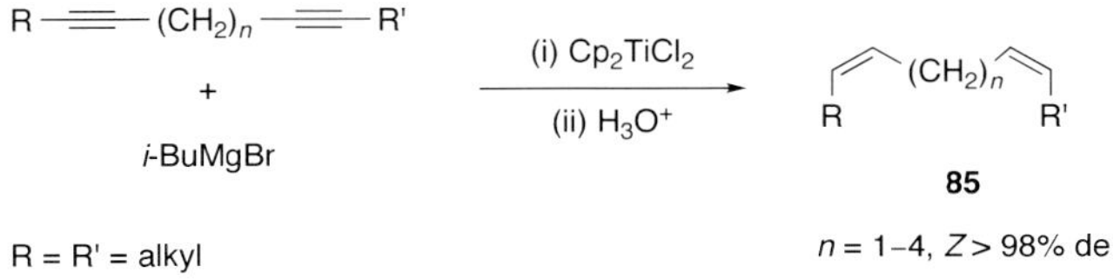

Scheme 18.87 Hydromagnesiation of diynes.

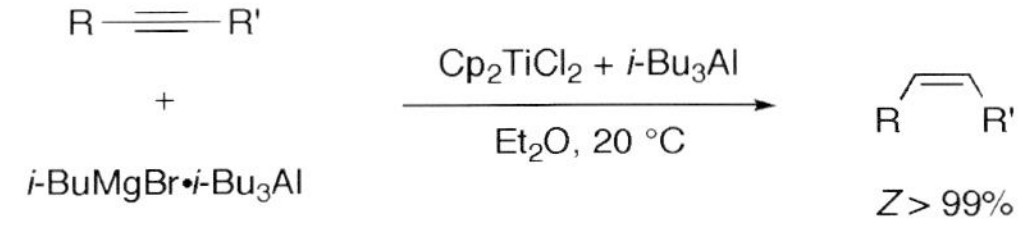

$n = 1$–4, $Z > 98\%$ de

Scheme 18.88 Stereoselective hydromagnesiation of alkynes with i-BuMgBr·i-Bu$_3$Al.

18.5
Summary

The material presented in this survey and the analysis of the published data for the preceding 15–20 years on hydrometallation of alkenes, dienes and alkynes with complex hydrides, and also magnesium and aluminum hydride derivatives, allow us to conclude that the reactions of thermal, catalytic hydroalumination and hydromagnesiation of unsaturated compounds have solidly entered into synthetic practice, being widely used both in the research laboratory and in industry.

Of special interest and practical value are the hydroalumination and hydromagnesiation reactions catalyzed by transition metal complexes. Application of these reactions allows the successful synthesis of the most important classes of natural compounds, including vitamins, antibiotics, alkaloids, steroids, carbohydrates, prostaglandins, and linear and cyclic isoprenoids.

We are convinced that the new highly effective homogeneous and supported metal complex catalysts, including enantioselective and also commercially available organomagnesium and organoaluminum reagents, that are capable under mild conditions of hydrometallating alkenes, acetylenes and conjugated dienes with high regio- and stereoselectivity to give practically important compounds will be successfully developed and realized in the very near future.

References

1. Ziegler, K., Gellert, H., Martin, H., Nagel, K. and Schneider, J. (1954) *Liebigs Annalen*, **589**, 91.
2. Ziegler, K. and Gellert, H. (1950) *Liebigs Annalen*, **567**, 184.
3. British Patent 763824 (1956) C.A. 1958, 52, 1203a.
4. Ziegler, K. (1952) *Angewandte Chemie*, **64**, 323.
5. Ziegler, K. (1952) *Brennstoff-Chemie*, **33**, 193.
6. Finholt, A.E., Bond, A.C. and Schlesinger, H.J. (1947) *Journal of the American Chemical Society*, **69**, 1199.
7. Braun, V.G. (1953) *Organitcheskie Reaktsii*, Inostrannaya literatura, Moskva, Sb.6 [Braun, V.G. (1953) *Organic Reactions*, Vol. 6, New York].
8. German Patent 917006 (1954).
9. USSR Patent 112349 (1958) Bull. Izobr. 1958, 3, 33.
10. US Patent 2794819 (1957) C.A. 1957, 51, 12961b.
11. German Patent 956956 (1957) C.A. 1959, 53, 14004d.
12. British Patent 757925 (1957)
13. US Patent 2872470 US (1959) C.A. 1959, 53, 7014g.
14. British Patent 789236 (1957) C.A. 1958, 52, 11893e.
15. Haiosh, A. (1971) *Kompleksnye Gidridy v Organitcheskoi Khimii*, Khimiya, Moskow [Hajós, A. (1966) *Komplexe Hydride und ihre Anwendung in der Organischen Chemie*, Veb Deutscher Verlag der Wissenschaften, Berlin].
16. Benn, H., Brandt, J. and Wilke, G. (1974) *Liebigs Annalen*, **2**, 189.
17. Ziegler, K., Kroll, W.-R., Larbig, W. and Stendel, O.-W. (1960) *Liebigs Annalen*, **629**, 53.
18. British Patent 926699 (1963) C.A. 1963, 59, 4059e.
19. Greker, G. and Egle, G. (1963) *Macromolecular Chemistry*, **64**, 68.
20. US Patent 2786860 (1957) C.A. 1957, 51, 12130a.
21. Ziegler, K., Martin, H. and Krupp, F. (1960) *Liebigs Annalen*, **629**, 14.
22. Netherlands Patent 6612261 (1967) C.A. 1967, 67, 64515n.
23. French Patent 1482790 (1967) C.A. 1968, 68, 49732h.
24. US Patent 3389161 (1968) C.A. 1968, 69, 67523r.
25. Asinger, F., Fell, B. and Theissen, P. (1967) *Chemische Berichte*, **100**, 937.
26. British Patent 1132377 (1968) C.A. 1969, 70, 19538y.
27. German patent 1270558 (1968) C.A. 1969, 70, 29045z.
28. French Patent 1505234 (1967) C.A. 1969, 70, 29044y.
29. Belgian Patent 630673 (1963) C.A. 1964, 61, 1757b.
30. Belgian Patent 660205 (1965) C.A. 1966, 64, 4936a.
31. Ziegler, K., Gellert, H.-G., Lehmkuhl, H., Pfohl, W. and Zosel, K. (1960) *Liebigs Annalen*, **629**, 1.
32. Ziegler, K., Gellert, H.-G., Zosel, K., Holzkamp, E., Schneider, J., Soll, M. and Kroll, W.-R. (1960) *Liebigs Annalen*, **629**, 121.
33. Zakharkin, L.I. and Gavrilenko, V.V. (1959) *Bulletin of the Academy of Sciences of the USSR, Division of Chemical Science*, **8**, 1460.
34. Natta, G., Pino, P., Mazzanti, G., Longi, P. and Bernardini, F. (1959) *Journal of the American Chemical Society*, **81**, 2561.
35. Eisch, J. and Husk, R. (1974) *Journal of Organometallic Chemistry*, **64**, 41.
36. Ziegler, K., and Lehmkuhl, H. (1970) *Methoden der Organischen Chemie*, **XIII/4**, 152.
37. Ziegler, K. (1956) *Angewandte Chemie*, **68**, 721.
38. Reinäcker, R. and Gothel, G. (1967) *Angewandte Chemie*, **79**, 862.
39. USSR Patent 237891 (1969) Bull. Izobr. 1969, 9, 25.
40. Reinäcker, R. and Ohloff, G. (1961) *Angewandte Chemie*, **73**, 240.
41. Schimpf, R. and Heimbach, P. (1970) *Chemische Berichte*, **103**, 2122.
42. Dolzine, T. and Oliver, J. (1974) *Journal of Organometallic Chemistry*, **78**, 165.
43. Tolstikov, G.A., Dzhemilev, U.M. and Shavanov, S.S. (1975) *Bulletin of the Academy of Sciences of the USSR, Division of Chemical Science*, **24**, 1754.

44 US Patent 2826598 (1958) C.A. 1958, 52, 7352i.
45 Zakharkin, L.I., Savina, L.A. and Antipin, L.M. (1962) *Bulletin of the Academy of Sciences of the USSR, Division of Chemical Science*, **11**, 931.
46 USSR Patent 406858 (1972) Bull. Izobr. 1973, 46, 73.
47 Hoberg, H., Martin, H., Reinäcker, R., Lozel, K. and Ziegler, K. (1969) *Brenstoff-Chemie*, **50**, 217.
48 Eisch, J. and Husk, R. (1965) *Journal of Organometallic Chemistry*, **4**, 415.
49 Montury, M. and Gore, J. (1980) *Tetrahedron Letters*, **21**, 51.
50 Kroll, W. and Hudson, B. (1971) *Journal of Organometallic Chemistry*, **28**, 205.
51 Gorobets, E.V., Kutchin, A.V. and Tolstikov, G.A. (1999) *Mendeleev Communications*, **6**, 252.
52 Kuchin, A.V., Nuruschev, R.A., Spivak, A.Yu. and Tolstikov, G.A. (1986) *Zhurnal Obshchei Khimii*, **56**, 2306.
53 Yur'ev, V.P., Salimgareeva, I.M., Kuchin, A.V., Tolstikov, G.A. and Rafikov, S.R. (1972) *Doklady AN SSSR*, **203**, 1097.
54 Yur'ev, V.P., Kuchin, A.V., Yakovleva, T.O., Romanova, T.Yu. and Tolstikov, G.A. (1978) *Zhurnal Obshchei Khimii*, **48**, 149.
55 Yur'ev, V.P., Kuchin, A.V., Yakovleva, T.O., Ivanova, T.Yu. and Tolstikov, G.A. (1976) *Zhurnal Obshchei Khimii*, **46**, 2559.
56 Kuchin, A.V., Akhmetov, L.I., Yur'ev, V.P. and Tolstikov, G.A. (1979) *Zhurnal Obshchei Khimii*, **49**, 401.
57 Kuchin, A.V., Akhmetov, L.I., Yur'ev, V.P. and Tolstikov, G.A. (1981) *Zhurnal Obshchei Khimii*, **51**, 1955.
58 Yur'ev, V.P., Gailyunas, G.A., Yusupova, F.G., Tolstikov, G.A. and Aminev, I.Kh. (1976) *Bulletin of the Academy of Sciences of the USSR, Division of Chemical Science*, **25**, 844.
59 Ziegler, K. (1955) *Experientia Supplementum*, **11**, 274.
60 Ziegler, K., Gellert, H., Lehmkuhl, H., Pfohl, W. and Zosel, K. (1960) *Liebigs Annalen*, **629**, 1.
61 Wilke, G. and Muller, H. (1960) *Liebigs Annalen*, **629**, 222.
62 Gavrilenko, V.V., Palei, B.A. and Zakharkin, L.I. (1968) *Bulletin of the Academy of Sciences of the USSR, Division of Chemical Science*, **17**, 872.
63 Surtess, J. (1965) *Australian Journal of Chemistry*, **18**, 14.
64 British Patent 831328 Brit. (1960) C.A. 1960, 54, 24560c.
65 Eisch, J. and Foxton, M. (1968) *Journal of Organometallic Chemistry*, **11**, P7.
66 Magoon, E. and Slaugh, L. (1967) *Tetrahedron*, **23**, 4509.
67 Eisch, J. and Foxton, M. (1971) *Journal of Organic Chemistry*, **36**, 3520.
68 Togni, A. and Grutzmacher, H. (2001) *Catalytic Heterofunctionalization*, Wiley-VCH Verlag GmbH, Weinheim.
69 Vostrikova, U.M., Dzhemilev, O.S. and Tolstikov, G.A. (1986) *Journal of Organometallic Chemistry*, **304**, 17.
70 Vostrikova, U.M., Dzhemilev, O.S. and Tolstikov, G.A. (1990) *Russian Chemical Reviews*, **59**, 1157.
71 Sato, F., Sato, S. and Sato, M. (1976) *Journal of Organometallic Chemistry*, **122**, C25.
72 Sato, F., Sato, S. and Sato, M. (1977) *Journal of Organometallic Chemistry*, **131**, C26.
73 Lee, H.S. and Lee, G.Y. (2005) *Journal of the Korean Chemical Society*, **49**, 321.
74 Lee, H.S. and Kim, C.E. (2003) *Journal of the Korean Chemical Society*, **47**, 297.
75 Isagawa, K., Tatsumi, K., Kosugi, H. and Otsuji, Y. (1977) *Chemistry Letters*, **6**, 1017.
76 Sato, F., Sato, S., Kodama, H. and Sato, M. (1977) *Journal of Organometallic Chemistry*, **142**, 71.
77 Japanese Patent 52 16102 (1978).
78 Ashby, E.C. and Lin, J.J. (1978) *Journal of Organic Chemistry*, **43**, 2567.
79 Sato, F. (1977) *Fundamental Research in Homogeneous Catalysis* (eds Y. Ishii and M. Tsutsui), Plenum Press, New York, p. 81.
80 Sato, F., Ishikawa, H., Ishikawa, H., Takahashi, Y., Miura, M. and Sato, M. (1979) *Tetrahedron Letters*, **20**, 3745.
81 Le Marechal, J.F., Epheritikhine, M. and Folcher, G. (1986) *Journal of Organometallic Chemistry*, **309**, C1.

82 Folcher, G., Le Marechal, J.F. and Marquet-Ellis, H. (1982) *Journal of the Chemical Society, Chemical Communications*, 323.
83 Sato, F., Tomuro, Y., Ishikawa, H. and Sato, M. (1980) *Chemistry Letters*, **9**, 99.
84 Tsuji, J. and Mandai, T. (1977) *Chemistry Letters*, **6**, 975.
85 Tsuji, J., Yamakawa, T. and Mandai, T. (1978) *Tetrahedron Letters*, **19**, 565.
86 Sato, F., Kodama, H. and Sato, M. (1978) *Journal of Organometallic Chemistry*, **157**, C30.
87 Sato, F., Oguro, K. and Sato, M. (1978) *Chemistry Letters*, **7**, 805.
88 Sato, F., Kodama, H. and Sato, M. (1978) *Chemistry Letters*, **7**, 789.
89 Sato, F., Kodama, H., Tomuro, Y. and Sato, M. (1979) *Chemistry Letters*, **8**, 623.
90 Sato, F., Oikawa, T. and Sato, M. (1979) *Chemistry Letters*, **8**, 167.
91 Sato, F., Mori, Y. and Sato, M. (1978) *Chemistry Letters*, **7**, 1337.
92 Sato, F., Mori, Y. and Sato, M. (1978) *Chemistry Letters*, **7**, 833.
93 Sato, F., Mori, Y. and Sato, M. (1979) *Tetrahedron Letters*, **20**, 1405.
94 Sato, F., Haga, S. and Sato, M. (1978) *Chemistry Letters*, **7**, 999.
95 Asinger, F., Fell, B. and Theissen, F. (1967) *Chemische Berichte*, **100**, 937.
96 Fell, B., Warwel, S. and Asinger, F. (1970) *Chemische Berichte*, **103**, 855.
97 Dzhemilev, U.M., Vostrikova, O.S. and Tolstikov, G.A. (1986) *Journal of Organometallic Chemistry*, **304**, 17.
98 Dzhemilev, U.M., Vostrikova, O.S., Ibragimov, A.G. and Tolstikov, G.A. (1980) *Izvestiy a Akademii Nauk SSSR, Seriya Khimicheskaya*, 2134.
99 Schwartz, J. (1980) *Pure and Applied Chemistry Chimie Pure et Appliquee*, **52**, 733.
100 Tolstikov, G.A., Dzhemilev, U.M., Vostrikova, O.S. and Tolstikov, A.G. (1982) *Bulletin of the Academy of Sciences of the USSR, Division of Chemical Science*, **31**, 596.
101 Ibragimov, A.G., Yusupov, Z.A., Akhmetov, M.F., Zagrebel'naya, I.V., Khalilov, L.M. and Dzhemilev, U.M. (1997) in The IWFAC' 97 "Fullerenes and Atomic Clusters" (Abstracts of Reports), St Petersburg, p. 121.
102 Vasil'ev, Yu.V., Tuktarov, R.F., Chegodaeva, M.F., Ibragimov, A.G., Khalilov, L.M. and Dzhemilev, U.M. (1997) ••, in The IWFAC' 97 "Fullerenes and Atomic Clusters" (Abstracts of Reports), St Petersburg, p. 216.
103 Lautens, M., Chiu, P., Ma, S. and Rovis, T. (1995) *Journal of the American Chemical Society*, **117**, 532.
104 Lautens, M. and Rovis, T. (1998) *Tetrahedron*, **54**, 1107.
105 Lautens, M. and Rovis, T. (1997) *Journal of the American Chemical Society*, **119**, 11090.
106 Lautens, M., Ma, S. and Chiu, P. (1997) *Journal of the American Chemical Society*, **119**, 6478.
107 Ashby, E.C. and Noding, S.R. (1977) *Tetrahedron Letters*, **18**, 4579.
108 Ashby, E.C. and Noding, S.R. (1979) *Journal of Organometallic Chemistry*, **177**, 117.
109 Maruoka, K., Sano, H., Shinoda, K., Nakai, S. and Yamamoto, H. (1986) *Journal of the American Chemical Society*, **108**, 6036.
110 Maruoka, K., Sano, H., Shinoda, K. and Yamamoto, H. (1987) *Chemistry Letters*, **16**, 73.
111 Maruoka, K., Shinoda, K. and Yamamoto, H. (1988) *Synthetic Communications*, **18**, 1029.
112 Maruoka, K. and Yamamoto, H. (1988) *Tetrahedron*, **44**, 5001.
113 Ibragimov, A.G., Zagrebel'naya, L.O., Khafizova, I.V., Parfenova, L.V., Sultanov, R.M., Khalilov, L.M. and Dzhemilev, U.M. (2001) *Russian Chemical Bulletin*, **50**, 292.
114 Asinger, F., Fell, B. and Osberghaus, R. (1971) *Chemische Berichte*, **104**, 1332.
115 Asinger, F., Fell, B. and Janssen, R. (1964) *Chemische Berichte*, **97**, 2515.
116 Negishi, E. and Yoshida, T. (1980) *Tetrahedron Letters*, **21**, 1501.
117 Wailes, P.C., Weigold, H. and Bell, A.P. (1972) *Journal of Organometallic Chemistry*, **43**, C32.
118 Weiles, P.C. and Weigold, H. (1970) *Journal of Organometallic Chemistry*, **24**, 405.

119 Barber, J.J., Willis, C. and Whitesides, G.M. (1979) *Journal of Organic Chemistry*, **44**, 3603.
120 Gagneur, S., Mokabe, H. and Negishi, E. (2001) *Tetrahedron Letters*, **42**, 785.
121 Giacomelli, G., Bertero, L. and Lardicci, L. (1981) *Tetrahedron Letters*, **22**, 883.
122 Fischer, K., Jones, K., Mollbach, P. and Wilke, G. (1984) *Zeitschrift fuer Naturforschung*, **39**, 1011.
123 Ibragimov, A.G., Zagrebel'naya, I.V., Satenov, K.G., Khalilov, L.M. and Dzhemilev, U.M. (1998) *Russian Chemical Bulletin*, **47**, 691.
124 Dzhemilev, U.M., Vostrikova, O.S., Ibragimov, A.G., Tolstikov, G.A. and Zelenova, L.M. (1981) *Bulletin of the Academy of Sciences of the USSR, Division of Chemical Science*, **30**, 281.
125 Dzhemilev, U.M., Ibragimov, A.G., Vostrikova, O.S., Vasil'eva, E.V. and Tolstikov, G.A. (1987) *Bulletin of the Academy of Sciences of the USSR, Division of Chemical Science*, **36**, 1004.
126 Ibragimov, A.G., Minsker, D.L., Berg, A.A., Schitikova, O.V., Lomakina, S.I. and Dzhemilev, U.M. (1992) *Bulletin of the Academy of Sciences of the USSR, Division of Chemical Science*, **41**, 2217.
127 Gavrilenko, V.V., Chekulaeva, L.A. and Zakharkin, L.I. (1984) *Bulletin of the Academy of Sciences of the USSR, Division of Chemical Science*, **33**, 1075.
128 Kuchin, A.V., Nuruschev, R.A., Umanskya, L.I. and Tolstikov, G.A. (1987) *Zhurnal Obshchei Khimii*, **57**, 1334.
129 Ziegler, K., Gellert, H.-G., Holzkamp, E., Wilke, G., Duck, E.W. and Kroll, W.-R. (1960) *Justus Liebigs Annalen der Chemie*, **629**, 172.
130 Eisch, J.J. and Foxton, M.W. (1968) *Journal of Organometallic Chemistry*, **12**, P33.
131 Lardicci, L., Giacornelli, G.P., Salvadori, P. and Pino, P. (1971) *Journal of the American Chemical Society*, **93**, 5794.
132 Fischer, K., Jonas, K., Misbach, P., Stabba, R. and Wilke, G. (1973) *Angewandte Chemie (International Ed in English)*, **12**, 943.
133 Fischer, K., Jonas, K., Mollbach, A. and Wilke, G. (1984) *Zeitschrift fuer Naturforschung*, **39**, 1011.
134 Eisch, J.J., Sexsmith, S.R. and Fichter, K.C. (1990) *Journal of Organometallic Chemistry*, **382**, 273.
135 Eisch, J.J., Ma, X., Singh, M. and Wilke, G. (1997) *Journal of Organometallic Chemistry*, **527**, 301.
136 Parfenova, L.V., Pechatkina, S.V., Khalilov, L.M. and Dzhemilev, U.M. (2005) *Russian Chemical Bulletin*, **54**, 316.
137 Shigeru, N., Tsutomu, U., Wataru, J. and Yoshihar, D. (1985) *Journal of the Chemical Society of Japan, Chemistry and Industrial Chemistry*, 2246.
138 Zaharkin, L.I., Vinnikova, M.I. and Gavrilenko, V.V. (1988) *Organometallic Chemical Reviews*, **1**, 101.
139 Ashby, E.C. and Noding, S.A. (1979) *Journal of Organic Chemistry*, **44**, 4364.
140 Dozzi, G., Cucinella, S. and Mazzei, A. (1979) *Journal of Organometallic Chemistry*, **164**, 1.
141 Nagahara, S., Maruoka, K. and Yamamoto, H. (1993) *Bulletin of the Chemical Society of Japan*, **66**, 3783.
142 Nagahara, S., Maruoka, K., Doi, Y. and Yamamoto, H. (1990) *Chemistry Letters*, **19**, 1595.
143 Tolstikov, G.A. and Yur'ev, V.P. (1979) *Aluminiiorganitcheski Sintez (Organoaluminium Synthesis)*, Nauka, Moskow.
144 Zweifel, G. and Arzoumanian, H. (1967) *Journal of the American Chemical Society*, **89**, 5085.
145 Eisch, J.J. and Amtmann, R. (1972) *Journal of Organic Chemistry*, **37**, 3410.
146 Eisch, J.J. and Damasevitz, G.A. (1976) *Journal of Organic Chemistry*, **41**, 2214.
147 Masure, D., Coutrot, P. and Normant, J.F. (1982) *Journal of Organometallic Chemistry*, **226**, C55.
148 Huang, Z. and Negishi, E. (2006) *Organic Letters*, **8**, 3675.
149 Ashby, E.C. and Noding, S.A. (1980) *Journal of Organic Chemistry*, **45**, 1035.
150 Rossi, R. and Carpita, A. (1977) *Synthesis*, **1977**, 561.
151 Parenty, A. and Campagne, J.-M. (2002) *Tetrahedron Letters*, **43**, 1231.
152 Kaskel, S., Schlichte, K. and Kratzke, T. (2004) *Journal of Molecular Catalysis A: Chemical*, **208**, 291.
153 Dzhemilev, U.M., Ibragimov, A.G., Ramazanov, I.R., Sultanov, R.M.,

Khalilov, L.M. and Muslukhov, R.R. (1996) *Russian Chemical Bulletin*, **45**, 2610.

154 Ibragimov, A.G., Ramazanov, I.R., Khalilov, L.M., Sultanov, R.M. and Dzhemilev, U.M. (1996) *Mendeleev Communications*, **6**, 231.

155 Musluhov, R.R., Khalilov, L.M., Ramazanov, I.R., Sharipova, A.Z., Ibragimov, A.G. and Dzhemilev, U.M. (1997) *Russian Chemical Bulletin*, **46**, 2082.

156 Negishi, E., Kondakov, D.Y., Choueiry, D., Kasai, K. and Takahashi, T. (1996) *Journal of the American Chemical Society*, **118**, 9577.

157 Makabe, H. and Negishi, E. (1999) *European Journal of Organic Chemistry*, **1999**, 969.

158 Caporusso, A.M., Giacomelli, G. and Lardicci, L. (1979) *Journal of the Chemical Society, Perkin Transactions I*, 3139.

159 Giacomelli, G., Caporusso, A.M. and Lardicci, L. (1979) *Journal of Organic Chemistry*, **44**, 231.

160 Langille, N.F. and Jamison, T.F. (2006) *Organic Letters*, **8**, 3761.

161 Zweifel, G. and Miller, R.L. (1970) *Journal of the American Chemical Society*, **92**, 6678.

162 Miller, J.A., Leong, W. and Zweifel, G. (1988) *Journal of Organic Chemistry*, **53**, 1839.

163 Zweifel, G. and Lewis, W. (1978) *Journal of Organic Chemistry*, **43**, 2739.

164 Eisch, J.J., Behrooz, M. and Dua, S.K. (1985) *Journal of Organometallic Chemistry*, **285**, 121.

165 Taapken, T. and Blechert, S. (1995) *Tetrahedron Letters*, **36**, 6659.

166 Zweifel, G., Clark, G.M. and Whitney, C.C. (1971) *Journal of the American Chemical Society*, **93**, 1305.

167 Andreeva, N.N., Kuchin, A.V., and Tolstikov, G.A. (1985) *Zhurnal Obshchei Khimii*, **55**, 1316.

168 Petry, N., Parenty, A. and Campagne, J.-M. (2004) *Tetrahedron: Asymmetry*, **15**, 1199.

169 Ioffe, S.T. and Nesmeyanov, A.N. (1967) *Methods of Elemento-Organic Chemistry. The Organic Compounds of Magnesium, Beryllium, Calcium, Strontium and Barium*, Vol. 2, North-Holland, Amsterdam.

170 Wilkinson, G., Stone, F.G.A. and Abel, E.W. (eds) (1982) *Comprehensive Organometallic Chemistry. The Synthesis, Reactions and Structures of Organometallic Compounds*, Vol. 1, Pergamon Press, Oxford, p. 155.

171 Sato, F. (1985) *Journal of Organometallic Chemistry*, **285**, 53.

172 Dzhemilev, U.M., Vostrikova, O.S. and Tolstikov, G.A. (1986) *Journal of Organometallic Chemistry*, **304**, 17.

173 Dzhemilev, U.M., Ibragimov, A.G. and Tolstikov, G.A. (1991) *Journal of Organometallic Chemistry*, **406**, 1.

174 Dzhemilev, U.M. (1995) *Tetrahedron*, **51**, 4333.

175 Hoveyda, A.N. and Heron, N.V. (1999) *Comprehensive Asymmetric Catalysis* (eds E.N. Yacobson, A. Pfaltz and H. Yamamoto), Springer Verlag, Berlin, p. 431.

176 Finkbeiner, H.L. and Cooper, G.D. (1961) *Journal of Organic Chemistry*, **26**, 4779.

177 Cooper, G.D. and Finkbeiner, H.L. (1962) *Journal of Organic Chemistry*, **27**, 1493.

178 Finkbeiner, H.L. and Cooper, G.D. (1962) *Journal of Organic Chemistry*, **27**, 3395.

179 Farady, L., Bencze, L. and Marco, L. (1967) *Journal of Organometallic Chemistry*, **10**, 505.

180 Farady, L. and Marco, L. (1971) *Journal of Organometallic Chemistry*, **28**, 159.

181 Felkin, H. and Swierczewski, G. (1975) *Tetrahedron*, **31**, 2735.

182 Ashby, E.C. and Ainslie, R.D. (1983) *Journal of Organometallic Chemistry*, **250**, 1.

183 Amin, Sk.R. and Crowe, W.E. (1997) *Tetrahedron Letters*, **38**, 7487.

184 Qian, Y., Li, G. and Huang, Y.-Z. (1989) *Journal of Molecular Catalysis*, **54**, L19.

185 Ashby, E.C. and Smith, T. (1978) *Journal of the Chemical Society, Chemical Communications*, **7**, 30b.

186 Bogdanovic, B., Schlichte, K. and Westeppe, U. (1985) Proceedings of the XII FECHEM Conference on Organometallic Chemistry. Abstracts of Reports, Vienna, p. 37.

187 Bogdanovic, B. and Schwickardi, M. (1984) *Zeitschrift fuer Naturforschung*, **39**, 1001.

188 Gubaidullin, L.Yu., Sultanov, R.M. and Dzhemilev, U.M. (1982) *Bulletin of the Academy of Sciences of the USSR, Division of Chemical Science*, **31**, 638.

189 Bogdanovic, B., Schwickardi, M. and Sikorsky, P. (1982) *Angewandte Chemie (International Ed in English)*, **21**, 457.

190 Bogdanovic, B. (1985) *Angewandte Chemie*, **97**, 253.

191 Bogdanovic, B., Bons, P., Konstantinovic, S., Schwickardi, M. and Westeppe, U. (1993) *Chemische Berichte*, **126**, 1371.

192 Gruter, G.M., Klink, G.P.M., Akkerman, O.S. and Bickelhaupt, F. (1995) *Chemical Reviews*, **95**, 2405.

193 Horeau, A., Menager, L. and Kagan, H. (1971) *Bulletin de la Societe Chimique de France*, **10**, 3571.

194 Eisch, J.J. and Galle, J.E. (1978) *Journal of Organometallic Chemistry*, **160**, C8.

195 Bogdanovic, B. and Maruthamuthu, M. (1984) *Journal of Organometallic Chemistry*, **272**, 115.

196 Bogdanovic, B., Liao, S.T., Schwickardi, M., Sikorsky, P. and Spliethoff, B. (1980) *Angewandte Chemie*, **92**, 845.

197 Sato, F., Ishikawa, H. and Sato, M. (1980) *Tetrahedron Letters*, **21**, 365.

198 Sato, F., Urabe, H. and Okamoto, S. (2000) *Chemical Reviews*, **100**, 2835.

199 Felkin, H., Kwart, L.D., Swierczewski, G.S. and Umpleby, J.D. (1975) *Journal of the Chemical Society Chemical Communications*, **7**, 242.

200 Viktorov, N.B., Zubritskii, L.M. and Petrov, A.A. (1993) *Zhurnal Obshchei Khimii*, **63**, 1601.

201 Viktorov, N.B. and Zubritskii, L.M. (1995) *Russian Journal of General Chemistry*, **65**, 267.

202 Victorov, N.B. and Zubritskii, L.M. (2001) *Russian Journal of General Chemistry*, **71**, 1773.

203 Victorov, N.B. and Zubritskii, L.M. (1996) *Russian Journal of General Chemistry*, **66**, 851.

204 Viktorov, N.B. and Zubritskii, L.M. (1997) *Russian Journal of Organic Chemistry*, **33**, 1706.

205 Viktorov, N.B. and Zubritskii, L.M. (1997) *Russian Journal of Organic Chemistry*, **33**, 305.

206 Viktorov, N.B. and Zubritskii, L.M. (1999) *Russian Journal of Organic Chemistry*, **35**, 1755.

207 Sato, F., Kusakabe, M. and Kobayashi, Y. (1984) *Journal of the Chemical Society, Chemical Communications*, **7**, 1130.

208 Sato, F., Takeda, J., Uchiyama, H. and Kobayashi, Y. (1984) *Journal of the Chemical Society Chemical Communications*, **7**, 1132.

209 Urabe, H., Yoshikawa, K. and Sato, F. (1995) *Tetrahedron Letters*, **36**, 5595.

210 Ibragimov, A.G., Gribanova, E.V., Khalilov, L.M., Zelenova, L.N. and Dzhemilev, U.M. (1985) *Zhurnal Organicheskoi Khimii*, **21**, 259.

211 Dzhemilev, U.M., Gribanova, E.V. and Ibragimov, A.G. (1987) *Bulletin of the Academy of Sciences of the USSR, Division of Chemical Science*, **36**, 369.

212 Ashby, E.C., Lin, J.J. and Goel, A.B. (1978) *Journal of Organic Chemistry*, **43**, 757.

213 Sato, F., Ishikawa, H. and Sato, M. (1981) *Tetrahedron Letters*, **22**, 85.

214 Hirao, T., Yamada, N., Oshiro, Y. and Agawa, T. (1982) *Chemistry Letters*, **11**, 1997.

215 Sato, F., Watanabe, H., Tanaka, Y., Yamaji, T. and Sato, M. (1983) *Tetrahedron Letters*, **24**, 1041.

216 Yamamoto, K., Kimura, T. and Tomo, Y. (1985) *Tetrahedron Letters*, **26**, 4505.

217 Takeda, Y., Matsumoto, T. and Sato, F. (1986) *Journal of Organic Chemistry*, **51**, 4728.

218 Isobe, M., Obeyama, J., Funabashi, Y. and Goto, T. (1988) *Tetrahedron Letters*, **29**, 4773.

219 Djadchenko, M.A., Pivnitsky, K.K., Spanig, J. and Schick, H. (1991) *Journal of Organometallic Chemistry*, **401**, 1.

220 Gao, Y. and Sato, F. (1995) *Journal of the Chemical Society Chemical Communications*, **6**, 659.

221 Sato, F., Watanabe, H., Tanaka, Y. and Sato, M. (1982) *Journal of the Chemical Society Chemical Communications*, **19**, 1126.

222 Dzhemilev, U.M., Vostrikova, O.S., Sultanov, R.M. and Gimaeva, A.R. (1988) *Bulletin of the Academy of Sciences of the*

USSR, Division of Chemical Science, **37**, 1936.
223 Sato, F. (2002) *Journal of Synthetic Organic Chemistry, Japan*, **60**, 891.
224 Zhao, H. and Cai, M. (2002) *Journal of Chemical Research S*, **12**, 608.
225 Sato, F., Ishikawa, H., Watanabe, H., Miyake, T. and Sato, M. (1981) *Journal of the Chemical Society Chemical Communications*, **14**, 718.
226 Cai, M., Peng, C., Zhao, H. and Hao, W. (2003) *Journal of Chemical Research S*, **5**, 296.
227 Cai, M., Zhou, Z. and Jiang, J. (2006) *European Journal of Organic Chemistry*, **2006**, 1400.
228 Zhao, H. and Cai, M. (2003) *Synthetic Communications*, **33**, 1643.
229 Sato, F. and Katsuno, H. (1983) *Tetrahedron Letters*, **24**, 1809.
230 Sato, F., Kanbara, H. and Tonaka, Y. (1984) *Tetrahedron Letters*, **25**, 5063.
231 Jefford, C.W. and Moulin, M.-C. (1991) *Helvetica Chimica Acta*, **74**, 336.
232 Tani, K., Sato, Y., Okamoto, S. and Sato, F. (1993) *Tetrahedron Letters*, **34**, 4975.
233 Kocienski, P., Love, C. and Whitby, R. (1988) *Tetrahedron Letters*, **29**, 2867.
234 Kocienski, P., Love, C., Whitby, R., Costello, G. and Roberts, D.A. (1989) *Tetrahedron*, **45**, 3839.
235 Ibragimov, U.M., Dzhemilev, A.G., Saraev, R.A. and Muslukhov, R.R. (1988) *Bulletin of the Academy of Sciences of the USSR, Division of Chemical Science*, **37**, 2150.
236 Cai, M., Hao, W., Zhao, H. and Song, C. (2003) *Journal of Organometallic Chemistry*, **679**, 14.
237 Cai, M., Hao, W., Zhao, H. and Xia, J. (2004) *Journal of Organometallic Chemistry*, **689**, 1714.
238 Cai, M., Hao, W., Zhao, H. and Xia, J. (2004) *Journal of Organometallic Chemistry*, **689**, 3593.
239 Cai, M., Xia, J. and Chen, G. (2004) *Journal of Organometallic Chemistry*, **689**, 2531.
240 Dzhemilev, U.M. and Ibragimov, A.G. (2005) *Russian Chemical Reviews*, **74**, 807.

Index

a
α-(acetamido) cinnamic acids 4, 5, 11
acetone drying 219–20
acetophenones 124–5, 128–30, 144, 167–8, 222, 351
acetylenes *see* alkynes
(±)-acorenone B 427
α-acyl imino esters 247
β-(acylamino) acrylates 16–18
alcohol dehydrogenase 211, 212
alcohols
  – C=C bond hydrogenation 25–8, 45
  – deoxygenation 437
  – diol synthesis 89–90
  – as hydrogen sources 212
aldehydes
  – DKR 229
  – enzyme-catalysed reductions 225, 229
  – organocatalysis
    – C=C bond reduction 344–5, 348–9
    – reductive amination 357–9
  – transfer hydrogenation 151–2, 344–5
  – *see also* reductive aldol reaction
aldimines 301, 312, 331, 335
aldiminoboranes 306
aldols *see* reductive aldol reaction
alkenes
  – dissolved metal reduction 426–7
  – hydroalumination
    – catalytic 456–66
    – thermal 448–52
  – hydroboration 65–7, 84
    – P,N ligands 74–82
    – P,P ligands 69–74, 82–3
    – regioselectivity 67–9
  – hydrogenation 3–4, 39
    – β-(acylamino) acrylates 16–18
    – alcohols 25–8
    – carboxylic acids 18–22
    – dehydroamino acids 4–13
    – enamides 13–16
    – Ir catalysis 13, 25–6, 28, 39–53, 62
    – ketones and lactones 23–5
    – metallocene catalysis 57–61
    – pharmaceutical syntheses 28–32
    – Rh catalysis 4–10, 13–18, 20, 21, 25–6, 53–5
    – Ru catalysis 10–13, 18–28, 54, 55–7
    – unfunctionalized alkenes 39–62
  – hydromagnesiation 472–5
  – hydrosilylation 87
    – aromatic compounds 87–93, 94
    – condensation reactions 102–3
    – Cu catalysts 99–102
    – 1,3 dienes 95–8
    – enynes 98–9
    – non-aromatic alkenes 93–4
    – Pd catalysts 87–93
    – α,β-unsaturated systems 99–103
  – organocatalysis 344–8
alkenols 469
alkoxides
  – in hydrosilylation 201–2
  – in transfer hydrogenation 135–6, 139
alkyl ketones
  – aryl alkyl
    – hydrogenation 124–7, 128–32
    – transfer hydrogenation 148–9
  – dialkyl 127, 149
  – heteroaryl alkyl 128, 130
alkyl-substituted alkenes 47–8

Modern Reduction Methods. Edited by Pher G. Andersson and Ian J. Munslow.
Copyright © 2008 WILEY-VCH Verlag GmbH & Co. KGaA, Weinheim
ISBN: 978-3-527-31862-9

alkylidene ketones/lactones 23–4
alkynes 363
  – dissolved metal reduction 430–1
  – hydroalumination
    – catalytic 468–71
    – thermal 456, 467
  – hydroboration 366–8
    – catalysts 368–71
    – mechanisms 371–3
  – hydrogenation 363–4
  – hydromagnesiation 479–82
  – hydrosilylation 89–90, 98–9, 373–4
    – catalysts 374–82
  – internal bonds 366, 371, 378–82
  – semi-hydrogenation 364–6
  – terminal bonds 366, 369–70, 371–3, 374–8, 431, 468, 471
allylic ethers/acetals 436–7
alpine-borane 163–4
aluminum (Al)
  – hydroalumination see hydroalumination
  – in transfer hydrogenation 135–6, 139, 280–1
Amberlite XAD resins 222–3
amine synthesis see imines
α-amino acids 3–13, 438
β-amino acids 16–18
β-amino alcohol ligands 145
amino alcohol-borane reagents 164–5
aminoalanes 470
aminophosphine-oxazoline ligands 243–4
ammonia 420
anions
  – in Ir catalysts 40, 41–4
  – in Rh catalysts 55
  – for transfer hydrogenation 144–5
anti-Markovnikov product of hydroboration 66, 68
arabinose 423
arborescidines 288–9
aryl iodides 406–7
aryl ketones
  – aryl alkyl
    – hydrogenation 124–7, 128–32
    – transfer hydrogenation 148–9
  – aryl heteroaryl 128, 129
  – diaryl 127–8
α-aryl-substituted acrylic acids 18–19
aryl-substituted alkenes 44–7
aziridines 289–90
azo compounds 425

b

(−)-bafilomycin A_1 83
baker's yeast 216–17, 218, 220, 221, 223, 224, 226–7, 229
BAr_F^- 40–1, 43, 240–1
BDPP ligand 239
benzil 151
benzimidazoline 343, 349
benzofurans 433
benzothiophenes 433
benzoylcystine derivatives 166
BINAP ligand
  – hydroboration 70–1
  – hydrogenation
    – alkenes 10, 13, 18–19, 21–2, 23–4, 55–6
    – carbonyl compounds 110, 120
  – hydrosilylation 92–3, 191
BINAPHANE ligand 239–40
BINOL ligand 280–1
biocatalysts see enzyme-catalyzed reductions
BIPHEMP ligand 10, 13
BIPNOR ligand 5
Birch reduction 431–3
bis(dialkylamino)alanes 460
bisferrocenyl ligands 96–7, 186
bisphosphine ligands
  – hydroboration 69–74, 82–3
  – hydrogenation of alkenes
    – β-(acylamino) acrylates 16–18
    – alcohols 26–8
    – carboxylic acids 18–22
    – dehydroamino acids 4–9, 10–13
    – enamides 13–15
    – ketones/lactones 23–5, 57
    – unfunctionalized 55–7
  – hydrogenation of imines 238–41
  – hydrogenation of ketones
    – activated 109–20
    – hydrogenation mechanisms 115–17, 121–4
    – unactivated 120–8, 132–3
  – hydrosilylation
    – alkenes 92–3
    – carbonyl compounds 186, 190–1
  – preparation methods 117
  – transfer hydrogenation 144
[bmim]PF_6 (1-N-butyl-3-methylimidazolium hexafluorophosphate) 224
boranes see hydroboration
borazines 306, 308
borohydrides 298

- carbonyl compounds 166–7, 171, 173
- imines 298, 307
- tosylhydrazones 304
boron-bis(oxazolinate) complexes 303–4
boron trifluoride etherate (BF$_3$.OEt$_2$) 198–9
Bouveault–Blanc reduction 423
Brintzinger's complexes 59
bromides 435
Brønsted acids 184, 197–8, 354–5, 356
butadiene 454
N-t-butanesulfinyl amines 302–3
(2,6-di-t-butyl-4-methylphenoxy)diisobutylaluminum 464

c
C—X bond reduction 434–41
camphene 450
candoxatril 21, 29–30
carbapenem 28, 120
carbenes 42, 281
carbon dioxide, supercritical
- carbonyl reductions 223–4
- imine reductions 255–7
carbonyl compounds
- dissolved metal reduction 420–2
- DKR 118–20, 227–9
- enzyme-catalyzed reductions 231
- biocatalyst characteristics 209–11, 215–18
- deracemization 230–1
- DKR 227–9
- hydrogen sources 211–15
- reaction conditions 210, 219–20
- solvents 221–4
- substrates 218–19, 225–7
- hydroboration
- diastereoselective 159–63, 173–7
- enantioselective 163–73
- hydrogenation 109
- diketones 112–14
- α-keto esters 112
- β-keto esters & their analogues 109–12, 114–20
- ketones 120–33
- hydrosilylation 183–4, 202–3
- σ-bond metathesis 184, 188–93
- Brønsted acids 184, 197–8
- high-valent oxo complexes 184, 193–7
- Lewis acids 184, 198–9
- Lewis bases 184, 200–2
- oxidative addition 183–4, 185–8
- reductive aldol reaction 387, 408–9

- catalysts 390–408
- cyclization reactions 393, 398, 399–401, 403, 404
- regioselectivity 387–9
- silanes 390–3, 398–9, 401, 403, 404, 407–8
- reductive amination 357–9
- transfer hydrogenation 135–7
- catalysts 138–9, 149–50
- donor molecules 137–8
- ligands 143–8
- mechanisms 139–43
- substrates 148–54
carboxylic acids
- aromatic ring reduction 432
- C=C bond reduction 18–22
- C=O bond reduction 423–4
catecholborane (HBcat)
- alkenes 65, 69
- carbonyl compounds 166
- imines 65, 301–2, 304, 308, 310–11
CATHy reduction system 277–8
Chalk–Harrod hydrosilylation 381
(+)-chatancin 27
chlorides 435
B-chlorodiisopinocamphenylborane ((–)-Ipc$_2$BCl) 164
chromium (Cr) catalysts 92
citronellol 26–7
Clemmensen reduction 421
cobalt (Co) catalysts
- hydroboration (Co-ketoiminato complexes) 166–7, 168, 174, 307
- hydrosilylation 380
- reductive aldol reaction 398–401
coenzymes 211–12
copper (Cu) catalysts
- alkenes 99–102, 458
- carbonyl compounds 139, 189, 190–2, 202–3, 403–6
- hydroalumination 458
- hydroboration 310–11
- hydrosilylation 99–102, 189, 190–2, 202–3, 328–9
- imines 310–11, 328–9
- reductive aldol reaction 403–6
- transfer hydrogenation 139
Crabtree's catalyst 25, 26, 39–40, 45
(R)-(+)-crispine A 289
Crixiran MK639 30–1
cyclization
- of 1,6-dienes 93–4
- reductive aldol reaction 393, 398, 399–400, 403, 404

N-(cyclobutylidene) amines 299
(1E,5Z)-cyclodecadiene 454
cyclodiborazanes 306–7
cyclohexanones 421–2
2-cyclohexylbutene 48–9
cyclopentanol 212
cyclopropanes 440

d

DDPPM ligand 240
DEAH (diethyl aluminum hydride) 449, 452
(E)-dec-3-en-1-ol ether 430
dehydroamino acids 4–13
dehydrogenases see enzyme-catalysed reductions
deoxygenation 424–5, 426, 436–8
deracemization 230–1
desulfurization 439
dialkyl ketones 127, 149
dialuminum compounds 458
diamine ligands 128–32
 – with diphosphines 120–8, 132–3
diaryl ketones 127–8
diazaborinanes 303
DIBAL (diisobutylaluminum hydride, i-Bu$_2$AlH) 449–50, 459, 460, 466, 468
cis-dibenzoquinone chlorofumarate 286
diborane 303
diboration 311–12
dichloroalane 460
dienes
 – 1,2-dienes 465, 467
 – 61,3-dienes 95–8, 467, 475–7
 – 1,6-dienes 93–4
 – α,ω-dienes 94, 454
 – dissolved metal reduction 429
 – hydroalumination 452–5, 467
 – hydrogenation 45–6
 – hydromagnesiation 475–8
 – hydrosilylation 93–8
diethyl aluminum hydride (DEAH) 449, 452
diethylzinc 400–1
dihydride mechanism 9, 54, 142–3, 273
dihydroisoquinolines 286, 291
diimines 303
diisobutylaluminum chloride (i-Bu$_2$AlCl) 464, 466, 470
diisobutylaluminum hydride (i-Bu$_2$AlH, DIBAL) 449–50, 459, 460, 466, 468
diketones
 – enzymatic reduction 228–9
 – hydroboration 173
 – hydrogenation 112–14
 – transfer hydrogenation 151

dimethyl sulfoxide (DMSO) 221
21,2-diols 89–90
DIOP ligand 70, 239
DIPAMP ligand 10, 72
diphenylphosphin(o)yl groups 321, 329
diphosphine ligands see bisphosphine ligands
diphosphinite ligands 241–2
diphosphite ligands 241–2
Diplogelasinospora grovesii reductase 215–16
direct hydrogen transfer 135–6, 139–41, 271–2
directed evolution 218
dissolving metals 419–20
 – aromatic compound (Birch) reduction 431–3
 – C=C bond reduction 426–9
 – C≡C bond reduction 430–1
 – C–X bond reduction 434–41
 – carbonyl groups 420–2
 – carboxylic acids 423–4
 – heteroaromatic compounds 433–4
 – imines 422–3
 – N–X bond reduction 425–6
 – nitro compounds 425
 – sulfoxides 424–5
diynes 481
DKR see dynamic kinetic resolution
duloxetine 309
Duphos ligand 6–7, 56–7, 247
dynamic kinetic resolution (DKR) 118
 – aldehydes 229
 – alkenes 100
 – in enzymatic reactions 227–9
 – imines 283, 290–1, 357
 – ketones 118–20, 227–9

e

electrochemical reactions 214–15
emetine 286–7
enamides 13–16
enol phosphinate esters 50–1
enolate synthesis see reductive aldol reaction
enones 101–2, 347, 388
environmentally-friendly approaches 210, 249–57, 441
enynes 98–9, 368
enzyme-catalyzed acylations 275
enzyme-catalyzed reductions 231
 – biocatalyst characteristics 209–11, 215–18
 – deracemization 230–1
 – DKR 227–9
 – hydrogen sources 211–15

– reaction conditions 210, 219–20
– solvents 221–4
– substrates 218–19, 225–7
(−)-epothilone 82–3
epoxides 437–8
Escherichia coli 217–18
esters
– unsaturated 23–5, 50–1
– *see also* Hantzsch esters; keto esters
ethyl-*S*-acetamido-8-nonenoate 29
ethyl chloroacetate 220
N-ethyl-*N*-isopropylaniline-borane 300–1
ethyl phenylpropiolate 430
eupomatilone-3 100

f

Felkin–Anh model of carbonyl hydroboration 159
ferrocene catalysts
– alkenes
– hydroboration 79–82
– hydrogenation 6
– hydrosilylation 90–2, 96–7, 98
– carbonyl hydrosilylation 186
– imine hydrosilylation 322
fluorinated olefins 51–2
fluoxetine 309
formate 213, 279–80
formic acid 137–8, 152, 274
N-formyl pyrrolidine 351
fullerenes 433, 459–60
– amino acid derivatives 308–9
furans 45, 433

g

genetic engineering 216–18
Geotrichum candidum reductase 210, 212, 219–20, 222, 224, 225–6
geranic acid 18
geraniol 26–7
glucose and its derivatives 212–13
gold (Au) catalysts, imines 249, 310–11

h

halogen bond reduction 434, 435
Halpern-Brown pathway 8–9
Hantzsch esters
– imine reductions 281–2, 354–6
– in organocatalysis 281–3, 342, 354–6
– transfer hydrogenation 138, 139, 154, 281–2
HBcat *see* catecholborane

heterocyclic molecules
– as hydrogen donors 342–3
– hydrogenation of 45
– as ligands 42, 281
hexa-1,5-dienes 458
51-hexene 93
Houk's model of carbonyl hydroboration 159
hydridic route of hydrogen transfer 141–3, 272–3, 278–9
hydroalumination
– catalytic
– alkenes 456–66
– alkynes 468–71
– dienes 467
– thermal
– alkenes 448–52
– alkynes 456, 467
– dienes 452–5
hydroamination
– hydrosilylation and 323–4, 327, 332, 335
– organocatalysis 357–9
hydroboration
– alkenes 65–7, 84
– P,N ligands 74–82
– P,P ligands 69–74, 82–3
– regioselectivity 67–9
– alkynes 366–8
– catalysts 368–71
– mechanisms 371–3
– carbonyl compounds
– diastereoselective 159–63, 173–7
– enantioselective 163–73
– diimines 303
– imines
– catalyzed 307–9, 310–11
– uncatalyzed 298–303
– nitriles
– catalyzed 309, 311
– uncatalyzed 305–7
– synthetic applications 82–3, 173–7
– tosylhydrazones 304–5
hydrogen, molecular
– in enzymatic reductions 213–14
– hydrogenation mechanisms 52, 258–9
– reductive aldol coupling 395–6
hydrogen transfer *see* transfer hydrogenation
hydromagnesiation
– alkenes 472–5
– alkynes 479–82

- dienes 475–8
- imines 474
hydrosilylation
- alkenes 87
- aromatic 87–93, 94
- condensation reactions 102–3
- Cu catalysts 99–102
- 1,3 dienes 95–8
- enynes 98–9
- non-aromatic 93–4
- Pd catalysts 87–93
- α,β-unsaturated systems 99–103
- alkynes 89–90, 373–4
- catalysts 374–82
- mechanisms 381–2
- carbonyl compounds 183–4, 202–3
- Brønsted acids 184, 197–8
- high-valent oxo complexes 184, 193–7
- Lewis acids 184, 198–9
- Lewis bases 184, 200–2
- oxidative addition 183–4, 185–8
- σ-bond metathesis 184, 188–93
- imines 321–2
- Lewis base catalysts 333–4
- metal catalysts 322–32
- organic catalysts 343, 351–4
- other catalysts 334–5, 343
- nitriles 323
- organocatalysis 343, 351–4
10-hydroxyasimicin 175
hydrozirconation 459

i

ibuprofen 18, 82
imidazolidinone 345, 347
imines 237–8, 297–8
- diboration 311–12
- dissolved metal reduction 422–3
- DKR 283, 290–1, 357
- hydroboration
- catalyzed 307–9, 310–11
- uncatalyzed 298–303
- hydrogenation
- Ir catalysts 238–47, 258–9, 262–3, 264–5
- mechanisms 257–66
- nitrogen-containing ligands 243–6, 248
- other catalysts 247–9
- recycling/reuse of catalysts 249–57
- hydromagnesiation 474

- hydrosilylation 321–2
- Lewis base catalysts 333–4
- metal catalysts 322–32
- organic catalysts 343, 351–4
- other catalysts 334–5, 343
- organocatalysis 281–3, 343, 351–6
- transfer hydrogenation 271–3, 291–2
- carbene catalysts 281
- DKR 283, 290–1
- Leuchart–Wallace reaction 279–80
- metal catalysts 273–9
- MPV reaction 271–2, 280–1
- organocatalysis 281–3
- synthetic applications 283–91
iminium ions 289, 345–7
immobilization of catalysts 252–4
indanones 125–7
indene 72, 76, 79
indium (In) catalysts
- reductive aldol reaction 407–8
- in solution
- C—X bond reduction 435
- imine reduction 422–3
- N—X bond reduction 426
- nitro reduction 425
indoles 433
iodine
- in imine hydrogenation 259
- in Ni-catalyzed reductive aldol reaction 406–7
(–)-Ipc$_2$BCl (B-chlorodiisopinocamphenylborane) 164
iridium (Ir) catalysts
- alkene hydrogenation 13, 25–6, 28, 39–53, 62
- alkyne hydrosilylation 377–8
- carbonyl hydrogenation 138, 142, 150, 151, 152
- imine hydrogenation 237, 238–47
- possible mechanisms 258–9, 262–3, 264–5
- imine hydrosilylation 322–4
- ligands 41, 238–47
- reductive aldol reaction 401
iron (Fe) catalysts
- alkenes
- hydroboration 79–82
- hydrogenation 6
- hydrosilylation 90–2, 96–7, 98
- carbonyl compounds
- hydrosilylation 186
- transfer hydrogenation 139
- imine hydrosilylation 322

isocyanides 439
isoprenoid alcohols 477
isopropanol 273
isoquinoline see tetrahydroisoquinoline
itaconic acids 20–1

j
Josiphos 6, 79, 128

k
K-glucoride 164
ketimines see imines
α-keto esters 112, 171
β-keto esters
– analogues of 114–15
– DKR 118–20, 227–8
– hydroboration 171–3
– hydrogenation 109–12, 114–20
β-keto nitriles 229, 309
ketones
– cyclic 57, 162–3
– diketones 112–14, 151, 173, 228–9
– dissolved metal reduction
– C=C bond 427–9
– C=O bond 420–2
– C–X bond 434–5
– DKR 118–20, 227–9
– enzyme-catalyzed reductions 211, 225–7, 225–31, 229
– hydrogen sources 211–15
– solvents 221–4
– stereochemical control 215–20
– halogenated 171, 434
– hydroboration
– diastereoselective 159–63, 173–7
– enantioselective 163–71
– hydrogenation 109, 120–33
– hydrosilylation 183–4, 202–3
– Brønsted acids 184, 197–8
– high-valent oxo complexes 184, 193–7
– Lewis acids 184, 198–9, 334
– Lewis bases 184, 200–2
– oxidative addition 183–4, 185–8
– σ-bond metathesis 184, 188–93
– organocatalysis
– C=C bond reduction 344–8
– condensation reactions 348–51
– reductive amination 357–9
– transfer hydrogenation 135–7
– catalysts 138–9, 149–50
– donor molecules 137–8
– ligands 143–8

– mechanisms 139–43
– synthetic applications 148–51, 152–4
– unsaturated
– C=C bond reduction 23–4, 57, 152–4, 427–9
– C=O bond reduction 132–3, 152–4, 169–71
ketoximines 323
Klebsiella pneumoniae reductase 216, 228
Kluyvromyces marxianus reductase 228
Knoevenagel condensation 348–9

l
lactones 23–4, 57, 100, 423
lanthanide catalysts 60–1, 330–1
Leuckart–Wallach reaction 274, 279–80
Lewis acids
– hydroboration 307–8
– hydrosilylation 184, 198–9, 334
Lewis bases, hydrosilylation 184, 200–2, 333–4
Lindlar's catalyst 364–5
lithium (Li)
– catalysts
– C=C bond reduction 428
– C=O bond reduction 421, 422
lithium aluminum hydride (LiAlH$_4$) 449, 451, 457, 469
lithium borohydride (LiBH$_4$) 166
lithium diisopropylamide (LDA) 387

m
magnesium see hydromagnesiation
magnesium hydride 474–5, 479
(–)-malyngolide 83
Mannich reaction 347–8, 349
Markovnikov product of hydroboration 66, 68, 70
Meerwein–Ponndorf–Verley (MPV) reduction (direct hydrogen transfer) 135–6, 139–41, 271–2, 280–1
(–)-menthone 421
MeO-BIPHEP ligand 21, 29, 110, 112
metallo-aldehyde enolates 397
metallocene catalysts
– alkenes
– hydroboration 79–82
– hydrogenation 6, 57–61
– hydrosilylation 90–2, 96–7, 98
– alkyne hydroboration 370–1, 379
– imines
– hydrogenation 248–9
– hydrosilylation 322, 324–5

– ketone hydrosilylation 186, 190
92-(4′-methoxyphenyl)-1-butene 44–5
methyl-2-benzamidomethylene-3-oxobutyrate 118–20
methyl vinyl ketone 220
methylene cyclobutane 451
methylene cyclopentane 451
(S)-metolachlor 237–8, 308
Mibefradil 19
Miniphos ligand 8
molybdenum (Mo) catalysts 195, 334
Monophos ligand 9, 245–6
monophosphine (MOP) ligands 88–9, 93, 96–7
monophosphorus ligands 9–10, 15
monoterpenes 453
montmorillonite 252
morphine and analogues 14, 286
MPV *see* Meerwein–Ponndorf–Verley (MPV) reduction
mutagenesis 218

n

N—X bond reduction 425–6, 438–9
NADH (nicotinamide–adenine dinucleotide) 211
nanoparticles 251, 365
naproxen 18–19, 82
NB-Enantrane 164
nickel (Ni) catalysts
    – hydroalumination 463, 465
    – hydrogenation 365
    – hydromagnesiation 473, 477
    – Ni(0) nanoparticles 365
    – reductive aldol reaction 406–7
nicotinamide–adenine dinucleotide (NADH) 211
nitriles
    – enzyme-catalyzed reduction 229
    – hydroboration 305–7, 309, 311
    – hydrosilylation 323
    – synthesis from alkynes 367–8
nitro compounds
    – dissolved metal reduction 425
    – hydrosilylation 101
norbornene
    – hydroalumination 464
    – hydroboration 70
    – hydrosilylation 93, 94

o

oct-1-yne 468
opioids 14, 286
Oppenauer oxidation 136, 271–2

organoboranes 366
organocatalysis
    – carbonyl compounds
        – condensation reactions 348–51
        – hydrogenation of C=C bond 344–8
        – reductive amination of C=O bond 357–9
    – hydrogen donors 342–3, 354
    – imines
        – Hantzsch esters 281–3, 354–6
        – hydrosilylation 343, 351–4
oxabicycloalkenes 460
oxazaborolidine
    – carbonyl hydroboration 165, 168, 170, 171
    – imine hydroboration 307–8
oxazoline-thioether ligands 244
α-oximino carbonyl compounds 423

p

palladium (Pd) catalysts
    – alkenes
        – hydroalumination 462
        – hydrosilylation 87–93
    – alkynes 365, 368
    – imines 247
    – reductive aldol reaction 403
paradisone 24–5
peloruside A 177
perfume industry
    – citronellol 26–7
    – paradisone 24–5
pharmaceutical syntheses
    – alkene hydroboration 82–3
    – alkene hydrogenation 28–32
    – carbonyl compound hydroboration 175–7
    – imine transfer hydrogenation 283–91
PHENAP ligand 76–7
2-phenyl-1-butene derivatives 55, 58
phenylethyne 368–9
phenylpropene 72, 79
phenylsilane 325
phosphinamine (P,N) ligands
    – alkenes 41, 74–82
    – hydroboration 74–82
    – hydrogenation 41, 243–5
    – hydrosilylation 187–8, 322–3
    – imines 243–5, 322–3
    – ketones 187–8
phosphine-imidazoline ligands 244

phosphine-oxazoline (phosphinooazoline) (PHOX) ligands 41, 243, 244–5, 323
phosphine–phosphite (P-OP) ligands 241
phosphines *see* bisphosphine ligands *and* monophosphine ligands
phosphinite–phosphite ligands 241–2
phosphoramidite ligands 90, 245–6
phosphoric acid 355
photochemical reactions 214
α-picoline-borane 301
pinacolborane 311
β-pinene 449–50
pipecolinic acid derivatives 353
platinum (Pt) catalysts
  – diboration of aldimines 312
  – hydrosilylation
    – alkenes 89, 90
    – alkynes 374, 378
polyenes 455, 478
poly(methylhydrosiloxane)/polymeric hydrosiloxane (PMHS) 100, 324
(*R*)-(−)-praziquantel 288
pregabalin 29
proline 348–9
prolinol 201, 347
42-propanol 137, 151, 212
n-propyl magnesium bromide (*n*-PrMgBr) 472, 473, 476, 477
pseudoephedrine-derived borane 69
psymberin 177
pyridine-borane 300
pyridines 434
pyrones 24, 57
pyrrole 433
pyrrolidines 322–3

q
QUINAP ligand 74–7
Quinazolinenap ligand 78–9
quinolines 356, 433

r
reductive aldol reaction 387, 408–9
  – catalysts
    – Co 398–401
    – Cu 403–6
    – In 407–8
    – Ir 401
    – Ni 406–7
    – Pd 403
    – Rh 390–8
    – Ru 401
  – cyclization reactions 393, 398, 399–401, 403, 404

  – regioselectivity 387–9
  – silanes 390–3, 398–9, 401, 403, 404, 407–8
reductive amination
  – hydrosilylation and 323–4, 327, 332, 335
  – organocatalysis 357–9
reductive deamination 439
rhenium (Re) catalysts 193–7, 330
rhodium (Rh) catalysts
  – alkenes
    – hydroboration 65–84
    – hydrogenation 4–10, 13–18, 20, 21, 25–6, 53–5
    – hydrosilylation 94
  – alkynes
    – hydroboration 368–70, 371–2
    – semi-hydrogenation 365–6
  – carbonyl compounds
    – hydrogenation 138, 142, 150
    – hydrosilylation 185–8
    – reductive aldol reaction 390–8
  – diboration 311–12
  – hydroboration 65–84, 311, 368–70, 371–2
  – hydrogenation
    – alkenes 4–10, 13–18, 20, 21, 25–6, 53–5
    – imines 247
    – ketones 138, 142, 150
  – hydrosilylation 94, 185–8
  – imines 247, 277–8, 279, 284, 311–12
  – transfer hydrogenation 277–8, 279, 284
ruthenium (Ru) catalysts
  – alkenes 10–13, 18–28, 54, 55–7
  – alkynes 374–7, 378–9
  – carbonyl compounds
    – hydrogenation 109–33
    – reductive aldol reaction 401
    – transfer hydrogenation 138, 142–3, 146–8, 149
  – hydrogenation
    – alkenes 10–13, 18–28, 54, 55–7
    – carbonyl compounds 109–33
    – imines 248, 260
  – hydrosilylation 322–3, 374–7, 378–9
  – imines
    – hydrogenation 248, 260
    – hydrosilylation 322–3
    – transfer hydrogenation 273–5, 278–9, 284
  – preparation methods 117

s

sabinene 450
Saccharomyces cerevisiae (baker's yeast) 216–17, 218, 220, 221, 223, 224, 226–7, 229
safety considerations
  – cyanide 298
  – enzymes 210
  – transfer hydrogenation 135
sertraline 82
Shvo's catalyst 54, 274, 278
silanes
  – hydrosilylation *see* hydrosilylation
  – in the reductive aldol reaction 390–3, 398–9, 401, 403, 404, 407–8
sodium (Na)
  – catalysis in solution
    – C=C bond reduction 426–9
    – C=O bond reduction 420, 422, 423
    – C≡C bond reduction 430
    – C–X bond reduction 435, 436
    – N–X bond reduction 426
sodium borohydride ($NaBH_4$) 162, 298, 304, 309
sodium cyanoborohydride ($NaNH_3CN$) 298, 304
sodium triacetoxyborohydride ($NaBH(OAc)_3$)(STAB-H) 298
solvents
  – for enzymatic reactions 221–4
  – for metals 420–1, 422, 423
  – for recycling catalysts 249–51, 255–7
  – steroid reactions 441
  – for transfer hydrogenation 137
spirophosphoramidite ligands 90, 93
(+)-spongistatin 175
steroids 161, 422, 441
stilbene 68–9, 79, 469
strained carbocycles 440
Stryker's reagent 403
styrene
  – hydroalumination 451
  – hydroboration 68, 70, 80, 81
  – hydrosilylation 87–9, 90–2
sulfides
  – desulfurization 439
  – hydromagnesiation 478
β-sulfido aldehydes 394–5
sulfinamides 353
sulfonated ligands
  – catalyst recovery and 249–50
  – imine reductions 284–5

β-*N*-sulfonylaminoalkylphosphine ligands 98
sulfoxamine derivative ligands 245
sulfoxides 424–5
sultams 283–5
Synphos 11, 110

t

tethered ligands in transfer hydrogenation 145–6
tetrahydro-β-carbolines 288–9
tetrahydroisoquinolines 275, 285–8
tetrahydroquinolines 14, 356
tetralin 432
tetralones 125–7
tetrasubstituted alkenes 17–18, 24, 39–40, 59
  – fluorinated 51–2
Thermoanaerobacter brockii dehydrogenase 212–13, 214
thiophenes 433
thiourea 359
TIBA (tri-isobutyl aluminum, *i*-Bu_3Al) 450, 462, 466, 471
tiglic acid 19, 22
tin (Sn) catalysts 331–2, 469
titanium (Ti) catalysts
  – alkenes
    – hydroalumination 457, 458, 461, 476, 478
    – hydrogenation 57–9
    – hydromagnesiation 472, 473–4
  – alkynes
    – hydroalumination 469, 470
    – hydroboration 371, 373
    – hydromagnesiation 479–80
    – hydrosilylation 378
  – carbonyl compounds
    – hydroboration 165, 168
    – hydrosilylation 188–90
  – imines
    – hydrogenation 248–9, 260, 265
    – hydrosilylation 324–7
tosyl diamine (TsDPEN) ligands 144–5, 275, 283–4
tosylhydrazones 304–5
transfer hydrogenation
  – carbonyl compounds 135–7
    – catalysts 138–9, 149–50
    – donor molecules 137–8
    – ligands 143–8
    – mechanisms 139–43
    – substrates 148–54

- imines 271–3, 291–2
    - carbene catalysts 281
    - DKR 283, 290–1
    - Leuchart–Wallace reaction 279–80
    - metal catalysts 273–9
    - MPV reaction 271–2, 280–1
    - organocatalysis 281–3
    - synthetic applications 283–91
- mechanisms
    - direct hydrogen transfer 135–6, 139–41, 271–2
    - hydridic route 141–3, 272–3, 278–9
- *see also* organocatalysis

TRAP ligands 14, 186
tributylstannane 404
trichlorosilane 333, 343, 351–4, 403
Trichosporon capitatum reductase 213
tricyclohexyltin chloride 464
tridentate ligands 187
triethyl aluminum (AlEt$_3$) 463, 470
triethylsilane 93, 407–8
trifluoroacetic acid 335
tri-isobutyl aluminum (*i*-Bu$_3$Al, TIBA) 450, 462, 466, 471
trisubstituted alkenes 28, 39–40, 51–2
trityl group 436, 438
TsDPEN (tosyl diamine) ligands 144–5, 275, 283–4
Tunaphos 112

u

uranium (U) catalysts 457

v

valine derivatives 351–3
vanadium (V) catalysts 457
vicinal diamines 439
vinyl arenes 75–6, 78
vinyl ketones 390–1
vinyl silanes 49, 373–8, 481

vinylic C–O bonds 437
vitamin E 27

w

Wilkinson's catalyst
    - hydroboration 65, 368–9
    - hydrogenation 25, 53
    - reductive aldol reaction 393
Wittig directed aldol condensation 387
Wolff–Kishner reduction 305

x

XAD resins 222–3
Xyliphos ligand 250–1, 266
xylyl-substituted catalysts 120

y

α,β-ynones 154, 170–1
ytterbium (Yb) catalysts 330–1
yttrium (Y) catalysts 94, 379

z

Zimmerman–Traxler transition state 395, 398, 408
zinc (Zn) catalysts
    - hydrosilylation
        - carbonyl compounds 189, 192–3
        - imines 328
    - in solution
        - C≡C bond reduction 430
        - C–X bond reduction 434
        - nitro reduction 425
        - sulfoxide reduction 424–5
    - *see also* diethylzinc
zirconium (Zr) catalysts
    - alkenes
        - hydroalumination 457, 459, 462, 464, 466
        - hydrogenation 45, 59
    - alkynes
        - hydroalumination 471
        - hydroboration 370–1, 372–3
    - imine hydrogenation 249